GENTIANACEAE: SYSTEMATICS AND NATURAL HISTORY

The family Gentianaceae is a diverse lineage of over 1600 angiosperm species, including many tropical and temperate trees, shrubs, and herbs with a wide range of floral types and colors. This volume provides the first comprehensive review of the family, covering phylogeny, classification, biogeography, palynology, phytochemistry, and morphology, and also presents the first classification of the entire family to be published for over 100 years, generated using modern molecular- and morphology-based phylogenetic data. The volume places the Gentianaceae in context with its relatives in the order Gentianales and subclass Asteridae; presents an updated, phylogenetic classification of tribes, subtribes, and genera; investigates the corroborative value of morphological features in phylogenetic diagnoses; and comprehensively summarizes palynology, seed morphology, and phytochemistry. Descriptions of each of the 87 gentian genera are provided, as are discussions on morphological evolution and biogeography for each major evolutionary lineage.

LENA STRUWE is Assistant Professor in Plant Systematics at Rutgers University, New Jersey. Her main research interests are the evolution and biogeography of the Gentianaceae, and also the order Gentianales, specifically the families Loganiaceae and Gelsemiaceae.

VICTOR A. ALBERT is Section Head of the Botanical Garden, The Natural History Museums and Botanical Garden, University of Oslo, Norway. His research is broadly concerned with plant evolutionary biology, spanning a wide range of studies from gentian systematics and island biogeography to understanding the molecular developmental basis for morphological novelty.

GENTIANACEAE
Systematics and Natural History

Edited by

LENA STRUWE
Rutgers University

VICTOR A. ALBERT
Universitetet i Oslo

PUBLISHED BY THE PRESS SYNDICATE OF THE UNIVERSITY OF CAMBRIDGE
The Pitt Building, Trumpington Street, Cambridge, United Kingdom

CAMBRIDGE UNIVERSITY PRESS
The Edinburgh Building, Cambridge CB2 2RU, UK
40 West 20th Street, New York, NY 10011-4211, USA
477 Williamstown Road, Port Melbourne, VIC 3207, Australia
Ruiz de Alarcón 13, 28014 Madrid, Spain
Dock House, The Waterfront, Cape Town 8001, South Africa

http://www.cambridge.org

© Cambridge University Press 2002

This book is in copyright. Subject to statutory exception
and to the provisions of relevant collective licensing agreements,
no reproduction of any part may take place without
the written permission of Cambridge University Press.

First published 2002

Printed in the United Kingdom at the University Press, Cambridge

Typeface Times NR 10/13pt   *System* QuarkXPress™ [SE]

A catalogue record for this book is available from the British Library

Library of Congress Cataloguing in Publication data
Gentianaceae: systematics and natural history / edited by Lena Struwe and Victor A. Albert.
p. cm.
Includes bibliographical references (p.).
ISBN 0 521 80999 1
1. Gentianaceae. 2. Gentianaceae – Classification. I. Struwe, Lena, 1967– . II. Albert, Victor A. (Victor Anthony), 1964– .

QK495.G35 G46 2002
583′.93–dc21   2001037404

ISBN 0 521 80999 1 hardback

Contents

List of contributors		*page* vii
Foreword		ix
1	Gentianaceae in context V. A. Albert and L. Struwe	1
2	Systematics, character evolution, and biogeography of Gentianaceae, including a new tribal and subtribal classification L. Struwe, J. W. Kadereit, J. Klackenberg, S. Nilsson, M. Thiv, K. B. von Hagen, and V. A. Albert	21
3	Cladistics of Gentianaceae: a morphological approach S. Mészáros, J. De Laet, V. Goethals, E. Smets, and S. Nilsson	310
4	Gentianaceae: a review of palynology S. Nilsson	377
5	The seeds of Gentianaceae F. Bouman, L. Cobb, N. Devente, V. Goethals, P. J. M. Maas, and E. Smets	498
6	Chemotaxonomy and pharmacology of Gentianaceae S. R. Jensen and J. Schripsema	573
Index		633

Contributors

Victor A. Albert
Universitetets naturhistoriske museer og Botanisk hage, Universitetet i Oslo, Sars' gate 1, 0562 Oslo, Norway

Ferry Bouman
Hugo de Vries-Laboratorium, Universiteit van Amsterdam, Kruislaan 318, 1098 SM Amsterdam, The Netherlands

Laura Cobb
Appelboom 22, 4101 VG Culemborg, The Netherlands

Jan De Laet
Department of Invertebrates, American Museum of Natural History, Central Park West at 79th Street, New York, NY 10024, USA. Formerly at: The Lewis B. and Dorothy Cullman Program for Molecular Systematics Studies, The New York Botanical Garden, Bronx, NY 10458-5126, USA

Nora Devente
Hugo de Vries-Laboratorium, Universiteit van Amsterdam, Kruislaan 318, 1098 SM Amsterdam, The Netherlands

Valérie Goethals
Instituut voor Natuurbehoud, Kliniekstraat 25, 1070 Brussel, Belgium. Formerly at: Laboratorium voor Systematiek, Instituut voor Plantkunde en Microbiologie, Katholieke Universiteit Leuven, Kasteel Arenbergpark 31, 3001 Leuven, Belgium

K. Bernhard von Hagen
Institut für Geobotanik, Martin-Luther-Universität Halle-Wittenberg, Neuwerk 21, 06099 Halle (Saale), Germany. Formerly at: Institut für Spezielle Botanik, Johannes Gutenberg-Universität, 55099 Mainz, Germany

List of contributors

Søren Rosendal Jensen
Kemisk Institut, Danmarks Tekniske Universitet, Kemitorvet, Bygn. 201, 2800 Lyngby, Denmark

Joachim W. Kadereit
Institut für Spezielle Botanik, Johannes Gutenberg-Universität, 55099 Mainz, Germany

Jens Klackenberg
Sektionen för Fanerogambotanik, Naturhistoriska Riksmuseet, Frescativägen 40, 104 05 Stockholm, Sweden

Paul J. M. Maas
Nationaal Herbarium Nederland, Vestiging Universiteit Utrecht, W.C. van Unnikgebouw, Heidelberglaan 2, 3584 CS Utrecht, The Netherlands

Sándor Mészáros
Agrárgazdasági Kutató és Informatikai Intézet, Postafiók 5, 1355 Budapest 55, Hungary

S. Nilsson
Palynologiska Laboratoriet, Naturhistoriska Riksmuseet, Roslagsvägen 101, 104 05 Stockholm, Sweden

Jan Schripsema
Setor de Quimica de Produtos Naturais, LCQUI/CCT, Universidade Estadual do Norte Fluminense, Av. Alberto Lamego 2000, 28015-620 Campos dos Goytacazes, RJ, Brazil

Erik Smets
Laboratorium voor Systematiek, Instituut voor Plantkunde en Microbiologie, Katholieke Universiteit Leuven, Kasteel Arenbergpark 31, 3001 Leuven, Belgium

Lena Struwe
Department of Ecology, Evolution, and Natural Resources, Rutgers University–Cook College, 237 Foran Hall, 59 Dudley Road, New Brunswick, NJ 08901-8551, USA. Formerly at: The Lewis B. and Dorothy Cullman Program for Molecular Systematics Studies, The New York Botanical Garden, Bronx, NY 10458-5126, USA

Mike Thiv
Botanischer Garten und Botanisches Museum Berlin-Dahlem, Koenigin-Luise-Str. 6–8, 14191 Berlin, Germany. Present address: Institut für Systematische Botanik, Universität Zürich, Zollikerstrasse 107, 8008 Zürich, Switzerland

Foreword

I am very pleased to introduce this important new contribution to systematics. This book is an excellent example of what can be achieved when a plant family is studied by an interdisciplinary group of researchers who each bring their different skills to answer the problems. The Gentianaceae is a good model for this kind of study because it is reasonably large, with its over 1600 species in 87 genera, and also because it contains several genera that were of dubious position before this work was carried out. The family has considerable morphological variation and has adapted to several different functional syndromes for pollination and dispersal. It is also a family that has not been studied on a complete worldwide basis at the tribal and subtribal level since the work of Gilg in 1895. This was a work waiting to be done. Many previous workers have speculated about the systematic position of *Fagraea* and *Potalia* and about *Saccifolium* with its extraordinary pouch-like leaves. These are now shown to fit well within the circumscription of the monophyletic Gentianaceae as defined here. I was also most interested to see where the strange saprophytic genera *Voyria* and *Voyriella* fit into the system, the latter near to *Saccifolium*. The authors have truly used all the available evidence to produce a phylogenetic framework that has yielded a monophyletic classification for the family.

I am also impressed with how the morphological work has combined so well with the molecular. The tribal system that has been produced here, using evidence from many different fields, also keys out well on morphology. This is something that has not always been true for systems based on molecular work. Here we have a well-integrated system that should be a model for other similar studies. This whole work is a most thorough study from the taxonomic history of the family to the molecular phylogeny. It also includes many interesting data on pollen and biogeography. The sources of data from the different contributors from around the world are well integrated. It was a delight and not a chore to read the manuscript in order to write this foreword, because it is a significant new contribution to systematics that has greatly expanded our knowledge of the Gentianaceae. I hope that this study will encourage other researchers to follow its example of the co-operative approach to systematic work for other plant families. I congratulate the authors and the editors for their good work.

The first chapter makes it quite clear that, even after a study of a family at this depth, much still remains to be done, especially at the species level and regarding the biogeographic history of Gentianaceae. This further work will certainly be stimulated and made easier by the study presented here. The data produced here also need to be applied to the conservation of some of the very interesting and rare plants that are included in this study.

Kalaheo, Hawaii,   *Ghillean T. Prance FRS, VMH*
June 2001

1
Gentianaceae in context

V. A. ALBERT AND L. STRUWE

INTRODUCTION

What does it take to recognize a family such as Gentianaceae? This is both an evolutionary biological question and one of perception and emphasis. Tracing back to the descriptor of Gentianaceae, Antoine Laurent de Jussieu (1789), gentians were distinguished as the Natural Order "Gentianeae", one of 15 such orders in Jussieu's class VIII, "*Dicotyledones monopetalae, corolla hypogyna*". According to Lindley (1853), Jussieu usually derived the names for his Natural Orders from genera deemed well representative in their general structure. We therefore have Gentianaceae from Linnaeus's *Gentiana*, defined by being dicotyledonous, sympetalous, and hypogynous. However, in the twenty-first century it is easy to see that many angiosperms, both those phylogenetically close and those phylogenetically distant from each other, could fit this bauplan. In one such example, Jussieu included *Mitreola* and *Spigelia* in Gentianaceae; in another, he included *Potalia*. These opinions were both pre-evolutionary (Darwin, 1859) and pre-phylogenetic (Hennig, 1966; Kluge & Farris, 1969), and were therefore based on different emphases of perceived morphological similarities and differences. In the first case, the hemi-apocarpous nature of the *Mitreola* and *Mitrasacme* gynoecium (Endress *et al.*, 1983; Conn *et al.*, 1996) matched the nascently apocarpous but postgenitally fused ovaries of many Gentianaceae (Padmanabhan *et al.*, 1978). *Spigelia* does not display this trait, but it does have generalized sympetalous and hypogynous flowers. *Potalia*, on the other hand, has a 4-lobed, decussate calyx, an 8–10-lobed sympetalous corolla with 8–10 adnate stamens, and a secondarily syncarpous gynoecium (Leeuwenberg & Leenhouts, 1980; L. Struwe & V. A. Albert, unpubl.). Moreover, all of these divergent taxa save *Gentiana* have been commonly assigned to Loganiaceae since Bentham (1857). To confuse matters still further, *Fagraea*, a non-polymerous but secondarily syncarpous relative of *Potalia* (Struwe & Albert, 1997; Struwe *et al.*, 1998), was included in Jussieu's original circumscription of yet another family, Apocynaceae.

To quote *Sesame Street*, "One of these things is not like the other, one of these things just doesn't belong". In Jussieu's classification of Gentianaceae, differences and similarities must have been given unlike emphases (weights), a condition that could describe most if not all decisions made by the classical system-builders. Modern biological data can come to the aid of artificial classifications, but reliably so only if these data are viewed in an explicit phylogenetic context. For example, the distributions of certain phytochemicals among the angiosperms can suggest natural groupings. In the case of Gentianaceae, the combination of secoiridoid and xanthone production is a defining feature for most taxa (Jensen, 1992; Jensen & Schripsema, 2002 (Chapter 6, this volume)); only *Exacum* and close relatives apparently lack xanthones (Saccifolieae have not yet been investigated). *Mitreola* lacks xanthones and *Spigelia* species lack secoiridoids and xanthones of the gentian types, whereas *Potalia* and *Fagraea* can be readily accommodated within Gentianaceae since completely comparable phytochemicals are produced (Jensen, 1992; Jensen & Schripsema, 2002). However, it is equally and ironically true, with reference to Jussieu, that the pollen of *Potalia* is more like that of some Apocynaceae rather than Gentianaceae, whereas *Fagraea* pollen is readily accommodated among that of the gentians (Nilsson, 2002 (Chapter 4, this volume)). Therefore, reliance on only single or few comparable traits cannot provide confidence in a natural classification. For example, the presence/absence of certain phytochemicals may result from duplication and divergence within single gene families (cf. Helariutta *et al.*, 1996; Eckermann *et al.*, 1998), and this could occur convergently among plants of different lineages (e.g., glucosinolates in Brassicales and Euphorbiaceae; Rodman *et al.*, 1998). Likewise, gross morphological similarities and differences can be phylogenetically deceptive (e.g., connation of petals into a tube in, for example, Asteridae, Cucurbitaceae, and Liliaceae, which are phylogenetically dispersed; APG, 1998; Soltis *et al.*, 2000).

The techniques of molecular phylogenetics have provided an opening to trace gentian phylogeny using traits of the genotype. Menyanthaceae, accepted as a subfamily of Gentianaceae by Gilg (1895), are now firmly nested inside Asterales from the standpoint of DNA data (Olmstead *et al.*, 1992; Gustafsson *et al.*, 1996; Gustafsson & Albert, 1999). Phenotypic characters are still of great importance to Gentianaceae circumscription

Gentianaceae in context 3

and classification, but it is the sum of all available evidence, cast in a phylogenetic framework, that most efficiently and objectively informs us about natural groupings of taxa. In the new, monophyletic classification of Gentianaceae presented in this book, *Mitreola*, *Spigelia*, and Menyanthaceae are excluded, while *Potalia* and *Fagraea* are included. These informed decisions are addressed below with respect to the gentian bauplan (architectural organization), its context, and its within-family variations, as well as the numerous structural gestalts (forms) shared among Gentianaceae, with other members of the gentian parent lineage (the order Gentianales and subclass Asteridae), and even with other angiosperms. We conclude with new, phylogenetically grounded perspectives on gentian diversity.

ASTERIDAE AND GENTIANALES: THE PARENT LINEAGE

The Asteridae comprises the numerous tube-flowered eudicot groups as well as some lineages of choripetalous plants that have sometimes been placed elsewhere (e.g., Cornales and Apiales; e.g., Cronquist, 1981). The subclass is strongly supported as monophyletic based on molecular systematic studies (APG, 1998; Soltis *et al.*, 2000). Two major subclades can be recognized in Asteridae, one including the Asterales (including Menyanthaceae), Dipsacales, and Apiales, and the second comprising the Lamiales, Oleales, Solanales, and Gentianales (Soltis *et al.*, 2000). Higher-order phylogenetic analyses identify the Gentianales as a large monophyletic group composed of three major and two minor subclades; these correspond to Gentianaceae (in the sense recognized in this book), Rubiaceae, Apocynaceae, Gelsemiaceae, and Loganiaceae (Struwe *et al.*, 1994, 1998; Backlund *et al.*, 2000). There are several important points to be made about this collection of taxa. First, pre-phylogenetic systems have commonly excluded the Rubiaceae based on their near ubiquitously epigynous flowers (e.g., Cronquist, 1981). This *ad hoc* emphasis placed on ovary position can be traced all the way back to Jussieu's system (1789), in which sympetalous angiosperms with hypogynous versus epigynous flowers were separated. Recent studies have shown that epigyny has evolved a number of times among otherwise hypogynous angiosperm groups (Gustafsson & Albert, 1999), and should therefore not necessarily be taken as a cardinal character at higher classificatory levels. Second, recent research has shown that the Asclepiadaceae, with their complex flowers, are merely an apomorphic derivation within Apocynaceae representing the epitome of floral synorganization (Endress *et al.*, 1996; Sennblad & Bremer, 1996; Civeyrel *et al.*, 1998;

Potgieter & Albert, 2002). Trends of character evolution leading up to the asclepiad condition can be clearly traced among various clades of Apocynaceae (MacFarlane, 1933; Endress *et al.*, 1996; Potgieter & Albert, 2002). Third, the heterogeneous Loganiaceae has been phylogenetically disassembled since Leeuwenberg and Leenhouts' 1980 treatment. *Buddleja* and Buddlejaceae have moved from Gentianales to Lamiales, near Verbenaceae (Oxelman *et al.*, 1999). Likewise, phylogenetic, morphological, and phytochemical evidence has moved *Desfontainia* to Dipsacales (Jensen, 1992; Struwe *et al.*, 1994; Backlund & Bremer, 1997), *Retzia* to Stilbaceae (Lamiales), and *Sanango* to Gesneriaceae (Bremer *et al.*, 1994; Jensen, 1994; Smith *et al.*, 1997; Oxelman *et al.*, 1999), to name just a few. Other former Loganiaceae (Struwe, 2002a, 2003) remain within Gentianales; *Potalia*, *Fagraea*, and *Anthocleista* (recall Jussieu!) form a monophyletic group nested far within the Gentianaceae (Struwe *et al.*, 1994, 1998; Struwe & Albert, 1997), and *Gelsemium* and *Mostuea* together form the sister group to Apocynaceae, now recognized as Gelsemiaceae (Struwe *et al.*, 1994; Backlund *et al.*, 2000; Struwe, 2003).

Despite the fact that many heterogeneous taxa were dispensed with as a result of phylogenetic consideration of Gentianales, such considerable diversity remains so as to make definition of an ordinal bauplan difficult. All recognized families except Rubiaceae have wood with intraxylary phloem (Struwe *et al.*, 1994). Solanaceae, which may be relatively closely related to Gentianales, also have this trait. Most Gentianales produce secoiridoid or complex indole alkaloid compounds, but several taxa of the strictly circumscribed Loganiaceae, as well as all investigated Gentianaceae, do not produce indole alkaloids (Jensen, 1992; Struwe *et al.*, 1994; Jensen & Schripsema, 2002). Nearly all Gentianales have opposite leaves with interpetiolar stipules or stipular lines, and colleters (small glandular structures) appear almost universally on the insides of leaf bases and/or calyces (Struwe *et al.*, 1994). Sympetaly is also nearly universal (save, e.g., *Theligonum* and *Dialypetalanthus*; Rutishauser *et al.*, 1998; Piesschaert *et al.*, 1997; Savolainen *et al.*, 2000), but then again, so is it in Gentianales' parent subclass, the Asteridae. The bottom line is that global morphological similarities among Gentianales families (i.e., synapomorphies) are not much to go on compared with such easily recognizable differences as inferior versus superior ovaries with placentation that varies both among and within the families of the order. Still, overlapping patterns of similarity and molecular phylogenetic evidence do identify Gentianales, and morphological evolution within the order's type family, Gentianaceae, should be evaluated from this context.

CHARACTERIZATION OF THE GENTIANACEAE LINEAGE

One of the historic problems with classifying Loganiaceae was that most taxa assigned to the family had rather generalized or plesiomorphic traits (see Leeuwenberg & Leenhouts, 1980). Thus, Loganiaceae became something of a grab-bag for taxa of Gentianales (and other orders) that didn't fit clearly anywhere else. As a whole, the same argument could be made for Gentianaceae, despite the fact that, based on molecular data, Gentianaceae are strongly supported both as monophyletic and as the second branching clade in Gentianales after Rubiaceae (Backlund et al., 2000). Gentians simply do not bear readily stereotyped (e.g., highly synorganized or otherwise jointly apomorphic) flowers, nor do they have a unified ovary or fruit structure or unique trait such as latex that would have pleased the early system-builders. The recent phylogenetic research that laid the foundation for the new classification of gentians presented in this book has in fact increased their already considerable morphological heterogeneity, legitimizing a perception of the family as becoming a sort of "natural grab-bag" in its own right. Three cases bear special mention.

Saccifolium

The addition to Gentianaceae of *Saccifolium* (Struwe et al., 1998; Thiv et al., 1999) alone adds saccate leaves (of still unknown functional relevance), a glandular disk inside the calyx possibly homologous to colleters, and a rare instance of imbricate corolla aestivation (most gentians are contort).

Saprophytes

In the same basal-most clade as *Saccifolium* lies *Voyriella*, a reduced, achlorophyllous mycotroph, as well as two to three genera that are heterostylous but that otherwise have generalized gentian flowers. Indeed, saprophytes (or parasites) have evolved at least four other times in Gentianaceae, in some *Sebaea* and *Cotylanthera* species (of Exaceae, the second-most basal clade of gentians; Struwe et al., 2002 (Chapter 2, this volume)), in *Bartonia* and in *Obolaria* (separately placed within the derived Gentianeae; Struwe et al., 2002), and in *Voyria*, which remains difficult to classify within the family (Albert & Struwe, 1997; Struwe & Albert, 2000).

Potaliinae

Another divergent and controversial gentian group comprises *Potalia*, *Anthocleista*, and *Fagraea*, which were for a long time assigned to the Loganiaceae as tribe Potalieae. Jussieu's early perceptions on *Potalia* (see above) were not heeded by later system-builders, including von Martius (1827), who erected Potaliaceae for all three genera. *Fagraea* was, ironically, the easiest to fit within the gentian bauplan, as was recognized by Bureau (1856) and Fosberg and Sachet (1974, 1980). Cladistic analyses of non-molecular data placed Potaliinae, now recognized at subtribal level, within Gentianaceae (Bremer & Struwe, 1992; Struwe *et al.*, 1994; Struwe & Albert, 1997), but owing to their shared possession of mainly generalized (plesiomorphic) phenotypic features only the combined presence of seco-iridoids and xanthones has rendered the Potaliinae decisively gentian. This relationship was strongly corroborated by molecular evidence (Downie & Palmer, 1992; Olmstead *et al.*, 1993; Struwe *et al.*, 1998; Thiv *et al.*, 1999). Appearances can be deceptive: Potaliinae, which often become large trees, are apparently closely related to small herbs (e.g., *Faroa* and *Neurotheca*) as well as to shrubs and small trees (*Lisianthius* spp.). Furthermore, the floral and palynological features of *Potalia* and *Anthocleista* are divergent (see above), and all three genera bear fleshy berries in contrast to the dry or leathery (in *Symbolanthus*) capsules of most other gentians. Fruit fleshiness might be linked with congenital syncarpy in the group (L. Struwe & V. A. Albert, unpubl.), another anomalous feature. By displaying a mosaic mainly of autapomorphic (uniquely derived) and plesiomorphic (generalized Gentianales) features, the Potaliinae had been a classic example of misclassification due to crypticity in shared derived (i.e., phylogenetically informative) traits that might otherwise have been diagnostic.

Summary

In conclusion, a Gentianaceae bauplan is difficult to identify. Other than in cryptic phytochemical features, Gentianaceae are not particularly divergent from basal Apocynaceae, Gelsemiaceae, or Loganiaceae *sensu stricto*. Opposite leaves with colleters and stipular lines to ocreas, 4–5-part calyces with colleters, 4–5-part corollas with contort aestivation, and postgenitally fused gynoecia characterize most taxa, but the exceptions are notable. The major clades within Gentianaceae, classified in this book as tribes and sub-tribes, are in some cases diagnosable by morphological features that could be taken to represent *unterbaupläne*. Indeed, these architectural variants in many cases hold much better within their respective lineages than do gentians among the various clades of Gentianales. For example, seeds with star-shaped testa cells, ovules positioned on entire inner surface of ovaries, helically twisting anthers, bilamellate stigmas, capitate stigmas, and stipitate gynoecia with sessile stigmas do well to identify most major gentian clades. Gentianaceae genera as classified in this book, however, are often more easily recognized as structural gestalts set off hierarchically by supporting molecular phylogenetic data. In other cases, it is species within genera that vary in this fashion. Since basic morphological, palynological, and phytochemical differences among Gentianaceae tribes, subtribes, and genera are covered in other chapters of this book, we will focus here on the patterns of occurrence of several common structural gestalts (themes) and their evolutionary correlates. Some of these will be seen to be perceptually related to emphases struck in previous, non-natural classifications of Gentianaceae and Gentianales. Our descriptions are not meant to be exhaustive treatises, but rather highlights of relevant themes.

GENTIANACEAE GESTALTS: EVOLUTION OF SHAPES AND OF CLASSIFICATIONS

Pollination syndromes

Features associated with pollination syndromes have long been recognized as structural gestalts that may or may not reflect phylogenetic relationships (see, e.g., Endress, 1994, for an in-depth treatment). For example, flowers with narrow corolla tubes and perpendicularly flattened corolla limbs (also known as salverform corollas) occur in several major gentian lineages, including *Voyria* (e.g., *Voyria caerulea*), Exaceae (e.g., *Tachiadenus carinatus*), Chironieae (*Centaurium erythraea*), and Gentianeae (e.g., *Gentiana verna*). Indeed, this flower form is not uncommon among other Gentianales (e.g., *Vinca* in Apocynaceae and *Ixora* in Rubiaceae), Asteridae (e.g., *Primula* in Primulaceae and *Syringa* in Oleaceae), and other angiosperms (e.g., *Dianthus* in Caryophyllaceae). Salverform flowers, which provide a landing platform for visiting insect pollinators such as butterflies, have therefore evolved convergently at several nested hierarchical levels, including during gentian diversification (i.e., they do not appear to be a phylogenetically primitive gentian trait).

Hawkmoth pollination, which involves longer and narrower, lightly colored corolla tubes, may occur in *Tachiadenus* (Exaceae) and the recently described *Aripuana* (Helieae). This syndrome is certainly present in other

Gentianales such as *Stephanotis* (Apocynaceae) and *Posoqueria* (Rubiaceae), and therefore, like salverform flowers, follows a similarly repetitive (homoplastic) phylogenetic distribution among Gentianaceae and Gentianales.

Features suggestive of bird pollination, not unexpectedly, also occur homoplastically within Gentianaceae and Gentianales. Although they are not common, tubular or salverform flowers with red-orange-yellow coloration are known among several apparently independent subclades of Helieae represented by *Calolisianthus, Celiantha, Lagenanthus, Neblinantha, Tachia*, and *Rogersonanthus* (*R. coccineus*, described by Struwe & Albert, 1998), and "Roraimaea" (ined., L. Struwe, S. Nilsson, & V. A. Albert, unpubl.). Additionally, some species of *Symbolanthus* (also Helieae), which typically have greenish-yellow or pink-red flowers (e.g., *S. elisabethae, S. pulcherrimus*, and *S.* "tetrapterus"), are known to be visited by hummingbirds (Struwe, 2000b). Bird pollination is also known among other families of otherwise insect-pollinated Gentianales, e.g., Loganiaceae (e.g., *Spigelia marilandica* and *Labordia waiolani*) and Rubiaceae (e.g., *Retiniphyllum*).

The aforementioned *Symbolanthus* case, however, becomes complicated since the gullet-shaped flowers of some yellow to green or white-flowered species (e.g., *Symbolanthus vasculosus*) are probably also visited by bats (cf. the case in *Chelonanthus alatus*; Machado et al., 1998). Moreover, not all *Calolisianthus, Rogersonanthus,* or *Tachia* species have red-yellow-colored flowers, and, returning to the issue of salverform corollas, other species of *Gentiana* (e.g., *Gentiana acaulis*) have bell-shaped flowers typical of bumblebee pollination. Therefore, form and phylogeny in Gentianaceae show a poor correlation with respect to pollination syndromes. Pollination syndromes do, however, correlate at one level with extensive intra-lineage diversification among genera, for example, within Helieae.

Fruit form

Most Gentianaceae have dry, capsular fruits, and this could be one main reason why the fleshy-fruited Potaliinae had been excluded from the family for so long. *Potalia, Anthocleista,* and *Fagraea* almost always bear fleshy, indehiscent berries (rarely dehiscent owing to the enormous size of the fruit in *Fagraea auriculata*; Leenhouts, 1962), as do some genera of Loganiaceae *sensu stricto* (*Gardneria* and *Strychnos*). The fleshiness of Potaliinae berries appears to correlate with congenital syncarpy, a trait only rarely seen among other gentians (i.e., dry-fruited *Aripuana*; Struwe et al., 1997), which characteristically have postgenitally fused ovaries. The state of ovary fusion in fleshy-fruited species of *Crawfurdia* and *Tripterospermum* (of

Gentianeae) has not been thoroughly investigated, but leathery-fruited *Symbolanthus* (Helieae) definitely has postgenital fusion (L. Struwe & K. Gould, unpubl.). In Potaliinae, the ovaries (but not always the mature fruits) of *Anthocleista* and *Potalia* have the same bilobed, inrolled placentas as do many other taxa of the family (e.g., Helieae), whereas *Potalia* placentas appear to be more axile, a finding that requires further study. Fruits of at least West African species of *Anthocleista* are known to be eaten by hammerheaded bats (Bradbury, 1984), and it is plausible that the fleshy, whitish, upright, and open flowers may be pollinated by the same animals. The animal dispersers of *Potalia* seeds are unknown, but their smaller flowers with small to tiny openings borne on electric-yellow cymes seem more likely bird pollinated. *Fagraea* species have many potential pollination syndromes from hawkmoth to bat pollination, but, at least in the widespread Pacific taxa *Fagraea berteroana* and *F. gracilipes*, seed dispersal by birds seems most likely, although some bat-pollinated species may also be bat dispersed (Ridley, 1930).

Not only is it *not* strange to find fleshy-fruitedness to be derived within a Gentianales clade, it is also not at all unexpected to find this feature associated with congenital syncarpy. Within Apocynaceae, *Carissa* and *Akocanthera* had previously been thought to occupy a rather basal position in this otherwise strongly apocarpous family (Leeuwenberg, 1994; Endress *et al.*, 1996). Instead, molecular data have proven them to be highly derived phylogenetically, occupying a position proximal to higher Apocynaceae, periplocoids, and asclepiadoids (Potgieter & Albert, 2002). Dry-fruited, apocarpous taxa such as *Aspidosperma* occupy the basal-most branches of the family, which suggests these features to be the ancestral conditions for the clade (Potgieter & Albert, 2002). After the overturning of such conventional wisdom in Apocynaceae, the derived status of Potaliinae in Gentianaceae seems hardly more controversial. Instead, it just reinforces an evolutionary hypothesis that congenital syncarpy may be a prerequisite to fleshiness in otherwise capsular clades. The fleshy, apocarpous fruits common in Apocynaceae (e.g., among the Alyxieae; Endress *et al.*, 2000) appear to have evolved on a different ontogenetic trajectory, directly from apocarpous, dry-fruited lineages. Regardless, one can easily see that fleshy-fruitedness is a homoplastic trait in both Gentianaceae and Apocynaceae, correlated more with fruit dispersal syndrome than with phylogeny. The homoplastic distribution of fruit fleshiness in Gentianales had been demonstrated already for Rubiaceae (e.g., Bremer & Eriksson, 1992), for which fruit features had previously served as cardinal characters for classification (Robbrecht, 1988). Similar cases in which fruit fleshiness is derived within an asterid clade come from taxa endemic to the Hawaiian Islands; the baccate-fruited genera *Clermontia* and *Cyanea* (Campanulaceae) are derived from dry-fruited lobelioid ancestors (Givnish *et al.*, 1995), and the mints with fleshy nutlets (*Phyllostegia* and *Stenogyne*, Lamiaceae) descend from a dry-fruited lineage characterized by several independent derivations of fruit fleshiness (Lindqvist *et al.*, 2000).

Plant habit

A further possible reason why Potaliinae were excluded from earlier Gentianaceae classifications could concern their woody habit. *Potalia* species range from sparsely stemmed shrubs to 12 m tall trees, *Anthocleista* species are also usually single-stemmed trees but are often much taller and considerably more branched, and *Fagraea* species range from lianas to shrubs to tall, branched trees up to 30 m in height. However, other Gentianaceae, for example members of neotropical Helieae, may also form small trees (*Macrocarpaea* spp. can be up to 10 m tall and 10 cm in diameter at the base; J. R. Grant & L. Struwe, unpubl.). Indeed, it was after examining some of these taxa that Fosberg and Sachet (1980) noted the strong similarity between *Fagraea* and genera such as *Macrocarpaea*. Although the primitive state for Gentianaceae may be woody (e.g., the Saccifolieae and Exaceae are largely suffrutescent perennials), short-lived, herbaceous plants with little secondary growth are derived in Chironieae, Faroinae, and Gentianeae, but so are trees in Potaliinae and Helieae. The latter pattern appears *not* to be the case for Apocynaceae, in which many of the basal-most taxa (e.g., *Aspidosperma*) are tall trees, whereas a number of derived taxa (such as *Asclepias*) are herbaceous. The same appears true of Loganiaceae *sensu stricto*, in which the basal lineages are woody and arborescent (e.g., *Antonia*), while some derived taxa such as *Mitreola* and *Mitrasacme* are herbaceous, and others are woody (e.g., *Geniostoma*). Woodiness and arborescence in Gentianales are obviously sequentially correlated, but the phylogenetic status of arborescence appears to be independent of the acquisition of woodiness in Gentianaceae versus Apocynaceae and Loganiaceae.

To conclude, taking a bauplan, adding various gestalts, and having multiple originations of features (i.e., parallelisms and reversals) is a simple way to parameterize diversity. A classification can be a model that employs these parameters while still retaining biological information for practical storage and retrieval. A phylogenetic system, such as that presented in this book, is precisely the sort that permits stable, well-grounded assessments of diversity (Stevens, 1997; Struwe & Albert, 1997) both within Gentianaceae (a lineage with sublineages) and between Gentianaceae and neighboring families (other lineages).

GENTIANACEAE DIVERSITY IN TAXONOMIC AND GEOGRAPHIC CONTEXT

Perception of diversity versus its measurement through phylogeny can surely differ depending on (1) whether monophyly is the core phylogenetic principle, and (2) which particular classification model is used, of which a great many will be intercompatible. The Gentianaceae, with its over 1600 species, is the third largest family of the Gentianales, with Rubiaceae and Apocynaceae being distinctly larger. Comparison of the number of species among monophyletic families of Gentianales, as well as among the monophyletic tribes of Gentianaceae, shows that sister clades are most often not of similar size in terms of species numbers. For example, at the ordinal level, Rubiaceae (c. 13000 species) is sister to all other families (Backlund et al., 2000), but contains approximately 66% of all Gentianales species. Gentianaceae (c. 1650 species) is then sister to Apocynaceae (c. 4600) plus Loganiaceae (c. 400) and Gelsemiaceae (11), but comprises only about 25% of all species in this clade. And for a final example, the very small family Gelsemiaceae is sister to the extremely large Apocynaceae clade, which has thousands of species.

Within the gentians themselves, we see a similar pattern of species skewness among clades (species numbers as approximated in Struwe et al., 2002). For example, tribe Saccifolieae is sister to the rest of Gentianaceae, but contains only 6% of the genera and 1% of the species of the family (Tables 1.1–1.2). For other gentianaceous tribes, the number of species and genera in a specific clade can be compared either (1) with all genera in the family, or (2) with only those taxa that form a monophylum with a particular tribe (as was done for the family-level examples, above). The difference is significant in that the comparison in the first case refers to an arbitrarily chosen circumscription of a group (i.e., all gentians except a certain tribe), which forms a paraphyletic unit. If a given tribe is compared with its sister clade, however, a better estimate of true (i.e., phylogenetically based) generic and species skewness results (see Tables 1.1 and 1.2). To illustrate this, when compared with the rest of Gentianaceae, Chironieae is the largest tribe in the family based on genera (23 genera, 27%), closely followed by Helieae (22 genera, 26%; Table 1.1). However, based on sister-group comparison, Chironieae represents only 30% of the genera of its monophylum, whereas Helieae holds 42% of the genera in its major clade.

Table 1.1. *The tribes of Gentianaceae, arranged in order by number of genera*

Tribe	No. of genera	Percentage of genera in family	Percentage of genera in tribe plus sister group
Chironieae	23	27	30[a]
Helieae	22	26	42[b]
Gentianeae	17	20	33[b]
Potalieae	13	15	25[b]
Exaceae	6	7	7[c]
Saccifolieae	5	6	6[d]

Notes:
[a] Based on generic counts of the monophylum Chironieae + Helieae + Gentianeae + Potalieae.
[b] Based on generic counts of the monophylum Helieae + Gentianeae + Potalieae.
[c] Based on generic counts of the monophylum Chironieae + Exaceae + Helieae + Gentianeae + Potalieae.
[d] Based on all genera included in the Gentianaceae.

Table 1.2. *The tribes of Gentianaceae, arranged in order by number of species*

Tribe	No. of species	Percentage of species in family[a]	Percentage of species in tribe plus sister group[a]
Gentianeae	939–968	58	74[b]
Helieae	184	11	14[c]
Chironieae	159	10	11[d]
Potalieae	154	10	12[e]
Exaceae	144–184	9	9[f]
Saccifolieae	16	1	1[g]

Notes:
[a] Based on minimum species counts.
[b] Based on species counts of the monophylum Helieae + Gentianeae + Potalieae.
[c] Based on species counts of the monophylum Helieae + Gentianeae + Potalieae.
[d] Based on species counts of the monophylum Chironieae + Helieae + Gentianeae + Potalieae.
[e] Based on species counts of the monophylum Helieae + Gentianeae + Potalieae.
[f] Based on species counts of the monophylum Chironieae + Exaceae + Helieae + Gentianeae + Potalieae.
[g] Based on all species included in the Gentianaceae.

Table 1.3. *The 10 most species-rich genera of Gentianaceae, arranged in order by number of species based on minimum number of species accepted*

Genus	No. of species	Percentage of no. of species in family[a]
Gentiana	360	22
Gentianella[b]	250	15
Swertia[c]	135	8
Macrocarpaea	90	6
Halenia	80	5
Fagraea	70	4
Exacum	65	4
Sebaea	60–100	4
Centaurium[d]	50	3
Symbolanthus	30	2
Total	1190	74

Notes:
[a] Based on minimum species counts.
[b] *Gentianella* in the current circumscription has been shown to be paraphyletic (P. Chassot, unpubl.; K. B. von Hagen, unpubl.).
[c] *Swertia* in the current circumscription has been shown to be paraphyletic (P. Chassot, unpubl.; K. B. von Hagen, unpubl.).
[d] *Centaurium* in the current circumscription has been shown to be paraphyletic (G. Mansion, unpubl.).

The values are even more divergent if we look at species diversity. Gentianeae is by far the largest tribe in the family, with 939–968 species, or 58% of total species diversity, but it contains 74% of the species in its monophylum (Table 1.2). It seems that the often more derived, species-rich clades originate within a family, the more these two values will differ (see Helieae for generic counts and Gentianeae for species counts, respectively, Tables 1.1 and 1.2).

The distribution of species within genera is also highly skewed among gentians, with a few large genera (30–360 species in 10 genera; Table 1.3) on the one side, and 39 genera with only one or two species (representing 45% of all genera in the family; Table 1.4) on the other. Many of the smaller genera are tropical and occur particularly in Helieae, Potalieae-Faroinae, and Saccifolieae. To some extent, the skewness observed reflects the preference of some specialists to split genera into smaller, distinct units (e.g., in Helieae and Potaliae), as well as the preference of others to lump segregated genera into larger, more polymorphic genera (e.g., in Chironieae and Gentianeae). Our message is this: in context, Gentianaceae and gentian subclades are small ("isolated") in comparison to their sister clades, but

Table 1.4. *Number of genera in Gentianaceae containing only one or two species*

	No. of genera	Percentage of no. of genera in family
1 species (monotypic)	27	31
2 species:	11	13

Table 1.5. *Number of indigenous genera in Gentianaceae for major geographic regions of the world*

Area	No. of genera	Percentage of no. of genera in family
South and Central America	47	54
Asia	26	29
Africa	21	24
North America	15	17
Europe	10	11
Madagascar	10	11
Australia, New Zealand, Pacific	7	8

that all such patterns, even when phylogenetically grounded, are subject to the non-objective interpretation of taxonomic circumscription.

Estimates of geographic diversity may also be distorted. For example, with tribe Gentianeae having 58% of all species in the family (Table 1.2), it might appear that the gentians are most diverse in temperate regions. However, a quick look at the phylogenetic trees and distributions of different tribes shows that four of the five largest genera (*Gentiana, Gentianella, Halenia*, and *Swertia*) belong to the same tribe, Gentianeae (Table 1.3). The more basal lineages in Gentianaceae are tropical to subtropical, with the exception of a few genera in the Chironieae that do enter the temperate zone. The Neotropics (South and Central America) contains an overwhelming mass of gentian diversity, not only from a phylogenetic viewpoint, but also from a taxonomic one. This area is home to over half of all gentian genera (54%, 47 genera; Table 1.5), and no fewer than 77% of these (36 genera; Table 1.6) are endemic to the area. Tropical Africa (52%) and Asia (44%) are also areas that have high generic endemicity, but the temperate areas of the world lag far behind (Table 1.6).

Table 1.6. *Number of endemic genera in Gentianaceae for major geographic regions of the world*

Area	No. of endemic genera	Percentage of no. of genera in family that are endemic to this area	Percentage of genera in the area that are endemic
South and Central America	36	41	77
Africa	11	13	52
Asia	11	13	44
Madagascar	3	3	30
North America	3	3	20
Europe	2	2	20
Australia, New Zealand, Pacific	0	0	0

However, temperate species radiations should not be discounted. As stated above, of the five largest genera, four are in the Gentianeae (*Gentiana, Gentianella, Halenia,* and *Swertia*; Table 1.3). These four genera alone account for 50% of all species in the family. Phylogenetic research has shown that *Gentianella* and *Swertia* are paraphyletic toward other taxa, but the fact that the large, temperate Gentianeae genera represent large and phylogenetically derived (later) radiations during gentian evolution cannot be ignored.

The fourth largest gentian genus is neotropical *Macrocarpaea*, with *c.* 90 species. Until recently this genus was thought to have only about 50 species; within the last three years, over 25 new species have been found (J. R. Grant, pers. comm.). Similarly, a revision of *Potalia* increased the number of well-defined species from one to seven (Struwe & Albert, 1998; L. Struwe & V. A. Albert, unpubl.), and studies on *Symbolanthus* have found more than 10 new species so far (Struwe & Albert, 1998; L. Struwe, unpubl.). New species of Gentianaceae can therefore be predicted to be found primarily in tropical areas, and by re-examining large and variable genera such as *Fagraea* and *Sebaea*.

One major goal of future Gentianaceae research should be to resolve species limits. Additionally, the monophyly of the genera accepted in this book must be either corroborated or rejected. Other future research could be aimed at detailed analysis of the biogeographic history of Gentianaceae, which is complicated by having two distinct clades with distributions around the Indian Ocean basin (Exaceae and Potaliinae), as well as other evidence (e.g., the strictly neotropical Saccifolieae and Helieae) that might suggest an austral (Gondwanic), and therefore heretically ancient, origin of modern lineages. The classification and palynological, phytochemical, and morphological data presented in this book, although comprehensive for a family of Asteridae, only scratch the surface of evolutionary knowledge that we may someday gain on *why* gentians have their particular bauplan, *why* genera or species look different (e.g., the genetic and selective basis for pollination syndrome evolution), and *what* are the important differences in terms of evolutionary history, evolutionary potential, and conservation of this fascinating group of plants.

ACKNOWLEDGMENTS

The authors thank Jason R. Grant and Katherine R. Gould for permission to cite unpublished results. V.A.A. thanks The College of Arts and Sciences, The University of Alabama, and L.S. thanks The Lewis B. and Dorothy Cullman Program for Molecular Systematics Studies, New York, and Rutgers University–Cook College for financial support.

LITERATURE CITED

Albert, V. A. & L. Struwe. 1997. Phylogeny and classification of *Voyria* (saprophytic Gentianaceae). *Brittonia* 49: 466–479.
APG (The Angiosperm Phylogeny Group). 1998. An ordinal classification for the families of flowering plants. *Ann. Missouri Bot. Gard.* 85: 531–553.
Backlund, A. & B. Bremer. 1997. Phylogeny of the Asteridae s. str. based on *rbc*L sequences, with particular reference to the Dipsacales. *Pl. Syst. Evol.* 207: 225–254.
Backlund, M., B. Oxelman, & B. Bremer. 2000. Phylogenetic relationships within the Gentianales based on *ndh*F and *rbc*L sequences, with particular reference to the Loganiaceae. *Amer. J. Bot.* 87: 1029–1043.
Bentham, G. 1857. Notes on Loganiaceae. *J. Linn. Soc. Lond., Bot.* 1: 52–115.
Bradbury, J. W. 1984. Buzzing bats: the lek mating system of hammer-headed bats. Pages 816–817 in: D. Macdonald, ed. *The encyclopedia of mammals.* Facts on File, New York.
Bremer, B. & O. Eriksson. 1992. Evolution of fruit characters and dispersal modes in the tropical family Rubiaceae. *Biol. J. Linn. Soc.* 47: 79–95.
Bremer, B. & L. Struwe. 1992. Phylogeny of the Rubiaceae and the Loganiaceae: congruence or conflict between morphological and molecular data? *Amer. J. Bot.* 79: 1171–1184.
Bremer, B., R. G. Olmstead, L. Struwe, & J. A. Sweere. 1994. *rbc*L sequences support exclusion of *Retzia*, *Desfontainia*, and *Nicodemia* from the Gentianales. *Pl. Syst. Evol.* 190: 213–230.

Bureau, L.-É. 1856. *De la famille des Loganiacées, et des plantes qu'elle fournit à la médecine*. Thèse pour le doctorat en médecine. Faculté de Médecine de Paris, Paris.

Civeyrel, L., A. le Thomas, K. Ferguson, & M. W. Chase. 1998. Critical reexamination of palynological characters used to delimit Asclepiadaceae in comparison to the molecular phylogeny obtained from plastid *mat*K sequences. *Mol. Phylogen. Evol.* 9: 517–527.

Conn, B. J., E. A. Brown, & C. R. Dunlop. 1996. Loganiaceae. Pages 29–62 in: *Flora of Australia*, vol. 28, *Gentianales*. CSIRO Australia, Melbourne.

Cronquist, A. 1981. *An integrated system of classification of flowering plants*. Columbia University Press, New York.

Darwin, C. 1859. *On the origin of species by means of natural selection*. John Murray, London.

Downie, S. R. & J. D. Palmer. 1992. Restriction site mapping of the chloroplast inverted repeat: a molecular phylogeny of the Asteridae. *Ann. Missouri Bot. Gard.* 79: 266–283.

Eckermann, S., G. Schröder, J. Schmidt, D. Strack, R. A. Edrada, Y. Helariutta, P. Elomaa *et al*. 1998. New pathway to polyketides in plants. *Nature* 396: 387–390.

Endress, M. E., B. Sennblad, S. Nilsson, L. Civeyrel, M. W. Chase, S. Huysmans, E. Grafström *et al*. 1996. A phylogenetic analysis of Apocynaceae s. str. and some related taxa in Gentianales: a multidisciplinary approach. *Opera Bot. Belg*. 7: 59–102.

Endress, M. E., B. Sennblad, R. W. J. M. Van der Ham, S. Nilsson, K. Potgieter, L. Civeyrel, M. Chase *et al*. 2000. Arils, wings, and other sneaky things: coming to terms with the Alyxieae (Apocynaceae). *Amer. J. Bot.* 87 (suppl.): 366.

Endress, P. K. 1994. *Diversity and evolutionary biology of tropical flowers*. Cambridge University Press, Cambridge.

Endress, P. K., M. Jenny, & M. E. Fallen. 1983. Convergent elaboration of apocarpous gynoecia in higher advanced dicotyledons (Sapindales, Malvales, Gentianales). *Nord. J. Bot.* 3: 293–300.

Fosberg, F. R. & M.-H. Sachet. 1974. A new variety of *Fagraea berteriana* (Gentianaceae). *Phytologia* 28: 470–472.

Fosberg, F. R. & M.-H. Sachet. 1980. Systematic studies of Micronesian plants. *Smithsonian Contr. Bot.* 45: 1–40.

Gilg, E. 1895. Gentianaceae. Pages 50–180 in: A. Engler & K. Prantl, eds. *Die natürlichen Pflanzenfamilien*, vol. 4(2). Verlag von Wilhelm Engelmann, Leipzig.

Givnish, T. J., K. J. Sytsma, J. F. Smith, & W. S. Hahn. 1995. Molecular evolution, adaptive radiation, and geographic speciation in *Cyanea* (Campanulaceae, Lobelioideae). Pages 288–337 in: W. L. Wagner & V. Funk, eds. *Hawaiian biogeography: evolution on a hot spot archipelago*. Smithsonian Institution Press, Washington, DC.

Gustafsson, M. H. G. & V. A. Albert. 1999. Inferior ovaries and angiosperm diversification. Pages 403–431 in: P. M. Hollingsworth, R. Bateman, & R. J. Gornall, eds. *Molecular systematics and plant evolution*. Taylor & Francis, London.

Gustafsson, M. H. G., A. Backlund, & B. Bremer. 1996. Phylogeny of the

Asterales sensu lato based on *rbcL* sequences with particular reference to the Goodeniaceae. *Pl. Syst. Evol.* 199: 217–242.

Helariutta, Y., M. Kotilainen, P. Elomaa, N. Kalkkinen, K. Bremer, T. H. Teeri, & V. A. Albert. 1996. Duplication and functional divergence in the chalcone synthase gene family of Asteraceae: evolution with substrate change and catalytic simplification. *Proc. Natl. Acad. Sci. USA* 93: 9033–9038.

Hennig, W. 1966. *Phylogenetic systematics.* University of Illinois Press, Urbana, IL.

Jensen, S. R. 1992. Systematic implications of the distribution of iridoids and other chemical compounds in the Loganiaceae and other families of the Asteridae. *Ann. Missouri Bot. Gard.* 79: 284–302.

Jensen, S. R. 1994. A reexamination of *Sanango racemosum.* 3. Chemotaxonomy. *Taxon* 43: 619–623.

Jensen, S. R. & J. Schripsema. 2002. Chemotaxonomy and pharmacology of Gentianaceae. Pages 573–631 in: L. Struwe & V. A. Albert, eds. *Gentianaceae: systematics and natural history.* Cambridge University Press, Cambridge.

Jussieu, A. L., de 1789. *Genera plantarum, secundum ordines naturales disposita, juxta methodum in Horto regio parisiensi exaratam, anno M.DCC.LXXIV.* Herissant, Paris.

Kluge, A. & J. S. Farris. 1969. Quantitative phyletics and the evolution of anurans. *Syst. Zool.* 18: 1–32.

Leenhouts, P. W. 1962 [1963]. Loganiaceae. Pages 293–387 in: C. G. G. J. van Steenis, ed. *Flora Malesiana,* ser. 1, vol. 6(2). Wolters-Noordhoff, Groningen.

Leeuwenberg, A. J. M. 1994. Taxa of the Apocynaceae above the genus level. Series of Apocynaceae XXXIII. *Wageningen Agric. Univ. Papers* 94(3): 47–60.

Leeuwenberg, A. J. M. & P. W. Leenhouts. 1980. Taxonomy. Pages 8–96 in: A. J. M. Leeuwenberg, ed. *Engler and Prantl's Die natürlichen Pflanzenfamilien, Angiospermae: Ordnung Gentianales, Fam. Loganiaceae,* vol. 28b (1). Duncker and Humblot, Berlin.

Lindley, J. 1853. *The vegetable kingdom: the structure, classification and uses of plants,* ed. 3. Bradbury and Evans, London.

Lindqvist, C., T. J. Motley, & V. A. Albert. 2000. A North American closest relative for the Hawaiian endemic mints (Lamiaceae): implications for pollination syndrome and fruit evolution. (http://www.ou.edu/cas/botany-micro/botany2000/section16/abstracts/29.shtml)*

Macfarlane, J. M. 1933. The Apocynaceae and Asclepiadaceae. Pages 1–181 in: *The Evolution and Distribution of Flowering Plants,* 1. Noel Printing Company, Philadelphia.

Machado, I. C. S., I. Sazima, & M. Sazima. 1998. Bat pollination of the terrestrial herb *Irlbachia alata* (Gentianaceae) in northeastern Brazil. *Pl. Syst. Evol.* 209: 231–237.

---

* The publisher has used its best endeavors to ensure that the URLs for external websites referred to in this book are correct and active at the time of going to press. However, the publisher has no responsibility for the websites and can make no guarantee that a site will remain live or that the content is or will remain appropriate.

Martius, C. F. P., von. 1827. *Nova genera et species plantarum quas in itinere par Brasiliam*, vol. 2. V. Wolf, München.
Nilsson, S. 2002. Gentianaceae: a review of palynology. Pages 377–497 in: L. Struwe & V. A. Albert, eds. *Gentianaceae: systematics and natural history.* Cambridge University Press, Cambridge.
Olmstead, R. G., H. J. Michaels, K. M. Scott, & J. D. Palmer. 1992. Monophyly of the Asteridae and identification of their major lineages inferred from DNA sequences of *rbc*L. *Ann. Missouri Bot. Gard.* 79: 249–265.
Olmstead, R. G., B. Bremer, K. M. Scott, & J. D. Palmer. 1993. A parsimony analysis of the Asteridae sensu lato based on *rbc*L sequences. *Ann. Missouri Bot. Gard.* 80: 700–722.
Oxelman, B., M. Backlund, & B. Bremer. 1999. Relationships of the Buddlejaceae s. l. investigated using branch support analysis of chloroplast *ndh*F and *rbc*L sequences. *Syst. Bot.* 24: 164–182.
Padmanabhan, D., D. Regupathy, & S. Pushpa Veni. 1978. Gynoecial ontogeny in *Enicostemma littorale* Blume. *Proc. Indian Acad. Sci.* 87B (Pl. Sci. – 2) 5: 83–92.
Piesschaert, F., E. Robbrecht, & E. Smets. 1997. *Dialypetalanthus fuscescens* Kuhlm. (Dialypetalanthaceae): the problematic taxonomic position of an Amazonian endemic. *Ann. Missouri Bot. Gard.* 84: 201–223.
Potgieter, K. & V. A. Albert. 2002. Phylogenetic relationships within Apocynaceae s. l. based on *trn*L intron and *trn*L–F spacer sequences and propagule characters. *Ann. Missouri Bot. Gard.*, in press.
Ridley, H. N. 1930. *The dispersal of plants throughout the world*. L. Reeve & Co., Ltd., Ashford, Kent.
Robbrecht, E. 1988. Tropical woody Rubiaceae. *Opera Bot. Belg.* 1: 1–271.
Rodman, J. E., P. S. Soltis, D. E. Soltis, K. J. Sytsma, & K. G. Karol. 1998. Parallel evolution of glucosinolate biosynthesis inferred from congruent nuclear and plastid gene phylogenies. *Amer. J. Bot.* 85: 997–1006.
Rutishauser, R., L.-P. Ronse Decraene, E. Smets, & I. Mendoza-Heuer. 1998. *Theligonum cynocrambe* – the developmental morphology of a peculiar rubiaceous herb. *Pl. Syst. Evol.* 210: 1–24.
Savolainen, V., M. F. Fay, D. C. Albach, A. Backlund, M. van der Bank, K. M. Cameron, S.A. Johnson *et al.* 2000. Phylogeny of the eudicots: a nearly complete familial analysis based on *rbc*L gene sequences. *Kew Bull.* 55: 257–309.
Sennblad, B. & B. Bremer. 1996. The familial and subfamilial relationships of Apocynaceae and Asclepiadaceae evaluated with *rbc*L data. *Pl. Syst. Evol.* 202: 153–175.
Smith, J. F., K. D. Brown, C. L. Carroll, & D. S. Denton. 1997. Familial placement of *Cyrtandromoea*, *Titanotrichum*, and *Sanango*: three problematic genera of the Lamiales. *Taxon* 40: 65–74.
Soltis, D. E., P. S. Soltis, M. W. Chase, M. E. Mort, D. C. Albach, M. Zanis, V. Savolainen *et al.* 2000. Angiosperm phylogeny inferred from a combined data set of 18S rDNA, *rbc*L and *atp*B sequences. *Bot. J. Linn. Soc.* 133: 381–461.
Stevens, P. F. 1997. What kind of classification should the practising taxonomist use to be saved? Pages 295–319 in: J. Dransfield, M. J. E. Coode, & D. A. Simpson, eds. *Plant diversity in Malesia*, III. Royal Botanic Gardens, Kew.

Struwe, L. 2002a. Loganiaceae (including Antoniaceae, Geniostomaceae, Spigeliaceae, and Strychnaceae), in: N. P. Smith, S. V. Heald, A. Henderson, S. A. Mori, & D. W. Stevenson, eds. *Flowering plant families of the American tropics*. Princeton University Press, Princeton, NJ/The New York Botanical Garden Press, Bronx, NY, in press.

Struwe, L. 2002b. Two new winged species of *Symbolanthus* (Gentianaceae: Helieae) from the western Colombian Andes. *Novon*, in press.

Struwe, L. 2003. Gelsemiaceae in: K. Kubitzki, ed. *Families and genera of flowering plants*, vol. *Asteridae* (J. W. Kadereit, vol. ed.). Springer-Verlag, Berlin, Heidelberg, and New York, in press.

Struwe, L. & V. A. Albert. 1997. Floristics, cladistics, and classification: three case studies in Gentianales. Pages 321–352 in: J. Dransfield, M. J. E. Coode, & D. A. Simpson, eds. *Plant diversity in Malesia*, III. Royal Botanic Gardens, Kew.

Struwe, L. & V. A. Albert. 1998. Six new species of Gentianaceae from the Guayana Shield. *Harvard Pap. Bot.* 3: 181–197.

Struwe, L. & V. A. Albert. 2000. Mycotrophic, non-chlorophyllous *Voyria* placed in Gentianaceae. *Amer. J. Bot.* 87 (suppl.): 161.

Struwe, L., V. A. Albert, & B. Bremer. 1994 [1995]. Cladistics and family level classification of the Gentianales. *Cladistics* 10: 175–206.

Struwe, L., P. J. M. Maas, & V. A. Albert. 1997. *Aripuana cullmaniorum*, a new genus and species of Gentianaceae from white sands of southeastern Amazonas, Brazil. *Harvard Pap. Bot.* 2: 235–253.

Struwe, L., M. Thiv, J. W. Kadereit, A. S.-R. Pepper, T. J. Motley, P. J. White, & J. H. E. Rova. 1998. *Saccifolium* (Saccifoliaceae), an endemic of Sierra de la Neblina on the Brazilian–Venezuelan frontier, is related to a temperate-alpine lineage of Gentianaceae. *Harvard Pap. Bot.* 3: 199–214.

Struwe, L., J. Kadereit, J. Klackenberg, S. Nilsson, M. Thiv, K. B. von Hagen, & V. A. Albert. 2002. Systematics, character evolution, and biogeography of Gentianaceae, including a new tribal and subtribal classification. Pages 21–309 in: L. Struwe & V. A. Albert, eds. *Gentianaceae: systematics and natural history*. Cambridge University Press, Cambridge.

Thiv, M., L. Struwe, V. A. Albert, & J. W. Kadereit. 1999. The phylogenetic relationships of *Saccifolium bandeirae* Maguire & Pires (Gentianaceae) reconsidered. *Harvard Pap. Bot.* 4: 519–526.

2

Systematics, character evolution, and biogeography of Gentianaceae, including a new tribal and subtribal classification

L. STRUWE, J. W. KADEREIT, J. KLACKENBERG,
S. NILSSON, M. THIV, K. B. VON HAGEN,
AND V. A. ALBERT

ABSTRACT

A new, monophyletic, genus-level classification of the Gentianaceae is presented based on cladistic analyses of *trn*L intron, *mat*K, and internal transcribed spacer (ITS) sequence data (key pp. 45–47; conspectus pp. 48–55). The family as presently circumscribed contains 87 genera and *c.* 1615–1688 species. Our analyses include nucleotide sequence data for 66 genera, for which no attempt has been made to test their monophyly. Some genera that could not be represented by DNA data are placed in higher categories based on morphological considerations. Our phylogenetic results suggest the recognition of six tribes: Exaceae, Chironieae, Gentianeae, Helieae, Potalieae, and Saccifolieae. The tribe Saccifolieae (Maguire & Pires) Struwe, Thiv, V. A. Albert, & Kadereit, *stat. nov.* is the most basally positioned lineage in the family and consists of five neotropical genera: *Curtia*, *Hockinia*, *Saccifolium*, *Tapeinostemon*, and *Voyriella*. The Exaceae, which is the next-most basal clade in the Gentianaceae, is paleotropical and includes the following genera: *Cotylanthera*, *Exacum*, *Gentianothamnus*, *Ornichia*, *Sebaea*, and *Tachiadenus*. Following Exaceae is tribe Chironieae, divided into three subtribes (which are resolved as a trichotomy): Chironiinae, Canscorinae Thiv & Kadereit, *subtrib. nov.*, and Coutoubeinae. Subtribe Chironiinae is pantropical to temperate and includes *Bisgoeppertia*, *Blackstonia*, *Centaurium*, *Chironia*, *Cicendia*, *Eustoma*, *Exaculum*, *Geniostemon*, *Ixanthus*, *Orphium*, *Sabatia*, and *Zygostigma*. The paleotropical subtribe Canscorinae contains *Canscora*, *Cracosna*, "Duplipetala", *Hoppea*, *Microrphium*, *Phyllocyclus*, and *Schinziella*. The third subtribe in Chironieae, the Coutoubeinae, is strictly neotropical with *Coutoubea*, *Deianira*, *Schultesia*, *Symphyllophyton*, and *Xestaea*. The last three tribes of the family, Gentianeae, Helieae, and Potalieae, are supported as monophyletic with the available data, but with no particular sister-group relationships well supported. The tribe Gentianeae is primarily temperate-alpine. It is divided into two subtribes. The first is Gentianinae, with *Crawfurdia*, *Gentiana*, and *Tripterospermum*. The second, larger subtribe is Swertiinae, including *Bartonia*, *Comastoma*, *Frasera*, *Gentianella*, *Gentianopsis*, *Halenia*, *Jaeschkea*, *Latouchea*, *Lomatogonium*, *Megacodon*, *Obolaria*, *Pterygocalyx*, *Swertia*, and *Veratrilla*. The only exclusively neotropical tribe is the Helieae, which contains the following genera: *Adenolisianthus*, *Aripuana*, *Calolisianthus*, *Celiantha*, *Chelonanthus*, *Chorisepalum*, *Helia*, *Irlbachia*, *Lagenanthus*, *Lehmanniella*, *Macrocarpaea*, *Neblinantha*, *Prepusa*, *Purdieanthus*, *Rogersonanthus*, "*Roraimaea*", *Senaea*, *Sipapoantha*, *Symbolanthus*, *Tachia*, *Tetrapollinia*, *Wurdackanthus*, and *Zonanthus*. The last tribe, Potalieae, is pantropical and includes three subtribes – Faroinae Struwe & V. A. Albert, *subtrib. nov.*, Lisianthiinae, and Potaliinae. The monotypic subtribe Lisianthiinae, based on *Lisianthius*, is restricted to the Caribbean and Central America. The pantropical subtribe Potaliinae, formerly Loganiaceae tribe Potalieae, includes *Anthocleista*, *Fagraea*, and *Potalia*. Finally, subtribe Faroinae is also pantropical and includes *Congolanthus*, *Djaloniella*, *Enicostema*, *Faroa*, *Karina*, *Neurotheca*, *Oreonesion*, *Pycnosphaera*, and *Urogentias*. The phylogenetic position of *Voyria*, an obligate mycotroph, is uncertain and this genus is therefore classified as *incertae sedis*. General morphology and palynology are discussed for each tribe and subtribe, and character evolution is discussed primarily with reference to gynoecial, staminal, and fruit morphology, woodiness, and pollination syndromes. A vicariance model based on the breakup of Gondwana is postulated to explain the branching patterns observed in the basal clades of the family; however, within the Chironieae, Helieae, Gentianeae, and Potalieae, Tertiary and Quaternary climatic changes and vicariance/dispersal events are suggested as possible causes of current distributions.

Keywords: anatomy, biogeography, classification, evolution, Gentianaceae, molecular systematics, morphology, palynology, phylogeny, taxonomy.

INTRODUCTION

The family Gentianaceae was described by Jussieu (1789: 141) and has been universally accepted in all subsequent major classifications of angiosperms. It is a family that shows great variation in habit, morphology, and anatomy, as well in geographic distribution. In October 1838, Grisebach published the only complete species- and genus-level treatment of all gentians, *Genera et species Gentianearum* (Grisebach, 1839). Although Grisebach intended to publish his book in 1839, which is the year printed on the first page, the work was validly published in 1838. The circumscription of Gentianaceae has not changed much since then, with the exception of the inclusion of tribe Potalieae (formerly in Loganiaceae; cf. Leeuwenberg & Leenhouts, 1980) and Saccifoliaceae (formerly a monotypic family; Maguire & Pires, 1978; Takhtajan, 1997; cf. Struwe et al., 1998, 1999; Thiv et al., 1999a). In this chapter we present the first complete tribal and subtribal classification of Gentianaceae since Gilg (1895a), a treatment primarily based on phylogenetic results derived from molecular data but with substantial morphological support (Tables 2.1, 2.2). Summaries are provided for all genera of Gentianaceae, and new and previously known palynological data are presented and reviewed. Character evolution and biogeography within each tribe are discussed using the available results from *trn*L intron and *mat*K data (Struwe et al., 1998; Thiv et al., 1999a,b; Figs. 2.1–2.3 (see also additional data for individual tribes)). These two loci reside in the plastid genome and have been used extensively in other plant phylogenetic studies (Taberlet et al., 1991; Endress et al., 1996; Gielly & Taberlet, 1996; Wallander & Albert, 2000).

The Gentianaceae are cosmopolitan in distribution (except continental Antarctica), and in the present circumscription contain 87 genera and c. 1615–1688 species in six tribes. This is a noteworthy increase in species numbers from the c. 1225 species estimated by Mabberley (1997). The Gentianaceae have been frequently treated in floristic works around the world, but never within a phylogenetic context (see discussion in Struwe & Albert, 1997). A summary of some of the available floristic treatments is presented in Appendix 2.1 to help researchers find keys, illustrations, and descriptions of species and genera from particular areas of the world.

For our discussion and classification of the large-scale relationships within the Gentianaceae we have used phylogenetic analyses based on previously published *trn*L intron data (Struwe et al., 1998; Thiv et al., 1999a; Fig. 2.1), new and previously published *mat*K sequences (Thiv et al., 1999a,b; Fig. 2.2), and a combined analysis including taxa for which both *trn*L and *mat*K sequences are available (Fig. 2.3). The three strict consensus trees are highly congruent with each other and are provided with jackknife values. Each recognized tribe has strong jackknife support (Farris et al., 1996), as do the subtribes and some internal clades. The Saccifolieae form the most basal clade, followed by the Exaceae, the Chironieae, and three tribes – Gentianeae, Helieae, and Potalieae – which have unresolved interrelationships (except in the combined analysis where Potalieae are sister to Gentianeae; Fig. 2.3).

This new classification of Gentianaceae is based on phylogenetic analyses of DNA sequences from individual accessions of exemplar species

Table 2.1. *The new tribal and subtribal classification of Gentianaceae, with genera arranged in tribal and subtribal order, from basal toward more apomorphic groups*

Tribe	Subtribe	Genus	No. of species
Saccifolieae		*Curtia* Cham. & Schltdl.	6–10
Saccifolieae		*Hockinia* Gardner	1
Saccifolieae		*Saccifolium* Maguire & Pires	1
Saccifolieae		*Tapeinostemon* Benth.	7
Saccifolieae		*Voyriella* (Miq.) Miq.	1
Exaceae		*Cotylanthera* Blume	4
Exaceae		*Exacum* L.	65
Exaceae		*Gentianothamnus* Humbert	1
Exaceae		*Ornichia* Klack.	3
Exaceae		*Sebaea* Sol. ex R. Br.	60–100
Exaceae		*Tachiadenus* Griseb.	11
Chironieae	Chironiinae	*Bisgoeppertia* Kuntze	2
Chironieae	Chironiinae	*Blackstonia* Huds.	4
Chironieae	Chironiinae	*Centaurium* Hill	50
Chironieae	Chironiinae	*Chironia* L.	15
Chironieae	Chironiinae	*Cicendia* Adans.	2
Chironieae	Chironiinae	*Eustoma* Salisb.	3
Chironieae	Chironiinae	*Exaculum* Caruel	1
Chironieae	Chironiinae	*Geniostemon* Engelm. & A. Gray	5
Chironieae	Chironiinae	*Ixanthus* Griseb.	1
Chironieae	Chironiinae	*Orphium* E. Mey.	2
Chironieae	Chironiinae	*Sabatia* Adans.	20
Chironieae	Chironiinae	*Zygostigma* Griseb.	1
Chironieae	Canscorinae	*Canscora* Lam.	9
Chironieae	Canscorinae	*Cracosna* Gagnep.	3
Chironieae	Canscorinae	"Duplipetala", ined.	2
Chironieae	Canscorinae	*Hoppea* Willd.	2
Chironieae	Canscorinae	*Microrphium* C. B. Clarke	2
Chironieae	Canscorinae	*Phyllocyclus* Kurz	5
Chironieae	Canscorinae	*Schinziella* Gilg	1
Chironieae	Coutoubeinae	*Coutoubea* Aubl.	5
Chironieae	Coutoubeinae	*Deianira* Cham. & Schltdl.	5
Chironieae	Coutoubeinae	*Schultesia* Mart.	15
Chironieae	Coutoubeinae	*Symphyllophyton* Gilg	1
Chironieae	Coutoubeinae	*Xestaea* Griseb.	1
Helieae		*Adenolisianthus* (Progel) Gilg	1
Helieae		*Aripuana* Struwe, Maas, & V. A. Albert	1
Helieae		*Calolisianthus* Gilg	6
Helieae		*Celiantha* Maguire	3
Helieae		*Chelonanthus* Gilg	7
Helieae		*Chorisepalum* Gleason & Wodehouse	5
Helieae		*Helia* Mart.	2
Helieae		*Irlbachia* Mart.	9
Helieae		*Lagenanthus* Gilg	1

Systematics, character evolution, and biogeography 25

Table 2.1. (cont.)

Tribe	Subtribe	Genus	No. of species
Helieae		*Lehmanniella* Gilg	2
Helieae		*Macrocarpaea* (Griseb.) Gilg	c. 90
Helieae		*Neblinantha* Maguire	2
Helieae		*Prepusa* Mart.	5
Helieae		*Purdieanthus* Gilg	1
Helieae		*Rogersonanthus* Maguire & B. M. Boom	3
Helieae		"Roraimaea", ined.	
Helieae		*Senaea* Taub.	1
Helieae		*Sipapoantha* Maguire & B. M. Boom	1
Helieae		*Symbolanthus* G. Don	30
Helieae		*Tachia* Aubl.	10
Helieae		*Tetrapollinia* Maguire & B. M. Boom	1
Helieae		*Wurdackanthus* Maguire	2
Helieae		*Zonanthus* Griseb.	1
Potalieae	Potaliinae	*Anthocleista* R. Br.	14
Potalieae	Potaliinae	*Fagraea* Thunb.	70
Potalieae	Potaliinae	*Potalia* Aubl.	9
Potalieae	Faroinae	*Congolanthus* A. Raynal	1
Potalieae	Faroinae	*Djaloniella* P. Taylor	1
Potalieae	Faroinae	*Enicostema* Blume	3
Potalieae	Faroinae	*Faroa* Welw.	19
Potalieae	Faroinae	*Karina* Boutique	1
Potalieae	Faroinae	*Neurotheca* Salisb. ex Benth. in Benth. & Hook.f.	3
Potalieae	Faroinae	*Oreonesion* A. Raynal	1
Potalieae	Faroinae	*Pycnosphaera* Gilg	1
Potalieae	Faroinae	*Urogentias* Gilg & Gilg-Ben.	1
Potalieae	Lisianthiinae	*Lisianthius* P. Browne	30
Gentianeae	Gentianinae	*Crawfurdia* Wall.	16–19
Gentianeae	Gentianinae	*Gentiana* L.	360
Gentianeae	Gentianinae	*Tripterospermum* Blume	24
Gentianeae	Swertiinae	*Bartonia* H. L. Mühl. ex Willd.	4
Gentianeae	Swertiinae	*Comastoma* (Wettst.) Toyok.	7–25
Gentianeae	Swertiinae	*Frasera* Walter	15
Gentianeae	Swertiinae	*Gentianella* Moench	250
Gentianeae	Swertiinae	*Gentianopsis* Ma	16–24
Gentianeae	Swertiinae	*Halenia* Borkh.	80
Gentianeae	Swertiinae	*Jaeschkea* Kurz	4
Gentianeae	Swertiinae	*Latouchea* Franch.	1
Gentianeae	Swertiinae	*Lomatogonium* A. Braun	21
Gentianeae	Swertiinae	*Megacodon* (Hemsl.) Harry Sm.	2
Gentianeae	Swertiinae	*Obolaria* L.	1
Gentianeae	Swertiinae	*Pterygocalyx* Maxim.	1
Gentianeae	Swertiinae	*Swertia* L.	135
Gentianeae	Swertiinae	*Veratrilla* Baill. ex Franch.	2
Incertae sedis		*Voyria* Aubl.	19

Table 2.2. *The new tribal and subtribal classification of Gentianaceae, with genera arranged in alphabetical order*

Genus	No. of species	Tribe	Subtribe
Adenolisianthus (Progel) Gilg	1	Helieae	
Anthocleista R. Br.	14	Potalieae	Potaliinae
Aripuana Struwe, Maas, & V. A. Albert	1	Helieae	
Bartonia H. L. Mühl. ex Willd.	4	Gentianeae	Swertiinae
Bisgoeppertia Kuntze	2	Chironieae	Chironiinae
Blackstonia Huds.	4	Chironieae	Chironiinae
Calolisianthus Gilg	6	Helieae	
Canscora Lam.	9	Chironieae	Canscorinae
Celiantha Maguire	3	Helieae	
Centaurium Hill	50	Chironieae	Chironiinae
Chelonanthus Gilg	7	Helieae	
Chironia L.	15	Chironieae	Chironiinae
Chorisepalum Gleason & Wodehouse	5	Helieae	
Cicendia Adans.	2	Chironieae	Chironiinae
Comastoma (Wettst.) Toyok.	7–25	Gentianeae	Swertiinae
Congolanthus A. Raynal	1	Potalieae	Faroinae
Cotylanthera Blume	4	Exaceae	
Coutoubea Aubl.	5	Chironieae	Coutoubeinae
Cracosna Gagnep.	3	Chironieae	Canscorinae
Crawfurdia Wall.	16–19	Gentianeae	Gentianinae
Curtia Cham. & Schltdl.	6–10	Saccifolieae	
Deianira Cham. & Schltdl.	5	Chironieae	Coutoubeinae
Djaloniella P. Taylor	1	Potalieae	Faroinae
"Duplipetala", ined.	2	Chironieae	Canscorinae
Enicostema Blume	3	Potalieae	Faroinae
Eustoma Salisb.	3	Chironieae	Chironiinae
Exaculum Caruel	1	Chironieae	Chironiinae
Exacum L.	65	Exaceae	
Fagraea Thunb.	70	Potalieae	Potaliinae
Faroa Welw.	19	Potalieae	Faroinae
Frasera Walter	15	Gentianeae	Swertiinae
Geniostemon Engelm. & A. Gray	5	Chironieae	Chironiinae
Gentiana L.	360	Gentianeae	Gentianinae
Gentianella Moench	250	Gentianeae	Swertiinae
Gentianopsis Ma	16–24	Gentianeae	Swertiinae
Gentianothamnus Humbert	1	Exaceae	
Halenia Borkh.	80	Gentianeae	Swertiinae
Helia Mart.	2	Helieae	
Hockinia Gardner	1	Saccifolieae	
Hoppea Willd.	2	Chironieae	Canscorinae
Irlbachia Mart.	9	Helieae	
Ixanthus Griseb.	1	Chironieae	Chironiinae
Jaeschkea Kurz	4	Gentianeae	Swertiinae
Karina Boutique	1	Potalieae	Faroinae

Systematics, character evolution, and biogeography 27

Table 2.2. (cont.)

Genus	No. of species	Tribe	Subtribe
Lageananthus Gilg	1	Helieae	
Latouchea Franch.	1	Gentianeae	Swertiinae
Lehmanniella Gilg	2	Helieae	
Lisianthius P. Browne	30	Potalieae	Lisianthiinae
Lomatogonium A. Braun	21	Gentianeae	Swertiinae
Macrocarpaea (Griseb.) Gilg	90	Helieae	
Megacodon (Hemsl.) Harry Sm.	2	Gentianeae	Swertiinae
Microrphium C. B. Clarke	2	Chironieae	Canscorinae
Neblinantha Maguire	2	Helieae	
Neurotheca Salisb. ex Benth. in Benth. & Hook.f.	3	Potalieae	Faroinae
Obolaria L.	1	Gentianeae	Swertiinae
Oreonesion A. Raynal	1	Potalieae	Faroinae
Ornichia Klack.	3	Exaceae	
Orphium E. Mey.	2	Chironieae	Chironiinae
Phyllocyclus Kurz	5	Chironieae	Canscorinae
Potalia Aubl.	9	Potalieae	Potaliinae
Prepusa Mart.	5	Helieae	
Pterygocalyx Maxim.	1	Gentianeae	Swertiinae
Purdieanthus Gilg	1	Helieae	
Pycnosphaera Gilg	1	Potalieae	Faroinae
Rogersonanthus Maguire & B. M. Boom	3	Helieae	
"Roraimaea", ined.		Helieae	
Sabatia Adans.	20	Chironieae	Chironiinae
Saccifolium Maguire & Pires	1	Saccifolieae	
Schinziella Gilg	1	Chironieae	Canscorinae
Schultesia Mart.	15	Chironieae	Coutoubeinae
Sebaea Sol. ex R. Br.	60–100	Exaceae	
Senaea Taub.	1	Helieae	
Sipapoantha Maguire & B. M. Boom	1	Helieae	
Swertia L.	135	Gentianeae	Swertiinae
Symbolanthus G. Don	30	Helieae	
Symphyllophyton Gilg	1	Chironieae	Coutoubeinae
Tachia Aubl.	10	Helieae	
Tachiadenus Griseb.	11	Exaceae	
Tapeinostemon Benth.	7	Saccifolieae	
Tetrapollinia Maguire & B. M. Boom	1	Helieae	
Tripterospermum Blume	24	Gentianeae	Gentianinae
Urogentias Gilg & Gilg-Ben.	1	Potalieae	Faroinae
Veratrilla Baill. ex Franch.	2	Gentianeae	Swertiinae
Voyria Aubl.	19	incertae sedis	
Voyriella (Miq.) Miq.	1	Saccifolieae	
Wurdackanthus Maguire	2	Helieae	
Xestaea Griseb.	1	Chironieae	Coutoubeinae
Zonanthus Griseb.	1	Helieae	
Zygostigma Griseb.	1	Chironieae	Chironiinae

assigned to particular genera. Only monophyletic tribes and subtribes within Gentianaceae have been recognized (Stevens, 1997), and we have made no assumption that any genus is monophyletic based on our data. Rather, we have attempted to maximize the diversity of taxa represented such that higher-order classificatory structure might be highlighted. Generic diversity was taken to be a placeholder for species diversity, considering the already fine-scaled generic circumscriptions accepted by Maguire (1981), Maguire and Boom (1989), and Struwe et al. (1999). Given this research focus, we have avoided the citation of genus-level synonymies in our proposed classification despite the fact that particular genera are implicitly recognized, leading to the equally implied rejection of others. This tacit recognition of sometimes poorly studied genera should be taken into account by all users of this classification. Additionally, some genera were placed to tribe or subtribe according to non-molecular evidence when DNA data were lacking. Accordingly, these assignments should be considered the most unstable points in our classification. Although it could not be attempted here, a detailed recircumscription of gentian species and genera is clearly in order. In some cases this is already underway (Helieae; *Centaurium*; *Gentianella*; *Swertia*; etc.), but for most genera discussed below, only estimates of species numbers could be provided.

Despite these caveats, the present treatment is the first comprehensive, suprageneric classification of Gentianaceae since Gilg (1895a). Its emphasis on monophyly and inclusion of maximum available gentian diversity should provide an anchor for future taxonomic research in the family. Our hope is that this effort will also stimulate detailed research on aspects of gentian character evolution, biogeography, and conservation. Gentianaceae are well suited for such issues considering their age, great morphological and anatomical variation, tendency toward narrow endemicity, and worldwide distribution in a variety of habitats.

MATERIALS, METHODS, AND NEW PHYLOGENETIC RESULTS

For the previously unpublished *mat*K sequences, DNA extraction, PCR, and sequencing were conducted as described by Thiv et al. (1999a). New sequences were added to those presented by Struwe et al. (1998) and Thiv et al. (1999a,b) and were used for the *trn*L intron-only, *mat*K-only, and *trn*L intron/*mat*K combined analyses. An overview of all sequences used in the three different analyses, with GenBank and voucher information, is given in Table 2.3. All sequences were aligned manually using Sequencher 3.0

Table 2.3. *Voucher information and Genbank accession numbers for trnL intron and matK DNA sequences presented and used in analyses in this chapter. Herbarium abbreviations follow Holmgren et al. (1990)*

Species	Family	matK Genbank no.	trnL Genbank no.	Voucher for new sequences
Alstonia boonei De Wild.	Apocynaceae		AF102374	
Anthocleista amplexicaulis Baker	Gentianaceae	AJ388137, AJ388206	AF102375	*B. Pettersson et al. 610* (UPS)
Anthocleista scandens Hook.f.	Gentianaceae	AJ388138, AJ388207	AF102376	*I. Friis et al. 4040* (UPS)
Anthocleista schweinfurthii Gilg	Gentianaceae	AJ388139, AJ388208	AF102377	*S. A. Thompson & J. E. Rawlins 1399* (NY)
Anthocleista vogelii Planch.	Gentianaceae			
Antirhea acutata (DC.) Urb.	Rubiaceae		AF102378	
Antonia ovata Pohl.	Loganiaceae	AJ388200, AJ388270	AF102379	*C. Gracie s.n.* (NY)
Apocynum cannabinum L.	Apocynaceae		AF102380	
Aripuana cullmaniorum Struwe, Maas, & V. A. Albert	Gentianaceae	AJ388140, AJ388209		*C. A. Cid Ferreira 5906* (NY)
Asclepias curassavica L.	Apocynaceae		AF102381	
Aspidosperma megalocarpon Muell. Arg.	Apocynaceae		AF102382	
Bartonia virginica (L.) Britton, Sterns, & Poggenb.	Gentianaceae	AJ388141, AJ388210	AF102383	*A. Kirschgessner 73* (NY)
Blackstonia imperfoliata (L.f.) Samp.	Gentianaceae	AJ010506, AJ011435	AF102384	
Bonyunia minor N. E. Br.	Loganiaceae		AF102385	
Bonyunia superba R. Schomb.	Loganiaceae		AF102386	
Calolisianthus pendulus (Mart.) Gilg	Gentianaceae		AF102387	
Calolisianthus pulcherrimus (Mart.) Gilg	Gentianaceae	AJ388142, AJ388211	AF102388	*R. M. Harley et al. 15674* (NY)
Canscora diffusa (Vahl) R. Br. ex Roem. & Schult.	Gentianaceae	AJ388143, AJ388212	AF102389	*Kokou s.n.* (TOGO)

Table 2.3. (cont.)

Species	Family	*matK* Genbank no.	*trnL* Genbank no.	Voucher for new sequences
Canscora pentanthera Clarke	Gentianaceae	AJ010507, AJ011436	AF102390	
Carissa bispinosa (L.) Desf. ex Brenan	Apocynaceae		AF102391	
Catharanthus roseus G. Don	Apocynaceae		AF102392	
Centaurium cf. *barrelieri* (Duf.) F. Q. & Rother.	Gentianaceae	AJ388144, AJ388213	AF102393	Mainz Botanical Garden, cult. no. 124 (562) (MJG)
Centaurium floribundum (Benth.) B. L. Rob.	Gentianaceae	AJ388145, AJ388214		*Roderick 710.439* (MJG)
Centaurium maritimum (L.) Fritsch	Gentianaceae	AJ010508, AJ011437	AF102394	
Cerberiopsis candelabra Vieill.	Apocynaceae		AF102395	
Chelonanthus alatus (Aubl.) Pulle	Gentianaceae	AJ010520, AJ011449	AF102396	
Chelonanthus albus (Spruce ex Progel) Badillo	Gentianaceae		AF102397	
Chelonanthus purpurascens (Aubl.) Struwe, S. Nilsson, & V. A. Albert	Gentianaceae	AJ388146, AJ388215		*G. Cremers 14561* (MJG)
Chelonanthus purpurascens (Aubl.) Struwe, S. Nilsson, & V. A. Albert	Gentianaceae		AF102398	
Chelonanthus viridiflorus (Mart.) Gilg	Gentianaceae		AF102399	
Chiococca alba (L.) Hitchc.	Rubiaceae	AJ010509, AJ011438	AF102400	
Chironia baccifera L.	Gentianaceae	AJ388147, AJ388216	AF102402	
Chorisepalum ovatum Gleason	Gentianaceae	AJ010510, AJ011439		*B. Maguire & L. Politi 27921* (NY)
Cicendia filiformis Delarbre	Gentianaceae		AF102403	
Cicendia quadrangularis (Lam.) Griseb.	Gentianaceae	AJ388148, AJ388217	AF102404	*P. J. M. Maas 8154* (U)

Species	Family			Voucher
Cinchona pubescens Vahl	Rubiaceae		Z70197	
Coffea arabica L.	Rubiaceae		AF102405	
Comastoma nana (Wulfen) N. M. Pritchard	Gentianaceae		X77890	
Comastoma tenellum (Rottb.) Toyok.	Gentianaceae	AJ388149, AJ388218	X77892	M. Thiv s.n. (MJG)
Condaminea corymbosa (R. & P.) DC.	Rubiaceae		AF102406	
Coutoubea minor H. B. K.	Gentianaceae	AJ010511, AJ011440	AF102407	
Coutoubea ramosa Aubl.	Gentianaceae	AJ388150, AJ388219	AF102408	
Coutoubea spicata Aubl.	Gentianaceae	AJ010512, AJ011441	AF102409	
Crawfurdia speciosa Wall.	Gentianaceae	AJ388151, AJ388220	AJ242606	
Curtia tenuifolia Knobl.	Gentianaceae	AJ388152, AJ388221		
Curtia verticillaris Knobl.	Gentianaceae	AJ388153, AJ388222	AF102410	W. A. Anderson 9385 (NY)
Deianira pallescens Cham. & Schltdl.	Gentianaceae			
Diocodendron dioicum (K. Schum. & Krause) Steyerm.	Rubiaceae		AF102411	
Ditassa sp.	Apocynaceae		AF102412	
Djaloniella ypsilostyla P. Taylor	Gentianaceae	AJ388154, AJ388223	AF102413	Morton SL2442 (K)
Echium vulgare L.	Boraginaceae		L33362	
Enicostema axillare (Lam.) A. Raynal	Gentianaceae	AJ010513, AJ011442		
Enicostema verticillatum (L.) Engl. ex Gilg	Gentianaceae	AJ388155, AJ388224	AF102414	J. Pruski & J. Steyermark 1473 (NY)
Eustoma exaltatum (L.) Salisb. ex G. Don	Gentianaceae		AF102415	
Eustoma grandiflorum (Raf.) Shinners	Gentianaceae	AJ010514, AJ011443	AF102416	
Exacum affine I. B. Balf. ex Regel	Gentianaceae	AJ010515, AJ011444	AF102417	K. Meyer 9644 (MJG)
Exacum tetragonum Roxb.	Gentianaceae	AJ388156, AJ388225	AF102418	

Table 2.3. (cont.)

Species	Family	matK Genbank no.	trnL Genbank no.	Voucher for new sequences
Fagraea berteroana A. Gray	Gentianaceae	AJ388157, AJ388226	AF102419	*L. Struwe 1219* (NY)
Fagraea elliptica Roxb.	Gentianaceae	AJ388158, AJ388227	AF102420	*W. Takeuchi 7122* (NY)
Fagraea fragrans Roxb.	Gentianaceae		AF102421	
Fagraea racemosa Jack ex Wall.	Gentianaceae	AJ010516, AJ011445		
Faramea multiflora A. Rich.	Rubiaceae		AF102422	*L. Andersson et al. 2041* (S)
Faroa axillaris Baker	Gentianaceae	AJ388159, AJ388228	AF102423	*M. Schaijes 5076* (BR)
Faroa schaijesiorum Bamps	Gentianaceae	AJ388160, AJ388229	AF102424	*M. Schaijes 3515* (BR)
Frasera albo-marginata S. Watson	Gentianaceae	AJ388230		*K. Gutsche 20* (MJG)
Frasera paniculata Torr.	Gentianaceae		AF102425	
Fraxinus americana L.	Oleaceae		X76812	
Gardenia taitensis DC.	Rubiaceae	AJ388201, AJ388271	AF102426	*L. Struwe & V. A. Albert 1208* (NY)
Gardenia thunbergia L.f.	Rubiaceae	Z70198		
Gardneria ovata Wall.	Strychnaceae		AF102427	
Gelsemium sempervirens (L.) Aiton	Gelsemiaceae	Z70195	AF102428	
Geniostemon gypsophilum B. L. Turner	Gentianaceae	AJ388161, AJ388231	AF102429	*G. Neson et al. 7621* (LL)
Geniostoma rupestre Forst.	Loganiaceae	Z70194	AF102430	
Gentiana acaulis L.	Gentianaceae		X77869	
Gentiana alpina Vill.	Gentianaceae		X77868	
Gentiana asclepiadea L.	Gentianaceae	AJ388165, AJ388235	X77871	Botanischer Garten München KG 37 (no voucher)
Gentiana bavarica L.	Gentianaceae		X77873	
Gentiana clusii Perr. & Song.	Gentianaceae		X77879	
Gentiana cruciata L.	Gentianaceae	AJ010519, AJ011448	AF102434	
Gentiana cruciata L.	Gentianaceae		X77880	

Gentiana cf. frigida Haenke	Gentianaceae	AJ388166, AJ388236	AF102435	K. Gutsche 42 (MJG)
Gentiana froelichii Jan	Gentianaceae		X77884	
Gentiana lutea L.	Gentianaceae		X75702	
Gentiana punctata L.	Gentianaceae		X77894	
Gentiana purpurea L.	Gentianaceae		X77893	
Gentiana pyrenaica L.	Gentianaceae		X77895	
Gentiana sedifolia H. B. K.	Gentianaceae	AJ388167, AJ388237	AF102436	J. Clarke et al. 1869 (MJG)
Gentiana terglouensis Hacq.	Gentianaceae		X77897	
Gentiana utriculosa L.	Gentianaceae		X77898	
Gentiana verna L.	Gentianaceae		X75704	
Gentianella anisodonta (Borbas) A. & D. Löve	Gentianaceae		X77870	
Gentianella aspera (Heg. & Heer) Dostal ex Skalicky, Chrtek, & Gill	Gentianaceae	AJ010517, AJ011446		
Gentianella aurea (L.) Harry Sm.	Gentianaceae	AJ388162, AJ388232	AF102431	
Gentianella bellidifolia (Hook.f.) Holub	Gentianaceae		AF102401	
Gentianella cerastioides (Kunth) Fabris	Gentianaceae	AJ010518, AJ011447		
Gentianella germanica (Willd.) Boerner	Gentianaceae		X77885	
Gentianella peruviana (Griseb.) Fabris	Gentianaceae	AJ388163, AJ388233	AF102432	Royal Botanic Garden Edinburgh, cult no 19932974 (no herbarium voucher)
Gentianopsis ciliata (L.) Ma	Gentianaceae	AJ388164, AJ388234	AF102433	Royal Botanic Garden Edinburgh, cult no 19950534 (no herbarium voucher)
Gentianopsis crinita (Froel.) Ma	Gentianaceae			K. Gutsche 52 (MJG)
Halenia corniculata (L.) Cornaz	Gentianaceae	AJ388168, AJ388238	AF102437	M. Togashi 518 (L)
Halenia palmeri A. Gray	Gentianaceae	AJ388169, AJ388239	AF102438	N. H. Holmgren & T. K. Lowrey 8073 (NY)
Halenia sp.	Gentianaceae		AF102439	
Hamelia papillosa Urb.	Rubiaceae		AF102440	
Hoppea dichotoma Willd.	Gentianaceae	AJ388170, AJ388240	X85796	C. D. K. Cook RHT307 (MJG)
Hydrangea paniculata Siebold	Hydrangeaceae			

Table 2.3. (cont.)

Species	Family	matK Genbank no.	trnL Genbank no.	Voucher for new sequences
Irlbachia poeppigii (Griseb.) L. Cobb & Maas	Gentianaceae		AF102441	
Irlbachia pratensis (H. B. K.) L. Cobb & Maas	Gentianaceae		AF102442	
Ixanthus viscosus (H. B. K.) Griseb.	Gentianaceae	AJ010521, AJ011450	AF102443	
Jaeschkea oligosperma (Griseb.) Knobl.	Gentianaceae	AJ388171, AJ388241	AF102444	R. McBeath 2300 (E)
Jasminum fluminense Vell.	Oleaceae	AJ388202, AJ388272	AF102445	L. Struwe & C. Specht 1098 (NY)
Labordia sp.	Loganiaceae		AF102446	
Labordia tinifolia A. Gray	Loganiaceae		AF102447	
Lehmanniella splendens (Hook.) Ewan	Gentianaceae	AJ388172, AJ388242		R. Callejas 8575 (NY)
Lisianthius jefensis A. Robyns & T. S. Elias	Gentianaceae	AJ010522, AJ011451	AF102448	
Lisianthius laxiflorus Urban	Gentianaceae	AJ388174	AF102449	
Lisianthius longifolius L.	Gentianaceae	AJ388173, AJ388243	AF102450	H. van der Werff 8690 (NY)
Lisianthius nigrescens Cham. & Schltdl.	Gentianaceae			Strybing Arboretum, San Francisco, 93-928 (no voucher)
Logania albiflora (Andrews & Jacks.) Druce	Loganiaceae	AJ388203, AJ388273	AF102451	H. Streimann 8916 (NY)
Lomatogonium carinthiacum (Wulf.) Rchb.	Gentianaceae		X77899	
Lomatogonium oreocharis (Diels) C. Marquand	Gentianaceae	AJ388174, AJ388244	AF102452	CLD -90 1106 (E)
Luculia cf. gratissima Sweet	Rubiaceae	Z70199	AF102453	

Species	Family			
Macrocarpaea domingensis Urb. & Ekman	Gentianaceae	AJ010523, AJ011452	AF102454	
Macrocarpaea cf. *glabra* (L.f.) Gilg	Gentianaceae		AF102455	
Macrocarpaea rubra Malme	Gentianaceae	AJ388175, AJ388245	AF102457	*A. C. Cervi 3825* (NY)
Macrocarpaea valerii Standley	Gentianaceae	AJ388176, AJ388246	AF102456	*S. R. Hill et al. 17751* (NY)
Megacodon stylophorus (C. B. Clarke) Harry Sm.	Gentianaceae	AJ388177, AJ388247	AF102458	*B. Alden et al. 1378* (E)
Microrphium pubescens C. B. Clarke	Gentianaceae	AJ388178, AJ388248		*C. E. Ridsdale SMHI 240A* (L)
Mitrasacme pilosa Labill.	Loganiaceae		AF102459	
Mitreola petiolata (Walt.) Torr. & Gray	Loganiaceae		AF102460	
Neblinantha parvifolia Maguire	Gentianaceae	AJ388179, AJ388249	AF102461	*B. Maguire et al. 42384* (NY)
Neuburgia corynocarpa (A. Gray) Leenh.	Loganiaceae		AF102462	
Neurotheca loeselioides (Spruce ex Progel) Baill.	Gentianaceae	AJ010524, AJ011453	AF102463	
Obolaria virginica L.	Gentianaceae	AJ388180, AJ388250	AF102464	*V. E. McNeilus 93-46* (NY)
Oreonesion testui A. Raynal	Gentianaceae	AJ388181		*J. M. G. Davies & J. Anton-Smith 242* (K)
Ornichia madagascariensis (Baker) Klack.	Gentianaceae	AJ388182, AJ388252	AF102465	*P. B. Philipson 3973* (BR)
Orphium frutescens (L.) E. Mey.	Gentianaceae	AJ010525, AJ011454	AF102466	
Pauridiantha sp.	Rubiaceae		AF102467	
Periploca graeca L.	Apocynaceae		AF102468	
Petunia × *hybrida* Hort. Vilm.	Solanaceae		X74572	
Portlandia platantha Hook.f.	Rubiaceae		AF102469	
Potalia amara Aubl.	Gentianaceae		AF102470	
Potalia amara Aubl.	Gentianaceae	AJ388183, AJ388253	AF102471	*S. Mori 24102* (NY)
Potalia resinifera Mart.	Gentianaceae	AJ388184, AJ388254	AF102472	*A. Henderson 2034* (NY)
Potalia resinifera Mart.	Gentianaceae			

Table 2.3. (cont.)

Species	Family	*mat*K Genbank no.	*trn*L Genbank no.	Voucher for new sequences
Pycnosphaera buchananii (Baker) N. E. Br.	Gentianaceae	AJ388185, AJ388255	AF102473	*Bingham MG9370* (K)
Rondeletia amoena (Planch.) Hemsl.	Rubiaceae		AF102474	
Rubia fruticosa Aiton	Rubiaceae		AF102475	
Sabatia angularis (L.) Pursh	Gentianaceae	AJ010526, AJ011455	AF102476	
Sabatia gentianoides Ell.	Gentianaceae	AJ388186, AJ388256	AF102477	*R. Dale Thomas 124383* (MO)
Saccifolium bandeirae Maguire & Pires	Gentianaceae	AJ388187, AJ388257	AJ242608	
Schinziella tetragona (Schinz) Gilg	Gentianaceae	AJ010527, AJ011456	AF102479	
Schultesia guianensis (Aubl.) Malme	Gentianaceae	AJ388188, AJ388258	AF102480	*C. C. Berg & A. Henderson BG 661* (NY)
Sebaea exacoides (L.) Schinz	Gentianaceae	AJ388189, AJ388259	AF102482	*Snijman 1562* (NBG, PRE)
Sebaea cf. *macrophylla* Gilg	Gentianaceae	AJ388190, AJ388260	AF102481	*R. D. A. Bayliss 8765* (NY)
Stapelia peglerae N. E. Br.	Apocynaceae		AF102483	
Strumpfia maritima Jacq.	Rubiaceae	AJ388204, AJ388274	AF102485	*L. Struwe 1101* (NY)
Strychnos nux-vomica L.	Loganiaceae	Z70193		
Swertia tomentosa Benth.	Loganiaceae	AJ388205, AJ388275	AF102484	*S. Mori 24166* (NY)
Swertia abyssinica Hochst.	Gentianaceae	AJ388191, AJ388261		*Carvalho 3654* (BR)
Swertia marginata Schrenk	Gentianaceae	AJ388192, AJ388262	AF102486	*H. Huss 252d* (M)
Swertia perennis L.	Gentianaceae	AJ010528, AJ011457	X75708	
Symbolanthus calygonus (Ruiz & Pav.) Griseb.	Gentianaceae	AJ010529, AJ011458		
Symbolanthus cf. *calygonus* (Ruiz & Pav.) Griseb.	Gentianaceae		AF102489	

Symbolanthus pulcherrimus Gilg	Gentianaceae		AF102488
Symbolanthus sp.	Gentianaceae		AF102487
Symphyllophyton caprifolioides Gilg	Gentianaceae	AJ010530, AJ011459	AF102490
Tachia grandifolia Maguire & Weaver	Gentianaceae	AJ388193, AJ388263		M. Nee 42892 (NY)
Tachia guianensis Aubl.	Gentianaceae	AJ011433, AJ011461	
Tachia loretensis Maguire & Weaver	Gentianaceae		AF102492
Tachiadenus carinatus (Desr.) Griseb.	Gentianaceae	AJ011434, AJ011460	AF102491
Telosma cordata Merr.	Apocynaceae		AF102493
Tetrapollinia caerulescens (Aubl.) Maguire & B. M. Boom	Gentianaceae		AF102494
Tetrapollinia caerulescens (Aubl.) Maguire & B. M. Boom	Gentianaceae	AJ388194, AJ388264		G. Cremers 14576 (MJG)
Thevetia peruviana (Pers.) K. Schum.	Apocynaceae	Z70188	
Urogentias ulugurensis Gilg & Gilg-Ben.	Gentianaceae	AJ388195, AJ388265	AF102495	D. J. Mabberley 1432 (K)
Usteria guineensis Willd.	Loganiaceae		AF102496
Valeriana officinalis L.	Valerianaceae		X85795
Veratrilla baillonii Franch.	Gentianaceae	AJ388196, AJ388266	AF102497	B. Alden et al. 1326 (E)
Voyriella parviflora Miq.	Gentianaceae	AJ388197, AJ388267	AJ242607
Wurdackanthus frigidus (Sw.) Maguire & B. M. Boom	Gentianaceae	AJ388198, AJ388268	AF102498	G. R. Cooley 8211 (NY)
Xestaea lisianthoides Griseb.	Gentianaceae	AJ388199, AJ388269		D'Arcy 14288 (U)

(GeneCodes), and alignments are available for viewing and downloading at http://www.rci.rutgers.edu/~deenr/LS.html. Gaps were coded as missing data. The matrices were analyzed using NONA (Goloboff, 1998), with the following options: "HOLD 10000" (the maximum number of trees to be kept), "MULT*25" (which specifies 25 random addition replicates with tree bisection-reconnection (TBR) branch swapping), followed by "MAX*" (which indicates TBR swapping on all shortest trees in memory). Some analyses were not run to completion (see below). Numbers of steps and consistency and retention indices were calculated using data from NONA. Parsimony jackknife support analysis (Farris *et al.*, 1996) was performed using the XAC application (J. S. Farris, unpubl.), specifying 1000 replicates, subtree pruning-regrafting (SPR) branch swapping, and five random entry orders per replicate. Groups supported by 50% or less of jackknife replicates are ambiguous, whereas those between 50% and *c.* 63% (actually, $1-e^{-1}$, e^{-1} being the character removal rate for each jackknife replicate) will show some robustness to extra steps (i.e., Bremer support; Bremer, 1988). Values at or above *c.* 63% (given sampling error related to number of replicates) represent support by the equivalent of one uncontradicted synapomorphy.

The first analysis, including only *trn*L intron sequences, was stopped after finding 3902 most-parsimonious trees (1290 steps, $C=0.55$, $R=0.82$). These resulted from swapping on shortest trees found among the initial 25 random addition sequence replicates. A strict consensus tree with jackknife values is shown in Fig. 2.1. Analysis of the second matrix, *mat*K sequences, resulted in 9817 most-parsimonious trees, and this search was also terminated prior to completion (1780 steps, $C=0.47$, $R=0.75$; consensus tree shown in Fig. 2.2). Analysis of the smaller, combined matrix of taxa for which *trn*L intron and *mat*K sequences were available gave 56 most-parsimonious trees, swapped to completion (2144 steps, $C=0.55$, $R=0.75$), and the resulting strict consensus tree is the most resolved of all the analyses (Fig. 2.3).

New palynological investigations were executed by S. Nilsson. Scanning electron microscopy (SEM) and light microscopy (LM) studies were performed following protocols detailed by Nilsson (2002 (Chapter 4, this volume)). For each newly investigated taxon, voucher information is provided under each pollen description, including country of collection, approximate locality, year of collection, collector and number, herbarium where voucher is deposited (abbreviations following Holmgren *et al.*, 1990), and slide accession number in the permanent collection of the Palynological Laboratory at the Swedish Museum of Natural History, Stockholm.

FAMILY GENTIANACEAE

Description of Gentianaceae

Most Gentianaceae are trees, shrubs, vines, or perennial or annual herbs. Saprophytes either occur without chlorophyll (*Cotylanthera*, *Voyria*, and *Voyriella*) or are weakly green (*Bartonia* and *Obolaria*). The plants are usually glabrous, have no latex, but often have colleters (finger-shaped multicellular glands) at the inner (adaxial) bases of the leaves and calyces. The stems and branches are terete or quadrangular and are often winged. The leaves are opposite (rarely alternate or verticillate) with entire margins (rarely dentate), and are simple, saccate (*Saccifolium*), or rarely reduced to scales (*Bartonia*, *Cotylanthera*, *Voyria*, and *Voyriella*). Leaf venation is usually characterized by a few pairs of secondary veins diverging close to the base of the leaf and arching toward the leaf tip (acrodromy), but pinnate venation with short and straight or curved secondary veins occurs as well (brochidodromy). No stipules are present, but interpetiolar lines or sheaths are often prominent.

The inflorescences are either terminal or axillary, and are usually cymose, but less often racemose, capitate, clustered, or spicate; the flowers are seldom solitary. The bisexual flowers (unisexual in *Veratrilla*) are hypogynous, 4- or 5-merous, or rarely 3- or 6–16-merous. Cleistogamous flowers are known only from *Sebaea oligantha*. The sepals are usually fused at their bases. The calyx is actinomorphic, rarely zygomorphic (e.g., *Exacum* spp.), 4- or 5-merous, or rarely with 2, 3, or 6–8 lobes. The calyx lobes are imbricate in bud (decussate in *Anthocleista*, *Chorisepalum*, and *Potalia*), and each lobe sometimes has a dorsal wing, keel, or glandular area. The sympetalous corollas are usually actinomorphic or less often slightly zygomorphic (e.g., *Chelonanthus* and *Symbolanthus* spp.), and are tubular, campanulate, to salver- or funnel-shaped. The corolla lobes are nearly always contorted in bud (valvate at the corolla lobe bases in *Aripuana* and imbricate in *Bartonia*, *Obolaria*, and *Saccifolium*). The corollas are sometimes nectariferous (e.g., *Frasera*, *Gentianella*, and *Swertia*).

The stamens are isomerous with the corolla lobes (reduced in number in *Canscora pro parte*, *Hoppea*, and *Schinziella*), and are inserted in the corolla tube or rarely in the sinuses of the corolla lobes (e.g., *Aripuana*, *Djaloniella*, *Faroa*, and *Swertia* spp.). The androecium as a whole may be either actinomorphic (the usual state) or zygomorphic (e.g., *Chelonanthus*, *Exacum* spp., *Orphium*, and *Symbolanthus*). The filaments of the stamens are nearly always free (at least partially fused in *Anthocleista*, *Exacum* spp., and

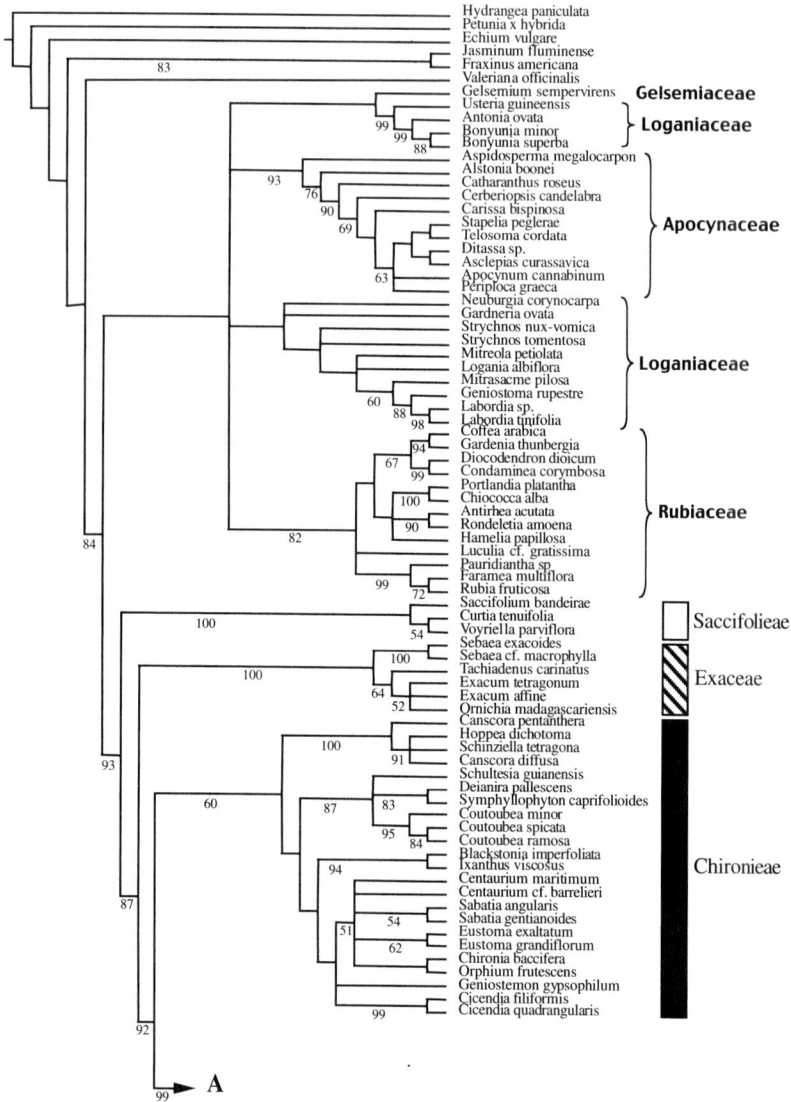

Figure 2.1. Strict consensus tree from an analysis of *trn*L intron sequences of Gentianales, particularly Gentianaceae, derived from 3902 most-parsimonious trees (1290 steps, $C=0.55$, $R=0.82$). Jackknife values ≥50% are shown above each branch. The data matrix is the same as in Thiv *et al.* (1999a).

Systematics, character evolution, and biogeography 41

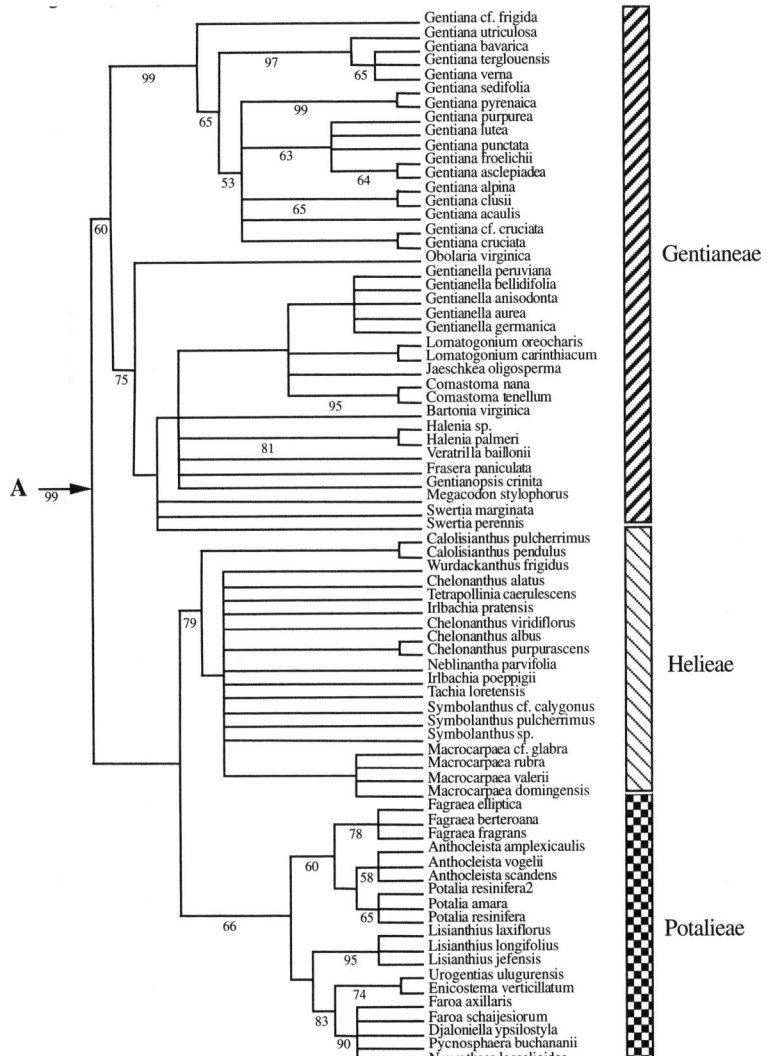

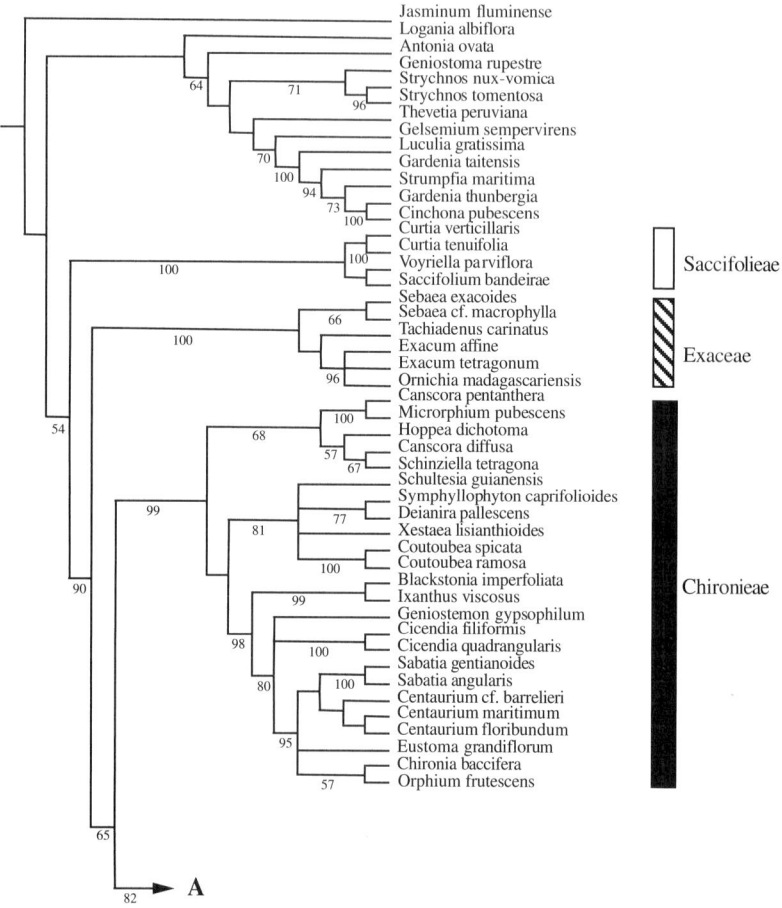

Figure 2.2. Strict consensus tree from an analysis of *mat*K sequences of Gentianales, particularly Gentianaceae, derived from 9817 most-parsimonious trees (1780 steps, $C=0.47$, $R=0.75$). Jackknife values $\geq 50\%$ are shown above each branch.

Potalia), and bear free or rarely connate anthers that are often sagittate. Anther dehiscence is longitudinal (porate in *Cotylanthera* and *Exacum* spp.) and usually introrse (latrorse in some *Voyriella* individuals and in *Saccifolium*). The bicarpellate gynoecium is syncarpous with a variety of placentation forms. There are often glands or a glandular disk at the base of the ovary. There is one terminal style and the stigmas can be simple, capitate, peltate, or bilamellate.

Systematics, character evolution, and biogeography 43

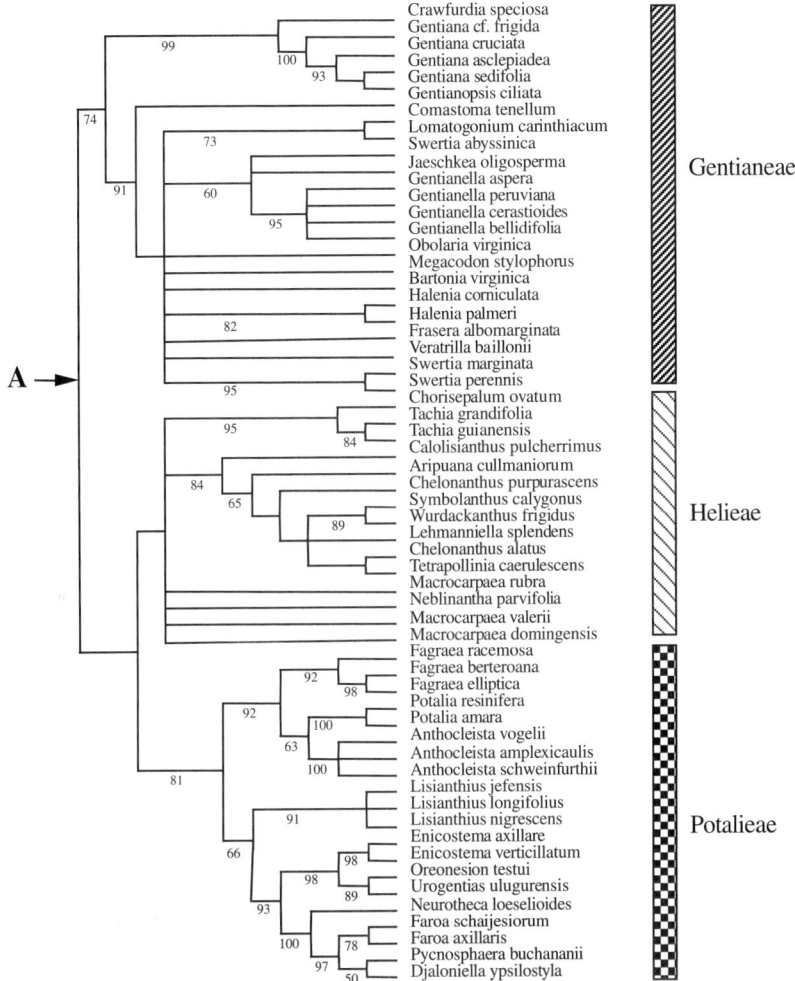

The fruits are usually dry and thin, less often woody or fibrous, and rarely leathery (*Symbolanthus*) or fleshy (*Anthocleista*, *Chironia* spp., *Fagraea*, *Potalia*, and *Tripterospermum* spp.). Dehiscence is medial or apical, or indehiscent (most species of the fleshy-fruited genera and *Voyria* spp.). The seeds are usually small, rounded or angular, winged or not, and without an aril (see Bouman *et al*., 2002 (Chapter 5, this volume), for more seed data). Extremely reduced seeds are found in *Voyria*.

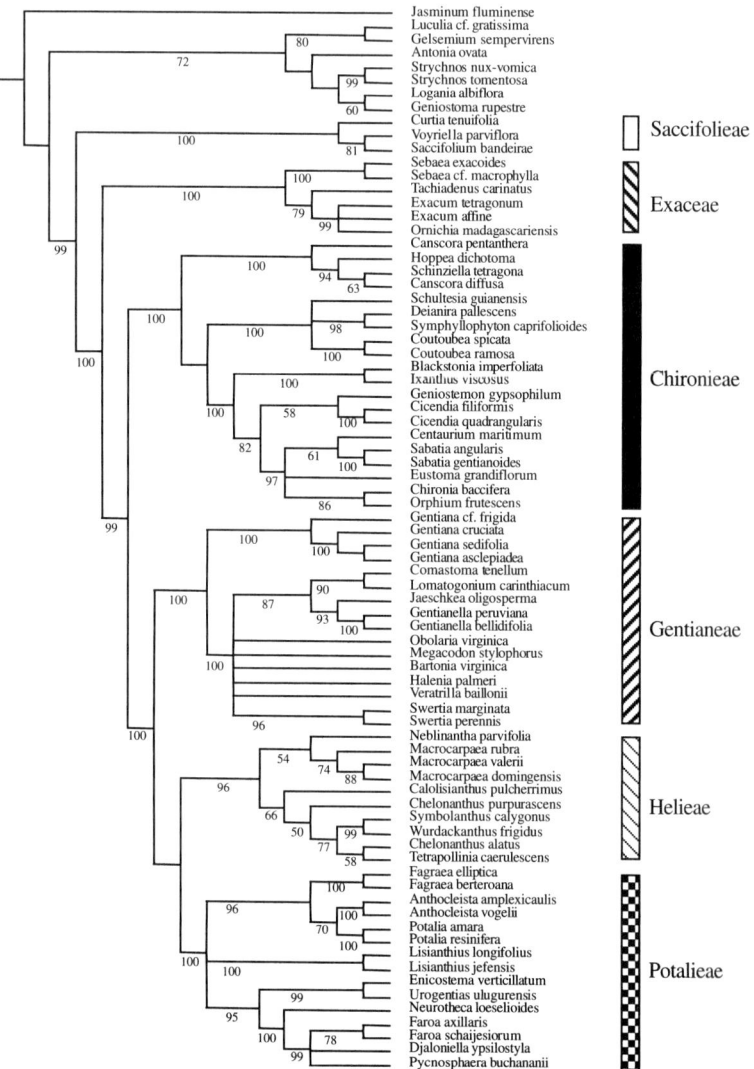

Figure 2.3. Strict consensus tree from an analysis of a combined matrix of taxa of the Gentianales for which both *mat*K and *trn*L intron sequences are available. The analysis resulted in 9817 most-parsimonious trees (1780 steps, $C=0.47$, $R=0.75$). Jackknife values $\geq 50\%$ are shown above each branch.

Systematics, character evolution, and biogeography 45

The pollen of Gentianaceae is very variable, especially in exine characteristics (Nilsson, 1968, 1970, 2002; see Appendix 2.2 for an explanation of palynological terms). The pollen grains are released as monads, tetrads (Chironieae-Coutoubeinae, some Helieae and *Potalia* spp.), or polyads (some Helieae), the last consisting of loosely to firmly united tetrads. Single grains are radially symmetrical, oblate to prolate, or rarely bilateral to asymmetric (*Voyria*). The pollen is usually 3-colporate (colpate to porate), or (2-)3(-6)-porate (*Anthocleista, Potalia*, and *Voyria*), with colpi and pores usually bordered by distinct margins or annuli. The ora (endopores) may have lateral extensions. The exine may be striate, striato-reticulate, reticulate, rugulate, foveolate, smooth, scabrate, perforate, imperforate, looped at the equatorial or polar areas, or vested with various surface processes, e.g., verrucae, pila, clavae, or granules (especially in tribe Helieae).

Compound pollen grains (i.e., tetrads or polyads) occur in Chironieae-Coutoubeinae, Helieae, and Potalieae. Striate to striato-reticulate and reticulate patterns are by far the most common exine features and occur in all tribes of Gentianaceae. Muri/lirae are either sharply crested to rounded in cross-section, parallel or oriented in various directions, of variable lengths, or criss-crossed and interlaced. At high magnification striation may dominate whereas a reticulate pattern due to transverse muri interconnecting the muri/lirae can appear at lower magnification (S. Nilsson, pers. obs.). Lateral extensions of the ora (endoapertures) are the result of ruptures and lack of nexine (endexine), and occur in all tribes.

KEY TO THE TRIBES AND SUBTRIBES OF GENTIANACEAE

1a Shrubs and trees always with fleshy fruits ... **Potalieae-Potaliinae** (p. 191)
1b Herbs with dry fruits (fleshy in *Chironia* spp. and *Tripterospermum* spp.), if trees or shrubs, then with dry fruits or rarely leathery fruits (*Symbolanthus*) 2

2a Plants without chlorophyll (and always with calyx tube clearly longer than calyx lobes); pollen usually reniform, rarely isodiametric, 2-6-porate (never in tetrads) .... *incertae sedis* (*Voyria*) (p. 266)
2b Plants with at least some chlorophyll, or if without chlorophyll (*Cotylanthera* and *Voyriella*), then with calyx tube shorter than or at most as long as calyx lobes; pollen never reniform, 3-colporate or 3-colpate (3-porate in some taxa with pollen in tetrads) 3

3a Calyx with dorsally thickened, glandular areas on lobes (rarely keeled); style flattened and twisted lengthwise after anthesis and in fruit; stigma bilamellate; staminal filaments strongly recurved after anthesis (except in *Aripuana* and "Roraimaea") .... **Helieae** (p. 137)

3b Calyx without dorsal glandular areas on lobes, but sometimes strongly keeled; style not flattened and not twisted lengthwise after anthesis and in fruit; stigma simple, capitate, peltate, to linear (bilamellate in *Eustoma* and *Centaurium* spp., decurrent along carpel sutures in *Bartonia* and *Lomatogonium* spp.); staminal filaments straight or only slightly bent (recurved in *Gentiana* sect. *Stenogyne* and *Tripterospermum*) 4

4a Gynoecium clearly bilocular for at least half of its length (pseudo-bilocular in *Gentianothamnus* and *Tachiadenus*); testa cells star- or puzzle-shaped (except *Cotylanthera*, *Exacum* spp., and *Sebaea* spp.); anthers often with apical or basal glands/knobs .... **Exaceae** (p. 66)

4b Gynoecium unilocular or pseudo-bilocular (i.e., two locules formed by parietal placentas intruding into center of ovary, e.g., *Centaurium*, *Halenia*, and *Saccifolium*); testa cells with straight or undulating cell walls; anthers never with glands/knobs, but sometimes with well-developed connectives 5

5a Anthers with broad and often extended connectives; plants heterostylous (unknown in *Saccifolium* and most *Tapeinostemon*) ..... **Saccifolieae** (p. 55)

5b Anthers without broad or extended connectives; plants never heterostylous 6

6a Corolla with plicae (elongated folds between corolla lobes, except *Gentiana lutea*); with membrane between calyx lobes (except *Crawfurdia* spp., *Gentiana* spp., and *Tripterospermum*) ... **Gentianeae-Gentianinae** (p. 222)

6b Corolla always without plicae; without membrane between calyx lobes (except *Gentianopsis*) 7

7a Ovules usually distributed over entire inner wall of gynoecium (except in *Halenia*, with large distinct placentas, and *Swertia* spp., with ovules not covering the entire inner wall); with one or two corolline nectaries per petal lobe (or nectaries on spurred petals), or when glands present at the base of the ovary then filaments inserted in sinuses of corolla lobes (nectaries at base of ovary missing in *Bartonia* and *Pterygocalyx*); seeds over 0.5 mm long (except *Bartonia* and *Obolaria*) .... **Gentianeae-Swertiinae** (p. 222)

Systematics, character evolution, and biogeography   47

7b Ovules always on distinct placentas or on indistinct placentas along carpel sutures only; corolla never with corolline glands or spurs, but nectariferous disk can be present at base of ovary and then stamens inserted in corolla tube (except *Djaloniella*, *Faroa*, and *Pycnosphaera*); seeds less than 0.5 mm long (rarely more, e.g., *Lisianthius* and *Urogentias*) 8

8a Pollen released as tetrads; corolla always 4-merous
.... **Chironieae-Coutoubeinae** (p. 108)
8b Pollen released as monads; corolla 3-merous or 5–13-merous (4-merous flowers present in *Canscora*, *Centaurium* spp., *Cicendia*, *Cracosna*, *Exaculum*, *Geniostemon*, *Hoppea*, rarely *Ixanthus*, *Sabatia* spp., and *Schinziella*) 9

9a Corolla with corolline fibers that are easily seen in dried flowers as thin white lines; with nectariferous disk at base of ovary (rarely absent in *Lisianthius* spp.) .... **Potalieae-Lisianthiinae** (p. 191)
9b Corolla without corolline fibers (except *Neurotheca*); without nectaries at base of ovary 10

10a Flowers sessile or subsessile, solitary or clustered in leaf axils or in capitate heads; appendages present in the corolla tube at the base of each stamen (absent in *Congolanthus*, *Neurotheca*, and possibly *Urogentias*) .... **Potalieae-Faroinae** (p. 191)
10b Flowers pedicellate (sessile in *Centaurium* spp.), in lax inflorescences (except *Canscora* spp.); staminal appendages never present in corolla tubes 11

11a Calyx tubes always longer than calyx lobes; corolla white or cream-colored, never purple or blue (yellow in *Phyllocyclus* spp. and *Schinziella*; pink to purple in *Canscora* spp.); anthers straight after anthesis, never spirally twisted .... **Chironieae-Canscorinae** (p. 108)
11b Calyx tubes shorter than calyx lobes (except *Centaurium* spp., *Chironia* spp., and *Cicendia*); corolla pink, yellow, purple, or blue (white or cream in *Centaurium* spp., *Chironia* spp., *Eustoma* spp., *Geniostemon* spp., and *Zygostigma*); anthers sometimes spirally twisted after anthesis (not twisted in *Bisgoeppertia*, *Blackstonia*, *Cicendia*, *Exaculum*, *Geniostemon*, *Ixanthus*, and *Zygostigma*) .... **Chironieae-Chironiinae** (p. 108)

A NEW TRIBAL AND SUBTRIBAL CLASSIFICATION OF THE GENTIANACEAE

Presented here is the first complete infrafamilial classification of the Gentianaceae since Gilg (1895a). This classification is based on the principle of monophyly and on current knowledge of phylogenetic relationships, which rests mainly on molecular data. The number of currently accepted species in each genus is indicated in parentheses after each genus. Genera that have not yet been analyzed with molecular markers (marked with *) have been placed according to morphological, anatomical, and palynological characteristics.

GENTIANACEAE Juss., Gen. Pl: 141, 1789, *nom. cons.* — subfamily Gentianoideae Gilg in Engler & Prantl, Nat. Pflanzf. 4(2): 62, 1895. TYPE: *Gentiana* L.
Coutoubeaceae Martinov, Tekhno-Bot. Slovar: 168, 1820. TYPE: *Coutoubea* Aubl.
Obolariaceae Martinov, Tekhno-Bot. Slovar: 427, 1820. TYPE: *Obolaria* L.
Potaliaceae Mart., Nov. Gen. Sp. Plant. 2: 89, 133, 1827 (as Potalieae). TYPE: *Potalia* Aubl.
Chironiaceae Horan., Tetractys: 27, 1843. TYPE: *Chironia* L.
Saccifoliaceae Maguire & Pires, Mem. New York Bot. Gard. 29: 242, 1978. — subfamily Saccifolioideae Thorne, Nord. J. Bot. 3: 108, 1983, *nom. nud.* TYPE: *Saccifolium* Maguire & Pires.
Number of tribes: 6.
Number of subtribes: 8 in three tribes.
Number of genera: 87.
Number of species: *c.* 1615–1688.

1. Tribe Saccifolieae (Maguire & Pires) Struwe, Thiv, V. A. Albert, & Kadereit, *stat. nov.*
Basionym: Saccifoliaceae Maguire & Pires, Mem. New York Bot. Gard. 29: 242, 1978. TYPE: *Saccifolium* Maguire & Pires.
Included genera:
Curtia Cham. & Schltdl. (*c.* 6–10)
*Hockinia** Gardner (1)
Saccifolium Maguire & Pires (1)
*Tapeinostemon** Benth. (7)
Voyriella (Miq.) Miq. (1)
Number of genera: 5.

Number of species: c. 16–20.

2. **Tribe Exaceae** Colla, Herb. Pedem., vol. 3: 174, 1834 [Jan–Feb 1835]. — subtribe Exacinae (Colla) Gilg in Engler & Prantl, Nat. Pflanzf. 4(2): 62, 1895. — family Exacaceae Daniel & Sabnis, Curr. Sci. 47: 111, 1978, *nom. inval.* TYPE: *Exacum* L.
subtribe Sebaeinae Rchb., Handb. Nat. Pflanz.: 210, 1837 (as Sebaearieae). — tribe Sebaeeae (Rchb.) Endl., Gen. Pl. 1: 604, 1838. TYPE: *Sebaea* Sol. ex R. Br.
Included genera:
*Cotylanthera** Blume (4)
Exacum L. (65)
*Gentianothamnus** Humbert (1)
Ornichia Klack. (3)
Sebaea Sol. ex R. Br. (c. 60–100)
Tachiadenus Griseb. (11)
Number of genera: 6.
Number of species: c. 144–184.

3. **Tribe Chironieae** (G. Don) Endl., Gen. Pl. 1: 600, Aug 1838. TYPE: *Chironia* L.
Number of subtribes: 3.
Number of genera: 23.
Number of species: c. 159.

3.1. **subtribe Chironiinae** G. Don, Gen. Syst. Dichl. Pl. 4: 173, Mar–Apr 1838 (as Chironiae). — tribe Chironieae (G. Don) Endl., Gen. Pl. 1: 600, Aug 1838. TYPE: *Chironia* L.
subtribe Erythraeinae Rchb., Handb. Nat. Pflanz.: 210, 1837 (as Erythraeariae), *nom. illeg.* — tribe Erythreae (Rchb.) Griseb., Gen. Sp. Gent.: 135–136, 1839 [Oct 1838] (as Erythraeaceae), *nom. illeg.* — series Erythraeeae (Rchb.) Griseb. in DC., Prodr. 9: 54, 1845 (as Erythraeaceae), *nom. illeg.* TYPE: *Erythraea* Borkh., *nom. illeg.* (= *Centaurium* Hill).
tribe Chloreae Griseb., Gen. Sp. Gent.: 115–116, 1839 [Oct 1838]. — subtribe Chlorinae (Griseb.) Griseb. in DC., Prodr. 9: 49, 1845 (as Chloreae). TYPE: *Chlora* Adans. (= *Blackstonia* Huds.).
series Teleiandrae Gilg-Ben., Notizbl. Bot. Gart. Mus. Berlin-Dahlem 14: 424, 1839, *nom. inval.*
series Sabatieae Griseb. in DC., Prodr. 9: 49, 1845 (as Sabbatieae). TYPE: *Sabatia* Adans.

Included genera:
Bisgoeppertia* Kuntze (2)
Blackstonia Huds. (4)
Centaurium Hill. (c. 50)
Chironia L. (15)
Cicendia Adans. (2)
Eustoma Salisb. (3)
Exaculum* Caruel (1)
Geniostemon Engelm. & A. Gray (5)
Ixanthus Griseb. (1)
Orphium E. Mey. (2)
Sabatia Adans. (20)
Zygostigma* Griseb. (1)
Number of genera: 12.
Number of species: c. 106.
Nomenclatural notes: A tribal name based on *Erythraea* was published before the name Chironieae. However, *Erythraea* shares the same type as *Centaurium* and is therefore *nomen superfluum* and not a legitimate name, and infrafamilial names cannot be based on illegitimate names. The series Teleiandrae was never validly published (since it was not based on a valid generic name), and so should not be typified.

3.2. subtribe **Canscorinae** Thiv & Kadereit, *subtrib. nov.* TYPE: *Canscora* Lam.

Subtribus Canscorinae Thiv & Kadereit, subtrib. nov., a subtribibus Coutoubeinis et Chironiinis staminibus specierum plurimarum minus numerosis et minoribus, ulterius a Coutoubeinis pollinum granis singulis (nec in tetradibus dispositis) tuboque calycino elongato (nec abbreviato) abstans.
series Ellipandrae Gilg-Ben., Notizbl. Bot. Gart. Mus. Berlin-Dahlem 14: 424–425, 1839, *nom. inval.*
Included genera:
Canscora Lam. (9)
Cracosna* Gagnep. (3)
"Duplipetala", ined. (M. Thiv, unpubl.) (2)
Hoppea Willd. (2)
Microrphium C. B. Clarke (2)
Phyllocyclus Kurz (5)
Schinziella Gilg (1)
Number of genera: 6.
Number of species: 24.

Systematics, character evolution, and biogeography 51

Nomenclatural note: The series Ellipandrae was never validly published (since it was not based on a valid generic name) and so should not be typified. "Duplipetala" is not yet described and is not included in the generic count. However, the two species of this genus are already described as *Canscora* species and are included in the count for that genus.

3.3. subtribe Coutoubeinae G. Don, Gen. Syst. Dichl. Pl. 4: 174, Mar–Apr 1838 (as Coutoubeae). TYPE: *Coutoubea* Aubl.
Included genera:
Coutoubea Aubl. (5)
Deianira Cham. & Schltdl. (7)
Schultesia Mart. (15)
Symphyllophyton Gilg (1)
Xestaea Griseb. (1)
Number of genera: 5.
Number of species: 29.

4. **Tribe Helieae** Gilg in Engler & Prantl, Nat. Pflanzf. 4(2): 95, 1895. TYPE: *Helia* Mart.
tribe Lisyantheae Griseb., Gen. Sp. Gent.: 172–173, 1839 [Oct 1838], *nom. inval.* (?). —subtribe Lisyanthinae (Griseb.) Griseb. in DC., Prodr. 9: 70, 1845 (as Lisiantheae), *nom. inval.* (?). TYPE: *Lisyanthus* Aubl., non *Lisianthius* P. Browne (= *Chelonanthus* Gilg; see Struwe & Albert, 1998a, for further discussion and explanation).
subtribe Tachiinae Gilg in Engler & Prantl, Nat. Pflanzf. 4(2): 90, 1895. TYPE: *Tachia* Aubl.
tribe Rusbyantheae Gilg in Engler & Prantl, Nat. Pflanzf. 4(2): 95, 1895. TYPE: *Rusbyanthus* Gilg (= *Macrocarpaea* (Griseb.) Gilg).
Included genera:
Adenolisianthus* (Progel) Gilg (1)
Aripuana Struwe, Maas, & V. A. Albert (1)
Calolisianthus Gilg (6)
Celiantha* Maguire (3)
Chelonanthus Gilg (c. 7)
Chorisepalum Gleason & Wodehouse (5)
Helia* Mart. (2)
Irlbachia Mart. (9)
Lagenanthus* Gilg (1)
Lehmanniella Gilg (2)
Macrocarpaea (Griseb.) Gilg (c. 90)

Neblinantha Maguire (2)
Prepusa Mart. (5)
*Purdieanthus** Gilg (1)
*Rogersonanthus** Maguire & B. M. Boom (3)
"Roraimaea", ined. (L. Struwe, S. Nilsson, & V. A. Albert, unpubl.) (1)
*Senaea** Taub. (1)
*Sipapoantha** Maguire & B. M. Boom (1)
Symbolanthus G. Don (*c.* 30)
Tachia Aubl. (10)
Tetrapollinia Maguire & B. M. Boom (1)
Wurdackanthus Maguire (2)
*Zonanthus** Griseb. (1)
Number of genera: 22.
Number of species: *c.* 184.
Nomenclatural note: The genus "Roraimaea" is not yet described and is not included in the genus or species counts.

5. **Tribe Potalieae** Rchb., Handb. Nat. Pfl.-Syst.: 211, 1–7 Oct 1837. TYPE: *Potalia* Aubl.
Number of subtribes: 3.
Number of genera: 13.
Number of species: 154.
Nomenclatural note: Reichenbach's name Potalieae has not been cited previously (to the best of our knowledge); instead the first publication of Potalieae has commonly been cited as Endlicher (1838: 576; cf. Leeuwenberg & Leenhouts, 1980). However, Reichenbach published Potalieae before Endlicher as an unranked name, and since we accept other Reichenbach names (e.g., Swertiinae) in this classification, we have chosen to accept Potalieae Rchb. as well (see also http://www.inform.umd.edu/PBIO/fam/rchbur.html).

5.1. **subtribe Faroinae** Struwe & V. A. Albert, *subtrib. nov.* TYPE: *Faroa* Welw.
A subtribu Potaliinis (Mart.) Progel habitu herbaceo fructibus capsularibus foliis que minoribus differt.
tribe Hippieae Griseb., Gen. Sp. Gent.: 129–130, 1839 [Oct 1838]. TYPE: *Hippion* Spreng. (= *Enicostema* Bl.).
Included genera:
*Congolanthus** A. Raynal (1)
Djaloniella P. Taylor (1)

Systematics, character evolution, and biogeography 53

Enicostema Bl. (3)
Faroa Welw. (19)
*Karina** Boutique (1)
Neurotheca Salisb. ex Benth. in Benth. & Hook.f. (3)
Oreonesion A. Raynal (1)
Pycnosphaera Gilg (1)
Urogentias Gilg & Gilg-Ben. (1)
Number of genera: 9.
Number of species: 31.

5.2. subtribe Lisianthiinae G. Don, Gen. Syst. Dichl. Pl. 4: 175, Mar–Apr 1838 (as Lisiantheae). TYPE: *Lisianthius* P. Browne, non *Lisyanthus* Aubl.
Included genus:
Lisianthius P. Browne (30)
Number of genera: 1.
Number of species: 30.

5.3. subtribe Potaliinae (Mart.) Progel in Mart., Fl. Bras. 6(1): 251, 267, 1 Dec 1865 (as Potalieae). — family Potaliaceae Mart., Nov. Gen. Sp. Plant. 2.: 89, 133, 1827 (as Potalieae). — subfamily Potalioideae (Mart.) Arn., Encycl. Brit., ed. 7, 5: 120, 9 Mar 1832 (as Potalieae). — tribe Potalieae Rchb., Handb. Nat. Pfl.-Syst.: 211, 1–7 Oct 1837. TYPE: *Potalia* Aubl.
tribe Fagraeeae Meisner, Pl. Vasc. Gen.: Table Diagn., 257, 259, Comm. 167, 5–11 April 1840 (as Fagraeaceae). — subtribe Fagraeinae Bureau, De la famille des Loganiacées, et des plantes qu'elle fournit a la médecine, Thèse, Faculté de Médecine de Paris: 69, 1856 (as Eufagraeacées). TYPE: *Fagraea* Thunb.
Included genera:
Anthocleista R. Br. (14)
Fagraea Thunb. (c. 70)
Potalia Aubl. (9)
Number of genera: 3.
Number of species: c. 93.

6. Tribe Gentianeae Colla, Herb. Pedem. 3: 157, 1834 [Jan–Feb 1835] (as Gentianaceae). TYPE: *Gentiana* L.
Number of subtribes: 2.
Number of genera: 17.
Number of species: c. 939–968.

6.1. subtribe Gentianinae G. Don, Gen. Syst. Dichl. Pl. 4: 173, Mar–Apr 1838 (as Gentianeae-verae). TYPE: *Gentiana* L.
subtribe Calathiinae Zuev, Bot. Zhurn. 75: 1301, 1990. TYPE: *Calathiana* (Froel.) Delarbre (= *Gentiana* L.)
Included genera:
  Crawfurdia Wall. (16–19)
  Gentiana L. (c. 360)
  Tripterospermum Blume (24)
Number of genera: 3.
Number of species: c. 400–403.

6.2. subtribe Swertiinae (Griseb.) Rchb., Handb. Nat. Pflanz.: 210, 1837 (as Swertieae). — tribe Swertieae Griseb., Gen. Sp. Gent.: 209–210, 1839 [Oct 1838]. TYPE: *Swertia* L.
subtribe Anagallidiinae Zuev, Bot. Zhurn. 75: 1304, 1990. TYPE: *Anagallidium* Griseb. (= *Swertia* L.)
Included genera:
  Bartonia H. L. Mühl. ex Willd. (4)
  Comastoma (Wettst.) Toyok. (7–25)
  Frasera Walter (15)
  Gentianella Moench (c. 250)
  Gentianopsis Ma (16–24)
  Halenia Borkh. (80)
  Jaeschkea Kurz (4)
  Latouchea* Franch. (1)
  Lomatogonium A. Braun (21)
  Megacodon (Hemsl.) Harry Sm. (2)
  Obolaria L. (1)
  Pterygocalyx Maxim. (1)
  Swertia L. (c. 135)
  Veratrilla Baill. ex Franch. (2)
Number of genera: 14.
Number of species: c. 539–565.

Genus et tribus incertae sedis
  Voyria* Aubl. (19)
tribe Voyrieae Gilg in Engler & Prantl, Nat. Pflanzf. 4(2): 102, 1895. TYPE: *Voyria* Aubl.
tribe Leiphaimeae Gilg in Engler & Prantl, Nat. Pflanzf. 4(2): 102, 1895. TYPE: *Leiphaimos* Cham. & Schltdl. (= *Voyria* Aubl.).

Number of genera: 1.
Number of species: 19.

Excluded taxa
Emblingia F. Muell. (1)
Note: *Emblingia* was suggested to be a gentian based on molecular phylogenetic results presented by Savolainen *et al.* (2000). However, this placement is strongly contradicted by other molecular data based on different plant samples (Chandler & Bayer, 2000) that support a position for *Emblingia* within the Brassicales, where the genus has been placed earlier (APG, 1998). Morphological characteristics also contradict a position in the Gentianaceae/Gentianales, since *Emblingia* has a labiate corolla (two unequal lobes), an androphore, two whorls of stamens, staminodes, a unicarpellate ovary, and no iridoids (Watson & Dallwitz, 1992–2001). *Emblingia* is the only member of the family Emblingiaceae, native to Australia.

TRIBE SACCIFOLIEAE

M. THIV, L. STRUWE, V. A. ALBERT, AND J. W. KADEREIT

The genera of this new tribe have not been placed together in any earlier classification. *Saccifolium*, the type genus for this tribe, has since its discovery in the 1960s and description in 1978 been the sole member of Saccifoliaceae (Maguire & Pires, 1978). Recent molecular work on freshly preserved material has shown that the aberrant genus *Saccifolium*, with its saccate leaves, is a member of the most basal lineage in the Gentianaceae together with *Curtia* and the achlorophyllous mycotroph *Voyriella* (Thiv *et al.*, 1999a). Earlier molecular results based on DNA extracted from a herbarium specimen had shown *Saccifolium* to be nested inside tribe Gentianeae (see Struwe *et al.*, 1998).

Curtia, *Voyriella*, and *Saccifolium* are all exclusively neotropical genera. In addition to these taxa, we have decided to place in Saccifolieae two additional genera for which molecular data are so far lacking. These are *Hockinia* and *Tapeinostemon*, and their inclusion in Saccifolieae is based on their morphological and palynological similarities with *Curtia* and *Voyriella*. *Hockinia*, *Saccifolium*, and *Voyriella* are all monotypic genera. *Curtia* and *Tapeinostemon* contain 6–10 and 7 species, respectively, which makes Saccifolieae the tribe of Gentianaceae with the fewest species.

Table 2.4. *Assignment of* Curtia, Hockinia, Tapeinostemon, *and* Voyriella *of tribe* Saccifolieae *to subgroups of the Gentianaceae in earlier classifications of the family.* Saccifolium *was not described until 1978 and is not included here*

Genus	Don (1837–1838)	Endlicher (1838)	Grisebach (1839)	Grisebach (1845)	Progel (1865)	Bentham (1876)	Knoblauch (1894)	Gilg (1895a)	Gilg (1939)
Curtia	Coutoubeaea[a]	Sebaeae	Erythraeaceae	Chloreae-Erythraeaceae	Chloreae	Chironieae-Euchironieae	Chironieae-Euchironieae	Gentianeae-Erythraeinae	Gentianeae-Erythraeinae-Ellipandrae
Hockinia	***[b]	***	***	Lisiantheae	Lisiantheae	Chironieae-Euchironieae	Chironieae-Euchironieae	Gentianeae-Tachiinae	(Not mentioned in classification)
Tapeinostemon	***	***	***	***	Chloreae	Chironieae-Euchironieae	Chironieae-Euchironieae	Gentianeae-Erythraeinae	Gentianeae-Erythraeinae-Ellipandrae
Voyriella	***	***	***	***	Lisiantheae	Chironieae-Euchironieae	Chironieae-Euchironieae	Leiphaimeae	(Not mentioned in classification)

Notes:
[a] Infrafamilial names are cited as they appear in each publication.
[b] Asterisks denote genera not described at the time of the listed publication.

Systematics, character evolution, and biogeography 57

Description of Saccifolieae

Saccifolieae are shrubs, robust perennial to small annual herbs, or fleshy, achlorophyllous, herbaceous plants. The stems are terete to angular and have alternate (*Saccifolium* and sometimes *Voyriella*), opposite, or whorled (*Curtia*) leaves. The leaves are normal or reduced to scales and are uniquely saccate in *Saccifolium*. The inflorescences are loose, terminal cymes or dense capitula (*Tapeinostemon sessiliflorum*), axillary cymes, or have solitary flowers in the leaf axils (*Saccifolium*). The actinomorphic flowers are (4–)5(–6)-merous, and heterostylous flowers have been reported from *Curtia*, *Hockinia*, *Tapeinostemon longiflorum*, and *Voyriella*. The calyx lobes are often free to the very base or deeply divided. The tubular to funnel-shaped corollas are mostly white or light-colored. The stamens are inserted in the corolla tube. Short-styled flowers have stamens with long filaments and free, non-sagittate anthers without apically elongate connectives. Long-styled flowers have stamens with short filaments and connate, sagittate (sometimes latrorse) anthers with elongate connectives. The ovary is sessile, without a disk at the base, and has a short or long style and a slightly bilobed stigma. The fruit is a capsule with numerous seeds.

Taxonomic history of Saccifolieae

In past classifications of the Gentianaceae, *Curtia*, *Hockinia*, *Tapeinostemon*, and *Voyriella* have been included in various tribes and subtribes (Table 2.4). They were included in Chironieae-"Euchironieae" by Bentham (1876) and Knoblauch (1894). A different view was presented in Gilg's (1895a) system, in which *Curtia* and *Tapeinostemon* were placed in Gentianeae-Erythraeinae, *Hockinia* was treated as a part of Gentianeae-Tachiinae, and *Voyriella* together with *Leiphaimos* (= *Voyria*; cf. Albert & Struwe, 1997) was included in tribe Leiphaimeae. The present classification marks the first time these four genera have been associated with *Saccifolium*.

Saccifolium was initially described as the only member of Saccifoliaceae, a family that has been associated mainly with Gentianales and with Gentianaceae in particular (Maguire & Pires, 1978). Thorne (1983) included *Saccifolium* in Gentianaceae as a member of a monotypic subfamily, the Saccifolioideae (which was never validly described), but he subsequently reversed this view, so that in a later system (Thorne, 1992) Saccifoliaceae again appeared as a separate family. Takhtajan (1997) treated *Saccifolium* as a distinct family but thought that it might not deserve to be recognized as a family separate from Gentianaceae.

Generic synopses for Saccifolieae

Curtia CHAM. & SCHLTDL.

This genus was described by von Chamisso and von Schlechtendal (1826: 209), based on *Curtia gentianoides*. The genus comprises *c.* 6–10 species distributed from Mexico to Argentina (Maguire, 1981; Struwe *et al.*, 1999) with a center of distribution in Brazil. Gilg (1939) observed heterostyly in *Curtia patula* and *C. tenuifolia*. The flower morphology of the remaining species is not well known. The scale-like leaves of *Curtia tenuifolia* suggest that this species may be mycotrophic. *Curtia* is a morphologically heterogeneous genus and may not be monophyletic.

Curtia plants are erect annual herbs (Struwe *et al.*, 1999). The stems are terete to quadrangular. Rosette leaves are absent, and the scale-like, linear to elliptic or ovate cauline leaves are opposite or sometimes arranged in whorls (up to 8 in each whorl). The 5-merous, pedicellate, and, in at least some species, heterostylous flowers are borne in lax monochasia. The calyx tube is shorter than the triangular calyx lobes. Calycine colleters are present. The white, yellow, pale lilac, pink, or purplish corollas (which may be hairy on the inside) are funnel- to salver-shaped with corolla tubes about as long as the corolla lobes. Short-styled flowers have long filaments and free, non-sagittate anthers without apically elongate connectives. Long-styled flowers have short filaments and connate, sagittate anthers with elongate connectives. The anthers are basifixed or rarely dorsifixed. The stamens are usually inserted lower in the corolla tube in long-styled flowers than in short-styled ones. The pollen is released as monads. The unilocular ovary with parietal placentas is narrowly ovoid and tapers into the distinct style. The stigmatic lobes are distinctly bilobed, or the lobes are closely adjacent, making the stigma appear club-shaped.

Hockinia GARDNER

Gardner (1843: 12–13) established this genus based on *Hockinia montana*. This monotypic genus is endemic to Brazil (Gilg, 1895a). Gilg (1895b) interpreted the flowers of *Hockinia* as heterostylous with three different flower types. In contrast, Knoblauch (1895) regarded the flowers as being dimorphic with somewhat variable long- and short-styled flowers.

Hockinia montana is an erect annual herb. The stems are terete to quadrangular. Rosette leaves are absent and the cauline leaves are elliptic. The 5-merous, pedicellate, heterostylous flowers are borne singly or are arranged in lax monochasia. The calyx tube is much shorter than the long, triangular calyx lobes. Calycine colleters are present. The pale lilac corollas are funnel-shaped with a corolla tube that is longer than the corolla lobes. Short-styled flowers have long filaments and free, sagittate anthers without apically elongate connectives. Long-styled flowers have short filaments and connate, non-sagittate anthers with elongate connectives. The anthers are basifixed. The filaments are often covered with hairs. The stamens are inserted in the lower part of the corolla tube. The pollen is released as monads. The unilocular ovary with parietal placentas is ovoid and tapers into the distinct style. The stigmatic lobes are distinctly bilobed, or the lobes are closely adjacent making the stigma appear clubshaped.

Saccifolium MAGUIRE & PIRES

The monotypic genus *Saccifolium* was only recently discovered and is known from only four collections. The type species is *Saccifolium bandeirae* (Maguire & Pires, 1978: 242), which was placed in the separate family Saccifoliaceae. This species is known only from the highest peaks of the Sierra de la Neblina tepui complex on the Brazilian–Venezuelan border.

Saccifolium is a shrub with corky-barked branches (Maguire & Pires, 1978; Struwe *et al.*, 1999; Fig. 2.4). The leaves are closely arranged, possibly alternate, pitcher-shaped, partly translucent, and possibly with secretory and/or absorptive glands on the inner surface that could aid in nutrient/water uptake (Struwe *et al.*, 1998). The 5-merous flowers are solitary in the axils of leaves. The calyces are rotate and divided almost to the base, and an intracalycine membrane is absent. The enlarged colleters found at the base of the calyx are presumably homologous to colleters (L. Struwe, pers. obs.), and are possibly nectariferous. The corollas are tubular, widened in the middle, and shortly lobed. The corolla aestivation is imbricate. The stamens are inserted in the lower part of the corolla tube. The ovary is sessile and unilocular with deeply protruding placentas (Struwe *et al.*, 1998; cf. Maguire & Pires, 1978). The style is long and slender. Mature fruits are unknown and immature seeds are unwinged.

Tapeinostemon BENTH.

This genus was described by Bentham (1854) based on *Tapeinostemon spenneroides*. The genus consists of seven species (Steyermark, 1951; Maguire, 1981; Struwe *et al.*, 1999) distributed in northern South America. The species grow in tropical rainforests and savannas from sea level up to 2200 m, most particularly in savanna and forest vegetation on tepui slopes and summits. Heterostylous flowers have been observed in herbarium

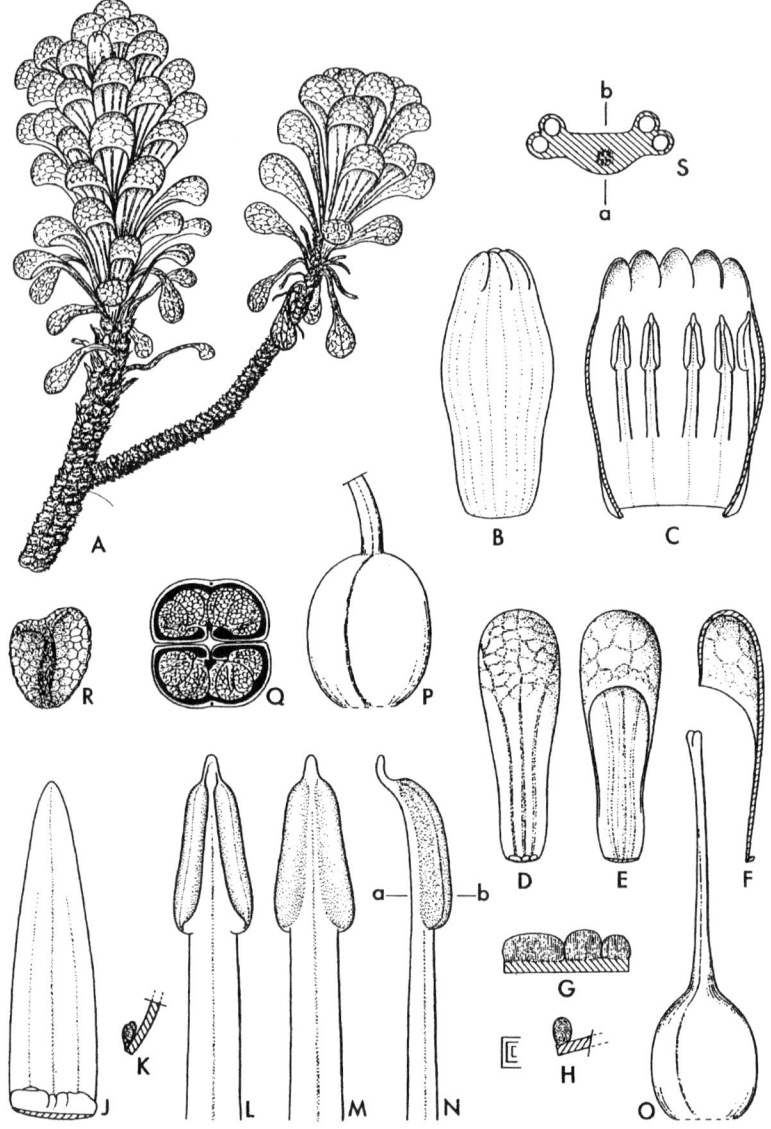

Figure 2.4. *Saccifolium bandeirae.* **A.** Habit. **B.** Corolla. **C.** Corolla, opened. **D–F.** Leaf. **G.** Colleters at inner leaf base. **H.** Cross-section of colleter and leaf. **J.** Inner side of sepal with glands. **K.** Cross-section of glands on sepal. **L–N.** Stamen. **O.** Pistil. **P.** Ovary. **Q.** Cross-section of ovary. **R.** Seeds (barely mature). **S.** Cross-section of anther. (Reprinted by permission of The New York Botanical Garden Press from Maguire *et al.*, 1978. The Botany of the Guayana Highland – Part X. *Memoirs of the New York Botanical Garden*, vol. 29.)

material of *Tapeinostemon longiflorum* (L. Struwe, pers. obs.). Free as well as connate anthers are present in this genus.

This genus comprises erect, slightly suffrutescent perennial herbs (Struwe et al., 1999). The stems are terete. Rosette leaves are absent and the petiolate cauline leaves are elliptic-lanceolate to elliptic-ovate. The 5-merous, pedicellate or sessile flowers are arranged in lax cymes or capitate inflorescences (*Tapeinostemon sessiliflorum*). The calyx tube is much shorter than the long, triangular or lanceolate calyx lobes. Calycine colleters are absent. The white or yellow to orange corollas are salver- to funnel-shaped or tubular with a corolla tube that is longer than the corolla lobes. The stamens are inserted in the middle of the corolla tube. The basifixed, sagittate, free or connate anthers have apically elongate anther connectives. The pollen is released as monads. The ovary is of ovoid shape and tapers into a distinct style of variable length. The placentas are strongly intruding and subdivide the ovary into two or four compartments. The stigmas are bilobed.

Voyriella (MIQ.) MIQ.

This genus was established by Miquel (1851) based on *Voyriella parviflora*. According to Maas and Ruyters (1986) the genus comprises one polymorphic species, *Voyriella parviflora*, which grows in tropical South America and Panama. Maguire and Boom (1989) accepted a second species, *Voyriella oxycarpha*, which differs from *V. parviflora* in the length of its corolla, filaments, and style, and in the morphology of the anther connective (Maas & Ruyters, 1986). This second species grows sympatrically with *Voyriella parviflora* (Maguire & Boom, 1989). Following Maas and Ruyters (1986), these differences may indicate the presence of heterostyly in *Voyriella*. Short-styled specimens are here interpreted as corresponding to *Voyriella parviflora* and long-styled specimens to *V. oxycarpha*.

Voyriella plants are erect, perennial herbs without chlorophyll (Maas & Ruyters, 1986; Struwe et al., 1999). The stems are quadrangular to slightly winged. Rosette leaves are absent and the cauline leaves are scale-like. The (4–)5(–6)-merous, shortly pedicellate flowers are borne in dense, mostly many-flowered dichasia. The calyx lobes are almost free. Calycine colleters are present. The white corollas are tubular to salver-shaped with a corolla tube that is longer than the corolla lobes. Short-styled flowers have short filaments and connate, non-sagittate anthers with apically elongate connectives. Long-styled flowers have long filaments and free or connate, sagittate anthers without elongate connectives. The anthers are basifixed. The stamens are usually inserted lower in the corolla tube in long-styled flowers than in short-styled ones. The pollen is released as monads. The unilocular ovary with parietal placentas is ovoid and tapers into a distinct style. The stigmatic lobes are bilobed or of variable shape. Reported chromosome numbers are $2n = 20-28$ (Oehler, 1927).

Phylogeny and character evolution in Saccifolieae

In the study based on *trn*L intron sequences of the Gentianales, Struwe *et al.* (1998) showed *Saccifolium* (based on DNA from herbarium material) to be nested inside the Gentianaceae close to the genus *Gentiana*. Further sequencing of freshly collected material of *Saccifolium* gave contradicting results, and *Saccifolium* is now accepted as a member of the most basal clade of the family. Separate analyses of *mat*K, *rbc*L, and *trn*L intron sequences show *Saccifolium* placed together with *Voyriella* and *Curtia* as the sister group to the rest of Gentianaceae (Thiv *et al.*, 1999a). Whereas a sister-group relationship with a clade consisting of *Voyriella* plus *Curtia* is only marginally supported with *trn*L intron data (jackknife support 54%; Fig. 2.1), *mat*K data alone (Fig. 2.2; Thiv *et al.*, 1999a) as well as the combined analysis of *mat*K and *trn*L sequences both suggest that *Saccifolium* and *Voyriella* are sister to each other with high support (jackknife value 81%; Fig. 2.3).

There is also morphological and palynological evidence for a close relationship among *Curtia*, *Hockinia*, *Saccifolium*, *Tapeinostemon*, and *Voyriella* (Thiv *et al.*, 1999a). *Curtia*, *Hockinia*, *Tapeinostemon*, and *Voyriella* have at least partly connate anthers with apically elongate anther connectives, and share an ovoid ovary that tapers into the style. Apically elongate anther connectives and a similar ovary morphology are also found in *Saccifolium*. *Curtia*, *Tapeinostemon*, *Saccifolium*, and *Voyriella* also share 3-colporate (or rarely 4-colporate) pollen grains with a perforate to finely reticulate exine (Maas & Ruyters, 1986; Maguire, 1981; Maguire & Boom, 1989; Nilsson, see below). Nilsson and Skvarla (1969) suggested a close relationship between *Curtia* and *Voyriella* on the basis of pollen morphological similarities. According to Nilsson (see under palynology, this chapter), *Hockinia* has the same type of pollen as *Curtia*. The heterostyly of *Curtia*, *Hockinia*, at least one species of *Tapeinostemon*, and possibly also *Voyriella* can be regarded as further evidence for a close relationship among these genera. Such a relationship had been postulated for *Curtia* and *Hockinia* by Grothe and Maas (1984), based on the fact that these two genera share similar seed coat characters. Some species of *Curtia* show seed-character similarities with *Tapeinostemon* and *Voyriella* (Bouman *et al.*, 2002). In contrast to the three perennial genera *Tapeinostemon* (except *T. sessiliflorum*), *Saccifolium*, and *Voyriella*, *Curtia* and *Hockinia* are characterized by an annual habit.

Of the characters listed above, only heterostyly is not found outside this group of gentians. Other characteristics that occur in the Saccifolieae are ovaries tapering into the style, a trait that is also found in Gentianeae, and connate anthers (also occurring in *Gentiana* and *Tachiadenus*; Duncan & Brown, 1954; Klackenberg, 1987a). The absence of a nectary disk at the base of the ovary has been reported for *Hockinia* and *Saccifolium* (Lindsey, 1940; Maguire & Pires, 1978). Additionally, the pollen features of Saccifolieae are found in various Gentianaceae, e.g., Chironieae, and apically elongate anther connectives are also characteristic for Helieae.

Within the Saccifolieae several highly apomorphic traits can be found, such as saccate leaves and imbricate corolla aestivation in *Saccifolium*, and loss of chlorophyll, presence of scale-like leaves, and reduction of seed and capsule size in *Voyriella*. This lineage of Gentianaceae stems from the oldest split in the family and it is therefore possible that the individual genera have themselves been separated for a very long time. The five genera, all monotypic or with rather few species, could also be the remnants or relicts of a formerly more diverse lineage in the Neotropics. This clade of Gentianaceae needs further study, especially regarding infrageneric relationships.

Palynology of Saccifolieae: Synopses of genera and general discussion
S. NILSSON

Curtia. *Curtia tenuifolia* ssp. *tenuifolia* (Aubl.) Knobl. The pollen grains are 3-colporate, subspheroidal to prolate, and *c.* 25×20 μm. The exine is smooth, perforate to foveolate, or finely reticulate with minute spines (microspinose). (Brazil: Mato Grosso, Serra da Chapada, 1903, *Malme s.n.* (S), slide no. 24402; Nilsson, 2002: Fig. 4.1.)

OTHER INVESTIGATIONS: Maguire (1981) reported smooth, perforate-finely reticulate to microspinose pollen grains in *Curtia tenuifolia*.

Hockinia. *Hockinia montana* Aubl. The pollen grains are 3-colporate, spheroidal to prolate, *c.* 30×28 μm. The exine is smooth, finely reticulate to perforate-foveolate, or microspinose. (Brazil: Minas Gerais, 1900, no collector (US 4718), slide no. 14718; Nilsson, 2002: Fig. 4.2.)

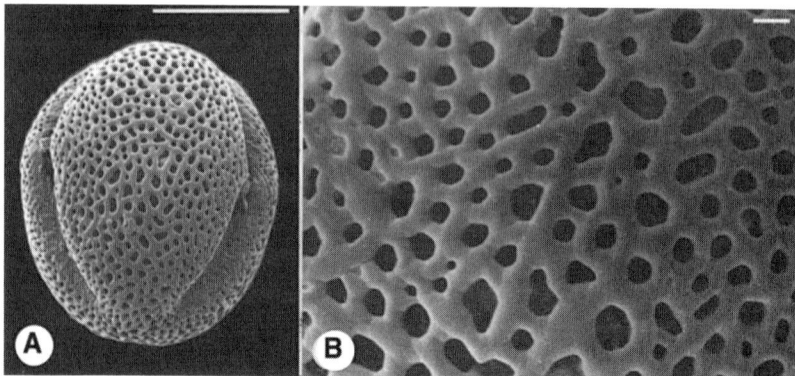

Figure 2.5. Pollen of Saccifolieae (*Saccifolium bandeirae*; *Maguire et al. 60532*). **A.** 3-colporate, perforate pollen in slightly oblique equatorial, apertural position. **B.** Detail of the perforate to finely reticulate exine. Scale bars: A = 10 μm; B = 1 μm.

Saccifolium. *Saccifolium bandeirae* Maguire & Pires. The pollen grains are 3-colporate, spheroidal, and *c.* 37 × 37 μm. The exine is perforate, less densely so at the polar areas and toward the relatively broad and smooth colpus margin. (Brazil–Venezuela: Amazonas, Sierra de la Neblina, 1965, *Maguire et al. 60532* (NY), slide no. 24885; Fig. 2.5.)

OTHER INVESTIGATIONS: Maguire and Pires (1978) described the pollen grains of *Saccifolium bandeirae* and provided LM and SEM pictures of the 3-colporate, perforate-foveolate pollen grains. Their results were confirmed by the present study.

Tapeinostemon. *Tapeinostemon zamoranum* Steyerm. The pollen grains are 3-colporate, subspheroidal to prolate, and *c.* 27 × 25 μm. The exine is perforate to microreticulate-reticulate. (Ecuador: Zamora, Chinchipe, 1988, *Gaarol 74940* (QCA); Nilsson, 2002: Fig. 4.3.)

OTHER INVESTIGATIONS: Maguire (1981) showed variability of exine patterns in two varieties of *Tapeinostemon longiflorum* and in two different collections of *T. spenneroides*, ranging from smooth, perforate, faintly striate-rugulate (low relief) to reticulate at the polar areas.

Voyriella. *Voyriella parviflora* (Miq.) Miq. The pollen grains are radially symmetrical, *c.* 15 × 15 μm in diameter, and 3–4-colporate. The exine is finely reticulate and distinctly stratified.

Systematics, character evolution, and biogeography 65

OTHER INVESTIGATIONS: Neither Gilg (1895a) nor Köhler (1905) noted the presence of colpi. Nilsson and Skvarla (1969), who studied pollen of *Voyriella oxycarpha*, *V. parviflora*, and other saprophytic taxa, concluded that *Voyriella* would better be associated with, for example, *Curtia* than with *Voyria/Leiphaimos*. This was later reconfirmed by Grothe and Maas (1984). Nilsson (in Maas & Ruyters, 1986: 5–9) described and illustrated pollen of *Voyriella parviflora* (the only recognized species). Maguire and Boom (1989) included LM and SEM pictures of pollen from the two species of *Voyriella* without comments on their possible relationships.

Discussion of palynological data in Saccifolieae. The perforate pollen grains of *Saccifolium* show similarities to pollen of *Curtia*, *Hockinia*, and *Voyriella* (all genera classified to Saccifolieae). They also differ significantly from pollen of the genera of Gentianeae-Gentianinae (cf. Struwe et al., 1998), but do resemble that of *Bartonia* in Gentianeae-Swertiinae. The current phylogenetic placement of *Saccifolium* and *Curtia* in Saccifolieae (cf. Thiv et al., 1999a) is therefore supported by palynological evidence.

The pollen grains of *Curtia* and *Hockinia* are similar. In both genera there are two different types of pollen. The first type has a smooth and finely reticulate to perforate exine, and the second type has a microspinose exine. This might be related to heterostyly (Nilsson, 2002: Figs. 4.1 vs. 4.2). Grothe and Maas (1984) suggested heterostyly and a close relationship between *Curtia* and *Hockinia* based on seed morphology, although the two genera were otherwise clearly distinct from each other.

Voyriella pollen differs in several respects from that of *Voyria*, the other neotropical achlorophyllous gentian genus, for example in number of apertures and exine pattern. Based on palynological data *Voyriella* may be best associated with *Curtia* and *Hockinia* (Nilsson & Skvarla, 1969), which is in agreement with molecular evidence and the current classification. *Curtia* and *Hockinia* both have two types of pollen, of which the microspinose type is similar to the pollen of *Bisgoeppertia* in the Chironieae (S. Nilsson, pers. obs.). The genus *Tapeinostemon* has pollen with variable exine patterns, from perforate to reticulate (microreticulate). Most pollen grains are relatively small (with polar axis exceptionally exceeding 40 μm in length). The pollen of the genera constituting Saccifolieae all show close similarities with regard to size, exine pattern, and the relatively broad and smooth colpus margin.

Biogeography of Saccifolieae

All genera of Saccifolieae are neotropical and the majority of the species occur on the Guayana Shield. *Saccifolium* is an extremely rare tepui endemic found only on Sierra de la Neblina, which straddles the Brazilian–Venezuelan border (Maguire & Pires, 1978). *Tapeinostemon* occurs primarily in rainforest or savannas of various elevations on or proximal to the Guayana Shield (Struwe et al., 1999), with one species endemic to the Andean foothills of northern Peru and Ecuador (Pringle, 1995). *Curtia* occurs mainly in savanna vegetation of the Amazon Basin and the Guayana and Brazilian Shields, also reaching Central America (Struwe et al., 1999), whereas *Hockinia* is a Brazilian Shield endemic from the Organ Mountains of Rio de Janeiro (J. E. Simonis, unpubl.). The saprophytic genus *Voyriella* occurs in lowland rainforests of Panama, on the Guayana Shield east to French Guiana, with one collection known from Mato Grosso (Brazilian Shield; Maas & Ruyters, 1986). Phylogenetic trees based on molecular data have so far included only three species of this tribe, so no particular biogeographic hypothesis for the group is well supported. A close relationship between *Curtia* and *Hockinia*, as suggested by morphological evidence, is also mirrored by their distribution in the sense that both genera are present in southeastern Brazil. Whereas the remainder of Saccifolieae is basically restricted to the Guayana Shield, *Curtia* is widespread in northern South America and Central America and *Hockinia* occurs exclusively on the Brazilian Shield. A more precise history of these genera on the South American shields must await greater taxonomic sampling and firmer resolution of intergeneric relationships.

TRIBE EXACEAE

J. KLACKENBERG

The Exaceae are a small tribe of six genera: *Cotylanthera*, *Exacum*, *Gentianothamnus*, *Ornichia*, *Sebaea*, and *Tachiadenus*. *Cotylanthera* and *Gentianothamnus*, not yet placed in molecular phylogenetic studies, are included here based on morphological criteria. The tribe includes c. 144–184 species, of which perhaps up to 165 belong to the genera *Exacum* and *Sebaea* (the species number for *Sebaea* is still highly uncertain). Taxa of Exaceae form a basal clade in Gentianaceae, with only Saccifolieae being more basally positioned. Although only six genera are included, the range of morphological variation is relatively great. The genera were placed in several different tribes during the nineteenth century, but the Exaceae have generally

Systematics, character evolution, and biogeography   67

been considered natural with uncontroversial circumscription during the twentieth century. *Tachiadenus*, however, has sometimes been excluded from the tribe based on palynological and gynoecial characters, and *Gentianothamnus* is here included in Exaceae for the first time, transferred from Chironieae.

Most authors have focused on the bilocular structure of the ovary as a diagnostic character for Exaceae. The bilocular ovary with axile placentation is the common state in plausible outgroups to Gentianaceae, e.g., Rubiaceae and Loganiaceae *sensu lato*. Consequently, this structure should be regarded as a symplesiomorphy and therefore cannot be used to characterize Exaceae (Klackenberg, 1985: 18). The testa cells of the seeds are isodiametric but mostly with wavy walls, usually rendering them star-shaped (Klackenberg, 1985, 1986, 1987a; Bouman et al., 2002). This feature has not been observed elsewhere in the family and is thought to be a synapomorphy for Exaceae. The shape of the petal epidermis cells, basically rounded and convex versus elongated and flat, has also been used as a supporting character for this clade (Klackenberg, 1985: 19).

According to the strict consensus tree derived from molecular data (*mat*K and *trn*L intron; Fig. 2.3), *Sebaea* is sister to *Exacum*, *Ornichia*, and *Tachiadenus*. The large and widely distributed genus *Exacum* is unresolved toward *Ornichia*, a small Malagasy endemic. The clade of *Exacum/Ornichia* is in turn sister to another small Malagasy endemic, *Tachiadenus*. Two species each of *Exacum* and *Sebaea* are included in the analyses. Owing to sampling considerations, the monophyletic origins of the genera are not refuted, although *Exacum* is not demonstrably monophyletic (Figs. 2.1–2.3). The monophyly of *Ornichia* and *Tachiadenus* cannot be evaluated, as only one representative of each is included in the analysis.

Description of Exaceae

Species of Exaceae are annual herbs to perennial subshrubs or tiny saprophytic achlorophyllous plants (*Cotylanthera* and *Sebaea oligantha*) between a few centimeters and a few meters high, erect and unbranched to much-branched, rarely with trailing branches (e.g., *Exacum walkeri*). Two species are prostrate and rooting at the nodes (*Exacum radicans* and *E. subverticillatum*). The stems are terete to quadrangular and usually furnished with four wings or lines. The leaves are herbaceous to rather succulent or somewhat coriaceous, rarely scale-like (*Cotylanthera* and *Sebaea oligantha*). The arrangement of the leaves is usually decussate but verticillate (*Exacum* spp.) or rosulate (*Exacum* spp. and *Sebaea* spp.) leaves are rarely present. The outline of the leaves varies from linear to orbicular, cordate to obovate, usually with a distinct petiole but sometimes amplexicaul (*Exacum* spp. and *Tachiadenus* spp.), glabrous to rarely hairy (*Ornichia*). The inflorescences are one (*Cotylanthera* spp. and *Gentianothamnus*), few or many-flowered mono- or dichasial terminal or axillary cymes but sometimes with the axes of the cymes suppressed, making the inflorescence umbel-shaped (*Exacum* spp. and *Tachiadenus* spp.), rarely with a basal underground inflorescence with cleistogamous flowers (*Sebaea oligantha*). The flowers are 4-merous (e.g., *Cotylanthera*, *Exacum* spp., and *Sebaea* spp.) to most commonly 5-merous and are actinomorphic to rarely zygomorphic by having the anthers connivent into a cone above a bent style (*Cotylanthera* spp. and *Exacum* spp.). The calyx is coalescent only at the very base or up to ¾ of its length, with or without a keel or wing; more rarely the calyx is zygomorphic with two well-developed and three more reduced wings (*Exacum* spp.), rarely with 10 wings (*Exacum decapterum* and *Tachiadenus longiflorus*; Fig. 2.6). The calyx lobes are linear and subulate to obovate and obtuse. The calyx wings, if present, are tapering to semicordate at the base and tapering to truncate at the apex, sometimes accrescent in fruit and then often with prominent veins. The corollas are small to large, white to blue/violet, or white to yellow (*Gentianothamnus* and *Sebaea*). The corolla tube is rather short and funnel-shaped to urceolate or long-cylindric (*Sebaea* spp. and *Tachiadenus*), rarely long and narrowly funnel-shaped (*Gentianothamnus*; Fig. 2.7). The corolla lobes are narrowly elliptic to orbicular, erect to most commonly spreading, sometimes long-persistent in fruit. The stamens are exserted on slender filaments inserted near the mouth and are usually connivent around or above the style forming a cone, or sometimes included in the tube on very short filaments (*Ornichia trinervis*, *Sebaea* spp., and *Tachiadenus*), rarely inserted on long filaments from the base of the tube (Fig. 2.8). The anthers, sometimes twisted backwards (*Sebaea* spp.), are thin to somewhat coriaceous (*Exacum* and *Gentianothamnus*) and are dehiscent by longitudinal slits or by one (*Cotylanthera*) or two apical pores (*Exacum*). The anthers are free or coherent with each other to form a ring (*Ornichia trinervis*, *Sebaea* spp., and *Tachiadenus*), and are often furnished with apical appendages (papilliform at the dorsal side (*Exacum* spp. and *Gentianothamnus*), filiform (*Tachiadenus*), or thick and rounded appendages that are sometimes also present at the bases of thecae (*Ornichia* spp., *Sebaea* spp., and *Tachiadenus tubiflorus*)). The style is filiform and straight or curved (*Cotylanthera paucisquama* and *Exacum* spp.) and sometimes has an additional lower stigma(s) (*Sebaea* spp.). The stigma (apical) is small and entire to faintly bilobed. The ovary, rarely placed on a disk

Systematics, character evolution, and biogeography

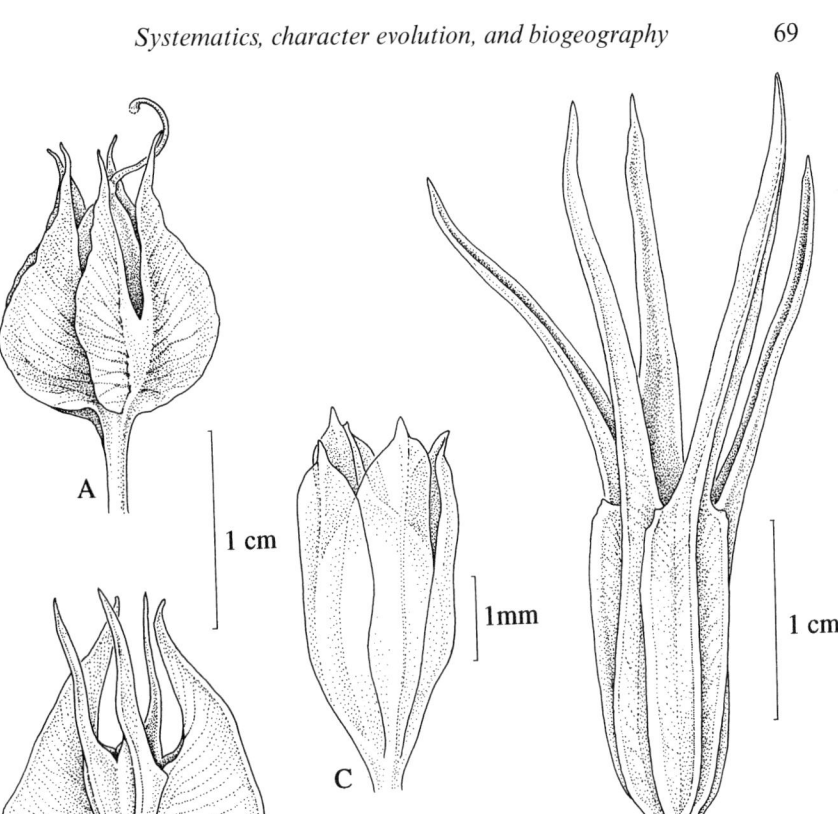

Figure 2.6. Calyces of Exaceae. **A.** *Exacum courtallense* (*Klackenberg & Lundin 527*), showing 5-merous calyx surrounding a mature capsule. Distinct wings are present, enlarged in fruit with prominent veins. **B.** *Exacum dipterum* (*Razafindrabe 44*), showing 5-merous calyx with three wings reduced, corolla removed. **C.** *Sebaea albens* (*Wiese 1832*), showing 4-merous calyx surrounding a mature capsule. Lobes with hyaline margins and without wings. **D.** *Tachiadenus longiflorus* (*Service Forestier de Madagascar 3240*), showing 5-merous calyx with 10 wings, corolla removed.

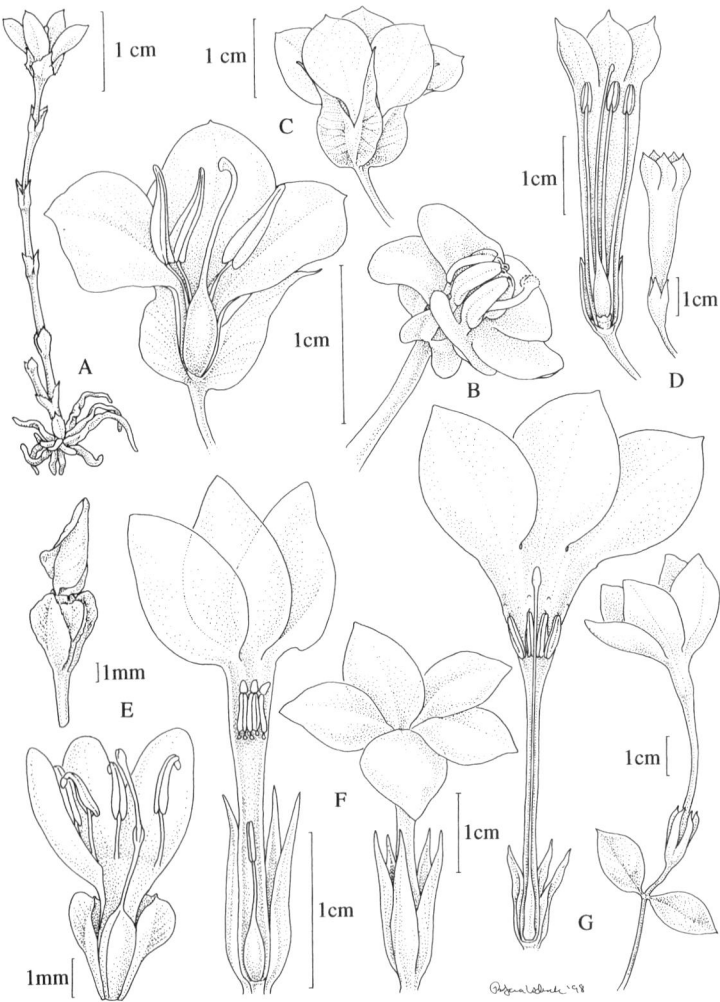

Figure 2.7. Flowers of Exaceae. **A.** *Cotylanthera tenuis* (*Kaudern 304*), whole plant. **B.** *Exacum axillare* (from greenhouse material), showing connivent anthers and bent style. **C–G.** Whole flower and flower with two calyx and corolla lobes removed, respectively. **C.** *Exacum courtallense* (*Klackenberg & Lundin 527*), with short corolla tube and free anthers inserted at tube mouth. **D.** *Gentianothamnus madagascariensis* (*Randrianasolo 136*), with long narrowly funnel-shaped corolla tube and free long-filamented anthers inserted at tube base just below a small 5-lobed disc. **E.** *Sebaea ambigua* (*Acock 1279*), with short corolla tube and free anthers inserted at tube mouth. Non-dissected flower in young fruit. **F.** *Sebaea thomasii* (*Hoener 2055*), with long cylindric corolla tube and anthers inserted in the tube and adhering to each other in a ring . **G.** *Tachiadenus carinatus* (*Lowry & Randrianasolo 4555*, whole flower; *Klackenberg 871124-5*, dissected flower), with long cylindric corolla tube and anthers inserted in the tube adhering to each other in a ring.

Systematics, character evolution, and biogeography 71

(*Gentianothamnus*), is bilocular with axile placentas or unilocular but becomes pseudo-bilocular from the centrally protruding parietal placentas (*Gentianothamnus* and *Tachiadenus*; Fig. 2.9). The fruit is a membranous to usually coriaceous septicidally bivalved capsule, rarely membranaceous and opening with a coriaceous lid (*Exacum* spp.). The seeds are usually angular with slightly to much-undulated testa cell walls that often become distinctly star-shaped (*Exacum* spp., *Ornichia*, and *Tachiadenus*; Fig. 2.10). More rarely the cells are isodiametric with minutely undulate (*Exacum* spp.) or straight walls (*Cotylanthera* and *Exacum* spp.). Alternatively, the seeds may be somewhat elliptic or oblong in outline, slightly flattened, with irregular margins, and having elongate to almost rectangular testa cells in distinct rows (*Sebaea* spp.). Rarely, the seeds are cup-shaped (*Exacum* spp.) or globular (*Tachiadenus tubiflorus*).

Taxonomic history of Exaceae

The tribe Exaceae was recognized in the first infrafamilial classification of Gentianaceae proposed by Colla (1834) in his description of the plants kept in the Turin herbarium. The family was divided into three tribes, viz. Gentianeae (as Gentianaceae), Exaceae, and Spigelieae, each characterized by the structure of the ovary. Bilocular ovaries, a rare feature in Gentianaceae, were recognized as an important character at this stage, and tribe Exaceae was erected for genera with this gynoecial structure. Four genera were included: *Exacum*, with two species, both now excluded from the genus; *Scoparia*, now in Scrophulariaceae, *Lisianthius*, now in Potalieae, and *Chironia*, now in Chironieae. No genus presently included in Exaceae was treated by Colla, although the name of course persists.

Three years later, Reichenbach (1837) presented his view of the classification of the family in *Handbuch des natürlichen Pflanzensystems*. Much emphasis was put on placentation in this system as well. Three subfamilies were accepted, spelled as Menyantheae, Gentianeae, and Loganieae. Gentianeae was subdivided into Sebaeariae, which included *Sebaea* but not *Exacum* and was characterized by a central placenta, Erythraeariae, which included, for example, *Exacum* and *Chironia*, and was characterized by parietal placentas, and Swertieae, which was characterized by having the ovules situated directly on the carpel margins in a parietal position. The material seen of *Exacum* was obviously not an *Exacum* species, or the gynoecium was wrongly interpreted.

A close relationship between *Exacum* and *Sebaea*, however, was

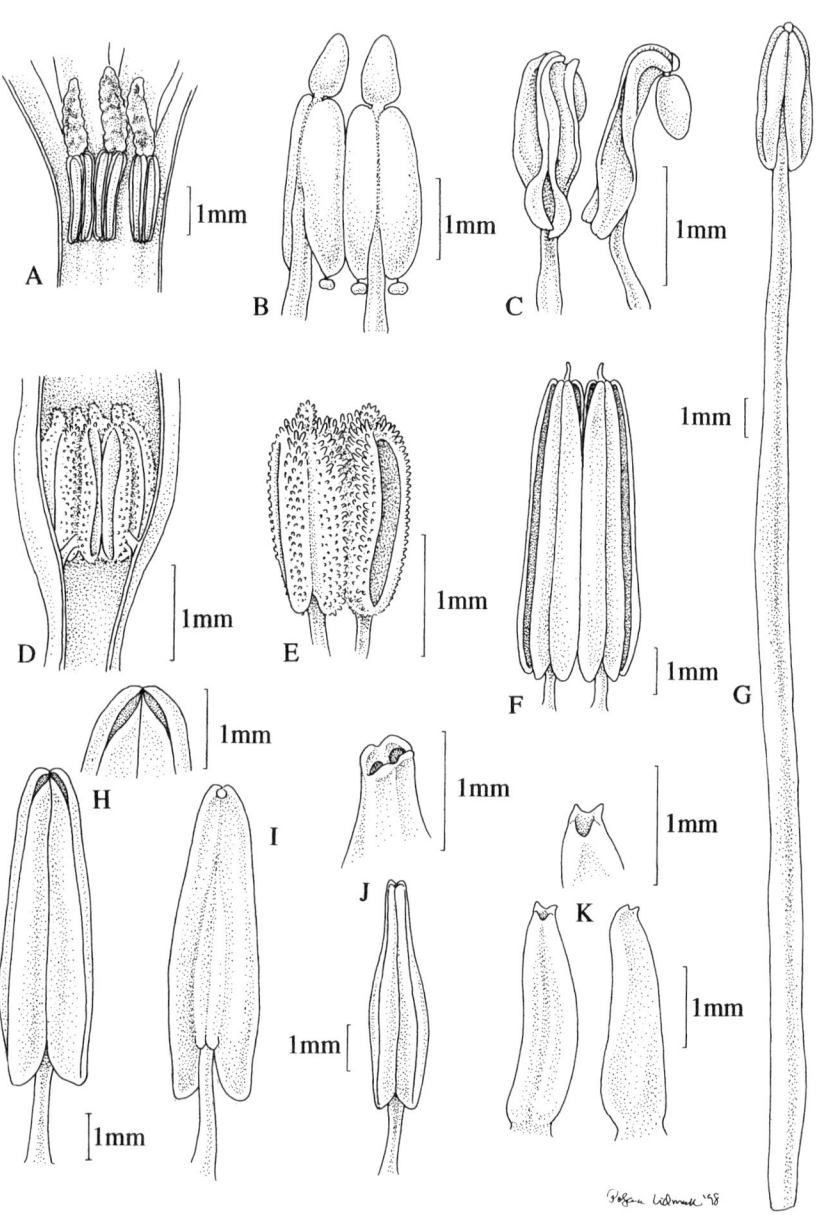

Systematics, character evolution, and biogeography

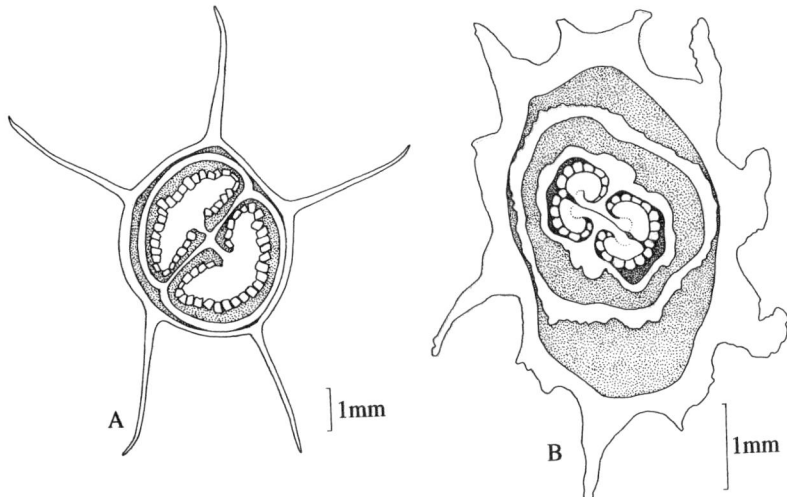

Figure 2.9. Cross-sections of ovaries of Exaceae. **A.** *Exacum courtallense* (*Klackenberg & Lundin 527*), showing axile placentation in bilocular capsule surrounded by a 5-winged calyx. **B.** *Tachiadenus longiflorus* (*Dorr 3681*), showing parietal placentation in pseudo-bilocular young ovary surrounded by the corolla and a 10-winged calyx.

Figure 2.8. (left) Stamens of Exaceae. **A–G.** Anthers opening by longitudinal slits. **A.** *Sebaea spathulata* (*Stewart 1901*), with anthers adhering to each other in a ring, exposing the pollen inwards. Each anther is furnished with a large apical appendage (gland). **B.** *Sebaea thomasii* (*Hoener 2055*), with anthers adhering to each other in a ring, seen from the dorsal side. Each anther is furnished with one apical and two basal appendages. **C.** *Sebaea albens* (*Wiese 1812*), with anthers free from each other, recurved when mature and furnished with apical appendages. **D, E.** *Ornichia trinervis* (*Randriamampionona 218*). **D.** Anthers adhering to each other in a ring, exposing the pollen outwards. Each anther is furnished with an apical appendage. **E.** Anthers from ventral side showing papillate margins of anthers adhering to each other. **F.** *Tachiadenus boivinii* (*Perrier 15532*), showing two anthers of an anther ring (seen from ventral side) adhering to each other along smooth margins, exposing the pollen outwards. Each anther has a small filiform apical appendage. **G.** *Gentianothamnus madagascariensis* (*Randrianasolo 136*), showing a free anther on long filament and with a small apical papilla. **H–K.** Anthers opening by pores, free. **H.** *Exacum divaricatum* (*Klackenberg 871028-1*), showing anther with two apical, oblique pores on ventral side. **I.** Same from dorsal side showing small apical papilla. **J.** *Exacum axillare* (*Bremer 848*), showing anther with two rounded apical pores. **K.** *Cotylanthera tenuis* (*Kaudern 304*), showing anthers with one apical pore from ventral (left and magnification) and dorsal (right) sides.

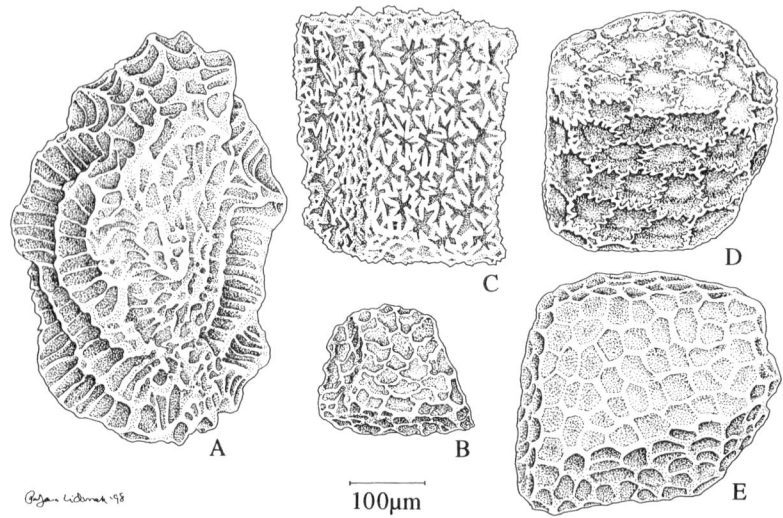

Figure 2.10. Seeds of Exaceae (redrawn from SEM micrographs). **A.** *Sebaea albens* (*Wall s.n.*), showing irregularly ellipsoid seeds with testa cells in distinct rows. **B.** *Sebaea grandis* (*Hilliard & Burtt 9911*), showing polyhedric seeds with shallowly undulated testa walls. **C.** *Exacum axillare* (*Bremer et al. 3*), showing polyhedric seeds with distinctly star-shaped testa cells. **D.** *Exacum stenopterum* (*Hildebrandt 3437*), showing seeds with finely undulated testa walls. **E.** *Exacum amplexicaule* (*Perrier 15959*), showing seeds with isodiametric testa cells with straight walls.

proposed the next year by Don (1837–1838), who divided Gentianaceae (as Gentianeae) into three tribes. Don also recognized the bilocular structure of the ovary as an important feature. Almost all genera (42) were put in tribe Gentianieae, and only *Desfontainia*, which was placed in tribe Desfontainieae (now Desfontainiaceae (Struwe & Albert, 1997; Struwe, 2002a) in Dipsacales (Backlund & Bremer, 1997)), and *Exacum*, *Microcala* (= *Cicendia*), and *Sebaea*, which were placed in tribe Exacieae, were separated.

The German botanists Endlicher (1838) and Grisebach (1839) nearly simultaneously presented tribal classifications, both without comments on earlier systems. Endlicher (1838: 600, 1841: 302) recognized two suborders within his order Gentianeae (= Gentianaceae), viz. Gentianeae-verae and Menyantheae, much in accordance with Reichenbach (1837). Gentianeae-verae was furthermore subdivided into two tribes – Chironieae, which included most of the genera, and Sebaeeae, which included only five genera. *Exacum* and *Tachiadenus* were placed in the former tribe, and *Belmontia, Lagenias* (both now included in *Sebaea*), and *Sebaea* were put in the latter. The diagnostic character used for the two tribes was also the structure of the gynoecium, i.e., a uni- or bilocular ovary. However, Endlicher obviously misinterpreted the gynoecium structure of *Exacum* as being unilocular, as had Reichenbach (1837). Consequently, he did not see a close affinity between *Exacum* and *Sebaea*. It is interesting to note, however, that Endlicher observed a close relationship between *Exacum* and *Tachiadenus*.

The same year, Grisebach (1839) published an elaborate account of Gentianaceae, *Genera et species Gentianearum*, in which the family was subdivided into seven tribes. Grisebach stressed the importance of the anther and style morphology, contrary to the earlier authors who focused on placentation. Also in this classification, *Exacum* and *Sebaea* were separated into different tribes (Chironieae and "Erythraeaceae", respectively), and neither of the earlier-erected tribes Exaceae and "Sebaeariae" was used. It should also be noted that *Exacum* was placed in Erythraeariae by Reichenbach (1837), with *Sebaea* outside, contrary to Grisebach, who placed *Sebaea* in "Erythraeaceae" with *Exacum* outside. Furthermore, *Tachiadenus* was referred to a third tribe, the Lisyantheae, because it was believed to have a persistent style, contrary to the presumably deciduous one of the former two tribes. *Exacum* (Chironieae) was said to lack a connective, in contrast to *Sebaea* (Erythraeaceae), which was characterized by anthers furnished with a connective.

Only a few years later, Grisebach (1845) revised his own account of the Gentianaceae for de Candolle's *Prodromus*, incorporating the classification presented by Endlicher (1838: 600, 1841: 302). The family was divided into the two tribes of Endlicher, Gentianeae (as Gentianeae-verae) and Menyantheae. Gentianeae was further subdivided into four subtribes. However, the genera today accepted in Exaceae were still referred to three different subtribes – *Exacum* in Chironieae, *Sebaea* in Chloreae, and *Tachiadenus* in Lisianthieae. Subtribe Chloreae was subdivided into two divisions – "Sabbatieae", with recurving anthers, and "Erythraeaceae", with anthers unchanged or becoming twisted. Unfortunately, because these differences in anther morphology were used as diagnostic traits for the divisions, *Sebaea sensu stricto* ("Sabbatieae"), *Belmontia*, and *Exochaenium* ("Erythraeaceae") – the last two genera both included in *Sebaea* by modern authors – were placed in different divisions. The key characters used for the subtribes Chironieae, Chloreae, and Lisianthieae were the same as in Grisebach's earlier classification (Grisebach, 1839, see above). *Tachiadenus*, endemic to Madagascar, was the only genus in Lisianthieae distributed outside the Americas. Regrettably, Grisebach's stress on certain style and anther structures created more confusion than clarification in the classification of Exaceae.

Bentham (1869) erected the tribe Eugentianeae for all gentians except the aquatic Menyantheae. Later, however, in his important work *Genera plantarum*, Bentham (1876) recognized four tribes. Grisebach's (1839, 1845) focus on anther morphology was abandoned, and Bentham put forward several new characters and again stressed the structure of the gynoecium. The genera with bilocular ovaries were grouped together in tribe Exaceae. *Tachiadenus*, although mistaken for having a true bilocular ovary, was also for the first time correctly placed in this tribe. Tribe Exaceae of Bentham very much corresponds to Exaceae as it is recognized today, i.e., including *Belmontia*, *Cotylanthera*, *Exacum*, *Sebaea*, and *Tachiadenus* (*Ornichia* and *Gentianothamnus* were not yet described). The classification of Bentham (1876) was followed in several large floristic works, e.g., *Flora of British India* by Clarke (1885) and *Flora of tropical Africa* (Baker & Brown, 1903). The French view of the nineteenth century, represented by Baillon (1888–1891) in *Histoire des plantes*, agreed with Bentham (1876) and included in Exaceae ("série des *Exacum*") *Exacum* (with *Cotylanthera* included as a section), *Sebaea*, and *Tachiadenus*, based on possession of bilocular ovaries.

The next overview of Gentianaceae was presented by Gilg (1895a). In his treatment for *Die natürlichen Pflanzenfamilien*, much emphasis was put on pollen morphology. The genera of the Exaceae were mostly kept together in the subtribe Exacinae, corresponding to Exaceae of Bentham (1876), with the exception of *Tachiadenus*, which was placed in subtribe Tachiinae together with *Lisianthius*, i.e., following the opinion of Grisebach (1839, 1845). The argument for this placement was pollen size and exine structure, both supposedly different from what is seen in *Exacum* and *Sebaea* and more similar to *Lisianthius* and related neotropical genera.

Engler followed the classification of Gilg (1895a) in subsequent editions of *Syllabus der Pflanzenfamilien*. In the latest edition, revised by Wagenitz (1964), subtribe Exacinae is represented by *Cotylanthera*, *Exacum*, and *Sebaea*. The recently described genus *Gentianothamnus* was added and placed in subtribe Erythraeinae owing to its large pollen grains and unilocular ovary. *Tachiadenus*, however, was left out of the classification.

There has been a general consensus in keeping Exaceae as a taxon separate but within the family. However, Daniel and Sabnis (1978) proposed, based mostly on their phytochemical analyses, that tribe Exacineae (Exaceae) should be raised to family level as Exacaceae. Molecular phylogenetic data (Figs. 2.1–2.3) contradict this standpoint, since the Exaceae clade is not the sister group to the rest of the family (tribe Saccifolieae is). Only if the most basal clade of gentians were to be accepted as a widely circumscribed Saccifoliaceae could Exacaceae also be recognized. However, morphological descriptions of such families would grossly overlap the description of the remaining Gentianaceae, and therefore no taxonomic clarity would be gained.

The most recent tribal classification with bearing on Gentianaceae was presented by Takhtajan (1997) in his *Diversity and classification of flowering plants*. Takhtajan (1997: 427), contrary to earlier studies, did not recognize the tribe Exaceae but included *Cotylanthera*, *Exacum*, and *Sebaea* in one of his four tribes, Gentianeae, together with, e.g., *Centaurium*, *Chironia*, *Gentiana*, *Halenia*, and *Swertia*. However, Exaceae do exist as a monophyletic group, and therefore the classification of Takhtajan does not improve our understanding of the tribe.

Generic synopses for Exaceae

Cotylanthera BLUME

This genus consists of small, almost leafless, saprophytic Southeast Asian herbs. Four species are recognized (Hara, 1975). Two of the species have a Malesian distribution: *Cotylanthera tenuis*, the type species, described from Java but ranging from Sumatra to New Guinea, and *C. loheri*, an endemic of the Philippines. The remaining two species occur in the Himalayas and the mountains of continental Southeast Asia; *Cotylanthera paucisquama* has been collected above 2000 m altitude in Sikkim, Bhutan, and Yunnan, and *C. caerulea* grows at slightly lower altitudes in Myanmar (formerly Burma), Thailand, Assam, and Nepal. All species grow among fallen leaves on shady ground in evergreen forests.

In the generic protologue, Blume (1826: 707) correctly described the characteristics of *Cotylanthera*, i.e., pale blue flowers with bilocular ovaries and anthers opening by single apical pores, although the genus was wrongly classified as closely related to Solanaceae. Endlicher (1838: 332) kept *Cotylanthera* as a dubious genus in Solanaceae, without any comment on its affinities. Miquel transferred the genus from Solanaceae and proposed a closer affinity to Gesneriaceae ("*Genus Cyrtandraceum dubium*"; Miquel, 1841: 734). Grisebach (1839, 1845) did not include *Cotylanthera* in his treatments of the family Gentianaceae. For de Candolle's *Prodromus* the genus was instead treated in connection with Solanaceae by Dunal (1852: 674) under dubious and insufficiently known genera. Dunal, however, was not satisfied with including *Cotylanthera* in Solanaceae. He cited Blume (1826: 707) in this regard, but added that perhaps the genus belongs to Cyrtandraceae (Dunal, 1852: 674). Bentham (1876: 800) was the first to transfer *Cotylanthera* to Gentianaceae, where it was placed in Exaceae next to *Exacum*. Later, Baillon (1888–1891: 114) discussed an even closer affinity with *Exacum* and reduced *Cotylanthera* to a section of this genus. Baillon's classification was never followed, however, and Gilg (1895a: 64) recognized three species of *Cotylanthera* and correctly placed them in his subtribe Exacinae.

The species of *Cotylanthera* are tiny, saprophytic, achlorophyllous herbs with simple or sparsely branched (*C. tenuis*) quadrangular stems. The leaves are scale-like. The inflorescences are few-flowered terminal cymes that sometimes are reduced to a single top flower. The flowers are 4-merous and are actinomorphic to slightly zygomorphic by having the anthers connivent into a cone above a bent style (*Cotylanthera paucisquama*). The calyx lobes are almost free, thin, and without wings. The blue to pale blue corolla is medium-sized (0.5–2.5 cm in diameter), shortly connate at the base, with oblong to narrowly oblong spreading lobes. The stamens are exserted from the tube, with the filaments inserted near the mouth. The anthers are thin and dehiscent through a small apical pore, rarely with a flattened apical appendix (*Cotylanthera loheri*). The filaments are slender and about as long as the anthers to broad and short (*Cotylanthera caerulea* and *C. tenuis*). The style is filiform and straight or slightly curved (*Cotylanthera paucisquama*). The stigma is small and capitate. The ovary is bilocular with axile placentas. The fruit appears to be a thin-walled capsule (J. Klackenberg, pers. obs.) but only poor herbarium material has been studied. The seeds are oblong with distinct testa cells, which have primarily straight walls.

Molecular phylogenetic information is still lacking for this genus. Nilsson and Skvarla (1969) did not find any palynological evidence to suggest a close relationship between *Cotylanthera* and *Exacum*, the former genus having a smooth exine in contrast to the striate to striato-reticulate pollen of the latter. However, it is most probable that *Cotylanthera* belongs to the same clade as *Exacum*, with which it shares two apomorphic anther characters, viz. anthers opening by pores and finely perforated endothecial walls. Autapomorphies for *Cotylanthera* are, e.g., anthers opening with one pore only (Fig. 2.8K) and saprophytism (if *Cotylanthera* and *Exacum* together form a monophyletic group). Furthermore, the corolla-androecium tube of *Cotylanthera* is completely fused with the ovary wall. The ovary is therefore semi-inferior with respect to the corolla-androecium tube, although the entire structure is very thin and membranaceous (Gopal

Systematics, character evolution, and biogeography 79

Krishna & Puri, 1962). No apomorphy is known for *Exacum* that excludes the species of *Cotylanthera*, and it is therefore possible that *Cotylanthera* is a derived lineage inside *Exacum*. A phylogenetic study of non-molecular characters including several species of *Exacum* and *Cotylanthera* (J. Klackenberg, in prep.) may help clarify the situation.

Exacum L.

There are at present 65 species of *Exacum* recognized (Klackenberg, 1985, 1990), a number rivaled in tribe Exaceae solely by *Sebaea*, the only other large genus. The genus *Exacum* has a wide paleotropical distribution, occurring from westernmost Africa to Madagascar and Socotra, and from India to New Guinea including the northernmost tip of Australia. The species, however, are very unevenly distributed over this large area; 59 of the 65 species occur in Madagascar (38 spp.), Socotra (4 spp.), and Sri Lanka (formerly Ceylon) plus the southern tip of India (17 spp.). Many species have small to local endemic distributions or have been collected only a few times and are therefore insufficiently known, but some are widely distributed and are sometimes common, e.g., *Exacum tetragonum* (India, Southeast Asia, northern Australia), *E. hamiltonii* and *E. teres* (the Himalayas and adjacent mountains), *E. petiolare* and *E. pedunculatum* (drier areas of India and Sri Lanka), *E. quinquenervium* (most of Madagascar and Mauritius), and *E. oldenlandoides* (all of tropical Africa). *Exacum* species have a wide spectrum of habitat preferences. Taxa are found from sea level up to the highest mountain tops in Madagascar (c. 2800 m elevation) and up to c. 2000 m in the Himalayas, South India, and New Guinea. Most species occur in lowland and montane rainforest areas, e.g., eastern Madagascar and the Western Ghats, although they usually grow in full sun. *Exacum* species are often found in grasslands and savannas, but they also occur in marshy or waterlogged areas and on wet rocks in ericoid mountain vegetation; many species are even found on road-cuts.

Exacum was introduced in 1747 by Linnaeus (1747a,b), and the genus was further treated in *Species plantarum* (Linnaeus, 1753) as one of four Linnaean genera (*Chironia, Exacum, Gentiana*, and *Swertia*) in today's Gentianaceae. Linnaeus included two species – *Exacum sessile* (the type species), and *E. pedunculatum*, which was described and illustrated in 1700 by Plukenet. Furthermore, one of Linnaeus's species of *Chironia* is now included in *Exacum, E. trinervium*. All three species were based on material from India and Sri Lanka. Not until the beginning of the nineteenth century were additional species described, also from India, by Roxburgh

(1814, 1820). More Asian species were added in the mid to late 1800s (Wallich, 1831; Don, 1837–1838; Arnott, 1839; Grisebach, 1839, 1845; Thwaites, 1860; Beddome, 1874), as were several species from the island of Socotra, including the well-known pot-plant *Exacum affine* (Regel, 1883; Balfour, 1884). The center of species diversity is now known to be Madagascar. However, before 1955 only six species were described from that island, the first being the widely distributed *Exacum quinquenervium*, whereas 38 are recognized there today (Klackenberg, 1990).

The species of *Exacum* are annual herbs to (probably) perennial subshrubs, between a few centimeters and a few meters high, erect and unbranched to usually moderately branched or rarely much-branched or even somewhat cushion-formed (*E. travancoricum*). Longer individuals sometimes have trailing branches (e.g., *Exacum walkeri*) that are rarely prostrate and rooting at the nodes (*E. radicans* and *E. subverticillatum*). The stems are terete to quadrangular and are usually furnished with four wings or lines. The leaves are herbaceous to rather succulent (e.g., *Exacum travancoricum*) or somewhat coriaceous (e.g., *E. fruticosum*, *E. lokohense*, and *E. marojejyense*). The arrangement of the leaves is decussate, but rarely verticillate (*Exacum radicans* and *E. subverticillatum*) or rosulate (*E. spathulatum*). The outline of the leaves varies from linear to orbicular, cordate to obovate, usually with a distinct petiole but sometimes amplexicaul. The inflorescences are usually ordinary mono- and dichasial cymes but sometimes the axes are shorter and the pedicels relatively longer toward the ends of the branches, sometimes even with the axes of the cymes totally suppressed, making the inflorescence umbel-shaped (15 spp.). The flowers are 4- or 5-merous and actinomorphic to often zygomorphic by having the anthers connivent into a cone above a bent style. The calyx is coalescent only at the very base or up to half of its length, and each lobe is furnished with a keel or broad wing that enlarges in fruit, although the wings and keels are sometimes lacking (*Exacum appendiculatum*, *E. gracile*, *E. lawii*, *E. linearifolium*, *E. naviculare*, and *E. sessile*). More rarely the calyx is zygomorphic with two well-developed and three more reduced wings (*Exacum anisopteris*, *E. bulbilliferum*, *E. dipterum*, *E. giganteum*, *E. penninerve*, and *E. subacaule*). The calyx wings are tapering to semicordate at the base and tapering to truncate at the apex, often with prominent veins. The corolla is white to pale blue to violet, small to medium-sized (a few millimeters to up to 7 cm in diameter), with a rather short and funnel-shaped to urceolate tube and erect to spreading, narrowly elliptic to orbicular lobes, rarely persistent in fruit and accrescent with hardening and prominent veins (*Exacum linearifolium* and *E. naviculare*). The anthers protrude from the corolla tube

Systematics, character evolution, and biogeography 81

on distinct filaments and are usually connivent around or above the style forming a cone, rarely with all anthers facing their dorsal sides upwards (the four Socotran species, *Exacum affine, E. caeruleum, E. gracilipes,* and *E. socotranum*). The anthers are somewhat coriaceous and are dehiscent by two apical pores. They are also often furnished with small apical papillae on their dorsal sides. The style is filiform and is straight or slightly curved. The stigma is small and entire to slightly bilobed. The ovary is bilocular with axile placentas. The fruit is a coriaceous, septicidally bivalved capsule, rarely membranaceous and opening with a coriaceous lid (*Exacum humbertii* group, 4 spp.). The seeds are angular to rarely cup-shaped (*Exacum radicans* and *E. subverticillatum*). The testa cell walls are usually undulate in a distinct star-shaped pattern, sometimes more shallowly so, or the cells are isodiametric with minutely undulate or straight walls.

Exacum is divided into two sections, section *Exacum* (21 spp. in India, Sri Lanka, the Himalayas, and Southeast Asia) and section *Africana* (44 spp. in Madagascar, Socotra, Oman, and tropical Africa; Klackenberg, 1985). Section *Exacum* is characterized by great variation in anther morphology (the thecae having thickened mid-walls of various shapes), but little diversity in seed morphology (mostly star-shaped testa cells) or in the structure of the cymes (axes as well as pedicels being well developed). On the other hand, section *Africana* is recognized by little variation in anther morphology (the thecae always having thin mid-walls), but great diversity in seed morphology (testa cells star-shaped to isodiametric with straight walls, sometimes minutely undulated) and in the structure of the cymes (which may be well developed, short, or reduced, i.e., umbelliform).

Apomorphies for *Exacum* are anthers opening by pores (Fig. 2.8H,J) and finely perforated endothecium walls (Klackenberg, 1985). These characters are also present in the small genus *Cotylanthera*. No synapomorphy for the genus *Exacum* which would exclude the four *Cotylanthera* species is presently known. Consequently, *Cotylanthera* might be an ingroup in *Exacum*, which would make *Exacum* paraphyletic.

Gentianothamnus HUMBERT

This monotypic genus is endemic to Madagascar. It was described by Humbert rather recently (1937a: 1747; 1937b: 388) and was placed in subtribe Erythraeinae, close to *Chironia* and *Orphium* because of unspecified details of pollen structure and other characters (also not stated). The only known species, *Gentianothamnus madagascariensis*, occurs in isolated populations from northern to southern Madagascar growing in the humid sclerophyllous montane forests above 1000 m altitude.

Gentianothamnus is an erect, sparsely branched subshrub up to 2 m in height. The stems are terete to subquadrangular with two close pairs of lines. The leaves are somewhat succulent and are usually elliptic with short petioles. The inflorescences each consist of single flowers borne atop suppressed axillary branches. The flowers are 5-merous and actinomorphic. The calyx is only shortly coalescent and the lobes are keeled to very narrowly winged. The corolla is large (2.5–4.0 cm long), yellow to orange, with a long, narrowly funnel-shaped tube and erect and elliptic to obovate lobes. The anthers are thick, dehiscent by longitudinal slits, and usually have small apical papillae on their dorsal sides. The long, slender filaments are inserted at the base of the corolla tube such that the anthers are free from each other and are slightly exserted from the tube. The style is filiform and long, usually becoming slightly exserted from the tube. The stigma is small and entire to faintly bilobed. The ovary is placed on a small, shallowly 5-lobed disk, apparently unilocular but in fact pseudo-bilocular with parietal placentas protruding to the center. The fruit is a coriaceous, septicidally bivalved capsule. The seeds are angular with testa cells having slightly undulate walls.

Gentianothamnus has not yet been characterized at the molecular level so its phylogenetic position is somewhat uncertain. However, it is here transferred from Chironieae and placed in Exaceae owing to several characters. The characteristic angular seeds with prominent undulate-walled testa cells, which together are interpreted as a synapomorphy for Exaceae, are present also in *Gentianothamnus*. Furthermore, the pollen grains show more similarity to *Tachiadenus* than to *Chironia* and *Orphium*. Additionally, the gynoecium, usually pseudo-bilocular but sometimes truly bilocular near the base, is similar to what is observed in other genera within Exaceae, and is particularly close to the condition seen in *Tachiadenus*. *Gentianothamnus* has, like *Tachiadenus*, a long corolla tube, although the shape of the tube differs in being narrowly funnel-shaped with essentially erect lobes (Fig. 2.7D). *Tachiadenus* has salver-shaped flowers. Furthermore, the corolla is yellow in *Gentianothamnus* and the anther filaments are long and inserted at the base of the tube. The thick-walled anthers, however, differ from the thin ones of *Sebaea* and are in general aspect more similar to the anthers of *Exacum* (Fig. 2.8G). Finely perforated endothecial cell walls, however – a synapomorphy for *Exacum* and *Cotylanthera* – are lacking. Thus, a more precise phylogenetic position of *Gentianothamnus* within Exaceae cannot be determined.

Ornichia KLACK.

This is a small Malagasy endemic genus of three species (Klackenberg, 1986) that occur in the eastern, more humid part of the island, where they grow mainly in evergreen forest up to 1500 m.

Ornichia was recognized as a separate taxon by Klackenberg (1986: 195). The genus consists of two species formerly placed in *Chironia* – the type species *Ornichia lancifolia* and *O. madagascariensis* – and one species formerly placed in *Tachiadenus*, *O. trinervis*. All species of *Ornichia* differ from *Chironia* and *Tachiadenus* by their bilocular ovaries and leaf pubescence. Furthermore, all species have angular seeds with conspicuously star-shaped testa cells, characteristic also for *Tachiadenus* and *Exacum*. Consequently, *Ornichia* was placed in Exaceae (Exacinae; Klackenberg, 1986).

The species of *Ornichia* are moderately to much-branched herbs or shrublets, at least finely hairy, and usually with terete but 4-lineolate stems. The leaves are herbaceous, narrow (*Ornichia lancifolia*) to elliptic or broadly ovate, mostly with rather distinct petioles. The inflorescences are terminal or axillary mono- and dichasial cymes. The flowers are 5-merous and actinomorphic. The calyx is coalescent only at the very base and the lobes are keeled to narrowly winged. The corolla, violet to blue, is medium-sized (1–2 cm in diameter) with a narrow and cylindric tube that is longer than the lobes (*Ornichia trinervis*) to usually narrowly funnel-shaped and slightly shorter than the lobes. The corolla lobes are elliptic to obovate and spreading. The anthers are exserted on slender filaments inserted near the mouth or are included (*Ornichia trinervis*) in the tube on very short filaments. The anthers are thin and dehiscent by longitudinal slits and have small apical appendices. They are free or cohere to each other in a ring (*Ornichia trinervis*) by their ventral theca walls, i.e., exposing the pollen outwards. The style is filiform and straight. The stigma is small and capitate to slightly bilobed. The ovary is bilocular with axile placentas. The fruit is a coriaceous, septicidally bivalved capsule. The seeds are angular with distinctly star-shaped testa cells.

Ornichia is similar to *Sebaea* in several characters, including the structure of membranaceous thecae that are furnished with an apical gland and open by lateral slits. Furthermore, one of the three species, *Ornichia trinervis*, has the anthers grouped and included together in a ring in a cylindric corolla tube (Fig. 2.8D), a flower structure very similar to that of several species of *Sebaea* (*Belmontia/Exochaenium*). The corolla, however, is blue, and although no morphological synapomorphy for the relationship is known at present, *Ornichia* is more closely related to *Exacum* according to molecular phylogenetic data (Fig. 2.3). *Ornichia* lacks, however, coriaceous anthers with finely perforated endothecium structure as well as dehiscence by apical pores, i.e., two apomorphies for *Exacum*. With its salver-shaped corolla, *Ornichia trinervis* shows affinity to the genus it was earlier placed in, viz. *Tachiadenus*, but the species is distinguished by the apomorphic character of having leaves with unicellular trichomes (Klackenberg, 1986). *Tachiadenus* differs in its pseudo-bilocular ovary, as well as by its anther connective appendices being small and filiform. *Ornichia* has a completely bilocular ovary and thick connective glands (Fig. 2.8D). A synapomorphy for *Ornichia* is hairy leaves (Klackenberg, 1986), a rare character in Gentianaceae not seen in any other species within tribe Exaceae.

Sebaea SOL. EX R. BR.

With its 60 (Dyer, 1975; Mabberley, 1997) to 100 (Wielgorskaya, 1995; Adams, 1996) species, *Sebaea* is probably the largest genus in Exaceae. In the latest monograph of *Sebaea sensu lato*, Schinz (1903, 1906) presented 95 species, but Boutique (1972: 38) estimated the species number to be as high as 150, whereas Paiva and Nogueira (1990: 3) postulated c. 159. The genus is distributed in Madagascar (4 spp.), India (1 sp.), Australia (2 spp.), New Zealand (1 sp.), and particularly Africa (remaining spp.), with a majority of the species in South Africa. The distributions of many species are insufficiently known, but a considerable number are endemic to the Cape region (e.g., *Sebaea albens*, *S. ambigua*, *S. exacoides*, and *S. pusilla*) or to the southeastern part of South Africa. Species from tropical Africa, however, usually have broad distributions, e.g., *Sebaea brachyphylla* (from Nigeria to Angola to tropical eastern Africa and Madagascar), *S. bojeri* (from the Cape Province to tropical eastern Africa and Madagascar), *S. aurea* (from the Congo to tropical eastern Africa and India), *S. grandis* (from Nigeria to Namibia and Tanzania to South Africa), *S. teuszii* (from Equatorial Guinea to Mozambique), *S. pumila* (from Nigeria to Mozambique), *S. oligantha* (from Ivory Coast to Uganda), *S. junodii* (from the Cape Province to Congo and Tanzania), and *S. leiostyla* (from the Cape Province to Congo and Ethiopia). The species of *Sebaea* occur from sea level up to 3000–3500 m altitude in the East African mountains (*S. brachyphylla* and *S. leiostyla*) and to 2000 m in the Himalayas (*S. aurea*). They grow in a wide variety of habitats, e.g., in dense tropical forests, grasslands, and savannas, in marshes or waterlogged areas, and on wet rocks and sandy riverbanks.

Solander erected the genus *Sebaea* in a manuscript that was published by

R. Brown (1810: 451). He transferred three yellow-flowered South African species (*Sebaea albens*, *S. aurea*, and *S. cordata*) and one Australian species (*S. ovata*, the type) from *Exacum* to the new genus, though he distinguished the new genus not on corolla color but on the anther opening mechanism (longitudinal slits) and on the form of the style (straight). Later, Rafinesque (1837: 78) accepted *Sebaea* but erected the genus *Parrasia* for the single species *S. cordata* (= *S. exacoides*). Rafinesque's genus *Parrasia* has been ignored by later authors. Likewise, Meyer (1838: 183) split *Sebaea* into three taxa when he transferred three South African species, among them *S. cordata*, to a separate genus, *Belmontia*. *Belmontia* was characterized by straight anthers included in the corolla tube but not exserted from the tube mouth or becoming recurved, as in *Sebaea*. In the same study Meyer (1838: 186) excluded another South African *Sebaea* species, for which he created the monotypic genus *Lagenias*. This genus was distinguished by having the staminal filaments inserted in the tube like *Belmontia*, but also by having the anthers becoming recurved, as in *Sebaea*. Only a few years later, Grisebach (1845: 55) further split *Belmontia* and erected one more monotypic genus, *Exochaenium*, which was based on the character of having the anthers, in addition to being included in the tube, also being connate to form a tube around the style. Consequently, in the mid-nineteenth century, *Sebaea* was split into four genera.

In *Genera plantarum* (Bentham, 1876), *Lagenias* and *Exochaenium* were included in *Belmontia*, which was kept distinct from *Sebaea*. Contrary to this classification, however, Schinz (1903) argued that it was impossible to discriminate between *Sebaea* and *Belmontia*. He therefore placed all species with anthers inserted in the tube, including *Belmontia*, *Exochaenium*, and *Lagenias*, into a section (*Belmontia*) of *Sebaea*, but nonetheless revised his own classification only three years later (Schinz, 1906: 744) by accepting *Exochaenium*, to which he transferred most species of *Belmontia*. *Exochaenium* was characterized by having included anthers and glands inside the calyx, characters seen also, for example, in *Tachiadenus* (Schinz, 1906: 744). Hill (1908: 317), when preparing *Flora Capensis*, followed Schinz's classification, adding one more character to *Exochaenium*, i.e., the style not having a biglandular swelling. Finally, however, Marais (1961), in preparing his account on Gentianaceae for *Flora of southern Africa*, argued that the three main characters mentioned above that were used to uphold *Belmontia/Exochaenium* as distinct from *Sebaea* were unreliable, and consequently all three satellite genera were synonymized under *Sebaea* (Marais & Verdoorn, 1963: 172).

The species of *Sebaea* are annual to perennial, erect or procumbent herbs, or rarely achlorophyllous saprophytes (*S. oligantha*). The stems are terete to quadrangular and are usually furnished with four wings or lines. The leaves are decussate and cauline or rarely rosulate, herbaceous and well developed or rarely scale-like (*Sebaea oligantha*). The outline of the leaves varies from linear to suborbicular, with a distinct petiole, but is often sessile and cordate at the base. The inflorescences are usually dichasial cymes, sometimes monochasial toward the end of the branches, rarely one-flowered or with a basal underground inflorescence with cleistogamous flowers (*Sebaea oligantha*). The flowers are 4- or 5-merous and are actinomorphic. The calyx is coalescent into a distinct tube only at the very base. The lobes are frequently thin with hyaline margins and are furnished with a keel or sometimes a broad, prominently veined wing that enlarges in fruit. The corolla is small to medium-sized (a few millimeters up to 4 cm in diameter), either with a short and funnel-shaped or urceolate tube with erect to spreading lobes or salver-shaped with a long (up to 2.5 cm) and cylindric tube. The lobes, white to cream or yellow, are ovate to obovate. The anthers are sometimes included in the tube on very short filaments but may also be exserted on slender filaments inserted near the mouth and connivent around or above the style forming a cone. The anthers are membranous, often becoming twisted backwards and dehiscent by longitudinal slits. Small but sometimes conspicuous appendices, one apical and often two basal (rarely only two basally), are often present on the anthers. The anthers are free or coherent to each other in a ring via their ventral or dorsal theca walls. The style is filiform and straight, exserted or included, often with an additional lower stigma(s), which is/are confluent with the apical stigma or usually placed distinctly below. The apical stigma is capitate or elongate, entire to bilobed. The ovary is bilocular with axile placentas. The fruit is a membranous or coriaceous septicidally bivalved capsule. The seeds may be angular with the walls of the more or less isodiametric testa cells slightly undulate, or more often somewhat elliptic or oblong in outline with irregular margins and slightly flattened. The testa cells are elongate to almost rectangular and form distinct rows.

The systematics of *Sebaea sensu lato* is confused, and no satisfactory classification of the genus currently exists, let alone any phylogenetic analysis. Two species of *Sebaea sensu lato* are included in the molecular analysis, viz. *S.* cf. *macrophylla* (*Sebaea sensu stricto*) and *S. exacoides* (*Belmontia/Exochaenium*; Figs. 2.1–2.3). In all of the phylogenetic results these two species form a distinct monophyletic group, and in the *trn*L intron and combined analysis (Figs. 2.1, 2.3), this group has a jackknife support value of 100%. Although a more inclusive phylogenetic analysis is lacking and necessary, *Belmontia, Exochaenium,* and *Lagenias* are here treated as synonymous with *Sebaea* based on morphological similarity.

Tachiadenus GRISEB.

This is a small, homogeneous genus of 11 species, all native to Madagascar. The species occur over a large part of the more humid East Malagasy phytogeographic region, except for the higher mountains. Some species are widely distributed along the eastern lowlands, growing in different habitats, e.g., forests, grasslands, and marshes (*Tachiadenus gracilis*) or coastal forests and dunes (*T. carinatus* and *T. tubiflorus*). Other common species are restricted to the central plateau above 800 m altitude and are often found in scrub or grassland (*T. longiflorus* and *T. platypterus*).

Tachiadenus was introduced by Grisebach (1839: 200), and was characterized by its very long and narrow corolla tubes. Three non-American species of *Lisianthius* were transferred to *Tachiadenus* (*T. carinatus* (type species), *T. trinervis*, and *T. tubiflorus*) and one new species (*T. longiflorus*) was added. *Tachiadenus trinervis*, a local endemic to southwesternmost Madagascar, was later transferred to the newly erected genus *Ornichia* (Klackenberg, 1986). A few more taxa were added during the twentieth century: Humbert (1963) added four, whereas Klackenberg (1987a) recognized 11 species in a revision of the group.

Grisebach placed *Tachiadenus* in tribe Lisyantheae, but the genus was later transferred to tribe Exaceae by Bentham (1876). However, owing to its pollen and gynoecium morphology, *Tachiadenus* was placed outside Exaceae in subtribe Tachiinae by Gilg (1895a), i.e., near the original position given to it by Grisebach in Lisyantheae (largely corresponding to tribe Helieae in this classification). Klackenberg (1985: 17) disagreed with this opinion and stated that the size and structure of the pollen grains of *Exacum* and *Tachiadenus* were similar; he therefore transferred the genus from Tachiinae back to Exaceae (Klackenberg, 1987a: 43).

The species of *Tachiadenus* are erect annual herbs with woody bases or perennial subshrubs from *c.* 0.1 m (several species) to 3 m in height (*T. tubiflorus*). The stems are terete to subquadrangular and are sometimes wingless (*Tachiadenus tubiflorus*), but they are usually furnished with four wings or lines. The leaves are herbaceous and are usually narrowly elliptic to narrowly ovate or ovate. The petioles are sometimes distinct (*Tachiadenus longiflorus* and *T. tubiflorus*) but are usually indistinct, and some species have semi-amplexicaul (*T. longiflorus, T. pervillei*, and *T. platypterus*) or fully amplexicaul (*T. antaisaka*) leaves. The inflorescences are ordinarily di- to monochasial cymes, but sometimes the axes of the cymes are suppressed and then the inflorescences are umbel-shaped (*Tachiadenus antaisaka*, *T. longifolius*, *T. tubiflorus*, and *T. umbellatus*). The flowers are 5-merous and actinomorphic. The calyx is coalescent only at the very base (*Tachiadenus gracilis* and *T. tubiflorus*) or up to ¾ of its length (*T. platypterus*). Each calyx lobe is furnished with a keel or a distinct wing (*Tachiadenus carinatus*, *T. longiflorus*, *T. pervillei*, and *T. vohimavensis*), but sometimes the calyx is completely wingless (*T. boivinii* and *T. tubiflorus*) or, more rarely, 10-winged (*T. longiflorus*). The calyx lobes are linear and subulate (*Tachiadenus boivinii* and *T. gracilis*) to obovate and obtuse (*T. platypterus*). The corolla, white to violet, is salver-shaped and large with an up to 2–20 cm long tube, usually with a small to minute coronula above the anthers. The anthers are thin and dehiscent by longitudinal slits and are usually furnished with a filiform appendix at the apex, rarely with a thicker apical appendix and two smaller basal ones (*Tachiadenus tubiflorus*). The anthers are included in the corolla tube and are inserted on very short filaments that cohere to each other in a ring by their ventral theca walls, i.e., exposing the pollen outwards. The style is filiform and long, usually becoming slightly exserted from the tube. The stigma is small and entire to faintly bilobed. The ovary is unilocular, becoming pseudo-bilocular via the parietal placentas that protrude to the center. The fruit is a coriaceous, septicidally bivalved capsule. The seeds are angular with distinctly star-shaped testa cells or rarely rounded with unevenly protruding testa walls (*Tachiadenus tubiflorus*).

Apomorphies for *Tachiadenus* are the large flowers with long corolla tubes (Fig. 2.7G), and the pseudo-bilocular ovary (Fig. 2.9B). All of these characters, however, are present in several other Gentianaceae genera of other tribes. However, in tribe Exaceae these traits are considered apomorphic, although also characteristic of the genus *Gentianothamnus*. Large flowers with long tubes also occur in some species of *Sebaea* (e.g., *S. exacoides*, *S. grandiflora*, and *S. thomasii*), probably as a parallelism.

Phylogeny and character evolution in Exaceae

Although the tribe Exaceae comprises only six genera, it nevertheless displays considerable character variation, much of which may be interpreted as having evolved in parallel with other Exaceae genera and taxa outside the tribe.

Vegetative structures

Exaceae are annual herbs or suffrutescent plants that can survive two or more years, a usual condition in Gentianaceae. *Cotylanthera* consists exclusively of small herbs, and *Gentianothamnus* is always woody at the base.

The small genus *Cotylanthera* comprises achlorophyllous saprophytes (Fig. 2.7A). Although the saprophytic life form is a synapomorphy for the four species of *Cotylanthera*, it is also known in one species of *Sebaea*, *S. oligantha* (Raynal, 1967b), similarly of tribe Exaceae. In Gentianaceae, saprophytism is also characteristic for *Voyria* and *Voyriella* (*incertae sedis* and Saccifolieae, respectively; see also Albert & Struwe, 1997). It is interesting to note that saprophytic taxa appear to have evolved twice within Exaceae, in Asia (*Cotylanthera*) and in Africa (*Sebaea oligantha*), respectively.

The leaves of Exaceae are opposite-decussate. However, in two species of Malagasy *Exacum*, viz. *E. radicans* and *E. subverticillatum*, the leaves are verticillate, usually with four leaves per whorl. This condition is due to an internodal reduction that places two to three nodes close together. Whorled leaves are an apomorphic character in Exaceae, but are not unique in Gentianaceae, being paralleled in, for example, *Curtia confusa* and *C. verticillaris*. In these two species of *Curtia* (Saccifolieae), the structural mechanism for verticillation is probably homologous with *Exacum* species, even though there are no external traces of shortened internodes. However, each leaf might subtend an axillary shoot, and the whorls do not seem to include stipules; there is also no splitting of the leaves.

The leaves in all genera of Exaceae are either parallel-nerved or one-nerved, with the lateral nerves, if present, diverging from the base. However, a dozen Malagasy species of *Exacum* are characterized by having apparently penninerved leaves (an apomorphic trait) owing to the lateral nerves being closely parallel to the mid-nerve near the base and diverging only after a distance, in some species up to about half the length of the lamina (Klackenberg, 1985, 1990). A similar type of venation, but usually with a more distinctly set mid-nerve (a truly penninerved condition), is also present in other tribes, e.g., many genera of Potalieae, as well as in *Tachia* sect. *Tachia* and in some *Macrocarpaea* in Helieae. Penninerved leaves are also present in possible outgroups to Gentianaceae, e.g., Rubiaceae or Loganiaceae. Being present in all genera of the basal clades of Exaceae and Saccifolieae, parallel-nerved leaves are considered plesiomorphic for Gentianaceae, and probably a synapomorphy for the family.

Inflorescences

The inflorescence is a monochasial or dichasial cyme, sometimes one-flowered. The cymes usually have a dichasial ground plan but often become monochasial toward their apices. In *Exacum* and *Tachiadenus*, umbel-like cymes are also observed, i.e., cymes with short to lacking internodes. This condition has evolved twice in *Exacum* (Klackenberg, 1985) and as a parallelism once in *Tachiadenus* (Klackenberg, 1987a). However, it has not been observed in the large genus *Sebaea*. Although in *Sebaea* the internodes of the cymes are often shortened, the resulting inflorescences are condensed, but not umbellate. In *Sebaea*, monochasia from the base have been observed, e.g., in *S. madagascariensis* and *S. condensata* (Klackenberg, 1987b). In these species, at each node only the top flower and one branch are present, the second branch being totally reduced.

Single flowers are sometimes observed, often in particular individuals of taxa with usually few-flowered, often umbelliform, cymes. However, *Gentianothamnus madagascariensis, Cotylanthera caerulea*, and *C. paucisquama* seem to have obligately one-flowered inflorescences, but they are not homologous. In *Cotylanthera* there is a solitary flower at the top of a much-reduced, unbranched stem. On the other hand, *Gentianothamnus* is a 2 m tall shrub that bears several flowers, but each inflorescence is reduced and carries a single flower in an axillary position.

A basal, underground inflorescence with cleistogamous flowers plus an ordinary chasmogamous aerial one has evolved in the saprophyte *Sebaea oligantha* (Raynal, 1967b: 217). This type of inflorescence is not present in other saprophytic taxa, i.e., *Cotylanthera* within Exaceae, nor in *Voyria* or *Voyriella*.

Flowers

In Gentianaceae, as well as in possible outgroups to the family, the flowers are usually 5-merous, more rarely 4-merous, which consequently is likely to be a derived state within Exaceae. Four-merous flowers, however, have evolved several times within the tribe. Klackenberg (1985) showed this character to be a parallelism in three different clades of *Exacum*, and 4-merous flowers are also found in c. 10 species of *Sebaea*. Furthermore, *Cotylanthera* is 4-merous.

Calyx

There is great morphological variation in calyx structure within Exaceae. In general, the five (four) calyx lobes are equal and each is furnished with a keel or wing along the mid-nerve (Figs. 2.6A, 2.7B–C,E,G), but several exceptions exist. In addition to the five ordinary wings, the calyx rarely has five extra wings in between them that are situated along the suture between the coalescent calyx lobes (Fig. 2.6D). This apomorphic trait has arisen independently in *Exacum* (*E. decapterum*) and *Tachiadenus* (*T. longiflorus*; Klackenberg, 1987a: 65; 1990: 32), but is not known from other genera. The usually actinomorphic calyx is zygomorphic in a clade of Malagasy

Exacum that displays two well-developed wings plus two smaller ones on one side and one smaller one on the opposite side (Klackenberg, 1985: 12). In this way, the calyx of *Exacum dipterum* becomes two-winged (Fig. 2.6B). This form of zygomorphy is not observed in the other genera within the tribe. Asymmetric calyx lobes are seen in *Tachiadenus tubiflorus* (Klackenberg, 1987a: 77). In several species of *Exacum*, *Sebaea*, and *Tachiadenus*, as well as in all species of *Ornichia* and *Gentianothamnus*, the calyx lobes are very narrowly winged or keeled. *Cotylanthera* and a few *Exacum* and *Sebaea* species totally lack wings on the calyx (Figs. 2.6C, 2.7A). In several species of *Exacum*, *Sebaea*, and *Tachiadenus* the calyx wings are accrescent in fruit with prominent veins (Figs. 2.6A, 2.7C), a characteristic which manifests itself as several parallelisms in Exaceae. Additionally, wingless, prominently veined calyx lobes are sometimes distinctly accrescent in fruit, e.g., in *Exacum lawii*. Distinctly hardening and prominent veins are also present outside Exaceae, e.g., in *Neurotheca* of the Potalieae (Raynal, 1968).

Corolla

The corolla is less variable than the calyx but shows large variation in tube length. All species of *Exacum*, two of the three *Ornichia* species, and most species of *Sebaea* have a short and often urceolate tube with the stamens exserted. In contrast, *Tachiadenus* has conspicuously long and narrow tubes (Fig. 2.7G), in some species up to 15–20 cm long, with the stamens included (Klackenberg, 1987a). Although the flowers are smaller in absolute dimensions, salver-shaped corollas are also present in several species of *Sebaea* (*Belmontia* and *Exochaenium*; Fig. 2.7F) and in *Ornichia trinervis* (Fig. 2.8D). At the present state of knowledge it is difficult to decide if long and narrow corolla tubes are apomorphic or plesiomorphic in Exaceae. In several genera of the possible outgroups to the tribe, salver-shaped corollas are present. *Sebaea*, being sister to the remainder of Exaceae, has both types of flowers. However, with corollas of one species of *Ornichia* also being salver-shaped, tube form must be an evolutionarily plastic character, one that shows several transformations within the tribe. The strikingly long, salver-shaped tubes of *Tachiadenus* are considered a synapomorphy for this genus. The corolla in *Gentianothamnus*, although furnished with a long tube, is not salver-shaped but instead narrowly funnel-shaped without spreading lobes.

There are two different patterns of corolla color variation within the tribe, white to usually blue or violet, and white to usually pale yellow or yellow. *Cotylanthera*, *Exacum*, *Ornichia* (only blue specimens known), and *Tachiadenus* belong to the white–blue pattern, and *Gentianothamnus* (only yellow specimens known) and *Sebaea* follow the white–yellow pattern. Yellow flowers are characteristic for individual genera and species in Gentianaceae, but the yellow corolla color in *Gentianothamnus* and *Sebaea* probably represents an apomorphy within the clade, possibly even parallelisms (the phylogenetic position of *Gentianothamnus* is unclear). In *Exacum linearifolia* and *E. naviculare* the corolla lobes are persistent and accrescent in fruit with hard, prominent veins (Klackenberg, 1985), probably assisting in the dispersal of seeds.

Androecium

The anthers are usually connivent, forming a cone or a ring; more rarely, they are arrayed at a distance from each other. The anthers open by apical pores (*Cotylanthera* and *Exacum*; Fig. 2.8H–K) or longitudinal slits (*Gentianothamnus, Ornichia, Sebaea*, and *Tachiadenus*; Fig. 2.8A–G). The connective in anthers of *Cotylanthera* is reduced in the upper part, and the thecae form a single apical pore (Fig. 2.8K). In *Exacum*, however, the connective is well developed and the anthers dehisce by two pores (Fig. 2.8H,J). Anthers opening by pores is considered a synapomorphy for *Cotylanthera* and *Exacum*, and anthers opening by one pore a synapomorphy for the *Cotylanthera* species. Thin-walled anthers opening by slits, present in *Sebaea* (the sister group to the rest of the tribe) but also in *Gentianothamnus, Ornichia*, and *Tachiadenus*, are considered plesiomorphic. Slit-opening anthers are prevalent in possible outgroups to Exaceae.

The apical part of the connective often bears projections of different shapes. The anthers of *Sebaea* are usually furnished with small apical appendages (glands, or Brown's bodies; Fig. 2.8A–C), which are thick structures that contain polysaccharides (Schinz, 1903: 130). They are eaten by the pollinators, i.e., different species of thrips (Marloth, 1909: 314). Often there is also a similar gland basal to each theca. In *Sebaea*, all three glands may be present, only the apical one, only the basal two, or none. Similar glands, although perhaps not homologous, are characteristic for *Ornichia* (apically; Fig. 2.8D) and are known in one species of *Tachiadenus*, *T. tubiflorus* (both apical and basal glands). The remaining species of *Tachiadenus* are instead characterized by a short, apical, filiform connectival appendix (Fig. 2.8F). On the other hand, in *Exacum* there is often a small, globular papilla situated apically on the dorsal side of the connective (Fig. 2.8I). This papilla seems not to differ in texture, however, from the rest of the anther. The function of these papillae is unclear. In *Cotylanthera loheri* the connectives are elongate and flattened (Hara, 1975), probably participating in the dispersal of pollen.

The ontogeny of the anthers of different taxa of Gentianaceae has been investigated. In Exaceae, *Cotylanthera tenuis* (Oehler, 1927: 670) and three species of *Exacum* (Maheswari Devi, 1962: 213) have been studied. In Gentianaceae, the subepidermal layer usually develops into a fibrous endothecium. This condition was not, however, observed in *Exacum* and *Cotylanthera*. Klackenberg (1985: 13, fig. 5) showed that the endothecial walls of all species of *Exacum* are in fact finely perforated and he proposed this to be a synapomorphy for the genus. Unlike the typically hard anthers of *Exacum*, the anther walls of *Cotylanthera* are very thin, but nevertheless exhibit the same finely perforated endothecial walls. The thick-walled anthers of *Gentianothamnus*, which are similar to *Exacum* anthers by being furnished with a small apical papilla on the dorsal side, lack these perforations and instead have the ribbed (fibrous) endothecial cell walls otherwise prevalent in the tribe. Such ribbed cells are also characteristic of the thin-walled anthers of *Ornichia*, *Sebaea*, and *Tachiadenus*, as well as of several other examined taxa of Gentianaceae (Klackenberg, 1985).

Cross-sections of anthers of all species of *Exacum* were studied by Klackenberg (1985: 13, fig. 4). Differently shaped thickenings and outgrowths of the septa between thecae were shown to be prevalent in section *Exacum*, but have not been observed elsewhere.

The staminal filaments are usually long and inserted in or near the lobe sinuses in *Exacum* (Fig. 2.7B–C) and *Cotylanthera*, or they may be long and inserted in the lower part or at the base of a rather open corolla tube in *Gentianothamnus* (Fig. 2.7D) and *Sebaea* (*Lagenias*) *pusilla*. In *Tachiadenus* the filaments are very short and are inserted in the upper part of a narrow, cylindric corolla tube (Fig. 2.7G). In *Sebaea* and *Ornichia*, both long-filamented anthers, as seen in *Exacum*, and short-filamented anthers, as seen in *Tachiadenus*, are present (Figs. 2.7E–F, 2.8A,C–D). The different filament lengths and insertions have apparently evolved several times within the tribe. In six species of *Exacum* (parallelisms in three different clades; Klackenberg, 1985) the filament bases are connate into a short tube.

All species of *Cotylanthera*, *Exacum* and the monotypic *Gentianothamnus* have exserted anthers on long filaments. In contrast, *Tachiadenus* has included anthers on very short filaments. However, this character, whichever state being plesiomorphic, very likely evolved several times, as both states are present in the sister clade to all other Exaceae, *Sebaea*, as well as in *Ornichia*, which is more closely related to *Exacum*. It should be noted that within the small genus *Ornichia*, one species has short-filamented, included stamens and two species have exserted stamens on long filaments.

In species with salver-shaped corollas the anthers adhere to each other in a ring around, below, or above the stigma. In *Tachiadenus* the anthers seem to be united to each other postgenitally by the ventral theca walls, and consequently the pollen grains are liberated outwards (Fig. 2.8F). Extrorse pollen dehiscence has also evolved in *Ornichia trinervis*. However, here the theca walls adhere to each other by marginal papillae (Fig. 2.8D–E). In *Sebaea*, species with anther rings opening inwards (Fig. 2.8A) or outwards are known. The introrse opening mechanism is functionally relevant for *Sebaea* species that have a secondary, lower stigmatic area ensuring self-pollination.

Palynology of Exaceae: Synopses of genera and general discussion

S. NILSSON

Exacum. Exacum affine I. B. Balf. ex Regel. The pollen grains are 3-colporate, prolate spheroidal, and $c.$ 20 × 17 μm. The exine is tectate-perforate, striate, with muri/lirae branched to unbranched, sharply keeled, and striae with rounded to oval-shaped perforations of varying size. (Oman: Dhufar Leje, 1984, *Miller 6201* (S), slide no. 23856; Fig. 2.11A–B.)

OTHER INVESTIGATIONS: Agababyan and Tumanyan (1976a) reported that most species of *Exacum* have relatively small pollen grains (polar axis not exceeding 25 μm). In his study of the genus *Exacum* in Sri Lanka, Klackenberg (1983) examined pollen of the eight species *Exacum axillare*, *E. macranthum*, *E. pallidum*, *E. pedunculatum*, *E. petiolare*, *E. sessile*, *E. trinervium*, and *E. walkeri*. Pollen size ranged between 13.5–35.0 μm (polar axis) and 9.5–26.0 μm (equatorial diameter). The exine pattern varied both inter- and intraspecifically from striate to striato-reticulate, insular, or with short, ridge-like elements. The striae were imperforate to perforate, or minutely reticulate. Each species had two types of pollen differing in shape. A remarkable diversity in exine pattern was noted and illustrated in *E. pallidum*, and this showed correlation to the geographic origin of the investigated material. In his monograph of *Exacum*, Klackenberg (1985) included pollen descriptions and illustrations for all species. A few species – *Exacum lawii*, *E. pumilum*, and *E. sessile* – have small pollen grains (polar axis less than 16 μm), while others, e.g., *E. anamallayanum*, *E. atropurpureum*, and *E. walkeri*, have larger pollen grains (polar axis $c.$ 30 μm). The exine pattern varies from striate, striato-reticulate to reticulate, or sometimes smooth with slits or perforations. The large variability of exine pattern seen within *Exacum* supports Klackenberg's (1985) conclusion that exine pattern of the same general type found in, for example, *Exacum*, *Gentiana*, and *Swertia* is

Systematics, character evolution, and biogeography 95

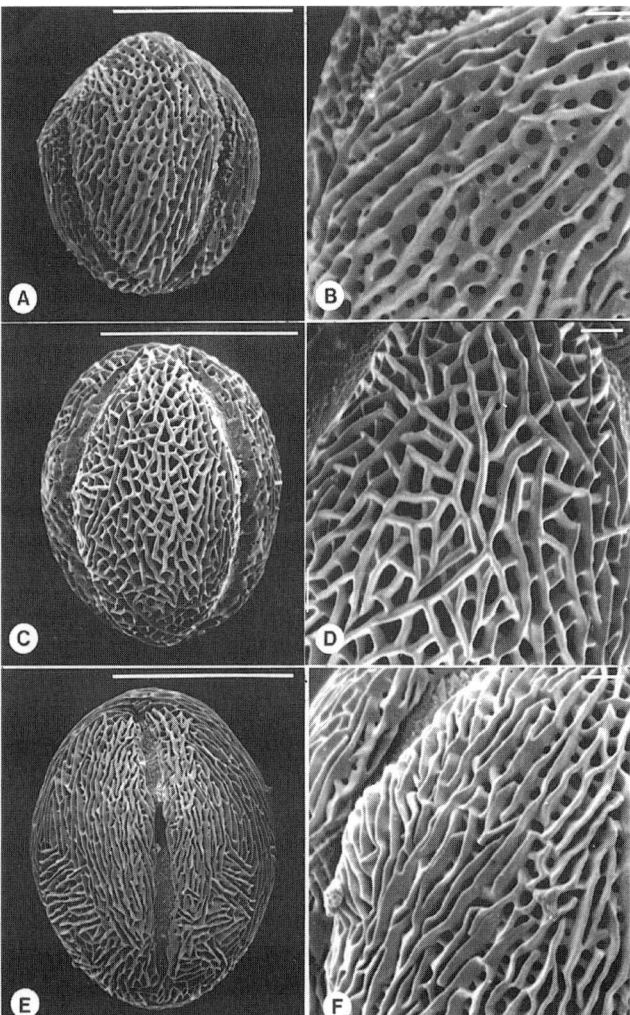

Figure 2.11. Pollen of Exaceae. **A, B.** *Exacum affine* (*Miller 62011*). **A.** 3-colporate, striate-perforate pollen in equatorial, mesocolpial view. **B.** Detail of tectate-perforate exine with parallel, often branched and keeled muri/lirae, separated by grooves (striae) with rounded to oval perforations up to the size of foveolae. **C, D.** *Sebaea leiostyla* (*Meeuse 10180*). **C.** 3-colporate, striato-reticulate pollen in equatorial, mesocolpial view. **D.** Detail of exine with strongly branched, partly interlaced lirae/muri, forming a striato-reticulate exine pattern. **E, F.** *Ornichia madagascariensis* ssp. *madagascariensis* (*Rakotovao s.n.*). **E.** 3-colporate, striate to striato-reticulate pollen in equatorial, apertural view, muri/lirae parallel to colpi but changing orientation at the colpus margin and toward the polar areas; colpus slightly constricted at equator. **F.** Detail of exine with mostly parallel and branched, relatively high and narrow muri/lirae interconnected by transverse muri at a lower level. Scale bars: A, C, E = 10 μm; B, D, F = 1 μm.

not indicative of a close relationship between these genera. Wang (1960) provided a description and illustration of the small, striato-reticulate pollen grains of *Exacum tetragonum*. Raj (1966) described pollen of *Exacum pedunculatum* as small (*c*. 22 × 15 µm), 3-colporate, and finely reticulate. Lienau *et al.* (1986) described a species of *Exacum* from Madagascar as having small (*c*. 20 × 19 µm) and rugulate pollen grains.

Gentianothamnus. *Gentianothamnus madagascariensis* Humbert. The pollen grains are 3-colporate and *c*. 22 × 18 µm. The exine is finely reticulate to rugulate and perforate. (Madagascar: 1949, *Humbert & Cours 23715* (P), slide no. 6933; Nilsson, 2002: Fig. 4.4.)

OTHER INVESTIGATIONS: Humbert (1937a) briefly described the pollen of his new genus, *Gentianothamnus*, as 3-colporate and perforate, whereas Lienau et al. (1986) described it as small and scabrate-perforate. The latter description has been confirmed in a recent study (Nilsson, 2002).

Ornichia. *Ornichia madagascariensis* ssp. *madagascariensis* (Baker) Klack. The pollen grains are 3-colporate, prolate spheroidal, and *c*. 21 × 19 µm. The exine is striate to striato-reticulate, with muri/lirae branched to unbranched, relatively high and narrow, slightly keeled, and parallel at the centers of mesocolpia but changing orientation abruptly at the colpus margin and toward the poles. The striae are perforate or with lumina. (Madagascar: Fianarantsoa, 1994, *Rakotovoa 113* (S), slide no. 24771; Fig. 2.11E–F.)

OTHER INVESTIGATIONS: Klackenberg (1986) described pollen from the genus *Ornichia*, based on all three species, as having 3-colporate and prolate pollen, being 17–22 × 15–20 µm (polar axis and equatorial diameter) and having striato-reticulate exine with perforate striae. This description is in accordance with *Ornichia madagascariensis* from the present study (Fig. 2.11E–F), although there is a tendency toward a partly rugulate and relatively open striato-reticulate exine pattern. Similar features have been found in some genera in the Chironieae, e.g., *Centaurium* and *Ixanthus*.

Sebaea. *Sebaea leiostyla* Gilg. The pollen grains are 3-colporate, prolate, and *c*. 23 × 17 µm. The exine is striato-reticulate with muri/lirae usually branched and interlaced and slightly keeled. The colpus margin is relatively broad, smooth, and sparsely perforated. (South Africa: Transvaal, 1957, *Meeuse 10180* (S), slide no. 18442; Fig. 2.11C–D.)

Systematics, character evolution, and biogeography 97

OTHER INVESTIGATIONS: The pollen grains of *Sebaea brachyphylla* from Madagascar were described as small (*c.* 23 × 22 μm) by Lienau *et al.* (1986). Moar (1993) included a detailed description of *Sebaea ovata* from New Zealand and described it as having 3-colporate and finely reticulate pollen grains (25–29 × 25–27 μm).

Tachiadenus. Tachiadenus carinatus (Desr.) Griseb. The pollen grains are 3-colporate, prolate spheroidal, and 30 × 27 μm. The exine is finely reticulate to perforate with muri of varying width. Colpus margins are distinct, thickened and smooth with few perforations. (Madagascar: Ambodimanga, 1987, *Klackenberg 871124-5* (S), slide no. 24144; Nilsson, 2002: Fig. 4.5A–C.)

OTHER INVESTIGATIONS: Apart from those of *Tachiadenus carinatus*, pollen grains of *T. antaisaka, T. gracilis, T. longiflorus,* and *T. platypterus* have been examined (Nilsson, 2002). *Tachiadenus tubiflorus* differs slightly from the other examined species by having a very finely reticulate-rugulate to almost smooth and perforate exine (Nilsson, 2002: Fig. 4.5D–F). The pollen grains of *Tachiadenus longiflorus, T. tubiflorus,* and *T.* sp. were described by Lienau *et al.* (1986) as being relatively large (polar axis exceeding 30 μm). The exine pattern of *Tachiadenus longiflorus* has been reported to be reticulate (Lienau *et al.,* 1986). In his revision of *Tachiadenus*, Klackenberg (1987a) investigated pollen from all 11 species in the genus.

Discussion of palynological data in Exaceae. The pollen grains of the Exaceae are 3-colporate with a striate to striato-reticulate to reticulate exine pattern, or, exceptionally, finely reticulate to rugulate-vermiculate, or mostly smooth. The average size (polar axis) is *c.* 20 μm or less, except for *Tachiadenus* and some *Exacum* spp., which have a polar axis that is *c.* 30 μm long.

Gilg's (1895a) description of Gentianeae-Exacinae as having small and smooth pollen grains is not supported. *Exacum* and *Sebaea* have similar types of pollen that are also similar to *Ornichia* (striate to striato-reticulate). Gilg placed *Tachiadenus* in Gentianeae-Tachiinae, together with genera such as *Lisianthius, Macrocarpaea,* and *Tachia*. This arrangement is not supported by recent pollen studies, but then again, neither is the position of *Tachiadenus* in the Exaceae entirely clear from a palynological standpoint. The exine pattern of *Gentianothamnus,* and to some extent *Ornichia,* also differs from *Exacum* and *Sebaea*. There are strong similarities in pollen exine pattern in *Tachiadenus* and *Gentianothamnus*

(J. Klackenberg & S. Nilsson, pers. obs.). Similarly, *Exacum* and *Ornichia* are very close in pollen morphology, and the infrageneric variation seen among pollen of *Exacum* species is substantially greater than that between *Ornichia* and *Exacum*.

Gynoecium

The ovary in *Exacum* and *Sebaea* is bilocular (Fig. 2.9A) to sometimes semi-unilocular, i.e., the placentation is axile over all to most of the ovary length. Sometimes, however, the septa disappear apically, leaving the ovary unilocular and the placentation free-central in the upper half of the gynoecium (Lindsey, 1940; Gopal Krishna & Puri, 1962; Klackenberg, 1985: 17). The same holds true for *Ornichia* (Klackenberg, 1986) and *Cotylanthera* (Oehler, 1927: 655; Lindsey, 1940; Gopal Krishna & Puri, 1962; Hara, 1975: 321). In *Tachiadenus* the ovary is pseudo-bilocular, i.e., the septal walls protrude substantially into the ovary and the placentas group close together centrally (Fig. 2.9B; Klackenberg, 1987a). *Gentianothamnus* has a similar though somewhat intermediate ovary structure. It is usually pseudo-bilocular but is sometimes truly bilocular near the base (Klackenberg, 1990). Bilocular ovaries, a structure historically used to characterize the tribe Exaceae, are, however, considered plesiomorphic within the entire Gentianaceae (Klackenberg, 1985: 18). This trait is present in most Exaceae, the next-most basal clade in the family, as well as in most of the closest outgroups to Gentianaceae, e.g., Loganiaceae *sensu lato* and Rubiaceae. It should be noted, however, that the possibly bilocular ovaries that are also present in *Anthocleista*, *Potalia*, and *Fagraea* must represent parallelism (see further discussion under Potalieae-Potaliinae). These genera, formerly placed in Loganiaceae, are now accepted as part of Gentianaceae (see below, subtribe Potaliinae, Fig. 2.3). Consequently, the bilocular condition is not unique to Exaceae within the family as was formerly thought. The most basally placed tribe Saccifolieae is poorly investigated with regard to ovary morphology, but *Saccifolium* has parietal placentation and unilocular ovaries with deeply protruding placentas that can form seemingly pseudo-bilocular ovaries (Maguire & Pires, 1978), and unilocular ovaries with parietal placentas have also been reported from the remaining genera of that tribe (see Saccifolieae, above). Nevertheless, the plesiomorphy of bilocular ovaries within Exaceae seems well assured. Within the tribe the presence of a pseudo-bilocular ovary is a possible synapomorphy for *Gentianothamnus* and *Tachiadenus*.

In a large group of Malagasy *Exacum* species the placentas are broad and replace part of the septal wall, and, in some species, almost the whole of the septum (Klackenberg, 1985: 16). In *Gentianothamnus* the ovary is placed on a small, shallowly lobed disk. A similar disk is also present in other Gentianaceae taxa, e.g., in *Potalia* (Potalieae-Potaliinae) and in many species of tribe Helieae, whereas it is absent from other genera of Exaceae.

Gopal Krishna and Puri (1962) studied the vascularization of the carpels in 26 species of Gentianaceae. Taxa from five tribes of the family were represented in the analysis (all but Saccifolieae). One species of *Exacum*, three of *Sebaea*, and one of *Cotylanthera* were selected from Exaceae. In *Exacum* and *Sebaea* each carpel was shown to be drained by five vascular bundles. The carpels of *Cotylanthera* were shown to be essentially three-traced, as were all other of the 21 species examined from the other tribes. There is a correlation, seen also in other families, between the type of placentation (closed carpels with axile placentation vs. open carpels with parietal placentation) and the number of carpellary traces (i.e., five vs. three; Gopal Krishna & Puri, 1962: 53); therefore, the five-traced state is considered possibly plesiomorphic in Gentianaceae-Exaceae with the three-traced carpel of *Cotylanthera* representing a unique character for this genus within Exaceae.

A specialized feature for most *Sebaea* species is the presence of two lateral, sometimes confluent, stigmatic swellings on the style below the level of the ordinary apical stigma (Fig. 2.7E). This unique trait, not being known from other plants, is considered a synapomorphy for *Sebaea*. It has been supposed that these stigmatic areas are ontogenetically connected with the apical stigmata by prolongation along the style, only later, in evolutionarily terms, to become severed from the apical part (Hill, 1913). The extra stigmatic swellings are situated below the anthers and become covered by self pollen, which obviously favors self-pollination (Marloth, 1909; Hill, 1913). This self-pollination, as well as cross-pollination, is, according to Marloth's (1909: 313) observations, performed by small insects belonging to the Thripsidae (Physopoda).

The style is declinate in most species of *Exacum*, as well as in at least one of the four species of *Cotylanthera*, *C. paucisquama* (Hara, 1975), i.e., both genera with pore-opening anthers forming a cone (Fig. 2.7B–C). This tendency toward zygomorphy, a rare condition in Gentianales, is considered apomorphic.

Seeds

In Exaceae, the seeds are small, generally cubic or polyhedric, and with distinct testa cells. Seeds that have isodiametric testa cells with straight walls are observed in many genera of Gentianaceae (Bouman *et al.*, 2002).

In Exaceae, however, the testa cells usually have finely (or in several genera, deeply) undulating or wavy walls that become star-shaped (Fig. 2.10B–D; see also Klackenberg, 1983: fig. 2; 1985: fig. 8; 1986: fig. 1b; 1987: fig. 1). Slightly wavy to distinctly star-shaped testa walls are characteristic for *Gentianothamnus*, *Ornichia*, and *Tachiadenus*, most *Exacum* and some species of *Sebaea*, viz. those that have been separated as *Belmontia* and *Exochaenium*. This seed structure is interpreted as a synapomorphy for tribe Exaceae. Consequently, the presence of ordinary isodiametric testa cells with straight walls in *Cotylanthera* as well as in some species of *Exacum* (Fig. 2.10E) is considered a derived character. The members of a clade of 15 small-flowered *Exacum* species from Madagascar and continental Africa are united by the synapomorphy of having finely undulated testa cell walls (Fig. 2.10D). Most species of *Sebaea* (*Sebaea sensu stricto*) have irregularly formed seeds, usually ovate or oblong in outline, but with the testa cells arranged in rows (Fig. 2.10A). This is also considered to be a derived condition. *Cotylanthera* has seeds of similar outline, but they lack the arrangement of testa cells into rows. The testa cells have thick, straight walls, similar to what is seen in several species of *Exacum*.

Karyology

Chromosome numbers have been determined for representatives of *Cotylanthera*, *Exacum*, and *Sebaea*. Two counts for *Sebaea*, the basal clade, are known, viz. *S. brachyphylla* from Africa with $2n=22$ (Thulin, 1970: 489) and *S. ovata* from Australia with $2n=c.\ 54$ (Beuzenberg & Hair, 1983: 14) – a large span without any specific base number. The karyology of several species of *Exacum* has been examined, all but one (*E. affine*, $2n=36$, from Socotra (Sugiura, 1936a,b; Post, 1967: 136)) distributed in India and belonging to section *Exacum*. Here also extensive variation is observed, viz. $2n=18$, 30, 34, 54, 56, 62, and 68 (Borgmann, 1964; Subramanian, 1980: 49; Mallikarjuna *et al.*, 1987: 766–767). No karyological information on African *Exacum*, *Gentianothamnus*, *Ornichia*, or *Tachiadenus* is available. *Cotylanthera* was stated by Oehler (1927: 724) to have $2n=32$–36. No particular karyological pattern can be discerned in the Exaceae clade, except that variation is great. This is seen even between closely related species, e.g., $2n=18$ in *Exacum tetragonum*, $2n=62$ in *E. tetragonum* (cited as *E. bicolor*; Borgmann, 1964), and $2n=68$ in *E. grande* (cited as *E. perrottetii*; Mallikarjuna *et al.*, 1987: 766), as well as within a single species, e.g., $2n=30$, 54, 56 in *E. pedunculatum* (Subramanian, 1980: 49).

Dispersal

The seeds of Exaceae are small and are discharged from xerochastic capsules by passing animals or by the wind. The prominent calyx wings that sometimes harden and enlarge in fruit, characteristic for many species of *Exacum*, *Sebaea*, and *Tachiadenus*, are thought to enhance this effect. The seeds are small, less than 0.5 mm in diameter, and probably readily adhere to animals. The dry capsules dehisce septicidally into two valves in all taxa except for a small group of Malagasy *Exacum* species (the *E. humbertii* group) in which the capsules open via a coriaceous lid that dehisces from a membranaceous bowl (Klackenberg, 1985: 16). Pringle (1979: 15) stated that fruits of *Gentiana* sect. *Chondrophyllae* function as splash-cups. A similar means of dispersal might be proposed for the seeds of the *Exacum humbertii* group. Dispersal by birds, directly or through earthworms subsequently eaten and spread by birds, has been suggested for *Cotylanthera tenuis* (Beccari, 1890: 325; Ridley, 1930: 530).

Chemistry

According to Daniel and Sabnis (1978) the tribe Exaceae differs considerably from other Gentianaceae in its phytochemistry. Flavone-O-glycosides are restricted to Exaceae. However, only two species of *Exacum* within the tribe were screened. Xanthones, glycoflavones, and L-(+)-bornesitol, abundant in the rest of the family, appear to be conspicuously absent in the Exaceae (Jensen & Schripsema, 2002 (Chapter 6, this volume)).

Uses

In Madagascar, *Tachiadenus* is used to produce a tonic against illnesses of the urinary tract, digestive system, respiratory organs, and blood circulation (Boiteau, 1986). In India, *Exacum tetragonum* extract is used as a stomachic and tonic for fevers. Powdered plants of *Exacum lawii* are reported to be used for kidney diseases, and plants boiled with oil for eye diseases. This species is also used as a laxative (Jain & DeFilipps, 1991).

Character synopsis

A synopsis of the distribution of the most consistent characters within Exaceae is presented in Fig. 2.12. The cladogram is based on the phylogenetic tree presented in Fig. 2.3 with *Gentianothamnus* and *Cotylanthera* added.

White-yellow flowers are regarded as a synapomorphy for *Sebaea*. Furthermore, most of the species are characterized by the unique double

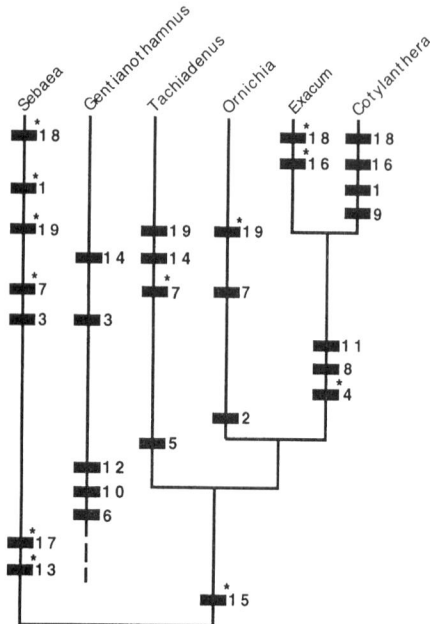

Figure 2.12. Distribution of some conspicuous characters in Exaceae. The tree is based on Fig. 2.3 with *Cotylanthera* and *Gentianothamnus* added. The placement of *Gentianothamnus* is uncertain. An asterisk indicates that not all species in the genus have that particular character.

stigma, the conspicuous apical and/or basal anther appendages, and seeds with testa cells in rows. The thick anther appendages seen in *Ornichia* and *Tachiadenus tubiflorus* might not be homologous to the ones present in *Sebaea*. In the sister group of *Sebaea*, consisting of *Tachiadenus*, *Ornichia*, *Exacum*, and *Cotylanthera*, seeds with distinctly star-shaped testa cells are predominant; however, they are lacking in *Cotylanthera* and some species of *Exacum*. *Tachiadenus* is characterized by its large flowers with very long corolla tubes as well as by its pseudo-bilocular ovary. No character is known, however, that unites *Ornichia*, *Exacum*, and *Cotylanthera*, the sister group of *Tachiadenus*. Hairy leaves are a synapomorphy for *Ornichia*. The position of *Cotylanthera* placed together with *Exacum* is based on anthers opening by pores and with finely perforated endothecia, both unique characters. Within this clade there is also a tendency toward zygomorphy, e.g., a declinate style in most species of *Exacum* and one species of *Cotylanthera*, and zygomorphic reduction of calyx wings in some *Exacum* species. *Cotylanthera* is presumably monophyletic, and is characterized by having anthers opening by one pore only and by being saprophytic. Saprophytism, however, has evolved once also in *Sebaea*. Straight testa walls, considered apomorphic within Exaceae, are observed in *Cotylanthera*, but are also present in some species of *Exacum*. The same holds true for 4-merous flowers, which are present also in some species of *Exacum* as well as in some of *Sebaea*. Another conspicuous example of a homoplastic character is included stamens with short filaments, which is observed in some *Sebaea*, all *Tachiadenus*, and one species of *Ornichia*. However, this feature can also be plesiomorphic in Exaceae, as it is present in several taxa within possible outgroups to Exaceae.

The position of *Gentianothamnus* within Exaceae is unclear. It is linked to *Sebaea* by the yellow-colored flowers. On the other hand it has a pseudo-bilocular ovary as does *Tachiadenus*. Furthermore, the pollen structure is similar to what is seen in *Tachiadenus*. Autapomorphies for *Gentianothamnus* are, for example, the corolla tube widening from the base, a basal disk inside the corolla, and very long filaments inserted at the corolla base.

Biogeography of Exaceae

The six genera included in tribe Exaceae have a principally paleotropical and southern African (temperate) distribution. In Africa, representatives of the tribe are found from Senegal to Kenya in the north to the Cape Province in the south. Exaceae are distributed in all phytogeographic regions in Madagascar, although very rarely so in the dry southwestern domain. They are also present on the adjacent Indian Ocean islands (Comores, Mauritius, La Réunion) as well as on Socotra and in the Dhufar mountains of the Arabian Peninsula. Exaceae are not known from the geologically old Seychelles islands. Remarkably, no native representatives of Gentianaceae are known from the Seychelles (Robertson, 1989; Friedmann, 1994). In Asia, tribe Exaceae is distributed from Sri Lanka and India to the Philippines and New Guinea, with a northern limit in the Himalayas and southern China. There are also a couple of species native to Australia and New Zealand. The center of diversity for Exaceae is Madagascar, where all genera except *Cotylanthera* are present. The great majority of the c. 60–100 species of *Sebaea* are distributed in continental Africa, mainly in South Africa, with only four species in Madagascar, two in Australia and New Zealand, and one in India. In contrast, only two of the 65 species of *Exacum* are found in continental Africa, the rest being divided between Madagascar (roughly two-thirds) and Asia (one-third). *Gentianothamnus*, *Ornichia*, and *Tachiadenus* are all endemic to Madagascar. *Cotylanthera* is distributed from the eastern Himalayas to the Philippines, Indonesia, and New Guinea.

Although seemingly paleotropical (and temperate South African) in general, tribe Exaceae shows a clear Gondwanic distribution pattern. In fact, only a few species are found in Southeast Asia, among which are all four species of *Cotylanthera*. No representatives of the tribe have been found in South America, the westernmost part of Gondwana. However, tribe Saccifolieae, which is the next-most basal clade below Exaceae, includes only neotropical genera.

Given their phylogenetic position, the Exaceae are relatively old by default. No fossils ascribable to the tribe have been found. The earliest megafossil evidence proposed to belong to Gentianaceae is given by flowers with *Pistillipollenites macgregorii* pollen from the lower Eocene of Texas (Crepet & Daghlian, 1981; Stockey & Manchester, 1986). It is uncertain, however, whether these specimens really belong to Gentianaceae, as the flowers are not similar to any extant genus in the family (L. Struwe, pers. comm.).

A model can be developed in which the ages and distributions of the basal-most clades of Gentianaceae correlate with known continental drift events. The split-up of the Gondwana supercontinent started in the Jurassic. Madagascar – with the now largely submerged (except for the Seychelles) Mascarene Plateau, India, and Antarctica and Australia – started to separate from Africa (Kenya/Somalia) by approximately 150 million years ago (Ma) and finished its southward movement during the mid-Cretaceous (*c.* 120 Ma; Rabinowitz *et al.*, 1983; Besse & Courtillot, 1988). The India/Madagascar/Mascarene plate separated from Antarctica/Australia in the Barremian–Aptian (*c.* 130–120 Ma). Later, during the late Cretaceous (*c.* 88 Ma), India separated from the Madagascar/Mascarene continent (Storey *et al.*, 1995) and moved rapidly northwards, establishing contact with southern Asia by Eocene–Paleocene time (*c.* 55 Ma (Klootwijk *et al.*, 1992, from paleomagnetic data; Beck *et al.*, 1995, from biostratigraphic data)). At this time, South America, Antarctica, and Australia were still joined, though Australia started to move northwards during the middle or late Eocene, separating totally from Antarctica by the late Oligocene (Wilford & Brown, 1994). Starting at the base of the Gentianaceae, it could be hypothesized that elements (lineages) of the family were already West Gondwanan (tribe Saccifolieae) and East Gondwanan (tribes Exaceae and Chironieae; see discussion of the Chironieae below) well before the split-up of the supercontinent, i.e., more than 100 Ma.

Some information on vicariance biogeography within tribe Exaceae can be extracted from the molecular phylogenetic results (Fig. 2.3). *Cotylanthera* and *Gentianothamnus*, not included in the cladogram, are discussed below. *Ornichia* and *Tachiadenus* are both endemic to Madagascar. They are sympatric with each other as well as with *Exacum*, and no vicariance events can be proposed on the generic level. It is interesting to note, however, that the large and widely distributed genus *Exacum*, with a distributional vicariance between India and Madagascar, is unresolved toward *Ornichia*, a small Malagasy endemic, and that these taxa are in turn the sister group of the small Malagasy endemic *Tachiadenus*. The first split within the Exaceae clade is that between *Sebaea* (Africa, Madagascar, India, and Australia/New Zealand) and the rest of the genera, viz. *Exacum, Ornichia,* and *Tachiadenus* (Africa, Madagascar, Socotra/Arabian Peninsula, India/Sri Lanka, Southeast Asia, and Australia).

The *Exacum/Ornichia/Tachiadenus* clade and *Sebaea* are sympatric over large areas in continental Africa, Madagascar, India, Southeast Asia, and Australia, i.e., over almost all of the distributional area of the tribe. A closer examination of the distribution pattern of the included taxa, however, reveals a possible old vicariance within Gondwana. Only two closely related species of *Exacum* are found in continental Africa: *E. oldenlandoides*, a weed found in almost all countries south of the Sahara, and *E. zombense*, which is locally endemic to Malawi. These species belong to a group of 15 small, weedy species, of which the remaining are Malagasy endemics. In Asia, only one species, *E. tetragonum*, is distributed widely outside the Indian subcontinent, being found from India throughout Southeast Asia to northern Australia. These distributions may be due to long-distance dispersal and/or migration. *Sebaea* has its species center in southern Africa (42 species, 32 of these endemic to South Africa; Marais & Verdoorn, 1963), but many species are also found in tropical continental Africa. However, only four of the *c.* 100 species of *Sebaea* are found exclusively outside continental Africa. Two closely related taxa of the *Exochaenium*-type, *Sebaea condensata* and *S. madagascariensis*, are endemic to Madagascar, and two species, *S. albidiflora* and *S. ovata*, are endemic to Australia/New Zealand. *Sebaea aurea* is widely distributed in Africa but also occurs in India; *S. bojeri* and *S. brachyphylla* are both found in Africa and Madagascar. Individual species of Exaceae, however, are most probably younger than the patterns proposed here to result from drifting continents, so species common to two or more areas are regarded as likely examples of long-distance dispersals and are therefore of limited value with regard to hypotheses of vicariance. Consequently, the *Exacum* clade including *Ornichia* and *Tachiadenus* is distributed principally around the eastern Indian Ocean, i.e., Madagascar, Socotra, India, and Sri Lanka, whereas *Sebaea* is centered in continental Africa. An old vicariance in accordance with the phylogeny (Fig. 2.3) can therefore be hypothesized between the Indian Ocean basin and Africa. However, the initial split-up of the Gondwana supercontinent, which began approximately 150 million years ago, may be too early to explain this vicariance. Tricolpate pollen has not been found earlier than c. 125 Ma (Crane *et al.*, 1995). Nevertheless, the phytogeographic pattern exists and must be explained (see also Struwe & Albert, 1997, and Potalieae, this chapter, for possible Gondwanan relationships among Potaliinae). It could be speculated that Madagascar/India had extended phytogeographic contact with continental Africa during its southwards movement along the East African coast until Madagascar reached its present position some 120 million years ago. Furthermore, Africa may have maintained floristic contact with Madagascar/India slightly longer through Antarctica. Whether the presence of *Sebaea albidiflora* and *S. ovata* in Australia and New Zealand manifests an old vicariance pattern through the Antarctic or is an example of long-distance dispersal cannot be established at present because of the lack of phylogenetic information within *Sebaea*. The Australian species *Sebaea albidiflora*, however, seems to be closely related to the African Cape species *S. minutiflora* (Marais & Verdoorn, 1963: 180).

The next major vicariance event affecting Exaceae could have been the split between Madagascar and India which occurred 90–80 Ma. This event is distinctly reflected in extant taxa of *Exacum*. *Exacum* is phylogenetically divided at the base into one India/Sri Lankan clade (section *Exacum*) and one Madagascar/Socotra–Arabian Peninsula/African clade (section *Africana*; Klackenberg, 1985). This vicariance hypothesis was based on a handmade cladogram (Klackenberg, 1985: figs. 10, 13, table 3). However, the same major clades are obtained from a PAUP (Swofford, 1993) analysis of the same data set. Consequently, the information available best supports the hypothesis that the *Exacum* lineage existed prior to 90–80 Ma, being subjected to the Madagascar/India vicariance event.

Within *Exacum* a close relationship between Socotra and Madagascar and an older vicariance between these two islands and India/Sri Lanka has been proposed (Klackenberg, 1985). As discussed above, geological data show that Madagascar/India started to separate from the African continent (Kenya/Somalia) by c. 150 Ma (Rabinowitz *et al.*, 1983; Besse & Courtillot, 1988). An early Cretaceous split within *Exacum* or any other modern angiosperm genus might seem improbable. However, in other genera of flowering plants we find the same distribution, i.e., continental Africa, Madagascar, India/Sri Lanka, and Socotra, with species endemic to each of these four regions. This is seen, for example, in *Secamone* (Apocynaceae *sensu lato*; Goyder, 1992; Klackenberg, 1992a,b) and *Neuracanthus* (Acanthaceae; Bidgood & Brummitt, 1998), genera with different types of dispersal mechanisms. However, no phylogenetic data for these genera are available. Schatz (1996: 78) stated that the xeric affinities between Madagascar and India (e.g., *Commiphora* (Burseraceae), *Delonix* (Fabaceae), *Moringa* (Moringaceae)) are best explained by overland migration through northeast Africa, the Arabian Peninsula and not via the "Lemuria land-bridge" (actually a Gondwanic vicariance) as Wild (1965) had suggested. However, here we also lack important data as no cladistic analyses have been presented for these groups either. Therefore, it is too early to decide if this repeated phytogeographic pattern is due to vicariance, a (xeric) migration, or long-distance dispersal.

Cotylanthera, not included in the present molecular phylogenetic studies, may be the sister taxon of *Exacum*, based on morphological synapomorphies (see above). The distribution patterns of *Exacum* and *Cotylanthera* by themselves suggest an old vicariance. The four species of *Cotylanthera* are distributed over a vast area from the Himalayas, throughout Southeast Asia, to New Guinea. *Exacum*, except for the widely distributed Indian and Southeast Asian species *E. tetragonum*, is concentrated in India, Sri Lanka, Socotra, and Madagascar, i.e., mostly to the south and east of *Cotylanthera*, with only slightly overlapping distributions in the eastern Himalayas, Myanmar, and Thailand. Here also we lack the necessary phylogenetic information to allow us to make a robust biogeographic hypothesis. Nonetheless, it is difficult to explain the Asian distribution of *Cotylanthera* as resulting from a vicariance event between Madagascar/India and Asia (i.e., between Gondwana and Laurasia). If it does indeed form a clade with *Exacum*, *Cotylanthera* and its broad distribution may be linked with migration from India after the collision of the Indian plate with Laurasia *c.* 55 Ma. It is interesting to note that such wide distributions are also seen in the similarly minute saprophytes *Sebaea oligantha* and *Voyria primuloides*, both of which are distributed over a large part of tropical Africa (Raynal, 1967b), as well as in *V. chionea*, which is broadly distributed from northwestern South America to eastern Brazil (Albert & Struwe, 1997). Ridley (1930: 531) noted that small saprophytes such as *Cotylanthera tenuis* often have curiously wide distributions. He proposed this to be an effect of dispersal by migrant birds such as ant-thrushes (*Pitta*), which feed on worms. In turn,

Ridley based his reasoning on Beccari's (1890: 325) proposition that seeds of minute tropical forest plants are eaten by worms.

Gentianothamnus is restricted to the highest mountain chains in Madagascar. Its precise phylogenetic position and biogeographic status are not yet known.

TRIBE CHIRONIEAE

M. THIV AND J. W. KADEREIT

The Chironieae as understood here comprise c. 159 species in 23 genera. Following the results of the *mat*K and *trn*L intron sequence phylogenetic analyses, as well as morphological evidence, these 23 genera fall into three major subgroups which are here classified as subtribes. These are subtribe Chironiinae, including *Bisgoeppertia*, *Blackstonia*, *Centaurium*, *Chironia*, *Cicendia*, *Eustoma*, *Exaculum**, *Geniostemon*, *Ixanthus*, *Orphium*, *Sabatia*, and *Zygostigma**, subtribe Canscorinae, including *Canscora*, *Cracosna**, "Duplipetala", *Hoppea*, *Microrphium*, *Phyllocyclus*, and *Schinziella*, and subtribe Coutoubeinae, including *Coutoubea*, *Deianira*, *Schultesia*, *Symphyllophyton*, and *Xestaea* (the asterisks indicate genera not included in the molecular investigations).

Description of Chironieae

The Chironieae comprise annual, biennial or short-lived perennial, erect, rarely climbing or creeping, glabrous, rarely bristly or papillose, sometimes suffrutescent herbs. The stems are terete to strongly winged. The leaves are simple, decussate, sessile or rarely petiolate, sometimes perfoliate, green, rarely glaucous, and of variable shape. Rosette leaves are present or absent. The flowers are single or are arranged in variously modified axillary cymose or rarely racemose inflorescences. The actinomorphic or zygomorphic flowers are sessile or pedicellate, without or rarely with bracteoles. The calyx consists of 4 or 5, rarely 2 or 3 or up to 12 fused sepals, and is usually unwinged and tubular. Calycine colleters may be present or absent.

The typically salver-shaped or tubular, rarely funnel-shaped or rotate corolla consists of 4 or 5, rarely up to 12 fused petals and is of white, cream, yellow, pink, or rarely of bluish or purple color. The corolla lobes are of equal or rarely unequal size (Fig. 2.13A). There are usually 4 or 5, rarely up to 12 stamens that are isomorphic or anisomorphic (Fig. 2.13A) and rarely staminodial. When anisomorphic, one or rarely two usually larger stamens are inserted at a higher level (Fig. 2.13A). The stamens are inserted in the

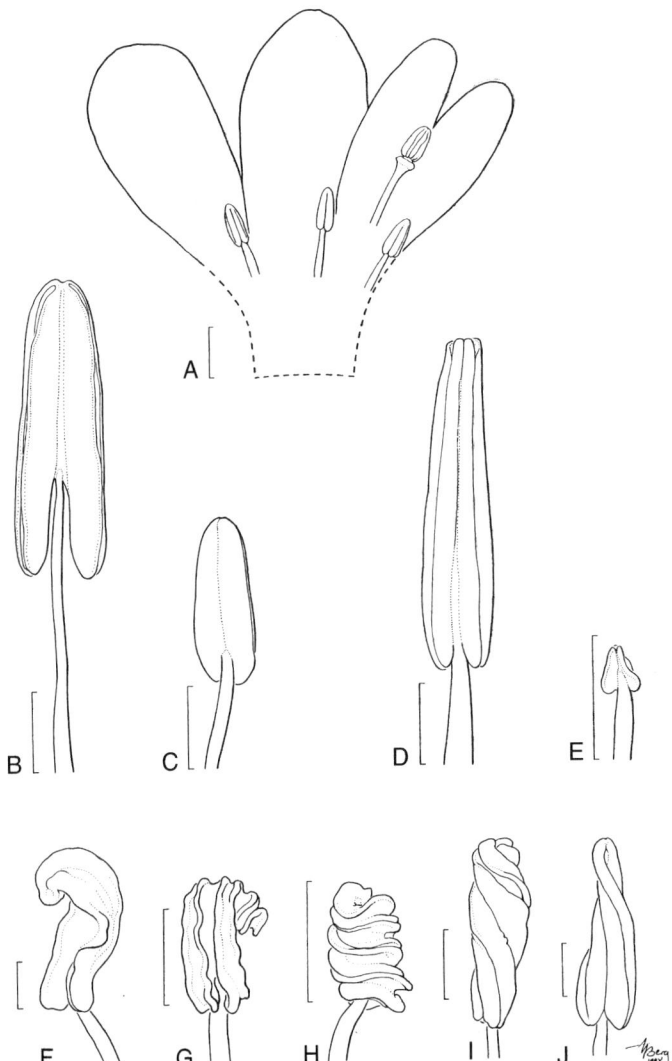

Figure 2.13. Corollas and stamens of Chironieae. **A.** Zygomorphic corolla and unequal insertion of anisomorphic stamens in *Canscora roxburghii* (*Cramer 4913*). **B–E.** Anther shape in Chironieae before anther dehiscence. **B.** Sagittate anther of *Coutoubea spicata* (*Billiet 1033*). **C.** Slightly sagittate anther of *Centaurium maritimum* (*Licht s.n.*). **D.** Non-sagittate anther of *Deianira erubescens* (*Irwin 13867*). **E.** Slightly "x"-shaped anther of *Cicendia quadrangularis* (*Halse 4461*). **F–J.** Anther shape in Chironieae after anther dehiscence. **F.** Slightly twisted anther of *Eustoma grandiflorum* (*Thiv s.n.*). **G.** Outwardly twisted anther of *Sabatia dodecandra* (*Frankenhäuser 106*). **H–J.** Helically twisted anthers. **H.** *Centaurium beyrichii* (*Kennedy 307*). **I.** *Chironia laxiflora* (*Msra Sumbi 16204*). **J.** *Orphium frutescens* in early stage (*Thiv s.n.*). Scale bars = 1 mm.

corolla tube or rarely in the sinuses between corolla lobes. The filaments are usually glabrous with basifixed, sagittate to non-sagittate anthers (Fig. 2.13B–E) that are sometimes helically or rarely outwardly twisted after dehiscence (Fig. 2.13F–J). Pollen is released in monads or tetrads. The unilocular ovary consists of two carpels with parietal placentation. The distinct, filiform style bears a bilobed stigma of variable shape. The fruit is usually a septicidal capsule or rarely a berry.

Taxonomic history of Chironieae

Don (1837–1838) established the "Chironiae" as a subtribe of his Gentianeae, based on the genus *Chironia*. The name of this genus is derived from Chiron, a centaur in Greek mythology. Endlicher (1838) divided the family into tribes Chironieae and Sebaeeae, of which Chironieae was a highly heterogeneous group as it contained the majority of genera known at that time. Grisebach (1839) initially treated the group at tribal rank but later (Grisebach, 1845) reduced it to a subtribe of Gentianeae. Bentham (1876), retaining Grisebach's earlier tribal rank, divided the group of genera into subtribes Euchironieae, Erythraeeae, and Lisiantheae. This treatment was closely followed by Knoblauch (1894).

The detailed taxonomic history of the genera included in Chironieae as understood here is shown in Table 2.5. As is evident from this table, the Chironieae are largely congruent with subtribes Erythraeinae and Chironiinae of Gilg's (1895a) Gentianeae. In addition to the genera of our Chironieae, Gilg (1895a) also included *Bartonia, Curtia, Enicostema, Faroa, Neurotheca, Obolaria,* and *Tapeinostemon* in his Erythraeinae. Following the molecular phylogenetic results presented here, *Bartonia* and *Obolaria* belong to Gentianeae, and *Enicostema, Faroa,* and *Neurotheca* should be reassigned to the Potalieae. *Curtia* and *Tapeinostemon* are placed in the new tribe Saccifolieae. Also included in Chironieae in this classification but placed by Gilg (1895a) elsewhere are *Ixanthus* (Gentianinae *sensu* Gilg), *Eustoma* and *Zygostigma* (Tachiinae *sensu* Gilg), and *Coutoubea, Deianira, Schultesia* (including *Xestaea*), and *Symphyllophyton* (Helieae *sensu* Gilg).

Generic synopses for Chironieae

Bisgoeppertia KUNTZE

This is a Caribbean genus described by Kuntze (1891) based on *Bisgoeppertia volubilis*. The genus now comprises two species endemic to the Greater Antilles (Liogier, 1989).

Systematics, character evolution, and biogeography   111

The species of *Bisgoeppertia* are annual climbing herbs. The rosette leaves are probably deciduous and the cauline leaves are lanceolate to linear. The 5-merous flowers are borne mostly in 1–3-flowered dichasia. They are slightly bent and have a pedicel with two small bracteoles. The calyx tube is shorter than the linear to lanceolate lobes. Calycine colleters are absent. The yellowish corolla is tubular and the tube is longer than the lobes. The stamens are inserted in the upper part of the corolla tube. The non-sagittate anthers do not coil after anthesis. The pollen is released as monads. The stigmatic lobes are obovate and bent downwards.

Blackstonia HUDS.

The genus was established by Hudson (1762: 146) based on *Blackstonia perfoliata*. According to Zeltner (1970) the genus comprises four species of Mediterranean and western to central European distribution. *Blackstonia perfoliata* has also been introduced to other parts of the world, e.g., Argentina and Australia (Fabris, 1953; Adams, 1996). The species of *Blackstonia* grow in at least temporarily humid patches in open places.

Blackstonia comprises glaucous, annual herbs. Whereas the elliptic rosette leaves are free, the cauline leaves are perfoliate. The 6–12-merous flowers are arranged in lax dichasia. The calyx tube is much shorter than the linear lobes. Calycine colleters are present. The yellow corolla is rotate to salver-shaped and the tube is slightly longer than the lobes. The stamens are inserted in the upper part of the corolla tube. The non-sagittate anthers do not coil after anthesis. The pollen is released as monads. The stigmatic lobes are apically divided. Reported chromosome numbers are $2n = 20$ and 40 (Zeltner, 1970).

Canscora LAM.

This genus was described by Lamarck (1785) based on *Canscora perfoliata*. *Canscora* contains nine species of paleotropical distribution, including tropical Africa, Madagascar, Oman, India, Sri Lanka, Nepal, Myanmar, Thailand, Cambodia, Vietnam, southern China, Malaysia, Java, the Philippines, and northern Australia. Most species grow in moist habitats, e.g., along the edges of rivers and on wet rocks, or in cultivated land.

Canscora as understood here (M. Thiv, in prep.) is the remainder of a formerly polyphyletic genus that was divided into four subgenera by Clarke (1885, 1905). Of these, subgenera *Canscora* and *Heterocanscora* form *Canscora* (M. Thiv, unpubl.). Both are characterized by anisomorphic stamens, a condition in which one larger stamen is inserted at a higher level than the remaining smaller stamens (Fig. 2.13). The species differ from each

Table 2.5. *Tribal and subtribal placement of Chironieae genera in selected previous classifications of the Gentianaceae*

Genus	Don (1837–1838)	Endlicher (1838)	Grisebach (1839)	Grisebach (1845)	Bentham (1876)	Knoblauch (1894)	Gilg (1895a)	Gilg (1939)
Bisgoeppertia	***[a]	***	Hippieae[b]	Chloreae-Erythraeaceae	Chironieae-Euchironieae	Chironieae-Euchironieae	Gentianeae-Erythraeinae	Gentianeae-Erythraeinae-Teleiandrae
Blackstonia	Gentianeae-verae	Chironieae	Chloreae	Chloreae-Erythraeaceae	Chloreae-Erythraeaceae	Chironieae-Erythraeeae	Gentianeae-Erythraeinae	Gentianeae-Erythraeinae-Teleiandrae
Canscora	Gentianeae-verae	Chironieae	Erythraeaceae	Chloreae-Erythraeaceae	Chironieae-Erythraeeae	Chironieae-Erythraeeae	Gentianeae-Erythraeinae	Gentianeae-Erythraeinae-Ellipandrae
Centaurium	Chironiae	Chironieae	Erythraeaceae	Chloreae-Erythraeaceae	Chironieae-Erythraeeae	Chironieae-Erythraeeae	Gentianeae-Erythraeinae	Gentianeae-Erythraeinae-Teleiandrae
Chironia	Chironiae	Chironieae	Chironieae	Chironieae	Chironieae-Euchironieae	Chironieae-Euchironieae	Gentianeae-Chironiinae	—[c]
Cicendia	Exacieae	Chironieae	Erythraeaceae	Chloreae-Erythraeaceae	Chironieae-Erythraeeae	Chironieae-Erythraeeae	Gentianeae-Erythraeinae	Gentianeae-Erythraeinae-Teleiandrae
Coutoubea	Coutoubeae	Chironieae	Hippieae	Chloreae-Erythraeaceae	Chironieae-Euchironieae	Chironieae-Erythraeeae	Helieae	—
Cracosma	***	***	***	***	***	***	***	—
Deianira	Gentianeae-verae	Chironieae	Chironieae	Chironieae	Chironieae-Erythraeeae	Chironieae-Erythraeeae	Helieae	—
"Duplipetala", ined.	***	***	***	***	***	***	***	Gentianeae-Erythraeinae-Ellipandrae
Eustoma	Lisiantheae			Chloreae	Chloreae-Sabbatieae	Chironieae-Lisiantheae		Gentianeae-Tachiinae	
Exaculum	Exacieae	Chironieae	Erythraeaceae	Chloreae-Erythraeaceae	Chironieae-Erythraeeae	Chironieae-Erythraeeae	Gentianeae-Erythraeinae	Gentianeae-Erythraeinae-Teleiandrae

Genus								
Geniostemon	***	***	***	***	***	Chironieae-Euchironieae	Gentianeae-Erythraeinae	Gentianeae-Erythraeinae-Teleiandrae
Hoppea	Gentianeae-verae	Chironieae	Erythraeaceae	Chloreae-Erythraeaceae	Chironieae-Euchironieae	Chironieae-Euchironieae	Gentianeae-Erythraeinae	Gentianeae-Erythraeinae-Ellipandrae
Ixanthus	Gentianeae-verae	Chironieae	Chloreae	Chloreae-Sabbatieae	Chironieae-Erythraeeae	***	Gentianeae-Gentianinae	—
Microrphium	***	***	***	***	***	***	***	—
Orphium	Chironiae	Chironieae	Chironieae	Chironieae	Chironieae-Euchironieae	Chironieae-Euchironieae	Gentianeae-Chironiinae	—
Phyllocyclus	***	***	***	***	Chironieae-Erythraeeae	Chironieae-Erythraeeae	Gentianeae-Erythraeinae	Gentianeae-Erythraeinae-Ellipandrae
Sabatia	Chironiae	Chironieae	Chloreae	Chloreae-Sabbatieae	Chironieae-Erythraeeae	***	Gentianeae-Erythraeinae	Gentianeae-Erythraeinae-Teleiandrae
Schinziella	***	***	***	***	***	***	Gentianeae-Erythraeinae	Gentianeae-Erythraeinae-Ellipandrae
Schultesia	Gentianeae-verae	Chironieae	Chloreae	Chironieae-Erythraeaceae	Chironieae-Erythraeeae	***	Helieae	—
Symphyllophyton	***	***	***	***	***	***	Helieae	—
Xestaea	***	***	***	***	Chironieae-Erythraeeae	***	Helieae	—
Zygostigma	Chironiae	Chironieae	Erythraeaceae	Chloreae-Sabbatieae	Chironieae-Erythraeeae	***	Gentianeae-Tachiinae	—

Notes:
[a] Asterisks denote genera not described at the time of the listed publication.
[b] Infrafamilial names are cited as they appear in each publication.
[c] Dash denotes genera not treated in publication.

other in inflorescence morphology. Subgenera *Phyllocyclus* and *Pentanthera* comprise species with mostly perfoliate leaves, a subglobose calyx, and stamens with even insertion. Molecular and morphological evidence suggests that both of these subgenera are more closely related to *Microrphium* than to *Canscora*. In consequence, they are here treated as separate entities (see *Phyllocyclus* and "Duplipetala", respectively) (M. Thiv & J. W. Kadereit, unpubl.).

Canscora species are annual herbs usually having winged stems. Rosette leaves are rarely present, and the free cauline leaves are of variable shape. The 4-merous flowers are arranged in cymes of variable structure. The sometimes winged calyx is tubular with a tube that is longer than the triangular to lanceolate lobes. Calycine colleters are sometimes present. The pinkish-purple or white corolla is tubular to salver-shaped and is mostly zygomorphic through the presence of two smaller and two larger lobes (Fig. 2.13A). The corolla lobes are shorter than the tube. The anisomorphic stamens are inserted in the upper part of the corolla tube. The level of stamen insertion is uneven. The sagittate anthers do coil after anthesis. The pollen is released as monads. The stigmatic lobes are of variable shape. A reported chromosome number is $2n=36$ (*Canscora diffusa*; Christopher, 1976).

Centaurium HILL

This genus was described by Hill (1756: 62), and was later typified by *Centaurium littorale* (Gillett, 1963). *Centaurium* contains *c.* 50 species (Zeltner, 1970). These are distributed mainly in the Mediterranean region, where both diploids and polyploids are found, and in North and Central America, where the species are mostly polyploids (Broome, 1976, 1978). A few species also occur in South America and temperate Asia and Australia. Some originally European species (e.g., *Centaurium erythraea*) have been introduced to, for example, Australia, and South, Central, and North America (Fabris, 1953; Broome, 1976; Wood & Weaver, 1982; Pringle, 1995; Adams, 1996). The ecology of *Centaurium* is similar to that of *Blackstonia* and *Cicendia*.

Centaurium comprises annual to biennial herbs. Rosette leaves are predominant and the cauline leaves are of variable shape. The (4–)5(–6)-merous flowers are borne in cymes of variable structure. The calyx tube is usually much shorter than the linear to lanceolate lobes. Calycine colleters are mostly absent. The pinkish-purple, rarely white or yellow corolla is salver-shaped and the tube is usually longer than the lobes. The stamens, which are bent toward the abaxial side of the flower, are inserted in the

Systematics, character evolution, and biogeography 115

upper part of the corolla tube. The weakly sagittate anthers (Fig. 2.13C) are helically twisted after dehiscence (Fig. 2.13H). The pollen is released as monads. The stigmatic lobes are of variable shape. Reported chromosome numbers are $2n = 20$, 34, 36, 40, 42, 44, 54, 56, 72, 80, 82, and 84 (Zeltner, 1970; Broome, 1978).

Chironia L.

Linnaeus (1753) established this genus based on *Chironia linoides*. Schoch's (1903) revision of *Chironia* recognized 36 species. Only 15 species are indicated by Mabberley (1997). The genus is found in grassy places in sub-Saharan Africa and Madagascar.

The species of *Chironia* are annual or perennial, occasionally suffrutescent herbs. The leaves are rarely covered with bristles or papillose hairs. Rosette leaves are sometimes present and the cauline leaves are of variable shape. The 5-merous flowers are arranged in cymose inflorescences of variable structure. The calyx tube is usually much shorter than the linear to lanceolate lobes (not in *Chironia baccifera*). Calycine colleters are present. The purple-pink or white corolla is salver-shaped and the tube is usually shorter than the lobes. The stamens are inserted close to the corolla sinuses. The weakly sagittate anthers are usually helically twisted after dehiscence (Fig. 2.13I). The pollen is released as monads. The stigmatic lobes are oblong to capitate. The fruit is usually a capsule or rarely a berry (*Chironia baccifera*). A reported chromosome number is $2n = 68$ (*Chironia baccifera*; Weaver & Rüdenberg, 1975).

Cicendia ADANS.

Adanson (1763) described this genus based on *Cicendia filiformis*. *Cicendia* contains two species (cf. Gilg, 1895a, as *Microcala*). These are *Cicendia filiformis*, a species of originally Mediterranean and western European distribution, and *C. quadrangularis*, which has a disjunct distribution in southern and western North America and South America from Ecuador to Argentina. Both species have been introduced to Australia (Adams, 1996). The species grow in humid and open, sometimes rocky places.

Cicendia is easily recognized by its small, annual, and filiform habit. The rosette leaves are deciduous and the lanceolate cauline leaves are very small. The 4-merous flowers are single or few and are arranged in lax cymes. The campanulate calyx tube is much longer than the triangular lobes. Calycine colleters are absent. The yellow corolla is salver-shaped and the tube is longer than the lobes. The stamens are inserted close to the corolla sinuses. The slightly x-shaped anthers (Fig. 2.13E) do not coil after anthesis. The pollen is released as monads. The stigmatic lobes are weakly constricted in the middle. Reported chromosome numbers are $2n = (24)$ 26 (Favarger, 1960; Löve & Löve, 1961; Post, 1967).

Coutoubea AUBL.

Aublet (1775) established this genus based on *Coutoubea spicata*. The genus contains five species distributed in northern South America (Guimarães & Klein, 1985). Only *Coutoubea spicata* extends into Central America. The species grow in savannas and weedy, open places.

The species of *Coutoubea* are annual or rarely short-lived perennial, sometimes suffrutescent herbs. Rosette leaves are absent and the cauline leaves are lanceolate. The 4-merous flowers with bracteoles at the base of the calyx are borne in dense spikes. The calyx tube is as long as or shorter than the triangular lobes. Calycine colleters are absent. The whitish corolla is salver-shaped and the tube is as long as the lobes. The stamens are inserted in the upper part of the corolla tube. The filaments have scale-like basal appendages. The sagittate anthers do not coil after anthesis (Fig. 2.13B). The pollen is arranged in tetrads. The small stigmatic lobes are oblong. A reported chromosome number is $2n = 30$ (*Coutoubea spicata*; Weaver & Rüdenberg, 1975).

Cracosna GAGNEP.

This genus was described by Gagnepain (1929) based on *Cracosna xyridiformis*. It consists of three species that occur in Thailand, Laos, and Cambodia (M. Thiv, unpubl.).

Cracosna xyridiformis is an annual herb with distinctly winged stems. Rosette leaves are absent and the cauline leaves are triangular to lanceolate. The 4-merous flowers (except for the calyx) are arranged in dense, capitate inflorescences. The calyx consists of two, three, or four sepals with a tube that is longer than the lanceolate lobes. Calycine colleters are absent. The cream-colored corolla is tubular to salver-shaped and the tube is longer than the lobes. The stamens are inserted in the upper part of the corolla tube. The sagittate anthers do not coil after dehiscence. The pollen is released as monads. The stigmatic lobes are linear.

Deianira CHAM. & SCHLTDL.

Von Chamisso and von Schlechtendal (1826) established this genus based on *Deianira erubescens*. Seven species from southern Brazil and Bolivia were recognized by Guimarães (1977). The species grow in open, sometimes rocky places.

The genus comprises annual or sometimes short-lived perennial, suffrutescent herbs. The rosette leaves are broadly elliptical and the cauline leaves are ovate to lanceolate and are sometimes slightly perfoliate. The 4-merous flowers are arranged in very dense dichasia. The calyx tube is shorter than the lanceolate lobes. Calycine colleters are absent. The purple-pink to white corolla is salver-shaped and the tube is shorter than the lobes. The stamens are inserted in the upper part of the corolla tube. The non-sagittate anthers are not coiled after dehiscence. The pollen is arranged in tetrads. The stigmatic lobes are small and oblong.

"Duplipetala", ined. (M. THIV, UNPUBL.)

This genus will be described by M. Thiv (in prep.) based on *Canscora pentanthera*. "Duplipetala" consists of two species with Southeast Asian distribution. The species grow in dipterocarp forests, savannas, or tropical forests, mostly on limestone rocks. This genus corresponds to Clarke's (1905) *Canscora* subgenus *Pentanthera*.

"Duplipetala" comprises annual and perennial herbs with unwinged stems. Rosette leaves are absent and the free cauline leaves are lanceolate to ovate in shape. The 5- or rarely 6-merous flowers are arranged in lax cymes. The calyx with unconnected wings is tubular, and the tube is longer than the triangular lobes. Calycine colleters are absent. The white- to cream-colored corolla is tubular to salver-shaped. The corolla lobes are shorter than the tube. The stamens are inserted in the upper part of the corolla tube. The level of stamen insertion is even. The sagittate anthers do not coil after anthesis. The pollen is released as monads. The stigmatic lobes are rounded.

Eustoma SALISB.

Eustoma was established by Salisbury (1806: t. 34) based on *Eustoma silenifolium*. A synopsis of the genus by Shinners (1957) recognized three species distributed in the southern United States, Mexico, and the Greater Antilles. The species grow on dry soil in disturbed places and in pine forests. *Eustoma grandiflorum* is a popular ornamental.

Eustoma consists of glaucous, annual or short-lived perennial herbs. The rosette and cauline leaves are elliptical. The 5-merous flowers are borne in lax, monochasial cymes. The calyx tube is much shorter than the narrowly triangular lobes. Calycine colleters are absent. The mostly bluish, sometimes purple, pink, or white corolla is funnel- to salver-shaped, and the tube is much shorter than the lobes. The stamens, which have sagittate anthers, are inserted between the corolla lobes. The anthers twist slightly after dehiscence (Fig. 2.13F). The pollen is released as monads. The stigmatic lobes are rounded. A reported chromosome number is $2n = 72$ (Rork, 1949).

Exaculum CARUEL

Exaculum was described by Caruel (1886) based on *E. pusillum*. The only species of the genus, *Exaculum pusillum* (cf. Gilg, 1895a, as *Cicendia*), is restricted to the western Mediterranean region and Atlantic region of western Europe, where it grows in humid places with sparse vegetation (Gilg, 1895a).

Exaculum pusillum is a small, annual, filiform herb. The deciduous rosette leaves are elliptical and the cauline leaves are lanceolate. The small, 4-merous flowers are borne in lax cymes. The calyx tube is shorter than the linear lobes. Calycine colleters are absent. The pink corolla is salver-shaped and the tube is longer than the lobes. The stamens are inserted close to the corolla sinuses. The sagittate anthers do not coil after dehiscence. The pollen is released as monads. The stigmatic lobes are weakly constricted in the middle. A reported chromosome number is $2n = 20$ (Favarger, 1960).

Geniostemon ENGELM. & A. GRAY

This genus was described by Engelmann and Gray (1881: 104–105) based on *Geniostemon coulteri*. *Geniostemon* comprises five species endemic to Mexico (Turner, 1994; Rzedowski & Calderon de Rzedowski, 1995), where they grow on gypsaceous soils.

The species of this genus are either small, erect, annual herbs or creeping, perennial herbs. Rosette leaves are absent and the cauline leaves are linear to ovate. The 4-merous flowers are single or arranged in frondose terminal dichasia. The calyx tube is shorter than the triangular to lanceolate lobes. Calycine colleters are absent. The pink or white corolla is salver-shaped and the tube is as long as the lobes. The stamens are inserted in the sinuses between corolla lobes. The filaments bear short glandular hairs at their middle. The sagittate anthers do not coil after dehiscence. The pollen is released as monads. The stigmatic lobes are minute.

Hoppea WILLD.

Hoppea was described by Willdenow (1801) based on *H. dichotoma*. The two species of *Hoppea* are distributed mainly in India, Sri Lanka, and Pakistan (Gilg, 1895a). One of them, *Hoppea dichotoma*, also occurs in Senegal. The species grow in open, moist places, and seem to prefer cultivated ground.

Hoppea comprises small, annual herbs. Rosette leaves are absent and the cauline leaves are broadly lanceolate. The small, 4-merous flowers are arranged in dense or lax dichasia. The calyx tube is as long as the triangular lobes. Calycine colleters are sometimes present. The whitish corolla is urceolate and the tube is as long as or longer than the lobes. The anisomorphic stamens are inserted unevenly in the corolla tube. The sagittate anthers do not coil after anthesis. The pollen is released as monads. The stigmatic lobes are small and oblong.

Ixanthus GRISEB.

Grisebach (1839) established this genus based on *Ixanthus viscosus*. The only species is endemic to the Canary Islands, where it grows in laurel forests (Gilg, 1895a). The biogeography and evolution of *Ixanthus* were discussed in detail by Thiv *et al.* (1999b), who found a close phylogenetic relationship between *Ixanthus* and the Mediterranean genus *Blackstonia*.

Ixanthus is a perennial suffrutescent herb. Gilg (1895a) noted the presence of glandular hairs, but we have not been able to confirm this from either living or dried plant material. However, the leaves and branches in the inflorescence are viscid, so sessile glands are probably present. Rosette leaves are absent. Whereas the lanceolate cauline leaves are free, the bracts are perfoliate. The usually 5-merous flowers are arranged in dichasia. The calyx tube is shorter than the triangular lobes. Calycine colleters are present. The yellow corolla is salver-shaped and the tube is as long as the lobes. The stamens are bent toward the abaxial side of the flower and are inserted close to the corolla sinuses. The sagittate anthers do not coil after anthesis. The pollen is released as monads. The stigma is capitate and slightly bilobed. A reported chromosome number is $2n=62$ (Thiv *et al.*, 1999b).

Microrphium C. B. CLARKE

Microrphium was established by Clarke (1905) based on *M. pubescens*. One of the two species of the genus (*Microrphium pubescens*) is distributed in the Malay Peninsula. Recently, *Microrphium elmerianum* was described from Palawan, the Philippines, but this species is probably a subspecies of *M. pubescens* (M. Thiv, unpubl.). Both species grow in basic soil, mostly on limestone rocks at low elevations (Regalado & Soejarto, 1997).

Microrphium comprises perennial herbs with bristles on all green parts except for the calyx. Rosette leaves are absent and the elliptical to lanceolate cauline leaves are petiolate. The 5-merous flowers are arranged in lax monochasia. The calyx consists of two lobes with a tube that is much longer than the broadly lanceolate lobes. Calycine colleters are absent. The cream- to white-colored corolla is salver-shaped and the tube is as long as the lobes. The stamens are inserted in the corolla tube. The sagittate anthers do not coil after dehiscence. The pollen is released as monads. The stigmatic lobes are pointed.

Orphium E. MEY.

Meyer (1838) described this genus based on *Orphium frutescens*. *Orphium* contains one (Gilg, 1895a) or probably two species (M. Thiv, unpubl.) that grow in the South African fynbos vegetation near the sea.

Orphium comprises perennial suffrutescent herbs with bristles on the upper green parts. Rosette leaves are absent and the cauline leaves are long-elliptical. The 5-merous flowers are borne in lax monochasia. The calyx tube is much shorter than the broadly lanceolate lobes. Calycine colleters are present. The purple-pink corolla is salver-shaped and the tube is shorter than the lobes. The stamens are bent toward the abaxial side of the flower and are inserted in the corolla tube. The weakly sagittate anthers are helically twisted after dehiscence (Fig. 2.13J). The pollen is released as monads. The stigmatic lobes are slightly bilobed and minute.

Phyllocyclus KURZ

This genus was described by Kurz (1873) based on *Phyllocyclus helferianus*. *Phyllocyclus* consists of five species of Southeast Asian distribution (M. Thiv, unpubl.). The species grow in tropical forests mostly on limestone rocks. This genus was formerly treated as *Canscora* subgenus *Phyllocyclus* (Clarke, 1885).

Phyllocyclus consists of annual herbs with unwinged stems. The rosette leaves are free at the base and the cauline leaves are perfoliate and orbicular in shape. The 5-merous flowers (except the calyx) are arranged in lax cymes of variable shape. The unwinged calyx is tubular to urceolate, and the tube is longer than the triangular lobes. Calycine colleters are absent. The white-, yellow-, or cream-colored corolla is urn-, funnel- to salver-shaped. The corolla lobes are shorter than the corolla tube. The stamens are inserted in the central or upper part of the corolla tube, and the level of stamen insertion is even. The sagittate anthers do not coil after anthesis. The pollen is released as monads. The stigmatic lobes are rounded.

Sabatia ADANS.

According to Wilbur (1989), the type species of this genus, which was described by Adanson (1763), is *Sabatia dodecandra*. *Sabatia* contains c. 20 species concentrated in eastern North America but extending to western

Systematics, character evolution, and biogeography 121

USA, Mexico, and the West Indies (Wilbur, 1955). They are found in a variety of habitats ranging from dry to wet near fresh or salt water. Three sections have been recognized (Perry, 1971). Species of section *Sabatia* and section *Campestria* have mostly 5-merous or rarely polymerous flowers on pedicels with scale-like bracts, whereas those of section *Dodecandrae* have polymerous flowers that are sessile and have large leaf-like bracts.

Sabatia contains annual or biennial herbs. The rosette and cauline leaves are of variable shape. The (4–)5 or 8–13-merous flowers are borne singly or are arranged in lax cymes. The calyx tube is much shorter than the triangular to lanceolate lobes. Calycine colleters are usually present. The pinkish-purple corolla is rotate to salver-shaped and the tube is much shorter than the lobes. The stamens are inserted between the corolla lobes. The sagittate or non-sagittate anthers are usually outwardly twisted after dehiscence (Fig. 2.13G). The pollen is released as monads. The stigmatic lobes are linear to oblong and usually start coiling during anthesis. Reported chromosome numbers are $2n = 26, 28, 32, 34, 36, 38, 40, 72$, and 76 (Wood & Weaver, 1982).

Schinziella GILG

This monotypic genus was described by Gilg (1895a) based on *Schinziella tetragona*. *Schinziella* is widespread in tropical Africa, where it grows in humid, grassy places (Gilg, 1895a; Paiva & Nogueira, 1990).

Schinziella tetragona is a perennial herb with strongly winged stems. Rosette leaves are absent and the cauline leaves are small and lanceolate. The 4-merous flowers are borne in dense, terminal, capitate cymes. The calyx tube is longer than the lanceolate lobes. Calycine colleters are present. The whitish-yellow corolla is tubular to salver-shaped and the tube is as long as the lobes. The anisomorphic stamens are inserted unevenly in the corolla tube. The sagittate anthers do not coil after dehiscence. The pollen is released as monads. The stigmatic lobes are small and rounded.

Schultesia MART.

Von Martius (1826–1827) established *Schultesia* based on *S. crenuliflora*. *Schultesia* contains c. 15 species (P. J. M. Maas, pers. comm.) distributed from Central America (including the West Indies) to Brazil. *Schultesia stenophylla* is also known from West Africa. The species of *Schultesia* grow in open, sometimes sandy places.

The species of *Schultesia* are annual herbs. Rosette leaves are absent and the cauline leaves are long-lanceolate. The 4-merous flowers are borne singly or are arranged in lax monochasia. The carinate to strongly winged calyx tube is as long as or longer than the lanceolate lobes. Calycine colleters are absent. The pink corolla is funnel-shaped and the tube is longer than the lobes. The stamens are inserted in the upper part of the corolla tube. The weakly sagittate anthers do not coil after dehiscence. The pollen is arranged in tetrads. The stigmatic lobes are rounded.

Symphyllophyton GILG

This monotypic genus was described by Gilg (1898) based on *Symphyllophyton caprifolioides*. The only species of the genus is a rare endemic of southern Brazil.

Symphyllophyton caprifolioides is a short-lived, perennial, slightly suffrutescent herb. Rosette leaves are absent and the cauline leaves are perfoliate. The 4-merous flowers are arranged in lax cymes. The calyx tube is shorter than the lanceolate lobes. Calycine colleters are absent. The pale yellow to cream corolla is salver-shaped and the tube is as long as the lobes. The stamens are inserted in the upper part of the corolla tube. The filaments are of different length. The non-sagittate anthers do not coil after dehiscence. The pollen is arranged in tetrads. The stigmatic lobes are small and oblong.

Xestaea GRISEB.

Xestaea was established by Grisebach (1849) based on *X. lisianthoides*. This monotypic genus is distributed in Central and northern South America, where it grows in humid thickets and fields. It has often been included in *Schultesia* (e.g., Elias & Robyns, 1975).

Xestaea is an annual herb. Rosette leaves are absent and the cauline leaves are long-lanceolate. The 4-merous flowers are arranged in lax dichasia. The calyx tube is shorter than the lanceolate lobes. Calycine colleters are absent. The pink to lavender corolla is funnel-shaped and the tube is longer than the lobes. The stamens are inserted unevenly in the corolla tube. The anthers are weakly sagittate and do not coil after dehiscence. The pollen is arranged in tetrads. The stigmatic lobes are oblong.

Zygostigma GRISEB.

Zygostigma was described by Grisebach (1839) based on *Z. australe*. This monotypic genus occurs in Brazil, Argentina, and Uruguay (P. J. M. Maas, unpubl.).

Zygostigma australe is an annual herb. Rosette leaves are absent and the cauline leaves are linear. The single flowers are 5-merous. The calyx tube is much shorter than the linear to lanceolate lobes. Calycine colleters are present. The white corolla is salver-shaped and the tube is as long as the lobes. The stamens are inserted in the upper part of the corolla tube. The sagittate anthers do not coil after dehiscence. The pollen is released as monads. The stigmatic lobes are oblong.

Phylogeny and character evolution in Chironieae

Among the molecular phylogenies shown in Figs. 2.1–2.3, the Chironieae are strongly supported in the *mat*K and combined analyses. Despite this, we have not been able to identify a unique morphological or anatomical synapomorphy for the group.

The tree shown in Fig. 2.14 summarizes a hypothetical phylogeny of the Chironieae based on combined *mat*K, *trn*L, and morphological characters. In comparison to Fig. 2.3, species of *Centaurium*, *Eustoma*, and *Microrphium* were added to the tree following their positions in the separate analyses, while *Cracosna*, *Exaculum*, *Phyllocyclus*, and *Schultesia* were placed according to morphological data (Thiv et al., 1999b). The more resolved topologies of the genera *Centaurium* and *Chironia* provided by internal transcribed spacer (ITS) data (Thiv et al., 1999b) was taken into consideration.

Within the tribe, two major lineages can be recognized (Fig. 2.14). Of these, the first (subtribe Chironiinae), which is well supported in the *mat*K and combined analyses, comprises taxa of mostly Northern Hemisphere distribution. A second clade of mostly tropical taxa, segregated into two well-supported subclades (subtribes Coutoubeinae and Canscorinae), is present only in the combined analysis and is not supported by parsimony jackknife resampling (Fig. 2.3). Additionally, this clade is not resolved in separate analyses of *mat*K and *trn*L intron sequence data. Because of their common tropical distribution (see below) and their mostly 4-merous flowers, we here wish to regard these two subclades together as a monophylum. However, they are here classified as different subtribes to reflect the low degree of support each receives among different phylogenetic analyses while retaining high individual support in all analyses.

Subtribe Coutoubeinae

Of the two subclades of the putative tropical lineage, the Coutoubeinae (*Coutoubea*, *Deianira*, *Schultesia*, *Symphyllophyton*, and *Xestaea*) have a neotropical distribution and are well characterized by the possession of pollen tetrads. *Xestaea* was not included in the *trn*L intron analysis (the original *trn*L intron sequence presented in Struwe et al. (1998: fig. 1) was misidentified; cf. Thiv et al., 1999a), so we have relied on the results of the

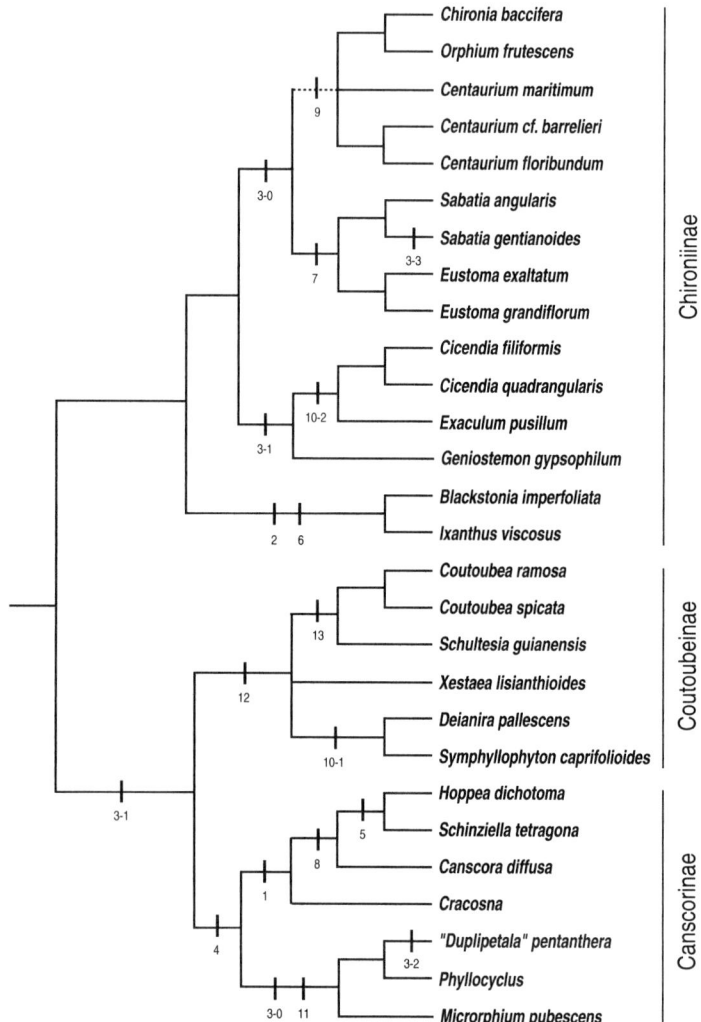

Figure 2.14. Hypothetical phylogeny of Chironieae based on results of combined *mat*K and *trn*L intron data. Compared to Fig. 2.3, *Centaurium floribundum*, *Eustoma exaltatum*, and *Microrphium pubescens* were added following the topologies of the separate *mat*K and *trn*L intron phylogenies (Figs. 2.1–2.2). The dashed line shows phylogenetic relationships as indicated by ITS sequence data (Thiv et al., 1999b). Bars indicate characters mapped onto the tree and discussed in the text as follows: 1: stems strongly winged; 2: perfoliate leaves; 3-0: 5-merous flowers; 3-1: 4-merous flowers; 3-2: 6-merous flowers; 3-3: polymerous flowers; 4: calyx tube longer than calyx lobes; 5: funnel-shaped calyces; 6: yellow petals; 7: stamen insertion in sinuses of corolla tube; 8: anisomorphic stamens inserted at different levels (e.g., Fig. 2.13A); 9: anthers after dehiscence strongly coiled (e.g., Fig. 2.13H–J); 10-1: long, basifixed, non-sagittate anthers; 10-2: slightly x-shaped anthers (Fig. 2.13E); 11: deciduous anthers; 12: pollen in tetrads; 13: reticulate to foveate exine structure.

*mat*K analysis (Fig. 2.2), the presence of pollen tetrads in this genus, and its distribution in the New World.

Although pollen tetrads appear to be a good synapomorphy for the Coutoubeinae, they are also found in the *Symbolanthus* clade of Helieae, where they are likely to have evolved in parallel. This had not been recognized by Gilg (1895a), who included all genera with pollen tetrads known to him in tribe Helieae. Within Coutoubeinae a reasonably well-supported sister group relationship is found between *Deianira* and *Symphyllophyton*. This relationship between two Brazilian taxa is supported particularly by the structure of their anthers, which are long and clearly non-sagittate, but also by their small and oblong stigmatic lobes and their partly perfoliate leaves. A sister-group relationship between *Coutoubea* and *Schultesia*, although not resolved in the molecular phylogenies, is indicated by the shared possession of pollen with a characteristically reticulate to foveate exine structure (Maguire & Boom, 1989; Nilsson, 2002). This exine pattern is clearly different from that of *Xestaea* (Elias & Robyns, 1975), which has often been regarded as congeneric with *Schultesia* (Gilg, 1895a; Elias & Robyns, 1975).

Subtribe Canscorinae

The second subclade (Fig. 2.14: subtribe Canscorinae) of the tropical lineage consists of *Canscora, Cracosna*, "Duplipetala", *Hoppea, Microrphium, Phyllocyclus*, and *Schinziella*. The group is of paleotropical distribution and is characterized by flowers with calyx tubes that are mainly much longer than the lobes as well as by its mostly cream-colored corollas. The phylogenetic relationships among all species of the Canscorinae were investigated in a cladistic analysis using morphological data (M. Thiv & J. W. Kadereit, unpubl.). This subtribe can be subdivided into two groups. The first comprises *Canscora, Cracosna, Hoppea*, and *Schinziella*, and is characterized by the possession of winged stems and a mostly annual habit. *Cracosna* forms the sister lineage to *Canscora, Hoppea*, and *Schinziella*. Of these, *Canscora, Hoppea*, and *Schinziella* share a 4-merous androecium in which one larger stamen is inserted at a higher level in the corolla tube than the three smaller ones (Fig. 2.14). The second group of the Canscorinae consists of "Duplipetala", *Phyllocyclus* (both formerly subgenera of *Canscora*), and *Microrphium*. These genera have an androecium with even stamen insertion, 5- or 6-merous flowers with a subglobose calyx, and a habitat preference for limestone rocks (Ubolcholaket, 1987; M. Thiv & J. W. Kadereit, unpubl.).

These morphology-based results are largely congruent with the molecular data. Although *Cracosna* and *Phyllocyclus* were not included in the molecular studies, the split into *Canscora*, *Hoppea*, and *Schinziella* on the one hand and "Duplipetala" (as *Canscora pentanthera*) and *Microrphium* on the other is corroborated. Relationships among *Canscora*, *Hoppea*, and *Schinziella* are resolved differently in the cpDNA analyses. According to the *mat*K data, *Canscora* is sister to *Schinziella*, but the *trn*L intron data suggest a sister-group relationship between *Schinziella* and *Hoppea*. The latter pair is supported by the presence of funnel-shaped calyces in *Schinziella* and *Hoppea*.

Subtribe Chironiinae

The predominantly northern-temperate subclade of Chironieae (Fig. 2.14: Chironiinae) is well supported in the *mat*K and combined molecular phylogenetic analyses, but not by the *trn*L intron data. Of the genera sampled, the group includes *Blackstonia*, *Centaurium*, *Chironia*, *Cicendia*, *Eustoma*, *Geniostemon*, *Ixanthus*, *Orphium*, and *Sabatia*. No unique morphological or anatomical synapomorphy could be identified to support this group, only its geographic distribution. *Ixanthus* and *Blackstonia*, which both have at least some perfoliate leaves and yellow petals, are sister to the remaining taxa in the *mat*K and combined analyses. The close relationship of the two genera has been discussed in detail by Thiv *et al.* (1999b). The clade of *Cicendia* and *Geniostemon*, which did not receive strong support in the combined analysis, is nonetheless marked by the shared presence of 4-merous flowers. Although not included in any of our molecular analyses, the filiform habit of *Exaculum*, its possession of 4-merous flowers, and the shapes of its anthers and stigmata all point to a sister-group relationship with *Cicendia*. In the past, some authors have treated *Exaculum* as part of *Cicendia* (e.g., Heß *et al.*, 1972). However, in the ITS analysis presented by G. Mansion (unpubl.), *Exaculum* was positioned close to some species of *Centaurium*.

The monophyly of the clade consisting of the remaining genera (*Centaurium*, *Chironia*, *Eustoma*, *Orphium*, and *Sabatia*) has very strong molecular support in the *mat*K and combined analyses. These genera share anthers that coil strongly after dehiscence (Fig. 2.13), pink, purple, or more rarely white or yellow corollas, and mostly 5-merous flowers. Within this clade, sister-group relationships are resolved with reasonably high support between *Chironia* and *Orphium*, and with lower support between *Centaurium* and *Sabatia*. In contrast to the *mat*K and *trn*L intron data, and more congruent with morphological characters, ITS sequence variation in this clade found by Thiv *et al.* (1999a) suggested a sister-group relationship between *Centaurium* and the *Chironia/Orphium* clade rather than one between

Centaurium and *Sabatia*. Such a relationship would be supported by the presence of anthers that very strongly and helically twist after dehiscence in *Centaurium, Chironia*, and *Orphium* (Fig. 2.13). These differing (and also poorly supported) relationships found for *Centaurium* were probably due to poor taxon sampling within this species-rich genus.

In contrast, in a larger ITS analysis by Mansion (2000; G. Mansion, unpubl.) concentrating on *Centaurium, Centaurium* is clearly shown as polyphyletic, with four monophyletic groups. These four clades are nested among *Chironia, Eustoma, Exaculum, Orphium*, and *Sabatia*, so if *Centaurium* were to be kept as a monophyletic genus, these five genera would have to be included in it. This seems a less practical solution than to identify the four *Centaurium* clades as separate genera, or to include them with their various sister groups outside the genus. Major generic rearrangements will therefore be necessary within the Chironiinae within the near future. One major clade of *Centaurium* (containing the type species) includes most of the Eurasian species, which are closely related to *Chironia* and *Orphium* from Africa. Nearly all of the North American *Centaurium* species form a separate monophyletic clade, and *Exaculum* is positioned in this clade as well. *Sabatia* and *Eustoma* form a clade with another *Centaurium* clade (approximately corresponding to *Centaurium* sect. *Gyrandra*). Within the *Centaurium* lineages found, there are also many examples of karyological changes within clades, for example polyploidy ($2x$ to $6x$), allopolyploidy (through hybridization), and dysploidy (G. Mansion, unpubl.).

Apart from the polyphyly of *Centaurium*, no other genera have so far been found to be poly- or paraphyletic in the Chironiinae. *Centaurium* is the largest genus in the subtribe, and *Chironia* and *Sabatia* also have comparatively large numbers of species (Table 2.1). It is not clear from the taxon sampling, however, whether or not *Chironia* is monophyletic. In particular, paraphyly in relation to *Orphium* should be considered as a possibility since, for example, a bristly or papillose indumentum is shared between species of *Chironia* and *Orphium*. *Eustoma* and *Sabatia*, both North to Central American genera, have stamens inserted in the sinuses between corolla lobes rather than in the corolla tube, a possible synapomorphy. Accordingly, the inclusion in our analysis of one 5-merous and one polymerous species representing two sections of *Sabatia* may indicate the monophyly of this morphologically variable genus. *Sabatia* was also found to be monophyletic in the study by G. Mansion (unpubl.). It seems likely to us that the annual *Bisgoeppertia* from the Greater Antilles and *Zygostigma* from South America – both of which have monad pollen, a calyx tube that is shorter than the calyx lobes, and 5-merous flowers – also belong in the

Chironiinae. Their precise relationships within this clade, however, remain unknown.

As is evident from the above discussion, several morphological characters provide additional support as either unique or homoplastic synapomorphies for the phylogenetic relationships found in the molecular analysis. Other characters useful for the identification of genera, however, are highly homoplastic. Thus, a suffrutescent habit is found in some species of *Chironia, Coutoubea, Deianira, Eustoma*, and *Symphyllophyton* as well as in *Ixanthus* and *Orphium*, and this habit appears to have evolved independently several times. Likewise, characters such as the presence or absence of a leaf rosette or presence/absence of bracteoles (*Bisgoeppertia/Coutoubea*) appear to contain little if any phylogenetic information. Perfoliate leaves are informative for some groups (within *Canscora, Blackstonia/Ixanthus*, and *Deianira/Symphyllophyton*) but have evolved repeatedly in the tribe. The presence of a bristly indumentum in this mostly glabrous tribe may support a relationship between *Chironia jasminoides* and *Orphium*, but this trait is likely to have arisen independently in *Microrphium*. Polymerous flowers have evolved in parallel in *Blackstonia* and *Sabatia* of the Chironieae, and they are also found in other Gentianaceae, such as *Anthocleista, Potalia*, and *Urogentias*. Calycine colleters, although consistently present or absent in some of the clades identified, are homoplastic in the Gentianaceae as a whole.

Palynology of Chironieae: Synopses of genera and general discussion

S. NILSSON

Centaurium. Centaurium maritimum (L.) Fritsch. The pollen grains are 3-colporate, prolate spheroidal, and *c.* 30×28 μm. The exine is tectate, densely striate to striato-rugulate, with straight to curved muri/lirae abruptly changing direction, and perforate. The perforations are often oval-shaped, relatively large, and of varying size. (Spain: Mallorca, 1995, *Klackenberg 950424-2* (S), slide no. 24639; Fig. 2.15A–B.)

OTHER INVESTIGATIONS: Gilg (1895a) described the pollen of Gentianeae-Erythraeinae (including *Centaurium*) as 3-colpate, smooth or finely granular, whereas Köhler (1905) described the pollen grains as 3-colporate and finely reticulate after studying 14 species of *Centaurium*. Huang (1972) also described one species of *Centaurium* (as *Erythraea spicata*) as having 3-colporate pollen with reticulate exine.

The pollen grains of *Centaurium pulchellum*-type, including *C. erythraea*,

C. littorale, C. maritimum, C. pulchellum, C. scilloides, C. spicatum, and *C. tenuiflorum,* have been described as 3-colporate, tectate, suprastriate, with lirae/muri abruptly changing direction, and striae perforate (Punt & Nienhuis, 1976). Agababyan and Tumanyan (1976b) described pollen of five species of *Centaurium*. The pollen grains of the Central American species *Centaurium strictum* and *C. quitense* have been described as 3-colporate, striate to striato-reticulate, rarely reticulate, and with lirae/muri parallel or not and interlaced and criss-crossed (Elias & Robyns, 1975). In a study of six *Centaurium* species Rao and Chinnappa (1983) found a significant variation, in particular regarding apertures (pericolporate conditions) and exine features related to the orientation of lirae and perforations in striae.

Chironia. *Chironia baccifera* L. The pollen grains are 3-colporate, prolate spheroidal, and *c.* 44×40 µm. The exine is striato-reticulate, with muri/lirae branched or unbranched, partly interlaced, and slightly keeled. (South Africa: Cape, Hermanus, 1972, *Bremer 507* (S), slide no. 24797; Fig. 2.16A–B.)

OTHER INVESTIGATIONS: Köhler (1905) described pollen of six species of *Chironia* as generally striate with finely reticulate pattern. Agababyan and Tumanyan (1976a) described pollen of four species of *Chironia*. In one specimen of *Chironia purpurascens* the striato-reticulate exine pattern was more open and distinct (cf. *Orphium*; S. Nilsson, pers. obs.). Pollen of *Chironia pubescens* (now *Ornichia madagascariensis* ssp. *pubescens* (see discussion under *Ornichia*); Klackenberg, 1986) was reported as small (21 × 19.5 µm) by Lienau *et al.* (1986).

Coutoubea. *Coutoubea ramosa* Aubl. The 3-colpate to colporate or porate pollen grains are united in tetrahedral tetrads that are *c.* 58 µm in diameter. The exine is reticulate with a coarser reticulum at the poles than at the equator. The lumina diminish in size toward the distinctly delimited aperture area. (Venezuela: Bolívar, 1948, *Tamayo 3460* (NY), slide no. 24093; Nilsson, 2002: Fig. 4.6B–C.)

Coutoubea spicata Aubl. The 3-colporate to porate pollen grains are united in tetrahedral tetrads that are *c.* 60 µm in diameter. The exine is as described for *Coutoubea ramosa* above. (Trinidad: Matura, 1985, *Nilsson 1985-4* (S), slide no. 19399.)

OTHER INVESTIGATIONS: *Coutoubea ramosa* and *C. spicata* have the same type of tetrads as *C. minor* and *C. reflexa* (Nilsson, 2002: Fig. 4.6A,D–F, respectively). The pollen tetrads are similar to those of *Schultesia*.

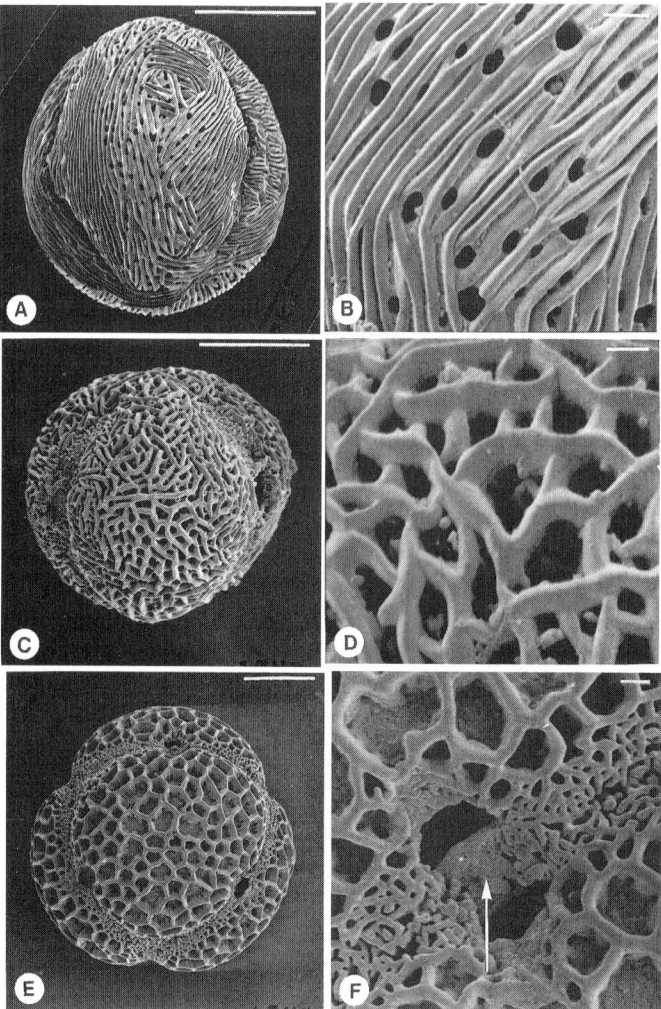

Figure 2.15. Pollen of Chironieae-Chironiinae. **A, B.** *Centaurium maritimum* (*Klackenberg 950425-2*). **A.** 3-colporate, striate pollen in equatorial, mesocolpial view. Muri/lirae oriented in various directions. **B.** Detail of tectate-perforate exine with densely spaced, straight and curved, often branched muri/lirae; striae perforate to almost foveolate. **C, D.** *Eustoma exaltatum* (*Palmer 14499*). **C.** 3-colporate, reticulate-striate pollen in equatorial, mesocolpial view. **D.** Detail of reticulate exine with a tendency toward striate pattern; muri straight, or sinuous, interconnecting at different levels. **E, F.** *Symphyllophyton* cf. *caprifolioides* (*Prance & Silva 58543*). **E.** Porate tetrad in tetrahedral position; reticulum relatively coarse and uniform, decreasing markedly at the immediate contact areas between grains. **F.** Detail at the aperture region showing pore(s) with part of the inner wall below (arrow), and the minutely reticulate pattern. Scale bars: A, C, E = 10 μm; B, D, F = 1 μm.

Systematics, character evolution, and biogeography   131

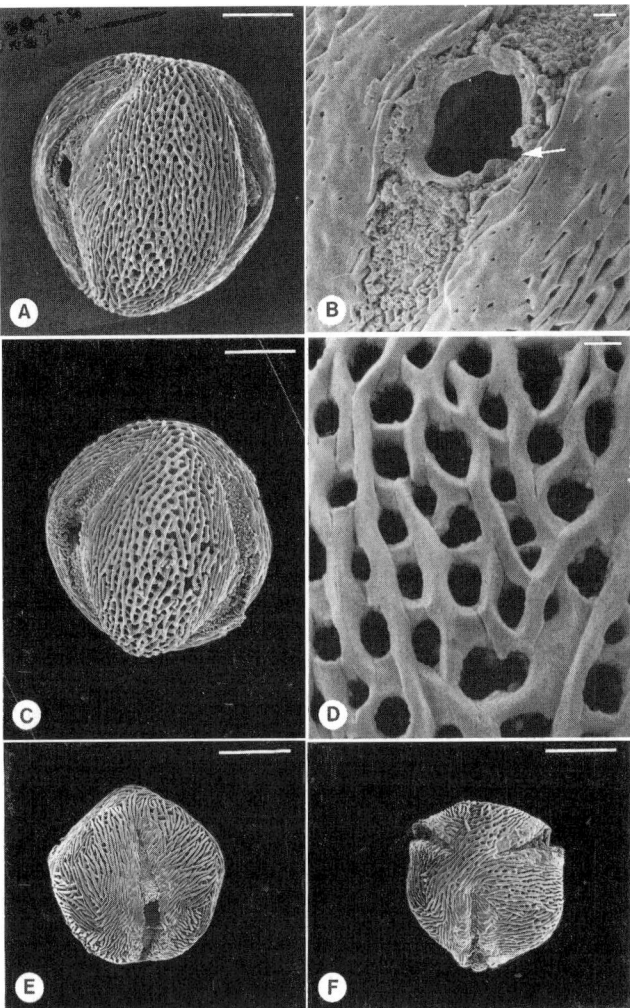

Figure 2.16. Pollen of Chironieae-Chironiinae. **A, B.** *Chironia baccifera* (*Bremer 507*). **A.** 3-colporate, striato-reticulate pollen in equatorial, mesocolpial view. **B.** Detail near the colpus and os region. The relatively broad colpus margin is smooth and sparsely perforated; the os is well delimited with a lateral extension on the right side (arrow). **C, D.** *Orphium frutescens* (*Barker 400*). **C.** 3-colporate, coarsely striato-reticulate pollen in equatorial, mesocolpial view. **D.** Detail of exine showing keeled, branched muri/lirae and transverse muri at different levels. **E, F.** *Ixanthus viscosus* (*Thiv s.n.*). **E.** 3-colporate, striate to striato-reticulate pollen in oblique equatorial, apertural view. The muri/lirae are branched, parallel near colpi, changing abruptly to oblique and transversely directed at mesocolpia; colpus margin narrow, smooth and sparsely perforated. **F.** Pollen in polar view; distinctly striate over the pole. Scale bars: A, C, E, F = 10 μm; B, D = 1 μm.

The tetrads of *Coutoubea* (placed in the tribe Helieae) were said to be indistinctly reticulate by Gilg (1895a). Köhler (1905) provided a more detailed description of *Coutoubea* (*C. ramosa* and *C. spicata*) and pointed out the clear net structure of the tetrads. Elias and Robyns (1975) also described tetrads of *Coutoubea spicata* in detail. *Coutoubea spicata* was also studied by Agababyan and Tumanyan (1977a), and more recently described and depicted by Roubik and Moreno (1991). Maguire and Boom (1989) treated three species of *Coutoubea* from the Guayana Highlands and reported that the pollen grains were released as tetrahedral tetrads. *Coutoubea spicata* was illustrated with SEM and LM micrographs in their publication.

Deianira. *Deianira nervosa* Cham. & Schltdl. The pollen grains are 3-colpate to colporate and are united in tetrahedral tetrads that are *c.* 56 µm in diameter. The exine is finely reticulate with muri provided with ring-like substructures. (Brazil: Minas Gerais, 1952, *Smith 6964* (NY), slide no. 2267; Nilsson, 2002: Fig. 4.7C–F.)

OTHER INVESTIGATIONS: Gilg (1895a) described the tetrads of *Deianira* (in his tribe Helieae) as finely reticulate to almost perforate. Köhler (1905) examined three species and pointed out that their tetrads were reticulate to finely reticulate with muri not always complete. Moreira (1961) examined pollen tetrads of *Deianira* and described them as having an average diameter of 30 µm, triporate to colpate (colpi short), and with reticulate-perforate exine. Agababyan and Tumanyan (1977a) described and illustrated tetrads of *Deianira nervosa* and *D. erubescens*. In addition to *Deianira nervosa* (Fig. 4.7C–F) the following species have been studied by Nilsson (2002): *D. cordifolia*, *D. cyathifolia* (Fig. 4.7), *D. erubescens*, *D. foliosa*, and *D. pallescens*. The tetrads of these species are all of the same type.

Eustoma. *Eustoma exaltatum* (L.) Griseb. The pollen grains are 3-colporate, oblate, spheroidal, and *c.* 29×32 µm. The exine is reticulate with a tendency toward a striato-reticulate pattern. (USA: Texas, 1918, *Palmer 14499* (MO), slide no. 24923; Fig. 2.15C–D.)

Eustoma russellianum (Hook.f.) Griseb. The pollen grains are 3-colporate, subprolate, and *c.* 30×26 µm. The exine is as in *Eustoma exaltatum*. (USA: Texas, 1961, *Ownbey & Baker 2980* (OKL), slide no. 24729.)

OTHER INVESTIGATIONS: The genus *Eustoma* was described as having 3-colporate and distinctly reticulate pollen by Gilg (1895a). Köhler (1905), who investigated pollen of *Eustoma russellianum*, confirmed Gilg's description. The investigated species in this study also have reticulate pollen grains, although *Eustoma grandiflora* has pollen grains with a stronger tendency toward a striato-reticulate pattern than *E. exaltatum* and *E. russellianum* (S. Nilsson, unpubl.).

Ixanthus. Ixanthus viscosus Griseb. The pollen grains are 3-colporate, prolate spheroidal, and c. 30×27 μm. The exine is striato-rugulate with muri/lirae differently oriented to indistinctly striato-reticulate. (Spain: Canary Islands, Tenerife, 1996, *Thiv s.n.* (MJG), slide no. 24346; Fig. 2.16E–F.)

OTHER INVESTIGATIONS: Pollen of *Ixanthus* was described as striate to indistinctly reticulate by Gilg (1895a), and as 3-colporate and reticulate by Köhler (1905). According to Nilsson (1967a) the pollen of *Ixanthus viscosus* is 3-colporate, striate to striato-reticulate with muri/lirae criss-crossing in varying directions from parallel to transverse.

Orphium. Orphium frutescens E. Mey. The pollen grains are 3-colporate, prolate, and c. 47×34 μm. The exine is striate to striato-reticulate with branched or unbranched, partly intertwisted, and sharply keeled muri/lirae. (South Africa: Riversdale, 1940, *Barker 400* (S), slide no. 6129; Fig. 2.16C–D.)

OTHER INVESTIGATIONS: A description of *Orphium frutescens* pollen was given by Agababyan and Tumanyan (1976a). The pollen grains of *Orphium* were described as large and striate-reticulate by Gilg (1895a). *Orphium* had earlier been placed in Gentianeae-Erythraeinae together with *Chironia* (Gilg, 1895a; Köhler, 1905).

Schultesia. Schultesia guianensis (Aubl.) Malme. The 3-porate pollen grains are united in tetrahedral tetrads that are c. 67 μm in diameter. The exine is coarsely reticulate with lumina markedly diminishing in size toward the extremely well-delimited apertural area. (Guyana: Rupununi savanna, 1987, *Jansen-Jacobs et al. s.n.* (U), slide no. 22343; Nilsson, 2002: Fig. 4.8E–F.)

OTHER INVESTIGATIONS: Gilg (1895a) provided the same pollen description for the genera *Schultesia* and *Coutoubea*, both of which he included in Helieae. Köhler (1905) described 11 species of *Schultesia*, including *Xestaea*.

He reported some non-significant differences in the exine structure between *Schultesia* and *Coutoubea*, but otherwise the two genera were similar.

Elias and Robyns (1975) described pollen of four species of *Schultesia* from Panama, including *Xestaea lisianthoides*. Differences between some species were noted, e.g., in shape and size of tetrads, reticulate pattern, and thickness of columellae. The similarity between the tetrads of *Coutoubea* and *Schultesia* was also noted.

Four species of *Schultesia* were described and depicted by Agababyan and Tumanyan (1977a; e.g., *S. angustifolia*). Pollen tetrads of *Schultesia guianensis* were illustrated by Maguire and Boom (1989). *Schultesia guianensis*, *S. benthamiana*, *S. brachyptera*, *S. pohliana*, *S. stenophylla*, and *S. subcrenata* all have tetrads of the same type (Nilsson, 2002: Figs. 4.8–4.9). Some features are subject to minor variations, for example, the shape and size of muri and the outline of tetrads; however, these characters are of unimportant diagnostic value. The tetrads are similar to those of *Coutoubea*.

Symphyllophyton. *Symphyllophyton* cf. *caprifolioides* Gilg. The 3-porate pollen grains are united in tetrahedral tetrads, which are *c.* 50 μm in diameter. The exine is reticulate and at the borderline between contiguous grains there is a distinct zone of minutely reticulate exine pattern. (Brazil: N. Goias, 1964, *Prance & Silva 58543* (NY), slide no. 24899; Fig. 2.15E–F.)

Xestaea. *Xestaea lisianthoides* Griseb. The pollen grains are 3-colporate to porate (with an apertural area that is not well defined) and united in tetrahedral tetrads, which are *c.* 46 μm in diameter. The exine is reticulate to finely reticulate and not clearly differentiated into coarser and finer areas as in *Coutoubea* and *Schultesia*. The borderline between contiguous grains is minutely reticulate. (Nicaragua: 1978, *Stevens & Krukoff 8224* (MO), slide no. 24691; Nilsson, 2002: Fig. 4.9A–C.)

OTHER INVESTIGATIONS: The relatively small tetrads of *Xestaea lisianthoides* were described and depicted by Roubik and Moreno (1991). Elias and Robyns (1975) mentioned the differences between *Schultesia* spp. and *S. lisianthoides* (= *Xestaea lisianthoides*) with reference to the size of the tetrads and the finer reticulate exine pattern of *X. lisianthoides*. This has been confirmed in this study.

Discussion of palynological data in Chironieae. The investigated genera of Chironieae as circumscribed here were distributed among most of Gilg's

(1895a) tribes and subtribes except the monotypic tribe Rusbyantheae and Exaceae, Gentianeae-Gentianinae, Leiphaimeae, and Voyrieae. *Chironia* and *Orphium* were placed in a separate subtribe (Gentianeae-Erythraeinae) by Gilg (1895a) because of the comparatively large size of their pollen.

Members of subtribe Chironiinae have striate to striato-reticulate (*Chironia* and *Orphium*) or striato-rugulate (*Ixanthus*) pollen grains. *Centaurium* has pollen with densely striate and distinctly rugulate exine pattern, whereas *Eustoma* differs by having clearly reticulate exine with a tendency toward a striato-reticulate patterning. The genus *Eustoma*, with its 3-colporate and distinctly reticulate pollen, was placed in Gentianeae-Tachiinae by Gilg (1895a). The pollen grains of *Eustoma* do not seem to be closely related morphologically to *Chironia* and *Orphium*.

In subtribe Coutoubeinae all genera bear pollen in tetrahedral tetrads. *Coutoubea* and *Schultesia* have very similar tetrads, but the exine structure of the tetrads of *Deianira* is different. The pollen of *Xestaea* differs slightly in size and exine details from *Schultesia* pollen, which supports segregation of this former *Schultesia* species into its own genus. Instead, *Symphyllophyton* resembles *Xestaea* by having minutely reticulate exine at the borders between individual grains. The placement of the genera of the Coutoubeinae as a subtribe in the Chironieae is not strongly supported by palynological data. However, their previous position in Gilg's (1895a) Helieae together with such genera as *Chelonanthus* and *Irlbachia* also appears unnatural. The basic tetrad type found in Helieae, i.e., tetrahedral tetrads with a zone of coarse reticulum around the equator and with enlarged lumina at the distal pole, differs from the pollen types of Coutoubeinae. The tetrads of *Coutoubea*, *Deianira*, *Schultesia*, *Symphyllophyton*, and *Xestaea* lack this differentiation of the exine.

Biogeography of Chironieae

When, as suggested above, Coutoubeinae and Canscorinae (Fig. 2.14) are considered to constitute a monophylum, the major subdivision of the tribe into tropical and temperate clades essentially corresponds to the supercontinents Gondwana and Laurasia. However, there is not much support for the assumption that this primary subdivision of the Chironieae is related to the separation of Laurasia and Gondwana during the early Cretaceous. Given the possible origin of angiosperms at the beginning of the Cretaceous (e.g., Doyle & Donoghue, 1993) it seems highly unlikely to us that the Chironieae, a subclade of a derived lineage of asterid eudicots, have the same age as early angiosperms. Also, the estimated minimum age of Gentianales (53 million years; Magallon et al., 1999) is considerably less than that of the first angiosperms. Thus, despite other arguments (see discussion under Exaceae and Potalieae, this chapter), we favor the view that the presence of the Chironiinae in Laurasia is secondary.

An assumption that the second split of the Chironieae into the neotropical Coutoubeinae and the paleotropical Canscorinae was caused by the separation of South America from Africa during the middle Cretaceous would also require a great age for this clade. Additionally, we argue that the Canscorinae are more likely of Laurasian than of Gondwanan origin (see below). In consequence, we contend that the present distribution of the Canscorinae and Coutoubeinae is best explained by long-distance dispersal or migration events.

The spread of some species (of *Coutoubea, Schultesia,* and *Xestaea*) of the Coutoubeinae into Central America may have taken place only after the closure of the Central American isthmus during the Pliocene (5 Ma; Winkler, 1990). Likewise, the occurrence of *Schultesia stenophylla* in West Africa (Berhaut, 1975) is probably of recent origin and may be the result of long-distance dispersal. A comparable disjunction within the Gentianaceae is known within *Voyria* (Raynal, 1967a; Albert & Struwe, 1997), and is also exemplified by other plant families such as Bromeliaceae (*Pitcairnia*; Smith & Till, 1998) and Cactaceae (*Rhipsalis*; Barthlott & Hunt, 1993). In the Coutoubeinae, the primary division appears to be between a northern group of genera (*Coutoubea, Schultesia,* and *Xestaea*, of which *Coutoubea* and *Schultesia* are largely limited to the Guayana Shield north of the Amazon Basin), and *Deianira* and *Symphyllophyton*, which are found in the Brazilian Shield, and the Paraná, São Francisco, and Parnaiba Basins, which are all located south of the Amazon Basin.

In the Old World Canscorinae, the primary subdivision is between *Cracosna, Canscora, Hoppea,* and *Schinziella,* from Africa, India and Southeast Asia, and "Duplipetala", *Microrphium,* and *Phyllocyclus*, from Indochina, the Malay Peninsula, and Palawan. Whereas Africa and India are ancient parts of Gondwana, Southeast Asia (except for the eastern Philippine islands, Sulawesi, and the Moluccas) belongs to Laurasia (Hall, 1996). According to M. Thiv & J. W. Kadereit (unpubl.), an origin of Canscorinae in the Laurasian parts of Southeast Asia is most likely. Thus, the distribution of *Canscora, Hoppea,* and *Schinziella* in Africa and India is best interpreted as secondary.

Within the Chironiinae, Old World/New World disjunctions are found on several levels. The *Blackstonia/Ixanthus* clade is of Old World distribution only. These two genera provide an example for the observation that

Macaronesian endemics (*Ixanthus*) often have their closest relatives (*Blackstonia*) in the Mediterranean area (Thiv et al., 1999b). In the next clade, *Geniostemon* is a New World taxon, whereas *Cicendia* has both an Old and New World distribution. The North/South American distribution of *Cicendia quadrangularis* is likely to be of recent origin. *Exaculum*, which is also likely to be a member of this clade, is distributed in the Old World. In the last group, *Eustoma* and *Sabatia* are of New World distribution, *Centaurium* is found in both the Old World and the New World, and *Chironia* and *Orphium* have an Old World range. The relationship of *Chironia* and *Orphium* as southern African genera related to one clade of *Centaurium* (G. Mansion, unpubl.), which has a partly Mediterranean distribution, exemplifies a distributional pattern known from many other plant groups (e.g., Thymeleaceae, Fumariaceae, Dipsacaceae; Burtt, 1971). The presence of *Centaurium* in America is probably of comparatively recent origin (Broome, 1978), and all but four species of North American *Centaurium* belong to one monophyletic clade – with the other American genera *Sabatia* and *Eustoma* – together representing a large radiation in the New World (Mansion, 2000; G. Mansion, unpubl.). Although both *Cicendia* and some *Centaurium* clades include Old and New World species, it is tempting to assume that the primary subdivision of their respective clades into Old and New World genera (*Cicendia* vs. *Geniostemon*, Eurasian *Centaurium* vs. *Eustoma*/*Sabatia*/North American *Centaurium*; G. Mansion, unpubl.) is related to the final separation of the Laurasian continents during the Tertiary. The New World species of *Cicendia* (and a few *Centaurium* species) represent a secondary dispersal. A Mediterranean origin for *Centaurium* has been postulated by Zeltner (1970) and Broome (1978) based on the absence of polyploid species in this area. Provided *Zygostigma* does belong in the Chironiinae, it would be the only genus of the subtribe endemic to South America.

TRIBE HELIEAE

L. STRUWE AND V. A. ALBERT

The genera of the strictly neotropical Helieae form a morphologically coherent group but also show extensive variation in vegetative, pollen, and floral morphological features. Generic circumscriptions in Helieae have traditionally been very complicated, especially in what has been called the *Irlbachia* complex, which originally included *Adenolisianthus*, *Aripuana*, *Calolisianthus*, *Chelonanthus*, *Helia*, *Irlbachia*, *Macrocarpaea*, *Rogersonanthus*, *Symbolanthus*, *Tachia*, *Tetrapollinia*, and *Wurdackanthus* (Struwe et al.,

Table 2.6. *A synopsis of the number of species in each genus of the Helieae, and their altitudinal range (elevation) and geographic distribution*

Genus	No. of species	Elevation[a]	Distribution
Adenolisianthus*[b]	1	Lowland	SE Colombia, SW Venezuela
Aripuana	1	Lowland	Central Brazil
Calolisianthus	6–10	Middle to highland	Brazil
Celiantha*	3	Highland	Brazil, Guyana, Venezuela
Chelonanthus	c. 8	Lowland to highland	Belize, Bolivia, Brazil, Colombia, Costa Rica, Ecuador, French Guiana, Grenada, Guatemala, Guyana, Honduras, Mexico, Nicaragua, Panama, Peru, Surinam, Trinidad, Venezuela
Chorisepalum	5	Highland	Guyana, Surinam, Venezuela
Helia*	2–3	Highland	Brazil
Irlbachia	9	Lowland to highland	Brazil, Guyana, Venezuela
Lagenanthus*	1	Highland	Colombia, Venezuela
Lehmanniella	2	Highland	Colombia
Macrocarpaea	c. 90	Highland	Bolivia, Brazil, Colombia, Costa Rica, Cuba, Dominican Republic, Ecuador, Guyana, Jamaica, Panama, Peru, Venezuela
Neblinantha	2	Highland	Brazil, Venezuela (Sierra de la Neblina only)
Prepusa	5	Highland	Brazil
Purdieanthus*	1	Highland	Colombia
Rogersonanthus*	3	Highland	Guyana, Trinidad, Venezuela
"Roraimaea"		Lowland	Brazil
Senaea*	1	Highland	Brazil
Sipapoantha*	1	Highland	Venezuela (Cerro Sipapo)
Symbolanthus	c. 30	(Lowland, 1 sp.), Highland	Bolivia, Brazil, Colombia, Costa Rica, Ecuador, Guyana, Panama, Peru, Venezuela
Tachia	10	Lowland to highland	Bolivia, Brazil, Colombia, Ecuador, French Guiana, Guyana, Peru, Surinam, Venezuela

Table 2.6. (cont.)

Genus	No. of species	Elevation[a]	Distribution
Tetrapollinia	1	Lowland to highland	Bolivia, Brazil, French Guiana, Guyana, Paraguay, Surinam, Venezuela
Wurdackanthus	2	Highland	Brazil, Dominica, Guadeloupe, St. Vincent, Venezuela
Zonanthus*	1	Highland	Cuba

Notes:
[a] Approximate elevations are as follows: lowland, 0–300 m above sea level; middle, 300–1000 m; highland, over 1000 m.
[b] An asterisk indicates genera that have not yet been examined using molecular data.

1997). All genera of the *Irlbachia* complex are also members of Helieae, and so the distinction is no longer a useful one. Other genera included in Helieae are *Celiantha*, *Chorisepalum*, *Lagenanthus*, *Lehmanniella*, *Neblinantha*, *Prepusa*, *Purdieanthus*, "*Roraimaea*", *Senaea*, *Sipapoantha*, and *Zonanthus*. Ongoing molecular, morphological, anatomical, and floristic studies will very likely result in taxonomic changes in this group of taxa. Nomenclatural issues add to the complexity. For example, the genus name "*Lisianthus*" has sometimes been used for many species of Helieae as well as for species belonging to *Lisianthius sensu* Weaver (1972) in tribe Potalieae (see Struwe & Albert, 1998a).

The Helieae are a rather species-rich group of Gentianaceae with c. 184 species in 22 genera. The largest genera are *Macrocarpaea* (comprising over half of all Helieae species) and *Symbolanthus*, both of which are most diverse in the Andes. A tree- or shrub-like perennial habit occurs in several genera, especially the montane ones. Pollination syndromes range from generalized insect-pollinated flowers (e.g., *Irlbachia* spp.) to bat and hummingbird pollination (e.g., *Chelonanthus* spp., *Macrocarpaea*, and *Symbolanthus*) and possibly hawkmoth-pollinated flowers in *Aripuana*.

Most species of Helieae have narrow distributions in particular habitats such as the mountains (tepuis) of the Guayana Highlands, *campo rupestre* areas in the highlands of southeastern Brazil, Andean valleys or ridges, and white-sand savannas on the Guayana and Brazilian Shields and in the Amazon Basin. Table 2.6 gives an overview of the number of species, altitudinal range, and geographic distribution for each genus.

Description of Helieae

Plants of the Helieae may be shrubs or subshrubs as well as perennial or annual herbs. The leaves are sessile or petiolated, of various shapes and textures, and often have arcuate, seldomly pinnate (e.g., *Tachia* sect. *Tachia*) secondary veins. The inflorescences are usually monochasially or dichasially branched compound cymes, although a few genera bear axillary and then usually solitary flowers. The bracts can be scale-like or leaf-like. The flowers are (4–)5(–6)-merous, horizontal or nodding, and are often large and showy (Fig. 2.17). The calyces are usually campanulate, less often tubular or with free lobes, thick and leathery to thin and membranaceous, and persistent in fruit (at least initially, although calyces fall off from old fruits in *Chorisepalum*, *Macrocarpaea*, and *Tachia*). Colleters are present on the inside (adaxial) surface of the calyx. The corollas are of various colors and textures and vary in shape from tubular, funnel- or salver-shaped, to campanulate. The stamens are inserted close to the base or higher up in the corolla tube (in the corolla lobe sinuses in *Aripuana*); the anthers are free and usually have sterile apical appendages. The filaments are sometimes sharply bent or curved in their upper part. The anthers are erect to strongly recurved after anthesis. The pollen grains are released as monads, tetrads, or polyads, and exine sculpturing is very variable. The stigmata are nearly always broadly bilamellate (rarely with narrow lobes, e.g., *Rogersonanthus coccineus* and "Roraimaea"). The fruit type is usually a dehiscing capsule with thin or woody pericarp (e.g., *Chorisepalum* and *Tachia*), rarely an indehiscent, leathery capsule (*Symbolanthus* spp.; Fig. 2.18). Seeds are numerous, angular to rounded (rarely flattened and winged in *Chorisepalum* and *Macrocarpaea* spp.).

Taxonomic history of Helieae

Genera of Helieae were primarily placed in tribe "Lisyantheae" in Grisebach's (1839) system, whereas they were divided among three different tribes in Gilg's (1895a) palynology-based classification. In the latter system, taxa with superficially similar pollen but distinct macromorphologies were sometimes assigned to the same tribes or subtribes. For example, *Chelonanthus* and *Coutoubea* were both placed in Helieae because of their tetrad pollen grains. A converse problem also occurred when Gilg placed morphologically closely related taxa in different tribes or subtribes solely because they had different pollen types. For example, *Macrocarpaea* and *Rusbyanthus* (now included in *Macrocarpaea*) were assigned to Gentianeae-Tachiinae and the monotypic Rusbyantheae, respectively.

Although palynology no doubt contributed to the artificiality of Gilg's (1895a) system, the importance of pollen characters to infratribal classification of Helieae, when balanced with other morphological and molecular data, cannot be ignored.

Generic synopses for Helieae

Adenolisianthus GILG

The monotypic genus *Adenolisianthus* was erected by Gilg (1895a: 98) when he gave Progel's (1865) "*Lisianthus*" sect. *Adenolisianthus* generic status. The sole species, *Adenolisianthus arboreus*, occurs only in lowland, white-sand areas on the western Guayana Shield around Rio Negro and Rio Vaupés in Brazil, Venezuela, and Colombia (Struwe *et al.*, 1997, 1999). *Adenolisianthus* was treated as "*Lisianthus*" *arboreus* in the floristic treatment by Progel in *Flora Brasiliensis* (1865), and as *Chelonanthus fruticosus* by Maguire and Boom (1989). No revision of this genus has been published.

Adenolisianthus plants are branched shrubs or subshrubs with leaves crowded at the branch apices (Struwe *et al.*, 1999). The leaves are sessile, broadly ovate to obovate, and are characteristically widest at or above the middle. The 5-merous, often horizontal flowers are arranged in long-stalked, monochasially branched, terminal inflorescences. The calyces are campanulate, thick and leathery, and are persistent in fruit. The corollas, thick and leathery, are broadly funnel-shaped with a narrow lower tube. The stamens are inserted close to the base of the corolla tube with filaments that are sometimes sharply bent at their apices. The anthers are slightly recurved after anthesis. The pollen is released as tetrads. The exine is reticulate. The stigma is broadly bilamellate and the fruits are nodding capsules.

Adenolisianthus was included in *Irlbachia sensu lato* by Maas (1985). Four years later it was transferred by Maguire and Boom (1989) into the palynologically and florally similar genus *Chelonanthus*. At the same time, *Adenolisianthus arboreus* was renamed *Chelonanthus fruticosus* since the name *Chelonanthus arboreus* had already been used for another species (now recognized as *Rogersonanthus arboreus*; Struwe *et al.*, 1999). In recent treatments, *Adenolisianthus* has been resurrected pending further phylogenetic studies that should clarify the relationships of this monotypic genus (Struwe *et al.*, 1997, 1999).

Aripuana STRUWE, MAAS, & V. A. ALBERT

This monotypic genus was recently described by Struwe *et al.* (1997: 236) based on *Aripuana cullmaniorum* from southeastern Amazonas, Brazil.

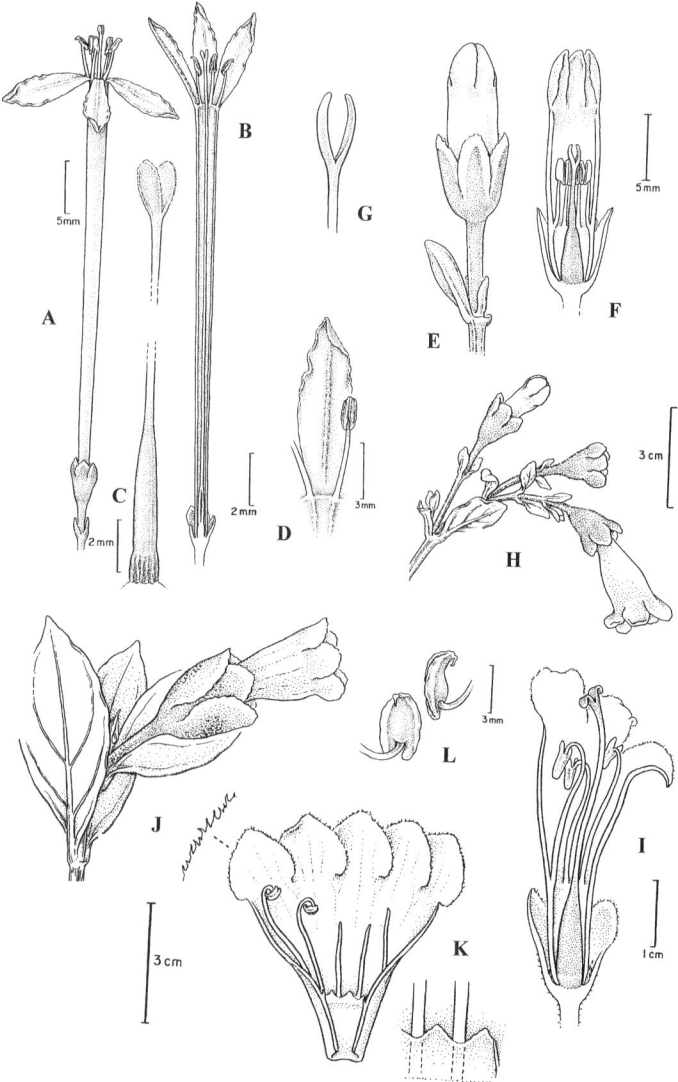

Figure 2.17. Flowers of Helieae. **A–D.** *Aripuana cullmaniorum* (*Cid Ferreira 5906*). **A.** Flower at anthesis. **B.** Opened flower (half). **C.** Gynoecium with basal glandular area and bilamellate stigma. **D.** Insertion of stamens in corolla lobe sinuses. **E–G.** *Rogersonanthus coccineus* (*Gentry & Stein 46708*). **E.** Flower in bud. **F.** Opened flower. **G.** Bilamellate stigma with linear lobes. **H–I.** *Macrocarpaea marahuacae* (*Liesner 24651*). **H.** Inflorescence with horizontal and slightly zygomorphic flowers. **I.** Opened flower. **J–L.** *Symbolanthus aureus* (J. *Liesner 23392*; K–L. *Maguire 33531*). **J.** Flower at anthesis. **K.** Opened flower with corona at the base of the recurved, hook-like filaments. **L.** Anthers after anthesis. (Redrawn from and reprinted with permission from *Harvard Papers of Botany*.)

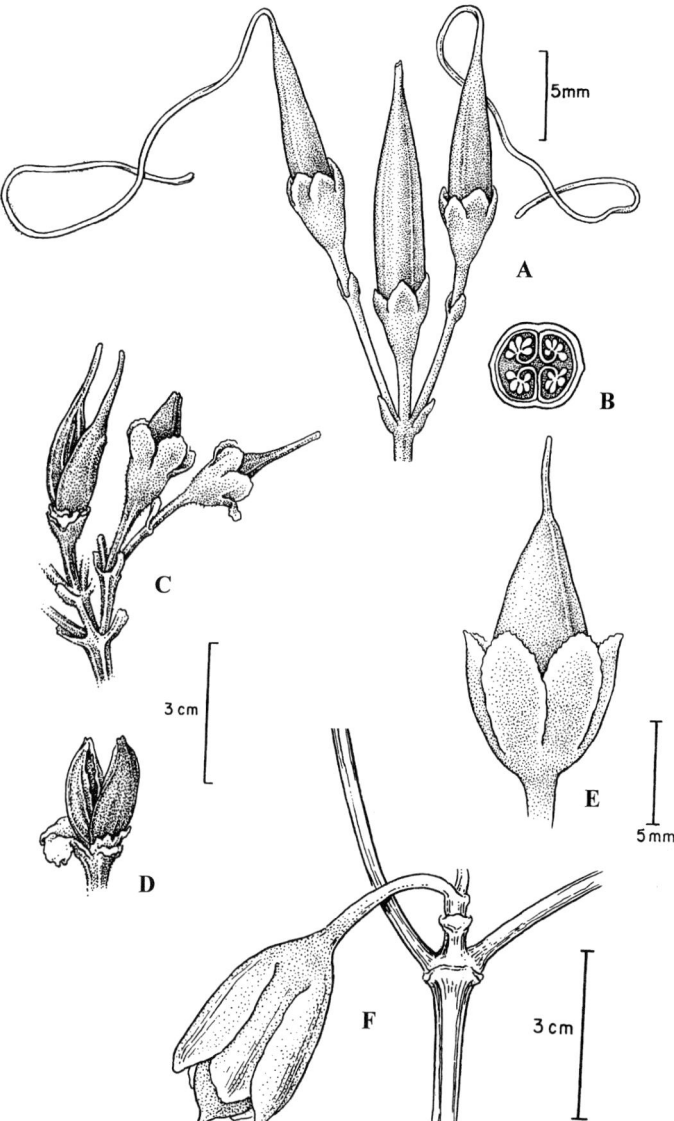

Figure 2.18. Fruits of Helieae. **A–B.** *Aripuana cullmaniorum* (*Cid Ferreira 5906*). **A.** Erect capsules with persistent styles; note bracts and floral bracts on side branches. **B.** Parietal placentation in mature fruit. **C–D.** *Macrocarpaea marahuacae* (*Liesner 24794*). **C.** Infructescence with partially nodding woody capsules. **D.** Older, dehisced capsule. **E.** *Rogersonanthus coccineus* (*Steyermark 103762*). Erect capsule with partly persistent style. **F.** *Symbolanthus aureus* (*Maguire 33531*). Leathery, nodding capsule, nearly completely enclosed by the persistent calyx. (Redrawn from and reprinted with permission from *Harvard Papers of Botany*.)

Aripuana is known by only a handful of collections from a small, white-sand savanna area along the Rio Aripuanã in the vicinity of Novo Aripuanã, Nova Prainha, and the Transamazônica highway. The description of the genus also included substantial data on vegetative and floral anatomy as well as palynology (Struwe *et al.*, 1997).

Aripuana cullmaniorum is a shrub or tree with elliptic leaves (Struwe *et al.*, 1997). The multi-flowered and terminal inflorescences are dichasial cymes that have scale-like bracts and bear long, showy, 5-merous flowers (Fig. 2.17A–D). The calyces, fused for more than half of their length, are small, campanulate, thick, and have rounded lobes with a thickened (probably nectariferous) keel on the outside (abaxial surface) of each lobe. The white corollas are salver-shaped with very narrow tubes. They have elliptic and spreading lobes and are deciduous in fruit. The corolla aestivation is unique among Gentianaceae by being valvate in the basal parts of the corolla lobes and contort at their apices. The long stamens are inserted at or directly below the corolla lobe sinuses on papillose ridges and are exserted and bent outwards at anthesis. The anthers are oblong, curving slightly backwards after anthesis, and have small, triangular, sterile appendages at their apices. The pollen is shed as tetrads and the exine is reticulate (Struwe *et al.*, 1997; Nilsson, 2002). The gynoecium has a prominent glandular area around the base and a 4-parted, deeply inrolled placenta that is initially axile, becoming secondarily (falsely) parietal in fruit (Fig. 2.18B; however, see discussion on placentation in Helieae, below). The long, exserted style has a bilamellate stigma with short, narrow, and oblong lobes. The fruits (known only as immature), containing many small and angular seeds, are slender, dry capsules with bent, twisted, and persistent styles (Fig. 2.18A).

Calolisianthus GILG

This genus was raised to generic rank by Gilg (1895a) from Grisebach's (1839) *Lisyanthus* sect. *Calolisyanthus*. The type species was designated by Pringle (1995) to be *Calolisianthus amplissimus*. *Calolisianthus* consists of 6–10 species distributed from 200 m to the highest elevations of the Brazilian Highlands in southeastern Brazil and Bolivia. The species grow mainly on rocky and sandy mountain summits, riversides, woodlands, and savannas on all types of soils and at lowland to highland elevations. H. Groen (unpubl.) wrote a revision of the genus.

Calolisianthus comprises usually unbranched, perennial herbs with showy flowers on terminal, few-flowered cymes (H. Groen, unpubl.). The leaves are often coriaceous, oblong-elliptic, and are always sessile. The flowers are 5-merous and horizontal to nodding. The calyces are campanulate, fused basally, with rounded or acute lobes and dorsal glandular areas. Colleters are present on the inside (adaxial) surface of the calyx. The color of the funnel- to salver-shaped corollas varies (often within species) from blue, to purple, to red. The corollas are persistent in fruit, abscising just before the fruits open. The stamens are inserted medially or lower in the corolla tube and have oblong anthers with sterile apical appendages. The pollen is shed as tetrads. The exine is unevenly thickened and reticulate. The gynoecium has a glandular disk at the base and is uni- to bilocular with inrolled, parietal placentas. The broadly bilamellate stigma terminates in a long style. The fruits are dry, multi-seeded capsules.

Celiantha MAGUIRE

This genus is endemic to the highlands of the Guayana Shield in northern South America (Struwe *et al.*, 1999). *Celiantha* was described by B. Maguire (1981: 382), and was named after his wife, Celia Maguire. The genus includes three species – the type species *Celiantha bella*, which occurs on Sierra de la Neblina at the Brazilian–Venezuelan border, and *C. imthurniana* and *C. chimantensis*, which are distributed on tepui summits in eastern Venezuela and Guyana. The species grow in open, rocky areas and savannas at high elevation. A revision of this genus is lacking.

Celiantha plants are erect and sometimes slightly woody herbs (Struwe *et al.*, 1999). The leaves are lanceolate to ovate and are often coriaceous. The inflorescences are multi-flowered, terminal, diffuse panicles or dichasia (*Celiantha bella*) or few-flowered terminal cymes (*C. chimantensis* and *C. imthurniana*). The flowers are 4- (*Celiantha chimantensis*) or 5-merous (*C. bella* and *C. imthurniana*) and nodding. The calyces are campanulate, thick, and persistent in fruit. The thin corollas are narrowly funnel-shaped and vary in color from purple, magenta, lilac or pink (*Celiantha bella* and *C. imthurniana*) to yellow (*C. chimantensis*). The stamens are inserted in the corolla tube with filaments of slightly unequal length. The anthers are linear to oblong and recurved after anthesis. The pollen is shed in polyads and the exine is beset with small globules. The stigma is bilamellate with ovate lobes. The fruits are nodding, elliptic capsules with many seeds.

Chelonanthus GILG

Chelonanthus was initially treated as a section of *Lisyanthus* by Grisebach (1839) and was raised to generic rank by Gilg (1895a: 98). *Lisyanthus* of Grisebach (1839, renamed "*Lisianthus*" in Grisebach, 1845) did not refer to *Lisianthius sensu* Browne, a member of the tribe Potalieae, but to *Lisyanthus sensu* Aublet (1775), a genus that should probably be treated as a homonym of *Lisianthius*. This nomenclatural problem was discussed by Struwe and Albert (1998a), who also lectotypified *Lisyanthus* Aubl. with *L. purpurascens*, a taxon that is now included in *Chelonanthus* as *C. purpurascens*. To add to the complexity, the type of *Chelonanthus*, *C. uliginosus*, is now treated as a synonym under *C. purpurascens*. Furthermore, Maas (1985) included *Chelonanthus* together with a few other Helieae genera in the genus *Irlbachia*. However, this was a nomenclaturally invalid move, as the name *Helia*, which was simultaneously published, should have been used instead. This is because Kuntze (1891) had already made the first specification that *Helia* should have priority over *Irlbachia* when synonymizing the two genera (see Struwe & Albert, 1998a). No complete monograph of *Chelonanthus* has been published. Many nomenclatural issues and species circumscription problems remain to be resolved, and, to compound matters, the genus is probably not monophyletic (see discussion below).

Chelonanthus has probably the widest distribution of all Helieae genera, with the widespread species *C. alatus* (previously known as *Irlbachia alata*) occurring from Mexico in the north, through Central America, the Andes, the Guayana Shield (including Trinidad), east to French Guiana, and south into Amazonas, southern Brazil and Bolivia (Struwe *et al.*, 1999). The highest species diversity occurs on the Guayana Shield, which harbors five of the total of *c.* eight accepted species (Struwe *et al.*, 1999). Two species are largely endemic to the Brazilian Shield, namely *Chelonanthus matogrossensis* and *C. viridiflorus* (P. J. M. Maas, unpubl.; Struwe & Albert, 1998a). Only *Chelonanthus alatus* and *C. purpurascens* occur both north and south of the Amazon river, bridging the gap between the two major geological shields.

The species of *Chelonanthus* are annual or perennial herbs (up to 3.5 m tall) sometimes woody at the base (Struwe *et al.*, 1999). The leaves are sessile or petiolate and of various shapes, sizes, and textures. The terminal inflorescences are characterized by being many-flowered with monochasial branches, or are rarely reduced, bearing only a few flowers and scale-like bracts. The flowers are 5-merous, often horizontal or nodding, and are actinomorphic to slightly zygomorphic. When zygomorphic, the flowers have a slightly bent corolla tube with the anthers clustered in the upper or lower part of the corolla mouth and a style bent toward the lower side of the mouth. The campanulate calyces are divided down to at least half of their length, are thick and coriaceous, and are persistent in fruit. The calyx lobes are elliptic with a dorsal glandular ridge. The campanulate to funnel-shaped corollas are blue, purple, green, yellowish, or white and have rounded lobes and rounded corolla bud apices. The stamens are inserted close to the base of the corolla tube with filaments of subequal to unequal length that are sometimes strongly bent close to their apices. The oblong to linear, sagittate anthers are erect or recurved (to 360°) after anthesis and are provided with sterile apical appendages. The pollen is released as tetrads or polyads (*Chelonanthus purpurascens* only) and has an unevenly reticulate exine, sometimes with additional loops and globose structures (*C. purpurascens* only; Nilsson, 2002). The ovary has a glandular disk at the base and the style is long, slender, and persistent in fruit. The bilamellate stigma has ovate-elliptic lobes. The elliptic capsules dehisce medially and are often nodding.

Chorisepalum GLEASON & WODEHOUSE

This genus was described by Gleason and Wodehouse (1931: 451) based on material collected during the first expedition to the Venezuelan tepui Cerro Duida. The most recent treatment of this genus recognized five species (Struwe *et al.*, 1999). The distribution of *Chorisepalum*, which is strictly confined to highland habitats, stretches from Cerro Sipapo (Amazonas, Venezuela) in the westernmost Guayana Highlands through the Highlands to a disjunct population on Tafelberg (Surinam) in the east. Ewan (1947) published a revision of *Chorisepalum*, and all described taxa were treated in the floristic works of Maguire (1981) and Struwe *et al.* (1999).

Chorisepalum consists of small trees and erect or scandent shrubs (Struwe *et al.*, 1999). The leaves are ovate to elliptic, glossy or rugose, and often very coriaceous. The dichasial terminal inflorescence is 1–9-flowered with leaf-like bracts. The flowers are actinomorphic and erect, with a 4-merous calyx that has free, decussate lobes, usually not keeled except for *Chorisepalum sipapoanum*, which has two dorsal keels (Struwe *et al.*, 1999; also described as *C. ovatum* var. *sipapoanum* by Maguire, 1981). The corollas are tubular to salver- or funnel-shaped, green, sometimes with purple margins, 6-merous with six sometimes aristate or acuminate lobes, apically tapering in bud. The stamens are attached near the base of the corolla tube with filaments of equal length. The anthers are basifixed, linear, and erect after anthesis. The pollen is released as monads with reticulate exine (Gleason & Wodehouse, 1931; Maguire, 1981; Nilsson, 2002). The ovary has a disk around its base and the style is long, slender, and deciduous in fruit. The bilamellate stigma has flattened and elliptic lobes. The capsules are elliptic, erect, with the pericarp of each fruit splitting from the base into four separate, deciduous parts, leaving fibrous traces of ovarian vasculature at the base. The seeds are elliptic, flattened, and winged, and the testa cells are papillose to rounded and smooth (*Chorisepalum acuminatum*).

The flowers of *Chorisepalum* are unique in the Gentianaceae by having the combination of a coriaceous, 4-merous, decussate calyx (like *Anthocleista* and *Potalia* in Potalieae), a 6(rarely 5)-merous corolla, a capsule that splits basally into four valves, and winged seeds. The nearly free, acuminate-lanceolate, coriaceous sepals from which the name of the genus is derived are also unusual features in Helieae and the rest of Gentianaceae. The affinities of *Chorisepalum* with Helieae are clear, not only from *mat*K and ITS phylogenetic results (Figs. 2.2, 2.19), but also from morphological and palynological data. The reticulate pollen monads of *Chorisepalum* very strongly resemble the pollen of *Tachia* and *Macrocarpaea* (Nilsson, 2002). The genus also has the broadly bilamellate stigmas characteristic of most Helieae, a woody habit, and calyx lobes with dorsal glandular areas, all characteristics that support its position in the tribe. The unique floral merosity of *Chorisepalum* must be viewed as an apomorphic trait within the Helieae, and it is interesting to note that the monophyletic grouping of *Tachia* and *Chorisepalum* in the *mat*K phylogenetic tree is supported by some structural characters, i.e., pollen in monads, and a perennial, woody habit. *Macrocarpaea* is unresolved toward *Tachia* and *Chorisepalum* in the *mat*K tree (Fig. 2.2), but a close relationship between *Chorisepalum* and *Tachia*/*Macrocarpaea* is found with ITS data (Fig. 2.19). Additionally, the only species of Guayanan Helieae with rugose leaves are members of *Chorisepalum* (2 spp.) or *Macrocarpaea* (4 spp.; Grant & Struwe, 2001).

Helia MART.

This herbaceous genus from wet savanna areas in the highlands of southeastern Brazil contains *c*. two species. It was described by von Martius (1826–1827) in his work on neotropical plants. *Helia* has never been revised and it is still uncertain if the described species should be considered conspecific (as in Maas, 1985) or as distinct entities (cf. Progel, 1865).

The simple-stemmed plants of *Helia* have a characteristic appearance. The stems have obovate to spathulate leaves aggregated at the base, and the inflorescence is a long-stalked terminal dichasium bearing flowers with salver-shaped corollas. The calyces are campanulate with rounded lobes. The white or yellow corollas are erect to horizontal with rounded lobes and narrow tubes. The stamens are inserted in the corolla tube, and the pollen is released in tetrads. The stigma is broadly bilamellate. The fruits are medially dehiscent capsules, often enclosed by the persistent corollas.

Although DNA data for *Helia* have not yet been obtained, the genus has definite morphological characteristics that place it in the Helieae (cf. Struwe *et al.*, 1997). For example, *Helia* has pollen tetrads very similar to

Systematics, character evolution, and biogeography 149

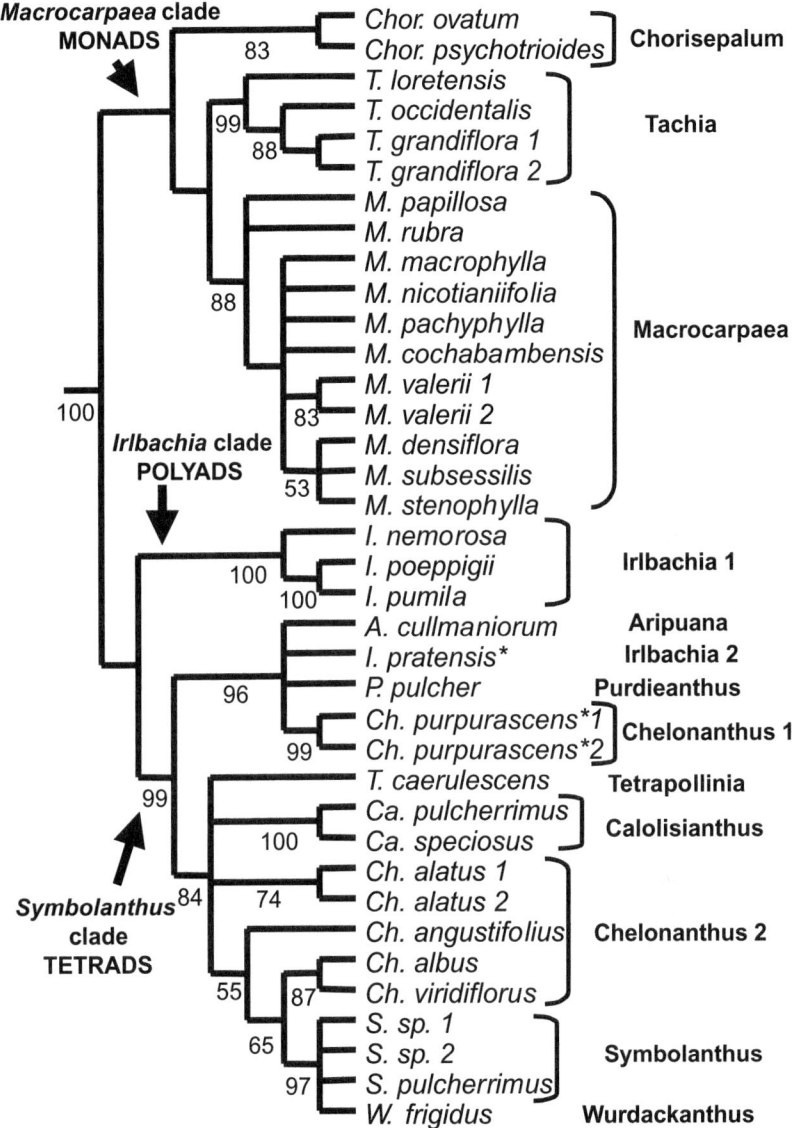

Figure 2.19. Phylogenetic relationships within Helieae. Strict consensus tree (showing only the Helieae part) from a preliminary analysis of Gentianeae, Helieae, and Potalieae ITS sequences (360 most-parsimonious trees, 908 steps, $C=0.61$, $R = 0.70$; L. Struwe, J. R. Grant, K. Gould, & V. A. Albert, unpubl.). Three major lineages are distinguished: the *Macrocarpaea*, *Irlbachia*, and *Symbolanthus* clades. Numbers below branches are jackknife values >50%. Asterisks signify species in the *Symbolanthus* clade with pollen in polyads.

those of *Chelonanthus* and *Aripuana*, rounded calyx lobes with a glandular ridge (as just described for *Chorisepalum*), and anthers that bear a sterile apical appendage and become recurved after anthesis (again, see *Chorisepalum*). *Helia, Irlbachia,* and *Prepusa,* which were published simultaneously, were the first genera described in Helieae apart from Aublet's (1775) *Lisyanthus* (a name that should probably be rejected; see Struwe & Albert, 1998a) and *Tachia*.

Irlbachia MART.

This herbaceous genus from the Guayana Shield in northern South America now comprises *c*. nine species. The highest species diversity is found in the upper Rio Negro basin and on the tepuis of Venezuela, with three species reaching down to Rio Amazonas in Brazil (Struwe *et al.*, 1999). *Irlbachia* was described by von Martius (1826–1827: 101) based on the only species known at the time, *I. elegans* from Brazil. Two subsequently described genera are now synonymized with *Irlbachia*, namely *Pagaea*, described by Grisebach (1845), and *Brachycodon*, described by Progel (1865). *Brachycodon* was initially treated as "*Lisianthus*" sect. *Brachycodon* by Bentham (1854).

Maas (1985) included *Adenolisianthus, Calolisianthus, Chelonanthus, Helia, Rogersonanthus, Tetrapollinia,* and *Wurdackanthus* under *Irlbachia*. However, this broad circumscription of *Irlbachia* used by Maas (1985) and some subsequent workers is nomenclaturally incorrect (Struwe & Albert, 1998a). *Helia*, which was described simultaneously with *Irlbachia* by von Martius (1826–1827), has priority over *Irlbachia* since Kuntze (1891) made the first priority choice between these names by synonymizing *Irlbachia* (and other members of tribe Helieae) under *Helia*. In any case, *Helia* in a wide sense (or *Irlbachia sensu* Maas) is most likely paraphyletic, with *Symbolanthus* at least embedded inside (Figs. 2.1–2.3, 2.19). Instead of placing species-rich *Symbolanthus* into an even more widely circumscribed genus *Helia*, we decided to accept more narrow generic circumscriptions within Helieae until its phylogeny has been further investigated and better hypotheses can be made regarding monophyly of genera. These circumscriptions also follow the tradition of Gilg (1895a), Maguire (1981), Maguire and Boom (1989), and Struwe *et al.* (1997, 1999). No revision covering the total distribution of *Irlbachia* exists.

Irlbachia plants are annual herbs that are sometimes slightly woody at the base (Maguire, 1981; Struwe *et al.*, 1999). The leaves are cauline, often most abundant close to the base or in a basal rosette (*Irlbachia plantagini-*

folia), and are of various shapes and textures. The inflorescences are terminal cymes usually with monochasial branches and with flowers often arranged along one of the sides or alternating on opposite sides of each branch. The bracts are scale-like in most species, but are leaf-like in *Irlbachia pumila* and *I. poeppigii*. The flowers are 5-merous and actinomorphic, often becoming nodding with age, and sometimes have zygomorphic stamens. The calyces are campanulate, divided ½–⅘ of their length, and are thick and persistent in fruit with elliptic lobes that have dorsal glandular areas or ridges. The corollas are salver-shaped, campanulate, or narrowly funnel-shaped, and may be white, pink, rose, lavender, blue, or purple. When in bud, the corolla apices are rounded (except for *Irlbachia pratensis*, in which they are tapering). The stamens are inserted in the lower part of the corolla tube with filaments of equal or unequal length. The anthers are erect to recurved after anthesis, slightly sagittate, and are equipped with apical sterile appendages. The pollen is released as polyads with an exine beset with spines or globules. The ovary has a glandular disk at the base (possibly absent in *Irlbachia pratensis*) and the style is slender and persistent in fruit and terminated by a bilamellate stigma. The capsule is ovoid to ellipsoid, dehiscing medially, and is often nodding.

Most species of *Irlbachia* are endemic to lowland white-sand areas (in Amazonas, Venezuela and Brazil) or tepui summits (Venezuela), making this an interesting group to study for evaluation of biogeographic relationships of the tepuis to the unique white-sand floras of the Amazon Basin (Pihlar *et al.*, 1998). DNA sequencing of ITS for *Irlbachia pratensis* has shown that this species is probably more closely related to *Chelonanthus purpurascens* (of the *Symbolanthus* clade) than to other *Irlbachia* species (Fig. 2.19). This relationship is also supported by *Irlbachia pratensis* having pollen tetrads of an intermediate type between the *Uliginosus*- and *Irlbachia*-types (Nilsson, 2002).

Lagenanthus GILG

When Lindley (1849: 261) first described *Lagenanthus princeps* (under the name "*Lisianthus*" *princeps*) he called it "one of the noblest plants in existence" (Ewan, 1952), while Steyermark, one of the world's most prolific plant collectors, has remarked on *Lagenanthus* as "one of the most beautiful and showy plants ever seen anywhere" (Ewan, 1952). Its hanging flowers are spectacular, being tubular and up to 15 cm long, with alternating colors of rich yellow, crimson-scarlet, and orange (Ewan, 1952). When it was first described it caused a sensation among horticulturists in Europe, and accounts and drawings of *Lagenanthus*, nicknamed the "king of genus *Lisianthus*", were published in several horticultural journals across the European continent (see Ewan, 1952, for references). *Lagenanthus princeps* is the only species of this monotypic genus, and is restricted to high elevations on the Colombian–Venezuelan border. The new generic name was provided by Gilg (1895a: 99), and was later included in *Lehmanniella* by Simonis in Maas (1985). Another species, *Lagenanthus parviflorus*, was described by Ewan (1952) from Panama, but later turned out to be synonymous with *Ravnia triflora* (Rubiaceae; Maas, 1981).

Lagenanthus princeps is a branched shrub with ovate to ovate-lanceolate leaves (Ewan, 1952; J. E. Simonis, unpubl.). The flowers are terminal, solitary or in few-flowered inflorescences. The calyces are broadly campanulate with broadly ovate lobes. The corollas are hanging, long-tubular and somewhat inflated, and have short, rounded corolla lobes. The corollas are yellow at the base of the tubes, followed by two crimson-scarlet and orange-vermilion bands higher up on the tube, which is tipped with green lobes (Ewan, 1952). The stamens are inserted very close to the base of the corolla tube and have long filaments. The fruits are thick-walled capsules.

Lehmanniella GILG

Lehmanniella consists of two species from Andean Colombia and Peru (Maas, 1985), and was described by Gilg (1895a: 101) based on *L. splendens*.

Plants of *Lehmanniella* are scandent, herbaceous, or slightly woody, with ovate to elliptic leaves (Ewan, 1948b; J. E. Simonis, unpubl.). The 5-merous flowers are borne in small, terminal umbels (reduced cymes) with 2–9 flowers and are pedicellate. The calyces are campanulate with elliptic lobes and obtuse apices. The corollas are red and tubular, constrained at the throat, and with short, rounded lobes. The stamens are inserted close to the base in the corolla tube and have sagittate anthers with small, sterile apices. The gynoecium has a glandular disk at the base and the long style has narrow stigma lobes. The fruits are woody capsules.

Macrocarpaea (GRISEB.) GILG

Gilg (1895a: 94) raised the genus *Macrocarpaea* to generic rank from Grisebach's (1839) *Lisyanthus* sect. *Macrocarpaea*. At the same time Gilg also described the monotypic genus *Rusbyanthus*, and its only species is now considered to belong to *Macrocarpaea* (Gilg, 1895a: 95; Weaver, 1974). *Macrocarpaea* is distributed in wet and usually montane forests in Central

Systematics, character evolution, and biogeography 153

America (Costa Rica and Panama), the Greater Antilles (Cuba, Dominican Republic, and Jamaica), and tropical South America (Andes, Guayana Shield, and southeastern Brazil; Table 2.6; Struwe *et al.*, 1999). The highest species diversity is found in the Andes of Peru, Ecuador, and Colombia (J. R. Grant, pers. comm.). Ewan (1948a) revised the genus over 50 years ago but since then many new species have been found and it is now estimated that the genus contains at least 90 species (Grant & Struwe, 2001; J. R. Grant, pers. comm.). The circumscription of the genus has also changed since Ewan's revision by the removal of the tetrad-bearing species to *Rogersonanthus* (Maguire & Boom, 1989).

The species of *Macrocarpaea* are branched shrubs (rarely epiphytic), small trees (up to 10 m), or perennial herbs (only *M. rubra*; Ewan, 1948a; Grant & Struwe, 2001). Leaf morphology shows great variation with leaves being sessile to petiolate, the shape linear, lanceolate, ovate, elliptic, to obovate, the bases cordate to acute, the apices obtuse to acute, the texture chartaceous to coriaceous, and typically with smooth to rarely scabrous or hairy surfaces. The inflorescences are terminal, composed of few- or many-flowered dichasia or cymes, and usually have leaf-like bracts below the rather large (2.0–7.5 cm long), 5-merous flowers (Fig. 2.17H–I). The flowers are often slightly zygomorphic with the anthers and style bent down toward the lower part of the corolla mouth. The calyces are campanulate, fused at least at the base, thick, and with much diversity in calyx lobe morphology, which is one of the most useful characters in delimiting species (the apices ranging from acute, acuminate, cuspidate, obtuse, to rounded). The yellow, white, to greenish funnel-shaped corollas are thick and fleshy, night-blooming and presumably bat pollinated, although hummingbirds have been observed to visit flowers diurnally (L. Struwe, pers. obs.). The stamens are inserted in the middle or lower part of the corolla tube and have linear to oblong anthers that are recurved after anthesis. The filaments are often also strongly recurved in their upper part. The pollen is released as monads. The gynoecium has a disk around the base, and the style is long, slender, and persistent in fruit, and tipped by a bilamellate stigma with broad lobes. The fruit is a dry, woody capsule with medial dehiscence that is erect to nodding at maturity (Fig. 2.18C–D). The angular seeds are sometimes flattened, but not winged. A reported chromosome number is $2n = 42$ (*Macrocarpaea thamnoides* (Griseb.) Gilg; Weaver, 1969).

Macrocarpaea belongs to the *Macrocarpaea* clade together with *Tachia* and *Chorisepalum* (and possibly *Zonanthus*), which forms the only monad pollen-bearing group in Helieae (Fig. 2.19). *Macrocarpaea* has interesting patterns of biogeographic distribution, with preliminary molecular phylogenetic results showing the most ancestral species to be located peripherally in the aggregate range of the genus (e.g., the Greater Antilles and southeastern Brazil), whereas the Andean species apparently represent more recent, species-rich radiations (Grant & Struwe, 2000; J. R. Grant, unpubl.).

Neblinantha MAGUIRE

Neblinantha is a small genus described by Maguire (1985: 311) consisting of only two species endemic to Sierra de la Neblina on the Brazilian–Venezuelan border. This genus has never been revised, and very few collections are known of the two species, *Neblinantha neblinae* (the type) and *N. parviflora* (Struwe *et al.*, 1999).

The species of *Neblinantha* are suffrutescent herbs or shrubs with rather small, sessile to subsessile, ovate-lanceolate leaves (Struwe *et al.*, 1999). The flowers are solitary, and in *Neblinantha neblinae* appear at first sight to be axillary, but probably represent terminal solitary flowers with one or two branches developing in the axils of the uppermost leaf pair; in *N. parvifolia* the flowers are terminal on short, leafy shoots. The flowers are pedicellate, 4–5-merous, actinomorphic, and erect or nodding. The thin and tubular calyces are divided about halfway down their length, and have narrowly triangular, long-acuminate lobes. The deep pink or coral-colored corollas are salver-shaped or narrowly funnel-shaped, thin, and with circular or triangular lobes. The stamens are inserted in the corolla tube, with the filaments sometimes prominently bent 180° at their apices. A small, sterile apical appendage is present on each ovate-oblong anther. The pollen is shed as tetrads. The gynoecium probably lacks a glandular disk at the base. The style is long and slender with a bilamellate stigma and narrow stigmatic lobes. The erect or nodding capsules dehisce apically and have angular, unwinged seeds.

Prepusa MART.

The genus *Prepusa* was described by von Martius (1826–1827: 120) based on *P. montana*, and now includes five narrowly distributed species (J. E. Simonis, unpubl.). *Prepusa* is restricted to mountainous areas in southeastern Brazil. No revision has been published.

The species of *Prepusa* are herbs, shrubs, or small trees, sometimes with a rhizome, and often with unbranched stems (J. E. Simonis, unpubl.). The leaves are characteristic in being very thick and coriaceous, sessile, and usually aggregated at the bases of the stems in the herbaceous species, but crowded at the branch apices in the arborescent species, *Prepusa montana*. The inflorescence is a terminal (or axillary?) dichasial cyme, or the flowers may be solitary. The 6-merous flowers are showy with a large, campanulate and membranaceous calyx that often has dorsal wings. The white or yellow corollas are much smaller in diameter than the calyces and are salver- to funnel-shaped. The stamens are inserted in the corolla tube. The style is long and tipped with a bilamellate stigma. The fruits are dry, elliptic capsules surrounded by the old calyx. *Prepusa* is classified here as a member of Helieae pending further phylogenetic investigations.

Purdieanthus GILG

In 1895, Gilg described *Purdieanthus* based on the species *P. pulcher* (Gilg, 1895a: 99). *Purdieanthus* grows at high elevations (2500–4000 m) in northeastern Colombia and southwestern Venezuela in montane forests and páramos. Maas (1985) included *Purdieanthus* in *Lehmanniella*, a view that is not followed here.

Purdieanthus plants are straggling vines with narrow branches and shortly petiolated leaves (Ewan, 1948b). The flowers are borne in a terminal, dichasial cyme. The small calyces have short, oblong calyx lobes. The corollas are long and tubular, and the fruits are dry capsules.

Rogersonanthus MAGUIRE & B. M. BOOM

Rogersonanthus was segregated by Maguire and Boom (1989: 3) to accommodate the tetrad-bearing species of *Macrocarpaea* (*M. arborea*, *M. cerronis*, *M. quelchii*, and *M. salicifolia*). The genus was named after Clark Rogerson, a long-time editor of *Memoirs of the New York Botanical Garden*, and was based on *Rogersonanthus quelchii*. *Rogersonanthus* originally included three species – *Rogersonanthus arboreus* from Trinidad, and the Guayana Highland endemics *R. tepuiensis* and *R. quelchii*. Four years earlier, Maas (1985) had included the tetrad-bearing *Macrocarpaea* species in his expanded genus *Irlbachia* as one species unit, *I. quelchii*. In the most recent treatment of these taxa, *Rogersonanthus tepuiensis* is synonymized with *R. arboreus* (Struwe et al., 1999). Actually, only geographic distribution was used to distinguish these latter two species in Maguire and Boom's (1989) original key to the species of the genus.

Struwe, Nilsson, and Albert described an additional species, *Rogersonanthus coccineus*, from Sierra de la Neblina on the Brazilian–Venezuelan border (Struwe & Albert, 1998b; Figs. 2.17E–G, 2.18E). This species differs in habit and floral details from *Rogersonanthus arboreus* and *R. quelchii*, but leaf and pollen morphology show clear similarities with the other species of *Rogersonanthus*. Therefore, three species are currently accepted in the genus. No revision of *Rogersonanthus* has been published.

The genus consists of small perennial herbs (*Rogersonanthus coccineus*) or shrubs and trees, and all species have coriaceous leaves with attenuate bases (Struwe et al., 1999). The inflorescence is a small, terminal dichasium, 1–12-flowered, with scale-like bracts. The 5-merous flowers have campanulate calyces that are fused at the base and are thick and leathery (except for *Rogersonanthus coccineus*). The corollas are green to yellow, funnel-shaped, and thick and leathery in *Rogersonanthus arboreus* and *R. quelchii*, but are narrowly salver-shaped and red to orange in *R. coccineus*. The corolla tubes have the stamens inserted in their lower halves, with lanceolate anthers that are recurved after anthesis. The pollen is shed as tetrads. A glandular area is present at the base of the bilocular ovary, which is tipped by a long style with a bilamellate stigma. The capsules are elliptic, leathery, dehiscing medially or apically, and are sometimes nodding. The seeds are angular, not winged, and the testa cells are concave with band-like thickenings (*Rogersonanthus quelchii*).

Rogersonanthus might belong to the *Symbolanthus* clade, and although formerly included in *Macrocarpaea*, differs consistently in having tetrad pollen and thick coriaceous leaves without conspicuous venation (Struwe et al., 1997). *Rogersonanthus* also shows palynological and morphological similarities, mainly in flower and inflorescence morphology, with the similarly tetrad-bearing genus *Chelonanthus*.

"Roraimaea", ined. (L. STRUWE, S. NILSSON, & V. A. ALBERT, UNPUBL.)

This undescribed genus is based on a similarly undescribed species with orange flowers that occurs in white-sand savannas of Roraima state, Brazil (collection *Cid Ferreira 9125*; L. Struwe, S. Nilsson, & V. A. Albert, unpubl.).

"Roraimaea" is a branched suffrutescent herb with terete, 4-winged stems. The leaves are narrowly ovate to lanceolate and the inflorescences are compound, terminal cymes. The bracts and floral bracts are leaf-like, gradually becoming smaller toward the apices of the inflorescences. The flowers are 5-merous and erect and have a campanulate calyx that is divided about half of its length. The calyx lobes are rounded and often ciliate. The orange corolla is tubular, 8–14 mm long. The corolla lobes are about $^{1}/_{10}$ of the length of the corolla tube. There is no thickened papillose ridge on the inside of the corolla. The stamens are inserted straight into the corolla tube, about $^{2}/_{3}$ of the way down, and have straight filaments and sagittate, oblong anthers with small, sterile tips. The stamens are not zygomorphic inside the corolla tube. The pollen is released as tetrads with a non-differentiated reticulate exine. The gynoecium is slender, with a long and narrow style tipped by a bilamellate stigma with narrow, c. 4 mm long lobes. A disk is present at the base of the gynoecium. The fruit is an erect, dry capsule with many seeds (for seed morphology see Bouman et al., 2002).

The orange corolla color of this taxon is unusual among Helieae, to which "Roraimaea" clearly belongs. The pollen tetrads are similar to the *Helia*-type, but the exine is less differentiated along the equatorial sector (S. Nilsson, unpubl.). Leaf-like bracts are also found in *Macrocarpaea*, but that genus differs in its monad pollen, generally woody habit, and larger flowers. No molecular or morphological phylogenetic data are available for "Roraimaea" so its precise relationships are not yet known. However, with its non-differentiated pollen tetrads and erect, non-zygomorphic corolla and stamens, "Roraimaea" shows several traits that are characteristic of basal members of the *Symbolanthus* clade (Fig. 2.19). As such, the undescribed genus might be another relictual, white-sand taxon of the Helieae, as *Aripuana* and *Irlbachia* appear to be.

Senaea TAUB.

The monotypic genus *Senaea*, described by Taubert (1893: 515–517) based on *S. coerulea*, is a narrow endemic from southeastern Brazil (J. E. Simonis, unpubl.). *Senaea* has a disjunct, high-elevation distribution between Rio de Janeiro and Mato Grosso (Brazil).

Senaea plants are shrubs with erect, terete or quadrangular branches (J. E. Simonis, unpubl.). The leaves are decussate, rather coriaceous, petiolate, and elliptic to ovate. The inflorescences have leaf-like bracts, and are terminal or axillary dichasia. The flowers are 6-merous. The calyces are campanulate. The corollas are blue and funnel-shaped to campanulate, with acute lobes about as long as the corolla tubes. The stamens are inserted in the corolla tube, and the anthers are sagittate and apiculate. It is unknown if a glandular disk is present at the base of the ovary. The style is long and slender with a bilobed stigma. The fruits are elliptic capsules.

Senaea and *Prepusa* both bear 6-merous flowers and have the same pollen type (Nilsson, 2002). These taxa are probably very closely related and might be congeneric. Like *Prepusa*, *Senaea* is classified here as a member of Helieae pending further phylogenetic investigations.

Sipapoantha MAGUIRE & B. M. BOOM

This genus contains only one species, *Sipapoantha ostrina*, an endemic of the Cerro Sipapo tepui in the western part of Amazonas (Venezuela). It was recently described by Maguire and Boom (1989: 23, 25), and occurs only on the open savannas of the tepui summit. *Sipapoantha* was treated in the floristic works of Maguire and Boom (1989) and Struwe *et al.* (1999), but no complete monograph has been published.

The only species of this genus is an unbranched, annual herb with characteristic thick and coriaceous, revolute, oblong to obovate leaves aggregated at the base of the stems (Maguire & Boom, 1989; Struwe *et al.*, 1999). The stems are quadrangular and often have four narrow wings. The inflorescences are terminal, monochasially branched cymes with 1–7 flowers, and the bracts are similar to very small leaves or are scale-like. The flowers are 5-merous, and are slightly zygomorphic with the anthers and bent style aggregated in the upper part of the corolla mouth. The calyces are campanulate, thick and coriaceous, and persistent in fruit, and the lobes are oblong and obtuse. The corollas are large and showy, funnel-shaped with broad, rounded lobes, dark blue to purple, thin, and deciduous in fruit. The stamens are inserted very close to the base of the corolla tube and have filaments of unequal lengths that are broadened at the base. The anthers are lanceolate with sterile appendages, sagittate, and straight, but are versatile after anthesis. The pollen is shed as tetrads with reticulate exine bearing small globules (Maguire & Boom, 1989; Nilsson, 2002). The ovary has a glandular disk below, and the style is long, slender, and deciduous in fruit. The stigma is bilamellate with broadly circular lobes. The fruits are woody and nodding capsules.

Sipapoantha ostrina has not yet been added to our molecular phylogenetic analyses, but its affiliation with Helieae is nonetheless clear. The taxon has many of the typical characteristics of Helieae, i.e., a glandular disk below the ovary, pollen shed in tetrads, sagittate anthers with sterile apical appendages, monochasially branched cymes, and a broadly bilamellate stigma. Within the tribe, *Sipapoantha* shows some vegetative similarities with *Helia* and *Prepusa* in its obovate to oblong, coriaceous leaves that are aggregated at the base of the plant. These other genera, however, occur on the Brazilian Highlands and are absent from the Guayana Shield where *Sipapoantha* grows. Both *Prepusa* and *Helia* have pollen shed in tetrads, as does *Sipapoantha*, but with reticulate exine not bearing globose processes. The flowers of *Sipapoantha* also show some resemblance to *Calolisianthus*, another genus present only on the Brazilian Shield, both in their size and by being broadly funnel-shaped.

Symbolanthus G. DON

Don (1837: 175) described this genus of primarily shrubby to tree-like gentians with large, conspicuous flowers. *Symbolanthus* was later included as a section of "*Lisianthus*" (Bentham, 1876: 814) Traditionally, this taxon has been considered a separate genus and is now one of the most species-rich genera in Helieae, with an estimated 30 species (Struwe, 2002b). However, no revision has yet been published. E. C. H. van Heusden wrote an unpublished revision in 1982 in which only three species were accepted (see the abstract by van Heusden in Maas *et al.*, 1983). However, we consider the actual number of species to be much larger, with, for example, seven distinct species occurring in the Guayana Highlands, the remaining 23 being native to the Andes and southern Central America (Struwe *et al.*, 1999; Struwe, 2002b). *Symbolanthus* ranges from Costa Rica and Panama through northern South America, mainly on mountain ranges such as the Andes (down to Bolivia), the Coastal Cordilleras of Venezuela and Colombia, and the tepuis of the Guayana Shield (K. Gould & L. Struwe, unpubl.). The species, absent from the Brazilian Shield, grow mainly in wet, montane forests or in highland savannas and páramos on rocky terrain. One undescribed species is known from lowland white-sand savannas in eastern Peru, distinctly outside the normal range and habitat of the genus (L. Struwe, unpubl.).

The species of *Symbolanthus* are branched shrubs to small trees, or seldom perennial, suffrutescent herbs (Struwe *et al.*, 1999). The shape, thickness, and texture of the leaves show great variation among species and are often useful as distinguishing characteristics. The flowers are borne in terminal, rarely axillary, few-flowered dichasia or are solitary (Fig. 2.17J–L). They are 5-merous, pedicellate and usually slightly zygomorphic (i.e., the stamens and the style are bent toward the lower side of the corolla mouth and the two lower corolla lobes are often strongly reflexed). The campanulate and coriaceous calyces are deeply divided, c. $^7/_8$–$^9/_{10}$ (seldom less) of their length, and are fused only close to the base. The showy corollas are among the largest in the tribe (e.g., sometimes over 10 cm long in *Symbolanthus vasculosus* and *S. pulcherrimus*), broadly salver- or funnel-shaped, and are usually pink, magenta, or red, rarely purple, white, yellow, or green, and often also with white or purple stripes or white areas on the inside of the corolla. The stamens are inserted in the lower third of the corolla tube, and a corona (or seldom only a ridge) is present at the base of the filaments (L. Struwe & K. Gould, unpubl.). The filaments are sometimes sharply bent at their apices and are tipped with oblong to linear anthers that bear sterile apical appendages and become recurved after anthesis. The pollen is shed as tetrads. The gynoecium has a glandular disk at the base with a long style terminated by a bilamellate stigma with oblong lobes. Only the base of the style is persistent in fruit. The fruits are leathery, nearly berry-like capsules, which are medially dehiscent (or possibly indehiscent) and usually nodding (Fig. 2.18F).

The number of species in *Symbolanthus* is uncertain. Twenty-nine species epithets have been named in the genus and at least *c.* 30 species should be accepted, probably more. Distinguishing characteristics for different species are leaf size, shape, and texture, calyx lobe morphology, and corolla shape and color. One example is *Symbolanthus vasculosus* from Venezuela and Colombia, considered a synonym of *S. calygonus* by E. C. H. van Heusden (unpubl.) although it differs from this taxon in having distinct, yellow-green, very broadly funnel-shaped corollas, as opposed to pink to red corollas with a narrower corolla tube, a non-keeled calyx, and larger leaves. Pinkish versus yellowish-greenish color types do not characterize monophyletic groups in *Symbolanthus*, but instead appear to have evolved several times independently (K. Gould & L. Struwe, unpubl.).

Symbolanthus shares extensive morphological and anatomical similarities with *Wurdackanthus*. These two genera have pollen tetrads of the unique *Symbolanthus*-type in common, as well as staminal pockets behind the bases of the stamens in the corolla tube (absent in *Wurdackanthus argyreus*; L. Struwe & K. Gould, unpubl.). Phylogenetic studies based on nuclear ribosomal 5S-NTS (non-transcribed spacer) sequences and morphological data support the inclusion of *Wurdackanthus* in *Symbolanthus* (K. Gould & L. Struwe, unpubl.); see further discussion under *Wurdackanthus* as well as Figs. 2.1, 2.19.

Tachia AUBL.

Tachia is an easily defined genus from northern South America and Panama, described by Aublet (1775: 75) based on *T. guianensis* from French Guiana. *Tachia* is distributed in lowland and middle-elevation forests from Colombia in the northwest, eastward to French Guiana, and through the Amazon Basin to Bolivia in the south (Maguire & Weaver, 1975). A majority of the species are found north or northwest of the Amazon. The genus was revised by Maguire and Weaver (1975), and a new key to *Tachia* was recently published (Cobb & Maas, 1998).

The species of *Tachia* are shrubs or small treelets, sometimes single-stemmed, often with hollow and conspicuously yellow branches (Maguire & Weaver, 1975; Struwe *et al.*, 1999). The leaves are petiolate, elliptic to obovate, coriaceous or chartaceous. The leaf venation is pinnate with many straight secondary veins in *Tachia* sect. *Tachia*, but with 1–3 pairs of basally divergent, arcuate secondary veins in sect. *Schomburgkiana*. The inflorescence structure is unique in Helieae with axillary, bractless flowers, borne solitarily or in pairs, (sub-)sessile on a very short branch/cushion in the upper leaf axils. The flowers are 5-merous, and are slightly zygomorphic because of the bent corolla tube, stamens, and style. The yellow, tubular calyces show great morphological variation, and are used to distinguish several of the species. The corollas vary from tubular to salver-shaped, and also vary in color from yellow, cream, orange, white, to greenish. The stamens are inserted close to the base in the corolla tube, with filaments of unequal length, sometimes sharply bent at their apices. The linear-oblong anthers with sterile apical appendages are recurved after anthesis, and the pollen is released as monads. The gynoecium has a glandular disk at the base, and a long, slender style with a broadly bilamellate stigma. The fruits are woody capsules, dehiscing apically, and have many globose seeds.

Tachia differs from all other genera of Helieae by having sessile and axillary flowers (Struwe *et al.*, 1997, 1999). The flowers of *Zonanthus* are also axillary, but are pedicellate. Ant nests are common in the hollow stems and the ants probably feed on the abundant secretions from the glands (colleters) in the leaf axils and on the calyx keels (Struwe *et al.*, 1999). *Tachia* is closely related to *Chorisepalum* and *Macrocarpaea* (see below).

Tetrapollinia MAGUIRE & B. M. BOOM

This monotypic genus, named after its unique, spinose pollen tetrads, was segregated from *Irlbachia* by Maguire and Boom (1989: 31). The type species, *Tetrapollinia caerulescens*, is a small annual herb widespread in wet and sandy savannas and similar areas on the Guayana Shield, in eastern Amazonia, and on the Brazilian Shield (Struwe *et al.*, 1999). It grows at both low and high altitudes, being one of the few species of Helieae that covers most altitudinal and major geographical areas in this region of the Neotropics. It has not been revised yet.

Tetrapollinia is a small, unbranched annual herb with usually linear, sessile leaves (Struwe *et al.*, 1999). The few- to many-flowered inflorescence is terminal, cymose with monochasial branches, and has scale-like bracts. The 5-merous flowers are erect to horizontal (rarely nodding). The calyx is campanulate, fused about halfway or more from the base, and has dorsal keels on the triangular and acute lobes. The color of the funnel-shaped corolla can vary considerably from light to dark blue, light rose to dark purple, to white. The stamens are inserted in the lower half of the corolla tube and have oblong anthers. The pollen grains are released as tetrads and have a spinose exine. The gynoecium is unusual in Helieae in lacking a disk at the base, and the bilamellate stigma has lobes that are linear to narrowly ovate. The fruit is a thin, erect to horizontal capsule that dehisces apically and bears many angular seeds.

Wurdackanthus MAGUIRE

This genus with two recognized species was described by Maguire (1985: 312) and named in honor of his colleague John J. Wurdack, who collected many gentians together with Maguire during their trips to the Guayana Shield. One of the species, *Wurdackanthus frigidus*, is endemic to Guadeloupe, St. Vincent, and Dominica in the Caribbean, whereas the type species, *W. argyreus*, is endemic to two tepui summits, Cerro Aracamuni and Sierra de la Neblina in the southwestern Guayana Highlands of Venezuela and Brazil (Struwe *et al.*, 1999). No revision of this genus has been published; however, the species have been treated in floristic works (Fournet, 1978; Maguire & Boom, 1989; Struwe *et al.*, 1999).

Wurdackanthus plants are perennial herbs or subshrubs with elliptic, lanceolate to ovate, coriaceous (*W. argyreus*) or chartaceous (*W. frigidus*) leaves (Struwe *et al.*, 1999). The 5-merous, nodding to horizontal flowers are borne in few-flowered, terminal monochasial cymes. The calyces are campanulate, fused at their bases, with ovate to oblong lobes. The funnel-shaped corollas are yellow or greenish-yellow (*Wurdackanthus frigidus*) or lavender-pink, pink, or whitish (*W. argyreus*), with broadly ovate and obtuse lobes. The stamens are inserted close to the base of the corolla tube and have staminal pockets at the insertion points of the filaments (*Wurdackanthus frigidus*). The filaments are often sharply bent near the apices and the anthers are lanceolate-oblong, recurved after anthesis, and are equipped with sterile apical appendages. The pollen, which is released in tetrads, has coarsely reticulate exine. The gynoecium has a glandular base and a long style that is persistent in fruit, the stigma being bilamellate. The fruits are woody and nodding capsules that dehisce medially, bearing many angular seeds.

Wurdackanthus is closely related to *Symbolanthus* (Figs. 2.1, 2.19), with which it shares not only molecular synapomorphies but also morphological ones. Both genera share pollen of *Symbolanthus*-type (pollen tetrads with coarse reticulum around the equatorial zone; Nilsson, 2002) and very similar seeds (Bouman *et al.*, 2002). A molecular study based on 5S-NTS sequences has shown that *Wurdackanthus* is probably non-monophyletic with the type species, *W. argyreus*, nested inside *Symbolanthus* (K. Gould & L. Struwe, unpubl.). *Wurdackanthus*, at least in part, might therefore be better included in *Symbolanthus* to preserve the monophyly of the latter. *Wurdackanthus* differs from *Symbolanthus* mainly in having a smaller calyx that is less deeply divided, the lobes usually more obtuse and no corona.

Systematics, character evolution, and biogeography 163

Zonanthus GRISEB.

This is an endemic Cuban genus described by Grisebach (1862: 145) comprising only one rare species, *Zonanthus cubensis*. Very few collections are known of this species, and the information on its morphology and anatomy is rather limited.

The habit of *Zonanthus* is woody and tree-like with leaves crowded at branch apices, fused at the base as a short interpetiolar sheath (Grisebach, 1862; Leon & Alain, 1957). The solitary 5-merous flowers are borne in the leaf axils. The campanulate calyx has rounded lobes with hyaline margins. The green corollas are funnel-shaped, with the stamens inserted in the middle of the corolla tube. The anthers have sterile apical tips. The gynoecium is sub-4-locular because of the strongly intruding parietal placentas and is tipped by a style with a bilamellate stigma. The fruit is a capsule bearing many angular seeds.

The precise relationships of *Zonanthus* are unknown. Not only are the collections of this taxon scarce, but the specimens we have seen are also old and not suitable for DNA extraction. *Zonanthus* is most probably a member of Helieae, having a bilamellate stigma, deeply inrolled placentas, a woody habit, and anthers with sterile apical appendages. The solitary, axillary flowers could indicate a close relationship with *Tachia*, from which *Zonanthus* differs in its arborescent habit. However, such an arborescent and branched habit is characteristic of *Chorisepalum* and some *Macrocarpaea* spp., and *Zonanthus* does bear pollen monads similar to these taxa. The evidence therefore suggests that *Zonanthus* belongs to the *Macrocarpaea* clade (Fig. 2.19).

Phylogeny and character evolution in Helieae

The tribe Helieae has some consistent characteristics but also shows large variation in features such as leaf morphology and venation, calyx morphology, corolla shape and color, palynology, and seed morphology. Some characters, e.g., rugose leaves, appear to have evolved several times independently in different genera, whereas others are present only in monophyletic groups, e.g., the corona at the base of the filaments in *Symbolanthus*. Limited phylogenetic information is presented in the trees based on either *trn*L intron or *mat*K sequences alone (Figs. 2.1–2.2). However, in the combined analysis of both *trn*L intron and *mat*K sequences (Fig. 2.3, an analysis that included only 10 species of the tribe), the relationships in Helieae are nearly completely resolved. Additionally, an analysis including 37 ITS sequences from Helieae yielded highly resolved clades inside most of the tribe (Fig. 2.19; Struwe, 1999; L. Struwe, J. R. Grant, K. Gould, & V. A. Albert, unpubl.). Despite this, the selection of taxa is different in the combined *trn*L intron/*mat*K (Fig. 2.3) and ITS (Fig. 2.19) analyses, but the resulting cladograms are nearly compatible with each other, differing mainly in nodes not supported by high jackknife values (i.e., where data are ambiguous). The ITS analysis has given the highest resolution and support for branching patterns inside the Helieae, and it includes more species; therefore, it will form the main basis for the following discussion. Note that *Neblinantha* (included in the *trn*L tree) has not been sequenced for ITS, so its position relative to the additional taxa sequenced for ITS is not known.

All phylogenetic results (Figs. 2.1–2.3, 2.19) can be summarized as identifying three major clades. *Macrocarpaea*, *Tachia*, and *Chorisepalum* form the first clade, here referred to as the *Macrocarpaea* clade. The *Macrocarpaea* clade (in the ITS result) is not supported by a jackknife value over 50%, owing to *Chorisepalum* sometimes forming a polytomy at the most basal branching point in the Helieae tree. The *Macrocarpaea* clade is supported by additional morphological data such as monad pollen of the *Glabra*- or cf. *Glabra*-type (Nilsson, 2002) and leaves with more than three secondary veins (pinnate venation, except *Tachia* sect. *Schomburgkiana*). All three genera in the *Macrocarpaea* clade are supported as monophyletic with jackknife values of 82% or more (Fig. 2.19). The sister group to *Macrocarpaea* appears to be *Tachia*, but this is not strongly supported by parsimony jackknife analysis.

A second major clade includes only *Irlbachia* (hereafter called the *Irlbachia* clade, jackknife support 100%; Fig. 2.19). Among mostparsimonious trees, the *Irlbachia* clade is positioned as sister to either the *Macrocarpaea* clade or the *Symbolanthus* clade, although neither position has any jackknife support. Therefore, the *Irlbachia* clade is best discussed as an independent basal clade in the Helieae. *Irlbachia* is distinguished from the other genera of Helieae by having pollen borne in polyads of the *Irlbachia*-type (Nilsson, 2002). The genus also has mostly small, nearactinomorphic, open corollas with very short corolla tubes (except *Irlbachia tatei*). One aberrant *Irlbachia* species has been found to be a member of the *Symbolanthus* clade, and this species, *I. pratensis*, will be discussed below.

The third major clade in Helieae includes tetrad-bearing genera such as *Aripuana*, *Calolisianthus*, *Chelonanthus*, *Purdieanthus*, *Symbolanthus*, *Tetrapollinia*, and *Wurdackanthus*. Two species included in this group have pollen shed in polyads (*Irlbachia pratensis* and *Chelonanthus uliginosus*). We refer to this lineage as the *Symbolanthus* clade, after its largest genus,

Symbolanthus, which is strongly supported by a jackknife value of 99% and is deeply nested inside the group. In the *Symbolanthus* clade, there is also a strongly supported dichotomy between two major subclades. One subclade is composed of *Aripuana, Chelonanthus purpurascens, Irlbachia pratensis*, and *Purdieanthus*, with all species positioned in a large polytomy. The other subclade includes *Calolisianthus*, the other species of *Chelonanthus, Tetrapollinia, Symbolanthus*, and *Wurdackanthus. Symbolanthus* and *Wurdackanthus* are strongly supported as a monophyletic group and confirm results obtained with 5S-NTS data (K. Gould & L. Struwe, unpubl.). Several nodes are unresolved in this second subclade, but the paraphyly of *Chelonanthus* toward *Symbolanthus* plus *Wurdackanthus* is clear, as is the exclusion of *Chelonanthus purpurascens* (the type of the genus) from this subclade. Also para-/polyphyletic is *Irlbachia*, but here the type of *Irlbachia* (*I. elegans*, not sequenced) belongs morphologically to the *Irlbachia* clade, not the *Symbolanthus* clade. *Irlbachia pratensis* has in fact been assigned to *Chelonanthus* previously, having pollen polyads that are intermediate between the *Uliginosus*-type found in *Chelonanthus purpurascens* and the *Irlbachia*-type found in the *Irlbachia* clade (Nilsson, 2002). It is clear from these results that new generic circumscriptions will be needed for *Chelonanthus* and *Irlbachia* at the least, and possibly also for more genera as new species are sequenced. Nevertheless, several hypotheses regarding phylogenetic relationships, character evolution, and biogeography can be supported.

Vegetative structures

The plants of Helieae are generally woody at least at the base, and many are suffrutescent herbs. A few annual herbs are known (e.g., *Irlbachia pumila* and *Tetrapollinia caerulescens*). The arborescent genera – *Aripuana, Chorisepalum* spp., *Macrocarpaea* spp., and *Symbolanthus* spp. – can be up to 10 m tall, and appear to hold derived phylogenetic positions, with suffrutescent, perennial herbs placed more basally (cf. *Calolisianthus, Chelonanthus*, and *Macrocarpaea rubra*). Lianas are also known in the Helieae, e.g., *Lehmanniella* and *Purdieanthus*. Development of secondary wood is widespread in the Gentianaceae and occurs in all tribes, and when interpreting the distribution of habit on the phylogenetic trees it seems probable that a perennial, suffrutescent, but herbaceous habit is plesiomorphic in the family. Shrubs have evolved rarely and independently in Chironieae (e.g., *Orphium* and *Ixanthus*; Thiv *et al.*, 1999b), Exaceae (e.g., *Gentianothamnus*; Klackenberg, 1990), and Saccifolieae (*Saccifolium*; Thiv *et al.*, 1999a). In Potalieae, subtribe Potaliinae is entirely woody and contains the largest trees of the family, whereas subtribe Lisianthiinae comprises primarily suffrutescent herbs but also shrubs and small trees, and plants of subtribe Faroinae are herbaceous.

Leaf morphology shows great variation in Helieae, ranging from taxa having tiny, thin, and linear leaves (e.g., *Irlbachia pumila*) to those having very large, obovate, and strongly rugose leaves (e.g., *Macrocarpaea marahuacae*; Struwe & Albert, 1998b; Struwe *et al.*, 1999). The leaves are always opposite, entire (except for *Irlbachia poeppigii* with slightly dentate leaves), sessile or petiolate, often with a thickened or recurved margin, or with a thin margin on chartaceous leaves. The texture of the leaves is very variable and no generalizations can be made regarding, for example, whether highland taxa have more coriaceous leaves than lowland taxa. However, there is rarely variation within species with regard to texture, venation, or margins.

The two types of leaf venation found within Gentianaceae are also found within Helieae. Gentianaceae are well known for having "pseudopalmate" venation, with 1–2 pairs of secondary veins branching off close to the base of the leaf and arcuating toward the leaf tip. This type is common in Exaceae and Gentianeae, and also occurs frequently in Helieae. The second common type is a pinnate venation of many, shorter and usually straight secondary veins that successively branch from the midvein toward the tip of the leaf. This venation pattern is found, for example, in *Tachia* sect. *Tachia* (Maguire & Weaver, 1975) as well as in subtribe Potaliinae (Potalieae). Intermediate forms occur as well, but are less common.

In several species of Helieae the veins are deeply impressed on the upper side and prominently raised below. This character also seems to be associated with the leaves having a rugose surface and turning brown when drying. It is interesting to note that this trait has so far been found only in some *Macrocarpaea* species from the Guayana Highlands, an area to which the two rugose-leaved *Chorisepalum* species are endemic. Rugose leaves with prominent venation are also known from the mainly Guayanan genus *Tapeinostemon*, here classified as Saccifolieae, and not hypothesized to be closely related to Helieae.

An interpetiolar line/sheath, which can be up to several millimeters high and conspicuous in some taxa (e.g., *Macrocarpaea* and *Tachia*), is usually present around the stem. At the inside of the base of each leaf are resin-secreting colleters, which are most prominent in *Tachia*. Branches of *Tachia* herbarium collections often stick to the paper during drying because of these secretions (L. Struwe, pers. obs.).

Inflorescence

The inflorescences of Helieae are compound cymes, sometimes reduced to solitary flowers. The upper branches of the inflorescences are mostly dichasial or monochasial cymes, traits that can be used to separate genera with multi-flowered inflorescences. Monochasial inflorescence branches occur in, for example, *Adenolisianthus, Chelonanthus, Irlbachia, Tetrapollinia*, and *Wurdackanthus frigidus*, whereas *Aripuana, Celiantha bella, Chorisepalum* spp., and *Macrocarpaea* have strictly dichasially branching cymes or thyrses. In the cases where only three or fewer flowers per inflorescence occur in a taxon, e.g., *Helia, Lagenanthus, Rogersonanthus*, and *Symbolanthus* spp., it is not possible to determine which of these two types they represent. Most genera have terminal inflorescences/flowers, but *Tachia* and *Zonanthus* have strictly axillary flowers, and since *Zonanthus* has not yet been molecularly characterized this trait remains a non-homoplastic synapomorphy for the genus *Tachia*. Otherwise, the type and position of the inflorescences do not appear to have any phylogenetic information value in Helieae as shown so far. It is also difficult to make a conclusion on whether multi-flowered inflorescences are plesiomorphic in the tribe.

The bracts and bracteoles of the cymes are paired. Since the bracteoles of the middle flower are the bracts of the side flowers, etc., in a dichasial cyme the terms "bract" and "bracteole" are not the most apt. No distinction or clear homology statement can be made between these two terms, and they are often used in confusing ways. Therefore, in Helieae it is better to use the term "bract" for all bracts and bracteoles except for the "floral bracts" that are situated below the side flowers and do not serve as bracts for other flowers (cf. Struwe *et al.*, 1999). In any case, the morphologies of bracts and floral bracts are usually the same among Helieae. Most genera have both types, with the exception of their absence in sessile-flowered *Tachia*. The bracts and floral bracts are generally small, scale-like, and triangular, but leafy bracts are a characteristic trait for *Aripuana* and "Roraimaea" (at least for lower parts of the inflorescence), *Irlbachia poeppigii*, and *Macrocarpaea* (Struwe *et al.*, 1997, 1999), all genera that have relatively basal positions in Helieae in different phylogenetic analyses.

Floral anatomy

The floral anatomy of Helieae has been studied by several researchers, and a summary of some of the available data was presented in Struwe *et al.* (1997). The tribe is characterized by having a star-shaped stele leading into the flower (Lindsey, 1940; van Heusden, 1986; Struwe et al., 1997: fig. 2b). Fused calyx lateral bundles (i.e., 10 vascular bundles to a 5-merous calyx in total) have been reported from seven genera in Helieae, including *Aripuana*, *Chelonanthus*, and *Irlbachia*, all placed in one subclade in the *mat*K and ITS results (Figs. 2.2, 2.19). Outside Helieae, this trait has also been reported from *Lisianthius* (Potalieae; Lindsey, 1940; Struwe et al., 1997). The receptacle and calyx also contain sclereids, a feature that has been reported only from Helieae (Lindsey, 1940; Struwe et al., 1997) and *Ixanthus* in tribe Chironieae (Thiv et al., 1999b), and therefore a case of parallel evolution. Colleters are common on the inside of the calyces throughout Helieae (and in the order Gentianales as a whole). Another typical character is the presence of "bridges" created by late repositioning of two staminal traces from the ovarian wall to the corolline/staminal tube (Lindsey, 1940; Struwe et al., 1997).

Helieae show the highest evolutionary degree of fusion and cohesion in floral anatomical traits in the entire Gentianaceae (Lindsey, 1940), a statement that is also supported by the relatively advanced position of the tribe in the phylogenetic trees. For further discussion on the floral anatomy of Helieae, see Struwe et al. (1997).

Calyx

Calyces in Helieae are most often distinctly fused at the base, sometimes up to $^9/_{10}$ of their lengths, with the least fused sepals found in *Symbolanthus* spp. and the most fused sepals occurring in *Tachia guianensis*. Free calyx lobes are known only from *Chorisepalum*, a genus that is unique in having a 4-merous, decussate calyx. All other taxa in the tribe have 4- or 5-merous calyces that are fused at the base and have imbricate aestivation. The calyx lobes are usually elliptic to ovate with obtuse apices (e.g., *Chelonanthus* and *Sipapoantha*), but acute apices occur as well (e.g., *Chorisepalum*, *Symbolanthus* spp., and *Tetrapollinia*).

A distinct trait for Helieae is the presence of dorsal glandular keels (or areas) on each calyx lobe, often decurrent down the calyx tube. These glands, the function of which is unknown, form an uneven, strongly safranin-staining and thickened surface area (Struwe et al., 1997: fig. 2d). These keels should not be confused with the carinate (dorsally keeled), but non-glandular, calyx lobes that are found in other tribes of the family, e.g., in *Exacum*, *Schultesia*, and *Tachiadenus*. Carinate calyx lobes are found only within Helieae in *Tachia schomburgkii* and *Symbolanthus* spp. (Struwe et al., 1997; Struwe, 2000b), and must be considered apomorphic in the tribe.

Systematics, character evolution, and biogeography 169

Macrocarpaea bracteata, from the Colombian–Venezuelan border mountains, shows an interesting feature in that the outer lobes of the calyx have a leaf-like appearance, possibly caused by a homeotic mutation in which some of the calyx lobes are expressing the genetic program and morphology of leafy bracteoles (Albert *et al.*, 1998). Most taxa have very coriaceous calyces, and all genera have calyces that are persistent at least initially in fruit. However, the dry calyx falls off before the fruit matures in *Chorisepalum*, *Macrocarpaea*, and *Tachia*.

Corolla

Helieae show great diversity in corolla color, shape, and morphology. The merosity of the corolla varies from 4-merous in *Tetrapollinia* and sometimes *Neblinantha neblinae*, 5-merous in most taxa, and uniquely 6-merous in *Chorisepalum*, *Prepusa*, and *Senaea*. Except for 6-merous corollas being a strong potential synapomorphy for *Chorisepalum*, merosity appears to have little value as a phylogenetic character in the tribe.

The corollas are tubular (e.g., *Lagenanthus* and "Roraimaea"), funnel-shaped (e.g., *Calolisianthus* spp., *Macrocarpaea* spp., *Rogersonanthus* spp., and *Symbolanthus* spp.), or salver-shaped (e.g., *Aripuana*, *Chorisepalum*, *Helia*, *Irlbachia*, and *Symbolanthus* spp.). Several shape classes often occur within single genera, apparently not characterizing phylogenetic groups. The same is also true for corolla color, with many genera having species with white, pink, purple, and blue flowers, sometimes also within the same species (e.g., *Tetrapollinia caerulescens*). Green and yellow, sometimes also whitish corollas are also common in the tribe and are characteristic for *Chelonanthus* (except for the type species, *C. purpurascens*, with blue-purple corollas), *Macrocarpaea*, and *Rogersonanthus* (except for *R. coccineus* with red-orange corollas; Struwe *et al.*, 1999). Red-orange corollas are also known from *Calolisianthus* (Progel, 1865), *Lagenanthus* (Ewan, 1948b), *Purdieanthus* (Ewan, 1948b), and "Roraimaea".

Aripuana is unique in the tribe in having valvate aestivation at the base of the corolla lobes and contort aestivation at the very apex of the corolla bud (Struwe *et al.*, 1997: fig. 3). All other genera have strongly contorted corolla buds. Since *Aripuana* is the only taxon in Helieae with the anthers inserted in the corolla lobe sinuses, Struwe *et al.* hypothesized that an incomplete postgenital fusion of the upper corolla tube above the anthers, i.e., where the valvate aestivation occurs, could explain this feature. The corollas of the Gentianaceae develop first as free lobe primordia with free stamen primordia on the inside, then after the lobes are nearly fully developed, the staminal-corolline tube begins to form (L. Struwe, pers. obs.). The development of the staminal-corolline tube occurs very late in development, directly before the bud opens, i.e., after the anthers and gynoecium are fully developed. Most Helieae have a long, well-developed corolla tube above the anthers, with the anthers situated rather low in the corolla tube as a whole. In *Aripuana*, however, the lower staminal-corolla tube is very long, and no upper corolla tube exists at all. If this hypothesis is true, then only the contorted part of the lobes of *Aripuana* are strictly homologous to the corolla lobes of other taxa in Helieae. Further studies of flower development in the tribe are necessary to clarify this.

The shape of the corolla bud apices is often characteristic for different genera in Helieae, but does not appear to have significant phylogenetic value. Acute to acuminate corolla lobe apices on tapering corolla buds are found in *Aripuana, Celiantha, Chorisepalum, Neblinantha, Symbolanthus, Tachia,* and *Tetrapollinia*. Rounded corolla buds, on the other hand, are characteristic for the genera with obtuse corolla lobes, e.g., *Calolisianthus, Chelonanthus, Macrocarpaea,* and *Rogersonanthus*.

Androecium

The stamens are generally of the same morphology throughout the entire Helieae. The filaments are filiform, long, not fused, and often of slightly to distinctly unequal length (Fig. 2.17). The anthers are oblong to linear, basifixed, sagittate, introrse, longitudinally dehiscent, and usually with small, sterile apical tips. After anthesis the anthers often become recurved backwards. In some genera such as *Symbolanthus* and *Calolisianthus*, the upper part of the filaments and/or the anthers can recurve so much that they form a full circle or more. Anther morphology does not appear to be phylogenetically informative either inside genera or in the tribe as a whole, with a few exceptions. All species of *Chorisepalum* have the longest and most linear anthers, forming a possible synapomorphy for this genus.

Palynology of Helieae: Synopses of genera and general discussion
S. NILSSON

Adenolisianthus. Adenolisianthus arboreus (Spruce ex Progel) Gilg. The pollen grains are 3-colporate to porate and are united in tetrahedral tetrads that are c. 63 μm in diameter. The exine is differentiated into a fine reticulum, but with a coarsely reticulate area at the poles and around the equator. The muri of the coarse reticulum are often fragmented into various elongated to rounded processes. (Venezuela: Amazonas, 1959,

Wurdack & Adderley 43722 (NY), slide no. 24089; Nilsson, 2002: Fig. 4.10A.)

OTHER INVESTIGATIONS: Gilg (1895a), Köhler (1905), and Nilsson (1970) described the pollen tetrads of *Adenolisianthus arboreus*, which was regarded as a synonym of *Chelonanthus fruticosus* by Maguire and Boom (1989; incorrectly listed as *C. arboreus* in their figure legend).

Aripuana. *Aripuana cullmaniorum* Struwe, Maas, & V. A. Albert. The pollen grains are 3-colporate and united in tetrahedral tetrads that are *c.* 58 µm in diameter. The exine is reticulate and slightly differentiated into a zone of coarser reticulum around the equator. (Brazil: Amazonas, 1985, *Ferreira et al. 5906* (NY), slide no. 24673.)

OTHER INVESTIGATIONS: The pollen tetrads of *Aripuana cullmaniorum* were described and illustrated in Struwe *et al.* (1997).

Calolisianthus. *Calolisianthus pulcherrimus* (Mart.) Gilg. The pollen grains are 3-colporate and united in tetrahedral tetrads that are *c.* 70 µm in diameter. The exine consists of reticulate islands. (Brazil: Minas Gerais, 1933, *Barreto 2786* (F), slide no. 5788.)

OTHER INVESTIGATIONS: Gilg (1895a) described the tetrads of some species of *Calolisianthus* (including *C. pulcherrimus*) and Köhler (1905) also investigated the same species. Nilsson (1970) discerned no fewer than five pollen types in the genus *Calolisianthus* as it was then circumscribed. The pollen of *Calolisianthus pulcherrimus* was classified as the *Speciosus*-type by Nilsson (1970) together with pollen from, for example, *C. pedunculatus*, *C. pendulus*, and *C. speciosus*.

Celiantha. *Celiantha bella* Maguire & Steyermark. The pollen grains are united in polyads that are *c.* 134 µm in diameter. The exine has globular and pila-like processes or elongated elements. (Venezuela: Amazonas, 1957, *Silva & Brazão 60921* (NY), slide nos. 24085 and 17831; Nilsson, 2002: Fig. 4.11.)

OTHER INVESTIGATIONS: The granular polyads of *Celiantha bella*, *C. chimantensis*, and *C. imthurniana* were described and illustrated by Maguire (1981). The pollen types of the latter two species had been earlier described and put in a new pollen type, the *Imthurnianus*-type of *Calolisianthus*, by Nilsson (1970).

Chelonanthus. *Chelonanthus alatus* (Aubl.) Pulle. The pollen grains are 3-colporate and united in tetrahedral tetrads that are *c.* 55 μm in diameter. The exine is differentiated, reticulate, and with an equatorial zone of coarser reticulum. (Brazil: Amazonas, 1971, *Prance et al. 14376* (NY), slide no. 24086; Nilsson, 2002: Fig. 4.12A–C.)

Chelonanthus purpurascens (Aubl.) Struwe, S. Nilsson, & V. A. Albert. The pollen grains are obscurely porate and united in polyads that are 115 μm in diameter. The exine is reticulate with polar loops. (French Guiana: Rochambeae, 1994, *Rova et al. 1902* (S), slide no. 24650; Nilsson, 2002: Fig. 4.13D–F.)

OTHER INVESTIGATIONS: Gilg (1895a) and Köhler (1905) described both tetrads and polyads in the genus *Chelonanthus*. These observations were later confirmed by Nilsson (1970), who distinguished the two types of pollen found in *Chelonanthus* as the *Chelonoides*-type (tetrads) and the *Uliginosus*-type (polyads).

Chorisepalum. *Chorisepalum carnosum* Ewan. The pollen grains are released as monads, 3-colporate, subprolate to prolate, *c.* 33×25 μm. The exine is reticulate to coarsely reticulate. (Venezuela: Amazonas, 1948, *Maguire & Politi 27921* (NY), slide 24091; Nilsson, 2002: Fig. 4.14.)

OTHER INVESTIGATIONS: The pollen grains of *Chorisepalum carnosum*, *C. ovatum*, *C. psychotrioides*, and *C. rotundifolium* are all similar (Maguire, 1981; Nilsson, 2002).

Helia. *Helia brevifolia* (Cham.) Gilg. The 3-colporate to porate pollen grains are united in tetrahedral tetrads that are *c.* 60 μm in diameter. The exine is differentiated into a finely perforate zone at the polar area, and a zone of coarse reticulum around the equator. (Paraguay: date unknown, *Jorgensen 4837* (S), slide no. 24652; Nilsson, 2002: Fig. 4.15A–C.)

OTHER INVESTIGATIONS: Gilg (1895a) mentioned that the tetrads of *Helia* were minutely reticulate to perforate, which is in agreement with the present study.

Irlbachia. *Irlbachia cardonae* (Gleason) Maguire. The pollen grains are united in obscurely porate polyads that are *c.* 130 μm in diameter. The exine

Systematics, character evolution, and biogeography 173

has polar spines in groups. (Venezuela: Bolívar, 1937, *Tate 1360* (NY), slide no. 6217; Nilsson, 2002: Fig. 4.16A–C.)

OTHER INVESTIGATIONS: Gilg (1895a) recognized *Irlbachia* as having spiny tetrads (*I. caerulescens* now = *Tetrapollinia caerulescens*) and *Pagaea* (including *Brachycodon*, now *Irlbachia*) as having spiny tetrads grouped into polyads. Köhler (1905) confirmed Gilg's descriptions. Nilsson (1970) also distinguished two pollen types in *Irlbachia*, namely, the *Irlbachia*-type (polyads) and the *Caerulescens*-type (tetrads). The genus *Pagaea* has pollen of the *Irlbachia*-type (Nilsson, 1970). Subsequently, Maguire (1981) joined *Irlbachia* (including *I. cardonae* above), *Pagaea*, and *Brachycodon* together in the genus *Irlbachia*, and moved the tetrad-bearing species *I. caerulescens* to the new genus *Tetrapollinia*.

Lagenanthus. Lagenanthus princeps (Lindl.) Gilg. The pollen grains are porate with well-delimited pores and are united in tetrahedral tetrads that are c. 69 μm in diameter. The exine is coarsely reticulate, only slightly differentiated, and with equal-sized lumina. (Colombia: Santander, 1969, *Murillo & Jaramillo s.n.* (COL), slide no. 24531; Nilsson, 2002: Fig. 4.18A–C.)

OTHER INVESTIGATIONS: Both Gilg (1895a) and Köhler (1905) described the reticulate tetrads of *Lagenanthus princeps*. Nilsson (1970) considered *Lagenanthus princeps* to have a unique pollen type.

Lehmanniella. Lehmanniella splendens (Hook.) Gilg ex Ewan. The pollen grains are united in tetrahedral tetrads that are c. 65 μm in diameter. The exine is coarsely to extremely coarsely reticulate. (Colombia: Antioquia, 1948, *Sandeman 6046* (K), slide no. 4810; Nilsson, 2002: Fig. 4.19A–B.)

OTHER INVESTIGATIONS: Gilg (1895a) described the tetrads of *Lehmanniella* as being coarsely reticulate with projecting muri. This was later confirmed by Nilsson (1970), who established a separate pollen type for *Lehmanniella splendens*, the *Lehmanniella*-type.

Macrocarpaea. Macrocarpaea bracteata Ewan. The pollen grains are released as monads, 3-colporate, prolate spheroidal, and c. 38×37 μm. The exine is reticulate to coarsely reticulate. (Venezuela: Trujillo, 1988, *Dorr et al. 4940* (NY), slide no. 24185; Nilsson, 2002: Fig. 4.20A–C.)

Macrocarpaea cinchonifolia (Gilg) Weaver (= *Rusbyanthus cinchonifo-*

lius Gilg). The pollen grains are released as monads, 3-porate to colporate, oblate spheroidal to spheroidal, and *c.* 32×33 μm. The exine has verrucae, various processes, and remnants of a reticulate pattern. (Bolivia: Mapiri, 1886, *Rusby 1173* (K), slide no. 4112; Nilsson, 2002: Fig. 4.20D–F.)

Macrocarpaea domingensis Urb. & Ekman. The pollen grains are released as monads, 3-colporate, oblate spheroidal, and *c.* 28×30 μm. The exine is coarsely reticulate to reticulate. (Dominican Republic: 1946, *Allard 18215* (S), slide no. 1164.)

Macrocarpaea rubra Malme. The pollen grains are released as monads, 3-colporate, spheroidal, and *c.* 28×28 μm. The exine is coarsely reticulate to reticulate. (Brazil: 1908, *Dusén 6965* (F), slide no. 1851.)

Macrocarpaea valerii Standl. The pollen grains are released as monads, 3-colporate, oblate spheroidal, and *c.* 29×34 μm. The exine is coarsely reticulate to reticulate. (Costa Rica: Alajuela, 1938, *Smith 1001* (F), slide no. 979.)

OTHER INVESTIGATIONS: Pollen from about 30 species of *Macrocarpaea* have been studied. The genus *Macrocarpaea* as circumscribed by Ewan (1948a) has pollen grains of three types: reticulate monads, verrucose monads, and tetrads (Nilsson, 1968, 1970; Robyns & Nilsson, 1970). The species with tetrads have been moved to a genus of their own, *Rogersonanthus* (Maguire & Boom, 1989), whereas the taxa with monads now form *Macrocarpaea sensu stricto*. Erdtman (1952) described the pollen of *Macrocarpaea rubra* as 3-colporate and coarsely reticulate. Maguire (1981) described three new species with reticulate pollen.

The exine features of the reticulate and verrucose pollen types in *Macrocarpaea* intergrade, and consequently the monotypic genus *Rusbyanthus* (Gilg, 1895a) with warty pollen grains was suppressed and its only species, *R. cinchonifolius*, was transferred to *Macrocarpaea* (Weaver, 1974). Erdtman (1952) made a description and drawing of the warty pollen of *Macrocarpaea cinchonifolia* (then *Rusbyanthus cinchonifolius*).

Neblinantha. *Neblinantha neblinae* Maguire. The pollen grains are united in tetrahedral tetrads that are *c.* 64 μm in diameter. The exine is coarsely reticulate with islands of loop-like, projecting muri. (Brazilian–Venezuelan border: Amazonas, 1970, *Steyermark 103862* (NY), slide no. 24097; Nilsson, 2002: Fig. 4.22.)

Systematics, character evolution, and biogeography   175

OTHER INVESTIGATIONS: Maguire and Boom (1989) described the pollen tetrads of *Neblinantha* as being basically of *Helia*-type, but this is not supported in the present study.

Prepusa. Prepusa connata Gard. The pollen grains are 3-colporate to porate and are united in tetrahedral tetrads that are c. 89 μm in diameter. The exine is verrucose to pilate with processes of different sizes. (Brazil: date unknown, *Gardner 541* (CGE), slide no. 17772; Nilsson, 2002: Fig. 4.23A-B.)

OTHER INVESTIGATIONS: The granular tetrads of *Prepusa* (including *P. connata* above) were described by Gilg (1895a) and Köhler (1905), both of whom pointed out the similarities between the tetrads of *Prepusa* and *Senaea*.

Purdieanthus. Purdieanthus pulcher (Hook.) Gilg. The pollen grains are 3-porate and united in tetrahedral tetrads that are c. 61 μm in diameter. The exine is coarsely reticulate and indistinctly differentiated. (Colombia: Mt. del Moro, date unknown, *Purdie s.n.* (K), slide no. 7357; Nilsson, 2002: Fig. 4.24D-F.)

OTHER INVESTIGATIONS: Gilg (1895a) described the minutely to coarsely reticulate pollen tetrads of *Purdieanthus pulcher*, and Köhler (1905) confirmed his observations. Nilsson (1970) put the pollen of *Purdieanthus pulcher* in a pollen type of its own, the *Purdieanthus*-type.

Rogersonanthus. Rogersonanthus arboreus (Britton) Maguire & Boom. The pollen grains are 3-colporate to porate and united in tetrahedral tetrads that are c. 68 μm in diameter. The finely reticulate exine is differentiated into a coarser reticulate zone around the equator and at the very distal pole. (Trinidad: 1984, *Baksh & Nilsson 1984-3* (S), slide no. 19409; Nilsson, 2002: Fig. 4.25.)

OTHER INVESTIGATIONS: Nilsson (1968, 1970) described the tetrads of four species of *Macrocarpaea* (*M. arborea, M. cerronii, M. quelchii,* and *M. salicifolia*). These four species were transferred to the new genus *Rogersonanthus* by Maguire and Boom (1989) based on having pollen released as tetrads instead of as monads (as in *Macrocarpaea sensu stricto*). *Rogersonanthus coccineus* tetrads were described by Struwe and Albert (1998b: fig. 5).

Senaea. Senaea janeirensis Brade (= *Senaea caerulea*). The 3-porate pollen grains are united in tetrahedral tetrads that are *c.* 82 μm in diameter. The exine is pilate with pila of different sizes. (Brazil: E. Rio de Janeiro, 1986, *Martinelli et al. 12003* (RB 272215), slide no. 21827; Nilsson, 2002: Fig. 4.26A–B.)

OTHER INVESTIGATIONS: Gilg (1895a) described the tetrads of *Senaea caerulea*. Köhler (1905) investigated *Senaea perpusilloides* and indicated the similarity that exists between the pollen of *Senaea* and *Prepusa*.

Sipapoantha. Sipapoantha ostrina Maguire & B. M. Boom. The 3-colporate to porate pollen grains have distinct aperture margins and are united in tetrahedral tetrads that are *c.* 71 μm in diameter. The exine is pilate to verrucose. (Venezuela: 1949, *Maguire & Politi 28092* (NY), slide no. 24098; Nilsson, 2002: Fig. 4.26D–F.)

OTHER INVESTIGATIONS: The tetrads of *Sipapoantha ostrina* were described and depicted by Maguire and Boom (1989), and similarities to *Rusbyanthus* (now *Macrocarpaea* spp.) pollen were suggested by these authors. However, the pollen grains of *Macrocarpaea* differ in being shed as monads (Nilsson, 1970).

Symbolanthus. Symbolanthus pulcherrimus Gilg. The pollen grains are 3-colporate and united in tetrahedral tetrads that are *c.* 82 μm in diameter. The exine is differentiated into a finely reticulate pattern with a zone of coarse reticulum with projecting muri around the equator. (Costa Rica: Cartago, 1968, *Weaver Jr. 1406* (S), slide no. 24636; Nilsson, 2002: Fig. 4.27.)

OTHER INVESTIGATIONS: The tetrads of some species of *Symbolanthus* were described by Gilg (1895a) and Köhler (1905). Nilsson (1970) investigated 18 species of *Symbolanthus* and Maguire and Boom (1989) included brief descriptions and illustrations of a few additional species of *Symbolanthus*.

Tachia. Tachia guianensis Aubl. The pollen grains are released as monads and are 3-colporate, prolate spheroidal, and 34×30 μm. The exine is coarsely reticulate. (Guyana: Mt. Mania, 1974, *Rova et al. 1963* (S), slide no. 24635; Nilsson, 2002: Fig. 4.28A–B.)

OTHER INVESTIGATIONS: Most species of *Tachia* have reticulate pollen grains according to Agababyan and Tumanyan (1977b), Maguire and Weaver (1975), and the present study. The exceptions are *Tachia loretensis*

Systematics, character evolution, and biogeography 177

with reticulate-rugulate pollen, and *T. occidentalis* and *T. schomburgkiana* with smooth pollen grains.

Tetrapollinia. *Tetrapollinia caerulescens* (Aubl.) Maguire & B. M. Boom. The 3-porate pollen grains are united in tetrahedral tetrads that are c. 66 μm in diameter. The exine is finely reticulate, suspended between concentrically or irregularly arranged spines. (Guyana: Rupununi, 1992, *Jansen-Jacobs et al. 2764* (NY), slide no. 24096; Nilsson, 2002: Fig. 4.29A–B.)

OTHER INVESTIGATIONS: Both Gilg (1895a) and Köhler (1905) described the spinose tetrads of *Tetrapollinia caerulescens* (as *Irlbachia caerulescens*) as having relatively numerous spines spread all over the surface (*Irlbachia* has only polar spines). Nilsson (1970) classified this as a special pollen type, the *Caerulescens*-type. Maguire and Boom's (1989) establishment of a new monotypic genus for this species, *Tetrapollinia*, is supported.

Wurdackanthus. *Wurdackanthus frigidus* (Sw.) Maguire & B. M. Boom. The pollen grains are 3-colporate to porate and united in tetrahedral tetrads that are c. 66 μm in diameter. The exine is differentiated into a finely reticulate pattern and an equatorial zone of coarse reticulum. (St. Vincent: 1942, *Beard 17* (S), slide no. 24715; Nilsson, 2002: Fig. 4.29D–F.)

OTHER INVESTIGATIONS: Maguire and Boom (1989) described the pollen tetrads of *Wurdackanthus argyreus* and pointed out the similarities between the two recognized species of *Wurdackanthus*, *W. argyreus* and *Wurdackanthus frigidus*, both having tetrads of *Symbolanthus*-type. Nilsson (1970) regarded *Wurdackanthus* (previously placed in *Calolisianthus*) as having pollen of *Symbolanthus*-type.

Zonanthus. *Zonanthus cubensis* Griseb. The pollen grains are released as monads and are 3-colporate, subprolate, and c. 48×37 μm. The exine is perforate to rugulate. (Cuba: Mt. Cristo, 1983, *Areas et al. PFC 49840* (HAJB), slide no. 24638; Nilsson, 2002: Fig. 4.30.)

OTHER INVESTIGATIONS: The pollen monads of *Zonanthus cubensis* were described as distinctly reticulate by both Gilg (1895a) and Köhler (1905).

Discussion of palynological data in Helieae. In the classification of Gilg (1895a), all genera with pollen united in tetrads or polyads were grouped together in the tribe Helieae. In the present classification, however, Helieae consists of taxa with monads, tetrads, and polyads. Gilg (1895a) keyed out the following genera of Helieae, mainly by means of the structure of the pollen tetrads: *Adenolisianthus, Calolisianthus, Chelonanthus, Helia, Irlbachia, Lagenanthus, Lehmanniella, Pagaea* (= *Irlbachia*), *Purdieanthus,* and *Symbolanthus*. The presence of both tetrads and polyads in *Chelonanthus* was mentioned. In addition, *Prepusa* and *Senaea* (placed together in the key) as well as *Coutoubea* and *Schultesia* were distinguished. *Deianira* was also included in the key and came out close to *Helia*. The removal to the Chironieae of *Coutoubea, Deianira, Schultesia, Symphyllophyton,* and *Xestaea* from Gilg's (1895a) Helieae, all with different types of tetrads from the current members of Helieae, has palynological support.

Köhler (1905) gave detailed and relevant morphological descriptions of all the above genera except *Helia* and *Lehmanniella*. In *Chelonanthus* he confirmed Gilg's (1895a) findings of both tetrads and polyads among the six species he examined. Furthermore he agreed that the tetrads of *Prepusa* and *Senaea* were similar to each other in having similar rounded processes and other exine features.

Most genera of Helieae, i.e., *Adenolisianthus, Calolisianthus, Celiantha, Chelonanthus, Helia, Irlbachia, Lagenanthus, Lehmanniella, Neblinantha, Purdieanthus, Rogersonanthus, Sipapoantha, Symbolanthus, Tetrapollinia* and *Wurdackanthus,* have been subject to extensive pollen morphological studies by Nilsson (1970), Maguire (1981), Maguire and Boom (1989), and the present study. In *Calolisianthus* there are two different pollen types (tetrads and polyads), and *Chelonanthus* has both tetrads and polyads. Different morphological trends among genera have been proposed. *Lehmanniella* (tetrads), *Calolisianthus* spp. (polyads) and *Celiantha* (polyads) may be regarded as endpoints in a proposed scheme of evolutionary trends (cf. Nilsson, 1970).

Of the included genera with monads, *Macrocarpaea* and *Tachia* were placed in another tribe, Gentianeae-Tachiinae, by Gilg (1895a). *Chorisepalum* was not described at the time (Gleason & Wodehouse, 1931). The genus *Chorisepalum* seems close to *Tachia* and *Macrocarpaea* palynologically. *Chorisepalum, Macrocarpaea,* and *Tachia* all have reticulate pollen grains of the same type (Nilsson, 2002), and a close relationship between these genera is also supported by *mat*K and ITS data (Figs. 2.2, 2.19). The relationship between reticulate monads and tetrads/polyads with a differentiated or undifferentiated net structure appears unclear. The position of *Zonanthus* in Helieae, also with monads but with a striato-reticulate exine, is even more enigmatic.

Tetrads with a differentiated reticulate exine, i.e., with a zone of coarse reticulum around the equator, and/or coarsely reticulate at the distal polar area, is a basic type from which a number of pollen types may be derived. A number of tetrads sticking together in aggregates and gradually losing their individual characteristics and reducing their aperture (e.g., becoming indistinctly porate) will lead to the specialized polyads of, for example, *Celiantha*.

The following genera have the basic type of tetrads with a coarsely reticulate zone around the equator and an area of coarse reticulum at the distal pole: *Adenolisianthus, Aripuana, Calolisianthus* spp., *Chelonanthus* spp., *Helia, Lagenanthus, Lehmanniella, Neblinantha, Purdieanthus, Rogersonanthus*, "Roraimaea", *Symbolanthus*, and *Wurdackanthus*. "Roraimaea" has pollen shed in tetrads but not with a distinct area of coarse reticulum, thereby being most similar to pollen of the *Helia*- and *Aripuana*-types (S. Nilsson, unpubl.). The polyads characteristic of *Calolisianthus* spp., *Celiantha, Chelonanthus* spp., and *Irlbachia* are regarded as being derived from the above tetrads.

Close affinity between *Chelonanthus* and *Tetrapollinia* has pollen morphological support, as does an affinity between *Symbolanthus* and *Wurdackanthus*. The delicate and lace-like reticulum of the pollen of *Tetrapollinia* shows some similarity with the polyads of *Irlbachia*, but this is not strongly supported by molecular data.

Sipapoantha, Prepusa, and *Senaea* appear more remote from the basic type of tetrads and lack the differentiated exine pattern typical for *Adenolisianthus*, etc., but do have globules or pila on the exine. The pollen of *Prepusa* and *Senaea* has certain similarities to the tetrads of the genera above, although they are easily separable on their tetrad morphology. A possible relationship of *Prepusa* and *Senaea* with *Sipapoantha* is not clearly understood.

Gynoecium and fruit

The gynoecia in Helieae are sessile, usually with a glandular disk or glandular area at the base, elliptic, and terminated by a usually long, slender style that ends with a bilamellate stigma (Struwe *et al.*, 1997). The stigma lobes are often characteristically circular, broad, and flattened, but more linear and narrow lobes occur in some taxa. No species has capitate or simple stigmas. These gynoecial characters could possibly be used as synapomorphies for the group, but homoplastic occurrences of several characters occur outside of Helieae. For example, bilamellate stigmas with broad lobes are also present in *Eustoma* (Chironieae) and *Fagraea berteroana* (Potalieae; Struwe & Albert, 1997). However, the presence of a long style that becomes twisted and flattened when dry might be unique for Helieae (L. Struwe, unpubl.). A glandular disk or nectaries at the base of the ovary is probably a plesiomorphic character for Helieae, since this trait also occurs in, for example, subtribe Potaliinae, tribe Gentianeae, and *Voyria* (*incertae sedis*).

The placentation in Helieae follows a common pattern with parietal placentas that are strongly inrolled, sometimes forming secondarily axile and bilocular ovaries (Struwe *et al.*, 1997). At the very base the ovaries are sometimes truly bilocular, but higher up this bilocular appearance is caused by a septum formed by the intruding parietal placentas. In anatomical cross-sections this can be seen as the incomplete fusion of the wall and septum (Struwe *et al.*, 1997: fig. 4). Presumably unusual in Helieae is the presence of a bilocular ovary in *Aripuana* (Struwe *et al.*, 1997). Ovaries with placentas that at first appear bilocular while in fact have strongly intruded parietal placentas are also found in many other genera of the Gentianaceae (Lindsey, 1940). They should not be confused with the true axile placentation found in *Exacum* and related genera (see Exaceae; Fig. 2.9A) as well as in the closely related gentianalean families Apocynaceae, Loganiaceae, Gelsemiaceae, and Rubiaceae. In Helieae, fruits might show unilocularity and/or bilocularity as a result of late development or inrolling of the placentas. Thus, further studies of the early stages are necessary to clarify the placentation patterns in this tribe (Struwe, 1999).

Helieae have capsular fruits, often woody or sometimes leathery (*Symbolanthus*; Fig. 2.18). There are some reports that the fruits of *Symbolanthus* are sometimes indehiscent; these fruits are up to 5 cm long, being the largest in the tribe. *Chorisepalum* has unique fruits that dehisce via four valves from the base upward, eventually leaving a central stalk and ruptured placental vascular bundles at the base (Struwe *et al.*, 1997). The styles of Helieae are often persistent (cf. Fig. 2.18A), and this causes the apically fused capsules to open medially and forcefully. Taxa with deciduous styles, e.g., *Macrocarpaea* spp. and *Tachia*, have apically dehiscent capsules.

Pollination syndromes

The flowers of Helieae show many pollination syndromes. The presence of a nectariferous disk aids in pollination, and all flowers can probably be assumed to be pollinated by animals (cleistogamous flowers are known only from one species of *Sebaea*). The most common type of flower is small or large, funnel-shaped with a basal narrow tube, and of white to green or yellowish color. *Chelonanthus alatus* is an example of this type, and bat pol-

Systematics, character evolution, and biogeography   181

lination has been reported in this species, which is interesting because it is unusual for bats to visit herbaceous plants (Vogel, 1969a,b; Machado *et al.*, 1998). Other species that show similar floral morphology and can be assumed to be bat pollinated are *Rogersonanthus arboreus* and *R. quelchii*, several *Macrocarpaea* species, as well as other *Chelonanthus* species.

Another large group of species are possibly hummingbird pollinated. These have large, tubular or funnel-shaped flowers, often of pink-red-orange corolla color, but yellow and green are also known. Hummingbirds have been reported to visit flowers of *Symbolanthus* and *Macrocarpaea*, and probably also visit the flowers of *Lagenanthus*, *Lehmanniella*, and *Purdieanthus*. Possibly hawkmoth-pollinated flowers are known only from *Aripuana* (Struwe *et al.*, 1997). The pollinators of the yellow and tubular corollas of *Tachia* are not known, and neither are there any pollinator reports for, e.g., *Adenolisianthus*, *Calolisianthus*, *Helia*, *Irlbachia*, *Prepusa*, *Senaea*, and *Tetrapollinia*.

Uses

Leaves of *Chelonanthus alatus* have been reported to be used against "athlete's foot" in Trinidad, and in Peru against worm-infested wounds of cattle (Morton, 1981, as *Chelonanthus chelonoides*). *Chelonanthus alatus* is commonly used in the Neotropics as a remedy against many illnesses (fevers, fungal infections, etc.).

Character synopsis

Typical characteristics for Helieae are calyces with low, dorsal glandular keels or glandular areas on the lobes. The stamens are inserted in the corolla tube (except *Aripuana*), the filaments are of unequal length (rarely not), and the anthers have sterile apical appendages and are recurved after anthesis. The gynoecia are sessile with a basal glandular area or disk (rarely not), and the placentation is parietal with strongly intruding and inrolled placentas, sometimes creating a falsely axile placentation. The taxa of Helieae also have long styles that become flattened and twisted when dry, bilamellate stigmas with flattened broad or narrow lobes, and usually medially dehiscent, dry fruits.

Biogeography of Helieae

Neotropical distributions

Helieae are found only in tropical South America, Central America, and some islands in the Caribbean (see Table 2.6 for an overview). Their distribution patterns can be classified into two general categories, i.e., narrow endemics and widespread taxa. Of the narrow endemics, *Calolisianthus, Helia, Prepusa,* and *Senaea* are restricted to the Brazilian Highlands in eastern South America, while *Celiantha, Chorisepalum, Neblinantha,* and *Sipapoantha* are endemic to the Guayana Highlands (Struwe *et al.*, 1999). *Lagenanthus* and *Purdieanthus* are endemics of the northern Andes (Colombia and Venezuela).

Chelonanthus and *Irlbachia* both have their center of distribution in the southwestern part of the Guayana Shield (lowland and highland), although the widespread species *C. alatus* reaches eastward to French Guiana, over Amazonia to Bolivia, the foothills of the Andes, and through Central America to Mexico (Struwe *et al.*, 1999). *Tachia* is a low- to middle-elevation genus that occurs from French Guiana in the northwest to the foothills of the Andes, but is absent from the Brazilian Shield.

Macrocarpaea and *Symbolanthus* are both rather widespread genera that are most species-rich in the Andes (Grant & Struwe, 2001; Struwe, 2002b). *Macrocarpaea* has three species on the Brazilian Shield (*M. glaziovii, M. obtusifolia,* and *M. rubra*; J. R. Grant, unpubl.), while *Symbolanthus* is not present on the Brazilian Shield at all. Both genera reach Central America (Panama and Costa Rica), but only *Macrocarpaea* is also distributed on the Greater Antilles (Cuba, Dominican Republic, and Jamaica). However, *Wurdackanthus*, which is closely related to and will likely be included in *Symbolanthus*, is known from the Guayana Highlands and the islands of Dominica, Guadeloupe, and St. Vincent, which are part of the Lesser Antilles and are more adjacent to the South American continent than the Greater Antilles. *Rogersonanthus* (*R. arboreus*) is found on Trinidad, the island closest to Venezuela, but otherwise occurs only on the tepuis of the Guayana Shield. The only genus endemic to the Caribbean is *Zonanthus*, which occurs on eastern Cuba.

Geographic origin of Helieae

It is plausible to assume that the ancestry of Helieae is both tropical and austral. All major clades of Gentianaceae except Gentianeae have tropical members occupying basal phylogenetic positions, supporting a tropical origin for the family. With major diversifications in South America, Africa, and around the Indian Ocean basin, an austral origin is supported for most of the major clades in the family. Further evidence within Helieae is the absence of North American relatives basally positioned within the tribe, which could have supported a more northern origin, but instead the presence of basal lineages in white-sand areas on Gondwanic crustal elements

Systematics, character evolution, and biogeography 183

in South America (i.e., the Guayana and Brazilian Shields). Furthermore, South America has the highest number of gentian genera of any continent, and the highest percentage of genera endemic to any major region as well (Albert & Struwe, 2002). Most of these endemic genera, primarily members of the Chironieae and Helieae, are restricted to the Guayana or Brazilian Shields.

Using the available phylogenetic information, we cannot yet clarify whether a distribution on the Brazilian Shield or on the Guayana Shield is most plesiomorphic for Helieae. If Helieae are older than the formation of the Amazon Basin and diversified while the shields were still connected, repeated biogeographic patterns could be expected to be seen within Helieae subclades and no particular shield might be supported as the most ancestral area. We evaluate this possibility below with respect to the *Macrocarpaea, Irlbachia,* and *Symbolanthus* subclades.

The *Macrocarpaea* clade

The *Macrocarpaea* clade consists of *Chorisepalum, Macrocarpaea,* and *Tachia* (*Neblinantha* was also included in this group in the *trn*L result, but this has not yet been confirmed with ITS data; cf. partial relationships from Figs. 2.1–2.3, 2.19). This group is centered on the Guayana Shield, and, as mentioned above, *Macrocarpaea* has three species endemic to the Brazilian Shield as well. Morphological support for this clade includes monads of the same pollen type present in *Chorisepalum, Macrocarpaea,* and *Tachia* (Nilsson, 2002). The only other genus in Helieae with monads is *Zonanthus*. This genus has not yet been evaluated with molecular data, but can be assumed to belong to this group as well because of its pollen morphology and axillary, sessile flowers (as in *Tachia*).

As one of the primary Helieae lineages, the *Macrocarpaea* clade is approximately as old as the tribe itself, and so may preserve evidence for ancestral distribution patterns. The longer a lineage has been present in an area, the more opportunities for dispersal have existed, but also the more threats of extinction due to geological and climatic changes that have occurred, all of which can give rise to (relictual) populations in small, isolated areas. *Zonanthus* and *Macrocarpaea* are the only genera in Helieae that are found in the Greater Antilles, Caribbean islands with a long geological and biogeographical history. However, species from these two genera are known only from restricted areas on Cuba, Jamaica, and the Dominican Republic, and no species occur on more than one island. This, as well as the fact that no *Macrocarpaea* species occur on younger islands closer to the

South American continent, could indicate that these species are comparatively old relicts that do not disperse easily. Supporting this interpretation is the fact that there are three possibly relictual species of *Macrocarpaea* present in the southeastern Brazilian Highlands and many narrow endemics in the Guayana Shield, Andes, and Greater Antilles (Grant & Struwe, 2000). Moreover, the six species of *Macrocarpaea* that are endemic to tepui summits in the Guayana Highlands do not show overlapping distributions (with one possible exception on Sierra de la Neblina, see Struwe & Albert, 1998b). *Macrocarpaea* has also reached Panama and Costa Rica, possibly rather recently, with *M. macrophylla* occurring in both Panama and Colombia (J. R. Grant, unpubl.). Biogeographic patterns found in phylogenetic analyses of *Macrocarpaea* based on ITS and 5S-NTS sequences have shown that species from the Dominican Republic, southeastern Brazil, and Central America are most basal in this genus (Grant & Struwe, 2000; J. R. Grant & L. Struwe, unpubl.). These species have the most peripheral distributions in the genus, which supports the hypothesis that these taxa are ancestral relicts. Within the rest of *Macrocarpaea* there appears to have been a high diversification rate among Andean taxa, producing two main clades, one primarily including species from Bolivia, Peru, and Ecuador, the other mainly including species with more northern distributions. Dispersal events have also occurred between the areas occupied by these two subclades, as well as into Central America from Colombian lineages.

Chorisepalum is also endemic to tepuis of the Guayana Shield and shows largely non-overlapping species distributions, although some species are known to be sympatric, e.g., in the Pakaraima Mountains at the Venezuela/Brazil/Guyana border. *Tachia* species also show a similarly non-overlapping pattern, despite the fact that the genus is a low- to middle-elevation rainforest element not constrained by the geographic limits of a Caribbean island or by altitudinal differences on the "virtual islands" of the tepuis. The opposite scenario is found in *Chelonanthus* and *Irlbachia*, for example, both more herbaceous and short-lived genera, in which several species have overlapping distributions at lowland and/or high elevations. A relevant difference may be that *Chorisepalum*, *Macrocarpaea* (except *M. rubra*), *Tachia*, and *Zonanthus* are long-lived, woody plants of primary rainforest vegetation and are not pioneer species in secondary vegetation and disturbed areas (as opposed to *Chelonanthus alatus*, for example).

To conclude, we would hypothesize that species of the *Macrocarpaea* clade and *Zonanthus* are not readily dispersed and that their distributions reflect comparatively old vicariance and/or dispersal events in the history of Helieae.

The *Irlbachia* clade

With only three species of the *Irlbachia* clade sampled for molecular data, not much can be said about biogeography within the group. Also, all species sequenced are lowland species, so the position of tepui species in *Irlbachia* unfortunately cannot be evaluated yet. *Irlbachia nemorosa* and *I. poeppigii* are both from lowland savannas and forests in northwestern Amazonas, especially along Rio Negro. Many *Irlbachia* species occur only or primarily on white-sand savannas, and the position of *Irlbachia* as one of three basic lineages of Helieae adds support to the hypothesis that white-sand areas may represent relictual habitats for the tribe (Struwe et al., 1997).

The *Symbolanthus* clade

The third major clade in Helieae found in the ITS phylogeny (Fig. 2.19) and the combined analysis of *trn*L and *mat*K data (Fig. 2.3) includes several genera: *Aripuana* (southeastern Amazonas, Brazil), *Chelonanthus* (widespread in tropical America), *Irlbachia pratensis* (northwestern Amazonas, Guayana Shield), *Purdieanthus* (Colombian Andes), *Tetrapollinia* (in Amazonia and on the Brazilian and Guayana Shields), *Symbolanthus* (Andes, Central America, and Guayana Highlands), and *Wurdackanthus frigidus* (the Lesser Antilles). From the results of the *mat*K analysis (Fig. 2.2) we can also conclude that *Lehmanniella splendens* (Colombian Andes) belongs to this clade. *Chelonanthus* and *Tetrapollinia* represent short-lived herbs that inhabit savannas (sometimes white-sand savannas), disturbed areas, riversides, and forests, whereas *Aripuana* is an endemic white-sand savanna tree or shrub. *Lehmanniella*, *Purdieanthus*, *Symbolanthus*, and *Wurdackanthus* are woody (at least at the base, except *W. frigidus*, which is sometimes herbaceous), and occur only in montane forests or on savannas or rocky areas on mountain summits.

The monophyletic group consisting of *Symbolanthus* and *Wurdackanthus* is deeply embedded within the *Symbolanthus* clade, and its origin may be younger than that of *Macrocarpaea*, despite its being the next-most species-rich genus in the tribe (with c. 30 species; *Macrocarpaea* has c. 90). Noteworthy is that no species from this lineage occur on the Greater Antilles in the Caribbean; however, *Wurdackanthus frigidus* occurs in the Lesser Antilles, and *Chelonanthus alatus* occurs on Trinidad, an island that has been connected with the South American continent relatively recently in geological time (Iturralde-Vinent & MacPhee, 1999).

To generalize, one lineage of Helieae seems to have predominantly woody, slowly-dispersed taxa with old, relictual distributions (the *Macrocarpaea* clade), whereas another (the *Symbolanthus* clade) has predominantly herbaceous representatives that have spread over vast areas in South and Central America, but have never managed to reach the Greater Antilles. No species in the *Macrocarpaea* or *Irlbachia* clades could be fairly termed a weed, whereas *Chelonanthus alatus* is indeed a weed in secondary vegetation and disturbed areas in large parts of Central and South America.

Climatic changes in the Quaternary and the refugia hypothesis

Severe changes in climate have affected tropical South America during the glaciations and inter-glaciations of the Quaternary (Gentry, 1982; Schubert, 1986, Haffer, 1987; Schubert & Salgado-Labouriau, 1987; Rull *et al.*, 1988; Irion, 1989; Rull, 1996). Drier periods have been interchanging with wetter ones. How reduced the forest cover was during the driest times is still not agreed upon. There is evidence that highland areas might have become too cold to support forests during glacial periods, but that forests remained in some lowland areas (the refugia hypothesis; cf. Haffer, 1969, 1987; Brown & Prance, 1987). Several partially opposing theories have been proposed for the presence of such refugia, while some authors have argued that some of the support for high-diversity refugial areas has been a collecting artifact (cf. Haffer, 1969, 1987; Prance, 1982, 1987; Steyermark, 1982; Kubitzki, 1989, 1990).

Adams (1997) and Adams and Faure (1997) presented an overview of the Quaternary history of South America (together with all other continents), focusing on the conditions prevailing during the last glacial maximum (LGM, *c.* 21000–17000 years ago) until the present. The forest cover was greatly reduced during the LGM to rather small areas, with a large rainforest refugium probably present in the upper Rio Negro basin (Colinvaux, 1996; Adams, 1997; Adams & Faure, 1997), and smaller refugia as gallery forests along the rivers. A broad savanna corridor was probably present along the Amazon Basin (Adams, 1997). In northwestern Venezuela, desert conditions were present and montane forests were lowered in altitude as a result of climatic cooling. Events like these must have affected the flora of South America severely, but apart from causing extinction, they could conceivably have favored the survival of some species. Several ice ages occurred during the Quaternary, so there has been a repeated pattern of drastic climatological change in South America during the last few million years.

Chelonanthus, especially *C. purpurascens* and *C. alatus*, and *Tetrapollinia* are probably easily dispersed or have relaxed environmental constraints, judging from their broad distributions. It is hard to hypothesize whether this distributional trait is due to some specific morphological character.

Systematics, character evolution, and biogeography 187

However, all of these taxa occur in a variety of habitats, and are especially common in savannas, grasslands, and other open areas, which should probably earn them classification as pioneer species. It is possible that the recurring cooling and drying of South America during the Quaternary could have favored the dispersal of these open-habitat taxa.

White-sand areas

It is interesting to note that the area in the upper Rio Negro basin that has been postulated by Colinvaux (1996) as a possible rainforest refugium during the LGM is also the area in South America with the highest number of lowland species in Helieae. Genera with endemic species in this area include *Adenolisianthus*, *Chelonanthus*, *Irlbachia*, and *Tachia* (Struwe *et al.*, 1999). *Curtia* (Saccifolieae) and *Potalia* (Potalieae) also have species restricted to this area (Struwe & Albert, 1998b; Struwe *et al.*, 1999). This region is also characterized by abundant white-sand areas, which are extremely nutrient-poor quartz-sand areas often with sparse vegetation and both very wet and very dry conditions depending on seasonal variations of the water table. The plants of these areas are subjected to nutrient- as well as water-related stress. Struwe *et al.* (1997) hypothesized that since many species of lowland Helieae are endemic to white-sand areas (e.g., *Aripuana* and genera mentioned above), and as these areas are very old and may have been continuously present (see below), the taxa involved might represent old lineages in the tribe (see also Pihlar *et al.*, 1998). The white sands are sedimentary deposits of the erosion products from the Guayana and Brazilian Shields, and are much older than the more recent sediments originating from the Andes that today cover most of the Amazon Basin and adjacent areas. The white-sand areas were probably larger before the younger, more clayish sediments of the Andes covered much of them (Struwe *et al.*, 1997). The hypothesis of whether the white-sand species are basally positioned in Helieae phylogeny and represent relictual lineages cannot be rigorously tested with current molecular data, but preliminary data from both Helieae and Potaliinae indicate that this might be the case (Fig. 2.19; Struwe, 1999; L. Struwe & V. A. Albert, unpubl.).

Highland and lowland taxa

The highlands (tepuis) of the Guayana Shield as well as the rocky, nutrient-poor mountains of the Brazilian Shield show a high endemicity of gentians and other plants (Berry *et al.*, 1995; Struwe, 1999; Struwe *et al.*, 1999). The region of the Guayana Highlands that harbors the vertical-cliffed mountain summits and plateaus has been called the Pantepui region (Maguire,

1970; Steyermark, 1986; Huber, 1995). Steyermark (1986) and Maguire (1970) regarded the Pantepui flora as representing ancient diversifications that had been isolated from other taxa through time and altitude. Steyermark (1986) also supported migration of taxa from the Andes and lowlands to the tepui areas, arguing for a mixed origin of the tepui flora. It has been suggested that both Maguire and Steyermark overestimated the endemicity rates for the Pantepui region, but in Gentianaceae the Pantepui endemicity rates are 26% for genera and 51% for species (Struwe, 1999). Furthermore, of gentians growing on the tepuis, 70% of the species are endemic to Pantepui, so the endemicity rate for these taxa is very high (Struwe, 1999). Correspondingly, the highlands of the southeastern part of the Brazilian Shield have several endemic gentian genera, e.g., *Calolisianthus*, *Helia*, *Hockinia*, *Prepusa*, and *Senaea*.

The highlands were also affected by the Quaternary cooling and drying during glacial maxima, but few specific data are known from the neotropical region (cf. Schubert, 1986; Schubert & Salgado-Labouriau, 1987). During the LGM, at least, it is hypothesized that the level of forest was lowered altitudinally as a result of lower temperatures, with some areas also receiving far less rainfall than today (Adams, 1997). It would be likely that humid forests on mountain slopes persisted during drier times because of increased rainfall associated with higher elevations (Haffer, 1987). This could have created more opportunities for both dispersal and radiation of species than are currently present. Prance (1987) presented a hypothesis on the importance of refugia following Pleistocene climate changes in South America. On the other hand, Kubitzki (1989) hypothesized that tepui endemics mainly represent not ancient lineages but rather *in situ* radiations of Andean elements that dispersed there following the Andean uplift. His ideas were based on affinities of tepui flora with Andean taxa. Dispersal and contact between mountain ranges and peaks in both the Pantepui area and the Andes could have been possible, as well as isolation of populations in, for example, forested valleys of the Andes (the latter would be relevant for *Macrocarpaea* and *Symbolanthus*, both species-rich in Andean montane forests). Preliminary results from independent phylogenetic analyses using 5S-NTS sequences in *Macrocarpaea* and *Symbolanthus* show that tepui species in both of these genera might have their closest relatives in the Ecuadorian Andes and so may be derived from Andean lineages (Grant & Struwe, 2000; J. Grant & L. Struwe, unpubl.; K. Gould & L. Struwe, unpubl.). Similar patterns of relationships between the tepuis and the Ecuadorian–northern Peruvian part of the Andes can also be found in other plant groups, for example the recently found disjunct distribution of the genus *Aratitiyopea* (Xyridaceae) and several orchids, which were thought to be endemic to the tepuis (L. Campbell

& G. Romero, pers. comm.). However, the basal position of *Chorisepalum* in the *Macrocarpaea* clade (and especially the broad, apparently vicariant distribution of *C. sipapoanum* from Cerro Sipapo on the western edge of the Guayana Shield to Tafelberg in Surinam) supports this genus as representative of ancient cladogenetic events and part of the truly endemic tepui flora *sensu* Maguire (1970). Similarly, *Wurdackanthus argyreus* (endemic to Cerro Aracamuni and Sierra de la Neblina in the southwest Guayana Highlands) has been found to be a basally positioned element of the *Symbolanthus* + *Wurdackanthus* clade, and so this species might also fit the "ancient tepui flora" hypothesis (K. Gould & L. Struwe, unpubl.).

The savannas on the tepui summits also occur further down on the inaccessible talus slopes, perhaps making it possible for lowland savanna species to come in contact with highland savanna habitats, and vice versa. Following this, several authors (Kubitzki, 1989; Givnish *et al.*, 1997; Struwe *et al.*, 1997) have suggested that some tepui taxa might have been derived from taxa inhabiting the lowland white-sand areas of the Guayana Shield. Preliminary data from ITS sequences show that some of the highland-inhabiting taxa in Helieae (e.g., *Symbolanthus*) are derived from within lineages of lowland taxa (Struwe, 1999; Fig. 2.19). Phylogenetic support for lowland-to-highland dispersal has also been shown for Bromeliaceae (*Brocchinia*; Givnish *et al.*, 1997) and Rapateaceae (Givnish *et al.*, 2000). In the Brazilian Highlands, which are generally drier than the Guayana Highlands, there are not such dramatic and abrupt altitudinal differences as in the Guayana Highlands, where there are also shorter horizontal distances to lowlands. This might have made it less possible for Brazilian Highland species to have contact with lowland savanna taxa during times of climatic changes. Only four species of Helieae occur both on the Brazilian Highlands and in Amazonia, i.e., three species of *Chelonanthus* and *Tetrapollinia*. On the other hand, the Brazilian Highlands have several endemic savanna (*campo rupestre*) or forest genera, i.e., *Calolisianthus*, *Helia*, *Prepusa*, and *Senaea*, that formerly may have had greater distributions.

In summary, the biogeography of Helieae is complex, and several patterns of different ages can be discerned in the group. The hypothesis that the Pantepui region was an area of ancient diversification and speciation is supported by the phylogenetic position of *Chorisepalum*, whereas other tepui species in *Macrocarpaea* and *Symbolanthus* are supported as having an Andean origin. Within the tribe as a whole, white-sand and lowland habitat appears to be an ancestral trait, with subsequent dispersal to montane regions. However, species number per genus is much larger in montane genera (*Macrocarpaea* and *Symbolanthus*), which could be due to a higher diversification rate, or large-scale extinction in the lowlands.

190   L. Struwe et al.

Large-scale distributions seen in Helieae might be the result of old vicariance or dispersal as well as geological events occurring during the Tertiary, whereas small-scale distributions within a particular area, e.g., the Amazon Basin, and mountain summits and lowlands of the Guayana and Brazilian Shields, might be the result of recurring climatic shifts and their consequent vegetational changes during the Quaternary.

Table 2.7. *A synopsis of the number of species in each genus of the Potalieae and their geographic distribution*

Genus	No. of species	Distribution
Anthocleista	14	Angola, Burundi, Cameroon, Congo, Ethiopia, Fernando Poo, Gabon, Gambia, Ghana, Guinea, Ivory Coast, Kenya, Liberia, Mozambique, Nigeria, Principe, Rwanda, Sao Tomé, Senegal, Sierra Leone, South Africa, Sudan, Tanzania (incl. Zanzibar), Togo, Uganda, Upper Volta; Madagascar, Comores
Congolanthus	1	Angola, Cameroon, Central African Republic, Congo, Gabon, Nigeria, Republic of Congo, Tanzania, Uganda, Zambia
Djaloniella	1	Guinea (Diaguissa Plateau)
Enicostema	3	Antigua, Costa Rica, Cuba, Dominica, Grenada, Guadeloupe, Guyana, Hispaniola, Jamaica, Martinique, Nevis, Panama, Puerto Rico, St. Kitts, St. Lucia, St. Vincent, Trinidad, Tobago, Venezuela; Angola, Botswana, Ethiopia, Kenya, Malawi, Mozambique, Somalia, South Africa, Sudan, Tanzania (incl. Zanzibar), Uganda, Zambia, Zimbabwe; Madagascar; India, Malaysia (Java, Madura, Lombok, Sumbawa, Sumba), Sri Lanka, Timor, Yemen
Fagraea	c. 70	Bismarck Islands, Cambodia, China, Fiji, Hong Kong, India, Indonesia, Laos, Malaysia (Malayan Peninsula, Sumatra, Java, Borneo), Micronesia, Moluccas, Myanmar, New Caledonia, New Guinea, Philippines, SE Polynesia, Samoa, Singapore, Solomon

Systematics, character evolution, and biogeography 191

Table 2.7. (cont.)

Genus	No. of species	Distribution
Fagraea (cont.)		Islands, Sri Lanka, Thailand, Tonga, Vanuatu, Vietnam; Australia (Northern Territories and Queensland)
Faroa	19	Angola, Burundi, Cameroon, Central African Republic, Chad, Congo, Ghana, Ivory Coast, Malawi, Mali, Mozambique, Nigeria, Rwanda, Senegal, South Africa, Sudan, Tanzania, Uganda, Zambia, Zimbabwe
Karina	1	Congo (Haut–Katanga District)
Lisianthius	30	Belize, Costa Rica, Cuba, Dominican Republic, Guatemala, Haiti, Honduras, Jamaica, Mexico, Nicaragua, Panama, Puerto Rico
Neurotheca	3	Brazil, French Guiana, Guyana, Surinam, Venezuela; Angola, Burundi, Cameroon, Central African Republic, Chad, Congo, Fernando Poo, Gabon, Ghana, Guinea, Ivory Coast, Liberia, Mali, Mozambique, Nigeria, Republic of Congo, Senegal, Sierra Leone, South Africa (Natal), Uganda; W Madagascar
Oreonesion	1	Gabon
Potalia	c. 9	Bolivia, Brazil, Colombia, Costa Rica, Ecuador, French Guiana, Guyana, Panama, Peru, Surinam, Venezuela
Pycnosphaera	1	Angola, Botswana, Burundi, Cameroon, Central African Republic, Congo, Guinea, Malawi, Republic of Congo, Rwanda, Tanzania, Uganda, Zaïre, Zambia, Zimbabwe
Urogentias	1	Tanzania (Uluguru Mountains)

TRIBE POTALIEAE

L. STRUWE AND V. A. ALBERT

The Potalieae contains 13 tropical genera of very diverse morphology and distribution. Flower merosity and habit are more variable than in most tribes of the gentians. Most genera have rather few species with the exceptions of *Fagraea* and *Lisianthius*, which have c. 70 and 30 species, respectively. An overview of species number and distribution for all genera of the Potalieae is presented in Table 2.7. The tribe is here subdivided into three

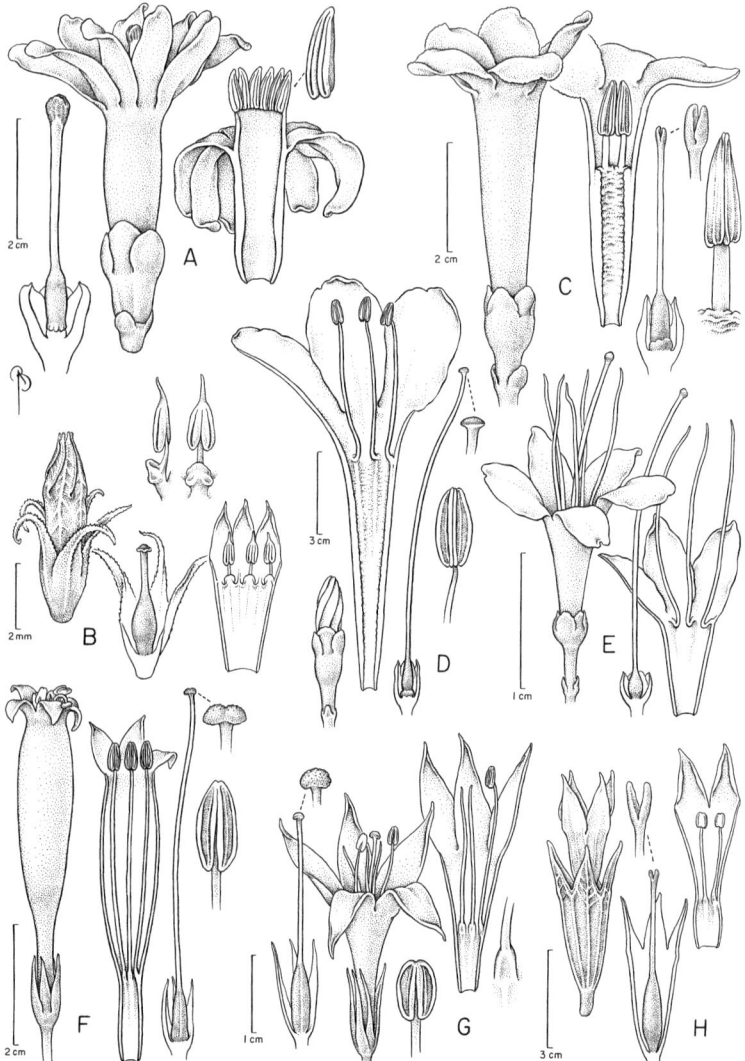

Figure 2.20. Flower morphology of Potalieae, showing insertion of stamens in corolla tube and calyx and style morphology. **A.** *Anthocleista grandiflora* (*Lindqvist s.n.*). **B.** *Enicostema verticillatum* (*Steyermark & Liesner 120963*). **C.** *Fagraea berteroana* (*Motley s.n.*). **D.** *Fagraea ceilanica* (*Struwe 1300*). **E.** *Fagraea fragrans* (*Worthington 12726*). **F.** *Lisianthius aurantiacus* (*Luteyn 14898*). **G.** *Lisianthius laxiflorus* (*Struwe 1153*). **H.** *Neurotheca loeselioides* (*Reitsma & Reitsma 3145*).

subtribes, the Faroinae, Lisianthiinae, and Potaliinae. The Central American and Caribbean genus *Lisianthius* is the only member of Lisianthiinae. The subtribes Faroinae and Potaliinae, however, include Old and New World genera. Subtribe Faroinae is mainly African, but both *Enicostema* and *Neurotheca* occur in South America and Madagascar, and one species of *Enicostema* occurs in Malesia. Subtribe Potaliinae includes three genera formerly assigned to the Loganiaceae (Leeuwenberg & Leenhouts, 1980), *Anthocleista*, *Fagraea*, and *Potalia* (the type genus), which are restricted to Africa–Madagascar, Australasia and the Pacific, and the Americas, respectively.

Description of Potalieae

Plants of the Potalieae are trees, lianas, shrubs, and herbs, with terete to quadrangular or winged stems. The leaves are opposite, usually petiolate, and of varying sizes, shapes, and textures. An interpetiolar sheath or prominent line is frequently present in *Anthocleista*, *Lisianthius*, *Potalia*, and most *Fagraea*. The flowers are borne in terminal or axillary inflorescences in dichasial cymes, clusters, or umbels, or are borne solitarily. The positioning of the flowers is either erect, horizontal, or nodding. The bracts can be scale- or leaf-like. The corollas are 3-merous (*Pycnosphaera*), 4-merous (*Congolanthus*, *Djaloniella*, *Faroa*, *Karina*, *Neurotheca*, and *Oreonesion*), 5-merous (*Enicostema* (usually, but rarely 3-, 4-, or 6-merous), *Fagraea*, and *Lisianthius*), 8-merous (*Urogentias*), 8-10-merous (*Potalia*), or 12-16(-24)-merous (*Anthocleista*). The calyces are isomerous with the corollas except in *Anthocleista* and *Potalia*, which have 4-merous calyces. The calyces are fused at the base, and are campanulate or tubular, thin or woody, and usually persistent in fruit. The corollas occur in many shapes, e.g., tubular, salver- or funnel-shaped, and vary in color from white, yellow, green, blue, lilac, red, or black. The stamens are inserted in the corolla tube or in the corolla lobe sinuses (*Faroa* and *Djaloniella*), often with various processes or appendages at the insertion points, and the filaments are at least partially fused in *Anthocleista* and *Potalia* (Fig. 2.20). The anthers are oblong to linear, and the pollen is shed as monads (tetrads in *Potalia maguireorum*). The gynoecium is sessile, surrounded by a disk in *Anthocleista*, *Fagraea* spp., *Lisianthius* spp., and *Potalia*, and is uni- to bilocular. The style is short or long and is tipped by a capitate, slightly bilobed, or broadly bilamellate stigma. The fruit is a capsule in the herbaceous genera and *Lisianthius*, and a fleshy berry in *Anthocleista*, *Fagraea*, and *Potalia*.

Taxonomic history of Potalieae

The present classification represents the first time the genera treated below have been placed together in the same tribe. As mentioned earlier, subtribe Potaliinae had commonly been included in Loganiaceae as tribe Potalieae (Leeuwenberg & Leenhouts, 1980), or had been treated as a separate family, the Potaliaceae (von Martius, 1826–1827; Lindley, 1836 (excluding *Fagraea*); Hutchinson, 1973). The affinity of Potaliinae genera with Gentianaceae is nothing new, however, since Jussieu (1791) considered *Potalia* to be a gentian. For an overview of the taxonomic history of the Potaliinae see Struwe *et al.* (1994). Relationships of Potaliinae to neotropical woody gentians were suggested by both Bentham (1857) and Gray (1859), who remarked on the similarity between *Fagraea* and *Lisianthius*. Bentham (1857) noted that the only large difference between these taxa was the more developed placentas of *Fagraea*, and Gray added that *Fagraea berteroana* had the bilamellate stigma of *"Lisianthus" sensu lato* (i.e., taxa assigned to Helieae; cf. Struwe *et al.*, 1997). In recent cladistic analyses of morphological, phytochemical (e.g., Jensen, 1992), and molecular data, representatives of the Potalieae are always positioned within the Gentianaceae (Downie & Palmer, 1992; Olmstead *et al.*, 1993; Bremer *et al.*, 1994; Struwe *et al.*, 1994, 1998). Although the transfer of tribe Potalieae (as "tribu des Fagraeacées") from Loganiaceae to Gentianaceae was accomplished by Bureau (1856), most subsequent workers on Loganiaceae have either ignored or overlooked Bureau in their taxonomic treatments. The reaffirmation of Bureau's view by Fosberg and Sachet (1974, 1980) was met with reluctance by Leeuwenberg and Leenhouts (1980), who advanced no evidence to support their alternative opinion. In fact, a cladistic re-examination of the same morphological data used by Leeuwenberg and Leenhouts (1980) indicates a placement in the Gentianaceae (Struwe *et al.*, 1994). Shared and diagnostic phytochemicals include the xanthones decussatin, gentianacaulin, and methyl-swertianin as well as unique secoiridoids and derivatives such as swertiamarin, gentiopicroside, and gentianine (Carpenter *et al.*, 1969; Rezende & Gottlieb, 1973; Bisset, 1980a; Jensen, 1992; Jensen & Schripsema, 2002 (Chapter 6, this volume)). The inclusion of *Anthocleista*, *Fagraea*, and *Potalia* in the Gentianaceae has been recently accepted in the systems presented by APG (1998), Struwe & Albert in Struwe *et al.* (1994), and Takhtajan (1987, 1997).

Lisianthius was placed by Gilg (1895a) in his Gentianeae-Tachiinae, together with genera now placed in the Chironieae, Exaceae, Helieae, and Saccifolieae. Only *Enicostema*, *Faroa*, and *Neurotheca* in the new subtribe

Faroinae were described at the time of Gilg's classification, and they were all placed in his heterogeneous Gentianeae-Erythraeinae. A close relationship between *Congolanthus, Djaloniella, Enicostema, Faroa, Neurotheca,* and *Oreonesion* was suggested by Raynal (1968). The monotypic genus *Karina* was not described until 1971, but was also considered similar to these African genera (Taylor, 1973). As far as we know, *Urogentias* has not been associated with the genera of Potalieae by earlier botanists.

Generic synopses for Potalieae

Anthocleista R. BR.

The genus *Anthocleista* was described by R. Brown (1818), who adopted the unpublished name inscribed by Afzelius on material collected in Sierra Leone. In the only revision of this genus, Leeuwenberg (1961) accepted 14 species. These are distributed in Madagascar (3 spp.), the Comores (1 sp.), and tropical sub-Saharan Africa (11 spp.; see Table 2.7). Flower and fruit size and shape as well as leaf morphology have been used as species-delimiting characters.

The species of *Anthocleista* are either tall, umbellate-branched trees with large (up to 2.5 m long) leaves crowded at the branch apices, or shrubs or lianas with smaller leaves more evenly distributed on the stems (Leeuwenberg, 1961). The leaves are often coriaceous and sometimes have paired spines at their bases. The inflorescences are usually very large, multi-flowered, dichotomously branched cymes with a few flowers open at a time. The flowers are unusual among Gentianaceae in having four decussate sepals and 11–16(–24)-merous corollas with an equal number of stamens (Fig. 2.20A). Colleters are present on the inside of the calyx (L. Struwe, pers. obs.). The filaments are at least partially fused into a staminal ring inside the corolla tube or close to the corolla mouth, and the anthers are linear and erect after anthesis. The floral characteristics are the same as those of the neotropical genus *Potalia*, but the number of corolla lobes is often greater (i.e., 11–16(–24) in *Anthocleista* vs. 8–10 in *Potalia*). The stigmas are capitate and slightly bilobed, and the fruit is a fleshy, green or yellow berry.

When Brown (1818) distinguished paleotropical *Anthocleista* from Aublet's (1775) neotropical *Potalia*, he based his conclusion on a woody versus herbaceous habit and 4- rather than bilocular ovaries. The first trait is simply erroneous, since all species of both genera are woody. The placentation of mature fruits of *Potalia* is axile. The investigated species of *Anthocleista* are not truly 4-celled as described by Brown (1818), but have parietal placentas that become deeply intruded and inrolled in the mature fruits (Leeuwenberg, 1961: fig. 9.10). The latter condition is a common feature in Gentianaceae and can function as a precursor state to secondary axile placentation (e.g., Lindsey, 1940; Struwe *et al.*, 1997; Thiv *et al.*, 1999b). Sectioning and scanning electron microscopy of ovaries in various developmental states of *Potalia* will show if the ovary is indeed primarily axile, or, as in most other gentians, initially parietal with later postgenital fusion of placentas and carpel edges. Rightly cautious in accepting the placentation difference, Bentham (1857: 75) stated that "had the genus *Anthocleista* not been already established, I should certainly have considered it as a second species of *Potalia*".

Some morphological phylogenetic data are available for *Anthocleista* and the closely related *Potalia*. The three Malagasy *Anthocleista* species are monophyletic in all most-parsimonious trees presented in Struwe and Albert (1997, and unpubl.) based on violet (rather than creamy or white) corollas and sterile apical appendages on the anthers. Additional features supporting this geographically distinct group include small flower size (comparable to *Potalia*) and corolla lobes broadly overlapping in bud (Leeuwenberg, 1961). Placed most basal in the above-mentioned study and sister to the arborescent species in *Anthocleista* was a group of lianous species, i.e., *A. laxiflora*, *A. obanensis*, and *A. scandens*. A derivation of *Potalia* within *Anthocleista*, sister to the Malagasy species, was the result in Struwe and Albert's (1997, and unpubl.) morphological cladistic analysis; however, this placement is not supported by the molecular trees presented in this chapter, which retain *Anthocleista* and *Potalia* as monophyletic sister taxa (Fig. 2.21). Additionally, molecular data suggest that the Malagasy species (*Anthocleista amplexicaulis*) are sister to all African *Anthocleista* species. Within the African clade there are two subclades, one composed of shrubby to liana-like, small-leaved species (e.g., *A. scandens*) and one composed of the tree-like, long-leaved species (e.g., *A. grandiflora* and *A. schweinfurthii*). In contrast, morphological data had suggested the small-leaved African species to be most basal in the genus (Struwe & Albert, 1997, and unpubl.).

Congolanthus A. RAYNAL

This monotypic genus was segregated from *Neurotheca* by Raynal (1968: 56) to make *Neurotheca* more homogeneous. *Congolanthus* occurs in disturbed areas, savannas, and swamp forest margins at up to 1600 m elevation in tropical Africa from Nigeria in the west to Tanzania in the east.

The only species, *Congolanthus longidens*, is an annual herb with linear to lanceolate leaves (Raynal, 1968; Boutique, 1972). The 4-merous flowers

Systematics, character evolution, and biogeography   197

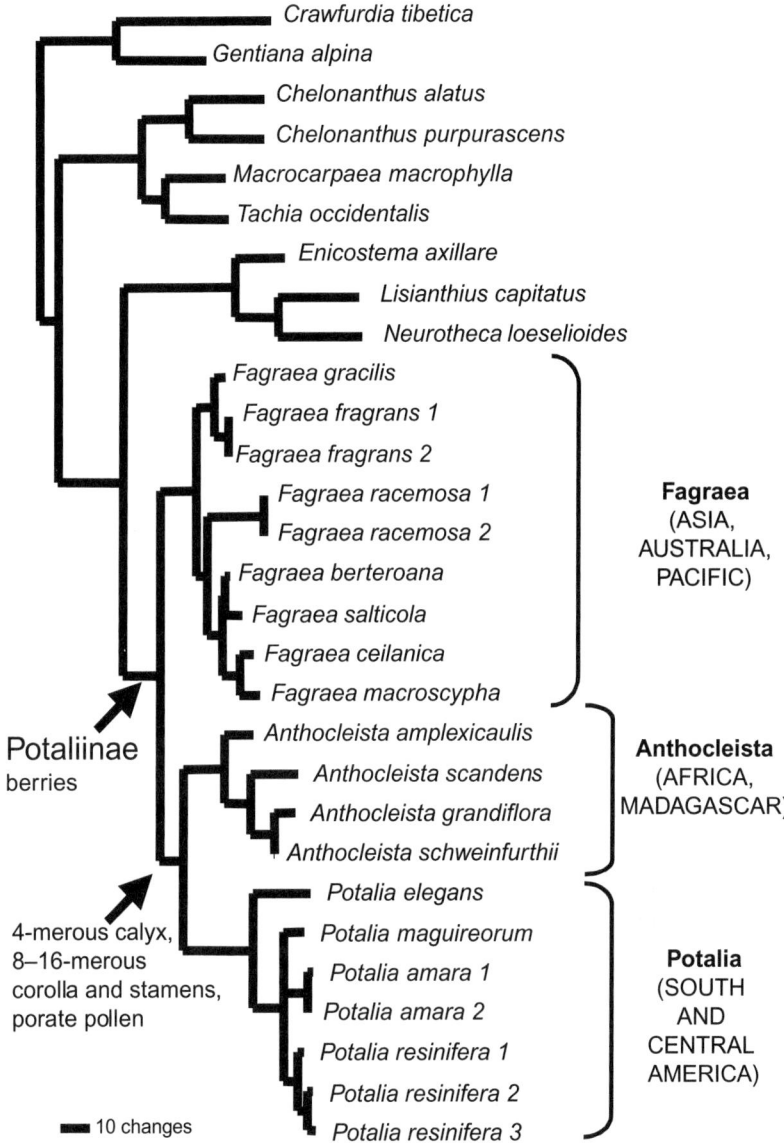

Figure 2.21. Phylogenetic relationships in Potalieae. Single most-parsimonious tree from a preliminary analysis of Gentianeae, Helieae, and Potalieae ITS sequences (shown as phylogram, scale bar = 10 changes, 706 steps, $C=0.70$, $R=0.75$; L. Struwe & V. A. Albert, unpubl.).

are axillary and solitary in the upper leaf axils. The calyces are divided to below the middle, and are thin with prominent veins. The lilac to mauve corollas are tubular and have slightly acute lobes. Inserted in the corolla tubes are the stamens, which are not equipped with staminal appendages. The anthers are oblong. The ovary is sessile, without a basal disk, unilocular, and has protruding placentas. On top of the ovary is a slim, rather long style with a slightly bilobed stigma. The fruits are apically dehiscent, dry capsules with coriaceous valves.

The prominent venation of the calyx of *Congolanthus* is very similar to the calyx morphology seen in *Neurotheca*, but differs in having no transverse veins on the calyx lobes (Raynal, 1968: figs. 2, 4). Furthermore, the genus differs from *Djaloniella*, *Faroa*, and *Oreonesion* in having solitary flowers in the leaf axils instead of multi-flowered fascicles (Raynal, 1968).

Djaloniella TAYLOR

Djaloniella was described by Taylor (1963) based on *Djaloniella ypsilostyla*, an endemic species from wet areas of the high-elevation Diaguissa Plateau in Guinea (Africa). This species was originally described as *Swertia caerulea* (Taylor, 1963).

The only species of *Djaloniella* is a very small annual herb with quadrangular, narrowly winged stems (Taylor, 1963, 1973). The acuminate to acute leaves are lanceolate to linear-lanceolate. The 4-merous flowers are borne in leafy, terminal, and congested dichotomous cymes. The calyces are very small, with linear-lanceolate sepals. The small, blue corollas have narrow and acuminate corolla lobes that are shorter than the corolla tube. There is a fringe of hairs in the corolla throat. The stamens are inserted in the corolla lobe sinuses. The filaments have fimbriate scales between their bases and the anthers are exserted from the corolla mouth. The stigma is bilobed.

Enicostema BLUME

This genus is comprised of three tropical species, *Enicostema verticillatum*, an endemic of northern South America, Central America, and the Caribbean, *E. axillare* (with three subspecies) from tropical Africa and Asia, and *E. elizabethae*, which is endemic to Madagascar (distribution map in Raynal, 1969). *Enicostema* was described by Blume (1826: 848) based on the species *E. littorale* (= *E. axillare* ssp. *littorale*) from Southeast Asia. A synopsis of *Enicostema* was published by Veldkamp (1968), and a full revision was presented by Raynal (1969). Both of these treatments provide a history of the complex taxonomy and nomenclature of *Enicostema*. The habitat preferences of *Enicostema* range from very wet to very dry, from savannas, sandy beaches, to open forests, and to radically different levels of salinity and humidity (Raynal, 1969).

The species of *Enicostema* are annual or perennial herbs with sessile and rather narrow leaves of various shapes (Raynal, 1969). The flowers are 5-merous (rarely 3-, 4-, or 6-merous), sessile, and borne in dense clusters or complex cymes in the leaf axils. The narrow, campanulate calyces are divided about halfway down, seldom more, are thin with white, hyaline margins, and are persistent in fruit (Fig. 2.20B). Colleters are present on the inside of the calyx (Lindsey, 1940). The small, white corollas are tubular to narrowly funnel-shaped. The stamens are inserted in the corolla tubes and have an unusual double hood-shaped appendix at the base of each filament, as well as small appendages between these. The anthers are erect after anthesis and have sterile apical appendages. The stigmas are capitate and slightly bilobed. The fruit is an obovoid capsule.

Fagraea THUNB.

This spectacularly flowered genus currently contains c. 70 species. In Leenhouts' (1962) treatment for *Flora Malesiana*, 31 species were accepted, but some of the polymorphic species accepted by Leenhouts are better considered as distinct species. An additional 20 species were recently described from Borneo by Wong and Sugau (1996). These authors accepted several species that were treated by Leenhouts as synonyms to polymorphic, widespread taxa such as *Fagraea ceilanica, F. fragrans,* and *F. racemosa*. Species circumscriptions have always been a problem in *Fagraea*, and further revisionary studies are needed. *Fagraea* was originally described from Sri Lanka by Thunberg (1782), based on the type species *F. ceilanica. Fagraea* is distributed from Sri Lanka and India in the west to the Pacific in the east, through the tropical areas of Southeast Asia, Malesia (where it is most species-rich, especially on Borneo and New Guinea), the Northern Territory and Queensland states of Australia, New Caledonia, and also on many island groups in Micronesia, Melanesia, and Polynesia (Table 2.7; Leenhouts, 1962; van Balgooy & Leenhouts, 1966).

Fagraea contains mainly terrestrial trees, large or small, but also shrubs and lianas, and some epiphytic shrubs and trees (Leenhouts, 1962). The leaves are often quite coriaceous and fleshy, and many species have well-developed auricles at their bases. The often very large and fragrant, 5-merous flowers are terminal (rarely axillary), and are borne either solitarily, in few- or many-flowered cymes, or seldom in umbels or nodding racemes. The calyces are usually very thick and woody. The corollas are white to cream-colored, tubular, funnel-shaped, or salver-shaped (Fig. 2.20C–E). One of the largest flowers in the world, up to 30 cm across, occurs in *Fagraea auriculata* (Leenhouts, 1962). The stamens are free and inserted in the corolla tube, and sometimes have filaments that are knee-bent at the insertion point. The *Fagraea berteroana* species group has linear anthers and a fleshy ring below the filaments, but the rest of the genus has shorter, oblong anthers and no ring (Struwe & Albert, 1997). The stigmas are capitate to broadly bilamellate, and the fruits are fleshy berries of many different colors (e.g., red, orange, green, white, and blue).

The generic and sectional nomenclature of *Fagraea* is quite complex. Leenhouts (1962) recognized three sections in *Fagraea*: section *Cyrtophyllum* (3 spp. in his classification), section *Racemosa* (1 sp.), and section *Fagraea* (27 spp.). Section *Cyrtophyllum* is based on the genus *Cyrtophyllum* described by Blume (1826), and also includes the type for the genus *Picrophloeus*, also described by Blume (1826). Section *Cyrtophyllum* is characterized by smaller flowers and berries that are often borne in large, thin-branched, multi-flowered inflorescences. Section *Racemosa* was treated as the separate genus *Utania* by Don (1837–1838), based on the illegitimate genus *Kuhlia* described by Blume (it is a homonym of a Kunth name described earlier). Both of these names were based on *Kuhlia morindaefolia*, which is now considered a synonym of *Fagraea racemosa*. *Fagraea racemosa* now represents a species complex that shares unique inflorescence and pollen morphology within the genus (Leenhouts, 1962; Punt, 1978).

A morphology-based cladistic study of *Anthocleista*, *Fagraea*, and *Potalia* showed *Fagraea* to be a paraphyletic grade toward *Anthocleista* (also paraphyletic) and *Potalia* (Struwe & Albert, 1997; L. Struwe & V. A. Albert, unpubl.). However, the rather limited molecular data that are so far available consistently show *Fagraea*, *Anthocleista*, and *Potalia* as three monophyletic genera, with *Fagraea* sister to the latter two genera (cf. Figs. 2.3 and 2.21; Struwe & Albert, 2000b). Further studies combining morphological and molecular data are needed to elucidate relationships within *Fagraea*. At present, the molecular evidence does not contradict sections *Cyrtophyllum* (*sensu lato*), *Racemosa*, or *Fagraea*.

Faroa WELW.

This exclusively tropical African genus has 19 species (Taylor, 1973; Bamps, 1982, 1987). A few of the species are widespread in western or southern tropical Africa, but most are narrow endemics in the Katanga–Zambia "copper belt" region (Taylor, 1973). A revision of *Faroa* was published by Taylor (1973), and an updated key including additionally described species was presented by Bamps (1987). The genus was described by Welwitsch (1869: 45) based on material of the type species *Faroa salutaris* from Angola. Among the vegetation and substrate types in which *Faroa* species occur are crevices on inselbergs, rocky areas, sandstone strata, sandy soils, deciduous woodlands, dry or wet grasslands, riverbanks, and marshes, where they are distributed from low elevations up to 2100 m.

The species of *Faroa* are small, annual or perennial herbs (Taylor, 1973). The leaves are linear to ovate and often connate. The inflorescences are axillary fascicles. The flowers are 4-merous. The campanulate calyces are fused at the base, the lobes of which often have a thickened dorsal nerve or prominent keel. The corolla is salver-shaped or nearly so (and only 1 mm long in *Faroa minutiflora*), and may be light blue, mauve, or cream-colored. The stamens are inserted in the sinuses of the corolla lobes, have ovate anthers, and have a smooth or minutely papillose scale below each filament. The ovary is shortly stipitate, unilocular with slightly intruded placentas, and the styles are often very long with capitate or bilobed stigmas. The fruit is a capsule with a relatively thick and sometimes fleshy wall.

Karina BOUTIQUE

The monotypic genus *Karina* was described from the Congo (then Zaïre, central Africa) by Boutique (1971: 261). It is known only from lateritic *Vellozia* steppes in the Haut–Katanga District (Boutique, 1972).

The only species, *Karina tayloriana*, is a small, annual herb with linear, acute leaves and sessile, 4-merous flowers in terminal spikes (Boutique, 1971, 1972). The calyces are divided nearly to their bases, and the calyx lobes are dorsally keeled. The white to light blue corollas are salver-shaped, with stamens inserted in the upper part of the corolla tubes. Below each filament is a hood-shaped appendix. The anthers are oblong and erect after anthesis. The gynoecium is slightly stipitate, lacking a disk at the base, and is unilocular; the style is short, with a small, bilobed stigma. The fruits are apically dehiscent, dry capsules.

Karina has not yet been treated phylogenetically, but the morphological characteristics of this genus clearly indicate a close relationship with *Enicostema* and related genera in the Faroinae. The presence of a hood-shaped appendix below each filament insertion in the corolla tube, axillary flowers, and linear leaves are common characteristics for *Karina* and most species in this group. The seeds of *Karina* are also very similar to the seeds of *Neurotheca* (Boutique, 1971). Taylor (1973) argued for close relationships among *Karina, Djaloniella, Faroa,* and *Oreonesion. Karina* differs from the latter three genera in having solitary flowers as opposed to flowers borne in fascicles or cymes.

Lisianthius P. BROWNE

This Caribbean and Central American genus was described from Jamaica by Browne (1756: 157) and has been typified with *Lisianthius longifolius*. *Lisianthius* now consists of 30 species (Weaver, 1972; Sytsma, 1988). Most of the species are restricted to specific islands of the Greater Antilles or have narrow distributions in Central America, but a few are more widespread. Their habitats range from pine forests, cliffs, roadsides, limestone thickets, and savannas to rain and cloud forests, where they occur from sea level up to 2100 m (Weaver, 1972). A monograph of *Lisianthius* was published by Weaver (1972), with an earlier revision by Perkins (1902). *Lisianthius* was divided into two sections (sects. *Omphalostigma* and *Lisianthius*), and the latter of these was further divided into two subsections and three series (Weaver, 1972).

Lisianthius consists of perennial herbs, shrubs, small trees, or rarely annual herbs (Weaver, 1972). The leaves are ovate, lanceolate, to elliptic, and a sheath is present around the stem. The inflorescences are either terminal or axillary, and the flowers are borne solitarily, in compound to reduced dichasial cymes, rarely appearing umbellate or capitate. The 5-merous flowers are usually nodding or horizontal. The calyces are campanulate to tubular and are divided down to varying lengths; they are often carinate or alate and usually have long-acuminate lobes (Fig. 2.20F–G). Colleters are present on the inside of the calyx. The corollas are yellow, sometimes red, black, orange and/or green, tubular, funnel-shaped, or less commonly salver-shaped, frequently persistent around the fruit, and most species have conspicuous fiber bundles (seen as white lines in dry corollas; Woodson, 1938). The stamens are most often inserted close to the base in the corolla tube and the filaments are thickened at the base and are knee-bent at the insertion point; however, no appendages are present in the corolla tube below the filaments. The anthers are oblong and often have small, sterile apical appendages. The ovary is unilocular or secondarily bilocular with protruding placentas, and is tipped by a long style with a capitate, peltate, or slightly bilobed stigma. The fruits are dry, apically dehiscent capsules with a whitish, thin, and dry placenta remnant on the inside of each valve. The seeds are small, angular, and numerous. Reported chromosome numbers are $2n = 36$ (nine species investigated; Weaver, 1969).

The name *Lisianthius* has sometimes been used in a broader generic concept including parts of the Helieae (see discussion in Struwe & Albert,

1998a, and above). It has also been confused with the probably illegitimate name *Lisyanthus* (Aublet, 1775), which represents taxa in *Chelonanthus* (Struwe & Albert, 1998a). In the present circumscription of *Lisianthius*, the genus forms a morphologically and, as far as we can tell from current data, molecularly homogeneous unit that is not related to Helieae, but should be included in the Potalieae.

Lisianthius is unresolved toward the rest of the Potalieae in the combined data analysis (Fig. 2.3), is supported as sister to Faroinae in the *mat*K analysis (Fig. 2.2), but is resolved without jackknife support in the *trn*L tree (Fig. 2.1). We have therefore decided to classify *Lisianthius* in a subtribe of its own, Lisianthiinae, since no particular sister-group relationship to either Potaliinae or Faroinae is strongly supported at this time.

Many interesting biogeographic and character evolutionary questions can be studied in *Lisianthius*, and morphological and molecular phylogenetic work at the species level is in progress (J. De Laet, unpubl.). For example, the relationships of flower morphologies associated with different pollination syndromes can be evaluated since moth-pollinated and presumably hummingbird-pollinated flowers both occur (cf. Weaver, 1972). Sytsma and Schaal (1985) showed in a population study based on restriction enzyme analysis that narrowly distributed highland species of *Lisianthius* in Panama had evolved at least two times from a more ancestral (paraphyletic) and widespread lowland species.

Neurotheca SALISB. EX BENTH. IN BENTH. & HOOK.F.

The genus *Neurotheca* was originally described under the genus name *Octopleura* by Progel (1865), using *Octopleura loeselioides* from South America as the type species. However, Grisebach had used the same genus name five years earlier for another taxon. Progel's generic name was therefore a homonym and had to be replaced. This was done by Bentham and Hooker (Bentham, 1876: 812) when they described *Neurotheca*; however, the combination *Neurotheca loeselioides* was not made until 10 years later for the type species (Baillon, 1888–1891).

This genus currently consists of three species, the annual *Neurotheca loeselioides* from northern South America, tropical Africa, and western Madagascar, and the perennial species *N. congolana* and *N. corymbosa* from tropical Africa (Table 2.7; Raynal, 1968). It is unclear if the morphologically variable and widespread *Neurotheca loeselioides* species complex might be better treated as several species, or, as is currently done, as subspecies (for discussion see Raynal, 1968). The species of *Neurotheca* grow in swamps, rocky places, and wet savannas from low to high elevation

(Paiva & Nogueira, 1990; Struwe et al., 1999). Raynal (1968) published a revisionary treatment, and in the same publication one former species of *Neurotheca* was made the type of the new genus *Congolanthus*.

The species of *Neurotheca* are low, annual or perennial herbs with linear or nearly linear leaves, sometimes in a basal rosette (Raynal, 1968; Paiva & Nogueira, 1990). The flowers are 4-merous and are borne in a lax or congested raceme (usually with a terminal flower), or rarely appearing 1–3 together in the leaf axils along the stem. The calyx has 8–12 prominent longitudinal veins as well as prominent transverse veins on the outsides of the calyx lobes, which gave the genus its Latin name (Fig. 2.20H). Colleters are present in the inside of the calyx (Lindsey, 1940). The corolla is tubular, and the corolla lobes have acute apices. The stamens are inserted in the corolla tube. There are no specialized appendages below the filaments; however, prominent nerves connect their bases (Raynal, 1968: fig. 1). The oblong anthers are erect after anthesis. The ovary is unilocular and has intrusive placentas. The stigma is slightly bilobed with small lobes (sometimes reported as capitate), and the fruit is an apically dehiscent capsule enclosed by a persistent calyx.

Oreonesion A. RAYNAL

Oreonesion, described by Raynal (1965: 271), is a rare, monotypic African genus known only from inselbergs in Gabon.

The only species, *Oreonesion testui*, is an annual herb with sessile, lanceolate leaves (Raynal, 1965). The 4-merous flowers are subsessile and are borne in multi-flowered axillary fascicles (congested cymes) in the upper leaf bases. The hyaline-margined calyces are shallowly divided and have a low keel on the outside of each lobe. The white corollas are salver-shaped with acute corolla lobes and are persistent in fruit. The stamens are inserted inside the upper part of the corolla tube, with a staminal appendage below each filament. The ovary, which has prominent veins on the outside, is sessile, without a disk at the base, unilocular, and has slightly intruded placentas. The style is long and tipped with a small bilobed stigma. The fruit is an apically dehiscent, dry capsule.

The position of *Oreonesion* in the Faroinae is clear based on its subfilamental appendages, linear-lanceolate leaves, and axillary flowers. In the first description of the genus, Raynal (1965) suggested a close relationship between *Enicostema* and *Oreonesion*. *Enicostema* differs from *Oreonesion* by having capitate instead of bilobed stigmas (Raynal, 1965). *Oreonesion* further differs from *Faroa* and *Djaloniella* by having stamens inserted in the corolla tube instead of in the corolla lobe sinuses (Taylor, 1973).

Potalia AUBL.

The genus *Potalia* is distributed in tropical America, from Bolivia in the south to Costa Rica in the north, and has nine species, two of these recently described (*P. elegans* and *P. maguireorum*), one recently resurrected (*P. resinifera*), and five as-yet undescribed by us. The genus and its type species, *Potalia amara*, were described by Aublet (1775: 394) based on material from French Guiana. Until recently, *Potalia* has usually been considered a monotypic genus. Most of the species accepted in the unpublished monograph by Struwe and Albert have rather narrow distributions, with the exception of *Potalia resinifera*, which is widespread in the Amazon Basin and Andean foothills. The species of *Potalia* differ prominently in leaf, inflorescence, and fruit characteristics, and the breadth of the morphological variation can be described as equivalent to that observed in *Anthocleista* (Struwe & Albert, 1998b; Struwe et al., 1999; cf. Leeuwenberg, 1961). The most polymorphic species, *Potalia resinifera*, is also the most widespread one (Struwe et al., 1999).

The species of *Potalia* are usually monopodial, unbranched shrubs or small trees with leaves crowded at the stem apex. The leaves are often very large, narrowly elliptic to obovate, with interpetiolar sheaths at the leaf bases. The terminal inflorescences are up to 50-flowered dichasial cymes, with conspicuously bright yellow or orange peduncles, pedicels, and calyces. The flowers are erect, with a 4-merous calyx and an 8–10-merous corolla. The coriaceous calyces are divided to the base, have decussate lobes, and are persistent in fruit. The white (to cream), yellow, orange, or green corollas are thick and fleshy, tubular with a narrower lower part and a wider upper part, and deciduous in fruit. The 8–10 stamens are inserted in the middle of the corolla tube with very short filaments that are completely fused into a staminal tube terminated by linear anthers. The pollen is released as monads (except *Potalia maguireorum*, which has tetrads; Nilsson, 2002) and has smooth and porate exine. Around the base of the ovary is a glandular disk, and the upper part of the ovary is fleshy and sterile. The style is short with a capitate stigma. The fruit is a fleshy, green berry, varying in shape between the different species from globose, turbinate, ovoid, to obovoid. The fruit wall often has a thickened horizontal ring (e.g., *Potalia resinifera*). The seeds are elliptic and flattened, but not winged.

The closest relative to *Potalia* is *Anthocleista* from tropical Africa and Madagascar. These two genera are supported as sister groups in the *mat*K, combined *mat*K/*trn*L intron, and ITS analyses (Figs. 2.2, 2.3, 2.21; Struwe

& Albert, 2000b). There are also several strong structural synapomorphies for a monophyletic group consisting of *Anthocleista* and *Potalia*, e.g., a 4-merous decussate calyx, supermerous corollas and stamens, completely or partially fused filaments, linear anthers, porate pollen, and large leaves crowded at the branch apices (the last character is not present in the lianous species of *Anthocleista*). Concerning relationships within *Potalia*, ITS data resolve *P. elegans* as sister to the rest of the genus, the exemplars of which follow current species boundaries (Fig. 2.21). *Potalia elegans* is a white-sand endemic from the western edge of the Guayana Shield in Venezuela; morphologically, this species and an undescribed entity from Colombia/Peru most closely resemble arborescent species of *Anthocleista* (L. Struwe & V. A. Albert, unpubl.). This suggests that the widespread *Potalia resinifera* and similar, undescribed species are evolutionarily derived, and that the current distribution of *P. elegans* is relictual, possibly even dating from Gondwanic continuity between South America, Africa, and Madagascar.

Pycnosphaera GILG

Pycnosphaera, a monotypic genus, described by Gilg (1903: 333), has a wide distribution in tropical Africa (Table 2.7). It is unique among Gentianaceae in having 3-merous flowers. *Pycnosphaera* grows in swamp forests and marshes, and on riverbanks (Paiva & Nogueira, 1990). No revision has been published for the genus, but the only species was treated in the floristic works of Boutique (1972) and Paiva and Nogueira (1990).

Pycnosphaera buchananii is an up to 1.5 m tall herb with narrow, oblong-lanceolate to oblong-elliptic leaves at the base and linear cauline leaves (Gilg, 1903; Boutique, 1972; Paiva & Nogueira, 1990). The terminal inflorescences are compact bracteate heads (sometimes in corymbs) with 3-merous, subsessile flowers. The calyx has unequally sized lobes that are carinate. The pink, mauve, blue-violet, to white corollas are salver-shaped with acute lobes. The three stamens are inserted in the sinuses of the corolla lobes with a small appendix at the base of each stamen; the anthers are oblong and minutely apiculate. The ovary is unilocular, with slightly intruding placentas, and with a short style terminated by a stigma with two long, thin lobes. The fruits are small, dry, apically dehiscent capsules.

Gilg (1903) indicated that *Pycnosphaera* was closely related to *Enicostema* and *Faroa*. One species of *Faroa*, *F. buchananii*, turned out to be conspecific with Gilg's (1903) type species of the genus, *Pycnosphaera trimera*, and so it was transferred into *Pycnosphaera*. Our molecular studies confirm the relationships suggested by Gilg (Figs. 2.1–2.3).

Urogentias GILG & GILG-BEN.

This spectacular genus is endemic to the Uluguru Mountains in Tanzania (Gilg & Gilg, 1933: 944), part of the Eastern Arc mountain chain in East Africa. The genus contains a single species, *Urogentias ulugurensis*, which at the time of the description of the genus was abundant at high elevations (c. 1800 m above sea level). No revision has been published of *Urogentias*, and not much is known about this taxon.

The habit of *Urogentias* is herbaceous, subligneous at the base, with oblong-lanceolate leaves and flowers borne in multi-flowered inflorescences in the upper leaf axils (Gilg & Gilg, 1933). The 8-merous flowers have calyces with nearly free, long-acuminate lobes. The corollas are thin, salver-shaped, and bright blue, with likewise long-acuminate lobes. The stamens are inserted in the corolla tube seemingly without appendages at the base of each filament, and the anthers are free. The sessile ovary is unilocular with protruding placentas, and is tipped by a style with a bilobed stigma. The fruit is a dehiscent, subligneous capsule with two-winged seeds.

Gilg and Gilg (1933) wrote that *Urogentias*, with its subligneous habit and unusual 8-merous flowers, was hard to place in the Gentianaceae classification of Gilg (1895a), showing no obvious relationship to any other specific gentian genus. However, these authors did indicate that one possibility could be a relationship with the Mediterranean genus *Chlora* (= *Blackstonia*), which can also have 8-merous flowers, but which differs in comprising small herbs with terminal, few-flowered cymes. The proposed relationship with *Blackstonia* has no support in recent phylogenetic analyses. Instead, *Urogentias* is shown to be a part of the Faroinae in the molecular studies presented here (Figs. 2.2, 2.3) and by Struwe et al. (1998; Fig. 2.1). Apart from the differences in floral merosity and seed wings, the morphological characters of *Urogentias* do not strongly disagree with a close relationship to the other genera of Faroinae. It is interesting to note that supermerous corollas occur also in subtribe Potaliinae, with *Anthocleista* being 11–16(–24)-merous and *Potalia* 8–10-merous.

Phylogeny and character evolution in Potalieae

The three subtribes of the Potalieae are placed in a trichotomy in the analysis based on combined *trn*L intron/*mat*K data (Fig. 2.3), and other, faster-evolving genes as well as combined analyses with morphology and anatomy will be necessary to resolve these relationships.

Within the subtribe Faroinae, *Enicostema* and *Urogentias* form the sister group to the remainder of the sequenced genera in the subtribe, among which relationships are completely resolved only in the *mat*K analysis (Fig. 2.2). Several morphological character complexes in the Faroinae were discussed and illustrated by Raynal (1968), who presented a phenetic study of all genera of the subtribe except *Karina, Pycnosphaera,* and *Urogentias.* Most similar in his study were *Djaloniella* and *Faroa,* and *Djaloniella* also showed many similarities to *Neurotheca* and *Oreonesion* (Figs. 2.1–2.3).

Molecular data from several sources show *Fagraea* to be sister to *Anthocleista* plus *Potalia* in the subtribe Potaliinae (Fig. 2.21). The morphological study of Potaliinae presented by Struwe and Albert (1997, 2000b; L. Struwe & V. A. Albert, unpubl.) gave resolution within the genera *Fagraea* and *Anthocleista,* but with very few characters supporting internal branches. The morphology-based positioning of *Potalia* plus *Anthocleista* as sister to the *Fagraea berteroana* group (Struwe & Albert, 1997) was supported by the presence of similar pollen characteristics and staminal tubes (in *Anthocleista* and *Potalia*) being homologous to the fleshy ring found below the stamens in the *Fagraea berteroana* group, characters that could also be interpreted as non-homologous parallelisms. Further sequencing and combining of molecular and morphological data are needed to study species-level relationships within and among these genera.

Woodiness

The genera of Potaliinae are invariably woody plants as are several species of *Lisianthius* in subtribe Lisianthiinae, and much has been said about the ancestral versus derived status of this feature among Gentianaceae. On various and sometimes conflicting grounds, Gilg (1895a), Lindsey (1940), Gopal Krishna and Puri (1962), and Wood and Weaver (1982) each advocated or implied the woody condition to be derived, although none of these authors included Potalieae in their considerations. Wood and Weaver (1982: 444–445) emphasized supposed evolutionary trends in floral glands (nectaries), considering "a well-developed glandular disc at the base of the ovary ... a specialized condition in the family". Species of "Lisiantheae" were those intended (e.g., *Lisianthius, Macrocarpaea,* and *Symbolanthus,* of tribe Helieae and Potalieae), but a nectary disk also characterizes *Potalia* and some species of *Anthocleista,* as well as several herbaceous genera in tribe Gentianeae. Additional characteristics used to argue for a derived and "natural" "Lisianthieae" were tendencies toward pollen aggregation (not commonly found in Potalieae, however), complex fusion of floral vascular bundles, and high chromosome numbers. Contrary to Wood and Weaver's

Systematics, character evolution, and biogeography 209

(1982) final assertion, not all woody gentians other than South African *Orphium* are neotropical. Aside from *Anthocleista* and *Fagraea* in Potaliinae, the Malagasy genus *Tachiadenus* (Grisebach, 1839; Gilg, 1895a) in the Exaceae contains annual species with woody bases as well as apparently perennial subshrubs (Klackenberg, 1987a, 1990). The subshrub *Gentianothamnus* is also a Malagasy member of Exaceae (Klackenberg, 1990; cf. Fosberg & Sachet, 1980). The recent placement of woody *Saccifolium* in Gentianaceae (Maguire & Pires, 1978; Thiv *et al.*, 1999a) further adds to the list of woody gentian taxa. Most genera of the Faroinae are herbaceous annuals, but perennial species that are subligneous at the base occur as well. The plesiomorphic habit in Potalieae might be woody, considering that apart from the subligneous genus *Urogentias* in subtribe Faroinae the subtribe Potaliinae is uniformly woody (Leeuwenberg & Leenhouts, 1980), and *Lisianthius* species (Lisianthiinae) are perennial, subligneous herbs (very rarely annual herbs) to shrubs and small trees (Weaver, 1972).

We now know from molecular phylogenetic studies that cases of stem and branch woodiness are spread out among all major clades of Gentianaceae (except Gentianeae), but that the ancestral trait appears to be a perennial, suffrutescent habit when this character is optimized onto the DNA-based cladograms (cf. Figs. 2.1–2.3). Mennega (1980) investigated the wood anatomy of *Anthocleista*, *Fagraea*, and *Potalia*, and concluded that they differed from woody taxa of Loganiaceae (with which they were included at the time) but that they were not similar to other investigated Gentianaceae (see Jansen & Smets, 1998, for a similar viewpoint). As it appears that arborescence has evolved in parallel several times among the gentians, it is not unlikely that wood anatomy from distantly related groups within the Gentianaceae might be different. Very few gentians have been investigated for wood anatomy (only a few taxa from each clade; e.g., Mennega, 1980; Carlquist, 1984; Jansen & Smets, 1998) and further, more comprehensive studies could illuminate the evolution of secondary growth in the family when taking new phylogenetic relationships into account.

True trees are found only in two groups of gentians, viz. the neotropical tribe Helieae (in *Chorisepalum* spp., *Macrocarpaea* spp., *Symbolanthus* spp., and *Zonanthus*; Grisebach, 1862; Ewan, 1948a; Maguire, 1981; Maguire & Boom, 1989; Pringle, 1995; Struwe *et al.*, 1999) and the Potalieae (*Anthocleista*, *Fagraea*, *Lisianthius*, and *Potalia*; Leenhouts, 1962; Weaver, 1972; Leeuwenberg & Leenhouts, 1980; Struwe & Albert, 1997).

Vegetative structures

Auricles at the base of the petioles are found in several species of *Fagraea* (Leenhouts, 1962). These are elaborated to various degrees from small, flat auricles to large, reflexed, bulbous-shaped structures (Leenhouts, 1962: fig. 2). Sometimes ants live in the reflexed auricles and feed on extrafloral nectaries (Leenhouts, 1962). These auricles are not stipules but appendages of the leaf lamina. Winged petioles are a common phenomenon in Potalieae and many other gentians, but auricles are uncommon.

Interpetiolar lines are generally present in all gentians (as well as in Apocynaceae and many Loganiaceae; Leeuwenberg & Leenhouts, 1980; Rosatti, 1989). More developed structures taking the shape of low, interpetiolar sheaths between the bases of the leaves or around the stem are found in *Fagraea*, *Lisianthius*, *Potalia*, and several Helieae (*Macrocarpaea* spp., *Tachia*, and *Zonanthus*; Leenhouts, 1962; Weaver, 1972; Struwe & Albert, 1998b; Struwe *et al.*, 1999). In *Fagraea*, the sheath often splits into two "axillary scales" during development (Leenhouts, 1962).

Inflorescence

Several types of inflorescence structure are found in Potalieae. Flowers or inflorescences in axillary positions are most common in subtribe Faroinae but are not found in Potaliinae. Weaver (1972: figs. 1–2) presented a hypothesis for the evolution of different inflorescence types in *Lisianthius*. He hypothesized evolutionary sequences from an ancestral, dichasial, terminal cyme (e.g., *Lisianthius* sect. *Omphalostigma*) either to axillary flowers (e.g., *L. axillaris*), umbels, or cymes, or to monochasially branched terminal cymes (e.g., *L. brevidentatus*) or reduced cymes (e.g., *L. longifolius*).

In *Karina*, a member of the subtribe Faroinae, the flowers are axillary in a terminal spike (Taylor, 1973) and *Pycnosphaera* also has a terminal inflorescence (Gilg, 1903). *Faroa*, *Oreonesion*, and *Urogentias*, however, have fascicles or dichasia in their leaf axils (Gilg & Gilg, 1933; Taylor, 1973). The fascicles of *Faroa* and *Oreonesion* are unusual in the Gentianaceae, but similar structures are found in *Swertia* spp. and *Gentiana lutea* (Taylor, 1973). *Djaloniella* has dichotomous cymes (Taylor, 1973), an inflorescence type that is very common in the family. Terminal and nearly perfect dichotomous cymes, albeit sometimes reduced, are also found in *Anthocleista*, *Potalia*, and most *Fagraea* species (Leeuwenberg, 1961; Leenhouts, 1962). Exceptions in *Fagraea* are the presence of axillary, umbellate or thyrsoid inflorescences in *F. fragrans* and *F. umbelliflora*, hanging and racemose inflorescences in *F. racemosa* and associated species, and solitary or few-flowered inflorescences in, for example, *F. auriculata* and *F. involucrata*.

Floral merosity

The Potalieae exhibit a general trend of relaxation in floral merosity. The other tribes of Gentianaceae generally have 4- to 5-merous flowers, but in this tribe the number varies the most, from 3 in *Pycnosphaera* up to 16(–24) corolline and staminal parts in *Anthocleista*. *Potalia* and *Anthocleista* are unique in having supermerous corollas and stamens that differ in number (8–16-merous, once reported as 24-merous) from the four calyx lobes (Struwe, 1999). Their closest relative, *Fagraea*, has 5-merous calyces, corollas, and androecia (Leenhouts, 1962). The increase is probably based on 4 (to 8, 12, 16, etc.) in *Anthocleista* and *Potalia*, which is supported by the initial formation of four corolla lobes, one each in the corners of a square-shaped floral primordium in *Anthocleista* (Struwe, 1999; L. Struwe, unpubl.). Additional corolla primordia later initiate medially on each side of the square, and primordia doubling is observed to an extent necessary to account for the final number of lobes seen in mature flowers. *Urogentias*, which is also 8-merous, differs in also having an 8-merous calyx, and is similar in this respect to the supermerous genera *Blackstonia* and *Sabatia* spp. of Chironieae, which also have calyx lobes isomerous to the corolla lobes.

Floral anatomy

Lisianthius is one of the few genera in Gentianaceae that has lignified metaxylary fibers in their corollas (Woodson, 1938; Lindsey, 1940), but these are not consistently present in all species in the genus (Weaver, 1972). The fibers are often seen with the naked eye as white lines in dried corollas, and they have also been found in *Enicostema*, *Faroa*, and *Neurotheca*, as well as in the following Chironieae genera: *Bisgoeppertia*, *Canscora*, *Cicendia*, *Coutoubea*, *Centaurium*, *Schinziella*, and *Zygostigma* (Woodson, 1938; Lindsey, 1940). They have not been found in any Helieae genera investigated (Woodson, 1938). These fibers are present not only in the corollas but also in the pedicel, calyx, staminal filaments, and rarely the bracts. The only difference in the morphology of these fibers between *Lisianthius brevidentatus* and *Coutoubea spicata* that was reported by Woodson (1938) is that the corolla fibers are longer in the former species.

Lisianthius differs from Helieae in the floral anatomy of the calyx. In Helieae the calyx laterals are initially fused with the calyx midvein and begin with one trace per lobe. This contrasts with the case found in *Lisianthius longifolius*, in which there is simultaneously one midrib plus two side traces per calyx lobe (van Heusden, 1986). *Lisianthius* also differs in the larger size of the calyx traces compared with the remaining vasculature of the corolla, stamens, and gynoecium. In general, taxa of Helieae have large calyx traces but smaller corolla traces. However, in *Lisianthius* and *Coutoubea* (and *Schultesia*), the corolla is heavily vascularized, whereas the calyx is much less so (Lindsey, 1940; van Heusden, 1986).

Floral anatomy of many species of *Lisianthius* has been investigated, primarily to study the corolline fibers and placentation patterns (Perrot, 1897; Lindsey, 1940; Weaver, 1972). Lindsey (1940) also investigated *Enicostema* and *Neurotheca*. The ovaries of *Lisianthius* can be either unilocular or bilocular, or both, and a surrounding glandular area is found at their bases (Weaver, 1972). One similarity between *Lisianthius*, *Enicostema*, and *Neurotheca* is the presence of fused calyx laterals (Lindsey, 1940). The floral anatomy of the Potaliinae and other Faroinae has not yet been thoroughly investigated.

Calyx

Calyx morphology is very variable in Potalieae, especially as regards merosity, texture, and calyx lobe shape. Dorsal keels are often prominent in *Faroa* spp. and *Karina* (Taylor, 1973), *Pycnosphaera* (Gilg, 1903), as well as in several *Lisianthius* spp. (Weaver, 1972). Several *Faroa* species do not have a dorsal keel but have instead a thickened midvein at the lobe apex (Taylor, 1973). The calyx lobes as well as the corolla lobes of *Urogentias* are unique in the tribe in being extremely long-acuminate. Subtribe Potaliinae never exhibits keeled or acuminate calyx lobes, and the calyces are often thick and leathery with calyx lobes of different sizes. Unequal calyx lobe size is also found in *Karina* (Taylor, 1973). The 4-merous calyces of *Anthocleista* and *Potalia* have free lobes, whereas those of *Fagraea* are 5-merous and campanulate to tubular (Leeuwenberg, 1961; Leenhouts, 1962). *Karina* and *Pycnosphaera* have nearly free calyx lobes as well, but their calyces are isomerous with the corollas and are thinner in texture (Gilg, 1903; Taylor, 1973).

Corolla

The color of the corolla varies greatly within Potalieae. White to cream-colored corollas are found in *Oreonesion* and *Anthocleista* spp. from continental Africa, as well as in *Fagraea* and *Enicostema* (Leeuwenberg, 1961; Leenhouts, 1962; Raynal, 1965). The Malagasy species of *Anthocleista* have purplish corollas (Leeuwenberg, 1961), whereas *Potalia* has white, cream, orange, yellow, and/or green corollas (Struwe *et al.*, 1999). *Lisianthius* has a range of corolla colors, from red, orange, yellow, to green or rarely black, sometimes with several colors present (Weaver, 1972). Blue, mauve, or lilac corollas are characteristic for *Congolanthus, Djaloniella, Faroa, Karina* (sometimes white), *Neurotheca*, and *Pycnosphaera* (both rarely pink or white; e.g., Raynal, 1968; Boutique, 1972). Corolla lobe shape is also a variable character, from the elliptic lobes with rounded apices in many *Fagraea* spp. to the long-acuminate lobes of *Urogentias*. The corollas are persistent after fruit maturity in *Lisianthius* and several Faroinae, but fall off before fruit maturity in the Potaliinae (Weaver, 1972; Leeuwenberg & Leenhouts, 1980).

Androecium

Staminal characters are abundant in the Potalieae and their variation identifies several groups. The stamens are inserted in the corolla tube, close to the corolla, or in the corolla lobe sinuses. The latter case is found in *Djaloniella, Faroa*, and *Pycnosphaera* (Gilg, 1903; Taylor, 1973), whereas the closely related genera *Enicostema, Congolanthus, Karina, Neurotheca, Oreonesion*, and *Urogentias* have their stamens inserted in the corolla tube (sometimes rather close to the sinuses; Gilg & Gilg, 1933; Raynal, 1968; Taylor, 1973). Stamens that are knee-bent at the insertion point in the tube are found in *Lisianthius* and *Fagraea* (Fig. 2.20E–F). *Anthocleista* and *Potalia* have stamens inserted close to the corolla lobe sinuses, with the filaments completely or partially fused into a staminal tube (Fig. 2.20A; Leeuwenberg, 1961).

There is a strong tendency toward the presence of staminal appendages at or directly below the insertion points of the stamens in the corolla tube. These can be present as semilunate papillose scales (*Faroa*), fimbriate scales (*Djaloniella*), hooded scales (*Enicostema* (Fig. 2.20B) and *Karina*), simple scales (*Neurotheca*; Fig. 2.20H), swelling at the bases of the filaments (*Lisianthius* (Fig. 2.20F–G) and *Oreonesion*), or papillate or glandular areas (*Pycnosphaera* and the *Fagraea berteroana* group (Fig. 2.20C); Gilg, 1903; Gilg & Gilg, 1933; Raynal, 1968; Taylor, 1973; Struwe & Albert, 1997). *Karina* and *Urogentias* appear to lack any special structures at the base of their stamens (Gilg & Gilg, 1933; Taylor, 1973).

Palynology of Potalieae: Synopses of genera and general discussion
S. NILSSON

Anthocleista. Anthocleista amplexicaulis Baker. The pollen grains are 3(–4)-porate, often with slightly protruding pore margins, oblate spheroidal, and

26×29 μm. The exine is smooth to scabrate. (Madagascar: 1994, *Pettersson 610* (UPS), slide no. 24160.)

OTHER INVESTIGATIONS: Punt and Leenhouts (1967) and Punt (1978, 1980) described the pollen of eight species of *Anthocleista*, which were included in the *Potalia* (later *Anthocleista nobilis*)-type. The pollen grains were said to be 3(–5)-porate, perforate, and with protruding pores with distinct annuli. *Anthocleista microphylla* agrees palynologically with the other species investigated in the genus (Nilsson, 2002: Fig. 4.31). Erdtman (1952) reported delicately reticulate pollen in *Anthocleista parviflora* (not seen by the present author), which he compared to *Potalia amara*.

Congolanthus. *Congolanthus longidens* (N. E. Br.) A. Raynal. The pollen grains are 3-colporate, oblate spheroidal, and *c.* 34×37 μm. The exine is finely reticulate. (Congo: 1936, *Louis 1885* (S), slide no. 24168; Nilsson, 2002: Fig. 4.32.)

Fagraea. *Fagraea berteroana* Gray. The pollen grains are 3-colporate to porate, oblate spheroidal, and *c.* 25×29 μm. The exine is finely reticulate. (USA: Hawaii, 1995, *Motley s.n.* (NY), slide no. 24161; Nilsson, 2002: Fig. 4.33A–C.)

Fagraea elliptica Roxb. The pollen grains are 3-colporate, oblate spheroidal to subprolate, and *c.* 20×18 μm. The exine is striato-reticulate and often distinctly striate over the polar areas. (Indonesia: Ceram, 1990, *Burley et al. 4335* (NY), slide no. 24178; Nilsson, 2002: Fig. 4.34D–F.)

Fagraea racemosa Jack. The pollen grains are 3-colporate to porate, prolate spheroidal, and 29×27 μm. The exine is striate with markedly sinuous muri. (Indonesia: Sumatra, 1932, *Rahmat Si Toroes 3193* (NY), slide no. 24179; Nilsson, 2002: Fig. 4.35A–C.)

OTHER INVESTIGATIONS: Pollen grains from almost 30 species of *Fagraea* were studied by Punt and Leenhouts (1967) and Punt (1978, 1980). The genus contains a number of pollen types, and a scheme of relationships of *Fagraea* pollen was presented by Punt (1978: fig. 1, p. 322). It was suggested that *Fagraea berteroana* (*F. berteroana*-type: pollen 3-porate, finely reticulate) was a link to *F. gardenioides* ssp. *borneensis* (*F. gardenioides*-type: pollen 3-porate, tectate, perforate). The latter type was considered to be connected with *Anthocleista* and *Potalia*. However, the pollen of *Fagraea berteroana* is 3-colporate rather than porate, and is finely reticulate, not smooth as in *Potalia* (Nilsson, 2002: Fig. 4.38). Erdtman

(1952) described pollen of two *Fagraea* species – *F. imperialis* and *F. morindaefolia* (= *F. racemosa*) – as 3-porate and reticulate and 3-colporate and striato-reticulate, respectively. Guinet (1962) described and illustrated pollen grains of *Fagraea obovata* (= *F. ceilanica*) as 3-porate and coarsely reticulate. Rao and Lee (1970) gave a short description of *Fagraea fragrans* as being 3-colpate, relatively large, and coarsely reticulate, whereas Punt and Leenhouts (1967) found the same species to have 3-colporate, relatively small, reticulate to finely reticulate and striato-reticulate pollen grains.

Lisianthius. Lisianthius cordifolius L. The pollen grains are 3-colporate, subprolate, and c. 36 × 30 μm. The exine is striato-reticulate to reticulate-finely reticulate. (Jamaica: St. Andrew, 1954, *Webster & Wilson 4860* (BM), slide no. 24380; Nilsson, 2002: Fig. 4.36.)

OTHER INVESTIGATIONS: Both Gilg (1895a) and Köhler (1905) described the pollen grains of *Lisianthius* as 3-colporate and reticulate to finely reticulate. Nilsson (1970) described c. 25 species and varieties of *Lisianthius*, all with a reticulate to finely reticulate exine pattern. Studies of four species from Panama by Elias and Robyns (1975) confirmed previous results. *Lisianthius cordifolius* pollen was described and illustrated by Agababyan and Tumanyan (1977a).

Neurotheca. Neurotheca loeselioides Oliver. The pollen grains are 3-colporate, prolate, and c. 51 × 39 μm. The exine is reticulate. (Burundi: Bururi, 1980, *Reekmans 9114* (S), slide no. 24156; Nilsson, 2002: Fig. 4.37A–D.)

OTHER INVESTIGATIONS: The pollen of both *Neurotheca congolana* and *N. corymbosa* is similar to that of *N. loeselioides* (Nilsson, 2002).

Potalia. Potalia amara Aubl. The pollen grains are 3-porate, with markedly protruding pore margins compared with pollen of *Anthocleista*, subprolate, and c. 39 × 32 μm. The exine is smooth to scabrate. (Brazil: Amapa, 1983, *Mori et al. 15785* (NY), slide no. 24162; Nilsson, 2002: Fig. 4.38A–B.)

OTHER INVESTIGATIONS: *Potalia amara* was studied by Punt and Leenhouts (1967) and Punt (1978, 1980). Erdtman (1952) pointed out the similarity between *Potalia amara* and *Anthocleista parviflora*. Nilsson (2002) investigated the additional species *Potalia* "coronata" (ined., L. Struwe & V. A. Albert, unpubl.), *P. elegans*, *P. maguireorum*, *P. resinifera*, and *P.* "turbinata" (ined., L. Struwe & V. A. Albert, unpubl.). *Potalia maguireorum* has pollen arranged in loose tetrads, but otherwise agrees closely with the other species of *Potalia* (Nilsson, 2002: Fig. 4.39).

Discussion of palynological data in Potalieae. *Congolanthus*, *Neurotheca*, *Lisianthius*, and *Fagraea* all have pollen grains that are 3-colporate and striato-reticulate to reticulate or finely reticulate. However, the genera are distinguishable from each other using minor exine features. *Fagraea* is the most heterogeneous genus, with a number of recognized pollen subtypes.

Anthocleista and *Potalia* have pollen grains that are similar *inter se* but totally different from those of all other gentianaceous taxa (Nilsson, 2002). Smooth and 3(–4)-porate pollen grains of this type are not found in the rest of the Gentianaceae, but more significantly are found in the Loganiaceae, e.g., *Geniostoma* (Endress *et al.*, 1996), and in genera of the Apocynoideae (Apocynaceae; e.g., *Baissea* and *Forsteronia*; Nilsson *et al.*, 1993). Acceptance of the molecular and morphological results presented in this chapter and elsewhere (e.g., Struwe *et al.*, 1994, 1998; Thiv *et al.*, 1999a; Backlund *et al.*, 2000) would indicate that this pollen type has arisen in Gentianales several times in parallel.

Gynoecium and fruit

The ovaries of Potalieae show considerable variation in development and placentation. Postgenital fusion of ovaries is commonly present throughout all clades in the family, but *Anthocleista* and *Fagraea* (as well as *Aripuana* in Helieae) apparently have congenitally fused ovaries (Struwe, 1999; L. Struwe & P. Endress, unpubl.). The gynoecial ontogeny of *Enicostema axillare* ssp. *littorale* (as *"Enicostemma" littorale*) has been investigated by Padmanabhan *et al.* (1978), who described the development of the postgenitally fused carpels in detail.

Locularity and placentation have been investigated by several authors. The presence of conflicting reports regarding uni- or bilocularity as well as axile versus parietal placentation is the result of interpretation problems associated with the highly plastic development of these traits in gentians (e.g., Lindsey, 1940; Struwe *et al.*, 1997). For example, in *Potalia* and its related genera *Anthocleista* and *Fagraea*, views on locularity and placentation have differed strongly among various authors. Mature *Potalia* fruits appear to have simple, axile placentation (a rarity among gentians) as opposed to the parietal and inrolled placentas of *Anthocleista* and *Fagraea*, but this requires confirmation (Struwe, 1999). The fruits of *Lisianthius* never have inrolled placentas, and are furthermore characterized by their white, erose margins formed from the old placentas of dehisced fruits (Weaver, 1972).

Most Potalieae have small, filiform or capitate stigmas that are sometimes slightly bilobed. Exceptions to this are the broadly bilamellate stigmas of the *Fagraea berteroana* species group, *Karina*, and *Urogentias* (Gilg & Gilg, 1933; Leenhouts, 1962; Taylor, 1973; Struwe & Albert, 1997), which are similar to the typical stigmas of Helieae and therefore possibly a derived trait in the Potalieae (Fig. 2.205C).

The fleshy to leathery indehiscent fruits with seeds embedded in fleshy pulp found in Potaliinae are unusual among Gentianaceae (Leenhouts, 1962; Leeuwenberg & Leenhouts, 1980; Struwe & Albert, 1997). Fleshy fruits are also known from, for example, *Chironia baccata* and *Tripterospermum* spp. (Chironieae and Gentianeae, respectively). In contrast, Faroinae and Lisianthiinae have dry capsules that dehisce apically to medially. *Potalia* fruits have unique and variable sterile apical parts, and therefore fruit morphology is a highly diagnostic trait at the species level (Struwe & Albert, 1998b).

Pollination syndromes and dispersal

Many species in the Potalieae have spectacular flowers, and a variety of pollination syndromes are discernible. The relatively open, mostly fleshy, and white to cream-colored flowers of African *Anthocleista* spp. are possibly bat pollinated, whereas the rather closed flowers of *Potalia elegans* have been seen being visited by hummingbirds (Struwe & Albert, 1998b). In *Fagraea*, several pollination types are found, e.g., smaller flowers visited by butterflies (e.g., *F. fragrans*), larger and more open flowers visited by birds or bats (e.g., *F. racemosa*; Leenhouts, 1962; Endress, 1994: 141, 209). Protandrous flowers have been reported from *Fagraea* and *Lisianthius* (Leenhouts, 1962; Weaver, 1972). The flowers of *Fagraea* open initially at night and last for about two days (Leenhouts, 1962). Also, in *Lisianthius* several flower types are found, i.e., presumably hummingbird-pollinated flowers (e.g., *L. aurantiacus*), and flowers visited by butterflies or by pollen- or nectar-gathering bees (Weaver, 1972). In *Lisianthius*, pollinator specialization has probably been a major evolutionary force leading to the development of diverse floral morphologies (Weaver, 1972).

The fleshy fruits of the Potaliinae are probably mostly animal dispersed (by birds, bats, flying foxes, etc.). Bradbury (1977, 1984) reported that hammerheaded flying foxes (*Hypsignathus*) in Gabon primarily ate *Anthocleista* fruits. Ants can apparently also aid seed dispersal in *Fagraea* (Leenhouts, 1962).

Uses

Humans have used Potalieae species extensively. *Anthocleista*, *Fagraea*, and *Potalia* have all been commonly used for pharmacological purposes, mainly against fevers, fungal diseases, and inflammations. The vernacular names of these plants often reflect their native uses as well, e.g., pao de cobra (= "snake stick") for *Potalia*, and fever tree for *Anthocleista*. *Potalia* is widely used in the Neotropics against snake bites, fevers, and inflammations (Bisset, 1980a,b; Schultes & Raffauf, 1990). The highest concentration of bioactive secoiridoids of any gentian species has been found in *Potalia amara* (S. R. Jensen, pers. comm.). Decoctions of *Anthocleista* leaves have been used against malaria, and chewing of the bark is said to help against diarrhea (Coates Palgrave, 1991). *Fagraea gigantea* can be up to 55 m in height, and is used for timber production in Malesia (Leenhouts, 1962).

The neotropical species *Enicostema verticillatum*, called quinine bush in the Caribbean, is used against malaria, dysentery, and fevers (Morton, 1981). It has also been used in Cuba as a substitute for gentian root, an herbal derived from the European species *Gentiana lutea* (Morton, 1981). The African–Asian species *Enicostema axillare* is sold at markets in India, and is commonly used as a stomachic and carminative (Morton, 1981). *Faroa salutaris* is extremely bitter, and has been used against digestive disorders in Angola (Welwitsch, 1869; cf. Taylor, 1973: 81).

Character diagnosis

A few potential synapomorphies can be distinguished for the subtribes of the Potalieae. The Faroinae are characterized by a generally herbaceous habit, usually small plant size, appendages of various morphology between the bases of the filaments (absent in a few genera; see above), and usually 3–4-merous flowers (although usually 5-merous in *Enicostema* and 8-merous in *Urogentias*).

Potential synapomorphies for the subtribe Potaliinae are usually a tree-like habit, interpetiolar sheaths, often coriaceous and large-sized leaves, terminal cymose inflorescences, fleshy calyx and corolla (thin in *Fagraea* sect. *Cyrtophyllum*), and fleshy berries. *Anthocleista* and *Fagraea* also have strongly inrolled placentas compared with *Potalia*, which has bilocular ovaries with simple, possibly axile placentas. The sister-group relationship of *Anthocleista* and *Potalia* is supported, however, by the presence of 4-merous calyces, supermerous flowers (generally 8–16 corolla lobes), linear anthers (also present in the *Fagraea berteroana* group), and porate pollen (Struwe & Albert, 1997; Nilsson, 2002). *Potalia* is distinguished from

Anthocleista based on the presence of 8–10-merous flowers versus 11–16-merous flowers, mostly green, orange, or yellow corollas versus white, cream, or lilac-colored corollas, and the presence of a sterile apex on the gynoecium and fruit (absent in *P. maguireorum* and *P. elegans*; Struwe *et al.*, 1999).

The genus *Lisianthius*, which is also the only member of subtribe Lisianthiinae, has characteristics that fit with both the Faroinae and Potaliinae. Similarities with the Potaliinae include a mostly woody habit, 5-merous flowers, and the absence of distinct appendages at the base of the anthers. Some taxa of *Lisianthius*, however, have membranous leaves and a more herbaceous habit (e.g., *L. laxiflorus*), or axillary flowers (e.g., *L. axillaris*), all traits that are more characteristic of the Faroinae than the Potaliinae. *Lisianthius* also has thin calyces and corollas and capsular, dry fruits, all traits that are not consistent with the Potaliinae. Therefore, no morphological support for a particular sister-group relationship for the Lisianthiinae can be specified.

Biogeography of Potalieae

The Potalieae is one of the two pantropical tribes in Gentianaceae, the other being Chironieae. Chironieae is primarily centered in subtropical, Mediterranean, and nearly temperate areas, while Potalieae is strictly tropical and distributed around the Equator. The subtribe Lisianthiinae and its sole genus *Lisianthius* is the only taxon present in the northern part of the Neotropics, whereas subtribes Potaliinae and Faroinae are distributed in both the New and Old World tropics. Repeated patterns of distribution occur in Potaliinae and Faroinae, suggesting a common vicariance event as an explanation. For example, Potaliinae has one African–Malagasy genus (*Anthocleista*), one American genus (*Potalia*), and one Australasian–Pacific one (*Fagraea*). In the Faroinae, most genera are endemic to the African continent, but both *Enicostema* and *Neurotheca* bridge the Atlantic and occur also in the Neotropics as well as on Madagascar (Klackenberg, 1990; Struwe *et al.*, 1999). In addition to this, *Enicostema* also occurs in Asia (Raynal, 1969).

Transatlantic disjuncts and pantropical patterns

Of the four gentian genera that occur on both sides of the Atlantic, two belong to the Potalieae, i.e., *Enicostema* and *Neurotheca*. These two genera are also rather basally positioned in their subclade and each is sister to African taxa (cf. Fig. 2.3). The other two transatlantic genera are *Schultesia*

(Chironieae) and *Voyria* (*incertae sedis*), both of which are primarily neotropical with single species also present in Africa. For *Schultesia*, long-distance dispersal is a plausible theory, considering that the same species is present in both eastern South America and West Africa and that *Schultesia* belongs to an otherwise completely neotropical subtribe, the Coutoubeinae. For *Voyria primuloides*, however, an endemic of West Africa, it might be possible that this species represents a remnant of a formerly larger subclade rendered disjunct by the breakup of the boreotropical continuum (Albert & Struwe, 1997).

The boreotropics was a northern, subtropical to tropical area that has been restricted to smaller refugia in today's tropical areas as a result of climatological changes that occurred at the Oligocene/Eocene boundary during the Tertiary, c. 38 million years ago (Wolfe, 1975; Tiffney, 1985a,b; Lavin & Luckow, 1993). Specifically, cooling restricted the distribution of tropical taxa, also causing extinction and dispersal, and typical boreotropical distributions can be found in taxa disjunct between Southeast Asia and Central America plus southern North America (e.g., *Gelsemium sempervirens* and *G. elegans* in Gelsemiaceae; Struwe & Albert, 1997). Following this scenario, northern taxa should be more basally placed than their more derived southern relatives in the phylogenies of particular groups (i.e., South American taxa derived from Central American, Caribbean, or southern North American taxa). Despite the lack of truly northern extant taxa in the Faroinae, Potalieae, and *Voyria*, this hypothesis should still be seriously considered to explain their biogeographic patterns (see below).

Enicostema and *Neurotheca* have species in the Caribbean and/or the Guayana Shield, tropical Africa, and Madagascar. Additionally, *Enicostema verticillatum* is endemic to the Guayana Shield, Central America, and the Caribbean, *E. elisabethae* occurs only on Madagascar, and the three subspecies of *E. axillare* are disjunct with one subspecies present only in Malesia, one endemic to East Africa, and the last disjunct between tropical Africa and India (including Sri Lanka and possibly also Thailand; Raynal, 1969; Struwe *et al.*, 1999). This distribution pattern in *Enicostema* almost completely overlaps with the Potaliinae (*Anthocleista*, *Fagraea*, and *Potalia*), the only difference being that the taxa with this distribution in the latter subtribe are genera, not species or subspecies as in *Enicostema* (Struwe, 1999).

The repeated pattern found in *Enicostema*/*Neurotheca* and Potaliinae could be explained by (1) the Boreotropics Hypothesis, or (2) the much earlier breakup of the Gondwana supercontinent (Struwe, 1999). Struwe and Albert (1997) suggested a Gondwanic explanation for the distribution patterns seen in *Anthocleista*, *Fagraea*, and *Potalia* based on a morpholog-

Systematics, character evolution, and biogeography 221

ical cladistic study, which was the hypothesis most compatible with the area relationships found. However, recent phylogenetic results based on molecular data show this clade to be deeply embedded inside Gentianaceae but with different internal phylogenetic relationships (Figs. 2.3, 2.21; Struwe & Albert, 2000b), results that were not known at the time of the morphological analysis. Divergence rates based on *rbc*L sequence variation and area relationships did not contradict either hypothesis (Struwe & Albert, 1997), but considering the placement of these taxa rather far inside the family the Boreotropics Hypothesis could provide a more plausible explanation for the distribution patterns seen in *Anthocleista/Fagraea/Potalia* and *Enicostema/Neurotheca*. However, Exacaeae are only a few nodes away on our molecular phylogenetic trees, and Klackenberg (this chapter, and 1985) has made a strong argument for a Gondwanan origin of biogeographic patterns in that tribe. At the very least, the cladogenesis leading to the three subtribes must be minimally of Tertiary age, since *Lisianthius* pollen is known from the Eocene in Panama (Graham, 1984).

Gentianaceae as a family could be much older than previously thought (see also Struwe & Albert, 1997, for a possible Gondwanic derivation of Strychnaceae (or Loganiaceae-Strychneae), a closely related group). However, one cannot discount the possibility of transatlantic dispersal after the Gondwanic breakup when sea levels were lowered, for example in the Oligocene. This hypothesis has been proposed to explain the age and derivation of New World monkeys from African ancestors (Aiello, 1993).

In conclusion, the causes behind the distributional patterns of Potalieae are still incompletely known, and heretical Gondwanic vicariant processes cannot yet be ruled out.

South and Central American patterns

Potalia is distributed in wet, tropical areas of South America east of the Andes, and also occurs in Chocó, Colombia, and the westernmost areas of Costa Rica and Panama in Central America. A morphological analysis by Struwe and Albert (unpubl.) of the taxa of *Potalia* shows that the Central American species *P.* "turbinata", ined., and taxa endemic to the Chocó, eastern Guayana Shield, and Amazonia are derived relative to white-sand taxa native to the western Guayana Shield. This is in agreement with the hypothesis presented by Struwe *et al.* (1997) that taxa endemic to white-sand areas represent not primarily ecologically specialized and divergent species, but rather relictual representatives of basally positioned evolutionary lineages. For further discussion see Struwe (1999) and under Helieae, within which similar distributional patterns have been found.

TRIBE GENTIANEAE

K. B. VON HAGEN AND J. W. KADEREIT

Tribe Gentianeae in our circumscription comprises c. 939–968 species in 17 genera and two subtribes. Subtribe Gentianinae contains *Crawfurdia*, *Gentiana*, and *Tripterospermum*, and subtribe Swertiinae includes *Bartonia*, *Comastoma*, *Frasera*, *Gentianella*, *Gentianopsis*, *Halenia*, *Jaeschkea*, *Latouchea*, *Lomatogonium*, *Megacodon*, *Obolaria*, *Pterygocalyx*, *Swertia*, and *Veratrilla*. With *Gentiana*, *Gentianella*, *Swertia*, and *Halenia*, four of the five largest genera of the Gentianaceae are part of Gentianeae (see Albert & Struwe, 2002). With few exceptions (see below) the generic composition of Gentianeae has been more or less the same for most authors. However, the delimitation of genera and segregate genera within tribe Gentianeae has been, and in part still is, very problematic. Examples of this are *Gentiana*, *Gentianella*, and *Swertia* and their closest relatives. Whereas the circumscription of *Gentiana* has recently been clarified (Smith, 1965; Yuan et al., 1996), and we here offer a possible solution for *Gentianella*, the circumscription of *Swertia* for the time being remains a problem because it is clearly polyphyletic (see below).

Most species of Gentianeae grow in alpine or temperate habitats. Except for subtropical and tropical lowlands, the tribe is of worldwide distribution. Whereas some genera (e.g., *Gentianella* and *Swertia*) occupy a subcosmopolitan range, others (e.g., *Latouchea*) are monotypic and restricted to small areas.

Description of Gentianeae

All taxa included in Gentianeae are annual, biennial, or perennial herbs. Most genera are autotrophic with mycorrhizae, but *Bartonia* and *Obolaria* are mycotrophic. The plants are usually glabrous. Sometimes they have glandular hairs in the leaf axils, and rarely also an indumentum of short stiff hairs. The stems are erect, decumbent or twining, terete, quadrangular or winged, sometimes hollow. Plants can be from 5 cm (*Gentiana*) to 3 m tall (*Frasera*); they are usually branched or rarely scapose. The leaves are often arranged in a distinct basal rosette; they are entire and opposite, or rarely whorled or alternate. Leaves can be sessile to long-petiolate, with a usually ovate, elliptical, or almost linear blade. In the mycotrophic genera leaves are scale-like. Leaf bases are decurrent or not and sometimes connate. The bracts are usually foliose. Flowers are arranged in dichasia or in few- to many-flowered axillary or terminal racemes or panicles, although sometimes the flowers are borne singly. Flowers can be almost sessile or pedicellate; they are 4- or 5-merous, hermaphroditic or rarely unisexual, and if so, then the plants are dioecious (*Gentianella* spp. and *Veratrilla*). The calyx tube is shortly lobed to divided to the base, sometimes split to the base on one side only (*Gentiana* spp. and *Gentianella* spp.) or consists of two free sepals only (*Obolaria*). The calyx lobes are ovate to filiform, obtuse, acute, or cuspidate, sometimes unequal, or rarely asymmetrical (*Frasera*). An intracalycine membrane is present in some genera at the base of the calyx lobes. The often showy corolla can be tubular, rotate, infundibuliform, hypocrateriform, or campanulate. It may be shallowly to deeply lobed. The corolla lobes are mostly entire but are sometimes ciliate or denticulate (*Gentianopsis* and *Pterygocalyx*). The color of the corolla varies and can be violet, blue, green, red, yellow or white. The aestivation is usually convolute or sometimes imbricate (*Bartonia* and *Obolaria*). Plicae, viz. small extra lobes of varying shape between the corolla lobes, can be absent or present. If present, they are symmetrical or asymmetrical and ciliate or not. The corolla throat is usually naked but sometimes has vascularized (*Gentianella* spp.) or non-vascularized (*Comastoma*, *Gentiana*, and *Gentianella* spp.) fimbriae at the base of the corolla lobes. The stamens are inserted at varying heights in the corolla tube or in the sinuses between the corolla lobes. The filaments are filiform, winged or thickened. All anthers are sagittate. They are basifixed and introrse, dorsifixed and introrse or extrorse, and versatile or not. Sometimes the anthers are connate (*Gentiana* spp.). The base of the corolla tube can have one or two (sometimes merging) nectaries per petal lobe, or the nectaries are at the base of the ovary or gynophore. In *Bartonia* and *Pterygocalyx*, flower nectaries are not known. When nectaries are present on the corolla, they are naked or have fimbriae or scale-like appendages, or are naked and spurred (in *Halenia*). Nectaries at the base of the ovary rarely form a collar-like disk (*Tripterospermum*). The ovary has a gynophore or is sessile. It is globose, ellipsoid or narrowly oblong, and is always unilocular with parietal placentation. The placentas sometimes protrude into the ovary (the pseudobilocular condition) or are appressed along the ovary walls. The style is sometimes long and filiform, but more often is very short or absent. The stigma usually consists of two free stigmatic lobes that are sometimes connate (*Gentiana* spp.) or decurrent along the carpel margins (*Bartonia* and *Lomatogonium pro parte*). The fruit is usually a septicidal capsule with two valves, but is rarely a berry (*Tripterospermum* spp.). The seeds are globose, fusiform or flattened. Circumferential or one-sided (semicircular) wings are present or absent. The seed coat is smooth to rugose.

Taxonomic history of Gentianeae

In 1838, Don (1837–1838) placed all Gentianeae genera (in the present circumscription) known to him in subtribe Gentianeae-verae of his tribe Gentianieae. This subtribe also contained (in modern nomenclature) *Blackstonia, Canscora, Deianira, Enicostema, Prepusa, Schultesia, Tachia,* and *Voyria,* which made this group a very heterogeneous assemblage. Through a new emphasis on calyx and ovary characters, this classification was much improved by Grisebach (1839) with the description of tribe Swertieae, which contained all genera of our Gentianeae known at that time and excluded all alien elements of Don's (1837–1838) classification. In 1845, Grisebach reduced this group to subtribal rank without changing its generic composition. Bentham's (1876) tribe Swertieae contained only genera that are also part of our Gentianeae. He divided the tribe into three informal groups on account of nectary position and flower aestivation.

In Huxley's (1888) classification, the basic subdivision of the Gentianaceae was between genera with nectaries at the base of the corolla, i.e., his series Perimelitae, and genera with nectaries on the gynophore or without nectaries, i.e., his large series Mesomelitae. All genera of his Perimelitae are part of our Gentianeae and roughly correspond to our Swertiinae. All other genera of our Gentianeae were placed in two of the subgroups of his series Mesomelitae, the Asteranthe and Ptychanthe. Although in retrospect Huxley's (1888) treatment was accurate in this respect, it had very little impact on later authors.

Gilg (1895a), as the last monographer of the Gentianaceae, included in his subtribe Gentianinae those genera contained in Grisebach's (1839) and Bentham's (1876) Swertieae and our Gentianeae, largely on the basis of pollen morphology. Gilg was incorrect, however, in removing *Bartonia* and *Obolaria* to his Erythraeinae, and in including *Ixanthus*, a member of tribe Chironieae, in his Gentianinae.

Generic synopses for Gentianeae

Bartonia WILLD.

This presumably mycotrophic genus was originally described by Willdenow (1801: 444). The type species is *Bartonia tenella* (= *Sagina virginica*; see Fernald & Weatherby, 1932, for nomenclatural history). Four species are included in the genus, which is distributed in eastern North America from Texas and Florida to Newfoundland (Gillett, 1959; Wood & Weaver, 1982).

The species of this genus are small, erect, annual, presumably mycotrophic herbs (Wood & Weaver, 1982). The stems are slender and quadrangular. The leaves are reduced to small, opposite or alternate scales with glandular hairs (colleters) in the axils. The leaf bases are decurrent. The 4-merous flowers are borne in terminal and axillary pedunculate dichasia, which sometimes are reduced to single flowers. The calyx is divided almost to the base, has lobes in unequal pairs, and lacks an intracalycine membrane. The campanulate corolla is deeply lobed and plicae are absent. The corolla aestivation is imbricate. The stamens are inserted in the sinuses between the corolla lobes. The filaments are short. Nectaries have not been reported from *Bartonia*. The ovary is sessile with an indistinct style. The stigmatic lobes are decurrent along the carpel sutures. The seeds are unwinged. Reported chromosome numbers are $2n=44$ and 52 (Wood & Weaver, 1982).

Comastoma TOYOK.

The species now included in *Comastoma* were once regarded as part of *Gentiana* subg. *Gentianella* sect. *Amarella* (Kusnezow, 1895), or as the separate section *Comastoma* (Wettstein, 1896). This group was raised to generic rank by Toyokuni (1961: 198) with *Comastoma tenellum* (= *Gentiana tenella*) as its type species. Following different authors, the genus contains between 7 and 25 species occurring in north temperate and alpine regions of Asia, Europe, and America (Gillett, 1957; Toyokuni, 1961; Yuan & Küpfer, 1993a; Ho & Pringle, 1995).

Comastoma consists of annual or perennial, branched or unbranched, erect herbs. The stems are slender and quadrangular. The 4- or 5-merous flowers are arranged in few-flowered axillary racemes, or the inflorescence is reduced to a single terminal flower. The calyx is divided almost to the base and an intracalycine membrane is absent. The corolla is tubular, infundibuliform or hypocrateriform, and plicae are absent. The aestivation is convolute. The base of the corolla lobes has a ring of non-vascularized linear fimbriae. The stamens are inserted in the corolla tube. There are two naked nectaries per corolla lobe at the base of the corolla tube. The ovary is sessile and a style is absent. The seeds are unwinged. Reported chromosome numbers are $2n=10$, 12, 16, 18, 28, and 30 (Löve, 1953; Yuan & Küpfer, 1993a).

Crawfurdia WALL.

Crawfurdia as circumscribed here is synonymous with *Crawfurdia* sect. *Dipterospermum* (Clarke, 1875) or *Gentiana* sect. *Dipterospermum* (Marquand,

1937). Two sections, *Protocrawfurdia* and *Crawfurdia*, were established by Smith (1965). Following the pollen morphological characters provided by Nilsson (1967a), sect. *Protocrawfurdia* is more similar to *Gentiana* sect. *Stenogyne* than to sect. *Crawfurdia*. *Tripterospermum* and *Pterygocalyx*, treated separately here, were included in *Crawfurdia* by some authors, e.g., Gilg (1895a). The genus and lectotype *Crawfurdia speciosa* (see Grossheim *et al.*, 1967) were described by Wallich (1826: 63). The genus in its present circumscription contains 16–19 species distributed in Bhutan, China, India, Myanmar, and Sikkim (Smith, 1965; Ho & Pringle, 1995).

The species of *Crawfurdia* are perennial and mostly twining or rarely erect herbs with stems that are usually terete. The leaves are mostly glabrous or have short stiff hairs (sect. *Protocrawfurdia*). The 5-merous flowers are arranged in terminal or axillary cymes. The calyx is tubular, 10-veined, and divided to the middle or less. The calyx lobes are narrow, often filiform, and separated by large sinuses. A small intracalycine membrane is present or rarely absent. The corolla is tubular, infundibuliform or campanulate, and has plicae. The aestivation is convolute. Corolla lobes are short. The stamens are inserted near the middle of the corolla tube. The filaments are winged and usually widen toward the base. Nectaries are found at the base of the ovary. The ovary is sessile or has an elongated gynophore. The style is filiform. The seeds are compressed and have discoid wings.

Frasera WALTER

This genus has been included in *Swertia* by some authors (e.g., St. John, 1941; Pringle, 1990). Its maintenance as a separate genus by us will be discussed below. The type species, *Frasera caroliniensis*, was described by Walter (1788: 87). *Frasera* consists of about 15 species distributed in eastern North America (Ontario to Louisiana) and mountains in western North America (Card, 1931; St. John, 1941; Toyokuni, 1965; Threadgill & Baskin, 1978; Wood & Weaver, 1982; Pringle, 1990).

All species of *Frasera* are biennial or perennial but always monocarpic herbs. The stems are simple, erect, and sometimes hollow. Some plants can be up to 3 m high. The leaves are opposite or whorled, and their bases are weakly connate and not decurrent, as in most species of *Swertia*. The 4-merous flowers form axillary panicles. The calyx is deeply lobed, the lobes form unequal pairs, and an intracalycine membrane is absent. The corolla tube is very short. The stamens are inserted in the sinuses between the corolla lobes. There are mostly one, rarely two (*Frasera speciosa*) nectaries per corolla lobe near the base of the corolla tube. The nectaries are sometimes bilobed. The upper margins of the nectaries are fimbriate and at their base there is usually a small, fimbriate scale. The ovary is sessile, the style filiform, and the stigma bilobed. The seeds are winged. A reported chromosome number is $2n = 78$ (Rork, 1949; Moore, 1973).

Gentiana L.

The lectotype of *Gentiana* is *G. lutea*, which was described by Linnaeus (1753: 227; lectotypification in Britton & Brown, 1913). *Gentiana* was treated very inclusively by Kusnezow (1895) and other authors, who included subgenus *Gentianella* and other taxa that were later segregated. On the other hand, even within a narrowly defined and more natural *Gentiana*, many segregate genera were established or re-established recently by, for example, Löve and Löve (1961, 1975, 1976), Holub (1973), Omer (1989), and Zuev (1984). Most other authors (e.g., Ho & Liu, 1990; Yuan *et al.*, 1996), however, use the same circumscription of *Gentiana* as applied here (see also the phylogenetic discussion of *Crawfurdia* and relatives). Ho and Liu (1990) and Ho and Pringle (1995) recognized 15 sections (see Table 2.8) within *Gentiana*. Recently used names of the very narrow segregate genera and their corresponding sectional names may be found in Yuan *et al.* (1996). The last complete revision of *Gentiana* was made by Kusnezow (1896). Other important treatments of different parts of *Gentiana* as well as nomenclatural issues were provided by Marquand (1937), Pringle (1977, 1978, 1979), Ho and Liu (1990), Ho and Pringle (1995), Ho *et al.* (1996), and Yuan *et al.* (1996). *Gentiana* in its current circumscription comprises about 360 species, with most species in Asia but also several dozen species in Europe, North and South America, northwest Africa, and east Australia.

All species of *Gentiana* are annual, biennial, or perennial herbs. The stem is usually erect, sometimes ascending, simple or branched, and mostly quadrangular. The leaves are opposite and very rarely whorled. The uppermost leaf pair sometimes forms an involucre around the flowers. The 5-merous flowers (rarely 4- to 8-merous) are arranged in simple dichasia, terminal clusters, axillary whorls, or are borne singly. The calyx is divided shallowly to up to the middle, and is sometimes split to the base on one side. The calyx is usually provided with an intracalycine membrane, which is rarely missing. The corolla is of various shapes and can be tubular, infundibuliform, obconical, hypocrateriform, campanulate, urniform, or rarely rotate (*Gentiana lutea*). The corolla tube is usually longer than the corolla lobes, and its throat is rarely provided with fimbriae. Symmetrical or asymmetrical plicae are present (except for *Gentiana lutea*), and sometimes these are fimbriate. The stamens are inserted in the corolla tube. The filaments

Table 2.8. *Sectional classification of* Gentiana *following Ho and Liu (1990) and Ho and Pringle (1995)*

Gentiana section	No. of species	Distribution
Sect. *Gentiana* L.	5	Europe, Turkey
Sect. *Ciminalis* (Adans.) Dum.	7	Europe
Sect. *Calathianae* Froel.	8	Europe, NE America, N and W Asia, NW Africa
Sect. *Chondrophyllae* Bunge	158	Europe, Asia, North and Central America, NW Africa, Australia
Sect. *Cruciata* Gaudin	21	Europe, Asia
Sect. *Kudoa* Masam. = sect. *Monopodiae* (Harry Sm.) T. N. Ho	37	Kashmir, China, E Asia, Malaysia, Indonesia
Sect. *Otophora* Kusn.	12	Himalayas, India, China, Myanmar
Sect. *Isomeria* Kusn.	18	E and NE Asia, Himalayas, NW North America
Sect. *Microsperma* T. N. Ho	10	Nepal, Bhutan, SW China
Sect. *Frigidae* Kusn.	18	Europe, Asia, North America
Sect. *Phyllocalyx* T. N. Ho	1	SW China
Sect. *Dolichocarpa* T. N. Ho	12	Europe, Asia, North, Central, and South America
Sect. *Fimbricorona* T. N. Ho	4	Himalayas, SW China
Sect. *Stenogyne* Franch.	14	Myanmar, China, Thailand

are usually straight, symmetrical, winged at the base, and filiform above. The anthers are sometimes connate. In *Gentiana* sect. *Stenogyne* the stamens are asymmetrical, of unequal length, and their tips are bent downwards. The ovary is sessile or has a gynophore, and there are nectary glands at the base. The style is usually short or absent, but sometimes it is elongated. The stigmatic lobes are often free, sometimes connate. Seeds are winged or unwinged. Reported chromosome numbers are $2n = 12, 14, 16, 18, 20, 22, 24, 26, 28, 30, 32, 34, 36, 38, 40, 42, 44, 46, 48, 52, 60$, and 96 (Favarger, 1949, 1952; Löve, 1953; Yuan & Küpfer, 1993b; Yuan *et al.*, 1998).

Gentianella MOENCH

Gentianella was described by Moench (1794: 482) with *G. tetranda*, *nomen illegitimum*, as its type species. *Gentianella tetranda* is a synonym of *G. campestris*. The generic name *Gentianella* was accepted as a *nomen conservandum*. The description of *Gentianella* given here is based on its circumscription by most authors (e.g., Wood & Weaver, 1982; Ho & Pringle, 1995; Pringle,

1995). Segregate genera such as *Pitygentias* (Gilg, 1916), *Aliopsis* (Omer & Qaiser, 1991), or *Chionogentias* (Adams, 1995) are based on regional material only. The Rafinesque name *Aloitis* was misapplied by Omer *et al.* (1988) to a group of predominantly binectariate species. This segregate genus was resurrected for the uninectariate *Aloitis aurea* by Löve and Löve (1986). The segregation of binectariate *Gentianella* from uninectariate *Gentianella*, and the possible division of uninectariate *Gentianella* into one group with vascularized fimbriae at the base of the corolla lobes and another group with non-vascularized fimbriae (or without them) will be discussed below (see also Fig. 2.22). The sectional classification of (subgenus) *Gentianella* by Kusnezow (1895) is not natural and is therefore not applied here. The highest species diversity of this genus is found in South America. In its current circumscription the genus contains about 250 species in New Zealand, Australia, Asia, Europe, northwest Africa, and North and South America (Litardiére & Maire, 1924; Gillett, 1957; Wood & Weaver, 1982; Adams, 1995; Pringle, 1995).

All species of *Gentianella* are annual, biennial, or perennial herbs; they are decumbent or erect. The stems are usually sparsely branched above and rarely there are many stems from the base. The stems are terete or quadrangular. The 4- or 5-merous flowers form terminal or axillary umbelliform cymes, or are borne singly. The calyx has a very short to well-developed tube, which sometimes is split to the base on one side. An intracalycine membrane is absent. The corolla is infundibuliform, cylindrical, hypocrateriform, rotate, or campanulate. The aestivation is convolute. The corolla is shallowly divided or divided almost to the base. The corolla tube can have vascularized fimbriae, non-vascularized fimbriae, or none at all. When fimbriae are present, they are scattered, united in a ring, or arranged in a row at the base of each corolla lobe. Plicae are absent. The stamens are inserted at or below the middle of the corolla tube. The filaments are filiform and occasionally barbate below. One or two naked nectaries per corolla lobe are present at the base of the corolla tube. The ovary is sessile or has a short gynophore. The style is indistinct or absent. The seeds are usually unwinged or rarely winged (*Gentianella tornezyana*). Reported chromosome numbers are $2n=(18, 26), 36, (48, 54)$ (Löve, 1953; Weaver & Rüdenberg, 1975; Wood & Weaver, 1982).

Gentianopsis MA

Before its segregation from *Gentiana/Gentianella* by Ma (1951), *Gentianopsis* was known as *Gentiana* sect. *Crossopetalum* (Grisebach, 1839) and *Gentiana* sect. *Imaicola* (Grisebach, 1845), or was treated as *Gentianella* subg.

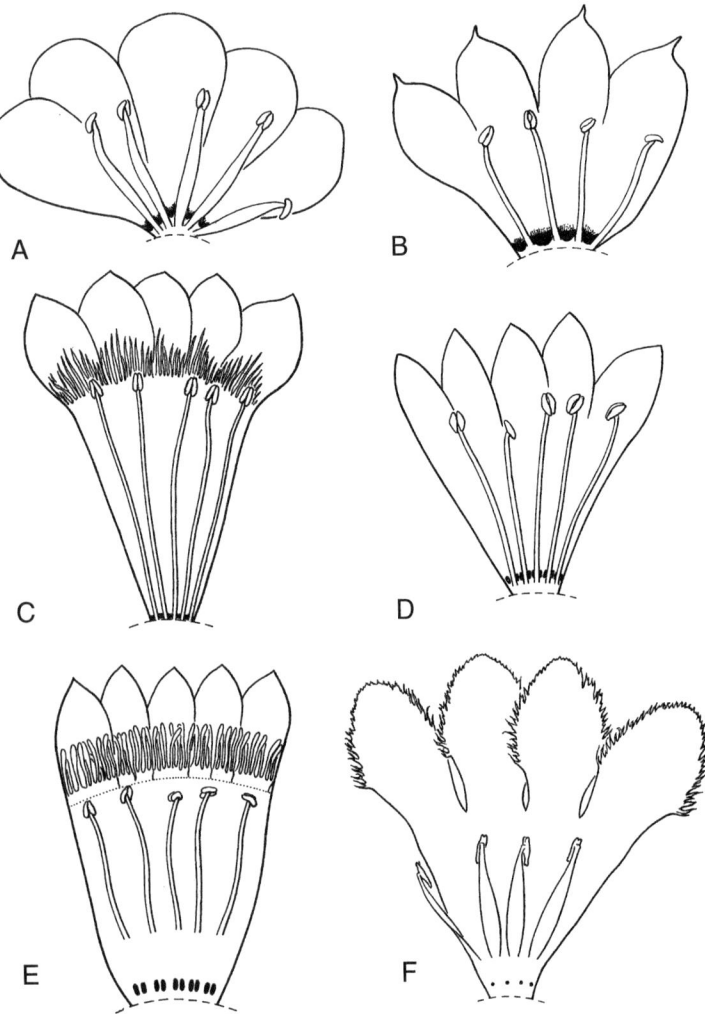

Figure 2.22. Flower types of *Gentianella* and species formerly placed in *Gentianella*. **A.** Uninectariate and efimbriate *Gentianella* from the Southern Hemisphere. **B.** Uninectariate and efimbriate *Gentianella* from the Northern Hemisphere. **C.** Uninectariate and fimbriate *Gentianella* with vascularized fimbriae in the corolla tube. **D.** Binectariate *Gentianella*. **E.** *Comastoma*, with two nectaries and non-vascularized fimbriae in the corolla tube. **F.** *Gentianopsis*, with one nectary and ciliate corolla lobes. According to our phylogeny (Fig. 2.24), *Gentianopsis* is not closely related to the other taxa. Binectariate *Gentianella* and *Comastoma* are related to each other and form a clade with *Lomatogonium*. Uninectariate *Gentianella* is monophyletic and contains two different groups: one with vascularized fimbriae and one with non-vascularized fimbriae (not shown) or without fimbriae.

Eublephis (Gillett, 1957). The type species is *Gentianopsis barbata*. Some 16–24 species are distributed in temperate and boreal Asia, Europe, and North America (Gillett, 1957; Iltis, 1965; Wood & Weaver, 1982).

The species of *Gentianopsis* are annual, biennial, or rarely perennial herbs. The stems are slender and quadrangular. The 4-merous flowers are arranged in few-flowered monochasia or are borne singly. The flower buds are flattened. The calyx is tubular, its lobes have hyaline margins, and there are small, discontinuous intracalycine membranes in the sinuses. Two opposite calyx lobes are usually conduplicate, and they are narrower and longer than the two other lobes. The corolla is campanulate to broadly infundibuliform and divided to the middle. Plicae are absent (but see Philipson, 1972). The aestivation is convolute. The margins of the corolla lobes are usually ciliate or denticulate. The stamens are inserted near or below the middle of the corolla tube. The filaments are often broadly winged. One nectary per corolla lobe is present near the base of the corolla tube. The ovary often has a gynophore; the style is short, and the bipartite stigma is comparatively large. The seeds are unwinged and strongly papillate. Reported chromosome numbers are $2n = 26$, 44, 52, and 78 (Favarger, 1949; Löve, 1953; Yuan & Küpfer, 1993a).

Halenia BORKH.

The genus *Halenia* was established by Borkhausen (1796: 25) with *H. sibirica, nomen illegitimum*, as its type (= *Swertia corniculata*; *Halenia corniculata* was accepted as *nomen conservandum*). The segregate genera *Exadenus* (Grisebach, 1839) and *Tetragonanthus* (Kuntze, 1891) are not used by modern authors. Gilg (1916) and Allen (1933) provided sectional classifications of *Halenia* based on the shape of the nectary spurs. These sections were not recognized by Wilbur (1984a) and Pringle (1995). There are three species in alpine, temperate, and boreal Asia, one species in temperate eastern North America, and 76 currently accepted species in alpine Central and South America (Allen, 1933; Wilbur, 1984a,b; Pringle, 1995).

The species of *Halenia* are annual, biennial, or perennial herbs, erect or decumbent. Stems are terete or narrowly winged. The 4-merous flowers form axillary and terminal panicles. The calyx is divided almost to the base, and often has colleters near the base of the calyx lobes; an intracalycine membrane is absent. The corolla is campanulate, and is lobed to the middle or almost to the base. Corolla aestivation is convolute. The stamens are inserted in the sinuses of the corolla lobes. The nectaries are on the corolla tube and form small protuberances or spurs of varying length and orientation. The spurs are often reduced in late flowers. The ovary is sessile and a style is absent or indistinct. The seeds are unwinged. A reported chromosome number is $2n = 22$ (Favarger, 1952; Moore, 1973; Weaver & Rüdenberg, 1975; Yuan & Küpfer, 1993a).

Jaeschkea KURZ

The genus *Jaeschkea* was described by Kurz (1870) with the type *J. gentianoides, nomen illegitimum* (a synonym of *J. oligosperma*). The genus *Kurramiana* (Omer & Qaiser, 1992) from Pakistan is probably synonymous with *Jaeschkea*, following Ho and Pringle (1995). There are four species in Pakistan, Kashmir, India, Sikkim, Tibet, Nepal, Bhutan, and Afghanistan (Omer & Qaiser, 1992; Ho & Pringle, 1995).

The species of *Jaeschkea* are annual, erect or decumbent herbs. Stems are branched and quadrangular. The 5-merous flowers form cymes or panicles, or are single and terminal. The calyx is divided to the base, and an intracalycine membrane is absent. The corolla is tubular or subcampanulate, and the tube is longer than the corolla lobes. The aestivation is convolute. Plicae are absent. The stamens are inserted in the sinuses between the corolla lobes, and the insertion point has hairs. The filaments are short. Two naked nectaries per corolla lobe occur near the base of the tube. The ovary is sessile or has a short gynophore. The style is indistinct. The seeds are unwinged. A reported chromosome number is $2n = 16$ (Yuan *et al.*, 1998).

Latouchea FRANCH.

This monotypic genus was described by Franchet (1899a: 212) with *Latouchea fokienensis* as its type. *Latouchea* is a rare and little-known genus. One species occurs in southeastern and southwestern China (Franchet, 1899a; Ho & Pringle, 1995).

Latouchea plants are perennial herbs. The rosette leaves are petiolate and much larger than the opposite and sessile cauline leaves. The 4-merous flowers form groups of three in the axils of bracts or are arranged in a terminal, umbel-like cluster. The calyx is divided to the base, and an intracalycine membrane is absent. The corolla is campanulate and divided to the middle. Corolla aestivation is convolute. Plicae are absent. The stamens are inserted in the sinuses between corolla lobes. The nectaries are at the base of the sessile ovary. The style is slender. The valves of the capsule have curved tips. The seeds are unwinged.

Lomatogonium A. BRAUN

The name *Lomatogonium* was first used by Braun (1830), but the type species *L. carinthiacum* was described by Reichenbach (1831–1832). The most important synonym of *Lomatogonium* frequently used in the older literature is *Pleurogyne* (Grisebach, 1839). As already suggested by Pringle (in Ho & Pringle, 1995), we do not maintain uninectariate *Lomatogoniopsis* (Ho & Liu, 1980) as separate from a more restricted binectariate *Lomatogonium*. The segregate genus *Pleurogynella* (Ikonnikov, 1970) was reduced to a section of *Lomatogonium* by Garg (1987). Discrimination of *Lomatogonium* from *Swertia* is not always easy because key characters can be inconspicuous or absent in some species, and because of this *Lomatogonium* was regarded as a subgenus of *Swertia* by Satake (1944). Nilsson (1967a) mentioned a species (*Lomatogonium bicoronatum*) that was considered to be intermediate between *Lomatogonium* and *Comastoma* in morphology and pollen characters. *Lomatogonium* was divided into three informal groups by Nilsson (1967a) on account of pollen morphological characters, but these groups are not congruent with the three main clades found by Liu and Ho (1992) in their morphological cladistic analysis. *Lomatogonium* has 21 species that occur in North America, temperate Asia, and Europe (Ho & Liu, 1980; Liu & Ho, 1992; Ho & Pringle, 1995).

All species of *Lomatogonium* are annual or perennial herbs. The stems are decumbent or erect, striate or angled. The 4- or 5-merous flowers form axillary or terminal cymes, or sometimes are borne singly. The calyx is lobed to the base or has a short tube, and an intracalycine membrane is absent. The corolla tube is very short. The aestivation is convolute. Stamens are inserted in the sinuses between the corolla lobes. The nectaries are found at the base of the corolla as one or two sometimes almost merging epipetalous glands per corolla lobe. The nectaries are apically lobulate or lamellate (lamellae are rarely absent). Nectaries sometimes are inconspicuous or absent. The ovary is sessile, a style is absent, and the stigmata are often but not always decurrent along the capsule. The seeds are unwinged. Reported chromosome numbers are $2n = 10$ and 16 (Löve, 1953; Yuan & Küpfer, 1993a).

Megacodon (HEMSL.) HARRY SM.

This small genus was erected by Smith (1936) with *Megacodon venosus* as its type species. *Megacodon* is synonymous with *Gentiana* sect. *Megacodon* (Hemsley, 1890) and *Gentiana* sect. *Stylophora* (Clarke, 1885). Two species occur in Bhutan, southwest China, India, Nepal, and Sikkim (Ho & Pringle, 1995; Aitken, 1999).

Both species of *Megacodon* are erect, perennial herbs with robust, terete, sometimes striate stems. The upper leaves are perfoliate. The large 5-merous flowers form few-flowered axillary cymes. The calyx tube is campanulate and divided to below the middle, and an intracalycine membrane is absent. The corolla is broadly campanulate and divided to below the middle. Corolla aestivation is convolute. Plicae are absent. The stamens are inserted in the sinuses between the corolla lobes. The nectaries are at the base of the gynophore or the sessile ovary. The style is stout or filiform. The seeds are unwinged.

Obolaria L.

This monotypic genus was described by Linnaeus (1753) with *Obolaria virginica* as its type. The only species grows in the southeastern United States from Texas and Florida to New Jersey and Ohio (Gillett, 1959; Wood & Weaver, 1982).

Obolaria virginica is a small, fleshy, perennial and mycotrophic herb. The lower leaves are scale-like, the upper leaves are spathulate, and all leaves are purple. Leaf bases are decurrent to the next node, and they have glandular hairs in their axils. The flowers usually form groups of three and are subtended by two bracts. The calyx consists of two free sepals that are usually foliose or sometimes scale-like. The interpretation of the two almost free foliose or bracteose leaves subtending the flower as a calyx has not always been accepted (Knoblauch, 1894). The corolla and the androecium are 4-merous. The corolla tube is campanulate, divided to the middle, and has fimbriate scales in the lower part of the tube. The aestivation is imbricate. Plicae are absent. The stamens are inserted in the sinuses between the corolla lobes. The nectary glands are probably at the base of the sessile ovary (see Lindsey, 1940; Wood & Weaver, 1982). The style is short. Seeds are unwinged. A reported chromosome number is $2n = 56$ (Wood & Weaver, 1982).

Pterygocalyx MAXIM.

This monotypic genus was described by Maximowicz (1859: 198–199) with *Pterygocalyx volubilis* as its type. *Pterygocalyx* was included in *Crawfurdia* by Bentham (1876) and Gilg (1895a), or was treated as part of *Gentianella* by Marquand (1931) or in *Gentianopsis* by Smith (in Nilsson, 1967a). Its one species occurs in China, Japan, Korea, and Russia (Toyokuni, 1963; Ho & Pringle, 1995).

Pterygocalyx plants are perennial herbs with long, twining or creeping stems. The 4-merous flowers are borne singly or are arranged in cymes. The calyx is campanulate to tubular and strongly winged, and an intracalycine membrane is absent. The calyx lobes are much shorter than the tube. The corolla is tubular, and the corolla lobes (with dentate upper margins) are shorter than the tube. Corolla aestivation is convolute. Plicae are absent.

The stamens are inserted in the upper half of the corolla tube. Nectaries are unknown. The ovary has a gynophore and the style is short. The seeds are winged.

Swertia L.

This genus was described by Linnaeus (1753: 226), and its lectotype is *Swertia perennis* (Hitchcock, 1929: 138). *Swertia* is a rather heterogeneous group of species with rotate flowers, one or two nectaries per petal lobe, and a variety of different (or sometimes missing) appendages around the nectaries. Many *Swertia*-like genera such as *Anagallidium* (Grisebach, 1839), *Agathotes* (Don, 1836), *Sczukinia* (Turczaninow, 1840), *Kingdon-Wardia* (Marquand, 1928–1931), and others have been described. Except for *Ophelia* (Don, 1836), *Frasera* (Walter, 1788), and *Veratrilla* (Baillon ex Franchet, 1899b), however, none of them are used by modern authors. For our opinion on *Frasera*, *Ophelia*, and *Veratrilla*, see the discussion on Swertiinae below. The latest sectional classification of *Swertia* into 11 sections by Ho and Liu (1980) and Ho et al. (1994a) was not used by Ho and Pringle (1995). A total of about 135 species occur in Africa (30 spp.), Madagascar (1 sp.), North America (1 sp.), Europe (3 spp.), and Asia (100 spp.; Pissjaukova, 1961; Grossheim et al., 1967; Karpati, 1970; Geesink, 1973; Ho & Liu, 1980; Garg, 1987; Shah, 1990, 1992; Zuev, 1990; Ho et al., 1994a; Ho & Pringle, 1995; Omer, 1995; Sileshi, 1998).

Swertia contains annual, biennial, and perennial herbs. The stems are ascending or erect, and are terete, striate or angled, and sometimes hollow. The leaves are usually opposite or rarely whorled or alternate. The leaf bases are decurrent or not, and sometimes the leaves are weakly perfoliate. The 4- or 5-merous flowers usually form foliose panicles or sometimes racemes, or are borne singly. The calyx is rotate and deeply divided, and an intracalycine membrane is absent. Corolla aestivation is convolute. The corolla tube is short, and there are one or two nectaries per corolla lobe. The nectaries can have fimbriae, small fimbriate basal scales, or are naked. When two nectaries are present, these are sometimes merging. The stamens are inserted in the sinuses between the corolla lobes; their insertion points are sometimes surrounded by long hairs. The ovary is sessile or has a gynophore. The style is sometimes slender and filiform, but more often robust or absent. The stigma is bilobed and not decurrent along the margins of the valves. The fruit is a flattened or oblong capsule. The seeds have wings or are unwinged. Reported chromosome numbers are $2n = 16, 18, 20, 24, 26, 28$, and 78 (Favarger, 1952; Khoshoo & Tandon, 1963; Wood & Weaver, 1982; Yuan & Küpfer, 1993a; Sileshi, 1998).

Tripterospermum BLUME

This genus was described by Blume (1826: 849). The type species is *Tripterospermum trinerve*. *Tripterospermum* was included in *Crawfurdia* (Clarke, 1875) or *Gentiana* (Marquand, 1931), but was resurrected as a separate genus by Smith (1965). Twenty-four species occur in India, Myanmar, China, Japan, Taiwan, Nepal, Bhutan, Korea, Vietnam, Thailand, Malaya, Sri Lanka, Indonesia, Indochina, the Philippines, and the Celebes (Smith, 1965; Murata, 1989).

Tripterospermum contains perennial herbs with mostly twining or prostrate, rarely erect stems. The 5-merous flowers form axillary and terminal cymes, or are borne singly. The calyx tube has five keeled ridges and is divided to the middle; an intracalycine membrane is inconspicuous or absent. The corolla is campanulate to infundibuliform, and its tube is much longer than the lobes. Corolla aestivation is convolute. Plicae are present. The stamens are inserted near the base of the corolla tube; they are unequal, and their tips are bent downwards. The nectaries form a collar-like disk at the base of the often long gynophore. The style is slender and slightly recurved. The fruit is sometimes a capsule but is more often a berry. The seeds are triquetrous to compressed, winged or not. A reported chromosome number is $2n=46$ (Moore, 1973).

Veratrilla BAILL. EX FRANCH.

Veratrilla was described by Franchet (1899b: 310) based on an unnamed species of Baillon (1888; see Smith, 1970). The type species is *Veratrilla baillonii*. *Veratrilla* is dioecious but in other characters rather similar to parts of *Swertia*. A better knowledge of *Swertia* is required to decide whether an enlarged *Veratrilla* should be maintained as a separate genus or whether *Veratrilla* should be included in *Swertia*. The two species of *Veratrilla* are distributed in Sikkim, India, Bhutan, and southwest China (Smith, 1970; Ho & Pringle, 1995).

Both species of *Veratrilla* are dioecious, erect, perennial herbs. The stems are angled or slightly winged. The male flowers form dense, many-flowered, sessile clusters, whereas the fewer female flowers are arranged in elongate inflorescences. The flowers are usually 4-merous, rarely 5-merous. The calyx is rotate and divided nearly to the base, and an intracalycine membrane is absent. The corolla has a short but distinct corolla tube. Corolla aestivation is convolute. Plicae are absent. The stamens are inserted in the sinuses between the corolla lobes. There are one or two nectaries per corolla lobe. A style is short or absent. The ovary is sessile. The seeds are commonly winged.

Phylogeny and character evolution in Gentianeae

The *mat*K and *trn*L intron data shown in Figs. 2.1–2.3 clearly resolve Gentianeae as a monophyletic group and partly elucidate the relationships of the tribe within the family (see below). However, these data are in many instances not sufficiently variable and do not include enough taxa for a conclusive discussion of the phylogeny of Gentianeae. For this reason we obtained and analyzed ITS sequences (Yuan & Küpfer, 1995; Yuan et al., 1996; K. B. von Hagen & J. W. Kadereit, unpubl.) of 93 species representing 16 of the 17 genera of the tribe. The results of this analysis will provide additional evidence for the following discussion (Figs. 2.23–2.24).

The Gentianeae are a well-supported clade, particularly in the combined analysis (Fig. 2.3). Gentianeae are sister to tribe Potalieae in the *trn*L intron analysis, to Potalieae plus Helieae in the *mat*K analysis, and part of a trichotomy with Helieae and Potalieae in the combined analysis (Figs. 2.1–2.3). Although the generic composition of Gentianeae has changed little since Grisebach (1839, as Swertieae), we cannot name an unambiguous morphological synapomorphy that supports the clade. In the past, characters such as sessile or nearly sessile stigmata and comparatively large and single pollen grains were used to define this group (Grisebach, 1839; Bentham, 1876; Gilg, 1895a). Although these characters are indeed found in most genera and species, many exceptions exist. These are the result both of the inclusion of *Bartonia* and *Obolaria* in Gentianeae, and of the increasingly detailed knowledge of morphological diversity within the group. One feature that does bear special mention is that the ovules are distributed over the entire ovary wall in most Gentianeae (except for *Halenia*, *Jaeschkea*, and *Swertia* spp.), and this character is not known from outside the tribe (Lindsey, 1940). However, further sampling is required to judge the apomorphic versus plesiomorphic nature of this character.

In the non-molecular cladistic analysis of Meszáros et al. (1996), all members of Gentianeae included are reported to have xanthones that are not oxygenated in position C6. This trait also differs from most other species of Gentianaceae. Although only few species from tribes other than Gentianeae have been examined for this character, it may well turn out to support the monophyly of the tribe.

Subtribe Gentianinae

All molecular data analyzed (*trn*L intron, *mat*K, ITS) show a clear subdivision of the Gentianeae into subtribes Gentianinae and Swertiinae. The genera of Gentianinae (*Crawfurdia*, *Gentiana*, and *Tripterospermum*) are characterized by the presence of plicae between the corolla lobes (Tables 2.9–2.10) and by usually having intracalycine membranes between the calyx lobes. The co-occurrence of these characters and their identical position in different floral whorls may suggest that their phenotypic expression is controlled by the same genetic or developmental pathway. An intracalycine membrane is also present in *Gentianopsis*, of subtribe Swertiinae. Because this structure in *Gentianopsis* is not continuous as in Gentianinae, these two features may not be homologous, an interpretation that has been suggested before (Gillett, 1957). Other characters that have been named to support Gentianinae as a monophyletic taxon and to separate it from Swertiinae are either more variable than initially thought or have been investigated in only a few species of Gentianeae. Characters that have been used are, for example, basifixed anthers, presence of three petal veins, and presence of calcium oxalate crystals in leaf mesophyll (Borodin, 1892; Gillett, 1957; Adams, 1995).

The relationships between and within *Crawfurdia*, *Gentiana*, *Tripterospermum*, and supposedly related genera have been discussed repeatedly. *Crawfurdia* and *Tripterospermum*, for example, have been combined into one genus (Clarke, 1875) or have been treated as sections of *Gentiana* (Marquand, 1931). Moreover, *Pterygocalyx* of subtribe Swertiinae has been postulated to be a close relative of *Crawfurdia* and *Tripterospermum* (Bentham, 1876; Gilg, 1895a) because of the shared twining habit, and *Gentiana* sect. *Stenogyne* has been regarded as being congeneric with *Tripterospermum* (Löve & Löve, 1976). These problems were solved almost completely in the morphological analysis by Smith (1965; see Table 2.10) and by the ITS DNA sequence analysis of Yuan et al. (1996). The ITS data demonstrate a clade of *Gentiana* sect. *Stenogyne*, *Crawfurdia*, and *Tripterospermum* (Fig. 2.23). This clade is sister to all other sections of *Gentiana*, and is characterized by the possession of compressed or triquetrous seeds. In this clade, *Gentiana* sect. *Stenogyne* is sister to *Crawfurdia* and *Tripterospermum*, themselves sister genera. With the exception of the monotypic *Crawfurdia* sect. *Protocrawfurdia*, *Crawfurdia* and *Tripterospermum* share a twining habit. The erect habit of *Crawfurdia* sect. *Protocrawfurdia*, its short stiff hairs (Smith, 1965), and its similarities in pollen morphology to *Gentiana* sect. *Stenogyne* (Nilsson, 1967a) may suggest that sect. *Protocrawfurdia* is more closely related to *Gentiana* sect. *Stenogyne* than to the remainder of *Crawfurdia*, which would then call for a reinvestigation of the characters given by Smith (1965) for sect. *Protocrawfurdia*. The presence of asymmetrical and curved stamens in both *Gentiana* sect. *Stenogyne* and *Tripterospermum* (Smith, 1965) must be

Table 2.9. An overview of important characters for generic delimitation in tribe Gentianeae

Genus-subtribe	Flower merosity	Intracalycine membrane	Plicae	Position of nectaries	No. of nectaries per petal lobe	Insertion of stamens[a]	Corolla aestivation
Crawfurdia-GEN[b]	5	+ (−)[c]	+	Ovary	—	Tube	Convolute
Gentiana-GEN	4 or 5	+ (−)	+ (−)	Ovary	—	Tube	Convolute
Tripterospermum-GEN	5	− (?)	+	Ovary	—	Tube	Convolute
Bartonia-SWE	4	—	—	− (?)	—	Sinus between lobes	Imbricate
Comastoma-SWE	4 or 5	—	—	Petals	2	Tube	Convolute
Frasera-SWE	4	—	—	Petals	1 or 2	*	Convolute
Gentianella-SWE	4 or 5	—	—	Petals	1 or 2	Tube	Convolute
Gentianopsis-SWE	4	+	− (?)	Petals	1	Tube	Convolute
Halenia-SWE	4	—	—	Petals/spurred	1	Sinus between lobes	Convolute
Jaeschkea-SWE	5	—	—	Petals	2	Sinus between lobes	Valvate
Latouchea-SWE	4	—	—	Ovary	—	Sinus between lobes	Convolute
Lomatogonium-SWE	(4) 5	—	—	Petals	1 or 2	*	Convolute
Megacodon-SWE	5	—	—	Ovary	—	Sinus between lobes	Convolute
Obolaria-SWE	4	—	—	Ovary	—	Sinus between lobes	Imbricate
Pterygocalyx-SWE	4	—	—	?	—	Tube	Convolute
Swertia-SWE	4 or 5	—	—	Petals	1 or 2	*	Convolute
Veratrilla-SWE	4	—	—	Petals	1 or 2	Sinus between lobes	Convolute

Notes:

[a] It is difficult to compare the insertion of stamens in the almost rotate genera *Lomatogonium*, *Swertia*, and *Frasera* (*) with the insertion of stamens in genera with a long corolla tube.

[b] GEN, subtribe Gentianinae; SWE, subtribe Swertiinae.

[c] "+" indicates presence of a character, "−" the absence of a character. For doubtful characters "?", see generic descriptions. Rare characters are noted in parentheses.

Table 2.10. *Important characters for generic delimitation in Gentiana and related taxa (after Smith, 1965)*

Character	*Gentiana* (except sect. *Stenogyne*)	*Gentiana* sect. *Stenogyne*	*Tripterospermum*	*Crawfurdia* sect. *Protocrawfurdia*	*Crawfurdia* sect. *Crawfurdia*
Habit	Perennial/annual	Annual (perennial)[a]	Perennial	Perennial	Perennial
Growth form	Erect	Erect	Twining	Erect	Twining
Short stiff hairs	Absent	Present/absent	Absent	Present	Absent
No. of vascular bundles in calyx tube	5	5	5	10	10
Nectariferous glands	Naked	Naked	With a collar-like disk	Naked	Naked
Stamen morphology	Symmetrical, straight	Asymmetrical, curved downwards	Asymmetrical, curved downwards	Symmetrical, straight	Symmetrical, straight
Staminal filaments	Winged	Filiform	Filiform	Thickened	Thickened
Fruit type	Capsular	Capsular	Baccate (capsular)	Capsular	Capsular

Note: [a] Rare characters are noted in parentheses.

Systematics, character evolution, and biogeography 241

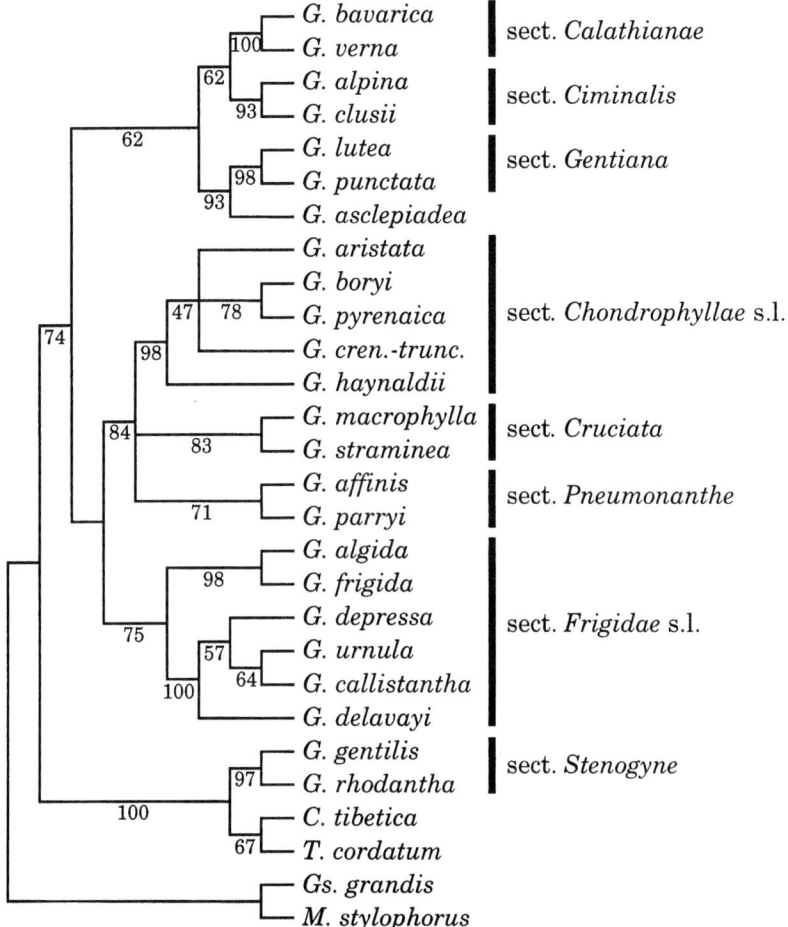

Figure 2.23. Phylogeny of *Gentiana* and related taxa, redrawn from Yuan *et al.* (1996). Strict consensus of six most-parsimonious trees of an ITS analysis. Numbers below branches are bootstrap values. G.= *Gentiana*, C.= *Crawfurdia*, T.= *Tripterospermum*, Gs.= *Gentianopsis*, M.= *Megacodon*, cren.-trunc.= *crenulato-truncata*. *Gentiana asclepiadea* was formerly included in the apparently unrelated sect. *Pneumonanthe*.

regarded either as the result of parallel evolution or as the plesiomorphic state for this clade. The phylogenetic relationships found in the Gentianinae suggest that *Gentiana* sect. *Stenogyne* should be recognized at generic rank. Within *Gentiana* (excluding sect. *Stenogyne*), two main clades (Fig. 2.23) are resolved in the ITS analysis. The first of these is European in distribution and comprises sects. *Gentiana*, *Ciminalis*, and *Calathianae*, as well as *Gentiana asclepiadea*. Except for *Gentiana asclepiadea*, this clade is supported by some morphological and karyological characters (Ho *et al.*, 1996) and phytochemical data (Meszáros, 1994). *Gentiana asclepiadea* had previously been placed in *Gentiana* sect. *Pneumonanthe* (Kusnezow, 1896), to which it appears not to be closely related. Because the species differs morphologically from *Gentiana* sect. *Gentiana* (its closest relative in the ITS topology), it perhaps should be allocated to a new monotypic section.

The second clade of *Gentiana* contains many widespread or exclusively Asian sections. The sections segregated from *Gentiana* sect. *Frigidae sensu lato* and sect. *Chondrophyllae sensu lato* by Ho and Liu (1990) and Ho and Pringle (1995) must be recircumscribed because they describe paraphyletic groups (Ho *et al.*, 1996; Yuan *et al.*, 1996; Yuan & Küpfer, 1997). Molecular, biogeographic, and karyological analyses of several sections of *Gentiana* have been published (Löve & Löve, 1972, 1986; Yuan & Küpfer, 1993b, 1997; Gielly & Taberlet, 1996; Hungerer & Kadereit, 1998; Yuan *et al.*, 1998; von Hagen & Kadereit, 2000); these are referenced here but will not be discussed further. For most authors (e.g., Tutin, 1972; Pringle, 1978; Ho & Liu, 1990; Yuan *et al.*, 1996) the morphological diversification of *Gentiana* (excluding *Gentiana* sect. *Stenogyne*) does not call for the establishment of segregate genera as suggested by Rydberg (1932), Holub (1973), Löve and Löve (e.g., 1976), Omer and Qaiser (e.g., 1992), and Omer (1995). Frequently the genera of these authors were sections or groups of sections sharing the same chromosome number. For example, this resulted in a polyphyletic *Gentiana* sect. *Ciminalis* that contained the perennial European species of sect. *Ciminalis* and entirely unrelated, annual species of *Gentiana* sect. *Chondrophyllae*, only because all share a diploid chromosome number of $2n = 36$.

Subtribe Swertiinae

No unique morphological synapomorphy exists in support of subtribe Swertiinae. Characters such as dorsifixed, versatile anthers, 5–9 petal veins, and the absence of calcium oxalate crystals in the leaf mesophyll were used by Borodin (1892), Gillett (1957), and Adams (1995) to distinguish their *Gentianella* lineage from *Gentiana*. However, these are either more variable than initially thought or are not yet known for every genus. It had been suggested that the position of the nectaries on the petals might differentiate Swertiinae from Gentianinae, which have nectaries at the base of the ovary (e.g., Gillett, 1957; Ho & Liu, 1990). The inclusion in Swertiinae of *Megacodon*, *Latouchea*, and *Obolaria*, which have nectaries at the base of the ovary, does not allow us to use petal nectaries as a synapomorphy of the entire subtribe, although they may characterize a subclade, as discussed below. Nectaries at the base of the ovary are rather common in Helieae and Potalieae and are present in all Gentianinae and in the probable basal genera of the Swertiinae (see below). As a consequence, nectaries at the base of the ovary are very likely to be plesiomorphic in Gentianeae and Swertiinae.

Although all our molecular data sets suffer from limited resolution of generic relationships within Swertiinae, and because the *mat*K and *trn*L intron data in particular are based on a very small taxon sample, we believe that the evidence available nevertheless permits several conclusions to be drawn about the phylogeny of the subtribe. The basal position in Swertiinae, although always without substantial support from resampling procedures, is occupied by *Gentianopsis* (*mat*K), *Megacodon* (ITS), or *Obolaria* (*trn*L intron). Morphological features in support of all three possibilities can be argued, viz. the presence of intracalycine membranes in *Gentianopsis* (which may not be homologous to those found in Gentianinae; see above) and the position of the nectaries at the base of the ovary for *Megacodon* and *Obolaria*, as found in Gentianinae and the closest relatives of Gentianeae.

The last two genera, and particularly *Obolaria*, share some characters with other genera of Swertiinae (Table 2.9). *Latouchea* also has nectaries at the base of the ovary (unknown for *Bartonia*). *Obolaria*, *Bartonia*, *Halenia*, *Latouchea*, *Megacodon*, and *Veratrilla* all have stamens inserted in the sinuses between the corolla lobes (similar only in *Jaeschkea*), and *Bartonia* and *Obolaria* share a strongly mycotrophic habit and imbricate corolla aestivation. *Bartonia* and *Obolaria* are sister genera in our ITS analysis but not in the *mat*K and *trn*L intron analyses. This relationship in the ITS analysis might result from spurious attraction between the long branches of these two genera, and is altered by even small changes in sequence alignment. Unfortunately, we could not include *Latouchea* in our molecular analysis. Taking all the evidence together, we believe that *Bartonia*, *Latouchea*, *Megacodon*, and *Obolaria* are basal within Swertiinae. Whether they constitute one clade or are paraphyletic in relation to the remainder of Swertiinae cannot yet be decided. A basal position for these four genera in

Swertiinae would imply that petal nectaries are a synapomorphy for the remaining genera.

The ITS tree provides interesting information on the circumscription and relationships of those taxa that were included in *Gentianella* (or *Gentiana* subg. *Gentianella*) *sensu* Kusnezow (1895) and Wettstein (1896; see Figs. 2.23–2.24). Three sections of *Gentianella* have since been segregated at generic rank. These are *Megacodon*, on account of the position of its nectaries (Smith, 1936), *Comastoma*, because of the presence of scales with non-vascularized fimbriae in the corolla tube (Toyokuni, 1961), and *Gentianopsis*, mainly because of its flattened flower bud, imbricate calyx lobes, and the discontinuous intracalycine membrane (Ma, 1951). For *Gentianopsis*, many more diagnostic characters such as papillate seeds, dentate corolla lobes, pattern of vascular supply of flower organs, and position of the ovules can be named (Lindsey, 1940; Iltis, 1965). *Comastoma* and *Gentianopsis*, however, have not been accepted by all authors (e.g., Gillett, 1957; Smith, 1967; Tutin, 1972; Aitken, 1999). In the ITS tree, *Megacodon* occupies a basal position in the Swertiinae, and *Gentianopsis* groups with *Frasera* and *Pterygocalyx*. Because *Megacodon* also shows no close relationship to *Gentianella* in the *trn*L intron data, and because the same is true for both *Megacodon* and *Gentianopsis* in the *mat*K tree, we believe that generic status for both genera is justified. *Comastoma* is well defined morphologically and is likely to be monophyletic and related to *Lomatogonium* and to those species of *Gentianella* with two nectaries per petal, which will be excluded from *Gentianella sensu stricto* (see below). Accordingly, generic rank also seems justified for *Comastoma*.

For the remainder of *Gentianella*, two major groups can be recognized in the ITS tree. The first group consists of species with two nectaries per petal (Fig. 2.22) and a naked corolla tube (see below). The second group is characterized by only one nectary per petal (Fig. 2.22), and is part of a polytomy with several other genera in our ITS tree. Besides having only one nectary per petal, this second group differs from the other taxa of the polytomy in having vascularized fimbriae, non-vascularized fimbriae, or no fimbriae at all in the corolla throat (scales with non-vascularized fimbriae in *Comastoma*), naked petal nectaries (usually fimbriate in *Swertia* and *Lomatogonium*), an always distinct but sometimes short corolla tube (mostly inconspicuously short in *Swertia* and *Lomatogonium*), and stamens inserted in the corolla tube (near sinuses between corolla lobes in *Jaeschkea*).

The monophyletic uninectariate clade of *Gentianella* is further subdivided in our ITS tree into those species with a long corolla tube usually with vascularized fimbriae, and those with an always naked and typically

Systematics, character evolution, and biogeography 245

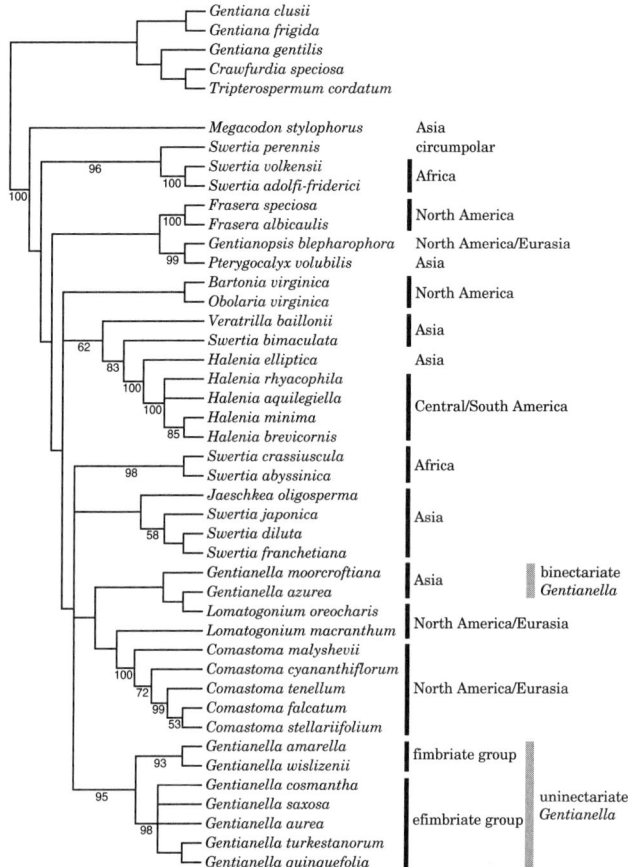

Figure 2.24. Preliminary phylogeny of subtribe Swertiinae (strict consensus of 312 most-parsimonious trees of the ITS analysis, 1042 steps, $C=0.48$, $R=0.60$, jackknife values (>50%) below branches; indels coded separately, ambiguous positions of the alignment and uninformative characters excluded; some of the sequences were obtained by Yuan & Küpfer, 1995, and were drawn from EMBL). Geographic distribution is given for genera or groups of species but not for species sampled. For a third binectariate *Gentianella*, *G. arenaria*, only ITS 1 could be obtained. When included in the data matrix, this species groups in a polytomy but not close to the uninectariate species of *Gentianella*. We have included a total of 44 uninectariate species of *Gentianella* in a separate analysis (not shown). The fimbriate group and the efimbriate group were both resolved as sister groups with ITS data, as in the analysis presented above. The efimbriate group in the larger analysis contains 32 species mostly from South America, Australia, and New Zealand but also from North America and Eurasia. The fimbriate group in the larger analysis contains 12 species from Eurasia and Central and North America. However, with the chloroplast data of a third analysis (not shown) most efimbriate species from the Northern Hemisphere, e.g., *Gentianella aurea*, *G. quinquefolia*, and *G. turkestanorum*, are basal to all other uninectariate *Gentianella*, including the fimbriate group.

short corolla tube. Whereas the first of these two subclades (the fimbriate group) is distributed largely in the Northern Hemisphere, the second (the efimbriate group) is found mostly in the Southern Hemisphere. However, in an analysis of sequence data from chloroplast genes (von Hagen & Kadereit, 2001) the efimbriate group became paraphyletic in relation to the fimbriate group, with species from the Northern Hemisphere basal to all other species, including the fimbriate group. In this analysis we were able to include *Gentianella ruizii*, one of the three species with non-vascularized fimbriae in the corolla throat that are endemic to South America. This species clearly grouped with species of the efimbriate group and not with other fimbriate species, which have vascularized fimbriae.

In any case, we feel that the phylogenetic relationships found and the distribution of morphological characters warrants a redefinition of *Gentianella* as containing species with one nectary per petal only, permitting us to exclude those species with two petal nectaries. If, within this group, generic rank were to be assigned to the two subclades, the group with vascularized fimbriae would have to be called *Gentianella*, because the type of the genus (*G. campestris*) clearly falls into this group on account of its flower morphology. For the efimbriate group at least three names are available. These are *Aloitis* (Rafinesque, 1837), *Pitygentias* (Gilg, 1916), and *Chionogentias* (Adams, 1995). These genera were segregated from *Gentianella* because of conspicuous differences from the type of the genus. Their circumscription, however, was based on regional observations only. Unfortunately, the oldest name, *Aloitis*, which would have to be used for uninectariate species of the efimbriate group, has recently been applied by Omer *et al.* (1988) to predominantly binectariate species from Pakistan.

The binectariate species of *Gentianella* sampled in our analysis group in one clade together with one species of a polyphyletic *Lomatogonium*. They are similar to *Comastoma* and *Lomatogonium* in having two nectaries per petal, but they differ from all binectariate genera in having a long corolla tube (short in *Frasera*, *Lomatogonium*, *Swertia*, and *Veratrilla*) without fimbriae (*Comastoma* having scales of non-vascularized fimbriae), naked nectaries (fimbriate in part of *Lomatogonium*, *Frasera*, *Swertia*, and *Veratrilla*), and stamens that are inserted in the corolla tube (in sinuses between corolla lobes in *Jaeschkea*). The future fate of the binectariate species of *Gentianella* must await the inclusion of further species sampling in *Lomatogonium* and *Comastoma*. A potential genus name for binectariate species of *Gentianella* is *Aliopsis* (Omer & Qaiser, 1991) which is based on *G. pygmaea*. However, it is not yet clear whether *Gentianella pygmaea* has one (Omer & Qaiser,

1991) or two (Grossheim et al., 1967; Ho & Pringle, 1995) nectaries per petal lobe. The two species presently sampled of *Lomatogonium* fall into two different but only weakly supported clades in the ITS tree, and are part of a polytomy in the *trn*L intron data. This problem requires further sampling in the different groups proposed by Nilsson (1967a) and Liu and Ho (1992). It will be particularly important to sample those uninectariate species of *Lomatogonium* that have been segregated as *Lomatogoniopsis* (Ho & Liu, 1980) because of our argument that the number of nectaries per petal is relevant for the redefinition of *Gentianella*. In the case that a close relationship of *Lomatogonium* to *Comastoma* should find further support, as suggested by both the ITS and the *mat*K data and as already suspected by Wettstein (1896), Löve and Löve (1956), and Nilsson (1967a,b), the fimbriae near the nectaries of *Lomatogonium* may well prove to be homologous to the nonvascularized fimbriae at the base of the corolla lobes of *Comastoma*. Because the decurrent stigmata of *Lomatogonium* sometimes cannot easily be distinguished from the regularly bilobed stigmata of *Swertia*, and because both genera have nectary appendages, a close relationship between the two genera has been postulated repeatedly (Gilg, 1895a; Satake, 1944; Liu & Ho, 1992). Such a relationship is not supported by any of our molecular data sets.

Besides *Gentianella*, the *Frasera/Swertia* complex is another taxonomically problematic group within Swertiinae. The first complete revision of *Swertia* after Grisebach (1845) was accomplished by Shah (1990, 1992). He recognized 35 informal groups of *Swertia* (32 excluding *Frasera*), defined mostly by flower morphological characters and geographic distribution. Shah made no attempt to clarify the phylogenetic position or evolution of the groups erected. It has now become clear that *Swertia* is highly polyphyletic and must be considered in the context of other genera of Swertiinae (Yuan & Küpfer, 1995; Chassot, 2000; K. B. von Hagen & J. W. Kadereit, unpubl.). Other work on *Swertia*, e.g., Grossheim et al. (1967), suffers from a regional perspective. The classification of *Swertia* into 11 sections by Ho and Liu (1980) and Ho et al. (1994a) was not used in the more recent account of the genus for the *Flora of China* (Ho & Pringle, 1995), probably because it was based on combinations of only very few characters. Classifications of African *Swertia* have been presented by, for example, Fries (1923), Hedberg (1957), and Sileshi (1998), mainly on the basis of pollen and flower morphology. A molecular phylogenetic analysis of most groups of *Frasera/Swertia* and a re-analysis of morphological characters in the context of the Swertiinae is now underway (Chassot, 2000; P. Chassot & P. Küpfer, pers. comm.).

Neither the *trn*L intron data nor the ITS analysis resolve *Swertia* and *Frasera* as a monophyletic group, and the different species sampled of *Swertia* group in different clades in both the *mat*K and ITS analyses. Extensive discussion both for and against the maintenance of *Frasera* as separate from *Swertia* can be found in, for example, St. John (1941), Wood and Weaver (1982), and Pringle (1990). In our ITS analysis, the two species of *Frasera* form one clade that is distinct from all included species of *Swertia*. Although representatives of *Swertia* are found in four different clades, and the inclusion of more species will very likely change their interrelationships, it seems improbable to us on the basis of our three molecular data sets (1) that the clade of *Swertia* containing the type of the genus (*S. perennis*) will move into a sister-group relationship to *Frasera*, or (2) that all five clades of *Swertia* and *Frasera* will eventually form a monophylum. Accordingly, *Frasera* may well be a monophyletic group that is also supported by the unique combination of 4-merous flowers, mostly one nectary per petal, weakly connate but not decurrent leaf bases, a filiform style, and some pollen characters (Nilsson, 1967a; Wood & Weaver, 1982). All these characters can also be found in Asian representatives of *Swertia*, but this combination of features is unknown outside *Frasera*. In consequence, the separation of *Frasera* from *Swertia* appears to be justified by our molecular data. Our present data set does not support the inclusion of annual *Swertia* from Asia in *Frasera* as suggested by Toyokuni (1965). In our ITS analysis, *Swertia bimaculata*, *S. japonica*, and *S. diluta*, as representatives of this group, are not closely related to *Frasera*.

The recognition of annual species of *Swertia* with often 4-merous flowers as sect. *Ophelia* (Gilg, 1895a) or even as a distinct genus (Toyokuni, 1963; Grossheim *et al.*, 1967) appears not to be justified because species of such morphology sampled by us (*S. abyssinica*, *S. adolfi-friderici*, *S. bimaculata*, *S. diluta*, *S. franchetiana*, and *S. japonica*) fall into four different clades. The two groups of African *Swertia* found in our analysis are congruent with those outlined by Sileshi (1998).

Dioecious *Veratrilla* is similar in part to the monoecious *Swertia* (e.g., large and naked nectary patches) but is usually maintained as a distinct genus. Our data suggest that the combination of *Veratrilla* with *Swertia* may eventually be necessary.

The monophyly of *Halenia* is not surprising. This large genus is well defined by its nectary spurs. Although it may be speculative on the basis of our taxon sample that *Swertia bimaculata* is the closest living relative of *Halenia* (as is found in our ITS phylogeny) it seems possible that *Halenia* is sister to one of the *Swertia* clades and evolved from a spurless, Asian *Swertia*-like progenitor. The basal position in *Halenia* of the long-spurred Asian *H. elliptica* may indicate that the group of short-spurred or spurless species in South America, exemplified by *H. brevicornis* in our taxon sample, are secondarily reduced forms that do not represent the ancestral type of the genus, as suggested by Allen (1933).

The close relationship of *Gentianopsis* and *Pterygocalyx* suggested by ITS data is supported by their similar pollen morphology (Nilsson, 1967a), the presence of a gynophore, and dentate margins on the corolla lobes. Yuan and Küpfer (1995) found that *Gentianopsis* is paraphyletic in relation to *Pterygocalyx*. In contrast to Smith (in Nilsson, 1967a), they did not merge the two genera because *Pterygocalyx* differs from *Gentianopsis* in its twining habit and winged seeds. There exist no obvious morphological characters supporting the sister-group relationship of both of these genera to *Frasera*, except perhaps for the presence of only one epipetalous nectary in *Gentianopsis* and most species of *Frasera*. Because *Frasera* and part of *Gentianopsis* occur in North America, however, this hypothetical clade is plausible from a geographical point of view as well.

Palynology of Gentianeae: Synopses of genera and general discussion

S. NILSSON

Bartonia. *Bartonia virginica* (L.) BSP. The pollen grains are 3-colporate, prolate spheroidal, and *c.* 14×13 μm. The exine is perforate to finely reticulate, with perforations usually rounded and sparsely spaced at the smooth colpus margin. (Canada: Nova Scotia, 1927, *Fernald et al. 390* (NO), slide no. 24929; Fig 2.29F.)

OTHER INVESTIGATIONS: Gilg (1895a) included *Bartonia* and *Obolaria* in Gentianeae-Erythraeinae, but did not describe any exine features for the two genera. According to Köhler (1905), the pollen grains of *Bartonia verna* are 3-colporate, small (14–16 μm), and reticulate, and those of *B. tenella* are small, 3-colporate, and smooth. In a study of saprophytic Gentianaceae (Nilsson & Skvarla, 1969), three species of *Bartonia* were described as having finely reticulate exine or exine with OL-pattern. Similarly, as shown in the present study, *Bartonia virginica* has small, perforate grains that are *c.* 15 μm in size, and thus has smaller pollen than the other genera of the subtribe Gentianinae.

Comastoma. *Comastoma polycladum* (Diels & Gilg) T. N. Ho. The pollen grains are 3-colporate, subprolate, and *c.* 41×32 μm. The exine is finely

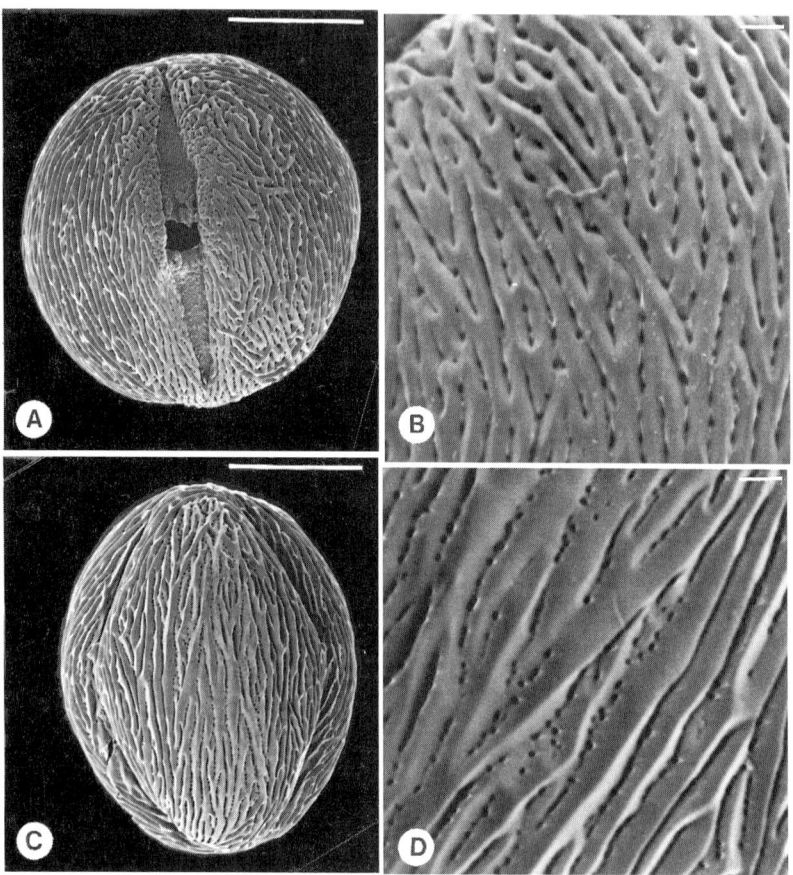

Figure 2.25. Pollen of Gentianeae-Gentianinae. **A, B.** *Crawfurdia bulleyana* (*Forrest 839*). **A.** 3-colporate, striate pollen grain in equatorial, apertural view; colpus margin narrow, smooth, perforate. **B.** Detail of relatively dense striate exine pattern with parallel, often branched muri/lirae separated by perforate striae. **C, D.** *Tripterospermum volubile* (*Ludlow et al. 20848*). **C.** 3-colporate, striate pollen in equatorial, mesocolpial view. **D.** Detail of tectate-perforate exine with sharply keeled, parallel, branched muri/lirae and striae with numerous perforations. Scale bars: A, C = 10 μm; B, D = 1 μm.

reticulate to perforate, with lumina rounded to oval-shaped. (China: Gansu, 1992, *Küpfer et al. 92290* (NEU), slide no. 23393; Fig. 2.27E–F.)

OTHER INVESTIGATIONS: As examined with light microscopy the pollen grains of *Comastoma* (formerly *Gentianella* sect. *Comastoma*) differed in some minor exine features from *Gentianella sensu stricto*, i.e., in not being typically striate or reticulate (as observed in, e.g., *G. aurea* and *G. foliosa*) or reticulate to striato-reticulate (as in *G. amarella* and *G. campestris*). Some species of *Comastoma* have pollen grains with a triangular amb, which is less common in *Gentianella*. Thus, there is palynological support to keep *Comastoma* as a separate section of *Gentianella* or to maintain *Gentianella* sect. *Comastoma* at generic rank.

Crawfurdia. Crawfurdia bulleyana G. Forrest. The pollen grains are 3-colporate, prolate spheroidal, and *c.* 32×30 μm. The exine is striate-perforate, with muri/lirae slightly keeled, branched, anastomosing, and separated by perforated striae. In the polar area the lirae/muri are oriented in different directions, and at the colpus border the muri/ lirae are fused to a relatively narrow, perforated margin. (China: *Forrest 839* (BM), slide no. 24470; Fig. 2.25A–B.)

OTHER INVESTIGATIONS: Gilg (1895a) described the exine of *Crawfurdia* pollen as striate or as irregularly reticulate, and Köhler (1905) noted two species of *Crawfurdia* to be striato-reticulate. The pollen grains of *Crawfurdia fasciculata* and *C. volubilis* appear to be striate to striato-reticulate (Wang, 1960). Huang (1972) investigated four species of *Crawfurdia*, all with striato-reticulate pollen grains. He also pointed out the narrow lateral slits of the ora. Pollen of *Crawfurdia angustata* was described and illustrated by Agababyan and Tumanyan (1977a). In a previous study (Nilsson, 1967a) 10 species of *Crawfurdia* were investigated and three different pollen types were distinguished. The newly investigated species *Crawfurdia bulleyana* resembles mostly *C. speciosa* in its exine features. Intraspecific variation of pollen of *Crawfurdia campanulacea* was noted by Nilsson (1967b).

Gentiana. Gentiana asclepiadea L. The pollen grains are 3-colporate, prolate spheroidal, and *c.* 30×25 μm. The exine is striate-perforate with straight muri/lirae that are branched at mesocolpia and over the poles (toward the poles sometimes running in various directions). (Switzerland: Tessin, 1929, *Ross s.n.* (S), slide no. 24798; Fig. 2.26A–B.)

Gentiana producta T. N. Ho. The pollen grains are 3-colporate, prolate, and 38×26 μm. The exine is striato-reticulate with transverse muri located at a lower level than the principal muri/lirae; the lumina are usually rounded to oval-shaped. (China: Sichuan, 1993, *Yuan & Zeltner s.n.* (NEU), slide no. 23402; Fig. 2.26C–D.)

Gentiana verna L. The pollen grains are 3-colporate, prolate, and

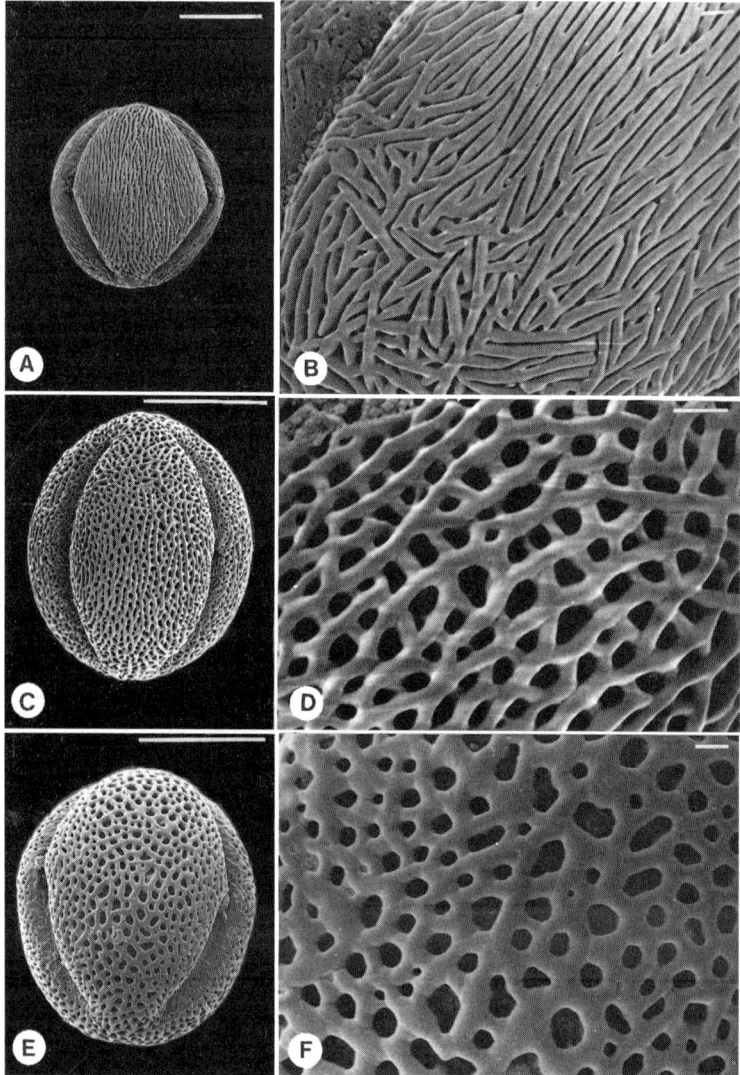

Figure 2.26. Pollen of Gentianeae-Gentianinae. **A, B.** *Gentiana asclepiadea* (*Ross s.n.*). **A.** 3-colporate, striate pollen grain in equatorial, mesocolpial view. **B.** Detail of exine with parallel and branched, densely spaced muri/lirae that become shorter, more ramified, and abruptly changing direction toward the poles. **C, D.** *Gentiana producta* (*Yuan & Zeltner s.n.*). **C.** 3-colporate, striato-reticulate pollen in equatorial, mesocolpial view. **D.** Detail of the striato-reticulate exine with parallel, branched muri/lirae, interconnected by muri at a lower level. **E, F.** *Gentiana verna* (*Treffer s.n.*). **E.** 3-colporate, reticulate pollen grain in equatorial, mesocolpial view. **F.** Detail of exine with rounded to angular lumina, and straight to sinuous muri. Scale bars: A, C, E = 10 μm; B, D, F = 1 μm.

c. 48×32 μm. The exine is reticulate to finely reticulate, with rounded to oval-shaped lumina. (Austria: Tirol, 1894, *Treffer s.n.* (S), slide no. 23341; Fig. 2.26E–F.)

OTHER INVESTIGATIONS: Kusnezow in Gilg (1895a) subdivided *Gentiana* into 19 sections, and *Gentianella* was recognized as a section, not as a separate genus. The whole subtribe of Gentianinae was said to have striate or reticulate pollen grains. Köhler investigated about 70 species of *Gentiana sensu lato* with 3-colporate or 3–5-porate, distinctly to indistinctly reticulate pollen grains. Pollen of *Gentiana pneumonanthe* was described as 3-colporate and striate (Erdtman *et al.*, 1961), and *G. nivalis* was reported to have similar 3-colporate and reticulate-striate pollen by Faegri and Iversen (1992).

Pollen from about 150 species of *Gentiana* was investigated by Nilsson (1967a). In *Gentiana*, 3-colporate (exceptionally 4-colporate or 6-aperturate) pollen grains with striate, striato-reticulate, and reticulate and transitional exine patterns (rarely smooth) were encountered. Infraspecific variation as to exine pattern was noticed in *Gentiana prostrata* and *G. utriculosa* (Nilsson, 1967b). The species of *Gentiana* described by Wang (1960) are striato-reticulate to reticulate. Seven species of *Gentiana* from Taiwan were all found to be 3-colporate and striato-reticulate (Huang, 1972).

Punt and Nienhuis (1976) described *Gentiana pneumonanthe* (tectate, finely suprastriate), and they added *G. cruciata, G. lutea, G. pneumonanthe,* and *G. purpurea* to the *G. pneumonanthe* type of pollen. Agababyan and Tumanyan (1977b) investigated 18 species of *Gentiana sensu lato*. Pollen from a few additional *Gentiana* species from China (Gansu) was described and depicted by Ma and Zhao (1991).

In two recently published works by Ho *et al.* (1994b, 1996) the pollen grains of 64 species of *Gentiana* were examined with special reference to their phylogenetic implications. *Crawfurdia* and *Tripterospermum* were selected as outgroups. The character states included exine ornamentation, shape of columellae, and relative thickness of sexine and nexine. Five pollen types and 14 subtypes were recognized. Several general evolutionary trends were proposed, including from striate-imperforate, striate-perforate, striate-foveolate, to striate-reticulate or reticulate. Apertures having long and wide colpi with distinctly thickened margins were regarded as more plesiomorphic than short colpi with indistinct margins. The evolutionary trend for shapes of pollen grains was proposed to be from spheroidal to prolate to perprolate.

However, some of the criteria used for evolutionary trends by Ho *et al.*

(1994b, 1996) are not sharply defined with reference to the variability of exine patterns and shapes. Transitions between character states are also common within species or even within accessions upon closer examination. Perforations are not easily traced in between closely spaced lirae, particularly at the mesocolpia. Perforations rarely exceed 1 μm in diameter (thus not foveolae). A general trend from striate (with low interconnecting muri between lirae) to striato-reticulate to reticulate seems plausible. A trend from spheroidal to prolate to perprolate is questionable and inconsistent as the shape variation is related to the chemical treatment (acetolysis), and its influence on variously expanded pollen grains in a sample can vary.

Gentianella. *Gentianella campestris* (L.) Börner. The pollen grains are 3-colporate, prolate, and *c.* 49×36 μm. The exine is striato-reticulate. The muri/lirae are unevenly wide and interconnected by tiny muri at a lower level. The lumina are rounded to angular-shaped, appearing as perforations toward the poles. (Sweden: Jämtland, 1995, *Nilsson s.n.* (S), slide no. 23882; Fig. 2.27A–B.)

Gentianella diffusa (H. B. K.) Fabris. The pollen grains are 3-colporate, subprolate, and *c.* 40×34 μm. The exine is striate-perforate, with muri/lirae relatively thin, parallel, and densely spaced. Toward the polar areas there is a more open pattern of irregularly shaped and oriented lirae/muri. (Colombia: Cauca, 1938, *von Sneidern 1886* (S), slide no. 23884; Fig. 2.27C.)

Gentianella montana G. Forst. The pollen grains are 3-colporate, oblate spheroidal, and *c.* 42×47 μm. The exine is coarsely and distinctly striate. The muri/lirae are unevenly wide, partly branched, anastomosing, and intertwined. At the center of mesocolpia, the muri/lirae are markedly prominent, but are fused to a narrow, smooth, and perforate margin at the colpus border. (New Zealand: Mt. Arthur, 1987, *Sneddon s.n.* (WELTU), slide no. 21407; Fig. 2.27D.)

OTHER INVESTIGATIONS: The pollen grains of *Gentianella baltica* were reported by Erdtman *et al.* (1961) to be 3-colporate and reticulate, a type that was also found in *G. amarella*, *G. campestris*, and *G. tenella* (Faegri & Iversen, 1992). Pollen grains of 60 species of *Gentianella* (including *Comastoma* and *Gentianopsis*) were studied by Nilsson (1967a). The pollen grains were found to be 3(–4)-colporate with an exine ornamentation varying from finely to coarsely striate, striato-reticulate, to reticulate.

Punt and Nienhuis (1976) described three pollen types under *Gentianella*: (1) *Gentianella campestris*-type (including *Gentiana nivalis* and *G. verna*,

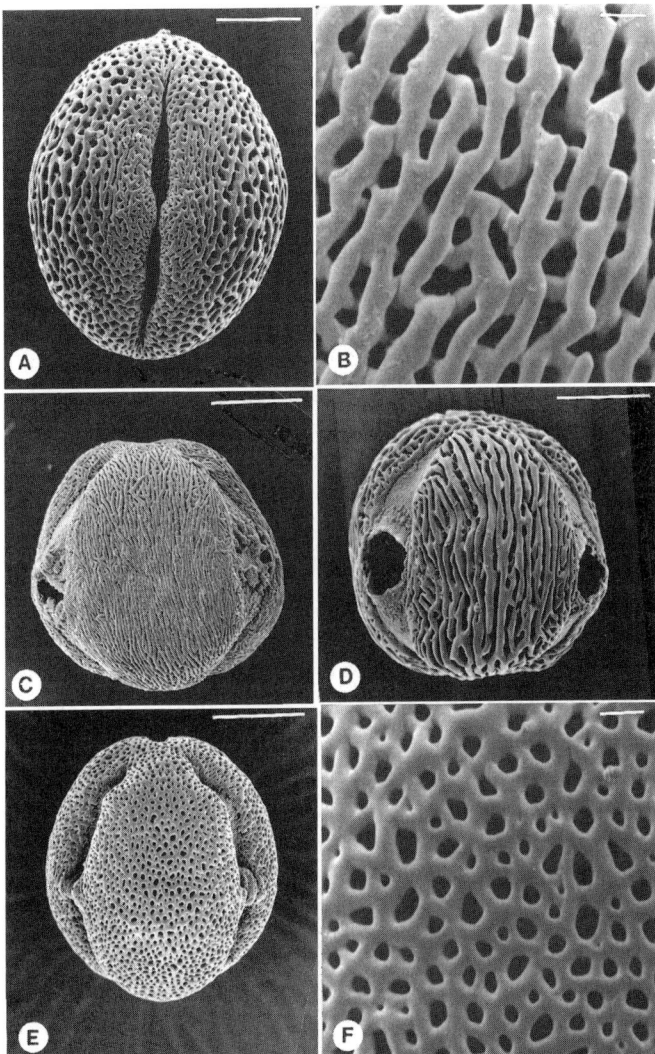

Figure 2.27. Pollen of Gentianeae-Swertiinae. **A, B.** *Gentianella campestris* (*Nilsson s.n.*). **A.** 3-colporate, coarsely striato-reticulate pollen grain in equatorial, apertural view; colpus equatorially constricted. **B.** Detail of exine with branched, anastomosing muri/lirae, interconnected by transverse muri at a lower level. **C.** *Gentianella diffusa* (*von Sneidern 1886*). 3-colporate, finely striate pollen grain with densely spaced, parallel, partly branched muri/lirae, in equatorial, mesocolpial view. **D.** *Gentianella montana* (*Sneddon s.n.*). 3-colporate, markedly coarse striate pattern with parallel, branched, and anastomosed muri/lirae. **E, F.** *Comastoma polycladum* (*Küpfer et al. s.n.*). **E.** 3-colporate, reticulate pollen grain in equatorial, mesocolpial view. **F.** Detail of reticulate exine; lumina rounded to oval in shape. Scale bars: A, C, D, E = 10 μm; B, F = 1 μm.

Gentianella anglica, *G. germanica*, and *G. uliginosus*), (2) *Gentianella tenella*-type (including *G. amarella*, *G. aurea*, *G. tenella* (= *Comastoma tenellum*), and *Swertia perennis*), and (3) *Gentianella detonsa*-type (including *G. detonsa* (= *Gentianopsis detonsa*); see below). Some additional Chinese species of *Gentianella* and *Gentianopsis* were studied by Ma and Zhao (1991). Wang (1960) investigated the pollen of one species of *Gentianella*, *G. trailliana*, which he described and illustrated as being reticulate.

Most of the Australian, New Zealand, and South American species, e.g., *Gentianella montana* (present study), have distinctly and coarsely striate pollen grains. They were transferred from *Gentiana* (sect. *Arctophila* and *Antarctophila*; Gilg, 1895a) to *Gentianella* by Ho and Liu (1993), or have even been treated as belonging to a new genus, *Chionogentias*, by Adams (1995). Moar (1993) investigated 23 New Zealand species of *Gentiana*, e.g., *G. bellidifolia* and other species with 3-colporate, coarsely striate-reticulate pollen grains, and still others with a finer exine pattern, from reticulate to finely reticulate to striate and perforate. He concluded that further taxonomic studies are needed. With reference to the coarsely striate to striato-reticulate pollen grains, these species could probably be kept as a separate group.

Gentianopsis. *Gentianopsis crinita* (Froel.) Ma. The pollen grains are 3–(4)-colpate-colporate, oblate spheroidal, and *c.* 49×50 μm. The exine is coarsely reticulate to reticulate. The muri are ring-shaped, without being fused with the contiguous muri delimiting the rounded lumina. The lumina are filled with granular to elongated processes. (USA: North Carolina, 1970, *H. & K. Clarke et al. 3909* (DUKE), slide no. 20622; Fig. 2.28A–B.)

OTHER INVESTIGATIONS: The species of *Gentiana* (*Gentianella*) sect. *Crossopetalum* (*sensu* Gilg, 1895a), i.e., *Gentianopsis*, were found to differ from all other sections of *Gentianella* by having apertures that are usually 4-colpor(oid)ate with relatively short colpi, and a deviating exine pattern, i.e., coarsely reticulate with numerous processes in the lumina (Nilsson, 1967a; see also *Gentianella* above). Pollen morphology is in favor of keeping species with this type of pollen in a separate section or genus.

Halenia. *Halenia brevicornis* (H. B. K.) G. Don. The pollen grains are 3-colporate, prolate spheroidal, and *c.* 30×28 μm. The exine is finely reticulate and perforate at the polar areas and finely striate toward the aperture margin. The colpus margin is distinct, smooth, sparsely perforate, and with markedly protruding overlappings above the ora. (Mexico: Tamascaltopoc, 1935, *Hinton 8311* (S), slide no. 5311; Fig. 2.28E–F.)

Systematics, character evolution, and biogeography 257

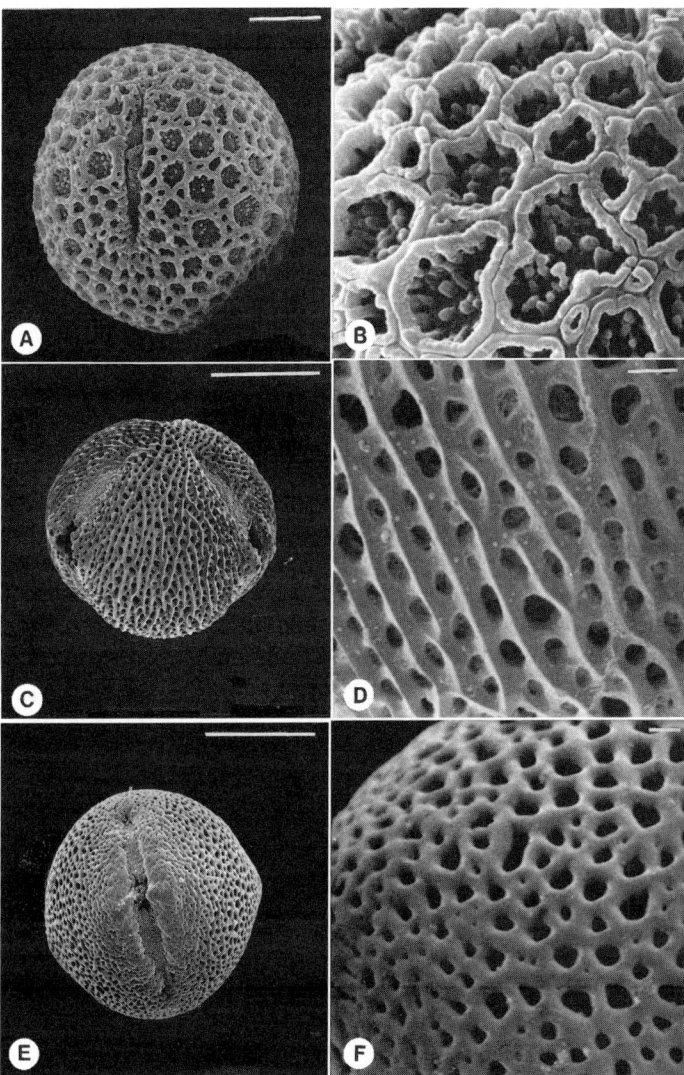

Figure 2.28. Pollen of Gentianeae-Swertiinae. **A, B.** *Gentianopsis crinita* (*Clarke et al. 3909*). **A.** 4(3)-colpor(oid)ate, reticulate pollen in equatorial, apertural view. **B.** Detail of exine with circular-shaped muri not fused with adjacent ones, and surrounding lumina with numerous processes inside. **C, D.** *Lomatogonium macranthum* (*Yuan & Zeltner s.n.*). **C.** 3-colporate, striate pollen in equatorial, mesocolpial view. **D.** Detail of tectate-perforate exine with sharply keeled muri/lirae and perforate to almost foveolate striae. **E, F.** *Halenia brevicornis* (*Hinton 8311*). **E.** 3-colporate, reticulate pollen in equatorial, apertural view; exine with protrusions above the os at equator. **F.** Detail of reticulate exine with rounded to oval-shaped lumina. Scale bars: A, C, E = 10 μm; B, D, F = 1 μm.

OTHER INVESTIGATIONS: Pollen of *Halenia* was described as striate to indistinctly reticulate (Gilg, 1895a). Köhler (1905) examined six species of *Halenia*, describing the pollen as 3-colporate, reticulate, with non-angular lumina. In a study of 16 species of *Halenia* by Nilsson (1967a), three types of pollen were distinguished using the shape of the grains and the shape of the columellae and their arrangements. The pollen grains studied were 3-colporate, usually with a triangular amb. In terms of pollen morphology, *Halenia* resembles *Lomatogonium* and *Comastoma*. One species of *Halenia* was included in the study of Ma and Zhao (1991).

Jaeschkea. Jaeschkea canaliculata (Royle ex Don) Knobl. The pollen grains are 3-colporate, spheroidal, and *c.* 37×37 μm. The exine is striate and tectate-perforate. The muri/lirae are sharply keeled and separated by rounded to oval-shaped perforations in the striae. (Kashmir: Bangar, 1943, *Ludlow & Sherriff 9283* (UPS), slide no. 661; Fig. 2.29A–B.)

OTHER INVESTIGATIONS: Pollen of *Jaeschkea* has been compared with that of *Crawfurdia*, i.e., both being striate to indistinctly reticulate (Gilg, 1895a). However, according to Köhler (1905), the pollen grains of *Jaeschkea* are 3-colporate and reticulate. *Jaeschkea oligosperma* has striate to striato-reticulate pollen grains similar to those of *J. canaliculata* (Nilsson, 1967a).

Lomatogonium. Lomatogonium macranthum (Diels & Gilg) Fern. The pollen grains are 3-colporate, oblate spheroidal, and *c.* 25×27 μm. The exine is tectate-perforate, striate, with muri/lirae sharply keeled, straight and branched, anastomosing, and with rounded to oval-shaped perforations in the striae. (China: Sichuan, 1993, *Yuan & Zeltner 93-91* (NEU), slide no. 23422; Fig. 2.28C–D.)

OTHER INVESTIGATIONS: According to Gilg (1895a), both *Lomatogonium* and *Crawfurdia* have pollen with striate or indistinctly reticulate exine. Köhler (1905) examined two species of *Lomatogonium* (*L. carinthiacum* and *L. rotatum*) with 3-colporate and reticulate pollen, also mentioning the projecting pores and triangular outline of the grains in polar view, features that have been confirmed in the present study. Wang (1960) described the pollen of *Lomatogonium rotatum* as having large ora and a striato-reticulate exine pattern.

Sixteen *Lomatogonium* species were studied palynologically by Nilsson (1964, 1967a). The 3-colporate pollen grains, usually with triangular amb,

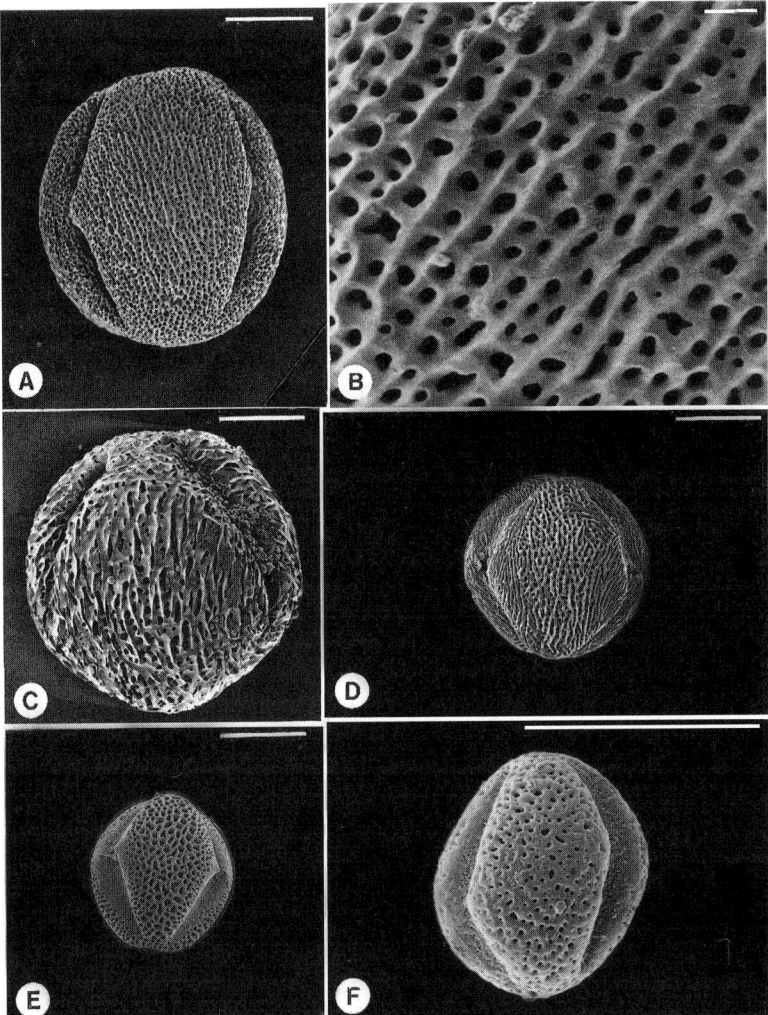

Figure 2.29. Pollen of Gentianeae-Swertiinae. **A, B.** *Jaeschkea canaliculata* (*Ludlow & Sherriff 9283*). **A.** 3-colporate, striate pollen grain in equatorial, mesocolpial view. **B.** Detail of the striate exine with sharply keeled muri/lirae and perforate to almost foveolate striae. **C.** *Megacodon venosus* (*Chu 3853*). 3-colporate, striate pollen grain in equatorial, mesocolpial view; muri/lirae sharply keeled, discontinuous. **D.** *Veratrilla baillonii* (*McClaren 140B*). 3-colporate, striato-reticulate pollen in equatorial, mesocolpial view; muri/lirae branched, anastomosing. **E.** *Obolaria virginica* (*Holthius s.n.*). 3-colporate, reticulate to finely reticulate pollen in equatorial, mesocolpial view. **F.** *Bartonia virginica* (*Fernald et al. 390*). 3-colporate, perforate to finely reticulate pollen in equatorial, mesocolpial view. Scale bars: A, C, D, E, F = 10 μm; B = 1 μm.

generally had large ora with prominent protrusions. Two main types of exine pattern were found: (1) exine striate to striato-reticulate, and (2) exine perforated, with spinules or shard-like processes. The latter type could be further subdivided using the shape of the pollen grains. The two different types of pollen were found in both annuals and perennials. Similarities between *Lomatogonium* and *Comastoma* pollen were noted by Nilsson (1967a).

Punt and Nienhuis (1976) described the pollen of the *Lomatogonium rotatum*-type (including *L. carinthiacum* and *L. rotatum*) as 3(–4)-colporate with large ora and striato-reticulate exine pattern. *Lomatogonium* was described as having 3-colporate, flattened pollen grains with striate exine by Faegri and Iversen (1992). Pollen from *Lomatogonium rotatum* from China (Gansu) was also described and illustrated by Ma and Zhao (1991).

Megacodon. *Megacodon venosus* (Hemsl.) Harry Sm. The pollen grains are 3-colporate, subprolate, and *c.* 48 × 38 μm. The exine is striate and tectate-perforate. The muri/lirae are sharply keeled, and not always continuous but cut off at places. The striae have rounded to oval-shaped perforations. (China: Pao-hsing-hsien, 1936, *Chu 3853* (BM), slide no. 24537; Fig. 2.29C.)

OTHER INVESTIGATIONS: The two species of *Megacodon* that were examined by Nilsson (1967a) had 3-colporate pollen, and showed differences in the apertures (ora) and exine features. Wang (1960) described and illustrated pollen of *Megacodon stylophorus*.

Obolaria. *Obolaria virginica* L. The pollen grains are 3-colporate, subprolate, and *c.* 24 × 18 μm. The exine is reticulate to finely reticulate, with a distinct smooth and sparsely perforate colpus margin. The lumina are rounded to angular in shape. (USA: Maryland, 1953, *Holthius s.n.* (L), slide no. 24336; Fig. 2.29E.)

OTHER INVESTIGATIONS: Gilg (1895a) compared pollen grains of *Obolaria* with those of *Enicostema* without mentioning the exine pattern specifically. The pollen grains of *Obolaria virginica* were described by Nilsson and Skvarla (1969) as being larger than those of *Bartonia* and clearly reticulate. This is confirmed in the present study.

Swertia. *Swertia franchetiana* Harry Sm. The pollen grains are 3-colporate, prolate spheroidal, and *c.* 31 × 29 μm. The exine is tectate-perforate with straight, branched and sharply keeled muri/lirae. The striae have rounded perforations of varying size. (China: Ganshu, 1993, *Yuan & Zeltner s.n.* (NEU), slide no. 23523.)

OTHER INVESTIGATIONS: Gilg (1895a) described the pollen grains of *Swertia* as striate to indistinctly reticulate. Köhler (1905) investigated 14 species of *Swertia*, the pollen of which he described as 3-colporate and clearly reticulate.

Among the more than 50 species of *Swertia* examined by Nilsson (1967a), the 3-colporate pollen grains were found to be both striate, striato-reticulate, reticulate, and spinuliferous. As in *Lomatogonium*, spinulose pollen grains occurred in both annuals and perennials. Huang (1972) described the pollen of four *Swertia* species from Taiwan as 3-colporate and striato-reticulate or reticulate. Punt and Nienhuis (1976) put *Swertia perennis* in the *Gentianella tenella*-type, characterized by 3-colporate and finely striate to microreticulate pollen grains. *Swertia bifolia*, with striato-reticulate pollen grains, was examined by Ma and Zhao (1991). Further studies are required to investigate if the genus *Frasera* is separable from *Swertia* with reference to pollen.

Tripterospermum. *Tripterospermum volubile* (D. Don) Harry Sm. The pollen grains are 3-colporate, subprolate, and *c.* 36×27 μm. The exine is tectate-perforate and striate. The muri/lirae are sharply keeled, branched, and run parallel at the mesocolpia and toward the poles. The densely perforate striae have perforations in one or more rows. (Bhutan: Tobrany, 1949, *Ludlow et al. 20848* (UPS), slide no. 25080; Fig. 2.25C–D.)

OTHER INVESTIGATIONS: *Tripterospermum* has striate to indistinctly reticulate pollen grains (Gilg, 1895a, as *Crawfurdia* subg. *Tripterospermum*).

Nine species of *Tripterospermum* were studied by Nilsson (1967a), who concluded that *T. caudatum* and *T. volubile* resembled *Crawfurdia delavayi* and *C. crawfurdioides* in certain exine features. *Tripterospermum affine* also varied infraspecifically (Nilsson, 1967b), with some specimens having distinctly striate pollen grains, whereas others were lacking the striate pattern.

Veratrilla. *Veratrilla baillonii* Franch. The pollen grains are 3-colporate, oblate spheroidal, and *c.* 25×26 μm. The exine is striato-reticulate, with branched and anastomosed muri/lirae interconnected by more narrow and lower mural bridges. The striato-reticulate exine pattern is more distinct at the center of the mesocolpia, with muri/lirae becoming progressively thinner and more densely spaced toward the aperture margin. (China: *McClaren 140 B* (BM), slide no. 24999; Fig. 2.29D.)

OTHER INVESTIGATIONS: The two species of *Veratrilla*, *V. baillonii* and *V. burkilliana*, both have 3-colporate and striate to striato-reticulate pollen (Nilsson, 1967a). *Veratrilla burkilliana* shows infraspecific palynological variation (Nilsson, 1967b).

Discussion of palynological data in Gentianeae. The pollen grains of Gentianeae are 3-colporate (exceptionally 3–4-colporoidate), striate to striato-reticulate, reticulate, and perforate. The polar size (P) of pollen ranges between c. 30 μm and 50 μm; however, *Lomatogonium*, *Obolaria*, and *Veratrilla* have pollen less than 30 μm in diameter, and *Bartonia* has pollen less than 20 μm in diameter. The striate to striato-reticulate exine pattern consists of densely arranged muri/lirae, with or without traceable perforations in the striae (tectate-perforate), to an open pattern with rounded to angular lumina.

Biogeography of Gentianeae

Reconstruction of the biogeographic history of Gentianeae encounters major difficulties because generic relationships are not yet well enough resolved. However, the distributions of extant taxa, and their interrelationships, so far as known, allow us to formulate some hypotheses for future testing. Except for *Gentiana*, *Gentianella*, *Halenia*, and *Swertia*, all genera of Gentianeae are found exclusively in the Northern Hemisphere. Most species of the tribe grow in temperate to alpine climates. For *Gentiana*, *Gentianella*, *Halenia*, and *Swertia*, it is likely that the genera colonized the Southern Hemisphere secondarily (see below).

Of the two major clades of Gentianeae, subtribe Gentianinae is clearly centered in the Old World. *Crawfurdia*, *Gentiana* sect. *Stenogyne*, and *Tripterospermum*, all members of one subclade, grow in Central to East Asia. Within *Gentiana* (excluding sect. *Stenogyne*), one subclade is primarily European, with a few species in northwest Africa, northeast North America, and Central Asia (Meusel *et al.*, 1978). The second clade contains Asian and widespread sections. The widespread sections are most diverse in eastern Asia (Ho & Liu, 1990). An exception to this is *Gentiana* sect. *Pneumonanthe*, which is most diverse in eastern North America. On the background of this distributional pattern and the phylogenetic relationships inferred, it seems most likely that the ancestor of Gentianinae occupied a temperate alpine range in the Old World, and that the New World and Southern Hemisphere were occupied secondarily (Yuan et al., 1996). The distribution of generic and specific diversity, if this is accepted as an argument, would even suggest an eastern Asian origin of the Gentianinae. Essentially nothing is known, however, about the timescale of this diversification. Only the European *Gentiana* sect. *Ciminalis* has been postulated to have begun to radiate 2 million years or less ago (Hungerer & Kadereit, 1998).

Whereas subtribe Gentianinae are likely to be of Old World origin, this is much more difficult to decide for subtribe Swertiinae. In this clade, *Jaeschkea, Megacodon, Latouchea, Pterygocalyx*, and *Veratrilla* are exclusively eastern Asian in distribution. In *Swertia*, only the circumpolar *S. perennis* grows in North America, whereas all other species are found in the Old World. Nine of the 11 sections of this genus (10, after the exclusion of *Frasera*) grow exclusively or partly in southwestern China and the Himalayas (Ho et al., 1994a). As a consequence, *Swertia* can be added to the above list of Asian taxa. The same applies to *Lomatogonium*, which contains 19 species in East Asia, one species widespread in Eurasia, and one circumpolar species, and perhaps also to *Comastoma*, with most of its species and its closest relatives (binectariate *Gentianella*, see below, and *Lomatogonium*) occurring in Asia. In contrast to these mostly or exclusively Asian genera, *Bartonia, Frasera*, and *Obolaria* are exclusively North American in distribution.

The remaining genera, i.e., *Gentianella, Gentianopsis*, and *Halenia*, are more complex in distribution. Binectariate species of *Gentianella* grow in Central to East Asia. The fimbriate group of *Gentianella*, as defined above, occurs with about an equal number of species in Eurasia and North and Central America, and the efimbriate group is widespread with some species in Eurasia and North America and most species in South America and Australia/New Zealand. The number of species of *Gentianopsis* is about equal in Eurasia and North America, and *Halenia* contains three Asian, one temperate North American, and many Central and South American species.

The phylogeny available for the Swertiinae does not permit a conclusive biogeographic interpretation of this subtribe, although three likely alternatives can be postulated. The first possibility is that the basic differentiation of an originally Laurasian tribe Gentianeae into subtribes Gentianinae and Swertiinae took place in Eurasia and North America, respectively. This would imply that Swertiinae are of North American origin and have spread into Asia. The second option is that Swertiinae, like Gentianinae, are of

Old World/Asian origin and have spread probably several times into the New World. The third alternative is that the original range of Swertiinae was Laurasia-wide, and a primary subdivision of the lineage was between the American genera *Bartonia* and *Obolaria* (but probably not *Frasera* and *Halenia*; see below) and the Eurasian genera. We prefer this third hypothesis because *Megacodon*, *Obolaria*, *Bartonia*, and *Latouchea* are very likely to be basal in Swertiinae, and these taxa are exclusively either American or Asian in distribution. The best explanation for this is a basal split of Swertiinae into an American and a Eurasian clade, although no direct support for this hypothesis is provided by our phylogeny. Moreover, this possibly ancient split in the subtribe has been masked by more recent migrations of several taxa between the continents.

In our ITS phylogeny, the Asian species *Halenia elliptica* is sister to the remainder of the genus (distributed in the Americas), and species of other genera in this clade also are Asian in distribution. This may indicate that *Halenia* originally was not of American distribution but arrived there secondarily. The presence of *Halenia* in South America is probably rather recent because alpine habitats in South America became available only about 3 million years ago (Simpson, 1975).

In our ITS data, the American genus *Frasera* is sister to *Gentianopsis* from America and Asia and *Pterygocalyx* from Asia. Whether this entire lineage originated in North America and spread to Asia, or whether an originally Laurasian-wide range of this lineage is more likely, cannot be decided at present.

Not considering the apparently unrelated Asian binectariate species, the distribution of *Gentianella* can be interpreted in several ways. The efimbriate group is either paraphyletic in relation to the fimbriate group (from chloroplast DNA data) or both are monophyletic and sister to each other (from nuclear ITS data). Considering the distribution of the close relatives of uninectariate *Gentianella*, one may assume an East Asian origin of the genus. Following the ITS data both groups spread from there across the Northern Hemisphere, and the efimbriate group also dispersed to South America. This subgroup strongly diversified there and also dispersed to Australia/New Zealand. However, following the chloroplast data, only the efimbriate group migrated to North America, where the fimbriate group originated. From there, Eurasia was secondarily colonized by the fimbriate group. Whereas in the first interpretation the Eurasian species of both groups are relictual in distribution, in the second interpretation the Northern Hemisphere species of the efimbriate group and the North American species of the fimbriate group are relictual. In view of the general distribution and ecology of *Gentianella* and its probable relatives, it seems very likely that the Southern Hemisphere parts of its range are secondary and possibly quite young. As already mentioned for *Halenia*, the alpine habitats of *Gentianella* in South America became available only about 3 million years ago (Simpson, 1975). This could also imply rather recent long-distance dispersal of *Gentianella* from South America to Australia/New Zealand, which is supported by short branches separating Australian/New Zealand species from South American species in the ITS tree (not shown).

The African and Malagasy groups of *Swertia* are likely to have dispersed there twice from the north, because the two African clades are not related to each other in our phylogeny. The occurrence of temperate Eurasian taxa in the African high mountains is a distribution pattern known from many other plant groups (Hedberg, 1957; Smith & Cleef, 1988). We do not have enough new data to interpret the biogeography of other groups of *Swertia*.

It is difficult to name specific geological or climatic events that may be responsible for the lineage splits seen in our phylogeny that could provide absolute dates for their cladogenesis. It could be suggested, however, that the split between New World *Bartonia* and *Obolaria* and other Swertiinae, and the split of the *Frasera–Gentianopsis–Pterygocalyx* lineage, were both caused by the increasing isolation between North America and Eurasia as a result of climatic cooling in northern latitudes. The easy exchange of boreotemperate taxa between North America and Eurasia was possible until 15 million years ago (Parks & Wendel, 1990), and alpine taxa could probably move between the continents until the end of the last ice age (Lang, 1994). Therefore, 15 million years may be the minimum age of the split between temperate *Bartonia/Obolaria* and the remainder of Swertiinae. The split between Gentianinae and Swertiinae and the origin of Gentianeae from tropical ancestors must have taken place earlier than this. In the absence of fossils, however, we currently cannot provide any estimate of absolute time. The other New World/Old World disjunctions (often between alpine taxa) are very likely to have arisen at many different times. For example, it seems possible that *Lomatogonium*, *Swertia*, and some groups of *Gentiana* migrated between Eurasia and North America only recently. North American/East Asian disjunct distributions are common in many plant groups (Li, 1952; Hong, 1993), and Gentianeae provide a good example for the very extensive exchange between the two continents. The migration of *Halenia* and *Gentianella* from North to South America and their subsequent extensive radiation has probably taken place since only 3 million years ago (see above). The long-distance dispersal of *Gentianella* to New Zealand/Australia may be even younger than this.

In summary, it is evident that, much in contrast to most other tribes of the family (except Chironieae-Chironiinae), the differentiation of Gentianeae appears to have been strongly related to geological and climatic changes in the Northern Hemisphere. Much of this differentiation is likely to have taken place in the Tertiary and the Quaternary.

TAXA *INCERTAE SEDIS*
L. STRUWE AND V. A. ALBERT

Voyria AUBL.

Voyria is a genus of pink, yellow, white, blue, or purple chlorophyll-less rainforest saprophytes with showy flowers. It was described by Aublet (1775: 208) from French Guiana based on *Voyria caerulea* Aubl. (lectotypified by Maas & Ruyters, 1986). Von Schlechtendal and von Chamisso (1831: 387) later described the genus *Leiphaimos*, which is now included in *Voyria* (Maas & Ruyters, 1986; Albert & Struwe, 1997). The plants often grow solitarily on the rainforest floor, never in abundance. There are 18 species in the Neotropics (centered on the Guayana Shield), and one additional species, *Voyria primuloides*, has a rather wide distribution in tropical West Africa. *Voyria aphylla* is the most common and widespread species in the genus, reaching several islands in the Caribbean, a distribution that has probably been facilitated by the filiform, wind-dispersed seeds. The larger-seeded species might be dispersed by rain-wash. *Voyria* grows in a wide variety of habitats (Maas in Struwe *et al.*, 1999). A monograph of *Voyria* was published by Maas and Ruyters (1986) that includes studies on wood, pollen, and seed anatomy. *Voyria primuloides* was treated in a floristic work by Boutique (1972).

The species of *Voyria* are saprophytic and perennial herbs (Maas & Ruyters, 1986; Albert & Struwe, 1997; Struwe *et al.*, 1999). The morphology of the roots and rhizomes is not well known, but can sometimes be coralloid or of the "bird's nest" type. The stem anatomy is highly reduced in several species with the vascular cylinder being replaced by a few individual vascular bundles and a sclerenchymatous ring (Maas & Ruyters, 1986). The small and scale-like leaves are slightly connate at the base. The terminal inflorescence is a few- to 30-flowered cyme or the flowers are borne singly. The flowers are (4–)5(–7)-merous, actinomorphic, and presumably not heterostylous (cf. *Voyriella*). The tubular to campanulate calyx is persistent in fruit. The actinomorphic and marcescent corolla is salverform to funnelform, and variously colored (pink, yellow, white, blue, or purple).

The stamens are inserted at varying levels in the corolla tube (in different species), with well-developed to nearly absent filaments. The bicarpellate ovary is unilocular with protruding parietal placentas. Two glands, sometimes stalked, are often present at the base of the ovary. The style is filiform and gradually widened toward the ovary, and is tipped with a funnelform, rotate, capitate, or slightly bilobed stigma. The fruit is a dry capsule that is septicidally dehiscent or indehiscent. The seeds are many, and can be globose to filiform.

The precise phylogenetic position of *Voyria* is not known. A reliable placement for the genus based on *trn*L intron data has not been possible so far because the sequences obtained (for 2 spp.) were strongly diverged compared with other Gentianales taxa (L. Struwe, M. Thiv, J. Kadereit, & V. A. Albert, unpubl.). Initial phylogenetic results placed *Voyria* with Solanaceae, with which it shares few morphological characters (Struwe & Albert, 2000a). An analysis of nuclear ribosomal 18S sequences showed *Voyria* to be placed among the basal clades of the Gentianaceae (Struwe & Albert 2000a). However, these data depicted *Voyria* as possibly non-monophyletic, with species instead grouping in two clades along current subgeneric designations (Albert & Struwe, 1997; L. Struwe & V. A. Albert, unpubl.). Nuclear ITS sequences were also produced, and whereas these differ markedly among presumably closely related taxa, the 18S data did not appear particularly sequence divergent relative to other gentians (Struwe & Albert, 2000a). Nevertheless, given the question about monophyly and our inability to resolve relationships using a plastid DNA marker, we do not consider the position of *Voyria* among gentians to be sufficiently supported to formally place them in our classification. Despite these uncertainties, *Voyria* flowers do fit the gentian bauplan, the plants have no latex (as would Apocynaceae), they are hypogynous (unlike virtually all Rubiaceae), and *Voyria* subg. *Voyria* has stem anatomy typical of gentians as well as clear embryological similarities.

Within the genus the phylogenetic relationships are better known. A morphological study identified two major clades in *Voyria*, which were subsequently classified as two subgenera, subgenus *Voyria* and subgenus *Leiphaimos* (Albert & Struwe, 1997). *Voyria* subgenus *Voyria* is characterized by several synapomorphies, e.g., urceolate, anatropous, and unitegmic seeds, corolla tubes gradually widened, and five calyx veins that each split into three veins, as opposed to subgenus *Leiphaimos*, which has several other unique characters, e.g., acuminate, atropous, and ategmic seeds, corolla tubes narrow and widened at staminal insertion level only, and five, unbranched calyx veins (Albert & Struwe, 1997). The only African species,

Voyria primuloides, is far nested inside American taxa, and has been hypothesized to be a boreotropical relict, possibly the remnant of a formerly larger paleotropical clade (Albert & Struwe, 1997). The Guayana Shield is supported as the ancestral distribution area for *Voyria* by the same morphological analysis.

Palynology of *Voyria*

S. NILSSON

Gilg (1895a) put the saprophytic genus *Voyria* in a tribe of its own, the Voyrieae, whereas *Leiphaimos* (= *Voyria*) and *Voyriella* were classified into another tribe, the Leiphaimeae. The shape of the pollen and number of pores were the main criteria to keep them as two separate tribes. The pollen grains of *Voyria* and *Leiphaimos* were later found to be basically the same (Nilsson & Skvarla, 1969), i.e., heteropolar, bilateral (reniform), radially symmetrical, or asymmetrical, with a maximum diameter not exceeding 20 μm, and 1-6-porate. The exine is smooth or scabrous and not stratified. The pollen of *Voyria* and *Voyriella* was further examined by Nilsson (in Maas & Ruyters, 1986) and later by Maguire and Boom (1989). Roubik and Moreno (1991) also described and illustrated pollen from three species of *Voyria*. The pollen grains of *Voyria* are different from all other taxa in the family Gentianaceae, and are distinct enough to justify *Voyria* being kept in a separate tribe.

ACKNOWLEDGMENTS

The authors thank all herbaria, botanical gardens, and institutions that have given access to their collections and libraries during the course of this work and sometimes given us permission to extract DNA or remove pollen from their collections (BC, BM, BR, DUKE, E, FR, HBG, JE, K, L, LL, M, MJG, MO, NBG, NEU, NO, NY, OKL, P, PRE, QCA, QCNE, S, TEX, TOGO, TSB, U, UPS, VER, WELTU, WU, and ZT; the botanical gardens in Edinburgh, Kyoto, Kew, Mainz, New York, Schachen, and Zürich). We also gratefully acknowledge the authors of other chapters in this book for sharing unpublished results, as well as Lisa Campbell, Philippe Chassot, Peter Endress, David Glenny, Katherine R. Gould, Jason R. Grant, Henk Groen, Philippe Küpfer, Paul Maas, Guilhem Mansion, Karsten Meyer, Tim Motley, Joachim Nerx, Gustavo Romero, Erik Simonis, and Noor van Heusden. Rupert Barneby provided the Latin diagnoses. We thank Stefan Beck, Alexander Berg, Paul Berry, Christiane Bittkau, Frank Blattner, Georges Cremers, Peter Endress, Tassilo Feuerer, Carol Gracie, Rainer

Greissl, Katja Gutsche, Andrew Henderson, Herbert Hurka, Klaus Kubitzki, Matt Lavin, Paul Maas, Scott Mori, Börge Pettersson, and Maximilian Weigend for collecting gentian material and/or providing helpful information. Philippe Küpfer and Jan Schlauer kindly put their private herbarium at our disposal. We thank James S. Farris of the Swedish Museum of Natural History for permission to use his unpublished parsimony jackknifing application, XAC. Jason R. Grant, Charlotte Lindqvist, Guilhem Mansion, Susan Pell, and Lowell Urbatsch are also thanked for help with floristic references. We also thank Elisabeth Grafström, Magnus Hellbom, and Wieslaw Smolenski (Palynological Laboratory, Swedish Museum of Natural History) for skillful technical help with pollen investigations. Anita Norrthon is acknowledged for help in preparing the palynological part of the chapter. Bobbi Angell prepared the Potalieae and Helieae drawings, Pollyanna von Knorring the Exaceae, and Anke Berg and Doris Franke the Chironieae and Gentianeae drawings.

This work was supported by The Lewis B. and Dorothy Cullman Foundation, the National Science Foundation (grant DBI-9601515 to V.A.A.), and a grant from the Deutsche Forschungs Gemeinschaft (to J.W.K.). L.S. thanks Rutgers University–Cook College and V.A.A. thanks The College of Arts and Sciences, The University of Alabama, for support during the final editing of the chapter.

Appendix 2.1
Overview of selected floristic works that include treatments of Gentianaceae

The list is arranged by geographical region. Also listed are treatments of Loganiaceae that include the three genera *Anthocleista*, *Fagraea*, and *Potalia* (see footnotes), which together now form subtribe Potaliinae in the Gentianaceae. The name of each flora is listed first, followed by the reference. Countries are indicated when it is not clear from the work's title which countries it includes. See Frodin's (2001) *Guide to standard floras of the world* for additional information on floristic research worldwide.

Africa and Madagascar
Flora Capensis [South Africa] Thistleton-Dyer (1909)
Flora of Libia Siddiqi (1977)
Flora of southern Africa Marais & Verdoorn (1963)
Flora of southern Africa[a] Verdoorn (1963)
Flora of tropical Africa Baker & Brown (1903)
Flora of tropical East Africa[a] Bruce & Lewis (1960)
Flora of west tropical Africa, 1st edn[a] Hutchinson & Dalziel (1931)
Flora of west tropical Africa, 2nd edn Taylor (1963)
Flora Zambesiaca[a] [Mozambique, Malawi, Zambia, Zimbabwe] Leeuwenberg (1983)
Flora Zambesiaca [Mozambique, Malawi, Zambia, Zimbabwe] Paiva & Nogueira (1990)
Flore d'Afrique centrale[a] [Congo, Rwanda, Burundi] Leeuwenberg & Bamps (1979)
Flore d'Afrique centrale [Congo, Rwanda, Burundi] Boutique (1972)
Flore de la Tunesie Pottier-Alapetite (1981)
Flore de Madagascar et des Comores Klackenberg (1990)
Flore de Madagascar et des Comores[a] Leeuwenberg (1984)
Flore du Cameroun[a] Leeuwenberg (1972)
Trees of southern Africa[a] Coates Palgrave (1991)

Australia and The Pacific
Flora of Australia[a] Conn et al. (1996)
Flora of Australia Adams (1996)
Flora Australiensis[a] Bentham (1869)
Flora of New South Wales [Australia] Harden (1992)

---

[a] Potaliinae treated.

Flora of Victoria [Australia]   Walsh & Entwisle (1993)
Flora vitensis nova[a] [Fiji]   Smith (1988)
Handbook of the New Zealand flora   Hooker (1867)
Manual of the flowering plants of   Wagner et al. (1990)
Hawai'i [USA]
Systematic studies of Micronesian   Fosberg & Sachet (1980)
plants[a]

Asia

A revised handbook to the flora of Ceylon   Cramer (1981)
Câyco Viêtnam[a]   Hô (1993)
Circumpolar arctic flora   Polunin (1959)
Conspectus gentianacearum japonicarum   Toyokuni (1963)
Flora Arctica URSS [Russia]   Tzvelev (1980)
Flora Armenii [Armenia]   Takhtajan (1980)
Flora Azerbajdzhana [Azerbaijan]   Sofieva (1957)
Flora Iranica   Schiman-Czeika (1967)
Flora Kavkaza [Armenia, Azerbaijan, Georgia, Russia]   Tamamshyan (1967)
Flora Kazakhstana   Semiotrocheva (1964)
Flora Koreana   Nakai (1911)
Flora Malesiana[a] [Indonesia, Malaysia, Singapore, Philippines, Brunei, Papua New Guinea]   Leenhouts (1962)
Flora of Bhutan   Aitken (1999)
Flora of China[a]   Li & Leeuwenberg (1996)
Flora of China   Ho & Pringle (1995)
Flora of Japan   Meyer & Walker (1965)
Flora of Java[a] [Malaysia]   Backer & Bakhuizen van den Brink (1965a)
Flora of Java [Malaysia]   Backer & Bakhuizen van den Brink (1965b)
Flora of lowland Iraq   Rechinger (1964)
Flora of Madhya Pradesh [India]   Srivastava (1997)
Flora of Rajasthan [India]   Parmar (1991)
Flora of Syria, Palestine and Sinai   Post (1933)
Flora of Taiwan   Liu & Kuo (1978)
Flora of Thailand[a]   Griffin & Parnell (1997)
Flora of Thailand   Ubolcholaket (1987)
Flora of the Malay Peninsula[a]   Ridley (1923)
Flora of the USSR   Grossheim et al. (1967)
Flora of Turkey and the East Aegean islands   Davis (1978)
Flora Sibiriae   Zuev (1997)
Flora Tadzhikskoy SSR [Tadjikistan]   Pissjaukova (1984)
Flora Turkmenii [Turkmenistan]   Nikitin (1954)

---

[a] Potaliinae treated.

Flora Uzbekistanica	Cherneva (1961)
Flore du Cambodge, du Laos et du Vietnam[a]	Tirel-Roudet (1972)
Flowers of the Himalaya	Polunin & Stainton (1984)
Key to the vascular plants of Mongolia	Grubov (2001)
Novelle flore du Liban et de la Syrie	Mouterde et al. (1978)
Opredelitel rasteniy Tuvinskoj ASSR [Russia]	Shaulo (1984)
Opredelitel sosudistykh rasteniy Kamchatskoy oblasti [Russia]	Probatova (1981)
The alpine flora of New Guinea[a]	van Royen (1983)
The flora of British India [India, Bangladesh, Burma, Bhutan, Ceylon, Pakistan, Nepal]	Hooker (1885)
The flora of Orissa [India]	Saxena & Brahman (1995)
The flora of the Tamilnadu Carnatic [India]	Matthew & Rani (1983)
Tree flora of Malaya[a] [Malaysia]	Kochummen (1973)
Wayside trees of Malaya[a] [Malaysia]	Corner (1988)

Europe

A Magyar Flóra és vegetáció rendszertani-növényföldrajzi kézikönyve [Hungary]	Soó (1966)
Circumpolar arctic flora	Polunin (1959)
Dansk feltflora [Denmark]	Hansen (1988)
Den nordiska floran [Denmark, Finland, Iceland, Norway, Sweden]	Mossberg et al. (1992)
Exkursionsflora von Deutschland [Germany]	Rothmaler (1990)
Flora Arctica URSS [Russia]	Tzvelev (1980)
Flora cr Srbije [Yugoslavia]	Jovanovich-Dunich (1973)
Flora d'Italia	Pignatti (1982)
Flora Ekskursioniste e Shqiperise [Albania]	Demiri (1983)
Flora Europaea	Tutin (1972)
Flora dels Països Catalans [Spain]	de Bolòs & Vigo (1995)
Flora de Mallorca [Spain]	Bonafè Barcelo (1979)
Flora der Schweiz und angrenzender Gebiete [Switzerland]	Heß et al. (1972)
Flora Helvetica [Switzerland]	Lauber & Wagner (1998a,b)
Flora of Cyprus	Meikle (1985)
Flora of Iceland	Löve (1983)
Flora of Russia [European part of Russia]	Tzvelev & Pissjaukova (2000)
Flora of the British Isles	Clapham et al. (1987)
Flora of the Baltic countries [Estonia, Latvia, Lithuania]	Jankeviciene et al. (1996)
Flora of the USSR	Grossheim et al. (1967)
Flora Polska	Jasiewicz (1971)

---

[a] Potaliinae treated.

Flora Portuguesa	Sampaio (1947)
Flora regionis boreali-orientalis territoriae Europaeae URSS	Laschenkova & Tolmatchev (1977)
Flora reipublicae popularis Bulgaricae	Jordanov (1982)
Flora republicii populare Romîne [Romania]	Savulescu (1961)
Flora URSR [Ukraine]	Visyulina (1957)
Flora vascular de Andalucía occidental [Spain]	Valdés *et al.* (1987)
Flora von Deutschland und angrenzender Länder [Germany]	Schmeil (2000)
Flora von Deutschland und angrenzender Länder [Germany]	Senghas & Seybold (1993)
Flore de la Suisse et des territoires limitrophes [Switzerland]	Aeschiman & Burdet (1994)
Flore descriptive et illustrée de la France	Coste (1990)
Illustrierte flora von Mittel-Europa [western and central Europe]	Hegi (1966)
Les quatre flores de France	Fournier (1961)
Mountain flora of Greece	Hartvig (1991)
New flora of the British Isles	Stace (1997)
Norsk, svensk, finsk flora [Norway, Sweden, Finland]	Lid (1985)
Nouvelle flore de la Belgique, du Grand-Duché de Luxembourg, du nord de la France et des régions voisines [Belgium, France, Luxembourg]	Lambinon *et al.* (1992)
Nuova flora Analitica d'Italia [Italy]	Fiori (1925)
Retkeilykasvio [Finland]	Hämet-Ahti *et al.* (1986)
Svensk flora [Sweden]	Krok & Almquist (1986)

North America

A flora of Arizona and New Mexico [USA]	Tidestrom & Kittel (1941)
A flora of New Mexico [USA]	Martin & Hutchins (1981)
A flora of tropical Florida [USA]	Long & Lakela (1971)
Anderson's flora of Alaska and adjacent parts of Canada [Canada, USA]	Welsh (1974)
An illustrated flora of the northern United States	Britton & Brown (1913)
An illustrated flora of the Pacific States: Washington, Oregon, and California [USA]	Abrams (1951)
A Utah flora [USA]	Welsh *et al.* (1993)
Budd's flora of the Canadian Prairie Provinces	Looman & West (1987)
Circumpolar arctic flora	Polunin (1959)

---

a Potaliinae treated.

Colorado flora: western slope/eastern slope [USA]	Weber & Wittmann (1996a,b)
Flora of Alaska and neighboring territories [Canada, USA]	Hultén (1968)
Flora of Alberta [Canada]	Moss (1983)
Flora of Baja California [Mexico]	Wiggins (1980)
Flora of Indiana [USA]	Deam (1970)
Flora of Maine [USA]	Haines & Vining (1998)
Flora of Manitoba [Canada]	Scoggan (1957)
Flora of Missouri [USA]	Steyermark (1963)
Flora of the Great Plains [USA]	McGregor et al. (1986)
Flora of the Northeast: a manual of the vascular plants of New England and adjacent New York [USA]	Magee & Ahles (1999)
Flora of the Pacific Northwest [USA]	Hitchcock & Cronquist (1976)
Flora of the Queen Charlotte Islands [Canada]	Calder & Taylor (1968)
Flora of the southeastern United States	Small (1903)
Flora of the Yukon Territory [Canada]	Cody (1996)
Flora of West Virginia [USA]	Strausbaugh & Cove (1978)
Flore Laurentienne [Canada]	Marie-Victorin (1995)
Gray's manual of botany [USA]	Fernald (1950)
Grønlands flora [Greenland (Denmark)]	Böcher et al. (1978)
Guide to the vascular flora of Illinois [USA]	Mohlenbrock (1986)
Guide to the vascular plants of central Florida [USA]	Wunderlin (1982)
Guide to the vascular plants of the Florida Panhandle [USA]	Clewell (1985)
Herbaceous plants of Maryland [USA]	Brown & Brown (1984)
Illustrated companion to Gleason and Cronquist's manual [USA]	Holmgren (1998)
Intermountain flora [USA]	Cronquist et al. (1984)
Manual of the vascular flora of the Carolinas [USA]	Radford et al. (1968)
Manual of vascular plants of northeastern United States and adjacent Canada	Gleason & Cronquist (1991)
Manual of the vascular plants of Texas [USA]	Correll & Johnston (1979)
Michigan flora [USA]	Voss (1996)
Plants of Iowa [USA]	Conard (1951)
The flora of Canada	Scoggan (1997)
The flora of New England [USA]	Seymour (1969)
The flora of Nova Scotia [Canada]	Roland & Smith (1969)
The genera of Gentianaceae in the southeastern United States	Wood & Weaver (1982)

---

[a] Potaliinae treated.

Systematics, character evolution, and biogeography 275

The gentians of Canada, Alaska and Greenland	Gillett (1963)
The Jepson manual – higher plants of California [USA]	Pringle (1993)
The new Britton and Brown illustrated flora of the northeastern United States and adjacent Canada	Gleason (1963)
The vascular flora of Ohio [USA]	Cooperrider (1995)
The vascular plants of South Dakota [USA]	Van Bruggen (1976)
Vascular plants of the Pacific Northwest [USA]	Hitchcock et al. (1959)
Vegetation and flora of the Sonoran Desert [Mexico, USA]	Shreve & Wiggins (1964)

Central America and the Caribbean

Descriptive flora of Puerto Rico and adjacent islands [USA]	Liogier (1995)
Flora of Barro Colorado Island [Panama]	Croat (1978)
Flora of Bermuda	Britton (1918)
Flora de Cuba	Leon & Alain (1957)
Flora of Costa Rica[a]	Standley (1938)
Flora of Guatemala	Williams (1969)
Flora of Panama	Elias & Robyns (1975)
Flora of Panama[a]	Blackwell (1968)
Flora of the Bahama archipelago	Correll & Correll (1982)
Flore illustrée des phanérogames de Guadeloupe et de Martinique	Fournet (1978)
Flowering plants of Jamaica	Adams (1972)

South America

A field guide to the families and genera of woody plants of northwest South America (Colombia, Ecuador, Peru)	Gentry (1993)
Catalogue of the flowering plants and gymnosperms of Peru	Zarucchi (1993)
Catalogue of the vascular plants of Ecuador	Pringle (1999)
Flora Brasiliensis	Progel (1865)
Flora Brasiliensis[a]	Progel (1868)
Flora da Reserva Ducke [Brazil]	Ribeiro et al. (1999)
Flora de Chile	Reiche (1910)
Flora illustrada Catarinense [Brazil]	Fabris & Klein (1971)
Flora Neotropica[b]	Maas & Ruyters (1986)
Flora of Ecuador	Pringle (1995)
Flora Patagonica	Correa (1999)

[a] Potaliinae treated.
[b] *Voyria* and *Voyriella* treated.

Flora of Peru[a]	MacBride (1959)
Flora of Suriname	Jonker (1936)
Flora of the Pico das Almas, Chapada Diamantina–Bahia, Brazil	Harvey (1995)
Flora of the Venezuelan Guayana	Struwe *et al.* (1999)
Flore de Guyane Française[a]	Lemée (1953)
Libro rojo de las plantas endémicas del Ecuador 2000	Montúfar (2000)
The botany of the Guayana Highland [Brazil, Venezuela]	Maguire (1981); Maguire & Boom (1989)
The botany of the Guayana Highland[c] [Brazil, Venezuela]	Maguire & Pires (1978)

[a] Potaliinae treated.
[c] *Saccifolium* treated.

Appendix 2.2
Glossary of selected palynological terms

The terminology follows Punt *et al.* (1994).

Amb. The outline of a pollen grain or spore seen in polar view.
Annulus. An area of the exine surrounding a pore that is sharply differentiated from the remainder of the exine, either in ornamentation or in thickness.
Colporate. Describing pollen with a compound aperture (colporus) consisting of an outer colpus (ectocolpus) with one or more inner aperture(s) (endoaperture, os).
Colpus (colpi). Elongate aperture with a length/width ratio greater than 2.
Columella(ae). A rod-like element of the ectexine/sexine, supporting either a tectum or a caput.
Ectexine. The outer part of the exine. It stains positively with basic fuchsin in optical microscopy, and has a higher electron density in conventionally prepared TEM sections than the endexine.
Endexine. The inner part of the exine. It remains relatively unstained with basic fuchsin in optical microscopy, and has a lower electron density in conventionally prepared TEM sections than the ectexine.
Exine. The outer layer of the wall of a palynomorph. It is highly resistant to strong acids and bases, and is composed primarily of sporopollenin.
Foveolate. With foveola(ae).
Foveola(ae). Ornamentation consisting of more or less rounded depressions or lumina more than 1 μm in diameter. The distance between foveolae is greater than their width.
Lira(ae). A narrow ridge which forms the murus in a striate pattern.
Lumen (lumina). The space enclosed by the muri.
Mesocolpium (mesocolpia). The area of a pollen grain surface delimited by lines between the apices of adjacent colpi.
Microspinose. With tapering, pointed elements less than 1 μm in length.
Murus (muri). A ridge that is part of the ornamentation and, for example, separates the lumina in a reticulate pollen grain or the striae in a striate pollen grain.
Nexine. The inner, non-sculptured layer of the exine.
Oblate. Describing the shape of a radially symmetrical pollen grain or spore in which the polar axis is shorter than the equatorial diameter.
OL-pattern. A pattern of ornamentation that appears as "dark islands" at high focus and becomes bright at low focus.
Os (ora). The inner aperture (endoaperture) of a compound aperture.
Perforate. With holes less than 1 μm in diameter and generally situated in the tectum.

Pilate. With pila.
Pilum (pila). A sexine element, usually standing directly on the nexine, consisting of a rod-like part (columellae) and a swollen apical part (caput).
Porate. With pore(s).
Prolate. Describing the shape of a radially symmetrical pollen grain or spore in which the polar axis is longer than the equatorial diameter.
Reniform. Kidney-shaped.
Reticulate. With reticulum.
Reticulum (reticula). A network-like pattern consisting of lumina or other spaces wider than 1 μm bordered by elements narrower than the lumina.
Rugulate. Describing a type of ornamentation consisting of elongated sexine elements more than 1 μm long, arranged in an irregular pattern that is intermediate between striate and reticulate.
Scabrate. With elements of ornamentation of any shape, smaller than 1 μm in all directions.
Sexine. The outer, sculptured layer of the exine.
Spinose. With long and tapering, pointed elements, exceeding 1 μm in length.
Stria(ae). A groove between elongated elements.
Striate. With elongated, generally parallel elements separated by grooves.
Striato-reticulate. Describing pollen in which parallel or subparallel muri (lirae) are cross-linked to a reticulum. The connections between the parallel muri (lirae) lie on different levels.
Tectate. With tectum.
Tectum. The layer of sexine which forms a roof over the columellae, granules, or other infratectal elements.
Tectum perforatum. Perforated tectum.
Verruca(ae). A wart-like sexine element.
Verrucose. With verruca(ae).

LITERATURE CITED

Abrams, L. 1951. Gentianaceae. Pages 350–365 in: *An illustrated flora of the Pacific States: Washington, Oregon, and California*, vol. 3, *Geraniaceae to Scrophulariaceae*. Stanford University Press, Stanford, CA.

Adams, C. D. 1972. Gentianaceae. Pages 582–586 in: *Flowering plants of Jamaica*. University of the West Indies, Mona, Jamaica.

Adams, J. M. 1997. Global land environments since the last interglacial. Oak Ridge National Laboratory, TN, USA. (http://www.esd.ornl.gov/ern/qen/adams1.html)

Adams, J. M. & H. Faure, eds., QEN members. 1997. Review and atlas of palaeovegetation: preliminary land ecosystem maps of the world since the last glacial maximum. Oak Ridge National Laboratory, TN, USA. http://www.esd.ornl.gov/ern/qen/adams1.html

Adams, L. G. 1995. *Chionogentias* (Gentianaceae), a new generic name for the Australasian "snow gentians", and a revision of the Australian species. *Austral. Syst. Bot.* 8: 935–1011.

Adams, L. G. 1996. Gentianaceae. Pages 72–103 in: A. Wilson, ed. *Flora of Australia*, vol. 28. CSIRO, Melbourne.
Adanson, M. 1763. *Familles de plantes*, vol. 2. Vincent, Paris.
Aeschiman, D. & H. M. Burdet. 1994. Gentianaceae. Pages 299–305 in: *Flore de la Suisse et des territoires limitrophes*, ed. 2. Editions du Griffon Neuchâtel, Neuchâtel.
Agababyan, V. S. & K. T. Tumanyan. 1976a. Materialy k palinomorfologicheskomu izocheniyu semeystva Gentianaceae 2 (Exacinae i Chironiinae). *Biol. Zh. Armenii* 29(7): 35–42.
Agababyan, V. S. & K. T. Tumanyan. 1976b. Materialy k palinomorfologicheskomu izocheniyu semeystva Gentianaceae 1 (Erythraeinae). *Biol. Zh. Armenii* 29(5): 26–38.
Agababyan, V. S. & K. T. Tumanyan. 1977a. Materialy k palinomorfologicheskomu izocheniyu semeystva Gentianaceae 4. *Biol. Zh. Armenii* 30(8): 43–53.
Agababyan, V. S. & K. T. Tumanyan. 1977b. Materialy k palinomorfologicheskomu izocheniyu semeystva Gentianaceae 3 (Gentianinae). *Biol. Zh. Armenii* 30(1): 40–47.
Aiello, L. C. 1993. The origin of the New World monkeys. Pages 100–118 in: W. George & R. Lavocat, eds. *The Africa–South America connection*. Oxford Monographs on Biogeography 7. Clarendon Press, Oxford.
Aitken, E. 1999. Family 160. Gentianaceae. Pages 602–656 in: A. J. C. Grierson & D. G. Long, eds. *Flora of Bhutan*. Royal Botanic Garden Edinburgh, Edinburgh.
Albert, V. A. & L. Struwe. 1997. Phylogeny and classification of *Voyria* (saprophytic Gentianaceae). *Brittonia* 49: 466–479.
Albert, V. A. & L. Struwe. 2002. Gentianaceae in context. Pages 1–20 in: L. Struwe & V. A. Albert, eds. *Gentianaceae: systematics and natural history*. Cambridge University Press, Cambridge.
Albert, V. A., M. H. G. Gustafsson, & L. Di Laurenzio. 1998. Ontogenetic systematics, molecular developmental genetics, and the angiosperm petal. Pages 349–374 in: P. Soltis, D. Soltis, & J. Doyle, eds. *Molecular Systematics of Plants* II. Kluwer Academic Publishers, Boston, MA.
Allen, C. K. 1933. A monograph of the American species of the genus *Halenia*. *Ann. Missouri Bot. Gard.* 20: 119–222.
APG (The Angiosperm Phylogeny Group). 1998. An ordinal classification for the families of flowering plants. *Ann. Missouri Bot. Gard.* 85: 531–553.
Arnott, G. A. W. 1839. Exaci species ex peninsula Indica ac ex insula Ceylano. *Ann. Sci. Nat. Bot.*, sér. 2, 11: 175–176.
Aublet, M. F. 1775. *Histoire des plantes de la Guiane Françoise*, vol. 1. P.-F. Didot, London and Paris.
Backer, C. A. & R. C. Bakhuizen van den Brink. 1965a. Loganiaceae. Pages 206–212 in: *Flora of Java*, vol. 2. N. V. P. Noordhoff, Groningen.
Backer, C. A. & R. C. Bakhuizen van den Brink. 1965b. Gentianaceae. Pages 437–441 in: *Flora of Java*, vol. 2. N. V. P. Noordhoff, Groningen.
Backlund, A. & B. Bremer. 1997. Phylogeny of the Asteridae s. str. based on *rbc*L sequences, with particular reference to the Dipsacales. *Pl. Syst. Evol.* 207: 225–254.

Backlund, M., B. Oxelman, & B. Bremer. 2000. Phylogenetic relationships within the Gentianales based on *ndh*F and *rbc*L sequences, with particular reference to the Loganiaceae. *Amer. J. Bot.* 87: 1029–1043.

Baillon, H. E. 1888–1891. *Histoire des plantes*, vol. 10. Librairie Hachette & Co., Paris.

Baillon, M. H. 1888. Observations sur le *Veratrilla*. *Bull. Mens. Soc. Linn. Paris* 1: 729–730.

Baker, J. G. & N. E. Brown. 1903. Gentianaceae. Pages 544–587 in: W. T. Thistleton-Dyer, ed. *Flora of tropical Africa*, vol. 4(1). L. Reeve & Co., London.

Balfour, I. B. 1884. Diagnoses plantarum novarum phanerogamarum Socotrensium, etc. *Proc. Roy. Soc. Edinburgh* 12: 76–98.

Balgooy, M. M. J. van & P. W. Leenhouts. 1966. *Fagraea* Thunb. Pages 168–169 in: C. G. G. J. van Steenis & M. M. J. van Balgooy, eds. *Pacific plant areas*, vol. 2. *Blumea*, suppl. 5.

Bamps, P. 1982. Une nouvelle espèce de *Faroa* (Gentianaceae) au Zaïre. *Bull. Jard. Bot. Nat. Belg.* 52: 486–487.

Bamps, P. 1987. Un nouveau *Faroa* (Gentianaceae) du Zaïre. *Bull. Jard. Bot. Nat. Belg.* 57: 479–480.

Barthlott, W. & D. R. Hunt. 1993. Cactaceae. Pages 161–197 in: K. Kubitzki, J. G. Rohwer, & V. Bittrich, eds. *The families and genera of vascular plants*, vol. 2. Springer, Berlin, Heidelberg, and New York.

Beccari, O. 1890. *Malesia*, vol. 3. Fratelli Bencini, Firenze-Roma.

Beck, R. A., D. W. Burbank, W. J. Sercombe, G. W. Riley, J. K. Barndt, J. R. Berry, J. Afzal et al. 1995. Stratigraphic evidence for an early collision between northwest India and Asia. *Nature* 373: 55–58.

Beddome, R. H. 1874. *Icones Plantarum Indiae Orientalis*. Gantz Brothers, Madras.

Bentham, G. 1854. Notes on north Brazilian Gentianeae, from the collections of Mr. Spruce and Sir Robert Schomburgk. *J. Bot. Kew Gard. Misc.* 6: 193–204.

Bentham, G. 1857. Notes on Loganiaceae. *J. Linn. Soc. Lond., Bot.* 1: 52–115.

Bentham, G. 1869. *Flora Australiensis*, vol. 4. L. Reeve & Co., London.

Bentham, G. 1876. Gentianeae. Pages 799–820 in: G. Bentham & J. Hooker, eds. *Genera plantarum*, vol. 2, part 2. L. Reeve & Co., Williams & Norgate, London.

Berhaut, J. 1975. Gentianaceae. Pages 55–73 in: *Flore illustrée du Sénégal*, vol. 4. Gouvernement du Sénégal, Ministère du Développement Rural et de l'Hydralique, Direction des Eaux de Forêts, Dakar.

Berry, P., O. Huber, & B. K. Holst. 1995. Floristic analysis and phytogeography. Pages 161–191 in: P. E. Berry, B. K. Holst, & K. Yatskievych, eds. *Flora of the Venezuelan Guayana*, vol. 1. Missouri Botanical Garden, St. Louis, MO and Timber Press, Portland, OR.

Besse, J. & V. Courtillot. 1988. Paleogeographic maps of the continents bordering the Indian Ocean since the early Jurassic. *J. Geophys. Res.* 93 (B10): 11791–11808.

Beuzenberg, E. J. & J. B. Hair. 1983. Contributions to chromosome atlas of the New Zealand flora – 25, miscellaneous species. *New Zealand J. Bot.* 21: 13–20.

Bidgood, S. & R. K. Brummitt. 1998. A revision of the genus *Neuracanthus* (Acanthaceae). *Kew Bull.* 53: 1–76.
Bisset, N. G. 1980a. Phytochemistry. Pages 211–233 in: A. J. M. Leeuwenberg, ed. *Engler and Prantl's Die natürlichen Pflanzenfamilien, Angiospermae: Ordnung Gentianales, Fam. Loganiaceae*, vol. 28b (1). Duncker and Humblot, Berlin.
Bisset, N. G. 1980b. Useful plants. Pages 238–244 in: A. J. M. Leeuwenberg, ed. *Engler and Prantl's Die natürlichen Pflanzenfamilien, Angiospermae: Ordnung Gentianales, Fam. Loganiaceae*, vol. 28b (1). Duncker and Humblot, Berlin.
Blackwell, W. H., Jr. 1968. Family 159. Loganiaceae in: R. E. Woodson, R. W. Schery, and collaborators. *Flora of Panama*, part VIII. *Ann. Missouri Bot. Gard.* 54: 393–413.
Blume, C. L. 1826. *Bijdragen tot de Flora van Nederlandsch Indië*. Lands Drukkerij, Batavia.
Böcher, T. W., B. Fredskild, K. Holmen, & K. Jakobsen. 1978. Gentianaeae. Pages 174–177 in: *Grønlands flora*, ed. 3. P. Haase & Søns Forlag, København.
Boiteau, P. 1986. *Médecine traditionelle et pharmacopée. Précis de matière médicale Malgache*. Agence de Cooperation Culturelle et Technique, Paris.
Bolòs, O., de & J. Vigo. 1995. Gentianàcies. Pages 111–130 in: *Flora dels Països Catalans*, vol. 3. Barcino, Barcelona.
Bonafè Barcelo, F. 1979. Gentianàcies. Pages 317–324 in: *Flora de Mallorca*, vol. 3. Moll, Mallorca.
Borgmann, E. 1964. Anteil der Polyploiden in der Flora des Bismarksgebirges von Ostneuguinea. *Z. Bot.* 52: 118–172.
Borkhausen, M. B. 1796. Über Linne's Gattung *Gentiana*. Pages 23–30 in: D. J. J. Roemer, ed. *Archiv für die Botanik*, vol. 1(1). In der Schäferschen Buchhandlung, Leipzig.
Borodin, J. 1892. [Paper title unavailable.] *Trav. Soc. Imp. Nat. St. Petersb.* 22: 131–137.
Bouman, F., L. Cobb, N. Devente, V. Goethals, P. J. M. Maas, & E. Smets. 2002. The seeds of Gentianaceae. Pages 498–572 in: L. Struwe & V. A. Albert, eds. *Gentianaceae: systematics and natural history*. Cambridge University Press, Cambridge.
Boutique, R. 1971. Deux Gentianacées nouvelles du Congo-Kinshasa. *Bull. Jard. Bot. Nat. Belg.* 41: 261–264.
Boutique, R. 1972. Gentianaceae. Pages 1–56 in: P. Bamps, ed. *Flore d'Afrique centrale (Zaire-Rwanda-Burundi)*. Jardin Botanique National de Belgique, Brussels.
Bradbury, J. W. 1977. Lek mating behavior in the hammer-headed bat. *Zeit. Tierpsych.* 45: 225–255.
Bradbury, J. W. 1984. Buzzing bats: the lek mating system of hammer-headed bats. Pages 816–817 in: D. Macdonald, ed. *The encyclopedia of mammals*. Facts on File, New York.
Braun, A. 1830. *Lomatogonium*; ein neues Genus für *Gentiana carinthiaca* Froehl. *Flora* 13: 293–297.
Bremer, B., R. G. Olmstead, L. Struwe, & J. A. Sweere. 1994. *rbc*L sequences support exclusion of *Retzia*, *Desfontainia*, and *Nicodemia* from the Gentianales. *Pl. Syst. Evol.* 190: 213–230.

Bremer, K. 1988. The limits of amino acid sequence data in angiosperm phylogenetic reconstruction. *Evolution* 42: 795–803.
Britton, N. L. 1918. Gentianaceae. Pages 291–292 in: *Flora of Bermuda*. Scribners, New York.
Britton, N. L. & A. Brown. 1913. *An illustrated flora of the northern United States*, ed. 2. Scribners, New York.
Broome, C. R. 1976. The Central American species of *Centaurium* (Gentianaceae). *Brittonia* 28: 413–426.
Broome, C. R. 1978. Chromosome numbers and meiosis in North and Central American species of *Centaurium* (Gentianaceae). *Syst. Bot.* 3: 299–312.
Brown, K. S., Jr. & G. T. Prance. 1987. Soils and vegetation. Pages 19–45 in: T. C. Whitmore & G. T. Prance, eds. *Biogeography and Quaternary history in tropical America*. Oxford Science Publications, Oxford.
Brown, M. L. & R. G. Brown. 1984. Gentianaceae. Pages 733–743. *Herbaceous plants of Maryland*. Book Center, University of Maryland, College Park, MD.
Brown, R. 1810. *Prodromus Florae Novae Hollandiae et Insulae van-Diemen*, vol. 1. R. Taylor, London.
Brown, R. 1818. *Observations systematical and geographical on the herbarium collected by Professor Christian Smith, in the vicinity of Congo*. W. Bulmer & Co., London.
Browne, P. 1756. *The civil and natural history of Jamaica in three parts*, ed. 1. T. Osborne & J. Shipton, London.
Bruce, E. A. & J. Lewis. 1960. Loganiaceae. Pages 1–47 in: C. E. Hubbard & E. Milne-Redhead, eds. *Flora of tropical East Africa*. Crown Agents for Oversea Governments and Administrations, London.
Bureau, L.-É. 1856. *De la famille des Loganiacées, et des plantes qu'elle fournit à la médecine*. Thèse pour le doctorat en médecine. Faculté de Médecine de Paris, Paris.
Burtt, B. L. 1971. From the South: an African view of the floras of western Asia. Pages 135–154 in: P. H. Davis, P. C. Harper, & I. C. Hedge, eds. *Plant life of south-west Africa*. Botanical Society of Edinburgh, Edinburgh.
Calder, J. A. & R. L. Taylor. 1968. Gentianaceae. Pages 479–482 in: *Flora of the Queen Charlotte Islands*, part 1, *Systematics of the vascular plants*. Roger Duhamel, Ottawa.
Card, H. H. 1931. A revision of the genus *Frasera*. *Ann. Missouri Bot. Gard.* 18: 245–282.
Carlquist, S. 1984. Wood anatomy of some Gentianaceae: systematic and ecological conclusions. *Aliso* 10: 573–582.
Carpenter, I., H. D. Locksley, & F. Scheinmann. 1969. Xanthones in higher plants: biogenetic proposals and a chemotaxonomic survey. *Phytochemistry* 8: 2013–2026.
Caruel, T. 1886. Gentianaceae in: F. Parlatore. *Flora Italiana continuata da Teodoro Caruel*, vol. 6. Le Monnier, Firenze.
Chamisso, L. K. A. von & D. F. L. von Schlechtendal. 1826. De plantis in expeditione speculatoria Romanoffiana observatis rationem dicunt Ad. de Chamisso et D. de Schlechtendal. *Linnaea* 1: 1–677.
Chandler, G. T. & R. J. Bayer. 2000. Phylogenetic placement of the enigmatic

western Australian genus *Emblingia* based on *rbc*L sequences. *Pl. Sp. Biol.* 15: 67–72.
Chassot, P. 2000. Phylogenetic position of the genus *Swertia* (Gentianaceae) in the subtribe Swertiinae. *Amer. J. Bot.* 87 (suppl.): 118–119.
Cherneva, O. V. 1961. Gentianaceae. Pages 95–109 in: A. I. Vvedensky, ed. *Flora Uzbekistanica*, vol. 5. Akademiya nauk Uzbekskoy SSR, Tashkent.
Christopher, J. 1976. Gentianaceae – *Canscora diffusa* R. Br., in: IOPB-chromosome number report 52. *Taxon* 25: 344.
Clapham, A. R., T. G. Tutin, & D. M. Moore. 1987. Gentianaceae. Pages 351–355 in: *Flora of the British Isles*, ed. 3. Cambridge University Press, Cambridge.
Clarke, C. B. 1875. Notes on Indian Gentianaceae. *J. Linn. Soc., Bot.* 14: 423–457.
Clarke, C. B. 1885. Gentianaceae. Pages 93–132 in: J. D. Hooker, ed. *The flora of British India*, vol. 4. L. Reeve & Co., London.
Clarke, C. B. 1905 [1906]. Flora of the Malayan peninsula. *J. Asiat. Soc. Bengal* 74 (II): 86–91.
Clewell, A. F. 1985. Gentianaceae. Pages 365–368 in: *Guide to the vascular plants of the Florida Panhandle*. Florida State University Press, Tallahassee, FL.
Coates Palgrave, K. 1991. *Trees of southern Africa*, ed. 2. Struik Publishers, Cape Town.
Cobb, L. & P. J. M. Maas. 1998. A new species of *Tachia* (Gentianaceae) from Suriname. *Brittonia* 50: 11–18.
Cody, W. J. 1996. Gentianaceae. Pages 483–487 in: *Flora of the Yukon Territory*. NRC Research Press, Ottawa.
Colinvaux, P. A. 1996. Quaternary environmental history and forest diversity in the Neotropics. Pages 359–405 in: J. B. C. Jackson, A. F. Budd, & A. G. Coates, eds. *Evolution and environment in tropical America*. University of Chicago Press, Chicago and London.
Colla, L. A. 1834. *Herbarium Pedemontanum*, vol. 3. Ex Typis Regis, Torino.
Conard, H. S. 1951. Gentianaceae in: *Plants of Iowa: keys for determining the names of the native trees, shrubs, flowers and ferns and most of the cultivated plants of Iowa (Tracheata)*, ed. 7. Grinnell, IA.
Conn, B. J., E. A. Brown, & C. R. Dunlop. 1996. Loganiaceae. Pages 1–72 in: A. Wilson, ed. *Flora of Australia*, vol. 28. CSIRO, Melbourne.
Cooperrider, T. S. 1995. Gentianaceae. Pages 294–304 in: *The vascular flora of Ohio*, vol. 2, *The dicotyledoneae*, part 2, *Linaceae through Campanulaceae*. Ohio State University Press, Columbus, OH.
Corner, E. J. H. 1988. *Wayside trees of Malaya*, vol. 1, ed. 3. Malayan Nature Society, Kuala Lumpur.
Correa, M. N. 1999. Gentianaceae. Pages 48–55 in: *Flora Patagonica*, part 6, *Dicotyledones Gamopétalas (Ericaceae a Calyceraceae)*. Coleccion Cientifica del INTA, Buenos Aires.
Correll, D. S. & H. B. Correll. 1982. Gentianaceae. Pages 1118–1124 in: D. S. Correll & H. B. Correll, eds. *Flora of the Bahama archipelago*. J. Cramer, Vaduz.
Correll, D. S. & M. C. Johnston. 1979. Gentianaceae. Pages 1204–1210 in: *Manual of the vascular plants of Texas*. University of Texas at Dallas, Richardson, TX.

Coste, H. 1990. Gentianaceae. Pages 557–565 in: *Flore descriptive et illustrée de la France, de la Corse et des contrées limitrophes.* Librairie Scientifique et Technique, Albert Blanchard, Paris.

Cramer, L. H. 1981. Gentianaceae. Pages 55–78 in: M. D. Dassanayake, ed. *A revised handbook to the flora of Ceylon*, vol. 3. Balkema, Rotterdam.

Crane, P. R., E. M. Friis, & K. R. Pedersen. 1995. The origin and early diversification of angiosperms. *Nature* 374: 27–33.

Crepet, W. L. & C. P. Daghlian. 1981. Lower Eocene and Paleocene Gentianaceae: floral and palynological evidence. *Science* 214: 75–77.

Croat, T. B. 1978. Gentianaceae. Pages 697–699 in: *Flora of Barro Colorado Island.* Stanford University Press, Stanford, CA.

Cronquist, A., A. H. Holmgren, N. H. Holmgren, J. R. Reveal, & P. K. Holmgren. 1984. Gentianaceae. Pages 4–23 in: *Intermountain flora*, vol. 4, *Subclass Asteridae* (*except Asteraceae*). The New York Botanical Garden, Bronx, NY.

Daniel, M. & S. D. Sabnis. 1978. Chemical systematics of family Gentianaceae. *Curr. Sci.* 47: 109–111.

Davis, P. H., ed. 1978. Gentianaceae (auctt. div.). Pages 176–195 in: *Flora of Turkey and the East Aegean islands*, vol. 6. Edinburgh University Press, Edinburgh.

Deam, C. C. 1970. Gentianaceae. Pages 755–760 in: *Flora of Indiana.* J. Cramer, Lehre.

Demiri, M. 1983. Gentianaceae. Pages 361–364 in: *Flora Ekskursioniste e Shqiperise.* Tirane.

Don, D. 1836. Descriptions of Indian Gentianaceae. *London Edinburgh Philos. Mag. & J. Sci.* 8: 75–78.

Don, G. 1837–1838. *A general history of the dichlamydeous plants*, vol. 4, *Corolliflorae.* J. G. & F. Rivington et al., London.

Downie, S. R. & J. D. Palmer. 1992. Restriction site mapping of the chloroplast inverted repeat: a molecular phylogeny of the Asteridae. *Ann. Missouri Bot. Gard.* 79: 266–283.

Doyle, J. J. & M. J. Donoghue. 1993. Phylogenies and angiosperm diversification. *Paleobiology* 19: 141–167.

Dunal, F. 1852. Solanaceae. Pages 1–690 in: A. P. de Candolle, ed. *Prodromus systematis naturalis regni vegetabilis*, vol. 13(1). Victoris Masson, Paris.

Duncan, W. H. & C. L. Brown. 1954. Connate anthers in *Gentiana* (Gentianaceae). *Rhodora* 56: 133–136.

Dyer, R. A. 1975. *The genera of southern African flowering plants*, vol. 1, *Dicotyledons.* Department of Agricultural Technical Services, Botanical Research Institute, Pretoria.

Elias, T. S. & A. Robyns. 1975. Family 160. Gentianaceae. Pages 61–101 in: R. E. Woodson, Jr., R. W. Schery, and collaborators, eds. *Flora of Panama*, part 8. *Ann. Missouri Bot. Gard.* 62.

Endlicher, S. L. 1838. *Genera plantarum secundum ordines naturales disposita.* F. Beck, Vienna.

Endlicher, S. L. 1841. *Enchiridion botanicum.* F. Beck, Vienna.

Endress, M. E., B. Sennblad, S. Nilsson, L. Civeyrel, M. W. Chase, S. Huysmans,

E. Grafström et al. 1996. A phylogenetic analysis of Apocynaceae s. str. and some related taxa in Gentianales: a multidisciplinary approach. *Opera Bot. Belg.* 7: 59–102.

Endress, P. K. 1994. *Diversity and evolutionary biology of tropical flowers.* Cambridge University Press, Cambridge.

Engelmann, G. & A. Gray. 1881 [1880]. A new genus of Gentianaceae, *Geniostemon* Engelm. & Gray in: A. Gray, ed. Contributions to North American botany. *Proc. Amer. Acad. Arts* 16: 78–108.

Erdtman, G. 1952. *Pollen morphology and plant taxonomy, Angiosperms.* Almqvist & Wiksell, Stockholm and Chronica Botanica Co., Waltham, MA.

Erdtman, G., B. Berglund, & J. Praglowski. 1961. *An introduction to a Scandinavian pollen flora.* Almqvist & Wiksell, Stockholm.

Ewan, J. 1947. A revision of *Chorisepalum,* an endemic genus of Venezuelan Gentianaceae. *J. Wash. Acad. Sci.* 37: 392–396.

Ewan, J. 1948a. A revision of *Macrocarpaea,* a neotropical genus of shrubby gentians. *Contr. U. S. Natl. Herb.* 29: 209–251.

Ewan, J. 1948b. A review of *Purdieanthus* and *Lehmanniella,* two endemic Colombian genera of Gentianaceae, and biographical notes on Purdie and Lehmann. *Caldasia* 5: 85–98.

Ewan, J. 1952. A review of the neotropical lisianthoid genus *Lagenanthus* (Gentianaceae). *Mutisia* 4: 1–5.

Fabris, H. A. 1953. Sinopsis preliminar de las Gencianaceas argentinas. *Bol. Soc. Argent. Bot.* 4: 233–259.

Fabris, H. A. & R. M. Klein. 1971. Gentianaceae. Pages 3–30 in: P. R. Reitz, ed. *Flora illustrada Catarinense,* part 1, fascicle GENC. Itajaí, Santa Catarina.

Faegri, K. & J. Iversen. 1992. *Textbook of pollen analysis,* ed. 4 by K. Faegri, P. E. Kaland & K. Krzywinski, 1989, reprinted. J. Wiley & Sons, Chichester, New York etc.

Farris, J. S., V. A. Albert, M. Källersjö, D. Lipscomb, & A. G. Kluge. 1996 [1997]. Parsimony jackknifing outperforms neighbor-joining. *Cladistics* 12: 99–124.

Favarger, C. 1949. Contribution à l'étude caryologique et biologique des Gentianacées. *Bull. Soc. Bot. Suisse* 59: 62–86.

Favarger, C. 1952. Contribution à l'étude caryologique et biologique des Gentianacées. II. *Bull. Soc. Bot. Suisse* 62: 244–257.

Favarger, C. 1960. Étude cytologique du *Cicendia filiformis* et du *Microcala pusilla* (Gentianaceae). *Bull. Soc. Bot. France* 197: 94–98.

Fernald, M. L. 1950. Gentianaceae. Pages 1153–1165 in: *Gray's manual of botany.* Dioscorides Press, Portland, OR.

Fernald, M. L. & C. A. Weatherby. 1932. *Bartonia*: a comedy of errors. *Rhodora* 23: 284–300.

Fiori, A. 1925. Gentianaceae. Pages 250–263 in: *Nuova flora analitica d'Italia.* M. Ricci, Firenze.

Fosberg, F. R. & M.-H. Sachet. 1974. A new variety of *Fagraea berteriana* (Gentianaceae). *Phytologia* 28: 470–472.

Fosberg, F. R. & M.-H. Sachet. 1980. Systematic studies of Micronesian plants. *Smithsonian Contr. Bot.* 45: 1–40.

Fournet, J. 1978. *Flore illustrée des phanérogames de Guadeloupe et de Martinique.* Institut National de la Recherche Agronomique, Paris.

Fournier, P. 1961. Gentianaceae. Pages 853–864 in: *Les quatre flores de France.* Editions Paul Lechevalier, Paris.

Franchet, M. A. 1899a. *Latouchea.* Page 212 in: M. M. Finet & M. A. Franchet, eds. Sur une collection de plantes réunie dans le fokien, par M. et Mme. de la Touche. *Bull. Soc. Bot. France* 46: 204–215.

Franchet, M. A. 1899b. Les *Swertia* et quelques autres Gentianées de la Chine. *Bull. Soc. Bot. France* 46: 302–324.

Friedmann, F. 1994. *Flore des Seychelles, Dicotylédones.* Editions de l'Orstom: Institut Français de Recherche Scientifique pour le Développement en Coopération, Paris.

Fries, T. C. E. 1923. Die *Swertia*-Arten der afrikanischen Hochgebirge. *Notizbl. Bot. Gart. Berlin-Dahlem* 77: 505–534.

Frodin, D. G. 2001. *Guide to standard floras of the world.*, ed. 2. Cambridge University Press, Cambridge.

Gagnepain, F. 1929. Un genre nouveau de Gentianacées. *Bull. Soc. Bot. France* 76: 776–777.

Gardner, G. 1843. Descriptions of four new genera of plants from the Organ Mountains. *Lond. J. Bot.* 2: 9–15.

Garg, S. 1987. *Gentianaceae of the northwest Himalaya (a revision).* Today and Tomorrow's Printers and Publishers, New Delhi.

Geesink, R. 1973. A synopsis of the genus *Swertia* (Gent.) in Malesia. *Blumea* 21: 179–183.

Gentry, A. H. 1982. Neotropical floristic diversity: Phytogeographical connections between Central and South America, Pleistocene climatic fluctuations, or an accident of the Andean orogeny? *Ann. Missouri Bot. Gard.* 69: 557–593.

Gentry, A. H. 1993. Gentianaceae. Pages 434–437 in: *A field guide to the families and genera of woody plants of northwest South America (Colombia, Ecuador, Peru) with supplementary notes on herbaceous taxa.* Conservation International, Washington, DC.

Gielly, L. & P. Taberlet. 1996. A phylogeny of the European gentians inferred from chloroplast *trn*L (UAA) intron sequences. *Bot. J. Linn. Soc.* 120: 57–75.

Gilg, C. 1939. Beiträge zur Kenntnis der Gentianaceen-Gattung *Curtia* Cham. & Schlecht. *Notizbl. Bot. Gart. Berlin-Dahlem* 121: 66–93.

Gilg, E. 1895a. Gentianaceae. Pages 50–108 in: A. Engler & K. Prantl, eds. *Die natürlichen Pflanzenfamilien*, vol. 4(2). Verlag von Wilhelm Engelmann, Leipzig.

Gilg, E. 1895b. Ueber die Blüthenverhältnisse der Gentianaceengattungen *Hockinia* Gardn. und *Halenia* Borckh. *Ber. Deutsch. Bot. Ges.* 13: 114–126.

Gilg, E. 1898. *Symphyllophyton. Beib. Bot. Jahrb.* 60(25): 43.

Gilg, E. 1903. Gentianaceae. Pages 331–335 in: H. Baum, ed. *Kunene-Sambesi-Expedition.* Kolonial-Wirtschaflisches Komitee, Berlin.

Gilg, E. 1916. Gentianaceae andinae. *Beib. Bot. Jahrb.* 118: 4–122.

Gilg, E. & C. Gilg. 1933. Gentianaceae. Pages 944–945 in: M. Mildbraed, ed.

XIV. Neue und seltene Arten aus Ostafrika (Tanganyika-Territ. Mandat) leg. H. J. Schlieben, IV. *Notizbl. Bot. Gart. Berlin-Dahlem* 11: 912–945.

Gillett, J. M. 1957. A revision of the North American species of *Gentianella* Moench. *Ann. Missouri Bot. Gard.* 44: 195–269.

Gillett, J. M. 1959. A revision of *Bartonia* and *Obolaria* (Gentianaceae). *Rhodora* 61: 43–62.

Gillett, J. M. 1963. *The gentians of Canada, Alaska and Greenland.* Research Branch, Canada Department of Agriculture, Ottawa.

Givnish, T. J., K. J. Sytsma, J. F. Smith, W. J. Hahn, D. H. Benzing, & E. M. Burkhart. 1997. Molecular evolution and adaptive radiation in *Brocchinia* (Bromeliaceae: Pitcairnioideae) atop tepuis of the Guayana Shield. Pages 259–311 in: T. J. Givnish & K. J. Sytsma, eds. *Molecular evolution and adaptive radiation.* Cambridge University Press, Cambridge.

Givnish, T. J., T. M. Evans, M. L. Zjhra, T. B. Patterson, P. E. Berry, & K. J. Sytsma. 2000. Molecular evolution, adaptive radiation, and geographic diversification in the amphiatlantic family Rapateaceae: evidence from *ndh*F sequences and morphology. *Evolution* 54: 1915–1937.

Gleason, H. A. 1963. Gentianaceae. Pages 386–392 in: *The new Britton and Brown illustrated flora of the northeastern United States and adjacent Canada.* The New York Botanical Garden, Bronx, NY.

Gleason, H. A. & A. Cronquist. 1991. Gentianaceae. Pages 386–392 in: *Manual of vascular plants of northeastern United States and adjacent Canada*, ed. 2. The New York Botanical Garden, Bronx, NY.

Gleason, H. A. & R. P. Wodehouse. 1931. *Chorisepalum* in: H. A. Gleason, ed. Botanical results of the Tyler–Duida Expedition. *Bull. Torrey Bot. Club* 58: 277–506.

Goloboff, P. A. 1998. *NONA.* Computer program and documentation. Published by the author, Tucumán, Argentina.

Gopal Krishna, G. & V. Puri. 1962. Morphology of the flower of some Gentianaceae with special reference to placentation. *Bot. Gaz.* 124: 42–57.

Goyder, D. 1992. *Secamone* (Asclepiadaceae subfam. Secamonoideae) in Africa. *Kew Bull.* 47: 437–474.

Graham, A. 1984. *Lisianthius* pollen from the Eocene of Panama. *Ann. Missouri Bot. Gard.* 71: 987–993.

Grant, J. R. & L. Struwe. 2000. Morphological evolution and neotropical biogeography in *Macrocarpaea* (Gentianaceae: Helieae). *Amer. J. Bot.* 87 (suppl.): 131.

Grant, J. R. & L. Struwe. 2001. Macrocarpaeae Grisebach (Gentianaceae) Species Novae seu notabiles Neotropicae I: An introduction to the genus *Macrocarpaea* and three new species from Colombia, Ecuador, and Guyana. *Harvard Pap. Bot.* 5: 521–530.

Gray, A. 1859. Notes upon some Polynesian Loganiaceae. *Proc. Amer. Acad. Arts Sci.* 4: 319–324.

Griffin, O. & J. Parnell. 1997. *Fagraea.* Pages 197–205 in: T. Santisuk & K. Larsen, eds. *Flora of Thailand*, vol. 6(3). Dimond Printing Co., Bangkok.

Grisebach, A. H. R. 1839 [1838]. *Genera et species Gentianearum.* J. G. Cotta, Stuttgart and Tübingen.

Grisebach, A. H. R. 1845. Gentianaceae. Pages 39–141 in: A. de Candolle, ed. *Prodromus systematis naturalis regni vegetabilis*, vol. 9. Fortin, Masson, et Sociorum, Paris.

Grisebach, A. H. R. 1849. Gentianeae in: J. F. Klotzsch, ed. Beiträge zu einer Flora der Aequinoctial-Gegenden der neuen Welt. *Linnaea* 22: 32–46.

Grisebach, A. H. R. 1862. Notes on *Coutoubea volubilis*, Mart., and some other Gentianeae from tropical America. *J. Proc. Linn. Soc., Bot.* 6: 140–146.

Grossheim, A. A. et al. 1967. Gentianaceae. Pages 387–470 in: B. K. Shishkin & E. G. Bobrov, eds. *Flora of the USSR*, vol. 18. S. Monson, Jerusalem. (English translation of the *Flora SSSR*, vol. 18 [1952].)

Grothe, E. H. M. & P. J. M. Maas. 1984. A scanning electron microscopic study of the seed coat structure of *Curtia* Chamisso & Schlechtendal and *Hockinia* Gardner (Gentianaceae). *Proc. Kon. Ned. Akad. Wetensch.*, ser. C, 87: 33–42.

Grubov, V. I. 2001. Gentianaceae. Pages 530–537 in: *Key to the vascular plants of Mongolia*, vol. 2. Science Publishers, Enfield, NH and Plymouth, UK. (English translation of *Opredelitel sosudistykh rasteniy Mongolii* [1982].)

Guimarães, E. F. 1977. Revisão taxonômica do gênero *Deianira* Chamisso et Schlechtendal (Gentianaceae). *Arq. Jard. Bot. Rio de Janeiro* 21: 46–123.

Guimarães, E. F. & V. L. G. Klein. 1985. Revisão taxonômica do gênero *Coutoubea* Aublet (Gentianaceae). *Rodriguésia (Rio de Janeiro)* 37: 21–45.

Guinet, P. 1962. Pollens d'Asie Tropicale. *Inst. Fr. Pond. Trav. Sect. Scient. Techn.* V(1).

Haffer, J. 1969. Speciation in Amazonian forest birds. *Science* 165: 131–137.

Haffer, J. 1987. Quaternary history of tropical America. Pages 1–18 in: T. C. Whitmore & G. T. Prance, eds. *Biogeography and Quaternary history in tropical America*. Oxford Science Publications, Oxford.

Hagen, K. B. von & J. W. Kadereit. 2000. Notes on the systematics and evolution of *Gentiana* sect. *Ciminalis. Bot. Jahrb. Syst.* 122: 305–339.

Hagen, K. B. von & J. W. Kadereit. 2001. The phylogeny of *Gentianella* (Gentianaceae) and its colonization of the southern hemisphere as revealed by nuclear and chloroplast DNA sequence variation. *Org. Divers. Evol.* 1: 61–79.

Haines, A. & T. F. Vining. 1998. Gentianaceae. Pages 655–658 in: *Flora of Maine: a manual for identification of native and naturalized vascular plants of Maine*. V. F. Thomas Co., Bar Harbor, ME.

Hall, R. 1996. Reconstructing Cenozoic SE Asia in: R. Hall & D. Blundell, eds. Tectonic evolution of SE Asia. *Geological Society Special Publications* 106: 153–184.

Hämet-Ahti, L., J. Suominen, T. Ulvinen, P. Uotila, & S. Vuokko. 1986. Gentianaceae. Pages 296–299 in: *Retkeilykasvio*. Suomen Luonnonsuojelun Tuki Oy, Helsinki.

Hansen, K. 1988. Gentianaceae. Pages 384–387 in: *Dansk feltflora*. Gyldendal, København.

Hara, H. 1975. A new species of *Cotylanthera* (Gentianaceae) from Philippines, with a conspectus of the genus. *J. Jap. Bot.* 50: 321–328.

Harden, G. J. 1992. Gentianaceae. Pages 508–512 in: *Flora of New South Wales*, vol. 3. New South Wales University Press, Kensington, NSW.

Hartvig, P. 1991. Gentianaceae. Pages 1–9 in: A. Strid & K. Tan, eds. *Mountain flora of Greece*, vol. 2. Edinburgh University Press, Edinburgh.
Harvey, Y. B. 1995. Gentianaceae. Pages 321–327 in: B. L. Stannard, ed. *Flora of the Pico das Almas, Chapada Diamantina–Bahia, Brazil*. Royal Botanic Gardens, Kew.
Hedberg, O. 1957. Afroalpine vascular plants; a taxonomic revision. *Symb. Bot. Upsal.* 15: 5–411.
Hegi, G. 1966. Gentianaceae. Pages 1953–2047 in: *Illustrierte flora von Mittel-Europa*, vol. 5, part 3. Carl Hanser Verlag, München.
Hemsley, W. B. 1890. *Gentiana* (§ *Megacodon*, Hemsl.) *venosa* in: F. B. Forbes & W. B. Hemsley. An enumeration of all the plants known from China proper, Formosa, Hainan, Corea, the Luchu archipelago, and the island of Hongkong, together with their distribution and synonymy. *J. Linn. Soc., Bot.* 26: 137, fig. 3.
Heß, H. E., E. Landolt, & R. Hirzel. 1972. Gentianaceae. Pages 15–39 in: *Flora der Schweiz und angrenzender Gebiete*, vol. 3, *Plumbaginaceae bis Compositae*. Birkhäuser Verlag, Basel.
Heusden, E. C. H., van. 1986. Floral anatomy of some neotropical Gentianaceae. *Proc. Kon. Ned. Akad. Wetensch.*, ser. C, 89: 45–59.
Hill, A. W. 1908. Notes on *Sebaea* and *Exochaenium*. *Bull. Misc. Inform.* 8: 317–341.
Hill, A. W. 1913. The floral morphology of the genus *Sebaea*. *Ann. Bot.* 27: 479–489.
Hill, J. 1756. *The British Herbal*. T. Osborne & J. Shipton, London.
Hitchcock, A. S. 1929. Proposal by A. S. Hitchcock (Washington) and M. L. Green (Kew) – standard-species of Linnean genera of Phanerogamae (1753–54), in: J. Ramsbottom, T. A. Sprague, A. J. Willmott, & E. M. Wakefield, eds. *International Botanical Congress, Cambridge (England), 1930, Nomenclature – Proposals by British botanists*. Wyman & Sons, London.
Hitchcock, C. L. & A. Cronquist. 1976. Gentianaceae. Pages 356–361 in: *Flora of the Pacific Northwest: an illustrated manual*. University of Washington Press, Seattle, WA.
Hitchcock, C. L., A. Cronquist, M. Ownby, & J. W. Thompson. 1959. Gentianaceae. Pages 57–76 in: *Vascular plants of the Pacific Northwest*, part 4, *Ericaceae through Campanulaceae*. University of Washington Press, Seattle, WA.
Hô, P.-H. 1993. *Fagraea* and Gentianaceae. Pages 846–856 in: *Câyco Viêtnam. An illustrated flora of Vietnam*, ed. 3, vol. 2(2). Mekong Printing, Santa Ana, CA.
Ho, T. N. & S. W. Liu. 1980. *Lomatogoniopsis* T. N. Ho et S. W. Liu – a new genus of Gentianaceae. *Acta Phytotax. Sin.* 18: 466–468.
Ho, T.-N. & S.-W. Liu. 1990. The infrageneric classification of *Gentiana* (Gentianaceae). *Bull. Brit. Mus. (Nat. Hist.), Bot.* 20: 169–192.
Ho, T.-N. & S.-W. Liu. 1993. New combinations, names and taxonomic notes on *Gentianella* (Gentianaceae) from South America and New Zealand. *Bull. Brit. Mus. (Nat. Hist.), Bot.* 23: 61–65.
Ho, T. N. & J. S. Pringle. 1995. Gentianaceae. Pages 1–139 in: Z.-Y. Wu & P. H. Raven, eds. *Flora of China*, vol. 16. Science Press, Beijing and Missouri Botanical Garden, St. Louis, MO.

Ho, T. N., C. Y. Xue, & W. Wang. 1994a. The origin, dispersal and formation of the distribution pattern of *Swertia* L. (Gentianaceae). *Acta Phytotax. Sin.* 32: 525–537.
Ho, T.-N., S.-W. Liu, Y.-Z. Xi, & J.-C. Ning. 1994b. Pollen morphology and phylogenetic analysis of *Gentiana*. *Cathaya* 6: 93–114.
Ho, T.-N., S.-W. Liu, & X.-F. Lu. 1996. A phylogenetic analysis of *Gentiana* (Gentianaceae). *Acta Phytotax. Sin.* 34: 505–530.
Holmgren, N. H. 1998. Gentianaceae. Pages 362–367 in: *Illustrated companion to Gleason and Cronquist's manual*. The New York Botanical Garden, Bronx, NY.
Holmgren, P. K., N. H. Holmgren, & L. C. Barnett. 1990. *Index Herbariorum*, part 1, *The herbaria of the world*, ed. 8. The New York Botanical Garden, Bronx, NY.
Holub, J. 1973. New names in Phanerogamae 2. *Folia Geobot. Phytotax.* 8: 155–179.
Hong, D.-Y. 1993. Eastern Asian–North American disjunctions and their biological significance. *Cathaya* 5: 1–39.
Hooker, J. D. 1867. Gentianaceae. Pages 189–191 in: *Handbook of the New Zealand flora*. L. Reeve & Co., London.
Hooker, J. D. 1885. Gentianaceae. Pages 93–132 in: *The flora of British India*, vol. 4, *Asclepiadeae to Amarantaceae*. L. Reeve & Co., London.
Huang, T.-C. 1972. *Pollen flora of Taiwan*. National Taiwan University, Botany Department Press, Taipei.
Huber, O. 1995. Geographical and physical features. Pages 1–61 in: P. E. Berry, B. K. Holst, & K. Yatskievych, eds. *Flora of the Venezuelan Guayana*, vol. 1. Missouri Botanical Garden, St. Louis, MO and Timber Press, Portland, OR.
Hudson, W. 1762. *Flora Anglica*. Impensis auctores, London.
Hultén, E. 1968. Gentianaceae. Pages 753–763 in: *Flora of Alaska and neighboring territories*. Stanford University Press, Stanford, CA.
Humbert, H. 1937a. *Gentianothamnus*, genre nouveau de Gentianacées de Madagascar. *Comptes Rend. Hebd. Séances Acad. Sci.* 204(23): 1747–1749.
Humbert, H. 1937b. Un genre nouveau de Gentianacées-Chironiinées de Madagascar. *Bull. Soc. Bot. France* 84: 386–390.
Humbert, H. 1963. Les Gentianacées de Madagascar. *Adansonia*, sér. 2, 3: 343–351.
Hungerer, B. K. & J. W. Kadereit. 1998. The phylogeny and biogeography of *Gentiana* L. sect. *Ciminalis* (Adans.) Dumort.: a historical interpretation of distribution ranges in the European high mountains. *Perspect. Pl. Ecol. Evol. Syst.* 1: 121–135.
Hutchinson, J. 1973. *The families of flowering plants*, ed. 3. Clarendon Press, Oxford.
Hutchinson, J. & J. M. Dalziel. 1931. *Flora of west tropical Africa*, ed. 1, vol. 2. Crown Agents for the Colonies, London.
Huxley, T. H. 1888. The Gentians: notes and queries. *J. Linn. Soc., Bot.* 24: 101–124.
Ikonnikov, S. S. 1970. Dopolnenie k flore Pamira (Additamenta ad floram Pamir). *Novit. Syst. Pl. Vasc. Acad. Sci. URSS* 6: 260–272.
Iltis, H. H. 1965. The genus *Gentianopsis* (Gentianaceae): transfers and phytogeographic comments. *Sida* 2(2): 129–154.

Irion, G. 1989. Quaternary geological history of the Amazon lowlands. Pages 23–34 in: L. B. Holm-Nielsen, I. C. Nielsen, & H. Balslev, eds. *Tropical forests: botanical dynamics, speciation and diversity*. Academic Press, London.
Iturralde-Vinent, M. A. & R. D. E. MacPhee. 1999. Paleogeography of the Caribbean Region: implications for Cenozoic biogeography. *Bull. Amer. Mus. Nat. Hist.*: 238.
Jain, S. K. & R. A. DeFilipps. 1991. *Medicinal plants of India*, vol. 1. Reference Publications, Algonac, MI.
Jankeviciene, R., H. Krall, & Z. Slangena. 1996. Gentianaceae. Pages 252–256 in: V. Kuusk, L. Tabaka, & R. Jankeviciene, eds. *Flora of the Baltic countries*. Eesti Loodusfoto AS, Tartu.
Jansen, S. & E. Smets. 1998. Vestured pits in some woody Gentianaceae. *IAWA J.* 19: 35–42.
Jasiewicz, A. 1971. Gentianaceae. Pages 7–42 in: B. Pawlowskiego & A. Jasiewicza, eds. *Flora Polska*, vol. 12. Polska Akademia Nauk, Krakow.
Jensen, S. R. 1992. Systematic implications of the distribution of iridoids and other chemical compounds in the Loganiaceae and other families of the Asteridae. *Ann. Missouri Bot. Gard.* 79: 284–302.
Jensen, S. R. & J. Schripsema. 2002. Chemotaxonomy and pharmacology of Gentianaceae. Pages 573–631 in: L. Struwe & V. A. Albert, eds. *Gentianaceae: systematics and natural history*. Cambridge University Press, Cambridge.
Jonker, F. P. 1936. Gentianaceae. Pages 400–427 in: A. Pulle, ed. *Flora of Suriname (Netherlands Guyana)*, vol. 4. J. H. de Bussy Ltd., Amsterdam.
Jordanov, D. 1982. Gentianaceae. Pages 389–418 in: *Flora Reipublicae Popularis Bulgaricae*, vol. 8. In Aedibus Academiae Scientiarum Bulgaricase, Serdicae.
Jovanovich-Dunich, R. 1973. Gentianaceae. Pages 403–433 in: M. Josifovich, ed. *Flora cr Srbije*, vol. 5. Akademija nauka i umetnosti, Belgrad.
Jussieu, A. L., de. 1789. *Genera plantarum secundum ordines naturales disposita*. Paris.
Jussieu, A. L., de. 1791. *Genera plantarum*. Ziegleri & Filiorum, Zürich.
Karpati, Z. 1970. Eine kritische-taxonomische Übersicht der Gattung *Swertia* in Europa. *Fragm. Florist. Geobot.* 16(1): 53–60.
Khoshoo, T. N. & S. R. Tandon. 1963. Cytological, morphological and pollination studies on some Himalayan species of *Swertia*. *Caryologia (Pisa)* 16: 445–477.
Klackenberg, J. 1983. A reevaluation of the genus *Exacum* (Gentianaceae) in Ceylon. *Nord. J. Bot.* 3: 355–370.
Klackenberg, J. 1985. The genus *Exacum* (Gentianaceae). *Opera Bot.* 84: 1–144.
Klackenberg, J. 1986. The new genus *Ornichia* (Gentianaceae) from Madagascar. *Bull. Mus. Nat. Hist. Nat., Paris*, sér. 4, 8, sect. B, *Adansonia* 2: 195–206.
Klackenberg, J. 1987a. Revision of the genus *Tachiadenus* (Gentianaceae). *Bull. Mus. Nat. Hist. Nat., Paris*, sér. 4, 9, sect. B, *Adansonia* 1: 43–80.
Klackenberg, J. 1987b. A new species of *Sebaea* (Gentianaceae) from Madagascar. *Bull. Mus. Nat. Hist. Nat., Paris*, sér. 4, sect. B, *Adansonia* 9: 133–136.
Klackenberg, J. 1990. Famille 168. Gentianacées. Famille 168 bis. Menyanthacées. Pages 1–185 in: P. Morat, ed. *Flore de Madagascar et des Comores*. Association de Botanique tropicale, Paris.

Klackenberg, J. 1992a. Taxonomy of *Secamone* (Asclepiadaceae) in Asia and Australia. *Kew Bull.* 47: 595–612.
Klackenberg, J. 1992b. Taxonomy of *Secamone* s. lat. (Asclepiadaceae) in the Madagascar Region. *Opera Bot.* 112: 1–127.
Klootwijk, C. T., J. S. Gee, J. W. Peirce, G. M. Smith, & P. L. McFadden. 1992. An early India–Asia contact: Paleomagnetic constraints from Ninetyeast Ridge, ODP Leg 121. *Geology* 20: 395–398.
Knoblauch, E. 1894. Beiträge zur Kenntniss der Gentianaceae. *Bot. Centralbl.* 60: 321–334, 353–363, 385–401.
Knoblauch, E. 1895. Ueber die dimorphen Blüthen von *Hockinia montana* und die Variabilität der Blüthenmerkmale bei den Gentianaceen. *Ber. Deutsch. Bot. Ges.* 13: 289–298.
Kochummen, K. M. 1973. Loganiaceae. Pages 267–275 in: T. C. Whitmore, ed. *Tree flora of Malaya*, vol. 2. Longman, London.
Köhler, A. 1905. Der Systematische Wert der Pollenbeschaffenheit bei den Gentianaceen. *Mitteil. Bot. Mus. Zürich* 25. (Dissertation)
Krok, T. O. B. N. & S. Almquist. 1986. Gentianaceae. Pages 424–427 in: *Svensk flora*, ed. 26. Esselte Studium, Uppsala.
Kubitzki, K. 1989. The ecogeographical differentiation of Amazonian inundation forests. *Pl. Syst. Evol.* 162: 285–304.
Kubitzki, K. 1990. The psammophilous flora of northern South America. *Mem. New York Bot. Gard.* 64: 248–253.
Kuntze, C. E. O. 1891. *Revisio generum plantarum*, vol. 2. Arthur Felix, Leipzig.
Kurz, S. 1870. *Gentiana Jaeschkei* re-established as a new genus of Gentianaceae. *J. Asiat. Soc. Bengal*, pt. 2, *Nat. Hist.* 39: 229–230.
Kurz, S. 1873. New Burmese plants – *Phyllocyclus*. *J. Asiat. Soc. Bengal* 42: 235–236.
Kusnezow, N. I. 1895. *Gentiana* Tournef. Pages 80–86 in: A. Engler & K. Prantl, eds. *Die natürlichen Pflanzenfamilien*, vol. 4(2). Wilhelm Engelmann, Leipzig.
Kusnezow, N. I. 1896[–1904]. Subgenus *Eugentiana* Kusnez. generis *Gentiana* Tournef. *Acta Hort. Petrop.* 15: 1–507. (German translation)
Lamarck, J. B. A. P. M. de. 1785. *Canscora*. Page 601 in: *Encyclopédie méthodique. Botanique*, vol. 1. Panckoucke, Paris.
Lambinon, J., J.-E. De Langhe, L. Delvosalle, & J. Duvigneaud. 1992. Gentianaceae. Pages 489–494 in: *Nouvelle flore de la Belgique, du Grand-Duché de Luxembourg, du nord de la France et des régions voisines (Pteridophytes et Spermatophytes)*. Editions du Patrimoine du Jardin botanique national de Belgique, Meise.
Lang, G. 1994. *Quartäre Vegetationsgeschichte Europas*. Fischer, Jena.
Laschenkova, A. N. & A. I. Tolmatchev. 1977. Gentianaceae. Pages 58–63 in: A. I. Tolmatchev, ed. *Flora regionis boreali-orientalis territoriae Europaeae URSS*, vol. 4. Nauka, Leningrad.
Lauber, K. & G. Wagner. 1998a. Gentianaceae. Pages 782–800 in: *Flora Helvetica, Flore illustrée de Suisse*. Editions Paul Haupt, Berne.
Lauber, K. & G. Wagner. 1998b. Gentianaceae. Pages 103–105 in: *Flora Helvetica, Flore illustrée de Suisse, Clef de détermination*. Editions Paul Haupt, Berne.

Lavin, M. & M. Luckow. 1993. Origins and relationships of tropical North America in the context of the boreotropics hypothesis. *Amer. J. Bot.* 80: 1–14.
Leenhouts, P. W. 1962 [1963]. Loganiaceae. Pages 293–387 in: C. G. G. J. van Steenis, ed. *Flora Malesiana*, ser. 1, vol. 6(2). Wolters-Noordhoff, Groningen.
Leeuwenberg, A. J. M. 1961. The Loganiaceae of Africa I. *Anthocleista. Acta Bot. Neerl.* 10: 1–53.
Leeuwenberg, A. J. M. 1972. Loganiacées. Pages 2–153 in: A. Aubréville & J.-F. Leroy, eds. *Flore du Cameroun*, vol. 12. Muséum national d'histoire naturelle, Paris.
Leeuwenberg, A. J. M. 1983. 109. Loganiaceae. Pages 327–374 in: E. Launert, ed. *Flora Zambesiaca*, vol. 7, part 1. Crown Agents for Oversea Governments and Administrations, London.
Leeuwenberg, A. J. M. 1984. Famille 167–Loganiacées. Pages 1–107 in: J.-F. Leroy, ed. *Flore de Madagascar et des Comores*. Muséum national d'histoire naturelle, Paris.
Leeuwenberg, A. J. M. & P. Bamps. 1979. Loganiaceae. Pages 1–149 in: *Flore d'Afrique centrale (Zaïre-Rwanda-Burundi), Spermatophytes*. Jardin Botanique National de Belgique, Brussels.
Leeuwenberg, A. J. M. & P. W. Leenhouts. 1980. Taxonomy. Pages 8–96 in: A. J. M. Leeuwenberg, ed. *Engler and Prantl's Die natürlichen Pflanzenfamilien, Angiospermae: Ordnung Gentianales, Fam. Loganiaceae*, vol. 28b (1). Duncker and Humblot, Berlin.
Lemée, A. 1953. Gentianaceae. Pages 277–288 in: *Flore de Guyane Francaise*, part 3. Lechevalier, Paris.
Leon, H. & H. Alain. 1957. Gentianaceae. Pages 158–168 in: *Flora de Cuba*, vol. 4. Contribuciones Ocasionales del Museo de Historia Natural del Colegio "De la Salle", no. 16. P. Fernandez, Habana.
Li, H.-L. 1952. Floristic relationships between eastern Asia and eastern North America. *Trans. Amer. Phil. Soc.* 42: 371–429.
Li, P.-T. & A. J. M. Leeuwenberg. 1996. Loganiaceae. Pages 320–338 in: Z.-Y. Wu & P. H. Raven, eds. *Flora of China*, vol. 15. Science Press, Beijing and Missouri Botanical Garden, St. Louis, MO.
Lid, J. 1985. Gentianaceae. Pages 379–382 in: *Norsk, Svensk, Finsk flora*. Det norske samlaget, Oslo.
Lienau, K., H. Straka, & B. Friedrich. 1986. Fam. 168. Gentianaceae. Pages 9–15 in: H. Straka, ed. *Palynologia Madagassica et Mascarenica, Familien 167 bis 181*. Tropische und Subtropische Pflanzenwelt 55. Akademie der Wissenschaften und der Literatur, Franz Steiner Verlag, Wiesbaden, Mainz.
Lindley, J. 1836. *A natural system of botany*, ed. 2. Longman, Rees, Orme, Brown, Green, and Longman, London.
Lindley, J. 1849. Memoranda concerning some new plants recently introduced into gardens otherwise than through the Horticultural Society, no. 1. *J. Hort. Soc. London* 4: 261–269.
Lindsey, A. A. 1940. Floral anatomy in the Gentianaceae. *Amer. J. Bot.* 27: 640–651.
Linnaeus, C. 1747a. *Nova plantarum genera*... [Dissertatio C. M. Dassow]. L. Salvii, Stockholm.
Linnaeus, C. 1747b. *Flora Zeylanica*. Sumtu & Literis L. Salvii, Stockholm.

Linnaeus, C. 1753. *Species plantarum*, vol. 1. L. Salvii, Stockholm.
Liogier, A. H. 1989. Gentianaceae. Pages 310–325 in: *La flora de la Española*, vol. 5. Universidad Central del Este, vol. 69, Serie Científica 26. San Pedro de Macorís.
Liogier, H. A. 1995. Gentianaceae. Pages 189–197 in: *Descriptive flora of Puerto Rico and adjacent islands*. Editorial de la Universidad de Puerto Rico. Río Pedras, Puerto Rico.
Litardiére, R., de & R. Maire. 1924. Contributions à l'Étude de la flore du Grand Atlas. *Mém. Soc. Sci. Nat. Maroc* 6, tome 4(1): 14–17.
Liu, S.-W. & T.-N. Ho. 1992. Systematic study on *Lomatogonium* A. Br. (Gentianaceae). *Acta Phytotax. Sin.* 30: 289–319.
Liu, T.-S. & C.-C. Kuo. 1978. Gentianaceae. Pages 161–201 in: H.-L. Li, T.-S. Liu, T.-C. Huang, T. Koyama, & C. E. DeVol, eds. *Flora of Taiwan*, vol. 4. Epoch Publishing Co., Taipei.
Long, R.W. & O. Lakela. 1971. Gentianaceae. Pages 692–697 in: *A flora of tropical Florida: a manual of the seed plants and ferns of southern peninsular Florida*. University of Miami Press, Coral Gables, FL.
Looman, J. & K. F. West. 1987. Gentianaceae. Pages 589–592 in: *Budd's flora of the Canadian Prairie Provinces*. Canadian Government Publishing Centre, Hull, Quebec.
Löve, A. 1983. Gentianaceae. Pages 308–311 in: *Flora of Iceland*. Almenna Bókafélagid, Reykjavík.
Löve, A. & D. Löve. 1956. Cytotaxonomical conspectus of the Icelandic flora. *Acta. Horti Gotob.* 20: 65–291.
Löve, A. & D. Löve. 1961. Some nomenclatural changes in the European flora. *Bot. Not.* 11: 33–47.
Löve, A. & D. Löve. 1972. *Favargera* and *Gentianodes*, two new genera of alpine Gentianaceae. *Bot. Not.* 125: 255–258.
Löve, A. & D. Löve. 1975. The Spanish gentians. *Anales Inst. Bot. Cavanilles* 32: 221–232.
Löve, A. & D. Löve. 1976. The natural genera of Gentianinae. Pages 205–222 in: P. Kachroo, ed. *Recent advances in botany*. Professor P. N. Mehra commemorative volume. Dehra Dun, Delhi.
Löve, A. & D. Löve. 1986. Gentianaceae-Gentianineae-Europaeae, in: A. Löve, ed. Chromosome number reports XCIII. *Taxon* 35: 897–899.
Löve, D. 1953. Cytotaxonomical remarks on the Gentianaceae. *Hereditas (Lund)* 39: 225–235.
Ma, J. & R. Zhao. 1991. Study on pollen morphology of Gentianaceae from Gansu. *Acta Bot. Boreal.-Occident. Sin.* 11: 251–257.
Ma, Y. C. 1951. *Gentianopsis*: a new genus of Chinese Gentianaceae. *Acta Phytotax. Sin.* 1: 5–19.
Maas, P. J. M. 1981. On the true identity of *Lagenanthus parviflorus* Ewan (Gentianaceae). *Ann. Missouri Bot. Gard.* 68: 685–688.
Maas, P. J. M. 1985. Nomenclatural notes on neotropical Lisyantheae (Gentianaceae). *Proc. Kon. Ned. Akad. Wetensch.*, ser. C, 88: 405–412.
Maas, P. J. M. & P. Ruyters, eds. 1986. *Voyria* and *Voyriella (saprophytic Gentianaceae)*. Flora Neotropica Monograph 41. The New York Botanical Garden, Bronx, NY.

Maas, P. J. M., S. Nilsson, A. M. C. Hollants, B. J. H. ter Welle, H. Persoon, & E. C. H. van Heusden. 1983. Systematic studies in neotropical Gentianaceae – the *Lisianthius* complex. *Acta Bot. Neerl.* 32: 371–374.

Mabberley, D. J. 1997. *The plant-book*, ed. 2. Cambridge University Press, Cambridge.

MacBride, J. F. 1959. Loganiaceae, Gentianaceae. Flora of Peru. *Field Museum of Natural History, Bot. Ser.* 13 (part 5, no. 1): 239–363.

Machado, I. C. S., I. Sazima, & M. Sazima. 1998. Bat pollination of the terrestrial herb *Irlbachia alata* (Gentianaceae) in northeastern Brazil. *Pl. Syst. Evol.* 209: 231–237.

Magallon, S., P. R. Crane, & P. S. Herendeen. 1999. Phylogenetic pattern, diversity, and diversification of eudicots. *Ann. Missouri Bot. Gard.* 86: 297–372.

Magee, D. W. & H. E. Ahles. 1999. Gentianaceae. Pages 834–839 in: *Flora of the Northeast: a manual of the vascular plants of New England and adjacent New York*. University of Massachusetts Press, Amherst, MA.

Maguire, B. 1970. On the flora of the Guayana Highland. *Biotropica* 2: 85–100.

Maguire, B. 1981. Gentianaceae. Pages 330–388 in: B. Maguire and collaborators, eds. The botany of the Guayana Highland – Part XI. *Mem. New York Bot. Gard.* 32.

Maguire, B. 1985. Gentianaceae – part 2. *Phytologia* 57: 311–312.

Maguire, B. & B. M. Boom. 1989. Gentianaceae, part 3. Pages 2–56 in: B. Maguire and collaborators, eds. The botany of the Guayana Highland – Part XIII. *Mem. New York Bot. Gard.* 51.

Maguire, B. & J. M. Pires. 1978. Saccifoliaceae – a new monotypic family of the Gentianales. Pages 230–245 in: B. Maguire and collaborators, eds. The botany of the Guayana Highland – Part X. *Mem. New York Bot. Gard.* 29.

Maguire, B. & R. E. Weaver, Jr. 1975. The neotropical genus *Tachia* (Gentianaceae). *J. Arnold Arbor* 56: 103–125.

Maheswari Devi, H. 1962. Embryological studies in Gentianaceae (Gentianoideae and Menyanthoideae). *Proc. Indian Acad. Sci.*, set. B, 56: 195–216.

Mallikarjuna, M. B., A. Sheriff, & D. G. Krishnappa. 1987. Chromosome number reports XCVII (ed. A. Löve). *Taxon* 36: 766–767.

Mansion, G. 2000. Phylogenetic position of the North and Central American species of *Centaurium* (Gentianaceae) based on molecular data and chromosome numbers; evidence of an old-world origin. *Amer. J. Bot.* 87 (suppl.): 416.

Marais, W. 1961. *Belmontia* and *Exochaenium* synonyms with *Sebaea*. *Bothalia* 7: 464.

Marais, W. & I. C. Verdoorn. 1963. Gentianaceae. Pages 171–243 in: R. A. Dyer, L. E. Codd, & H. B. Rycroft, eds. *Flora of southern Africa*, vol. 26. Department of Agricultural Technical Services, Republic of South Africa, Pretoria.

Marie-Victorin, F. 1995. Gentianaceae. Pages 510–516 in: *Flore Laurentienne*, ed. 3 [mise à jour et annotée par L. Brouillet, S. G. Hay, I. Gulet, M. Blondeau, J. Cayouette, & J. Labrecque]. Les Presses de l'Université de Montréal, Montréal.

Marloth, R. 1909. A diplostigmatic plant, *Sebaea exacoides* (L.) Schinz (*Belmontia cordata* L.). *Trans. Roy. Soc. South Africa* 1: 311–314.
Marquand, C. V. B. 1928–1931. The botanical collection made by Captain F. Kingdon Ward in the Eastern Himalaya and Tibet in 1924–25. *J. Linn. Soc., Bot.* 48: 149–229.
Marquand, C. V. B. 1931. New Asiatic gentians II. *Bull. Misc. Inform. Kew* [1931]: 68–88.
Marquand, C. V. B. 1937. The gentians of China. *Bull. Misc. Inform. Kew* [1937]: 134–180.
Martin, W. C. & C. R. Hutchins. 1981. Gentianaceae. Pages 1499–1515 in: *A flora of New Mexico*. J. Cramer, Vaduz.
Martius, C. F. P., von. 1826–1827. *Nova genera et species plantarum quas in itinere per Brasiliam*, vol. 2. V. Wolf, München.
Matthew, K. M. & N. Rani. 1983. Gentianaceae. Pages 970–978 in: K. M. Matthew, ed. *The flora of the Tamilnadu Carnatic*, vol. 3, part 2. Diocesan Press, Madras.
Maximowicz, C. J. 1859. Primitiae Florae Amurensis – Versuch einer Flora des Amur-landes. *Mém. Acad. Imp. Sci. St.-Pétersbourg Divers Savans* 9: 1–467. (Reprinted by Buchdrukerei der kaiserlichen Akademie der Wissenschaften, St. Petersburg, 1859.)
McGregor, R. L., T. M. Barkley, R. E. Brooks, & E. K. Schofield. 1986. Gentianaceae. Pages 604–609 in: *Flora of the Great Plains*. University Press of Kansas, Lawrence, KS.
Meikle, R. D. 1985. Gentianaceae. Pages 1108–1118 in: *Flora of Cyprus*. Bentham-Moxton Trust, Royal Botanic Gardens, Kew.
Mennega, A. M. W. 1980. Anatomy of the secondary xylem. Pages 112–161 in: A. J. M. Leeuwenberg, ed. *Engler and Prantl's Die natürlichen Pflanzenfamilien, Angiospermae: Ordnung Gentianales, Fam. Loganiaceae*, vol. 28b (1). Duncker and Humblot, Berlin.
Mészáros, S. 1994. Evolutionary significance of xanthones in Gentianaceae: a reappraisal. *Biochem. Syst. Ecol.* 22: 85–94.
Mészáros, S., J. De Laet, & E. Smets. 1996. Phylogeny of temperate Gentianaceae: a morphological approach. *Syst. Bot.* 21: 153–168.
Meusel, H., E. Jäger, S. Rauschert, & E. Weinert. 1978. *Vergleichende Chorologie der zentraleuropäischen Flora*. Fischer, Jena.
Meyer, E. H. F. 1838. *Commentariorum de plantis Africae Australioris*. Leopold Voss, Leipzig.
Meyer, F. G. & E. H. Walker. 1965. Gentianaceae. Pages 735–742 in: *Flora of Japan*. Smithsonian Institution, Washington, DC.
Miquel, F. A. G. 1841. *Flora van Nederlandsch Indië*, vol. 2. C. G. van der Post, Amsterdam.
Miquel, F. A. W. 1851. *Stirpes surinamensis selectae*. Arnz & Soc., Leiden.
Moar, N. T. 1993. *Pollen grains of New Zealand dicotyledonous plants*. Manaaki Whenua Press, Lincoln, Canterbury.
Moench, C. 1794. *Methodus plantas horti botanici et agri marburgensis, a staminum situ describendi*. In officina nova libraria academiae, Maburgi Cattorum.
Mohlenbrock, R. H. 1986. Gentianaceae. Pages 356–358 in: *Guide to the vascular flora of Illinois*, ed. 2. Southern Illinois University Press, Carbondale, IL.

Montúfar, R. 2000. Gentianaceae. Pages 202–204 in: R. Valencia, N. Pitman, S. León-Yánez, & P. M. Jørgensen, eds. *Libro rojo de las plantas endémicas del Ecuador 2000*. Publicaciones del Herbario QCA, Pontifica Universidad Católica del Ecuador, Quito.
Moore, R. J., ed. 1973. *Loganiaceae, Apocynaceae, Asclepiadaceae, Gentianaceae. Index to plant chromosome numbers 1967–1971*. Oosthoek's Uitgeversmaatschappij b.V., Utrecht.
Moreira, A. X. 1961. Sõbre o pôlen de Gentianaceae. *An. Acad. Brasileira Ciên.* 33: 2–3.
Morton, J. F. 1981. *Atlas of medicinal plants of middle America: Bahamas to Yucatan*. Charles C. Thomas, Springfield, IL.
Moss, E. H. 1983. Gentianaceae. Pages 455–458 in: *Flora of Alberta*, ed. 2, revised by J. G. Packer. University of Toronto Press, Toronto.
Mossberg, B., L. Stenberg, & S. Ericsson. 1992. Gentianaväxter. Pages 348–351 in: *Den nordiska floran*. Wahlström & Widstrand, Stockholm.
Mouterde, S. J. P., A. Charpin, & W. Greuter. 1978. Gentianaceae. Pages 24–28 in: *Nouvelle flore du Liban et de la Syrie*, vol. 3. Dar El-Machreq éditeurs, Beyrouth.
Murata, J. 1989. A synopsis of *Tripterospermum* (Gentianaceae). *J. Fac. Sci., Univ. Tokyo*, sect. 3, *Bot.* 14: 273–339.
Nakai, T. 1911. Gentianaceae. Pages 96–101 in: *Flora Koreana*. Imperial University of Tokyo, Japan.
Nikitin, V. V. 1954. Gentianaceae. Pages 40–46 in: B. K. Shishkin, ed. *Flora Turkmenii*, vol. 6. Akademiya nauk Turkmenskoy SSR, Ashkhabad.
Nilsson, S. 1964. On the pollen morphology in *Lomatogonium* A. Br. *Grana Palynol.* 5: 298–329.
Nilsson, S. 1967a. Pollen morphological studies in the Gentianaceae-Gentianinae. *Grana Palynol.* 7: 46–145.
Nilsson, S. 1967b. Notes on pollen morphological variation in Gentianaceae-Gentianinae. *Pollen Spores* 9: 49–58.
Nilsson, S. 1968. Pollen morphology in the genus *Macrocarpaea* (Gentianaceae) and its taxonomical significance. *Svensk Bot. Tidskr.* 62: 338–364.
Nilsson, S. 1970. Pollen morphological contributions to the taxonomy of *Lisianthus* L. s. lat. (Gentianaceae). *Svensk Bot. Tidskr.* 64: 1–43.
Nilsson, S. 2002. Gentianaceae: a review of palynology. Pages 377–497 in: L. Struwe & V. A. Albert, eds. *Gentianaceae: systematics and natural history*. Cambridge University Press, Cambridge.
Nilsson, S. & J. J. Skvarla. 1969. Pollen morphology of saprophytic taxa in the Gentianaceae. *Ann. Missouri Bot. Gard.* 56: 420–438.
Nilsson, S., M. E. Endress, & E. Grafström. 1993. On the relationship of the Apocynaceae and Periplocaceae. *Grana*, suppl. 2: 3–20.
Oehler, E. 1927. Entwicklungsgeschichtliche-zytologische Untersuchungen an einigen saprophytischen Gentianaceen. *Planta* 3: 641–733.
Olmstead, R. G., B. Bremer, K. M. Scott, & J. D. Palmer. 1993. A parsimony analysis of the Asteridae sensu lato based on *rbc*L sequences. *Ann. Missouri Bot. Gard.* 80: 700–722.
Omer, S. 1989. *Quaisera* Omer, a new genus of Gentianaceae. *Bot. Jahrb.* 111: 205–212.

Omer, S. 1995. No. 197, Gentianaceae. Pages 1–172 in: S. I. Ali & M. Qaiser, eds. *Flora of Pakistan*. Department of Botany, University of Karachi, Karachi.

Omer, S. & M. Qaiser. 1991. *Aliopsis*, a new genus of Gentianaceae from C. Asia. *Willdenowia* 21: 189–194.

Omer, S. & M. Qaiser. 1992. Generic limits in *Gentiana* (Gentianaceae) and related genera in Pakistan and adjoining areas along with a new genus *Kurramiana*. *Pakistan J. Bot.* 24: 95–106.

Omer, S., M. Qaiser, & S. I. Ali. 1988. Studies in the family Gentianaceae. The genus *Aloitis* Rafin. from Pakistan and Kashmir. *Pakistan J. Bot.* 20: 153–160.

Padmanabhan, D., D. Regupathy, & S. Pushpa Veni. 1978. Gynoecial ontogeny in *Enicostemma littorale* Blume. *Proc. Indian Acad. Sci.* 87B (Pl. Sci.–2)5: 83–92.

Paiva, J. & I. Nogueira. 1990. Gentianaceae. Pages 3–51 in: E. Launert & G. V. Pope, eds. *Flora Zambesiaca*, vol. 7, part 4. Flora Zambesiaca Managing Committee, Kew.

Parks, C. R. & J. F. Wendel. 1990. Molecular divergence between Asian and North American species of *Liriodendron* (Magnoliaceae) with implications for interpretation of fossil floras. *Amer. J. Bot.* 77: 1243–1256.

Parmar, P. J. 1991. Gentianaceae. Pages 492–497 in: B. V. Shetty & V. Singh, eds. *Flora of Rajasthan*, vol. 2. Botanical Survey of India, Calcutta.

Perkins, J. 1902. Monographische Übersicht der Arten der Gattung *Lisianthus*. *Bot. Jahrb.* 31: 489–494.

Perrot, M. E. 1897 [1899]. Anatomie comparée des Gentianacées. *Ann. Sci. Nat. Bot.*, sér. viii, 7: 105–292.

Perry, J. D. 1971. Biosystematic studies on the North American genus *Sabatia* (Gentianaceae). *Rhodora* 73: 309–369.

Philipson, W. R. 1972. The generic status of the southern hemisphere gentians. Pages 417–422 in: Y. S. Murty, B. M. Johri, H. Y. Mohan Ram, & T. M. Varghese, eds. *Advances in plant morphology*. Professor V. Puri commemorative volume. Sarita Prakashan, Meerut.

Pignatti, S. 1982. Gentianaceae. Pages 326–346 in: *Flora d'Italia*, vol. 2. Edagricole, Bologna.

Pihlar, O., L. Struwe, & V. A. Albert. 1998. Neotropical Gentianaceae and whitesands: biogeography and character evolution. *Amer. J. Bot.* 85 (suppl.): 150–151.

Pissjaukova, V. 1961. Notulae de genere *Swertia* L. *Bot. Mater. Gerb. Bot. Inst. Komarova Akad. Nauk SSSR* 21: 292–313.

Pissjaukova, V. V. 1984. Gentianaceae. Pages 278–314 in: P. N. Ovchinnikova, ed. *Flora Tadzhikskoy SSR*, vol. 7. Nauka, Leningrad.

Polunin, N. 1959. Gentianaceae. Pages 356–363 in: *Circumpolar arctic flora*. Clarendon Press, Oxford.

Polunin, O. & A. Stainton. 1984. Gentianaceae. Pages 265–275 in: *Flowers of the Himalaya*. Oxford University Press, Oxford.

Post, D. M. 1967. Documented chromosome numbers of plants. *Madroño* 19: 134–136.

Post, G. E. 1933. Gentianaceae. Pages 197–199 in: J. E. Dinsmore, ed. *Flora of Syria, Palestine and Sinai*, vol. 2, ed. 2. American Press, Beirut.

Pottier-Alapetite, G. 1981. Gentianacées. Pages 699–704 in: *Flore de la Tunesie*, vol. 1. Ministère de l'Enseignement Superieur.
Prance, G. T. 1982. Forest refuges: evidence from woody angiosperms. Pages 137–157 in: G. T. Prance, ed. *Biological diversification in the tropics*. Columbia University Press, New York.
Prance, G. T. 1987. Biogeography of neotropical plants. Pages 46–65 in: T. C. Whitmore & G. T. Prance, eds. *Biogeography and quaternary history in tropical America*. Oxford Science Publications, Oxford.
Pringle, J. S. 1977. Taxonomy and distribution of *Gentiana* (Gentianaceae) in Mexico and Central America: 1, sect. *Pneumonanthe*. *Sida* 7: 174–217.
Pringle, J. S. 1978. Sectional and subgeneric names in *Gentiana* (Gentianaceae). *Sida* 7: 232–247.
Pringle, J. 1979. Taxonomy and distribution of *Gentiana* (Gentianaceae) in Mexico and Central America: 2, sect. *Chondrophyllae*. *Sida* 8: 14–33.
Pringle, J. S. 1990. Taxonomic notes on western American Gentianaceae. *Sida* 14: 179–187.
Pringle, J. S. 1993. Gentianaceae. Pages 666–672 in: J. C. Hickman, ed. *The Jepson manual: higher plants of California*. University of California Press, Berkeley, CA.
Pringle, J. S. 1995. Family 159A. Gentianaceae. Pages 1–131 in: G. Harling & L. Andersson, eds. *Flora of Ecuador*, vol. 53. Department of Systematic Botany, Gothenburg University, Göteborg.
Pringle, J. S. 1999. Gentianaceae. Pages 487–491 in: P. M. Jørgensen & S. Léon-Yánez, eds. *Catalogue of the vascular plants of Ecuador*. Missouri Botanical Garden Press, St. Louis, MO.
Probatova, N. S. 1981. Gentianaceae. Pages 215–218 in: S. S. Kharkevich & S. K. Cherepanov. *Opredelitel sosudistykh rasteniy Kamchatskoy oblasti*. Nauka, Moscow.
Progel, A. 1865. Gentianaceae. Pages 197–248 in: C. F. P. von Martius, ed. *Flora Brasiliensis*, vol. 6(1). Frid. Fleischer, Munich.
Progel, A. 1868. Loganiaceae. Pages 251–300 in: C. F. P. von Martius, ed. *Flora Brasiliensis*, vol. 6(1). Frid. Fleischer, Munich.
Punt, W. 1978. Evolutionary trends in the Potalieae (Loganiaceae). *Rev. Palaeobot. Palynol.* 26: 313–335.
Punt, W. 1980. Pollen morphology. Pages 162–191 in: A. J. M. Leeuwenberg, ed. *Engler and Prantl's Die natürlichen Pflanzenfamilien, Angiospermae: Ordnung Gentianales, Fam. Loganiaceae*, vol. 28b (1). Duncker and Humblot, Berlin.
Punt, W. & P. W. Leenhouts. 1967. Pollen morphology and taxonomy in the Loganiaceae. *Grana Palynol.* 7: 469–516.
Punt, W. & W. Nienhuis. 1976. The northwest European pollen flora. 6. Gentianaceae. *Rev. Palaeobot. Palynol.* 21: NEPF 89–123.
Punt, W., S. Blackmore, S. Nilsson, & A. Le Thomas. 1994. *Glossary of pollen and spore terminology*. LPP Contrib. Ser. No. 1. LPP Foundation, Utrecht.
Rabinowitz, P. D., M. F. Coffin, & D. Falvey. 1983. The separation of Madagascar and Africa. *Science* 220: 67–69.
Radford, A. E., H. E. Ahles, & C. R. Bell. 1968. Gentianaceae. Pages 835–845 in:

Manual of the vascular flora of the Carolinas. University of North Carolina Press, Chapel Hill, NC.

Rafinesque, C. S. 1837. *Flora telluriana.* Probasco, Philadelphia.

Raj, B. 1966. Palynological studies in some South Indian weeds – II. *J. Osmania Univ.* 3: 61–69.

Rao, A. N. & Y. K. Lee. 1970. Studies on Singapore pollen. *Pacific Sci.* 24: 255–268.

Rao, K. S. & C. C. Chinnappa. 1983. Pericolporate pollen in Gentianaceae. *Can. J. Bot.* 61: 174–178.

Raynal, A. 1965. Un nouveau genre Africain *Oreonesion* A. Rayn. (Gentianaceae). *Adansonia,* sér. nov., 5: 271–275.

Raynal, A. 1967a. Étude critique des genres *Voyria* et *Leiphaimos* (Gentianaceae) et révision des *Voyria* d'Afrique. *Adansonia,* sér. 2, 7: 53–71.

Raynal, A. 1967b. Sur un *Sebaea* Africain saprophyte (Gentianaceae). *Adansonia,* sér. 2, 7: 207–219.

Raynal, A. 1968. Les genres *Neurotheca* Benth. et Hook. et *Congolanthus* A. Raynal, *gen. nov.* (Gentianaceae). *Adansonia,* sér. 2, 8: 45–68.

Raynal, A. 1969. Révision du genre *Enicostema* Blume (Gentianaceae). *Adansonia,* sér. 2, 9: 57–83.

Rechinger, K. H. 1964. Gentianaceae. Pages 476–479 in: *Flora of lowland Iraq.* Cramer, Weinheim.

Regalado, J. C. & D. D. Soejarto. 1997. The genus *Microrphium* (Gentianaceae) in the Philippines. *Novon* 7: 77–80.

Regel, E. 1883. Orginalabhandlungen. 1) Abgebildete Pflanzen B. *Exacum affine* Balfour. *Gartenflora* 32: 34–36.

Reiche, C. 1910. Gentianaceae. Pages 120–134 in: *Flora de Chile,* vol. 5. Imprensa Cervantes, Santiago de Chile.

Reichenbach, H. G. L. 1831–1832. *Flora germanica excursoria.* Apud Carolum Cnobloch, Lipsia.

Reichenbach, H. G. L. 1837. *Handbuch des natürlichen Pflanzensystems.* Arnold, Dresden and Leipzig.

Rezende, C. M. A. D. M. & O. R. Gottlieb. 1973. Xanthones as systematic markers. *Biochem. Syst.* 1: 111–118.

Ribeiro, J. E. da S. and collaborators. 1999. Gentianaceae. Pages 566–567 in: *Flora da Reserva Ducke: Guia de identificação das plantas vasculares de uma floresta de terra-firme na Amazônia Central.* INPA, Manaus.

Ridley, H. N. 1923. Fagraea, Gentianaceae. Pages 415–420 and 432–437 in: *Flora of the Malay Peninsula.* L. Reeve & Co., London.

Ridley, H. N. 1930. *The dispersal of plants throughout the world.* L. Reeve & Co., Ltd., Ashford, Kent.

Robertson, S. A. 1989. *Flowering plants of Seychelles.* Royal Botanic Gardens, Kew.

Robyns, A. & S. Nilsson. 1970. *Macrocarpaea browallioides* (Ewan) A. Robyns & S. Nilsson, *comb. nov.* (Gentianaceae). *Bull. Jard. Bot. Natl. Belg.* 40: 13–15.

Roland, A. E. & E. C. Smith. 1969. Gentianaceae. Pages 579–582 in: *The flora of Nova Scotia.* Nova Scotia Museum, Halifax.

Rork, C. L. 1949. Cytological studies in the Gentianaceae. *Amer. J. Bot.* 36: 687–701.

Rosatti, T. J. 1989. Apocynaceae. Flora of the southeastern United States. *J. Arnold Arbor.* 70: 307–401.
Rothmaler, W. 1990. Gentianaceae. Pages 409–416 in: *Exkursionsflora von Deutschland*, Band 4, kritischer Band, ed. 8. Volk und Wissen Verlag, Berlin.
Roubik, D.W. & J. E. Moreno. 1991. *Pollen and spores of Barro Colorado Island.* Monogr. Syst. Bot. Missouri Bot. Gard. 36. Missouri Botanical Garden, St. Louis, MO.
Roxburgh, W. 1814. *Hortus bengalensis.* Mission Press, Serampore.
Roxburgh, W. 1820. *Flora Indica*, vol. 1. Mission Press, Serampore.
Royen, P., van. 1983. *Fagraea* and Gentianaceae. Pages 2648–2651 and 2793–2846 in: *The alpine flora of New Guinea*, vol. 4, *Taxonomic part, Casuarinaceae to Asteraceae.* Strauss & Cramer, Vaduz.
Rull, V. 1996. Holocene vegetational succession on the Guaiquinima and Chimanta massifs (SE-Venezuela). *Interciencia* 21: 7–20.
Rull, V., C. Schubert, & R. Aravena. 1988. Palynological studies in the Venezuelan Guayana Shield: preliminary results. *Curr. Res. Pleistocene* 5: 54–56.
Rydberg, A. 1932. *Flora of the prairies and plains of central North America.* Hafner Publishing Company, New York and London. (Facsimile from 1965)
Rzedowski, J. & G. Calderon de Rzedowski. 1995. Tres adiciones a la flora fanerogamica de Mexico. *Acta Bot. Mexicana* 32: 1–10.
Salisbury, R. A. 1806. *The paradisus londinensis.* D. N. Shury & W. Hooker, London.
Sampaio, G. 1947. Gentianáceae. Pages 458–462 in: *Flora Portuguesa*, ed. 2. Imprensa Moderna, Porto.
Satake, Y. 1944. Species Swertiae nipponenses. *J. Jap. Bot.* 20: 334–344.
Savolainen, V., M. F. Fay, D. C. Albach, A. Backlund, M. van der Bank, K. M. Cameron, S. A. Johnson et al. 2000. Phylogeny of the eudicots: a nearly complete familial analysis based on *rbc*L gene sequences. *Kew Bull.* 55: 257–309.
Savulescu, T. 1961. Gentianaceae. Pages 430–479 in: E. I. Nyárády, ed. *Flora Republicii Populare Romîne*, vol. 8, ed. 10. Editura Academiei Republicii Populare Romîne.
Saxena, H. O. & M. Brahman. 1995. *The flora of Orissa*, vol. 2. Orissa Forest Development Corporation Ltd., Bhubaneswar.
Schatz, G. E. 1996. Malagasy/Indo-Australo-Malesian phytogeographic connections. Pages 73–83 in: W. R. Lourenço, ed. *Biogéographie de Madagascar.* Orstom Editions, Paris.
Schiman-Czeika, H. 1967. Gentianaceae. Pages 1–28 in: K. H. Rechinger, ed. *Flora Iranica*, part 41. Akademische Druck-u. Verlagsanstalt, Graz.
Schinz, H. 1903. Versuch einer monographischen Übersicht der Gattung *Sebaea* R. Br. 1. Die Sektion *Eusebaea* Griseb. *Mitt. Geogr. Ges. Naturhist. Mus. Lübeck* 17: 3–55.
Schinz, H. 1906. Gentianaceae. Versuch einer monographischen Uebersicht: 1. der Gattung *Sebaea* R. Br. 2. der Gattung *Exochaenium* Griseb. *Bull. Herb. Boissier*, sér. 2, 6: 714–746, 801–823.

Schlechtendal, D. F. L., von & L. K. A. von Chamisso. 1831. Plantarum mexicanarum a cel. viris Schiede et Deppe collectarum recensio brevis, Addenda. *Linnaea* 6: 385–430.
Schmeil, O. 2000. Gentianaceae. Pages 459–467 in: O. Schmeil & J. Fitschen, eds. *Flora von Deutschland und angrenzender Länder*, ed. 91. Quelle & Meyer, Wiebelsheim.
Schoch, E. 1903. Monographie der Gattung *Chironia* L. *Beih. Bot. Centralbl.* 14: 177–242.
Schubert, C. 1986. Paleoenvironmental studies in the Guayana region, southeast Venezuela. *Curr. Res. Pleistocene* 3: 88–90.
Schubert, C. 1988. Climatic changes during the last glacial maximum in northern South America and the Caribbean: a review. *Interciencia* 13: 128–137.
Schubert, C. & M. L. Salgado-Labouriau. 1987. Alluvial and palynological studies in the Venezuelan Guayana Shield. *Cur. Res. Pleistocene* 4: 54–56.
Schultes, R. E. & R. F. Raffauf. 1990. *The healing forest: Medical and toxic plants of the northwest Amazonia*. Dioscorides Press, Portland, OR.
Scoggan, H. J. 1957. Gentianaceae. Pages 443–447 in: *Flora of Manitoba*. Bulletin (National Museum of Canada), Biol. Ser. 47. Department of Northern Affairs and National Resources, Ottawa.
Scoggan, H. J. 1997. Gentianaceae. Pages 1234–1246 in: *The flora of Canada*, vol. 4, *Dicotyledoneae (Loasaceae to Compositae)*. National Museum of Canada, Ottawa.
Semiotrocheva, N. L. 1964. Gentianaceae. Pages 94–119 in: N. V. Pavlov, ed. *Flora Kazakhstana*, vol. 7. Akademiya nauk Kazakhskoy SSR, Alma-Ata.
Senghas, K. & S. Seybold. 1993. Gentianaceae. Pages 438–445 in: *Flora von Deutschland und angrenzender Länder*. Quelle & Meyers Verlag, Heidelberg.
Seymour, F. C. 1969. Gentianaceae. Pages 441–444 in: *The flora of New England*. Charles E. Tuttle Company Publishers, Rutland, VT.
Shah, J. 1990. Taxonomic studies in the genus *Swertia* L. Gentianaceae monograph (part 1). *Sci. Khyber* 3: 17–114.
Shah, J. 1992. Taxonomic studies in the genus *Swertia* L. Gentianaceae monograph (part 2). *Sci. Khyber* 5: 117–231.
Shaulo, D. N. 1984. Gentianaceae. Pages 179–183 in: I. M. Krasnoborov, ed. *Opredelitel rasteniy Tuvinskoj ASSR*. Nauka, Novosibirsk.
Shinners, L. H. 1957. Synopsis of the genus *Eustoma* (Gentianaceae). *Southwest. Naturalist* 2: 38–43.
Shreve, F. & I. L. Wiggins. 1964. Gentianaceae. Pages 1092–1095 in: *Vegetation and flora of the Sonoran Desert*. Stanford University Press, Stanford, CA.
Siddiqi, M. A. 1977. Gentianaceae. Pages 1–10 in: S. I. Ali & S. M. H. Jafri, eds. *Flora of Libia*, vol. 22. Al Faateh University, Tripoli.
Sileshi, N. 1998. A synopsis of *Swertia* (Gentianaceae) in east and northeast tropical Africa. *Kew Bull.* 53: 419–436.
Simpson, B. B. 1975. Pleistocene changes in the flora of the high tropical Andes. *Palaeobiology* 1: 273–294.
Small, J. K. 1903. Gentianaceae. Pages 924–932 in: *Flora of the southeastern United States*. New Era Printing Co., Lancaster, PA.

Smith, A. C. 1988. *Flora vitensis nova*, vol. 4. Pacific Tropical Botanical Garden, Lawai.
Smith, H. 1936. Gentianaceae. Pages 948–988 in: H. R. E. Handel-Mazzetti, ed. *Symbolae Sinicae*, vol. VII. Verlag von Julius Springer, Wien.
Smith, H. 1965. Notes on Gentianaceae. I. The status of *Crawfurdia* and *Tripterospermum*. II. New species of Gentianaceae. *Notes Roy. Bot. Gard. Edinburgh* 26: 237–258.
Smith, H. 1967. Appendix. Pages 144–145 in: S. Nilsson, Pollen morphological studies in the Gentinaceae-Gentianinae. *Grana Palynol.* 7: 46–145.
Smith, H. 1970. New or little known Himalayan species of *Swertia* and *Veratrilla* (Gentianaceae). *Bull. Br. Mus. (Nat. Hist.), Bot.* 4: 239–258.
Smith, J. M. B. & A. M. Cleef. 1988. Composition and origins of the world's tropicalpine floras. *J. Biogeogr.* 15: 631–645.
Smith, L. B. & W. Till. 1998. Bromeliaceae. Pages 74–99 in: K. Kubitzki, ed. *The families and genera of vascular plants*, vol. 4. Springer-Verlag, Berlin, Heidelberg, and New York.
Srivastava, R. C. 1997. Gentianaceae. Pages 98–109 in: V. Mugdal, K. K. Khanna, & P. K. Hajra, eds. *Flora of Madhya Pradesh*, vol. 2. Botanical Survey of India, Calcutta.
Sofieva, R. M. 1957. Gentianaceae. Pages 80–103 in: I. I. Karyagin, ed. *Flora Azerbajdzhana*, vol. 7. Akademiya nauk Azerbajdzhanskoy SSR, Baku. (In Russian)
Soó, R. 1966. Gentianaceae. Pages 616–624 in: *A Magyar Flóra és vegetáció rendszertani-növényföldrajzi kézikönyve*, vol. 2. Akadémiai Kiadó, Budapest.
St. John, H. 1941. Revision of the genus *Swertia* (Gentianaceae) of the Americas and the reduction of *Frasera*. *Amer. Midl. Naturalist* 26: 1–29.
Stace, C. 1997. Gentianaceae. Pages 518–524 in: *New flora of the British Isles*, ed. 2. Cambridge University Press, Cambridge.
Standley, P. C. 1938. Flora of Costa Rica: Loganiaceae, Gentianaceae. *Field Mus. Nat. Hist., Bot. Ser.* 18: 919–929.
Stevens, P. F. 1997. What kind of classification should the practising taxonomist use to be saved? Pages 295–319 in: J. Dransfield, M. J. E. Coode, & D. A. Simpson, eds. *Plant diversity in Malesia*, III. Royal Botanic Gardens, Kew.
Steyermark, J. A. 1951. The genus *Tapeinostemon* (Gentianaceae). *Lloydia* 14: 58–64.
Steyermark, J. A. 1963. Gentianaceae. Pages 1184–1194 in: *Flora of Missouri*. Iowa State University Press, Ames, IA.
Steyermark, J. A. 1982. Relationships of some Venezuelan forest refuges with lowland tropical floras. Pages 182–220 in: G. T. Prance, ed. *Biological diversification in the tropics*. Columbia University Press, New York.
Steyermark, J. A. 1986. Speciation and endemism in the flora of the Venezuelan tepuis. Pages 317–373 in: F. Vuilleumier & M. Monasterio, eds. *High altitude tropical biogeography*. Oxford University Press, New York.
Stockey, R. A. & S. R. Manchester. 1986. A fossil flower with in situ *Pistillipollenites* from the Eocene of British Columbia. *Can. J. Bot.* 66: 313–318.

Storey, M., J. J. Mahoney, A. D. Saunders, R. A. Duncan, S. P. Kelley, & M. F. Coffin. 1995. Timing of hot spot-related volcanism and the breakup of Madagascar and India. *Science* 267: 852–855.
Strausbaugh, P. D. & E. L. Cove. 1978. Gentianaceae. Pages 742–748 in: *Flora of West Virginia*, ed. 2. Seneca Books, Inc., Morgantown, WV.
Struwe, L. 1999. *Morphological and molecular phylogenetic studies in neotropical Gentianaceae.* Dissertation. Stockholm University, Stockholm.
Struwe, L. 2002a. Desfontainiaceae in: N. P. Smith, S. V. Heald, A. Henderson, S. A. Mori, & D. W. Stevenson, eds. *Flowering plant families of the American tropics.* Princeton University Press, Princeton, NJ/The New York Botanical Garden Press, Bronx, NY, in press.
Struwe, L. 2002b. Two new winged species of *Symbolanthus* (Gentianaceae: Helieae) from the western Colombian Andes. *Novon*, in press.
Struwe, L. & V. A. Albert. 1997. Floristics, cladistics, and classification: three case studies in Gentianales. Pages 321–352 in: J. Dransfield, M. J. E. Coode, & D. A. Simpson, eds. *Plant diversity in Malesia*, III. Royal Botanic Gardens, Kew.
Struwe, L. & V. A. Albert. 1998a. *Lisianthius* (Gentianaceae), its probable homonym *Lisyanthus*, and the priority of *Helia* over *Irlbachia* as its substitute. *Harvard Pap. Bot.* 3: 63–71.
Struwe, L. & V. A. Albert. 1998b. Six new species of Gentianaceae from the Guayana Shield. *Harvard Pap. Bot.* 3: 181–197.
Struwe, L. & V. A. Albert. 2000a. Mycotrophic, non-chlorophyllous *Voyria* placed in Gentianaceae. *Amer. J. Bot.* 87 (suppl.): 161.
Struwe, L. & V. A. Albert. 2000b. Supermerous corollas, fleshy fruits, and pantropical biogeography in *Anthocleista*, *Fagraea*, and *Potalia* (Gentianaceae). *Amer. J. Bot.* 87 (suppl.): 161.
Struwe, L., V. A. Albert, & B. Bremer. 1994 [1995]. Cladistics and family level classification of the Gentianales. *Cladistics* 10: 175–206.
Struwe, L., P. J. M. Maas, & V. A. Albert. 1997. *Aripuana cullmaniorum*, a new genus and species of Gentianaceae from white-sands of southeastern Amazonas, Brazil. *Harvard Pap. Bot.* 2: 235–253.
Struwe, L., M. Thiv, J. W. Kadereit, A. S.-R. Pepper, T. J. Motley, P. J. White, J. H. E. Rova et al. 1998. *Saccifolium* (Saccifoliaceae), an endemic of Sierra de la Neblina on the Brazilian–Venezuelan frontier, is related to a temperate-alpine lineage of Gentianaceae. *Harvard Pap. Bot.* 3: 199–214.
Struwe, L., P. J. M. Maas, O. Pihlar, & V. A. Albert. 1999. Gentianaceae. Pages 474–542 in: P. E. Berry, K. Yatskievych, & B. K. Holst, eds. *Flora of the Venezuelan Guayana*, vol. 5. Missouri Botanical Garden, St. Louis, MO.
Subramanian, D. 1980. Cyto-taxonomical studies in South Indian Gentianaceae. *Proc. Indian Sci. Congr. Assoc.* (III, C) 67: 49.
Sugiura, T. 1936a. A list of chromosome numbers in angiospermous plants, 2. *Proc. Imp. Acad. Japan* 12: 144–146.
Sugiura, T. 1936b. Studies on the chromosome numbers in higher plants, with special reference to cytokinesis, 1. *Cytologia* 7: 544–595.
Swofford, D. L. 1993. *PAUP – Phylogenetic Analysis Using Parsimony*, vers. 3.1.1. Computer program and manual distributed by Illinois Natural History Survey, Champaign, IL.

Sytsma, K. J. 1988. Taxonomic revision of the Central American *Lisianthius skinneri* species complex (Gentianaceae). *Ann. Missouri Bot. Gard.* 75: 1587–1602.
Sytsma, K. J. & B. A. Schaal. 1985. Phylogenetics of the *Lisianthius skinneri* (Gentianaceae) species complex in Panama utilizing DNA restriction fragment analysis. *Evolution* 39: 594–608.
Taberlet, P., L. Gielly, G. Patou, & J. Bouvet. 1991. Universal primers for amplification of three non-coding regions of chloroplast DNA. *Plant Mol. Biol.* 17: 1105–1109.
Takhtajan, A. A., ed. 1980. Gentianaceae. Pages 28–48 in: *Flora Armenii*, vol. 7. Akademiya nauk Armyanskoy SSR, Erevan.
Takhtajan, A. L. 1987. *Systema magnoliophytorum*. Nauka, Leningrad.
Takhtajan, A. 1997. *Diversity and classification of flowering plants*. Columbia University Press, New York.
Tamamshyan, S. G. 1967. Gentianaceae. Pages 202–217 in: A. A. Grossheim, ed. *Flora Kavkaza*. Nauka, Leningrad.
Taubert, P. 1893. Plantae Glaziovianae novae vel minus cognitae, IV. *Bot. Jahrb.* 17: 502–526.
Taylor, P. 1963. Gentianaceae. Pages 271–275 in: J. Hutchinson & M. D. Dalziel, eds. *Flora of west tropical Africa*, vol. 2, ed. 2. Crown Agents for Oversea Governments and Administrations, London.
Taylor, P. 1973. A revision of the genus *Faroa* Welwitsch. *Garcia de Orta, Bot.* 1: 69–82.
Thistleton-Dyer, W. T. 1909. Gentianaceae. Pages 1056–1121 in: A. W. Hill & D. Prain, eds. *Flora Capensis: being a systematic description of the plants of the Cape Colony, Caffraria, & Port Natal (and neighboring territories)*. L. Reeve & Co., London.
Thiv, M., L. Struwe, V. A. Albert, & J. W. Kadereit. 1999a. The phylogenetic relationships of *Saccifolium bandeirae* Maguire & Pires (Gentianaceae) reconsidered. *Harvard Pap. Bot.* 4: 519–526.
Thiv, M., L. Struwe, & J. W. Kadereit. 1999b [2000]. The phylogenetic relationships and evolution of the Canarian laurel forest endemic *Ixanthus viscosus* (Alt.) Griseb. (Gentianaceae): evidence from *mat*K and ITS sequence variation, and floral morphology and anatomy. *Pl. Syst. Evol.* 218: 299–317.
Thorne, R. F. 1983. Proposed new realignments in the angiosperms. *Nord. J. Bot.* 3: 85–117.
Thorne, R. F. 1992. An updated phylogenetic classification of the flowering plants. *Aliso* 13: 365–389.
Threadgill, P. F. & J. M. Baskin. 1978. *Swertia caroliniensis* or *Frasera caroliniensis*. *Castanea* 43: 20–22.
Thulin, M. 1970. Chromosome numbers of some vascular plants from East Africa. *Bot. Not.* 123: 488–494.
Thunberg, C. P. 1782. Beskrifning på et nytt och vackert örte-genus, kalladt *Fagraea ceilanica*. *Kongl. Vetenskapsakad. Nya Handl.* 3: 132–134.
Thwaites, G. H. K. 1860. *Enumeratio plantarum Zeylaniae*, vol. 3. Dulau, London.

Tidestrom, I. & T. Kittel. 1941. Gentianaceae. Pages 531–538 in: *A flora of Arizona and New Mexico*. Catholic University of America Press, Washington, DC.
Tiffney, B. H. 1985a. Perspectives on the origin of the floristic similarity between eastern Asia and eastern North America. *J. Arnold Arbor.* 66: 73–94.
Tiffney, B. H. 1985b. The Eocene North Atlantic land bridge: its importance in Tertiary and modern phytogeography of the northern hemisphere. *J. Arnold Arbor.* 66: 243–273.
Tirel-Roudet, C. 1972. Loganiaceae. Pages 3–89 in: A. Aubréville & J.-F. Leroy, eds. *Flore du Cambodge, du Laos et du Vietnam*, vol. 13. Muséum national d'histoire naturelle, Paris.
Toyokuni, H. 1961. Séparation de *Comastoma*, genre nouveau, d'avec *Gentianella*. *Bot. Mag. (Tokyo)* 74: 198.
Toyokuni, H. 1963. Conspectus gentianacearum japonicarum. *J. Fac. Sci. Hokkaido Univ.*, ser. V, 7: 137–259.
Toyokuni, H. 1965. Systema Gentianinarum Novissimum. *Symb. Asahikawensis* 1: 147–158.
Turczaninow, N. 1840. Description de deux nouveaux genres de la familie des Gentianées. *Bull. Soc. Imp. Nat. Moscow* 22: 313–342.
Turner, B. L. 1994. Taxonomic revision of *Geniostemon* (Gentianaceae). *Phytologia* 76: 8–13.
Tutin, T. G. 1972. Gentianaceae. Pages 56–67 in: T. G. Tutin, V. H. Heywood, N. A. Burges, D. M. Moore, D. H. Valentine, S. M. Walters, & D. A. Webb, eds. *Flora Europaea*, vol. 3. Cambridge University Press, Cambridge.
Tzvelev, N. N. 1980. Gentianaceae. Pages 195–213 in: A. I. Tolmatchev & B. A. Jurtzev, eds. *Flora Arctica URSS*, vol. 8. Nauka, Leningrad.
Tzvelev, N. N. & V. V. Pissjaukova. 2000. Gentianaceae. Pages 70–111 in: A. A. Fedorov, ed. *Flora of Russia, the European part and bordering regions*, vol. 3. Balkema, Rotterdam. (English translation of *Flora Evropeiskoi chasti SSSR* [1978].)
Ubolcholaket, A. 1987. Gentianaceae. Pages 72–92 in: T. Smitinand & K. Larsen, eds. *Flora of Thailand*, vol. 5(1). Chutima Press, Bangkok.
Valdés, B., S. Talavera, & E. Fernández-Galiano. 1987. Gentianaceae. Pages 339–345 in: *Flora vascular de Andalucía occidental*, vol. 2. Ketres Editora, S.A., Barcelona.
Van Bruggen, T. 1976. Gentianaceae. Pages 347–349 in: *The vascular plants of South Dakota*. Iowa State University Press, Ames, IA.
Veldkamp, J. F. 1968. A synopsis of the genus *Enicostema* Bl., *nom. cons.* (Gentianaceae). *Blumea* 41: 133–136.
Verdoorn, I. C. 1963. Loganiaceae. Pages 134–171 in: R. A. Dyer, L. E. Codd, & H. B. Rycroft, eds. *Flora of southern Africa*, vol. 26. Department of Agricultural Technical Services, Republic of South Africa, Pretoria.
Visyulina, O. D. 1957. Gentianaceae. Pages 221–260 in: M. I. Kotov & A. I. Barbarich, eds. *Flora URSR*, vol. 8. Akademiya nauk Ukraynskoy RSR, Kiyev.
Vogel, S. 1969a. Chiropterophilie in der neotropischen Flora. Neue Mitteilungen II, II. Spezieller Teil (Fortsetzung). *Flora, Abt. B* 158: 185–222.

Vogel, S. 1969b. Chiropterophilie in der neotropischen Flora. Neue Mitteilungen III, II. Spezieller Teil (Fortsetzung). *Flora, Abt. B* 158: 289–323.
Voss, E. G. 1996. Gentianaceae. Pages 71–81 in: *Michigan flora, Dicots*, part III. Cranbrook Institute of Science and University of Michigan Herbarium, Ann Arbor, MI.
Wagenitz, G. 1964. Gentianales. Pages 405–424 in: H. A. Melchior. *Engler's Syllabus der Pflanzenfamilien*, vol. 2, ed. 12. Gebr. Börntraeger, Berlin.
Wagner, W. L., D. R. Herbst, & S. H. Sohmer. 1990. Gentianaceae. Pages 724–725 in: *Manual of the flowering plants of Hawai'i*, vol. 1. University of Hawaii Press, Honolulu.
Wallander, E. & V. A. Albert. 2000. Phylogeny and classification of Oleaceae based on *rps*16 and *trn*L–F sequence data. *Amer. J. Bot.* 87: 1827–1841.
Wallich, N. 1826. *Tentamen florae napalensis illustratae*. Calcutta and Serampore.
Wallich, N. 1831. *A numerical list of the dried specimens of plants in East India Company's Museum (Wallich's Catalogue)*. London.
Walsh, N. G. & T. J. Entwisle. 1993. Gentianaceae. Pages 310–321 in: N. G. Walsh, ed. *Flora of Victoria*, vol. 4, *Dicotyledons, Cornaceae to Asteraceae*. Inkata Press, Melbourne.
Walter, T. 1788. *Flora caroliniana*. Sumptibus J. Fraser, London.
Wang, F. H. 1960. *Pollen grains of China*. Science Press, Beijing.
Watson, L. & M. J. Dallwitz. 1992–2001. The families of flowering plants: descriptions, illustrations, identification, and information retrieval, version 14 December 2000. (http://biodiversity.uno.edu/delta/)
Weaver, R. E., Jr. 1969. Cytotaxonomic notes on some neotropical Gentianaceae. *Ann. Missouri Bot. Gard.* 56: 439–443.
Weaver, R. E., Jr. 1972. A revision of the neotropical genus *Lisianthius* (Gentianaceae). *J. Arnold Arbor.* 53: 76–100, 234–272, 273–311.
Weaver, R. E., Jr. 1974. The reduction of *Rusbyanthus* and the tribe Rusbyantheae (Gentianaceae). *J. Arnold Arbor.* 55: 300–302.
Weaver, R. E., Jr. & L. Rüdenberg. 1975. Cytotaxonomic notes on some Gentianaceae. *J. Arnold Arbor.* 56: 211–222.
Weber, W. A. & R. C. Wittmann. 1996a. Gentianaceae. Pages 196–199 in: *Colorado flora: western slope*, rev. ed. University Press of Colorado, Niwot, CO.
Weber, W. A. & R. C. Wittmann. 1996b. Gentianaceae. Pages 210–213 in: *Colorado flora: eastern slope*, rev. ed. University Press of Colorado, Niwot, CO.
Welsh, S. L., ed. 1974. Gentianaceae. Pages 239–245 in: *Anderson's flora of Alaska and adjacent parts of Canada*. Brigham Young University Press, Provo, UT.
Welsh, S. L., N. D. Atwood, S. Goodrich, & L. C. Higgins, eds. 1993. Gentianaceae. Pages 346–350 in: *A Utah flora*, rev. ed. 2. Brigham Young University Press, Provo, UT.
Welwitsch, F. 1869. Sertum Angolense. *Trans. Linn. Soc.* 27: 1–94, 25 figs.
Wettstein, R. 1896. Die Gattungszugehörigkeit und systematische Stellung der *Gentiana tenella* Rottb. und *G. nana* Wulf. *Österr. Bot. Zeitschr.* 46: 121–128, 172–176.

Wielgorskaya, T. 1995. *Dictionary of generic names of seed plants.* Columbia University Press, New York.
Wiggins, I. L. 1980. Gentianaceae. Pages 404–405 in: *Flora of Baja California.* Stanford University Press, Stanford, CA.
Wilbur, R. L. 1955. A revision of the North American genus *Sabatia* (Gentianaceae). I. *Rhodora* 57: 1–33.
Wilbur, R. L. 1984a. A synopsis of the genus *Halenia* (Gentianaceae) in Mexico. *Rhodora* 86: 311–337.
Wilbur, R. L. 1984b. A synopsis of the genus *Halenia* (Gentianaceae) in Central America. *Bull. Torrey Bot. Club* 111: 366–374.
Wilbur, R. L. 1989. The type species of the genus *Sabatia* Adans. (Gentianaceae). *Rhodora* 91: 167–171.
Wild, H. 1965. Additional evidence for the Africa–Madagascar–India–Ceylon land-bridge theory with special reference to the genera *Anisopappus* and *Commiphora. Webbia* 19: 497–505.
Wilford, G. E. & P. J. Brown. 1994. Maps of late Mesozoic–Cenozoic Gondwana break-up: some palaeogeographical implications. Pages 5–13 in: R. S. Hill, ed. *History of the Australian vegetation: Cretaceous to recent.* Cambridge University Press, Cambridge and New York.
Willdenow, C. L. 1801. *Species plantarum.* Impensis G. C. Nauk, Berolini.
Williams, L. O. 1969. Gentianaceae in: P. Standley & L. O. Williams, eds. Flora of Guatemala. *Fieldiana Bot.* 24: 302–344.
Winkler, S. 1990. Zur Evolution der Gattung *Tillandsia* L. *Bot. Jahrb.* 122: 43–77.
Wolfe, J. A. 1975. Some aspects of plant geography of the northern hemisphere during late Cretaceous and Tertiary. *Ann. Missouri Bot. Gard.* 62: 264–279.
Wong, K. M. & J. B. Sugau. 1996. A revision of *Fagraea* (Loganiaceae) in Borneo, with notes on related Malesian species and 21 new species. *Sandakania* 8: 1–93.
Wood, C. E., Jr. & R. E. Weaver, Jr. 1982. The genera of Gentianaceae in the southeastern United States. *J. Arnold Arbor.* 63: 441–487.
Woodson, R. E. 1938. Observations on the floral fibres of certain Gentianaceae. *Ann. Bot. (London)* 50: 759–766.
Wunderlin, R. P. 1982. Gentianaceae. Pages 295–297 in: *Guide to the vascular plants of central Florida.* University Presses of Florida, Tampa, FL.
Yuan, Y.-M. & P. Küpfer. 1993a. Karyological studies of *Gentianopsis* Ma and some related genera of Gentianaceae from China. *Cytologia* 58: 115–123.
Yuan, Y.-M. & P. Küpfer. 1993b. Karyological studies on *Gentiana* sect. *Frigida* s. l. and sect. *Stenogyne* (Gentianaceae) from China. *Bull. Soc. Neuchâteloise Sci. Nat.* 116: 65–78.
Yuan, Y.-M. & P. Küpfer. 1995. Molecular phylogenetics of the subtribe Gentianinae (Gentianaceae) inferred from the sequences of internal transcribed spacers (ITS) of nuclear ribosomal DNA. *Pl. Syst. Evol.* 196: 207–226.
Yuan, Y.-M. & P. Küpfer. 1997. The monophyly and rapid evolution of *Gentiana* sect. *Chondrophyllae* s. l. (Gentianaceae): evidence from the nucleotide sequences of the internal transcribed spacers (ITS) of nuclear ribosomal DNA. *Bot. J. Linn. Soc.* 123: 25–43.

Yuan, Y.-M., P. Küpfer, & J. J. Doyle. 1996. Infrageneric phylogeny of the genus *Gentiana* (Gentianaceae) inferred from nucleotide sequences of the internal transcribed spacers (ITS) of nuclear ribosomal DNA. *Amer. J. Bot.* 83: 641–652.

Yuan, Y.-M., P. Küpfer, & L. Zeltner. 1998. Chromosomal evolution of *Gentiana* and *Jaeschkea* (Gentianaceae), with further documentation of chromosome data for 35 species from western China. *Pl. Syst. Evol.* 210: 231–247.

Zarucchi, J. L. 1993. Gentianaceae. Pages 534–545 in: L. Brako & J. L. Zarucchi, eds. *Catalogue of the flowering plants and gymnosperms of Peru.* Missouri Botanical Garden Press, St. Louis, MO.

Zeltner, L. 1970. Recherches de biosystématique sur les genres *Blackstonia* Huds. et *Centaurium* Hill (Gentianacées). *Bull. Soc. Neuchâteloise Sci. Nat.* 93: 1–164.

Zuev, V. V. 1984. On the systematics of the representatives of the Siberian genus *Gentiana* s. l. (Gentianaceae). *Bot. Zhurn. (Moscow & Leningrad)* 70: 916–923.

Zuev, V. V. 1990. On the systematics of the Gentianaceae family in Siberia. *Bot. Zhurn. (Moscow & Leningrad)* 75: 1296–1305.

Zuev, V. V. 1997. Gentianaceae. Pages 56–85 in: L. I. Malyschev, ed. *Flora Sibiriae*, vol. 11. Nauka, Novosibirsk.

3
Cladistics of Gentianaceae: a morphological approach

S. MÉSZÁROS, J. DE LAET, V. GOETHALS, E. SMETS, AND S. NILSSON

ABSTRACT

The infrafamilial relationships of the Gentianaceae are investigated by means of a cladistic analysis of 84 phenotypic characters, based mainly on data from the literature. The 41 genera that were selected for the analysis, including the formerly loganiaceous genera *Anthocleista* and *Fagraea* and the monotypic genus *Saccifolium*, are a fair representation of the character diversity in the family. The diverse genus *Gentiana* is represented by six of its sections. As outgroups we used *Strychnos* and *Geniostoma* (Loganiaceae), *Gelsemium* (Gelsemiaceae), and two genera each of Apocynaceae and Rubiaceae.

In the strict consensus cladogram of all most-parsimonious trees Gentianaceae has an unresolved basal trichotomy between *Saccifolium* (of tribe Saccifolieae), Potaliinae, and a major clade including all other genera. In this clade only tribe Gentianeae and subtribe Chroniinae of tribe Chronieae (*Ixanthus* excepted) are recognized as monophyletic groups. Within tribe Gentianeae, subtribe Gentianinae is nested in a paraphyletic subtribe Swertiinae. The relationships between the representatives of Exaceae, Canscorinae and Coutoubeinae (Chronieae), and Helieae are almost completely unresolved. An interesting exception is the sister-group relationship between *Exacum* and *Cotylanthera*.

Two complementary explanations for the lack of resolution in most parts of the cladogram are discussed: (1) the morphological characters of most tropical members of the family are insufficiently known, and (2) morphological characters are not well suited to resolve the more basal relationships in Gentianaceae.

Keywords: Gentianaceae, infrafamilial classification, morphology, phylogeny, phytochemistry.

INTRODUCTION

The family Gentianaceae is a (sub)cosmopolitan group of 87 genera and more than 1600 species (Struwe *et al.*, 2002 (Chapter 2, this volume)). There is strong evidence from cladistic analyses of both morphological and molecular data that its closest relatives are Apocynaceae *sensu lato*, Rubiaceae, and parts of the paraphyletic assemblage Loganiaceae *sensu lato* (e.g., Downie & Palmer, 1992; Olmstead *et al.*, 1993; Bremer *et al.*, 1994; Struwe *et al.*, 1994, 1998, 2002; Bremer, 1996; De Laet & Smets, 1996; Struwe & Albert, 1997). Compared with the classification of Gilg (1895), the most recent worldwide treatment of Gentianaceae, the delimitation of the family has been changed in two important ways: (1) Gilg's subfamily Menyanthoideae was raised to family level by Wagenitz (1964), and Menyanthaceae is now generally considered to be related to the Campanulales–Asterales complex (Downie & Palmer, 1992; Lammers, 1992; Olmstead *et al.*, 1992, 1993; Cosner *et al.*, 1994; Gustafsson *et al.*, 1996; Erbar, 1997), and (2) the loganiaceous tribe Potalieae (including *Potalia, Fagraea,* and *Anthocleista*) was transferred from Loganiaceae to Gentianaceae (Struwe & Albert in Struwe *et al.*, 1994, following Bureau, 1856; see also Fosberg & Sachet, 1980; Jensen, 1992). We also include *Saccifolium*, described as the monotypic family Saccifoliaceae (Maguire & Pires, 1978), in Gentianaceae (Thiv *et al.*, 1999a; Struwe *et al.*, 2002). Within Gentianales, Gentianaceae are characterized by the presence of internal phloem, contort corolla aestivation, superior ovary, and xanthones, and by the absence of laticifers, interpetiolar stipules, cardenolides, and indole alkaloids.

Other attempts at infrafamilial classification covering the whole family are Grisebach (1839, 1845) and Bentham (1876). Both Grisebach and Bentham based their classifications on a broader array of floral characters (mainly from anthers, styles, stigmas, and ovaries) than Gilg, who almost exclusively used pollen features. Even though Gilg's system has often been criticized (see Mészáros *et al.*, 1996, for details), it is the most used and best known of the three. Recent authors have used Gilg's classification (Ho *et al.*, 1988), partly returned to that of Bentham (Garg, 1987), or have not used any infrafamilial classification at all (e.g., Hutchinson, 1959; Wood & Weaver, 1982; Ho & Pringle, 1995). The different infrafamilial classifications are compared in Table 3.1, using Gilg's (1895) subfamily Gentianoideae as the point of reference. With some generalization, Grisebach's Lisyantheae and Bentham's subtribes Erythraeinae and Lisiantheae of tribe Chironieae correspond in outline to the ensemble of

Table 3.1. *Selected infrafamilial classifications of Gentianaceae: tribes (bold) and subtribes (not bold) of Gilg's (1895) subfamily Gentianoideae of Gentianaceae, and their (partial) correspondence to the (sub)tribes of Grisebach (1845), Bentham (1876), and Garg (1987)*

Gilg (1895)	Grisebach (1845)	Bentham (1876)	Garg (1987)
Gentianeae Exacinae	Chironieae	Exaceae	Exaceae
Gentianeae Erythraeinae	Chloreae	Chironieae Euchironieae Erythraeae Swertieae	Chironieae
Gentianeae Chironiinae	Chironieae	Chironieae Euchironieae	—[a]
Gentianeae Gentianinae	Swertieae	Swertieae	Gentianeae Swertieae
Gentianeae Tachiinae	Lisyantheae	Chironieae Euchironieae Lisiantheae	—
Rusbyantheae	Lisyantheae	Chironieae Lisiantheae	—
Helieae	Lisyantheae	Chironieae Erythraeae Lisiantheae	—
Voyrieae	Lisyantheae	Chironieae Euchironieae	—
Leiphaimeae	Lisyantheae	Chironieae Euchironieae	—

Note: [a] A dash indicates (sub)tribes that are absent from northwest Himalaya, the scope of Garg's regional treatment.

Gilg's subtribe Tachiinae of tribe Gentianeae and his tribes Rusbyantheae and Helieae. Considering that the only species of Gilg's tribe Rusbyantheae, *Rusbyanthus cinchonifolius*, is now included in *Macrocarpaea* (Weaver, 1974; Maas *et al.*, 1983), the main differences between Grisebach, Bentham, and Gilg center around the genera that are included in Gilg's much criticized neotropical (sub)tribes Helieae and Tachiinae; for example, Wood and Weaver (1982) proposed to merge Helieae and Tachiinae, thereby echoing Grisebach (1845). Gilg's tribe Rusbyantheae itself is a prime example of the artificial nature of a system that is too exclusively based on few characters.

Within Gentianaceae, several phylogenetic studies on the generic and tribal level have been published. Morphological cladistic analyses exist for *Exacum* (Klackenberg, 1985), *Tachiadenus* (Klackenberg, 1987), *Lomatogonium* (Liu & Ho, 1992), *Potalia* (L. Struwe & V. A. Albert, unpubl.), and *Voyria* (Albert & Struwe, 1997), while molecular studies have been published for part of *Lisianthius* (Sytsma & Schaal, 1985) and *Gentiana* (Gielly & Taberlet, 1996; Gielly *et al.*, 1996; Yuan *et al.*, 1996; Hungerer & Kadereit, 1997; Yuan & Küpfer, 1997). Studies at the tribal level cover Gentianinae (Yuan & Küpfer, 1995; Gutsche *et al.*, 1997), Erythraeinae (Thiv & Kadereit, 1997; Thiv *et al.*, 1999b), Helieae (Pihlar *et al.*, 1998; Struwe, 1999), and Potalieae (Struwe & Albert, 1997). Morphological cladistic analyses covering several tribes can be found in Mészáros (1994) and Mészáros *et al.* (1996). Lastly, cladistic studies of more than 150 *trn*L intron sequences and over 100 *mat*K sequences cover the whole family (Struwe *et al.*, 1998, 2002; Thiv *et al.*, 1999a) and form the basis of the infrafamilial classification that is proposed in Chapter 2 of this volume (Struwe *et al.*, 2002).

In this chapter we continue our principally morphological approach, broadening the scope from mainly temperate Gentianaceae (Mészáros *et al.*, 1996) to a more even sampling across the entire family. In order to contribute to an improved knowledge of the Gentianaceae, we aim to extend the documentation of character state distributions in the family and to present a cladistic analysis of the enlarged data set.

MATERIALS AND METHODS
Taxa

In Table 3.2 we present a survey of the genera of Gentianaceae, using Gilg's (1895) classification as a point of reference (but with Menyanthoideae excluded, Potalieae and Saccifoliaceae included, and Rusbyantheae reduced to Tachiinae), including the many new species and genera that have been described since Gilg presented his classification. We want to stress that the only purpose of the table is to provide a baseline against which new findings from phylogenetic analyses can be evaluated. Therefore, in compiling the table we followed Gilg (1895, 1897, 1908) for all the genera that were known to him, even though some transfers have subsequently been proposed (e.g., *Hockinia* to Erythraeinae (Maas & Ruyters, 1986), *Tachiadenus* to Exacinae (Klackenberg, 1987), *Eustoma* and *Coutoubea* to Erythraeinae (Kaouadji, 1990)); the remaining genera were accommodated using

Table 3.2. *Genera of Gentianaceae according to Gilg's (1895) classification. Accepted genera as in Struwe et al. (2002) except for* Xestaea *and* Frasera, *which are here included in* Schultesia *and* Swertia, *respectively. Figures in parentheses are number of species*

Genera included in the current data set	Genera not included in this study
Gentianeae-Exacinae	
Cotylanthera Blume (4)	*Microrphium* C. B. Clarke (2)[a1]
Exacum L. (65)	*Ornichia* Klack. (3)[1]
Sebaea Sol. ex R. Br. (60–100)	
Gentianeae-Erythraeinae	
Bartonia H. L. Mühl. ex Willd. (4)	*Bisgoeppertia* Kuntze (2)
Blackstonia Huds. (4)	*Cicendia* Adans. (2)
Canscora Lam. (9)	*Congolanthus* A. Raynal (1)[b2]
Centaurium Hill (50)	*Cracosna* Gagnep. (3)[c3]
Curtia Cham. & Schltdl. (6–10)	*Djaloniella* P. Taylor (1)[d4]
Enicostema Blume (3)	*Exaculum* Caruel (1)
Faroa Welw. (19)	*Geniostemon* Engelm. & A. Gray (5)
Hoppea Willd. (2)	*Karina* Boutique (1)[4]
Obolaria L. (1)	*Neurotheca* Salisb. ex Benth. (3)
Sabatia Adans. (20)	*Oreonesion* A. Raynal (1)
	Phyllocyclus Kurz. (5)[3]
	Pycnosphaera Gilg (1)
	Schinziella Gilg (2)
	Tapeinostemon Benth. (7)
	Urogentias Gilg & Gilg-Ben. (1)
Gentianeae-Chironiinae	
Chironia L. (15)	*Gentianothamnus* Humbert (1)[e5]
Orphium E. Mey. (2)	
Gentianeae-Gentianinae	
Crawfurdia Wall. (16–19)	*Comastoma* (Wettst.) Toyok. (7–25)[f6]
Gentiana L. (360)	*Jaeschkea* Kurz (4)
Gentianella Moench. (250)	*Latouchea* Franch. (1)
Gentianopsis Ma (16–24)[6]	*Megacodon* (Hemsl.) Harry Sm. (2)[6]
Halenia Borkh. (80)	*Pterygocalyx* Maxim. (1)
Ixanthus Griseb. (1)	*Veratrilla* Baill. ex Franch. (2)
Lomatogonium A. Braun (21)	
Swertia L. (150) (including *Frasera* Walter)	
Tripterospermum Blume (24)	
Gentianeae-Tachiinae	
Chorisepalum Gleason & Wodehouse (5)[g7]	*Hockinia* Gardn. (1)
Eustoma Salisb. (3)	*Zonanthus* Griseb. (1)
Lisianthius P. Browne (30)	*Zygostigma* Griseb. (2)
Macrocarpaea (Griseb.) Gilg (90)[h]	
Tachia Aubl. (10)	
Tachiadenus Griseb. (11)	

Cladistics: a morphological approach 315

Table 3.2. (*cont.*)

Genera included in the current data set	Genera not included in this study
	Helieae
Celiantha Maguire (3)[7]	*Adenolisianthus* Gilg (1)
Coutoubea Aubl. (5)	*Aripuana* Struwe, Maas, &
Deianira Cham. & Schltdl. (5)	V. A. Albert (1)[7]
Irlbachia Mart. (9)	*Calolisianthus* Gilg (6)
Schultesia Mart. (16)	*Chelonanthus* Gilg (7)
(including *Xestaea* Griseb.)	*Helia* Mart. (2)
Symbolanthus G. Don (30)	*Lagenanthus* Gilg (1)
	Lehmanniella Gilg (2)
	Neblinantha Maguire (2)[7]
	Prepusa Mart. (5)
	Purdieanthus Gilg (1)
	Rogersonanthus Maguire & B. M. Boom (3)[7]
	Senaea Taub. (1)
	Sipapoantha Maguire & B. M. Boom (1)[7]
	Symphyllophyton Gilg (1)
	Tetrapollinia Maguire & B. M. Boom (1)[7]
	Wurdackanthus Maguire (2)[7]
	Voyrieae
Voyria Aubl. (19)	
	Leiphaimeae
Voyriella (Miq.) Miq. (1)	
	Potalieae *sensu* Leeuwenberg & Leenhouts[i]
Anthocleista R. Br. (14)	*Potalia* Aubl. (9)
Fagraea Thunb. (70)	
	Saccifoliaceae *sensu* Maguire & Pires[j]
Saccifolium Maguire & Pires (1)	

Notes:
[a] Genera not known or excluded by Gilg (1895) were classified on the basis of Gilg (1897, 1908) and Pilger and Krause (1915) and are marked with a superscript "1".
[b] Genera not known or excluded by Gilg (1895) were classified on the basis of Raynal (1968) and are marked with a superscript "2".
[c] Genera not known or excluded by Gilg (1895) were classified on the basis of Struwe *et al.* (2002) and are marked with a superscript "3".
[d] Genera not known or excluded by Gilg (1895) were classified on the basis of Taylor (1973; as relatives of *Faroa*) and are marked with a superscript "4".
[e] Genera not known or excluded by Gilg (1895) were classified on the basis of Humbert (1937) and are marked with a superscript "5".
[f] These genera were classified by the taxonomic position of the broader genus from which the new genus was segregated and are marked with a superscript "6".
[g] Genera not known or excluded by Gilg (1895) were classified on the basis of Gilg's (1895: 62) key (pollen in monads (=Tachiinae) vs. pollen in tetrads or polyads (=Helieae)) and Struwe and Albert (1998) and are marked with a superscript "7".

cont.

Notes to Table 3.2 (*cont.*)

[h] *Macrocarpaea* includes *Rusbyanthus*, the only genus of Gilg's tribe Rusbyantheae (Weaver, 1974).
[i] Potalieae *sensu* Leeuwenberg and Leenhouts (1980) are gentians (Bureau, 1856; Fosberg & Sachet, 1980; Jensen, 1992; Struwe *et al.*, 1994, 1998, 2002; Struwe & Albert, 1997; Thiv *et al.*, 1999a) but do not fit into Gilg's classification.
[j] Saccifoliaceae *sensu* Maguire and Pires (1978) are gentians (Thiv *et al.*, 1999a; Struwe *et al.*, 2002) but do not fit into Gilg's classification.

information from various sources, indicated by footnotes. Note that, unless indicated otherwise, we follow the classification of Struwe *et al.* (2002) in the remainder of the text.

Forty-one genera were selected for the current analysis (those in the left-hand column of Table 3.2; note that we include *Frasera* in *Swertia* and *Xestaea* in *Schultesia*). This amounts to about half of the total number of genera and over 90% of all species. These numbers are an indication of the covered diversity within the family rather than an assessment of the phylogenetic significance of the unsampled genera. To search for systematic affinities of the mycotrophic Gentianaceae (*Cotylanthera*, *Bartonia*, *Obolaria*, *Voyria*, and *Voyriella*), all these genera were included, and so was the enigmatic monotypic genus *Saccifolium* (Maguire & Pires, 1978). In order to reduce problems with polymorphism (Nixon & Davis, 1991), the diverse genus *Gentiana* was represented by six of its sections.

As outgroups we included *Geniostoma* and *Strychnos* (Loganiaceae), *Gelsemium* (Gelsemiaceae), *Plumeria* and *Rauwolfia* (Apocynaceae), and *Danais* and *Exostema* (Rubiaceae). In the selection of the outgroups we took into account the availability of recent cladistic studies (Bremer, 1996; Endress *et al.*, 1996), the availability of data from modern revisions (e.g., Conn, 1980; Buchner & Puff, 1993), and suggestions from B. Bremer (pers. comm.).

Characters

The data set (Table 3.3 and Appendix 3.1) contains 84 phenotypic characters, predominantly from morphology (including palynology and seed micromorphology) but supplemented with anatomical, embryological, karyological, and phytochemical characters. The data are mostly compiled from the literature, in some cases are supplemented with observations of herbarium (BP, BR, and DBN) and living plant material (*Centaurium*,

Cladistics: a morphological approach 317

Table 3.3. *Characters and character states*

1. Heterotrophic syndrome: absent (0); present (1)
2. Life form: trees or shrubs (0); perennial herbs (1); biennial and annual herbs (2)
3. Cross-section of main stem: terete (0); quadrangular (1); winged (2)
4. Interxylary phloem in stem: absent (0); present (1)
5. Xylem rays: multi- and uniseriate (0); bi- and uniseriate (1); rayless (2)
6. Nodal anatomy: uni(tri)lacunar (0); multilacunar (1)
7. Stolons and runners: absent (0); present (1)
8. Stem: erect (0); twining (1)
9. Vessels: solitary (0); in chains or in clusters (1)
10. Vessel perforation plates: scalariform (0); simple (1)
11. Axial parenchyma: apotracheal (0); paratracheal (1)
12. Lacticifers in stems: absent (0); present (1)
13. Extrafloral nectaries: absent (0); present (1)
14. True interpetiolar stipules: absent (0); present (1)
15. Leaf venation: pinnate, brochidodromous (0); acrodromous (1)
16. Mesophyll anatomy: bifacial (heterogeneous) (0); homogeneous (1)
17. Mature stomata: anomocytic (0); paracytic (1); anisocytic (2); diacytic (3)
18. Calcium oxalate crystals in mesophyll: absent (0); present (1)
19. Inflorescence: dichasium (0); monochasium (1); flowers in clusters (2); solitary flowers (3)
20. Flower color: white, green, or yellow (0); pink, red, blue, lilac, or brown (1)
21. Calyx: polymerous (0); 5-merous (1); 4-merous (2); 2-merous (3)
22. Size of calyx lobes: equal (0); unequal (1)
23. Fusion of sepals: scarcely (0); half (1); almost completely (2)
24. Abaxial side of calyx lobes: smooth (0); keeled (1); winged (2)
25. Intracalycine membrane: absent (0); present (1)
26. Colleters or squamellae on adaxial side of calyx tube: absent (0); present (1)
27. Glandular area on top of calyx: absent (0); present (1)
28. Lateral traces of calyx: free (0); fused at origin (1); fused throughout (2)
29. Metaxylary fibers in calyx: absent (0); present (1)
30. Corolla aestivation: valvate (0); imbricate (1); contorted (2); plicate (3)
31. Corolla: polymerous (0); 5-merous (1); 4-merous (2)
32. Corolla shape: rotate (0); salver-shaped (hypocrateriform) (1); funnel-shaped (infundibular) (2); campanulate (3)
33. Petal fusion: scarcely (0); half (1); almost completely (2)
34. Nectar guide: absent (0); present (1)
35. Floral nectaries: none or rudimentary (0); on the corolla (1); gynoecial (2)
36. Anther shape: long, non-sagittate (0); long, sagittate (1); short (2)
37. Anther size: normal (0); enlarged (1)
38. Anther fixation: dorsifixed and non-versatile or basifixed (0); dorsifixed and versatile (1)
39. Anther dehiscence: longitudinal slits (0); apical pores (1)
40. Anther twisting after ripening: absent (0); present (1)
41. Anther cohesion: free (0); connate (in at least one floral type in heterostylous taxa) (1)
42. Anther abortion: none (0); 1–3 aborted stamina (1); one fertile stamen (2)
43. Anther appendix: absent (0); present (1)
44. Filament bases: not united (0); united by a membrane (1)

Table 3.3. (*cont.*)

45. Endothecium: fibrous (0); non-fibrous (1)
46. Heterostyly: absent (0); present (1)
47. Stamen insertion: near the base of the corolla (0); between the base and the mouth of the corolla (1); near the mouth of the corolla (2)
48. Ovary: superior (0); inferior (1)
49. Ovary: syncarpous (0); apocarpous (1)
50. Ovary shape: globular (0); oval (1); long, at least three times as long as wide (2)
51. Ovary position: sessile (0); stipitate (1)
52. Stigma: simple (0); capitate (1); lobed (2); decurrent (3); dichotomously lobed (4)
53. Carpel ventral traces: free (0); fused at origin (1); fused throughout (2)
54. Fruit dehiscence: septicidal (0); loculicidal (1); indehiscent (2); irregular (3)
55. Fruit type: capsule (0); baccate (1); drupe (2)
56. Ovule type: hemianatropous (0); anatropous (1); orthotropous (2); campylotropous (3)
57. Integuments: normal (0); absent or rudimentary (1)
58. Antipodal number: three (0); 8–12 (1)
59. Antipodals: ephemeral (0); persistent (1)
60. Antipodals: non-haustorial (0); haustorial (1)
61. Endosperm development: *ab initio* nuclear (0); *ab initio* cellular (1)
62. Embryo suspensor: uniseriate (0); 2–4-seriate (1)
63. Seed shape: angular or cubical (to irregular: e.g., *Schultesia* and *Celiantha*) (0); globular (1); oval (2); elongated (3)
64. Seed wing: absent (0); present (1)
65. Seed testa cell shape (away from hilum): not elongated (0); elongated (1)
66. Radial cell walls of seed testa cells: straight (to slightly undulated: e.g., *Celiantha*) (0); with clear undulations (1)
67. Inner tangential cell walls of seed testa cells: smooth (0); pitted (1); with papillae (2); with reticulum (3); multiply-pitted (*Sebaea*) (4)
68. Pollen germination: outside the thecae (0); within the thecae (1)
69. Pollen unit: monads (0); tetrads (1); polyads (2)
70. Pollen apertures: colpi (0); colpori (1); pori (2)
71. Exine structure: atectate (0); semitectate (1); tectate (2)
72. Supratectal processes: absent (0); present (1)
73. Haploid chromosome number n: below 15 (0); between 15 and 29 (1); 30 or above (2)
74. L-(+)-bornesitol (a cyclitol): absent (0); present (1)
75. Sugars: simple (glucose, primverose, rhamnose, galactose) (0); compound (gentianose, gentiobiose) (1)
76. End-product of secoiridoid biosynthesis: sweroside, including its derivatives (0); swertiamarin, including its derivatives (1); gentiopicroside (2); indole alkaloids (3)
77. Flavonoids: flavonols (O-glycosides) (0); flavone-O-glycosides (1); flavone-C-glycosides (2); flavonons (3)
78. Xanthones: absent (0); xanthone-C-glycosides (1); free xanthones and xanthone-O-glycosides (2)
79. Oxygenation of xanthone position C2: absent (0); present (1)
80. Oxygenation of xanthone position C4: absent (0); present (1)

Table 3.3. (cont.)

81. Oxygenation of xanthone position C5: absent (0); present (1)
82. Oxygenation of xanthone position C6: absent (0); present (1)
83. Oxygenation of xanthone position C7: absent (0); present (1)
84. Oxygenation of xanthone position C8: absent (0); present (1)

Eustoma, Exacum, Gentiana, Ixanthus, and *Plumeria*). The pollen characters and state distributions are mostly from personal observations of S. Nilsson; characters and character states of the seed characters are mostly from the work of V. Goethals. Of the cells in the data set, 24.4% are scored as question marks (designating either missing information or inapplicable characters); 9.8% of the data cells represent polymorphisms. Below we review the character state distributions and sources according to character groups.

Habit, duration of life cycle, and trophy

Gentianaceae cover a wide spectrum of habit and duration of the life cycle, from trees and shrubs to perennial, biennial, and annual herbs. Stebbins (1974) described some trends in the evolution of growth habits: (1) modern dicotyledonous trees have evolved from shrubby ancestors; (2) shrubs transformed to perennial herbs; and (3) annuals and biennials evolved from perennial herbs. In the first and second trend many examples of reversals are known, but the trend from perennial to monocarpic life cycle seems to be almost irreversible (Kremer & van Andel, 1995).

In our study, the outgroups are trees and shrubs but at the same time there are some indications for a reverse infrageneric trend in *Lisianthius*, *Exacum*, and *Tachiadenus* (Weaver, 1972; Klackenberg, 1985, 1987). Based on a comparison of families, Kremer and van Andel (1995) also argue that biennials emerged from annuals. Infrageneric data seem to contradict this trend; for example, within European species of *Centaurium* sect. *Centaurium* biennial species are diploids ($n=10$) while annuals are the tetraploid ones ($n=18, 20$; see Melderis, 1972). We therefore grouped biennials and annuals to one character state, as we did with trees and shrubs.

Another characteristic of Gentianaceae is a special type of arbuscular mycorrhiza, which differs from normal mycorrhizae in the structure of the endophyte and the way it spreads (Demuth, 1993). This type of mycorrhiza has also been reported for *Gelsemium* (Tiemann *et al.*, 1993), Rubiaceae (Rath, 1993), and Apocynaceae (Klahr, 1993), in the last case together with normal and transitional forms of arbuscular mycorrhizae. The family

Gentianaceae is also unusual in that, in tandem with the reduction in chlorophyll content, a phylogenetic transition from facultative to obligate mycotrophy has occurred. *Bartonia* and *Obolaria* have a low chlorophyll content while *Cotylanthera, Voyria*, and *Voyriella* seemingly have no chlorophyll at all, becoming endoparasites. Another conspicuous tendency in these genera is a reduction of the root system, changing to a coralloid or morning-star type (Furman & Trappe, 1971; Weber, 1992; Imhof *et al.*, 1994). The co-occurrence of these characteristics is coded as the presence of a heterotrophic syndrome.

Stem

When characterizing the stem of the Gentianaceae, anatomical characters related to secondary growth seem to have the most phylogenetic significance (note that according to Dickison, 1975, trends of specialization in the secondary xylem elements of dicotyledons tend to be paralleled, with an evolutionary lag, in the primary xylem). Wood anatomy of *Symbolanthus, Chelonanthus*, and *Ixanthus* was studied by Carlquist (1984), while the general anatomy of the family was described by Perrot (1897). In addition anatomical data for some genera can be found in Solereder (1885, 1899), Figdor (1897), Holm (1897, 1906), Metcalfe and Chalk (1950), Szujkó-Lacza and Sen (1979), Szujkó-Lacza and Gondar (1983), and ter Welle (1986).

Based on Carlquist (1984), general characteristics of the wood of Gentianaceae are absence of storied structures, absence of crystals, vessels round in transection and standing in radial chains, and presence of intraxylary (internal or medullary) phloem. The perforation plates of vessels are generally simple but the scalariform state is reported to occur in some of the outgroups (*Geniostoma* and *Rauwolfia*) and in *Saccifolium* (Maguire & Pires, 1978); both states co-occur in *Lisianthius* (Solereder, 1885), *Chironia, Coutoubea*, and *Orphium* (Solereder, 1899).

The imperforate tracheary elements are predominantly fiber-tracheids with bordered pits, but primitive tracheids occur in some of the outgroups (*Strychnos* and *Rauwolfia*). Libriform fibers with simple pits, considered the most advanced type, were reported for *Anthocleista, Fagraea* (Mennega, 1980), and *Ixanthus* (Carlquist, 1984).

In most of the outgroups and in *Symbolanthus* both multi- and uniseriate (heterogeneous) rays are present; the presence only of uniseriate (homogeneous) rays is characteristic for some other genera, and raylessness is reported for, or can be observed in, *Saccifolium* (Maguire & Pires, 1978) and in herbaceous genera: *Blackstonia, Centaurium*, and *Exacum* (Metcalfe

Cladistics: a morphological approach 321

& Chalk, 1950), *Schultesia* (Solereder, 1899), *Swertia* (Perrot, 1897), *Bartonia* and *Obolaria* (Holm, 1897, 1906), and *Gentiana* sect. *Gentiana* (*G. asclepiadea*; Szujkó-Lacza & Sen, 1979). Absence of interxylary (included) phloem is reported for *Anthocleista* and *Fagraea* (Mennega, 1980), and *Symbolanthus* and *Irlbachia* (Carlquist, 1984); its presence was documented for *Ixanthus* (Carlquist, 1984), *Chironia* (Vesque, 1875), *Orphium* (Solereder, 1885), *Crawfurdia*, *Schultesia*, *Swertia*, and *Tripterospermum* (Metcalfe & Chalk, 1950), and *Gentiana* sect. *Gentiana* (Szujkó-Lacza & Sen, 1979); furthermore it can be observed on Perrot's (1897) drawings of *Centaurium* and *Exacum*.

Leaves

In the autotrophic genera the leaves are simple, entire, and opposite (symplesiomorphies with other families of Gentianales), rarely verticillate (*Curtia*) or alternate (*Swertia*); in the heterotrophic genera they are reduced to scales. The principal venation pattern is acrodromous according to the terminology of Hickey (1979): two or more secondary veins run in convergent arches toward the leaf apex. This type is reported for *Canscora*, *Centaurium*, *Enicostema*, *Exacum*, and *Hoppea* by Mohan et al. (1989) and was observed in many other genera. Pinnate, brochidodromous venation (with a single primary midvein and secondaries joined together), characteristic for Gentianales (Hickey & Wolfe, 1975), occurs in some woody Gentianaceae (*Anthocleista*, *Fagraea*, *Macrocarpaea*, and some species of *Chorisepalum*). It is interesting to note that in the outgroup genus *Strychnos* both states exist (Leenhouts, 1962).

Types of mature stomata were reported in a series of papers (Pant & Kidway, 1969; Patel et al., 1981; Trivedi & Upadhyay, 1983; Gill & Nyawuame, 1990). Two types dominate: the anomocytic (ranunculaceous) type, without subsidiary cells; and the paracytic (rubiaceous) one with two subsidiary cells beside the two guard cells (following the definitions of van Cotthem, 1970). The latter type occurs mainly in the genus *Gentiana*. Gill and Nyawuame (1990) tried to define the phylogenetic sequence of stomatal types based on the distribution of the types in 320 taxa of Bentham and Hooker's Bicarpellatae, and considered the anomocytic type to be the primitive one.

Calyx

The calyx in Gentianaceae is persistent, often gamosepalous and isomerous with the corolla lobes; in other respects it is very variable, resulting in seven characters for the analysis. We subdivided the considerable variation in the degree of sepal fusion into three states: scarcely, half, and almost completely. The state "scarcely" is found in the outgroups, in many tropical genera, in Gilg's (1895) tribe Exacinae and subtribe Erythraeinae, and in *Halenia, Lomatogonium,* and *Swertia* (polymorphic) of tribe Gentianeae. The other two states occur in the remaining part of the tribe Gentianeae as well as in some other genera (e.g., in *Canscora, Faroa,* and *Hoppea*). In addition to gamosepaly, the sepals of *Crawfurdia, Gentiana,* and *Gentianopsis* are connected by a *"membrana intracalycina"* (Grisebach, 1845). Kusnezow (1896–1904: 38–44) described the character in detail and found it in all species of (*Eu*)*Gentiana* that he investigated.

Vascular bundles to the calyx originate in whorls with one trace to each sepal; each of these traces then branches into three. According to Wood and Weaver (1982) specialization has tended toward fusion of the lateral traces of adjacent sepals. Lindsey (1940) demonstrated such fused calyx laterals in *Lisianthius* and in seven investigated genera of Gilg's Helieae. It occurs only sporadically in other parts of the family.

An interesting character is the squamellae or colleters that develop on the adaxial surface of the calyx tube and degenerate during anthesis. McCoy (1940) described the details of their structure in *Swertia* (*Frasera*) *carolinensis*, as did Vijayaraghavan and Padmanaban (1969) in *Centaurium ramosissimum*. The presence of this structure has been documented in many of the taxa included in our analysis; it is absent in, for example, *Gelsemium, Plumeria,* and *Rauwolfia* among the outgroups as well as in *Celiantha, Curtia, Coutoubea,* and *Lisianthius* within the ingroup; *Schultesia* and *Voyria* are polymorphic in this aspect.

Corolla

Corollas are sympetalous and generally actinomorphic or rarely slightly zygomorphic. Variable characters are aestivation, merosity, and corolla shape. Contort aestivation is reported to be characteristic for the family (including *Anthocleista* and *Fagraea*); imbricate aestivation, considered the most primitive by Takhtajan (1991), occurs in *Bartonia* and *Obolaria* (Wood & Weaver, 1982) and is also found in the outgroups *Exostema, Gelsemium,* and *Geniostoma*. Plicate aestivation, a special form of contort aestivation in which folds are alternating to lobes, is characteristic for *Crawfurdia, Gentiana,* and *Tripterospermum*.

Pentamery is the common and probably ancestral state for the family, but constant or occasional tetramery, presumably as a reduction, also occurs in many genera. More interesting are the cases exceeding pentamery: constant 6 in *Chorisepalum* (Maguire, 1981; with four sepals), 6–12 in *Blackstonia*

(Tutin, 1972), 5–12 in *Sabatia* subsect. *Dodecandrae* (Wilbur, 1955), 8–16 in *Anthocleista* and *Potalia* (Leeuwenberg & Leenhouts, 1980; L. Struwe & V. A. Albert, unpubl.; with four sepals), and 5–9 in *Gentiana* sect. *Gentiana* (Tutin, 1972).

Main corolla shapes are rotate, salver-shaped (hypocrateriform), funnel-shaped (infundibular), and campanulate. Rotate flowers are characteristic for tribe Exacinae and Gilg's (1895) subtribe Erythraeinae, with some exceptions. In tribe Gentianeae the other three corolla forms predominate, but *Lomatogonium* and *Swertia* have rotate flowers. Variations are often infrageneric (e.g., *Gentiana* and *Gentianella*).

Based on our own observations there appear to be three mechanisms for constricting the corolla tube to form a nectar guide for pollinators with long proboscides: (1) developing a salver-shaped corolla (typical for *Gentiana* sect. *Calathianae*), (2) growing fimbriae in the corolla throat (e.g., *Gentianella*), or (3) stamens adnating to the style, sometimes called "revolver-flowers" (*Gentiana* sects. *Gentiana*, *Ciminalis*, and *Pneumonanthe*).

Androecium

Stamens are generally isomerous, epipetalous, alternating with the corolla lobes, representing haplostemony according to the definition of Ronse Decraene & Smets (1995). Anthers are dithecal, tetrasporangiate, and mostly introrse. Reductions of the androecium shown to be typical for Asteridae (Ronse Decraene & Smets, 1995) are rare in Gentianaceae, occurring only in *Canscora*, *Hoppea*, and *Schinziella*.

Anthers are typically basifixed, the original configuration for the angiosperm stamen according to Baum and Leinfellner (1953). Dorsifixed and versatile anthers occur in *Gentiana* sect. *Otophora*, *Gentianella*, *Gentianopsis*, *Halenia*, and *Swertia*, as well as in *Bartonia* and *Obolaria*. This specialization is connected with a pollination mechanism where stamens rather than the stalk of the ovary are moving during anthesis (Philipson, 1972).

Another interesting specialization is twisting of anthers after ripening. This phenomenon is well known in *Centaurium* (drawn in Wagenitz, 1964) but it is also documented in *Orphium* (Gilg, 1895), *Chironia* (Schoch, 1903; Boutique, 1972; Paiva & Nogueira, 1990), *Sabatia* (Wood & Weaver, 1982), *Blackstonia* (Tutin, 1972), and *Bartonia* (Gillett, 1959), and was seen on living plants of *Eustoma*.

Several characters are found in only a few taxa. Anther appendices, called "Brown's bodies" by Schinz (1903), were observed in *Sebaea* (Marais & Verdoorn, 1963) and *Tachiadenus* (Klackenberg, 1987). Klackenberg (1985) considered non-fibrous (finely perforated) endothecium cell walls as a generic attribute of *Exacum*, but they are also characteristic for *Cotylanthera* (Figdor, 1897; Oehler, 1927).

Pollen

Pollen grains are generally radially symmetrical, tricolporate, two- or three-celled at the time of shedding, and with the longest axis varying from about 20 μm to 35 μm. Several states that are generally considered to be advanced occur mostly in neotropical genera: (1) pollen units are tetrads in *Coutoubea, Deianira, Schultesia,* and *Symbolanthus,* and polyads in *Celiantha* and *Irlbachia*; (2) besides the Rubiaceae outgroups, only *Celiantha* and *Irlbachia* have supratectal processes; and (3) porate ectoapertures occur in the neotropical genera *Celiantha, Coutoubea, Irlbachia, Schultesia,* and *Voyria,* in the paleotropical genera *Anthocleista* and *Fagraea,* and in the outgroup *Geniostoma*. Exine sculpturing varies throughout the family, with some genera even being polymorphic.

Pollen characters and character states were established by S. Nilsson (see Nilsson, 1964, 1967a,b, 1968, 1970, 2002 (Chapter 4, this volume), and Nilsson and Skvarla, 1969, for documentation of these characters); data for several other neotropical genera are documented in Elias and Robyns (1975). Walker and Doyle (1975) and Punt (1978) discussed phylogenetic trends.

Pollination

Pollination syndromes in the family are rather diverse. Melittophily is considered as most common and probably ancestral. Chiropterophily was reported for several neotropical genera such as *Symbolanthus, Irlbachia, Lisianthius,* and *Macrocarpaea,* together with ornithophily (*Symbolanthus*) or melittophily and sphingophily (*Irlbachia*) for some species (Vogel, 1958, 1969). Pollen flowers, with pollen as the main reward, were observed in *Chironia, Exacum, Orphium, Sabatia* (Vogel, 1978), and *Eustoma* (Vogel, 1993) as well as in *Centaurium* and *Deianira* (S. Vogel, pers. comm.).

Several floral characteristics are correlated with the mode of pollination. Genera with pollen flowers are generally nectarless or have rudimentary nectaries; in other cases the flowers are nectariferous. The principal nectary type is a gynoecial nectary but another type, situated on the corolla, also frequently occurs (*Gentianella, Gentianopsis, Halenia, Lomatogonium,* and *Swertia*). All these genera have rotate flowers that supply free nectar for a large array of pollinators (Beattie *et al.*, 1973). The homology of the gynoecial glands of *Voyria* (Maas & Ruyters, 1986; see also Albert & Struwe,

Cladistics: a morphological approach   325

1997) is difficult to interpret; we left the question open and coded *Voyria* as unknown for presence of floral nectaries.

The pollen flowers are of the *Solanum* type (Vogel, 1978): melittophilous, oligandrous, with shortened filaments and with enlarged anthers capable of producing excess pollen. In *Exacum, Cotylanthera,* and *Deianira* poricidal anther dehiscence has been reported (Figdor, 1897; Guimarães, 1977; Klackenberg, 1985), pointing to buzz pollination, which also occurs in Rubiaceae. Buzz-pollinated flowers probably developed secondarily from nectariferous flowers (Dukas & Dafni, 1990).

With the exception of *Gentianella* and *Veratrilla* (sometimes included in *Swertia*), where dioecy occurs, flowers in Gentianaceae are hermaphroditic. Protandry is the general form of dichogamy but other forms occur as well (e.g., approach herkogamy; Webb & Pearson, 1993). As these developments are infrageneric, characters of the breeding system could not be used for phylogenetic inference on the family level.

Gynoecium and fruit

The gynoecium is bicarpellate and syncarpous or paracarpous (Shamrov, 1996) with a superior ovary. Varying characters are shape, position, placentation of the ovary and number of locules, degree of fusion of carpel ventral traces, stigma morphology, and type of fruit dehiscence.

The ovary in Gentianaceae is unilocular or bilocular except in *Anthocleista*, where it seems to be 4-locular. The polarity of this character has been much discussed in the past (Lindsey, 1940; Krishna & Puri, 1962), but nowadays the bilocular condition, prevailing in six of the seven outgroups, is generally considered to be primitive in the family.

Axile placentation is associated with the bilocular state of the ovary. It prevails in the outgroups, in Gilg's tribe Exacinae, and in some woody neotropical genera. Parietal and superficial placentation (Krishna & Puri, 1962) are correlated with the unilocular ovary, the latter being characteristic for *Crawfurdia, Gentiana, Gentianella, Gentianopsis, Lomatogonium,* and *Tripterospermum*. Even if transitional states do occur, mainly for locule number (e.g., in *Lisianthius*), the distinction of ovary zones made by Leinfellner (1950) could not be used as characters because the detailed data required were available only for some neotropical genera (van Heusden, 1986; Struwe *et al.*, 1997).

The fruit is generally capsular, which is considered to be primitive in Gentianales. Berries are less widespread than in Rubiaceae; they are characteristic only for *Anthocleista* and *Fagraea* and occur only sporadically elsewhere (one section of *Tripterospermum* and one species of *Chironia* and

Symbolanthus). Septicidal fruit dehiscence is the common state for the ingroup; *Voyriella*, with an indehiscent fruit, and *Voyria*, with dehiscent, indehiscent, and transitional types, are the exceptions (Maas & Ruyters, 1986).

Embryology

Many embryological characters are constant throughout the family: microsporangial development is of the dicotyledonous type, ovules are unitegmic and tenuinucellar, megagametogenesis is of the *Polygonum* type, and embryogeny is of the Solanad type (Rao & Nagaraj, 1982). These character states constitute symplesiomorphies within the dicotyledons or only among some families of the Asteridae.

Characters varying within the family are ovule type and integument development, antipodal characteristics, endosperm development, and specializations of the nucellus and embryo. These characters are distinctive partly between autotrophic and heterotrophic genera, and partly between subtribe Gentianinae and the other (sub)tribes of the family.

Ovules are commonly anatropous. In monotypic *Voyriella* the ovule is orthotropous, and orthotropous ovules also occur in *Cotylanthera* and some *Voyria*; other ovule types were reported (Stolt, 1921; Oehler, 1927; Shamrov, 1988, 1991) for *Swertia* (campylotropous), *Halenia* (orthotropous), and *Gentianella* (hemitropous). *Voyriella* and *Voyria* also deviate from the common state of (nuclear) endosperm development: in *Voyriella* and *Voyria caerulea* (Oehler, 1927) endosperm development is *ab initio* cellular; in five other species of *Voyria* nuclear endosperm was recorded (Maas & Ruyters, 1986).

Antipodal variation is stated to be important within the family (Stolt, 1927; Rao & Chinnappa, 1983). Their number is generally three, but in *Swertia* and *Gentianella* there may be 8–12 antipodals. They may be haustorial or non-haustorial, and either ephemeral, degenerating before fertilization, or persistent. Rao and Nagaraj (1982) proposed a distinction between Gilg's (1895) subtribe Gentianinae and the other Gentianaceae, the latter characterized by three ephemeral, non-haustorial antipodal cells. This statement seems to be correct for *Cotylanthera*, *Exacum*, *Canscora*, *Hoppea*, *Voyria*, and *Voyriella*, but a wider variation has been reported for *Blackstonia*, and *Centaurium* and for the genera of subtribe Gentianinae (Stolt, 1927; Arekal, 1961; Vijayaraghavan & Padmanaban, 1969; Drexler & Hakki, 1979; Rao & Nagaraj, 1982; Shamrov, 1988).

Embryological reports are scarce (e.g., an integumentary tapetum – endothelium – has been reported only for *Exacum*; Maheswari Devi, 1962) or contradictory, as in reports of endosperm type and in the terminology of ovule types. It is also problematic that no reports exist for the neotropical autotrophic genera.

Seed

The seeds mostly develop from unitegmic and tenuinucellate ovules. The outer layer of the integument develops into a mechanical layer that gives the seed hardness and strength. Since only the outer layer of the integument contributes to the formation of the seed coat (Bouman & Schrier, 1979), the seeds are exotestal according to Corner's (1976) terminology. The remaining tissues of the testa are usually compressed or resorbed by the endosperm or the embryo. The seed coat fits tightly to the endosperm (when present).

Enlarged exotestal cells and secondary thickenings of radial (anticlinal) cell walls make up the reticulations of the mature seed coat, generally without intercellular gaps. These elaborations of the seed coat facilitate anemochory, or, in the case of tropical mycotrophs, ombrohydrochory (Bouman & Devente, 1986). The exotesta exhibits a great diversity in cell shape and especially in cell wall thickenings, as demonstrated by Guérin (1904). Varying characters are seed shape, presence or absence of seed wings, testa cell shape, anticlinal wall undulations, and inner tangential (periclinal) wall sculpturing (terminology as in Barthlott, 1981).

For seed shape we distinguished four types. The angular (cubical) type is best documented for *Exacum* (Klackenberg, 1985) but it seems to dominate among neotropical woody genera as well. The globular shape is typical for Gilg's subtribes Erythraeinae and Chironiinae. The transitional oval type occurs in tribe Gentianeae while long seed is typical for *Gentiana*, well documented in papers of Miège and Wüest (1984), Ho and Liu (1990), and Yuan (1993b).

In a number of taxa of tribe Gentianeae (sections *Gentiana, Otophora, Stenogyne,* and *Pneumonanthe* of *Gentiana, Crawfurdia, Tripterospermum,* the Asian species of *Swertia,* and some American species of *Frasera*) and in one genus of the Potalieae (*Urogentias*) the seeds have flat, marginal outgrowths of the seed coat, called seed wings. The presence of wings may be considered an advanced feature and it is seemingly correlated with seed size; the mentioned genera all have some of the largest seeds in Gentianaceae. The morphology of the seed wings is very diverse. They may be (1) regular and more or less equal all around the edge of the seed (e.g., *Gentiana, Swertia,* and *Frasera*), or (2) unequal or asymmetric (e.g., *Crawfurdia, Tripterospermum,* and *Urogentias*). In *Crawfurdia* and *Tripterospermum* the seeds have three wings. *Urogentias* has striate membranous wings that are extensions of the chalazal end of the seed coat.

The *Voyria aphylla* species group is characterized by fusiform to filiform seeds (Bouman & Devente, 1986) that are adapted to wind dispersal. In this group the seeds have long projections that show reticulate secondary thickenings on the radial and inner tangential cell walls and that are air-filled in the dry, mature state.

The outline of the exotestal cell can be isodiametric or elongated, the latter state being typical in *Gentiana*. The isodiametric state can be considered as primitive; it is interesting to note that African species of *Exacum* (Klackenberg, 1985) and the *Voyria truncata* species group, thought to be the most primitive in *Voyria* (Bouman & Devente, 1986; subgenus *Voyria* of Albert & Struwe, 1997), have isodiametric testa cells.

Straight anticlinal cell walls, the most common state, are thought to be plesiomorphic. Undulate (sinuated) anticlinal walls were shown for *Exacum* (Guérin, 1904; Klackenberg, 1985), *Irlbachia* (Cobb & Maas, 1983), *Curtia* (Grothe & Maas, 1984), *Tachiadenus* (Klackenberg, 1987), *Centaurium*, and *Faroa* (Goethals & Smets, 1995). The inner tangential cell walls of the exotesta often have sculpturings, as do sometimes the radial walls. The partial or reticulate thickenings of exotestal cell walls combine strength with low seed weight, thus advancing both seed dispersal and survival. We distinguished four types of inner tangential cell wall sculpturings: pitted, papillate, reticulate, and multiple pitted; some genera are polymorphic.

Cytology

Chromosome numbers are partially known for 36 of the 46 ingroup terminals (78.2%). There are genera with constant chromosome numbers, e.g., *Halenia* and *Lisianthius* (the latter documented for 10 species by Weaver, 1969), but in the majority of taxa chromosome numbers are variable because of euploidy, dysploidy, or aneuploidy. Infrageneric variation is best documented for *Centaurium* (Zeltner, 1970; Broome, 1978), *Sabatia* (Perry, 1971), *Swertia* (Khoshoo & Tandon, 1963; Shigenobu, 1983; Pringle, 1990), and *Gentiana* sect. *Calathianae* (Müller, 1982).

We used haploid chromosome number as a character in our analysis, even though it has a wide and almost continuous range from $n=5$ to $n=42$ in the family (the exceptions are $n=25$, 29, 35, and 37; in one variety of *Gentiana nipponica* even $n=48$ and 49 was observed; Shigenobu, 1984). However, on a generic level the distribution of haploid chromosome numbers is bimodal, with a local maximum at 9–11 and a global maximum at 18–21. We used $n=15$ as a demarcation between these two modi. Next, the distribution has a long right tail in which, among others, the woody genera *Anthocleista* ($n=30$), *Symbolanthus* ($n=40$), and *Fagraea* ($n=$ up to 42) are present. Considering the long right tail, $n=30$ was (arbitrarily) chosen as a second demarcation point. With maxima at $n=9$–11 and $n=$ 18–21 in the distribution of haploid chromosome numbers, $x=9$, 10, 11 may be frequent base numbers ($x=10$ or 11 is also found in many Apocynaceae and Rubiaceae). Zeltner (1970) documented different ploidy levels for *Centaurium* (based on $x=9$–11; see also Ubsdell, 1979) and *Blackstonia* ($x=10$). Different ploidy levels are also documented for *Swertia* ($x=10, 13$), *Gentianella* ($x=9$), *Gentianopsis* ($x=13$), and *Gentiana* sect. *Cruciata* ($x=13$) and sect. *Frigidae* ($x=12$) (Shigenobu, 1983; Yuan, 1993a; Yuan & Küpfer, 1993a,b). Therefore, ploidy levels are fairly well assessable based on $x=9$–13, at least for these genera.

Dysploidy or aneuploidy has been reported for *Sabatia* (Perry, 1971), *Swertia* (Vasudevan, 1975), the American species of *Centaurium* (Broome, 1978), and for *Gentiana* sect. *Calathianae* (Müller, 1982).

Chemistry

From the various biochemical compounds that are found in the family only iridoids, secoiridoids, xanthones, flavonoids, and carbohydrates are used as a source for characters in this analysis. Other interesting compounds, such as pseudo-alkaloids and triterpenes, do occur in the family, but there are insufficient data.

Regarding iridoids and secoiridoids, the biosynthetic route of mevalonate → loganin (loganic acid) → secologanin (secologanic acid) → sweroside → swertiamarin → gentiopicrine can be considered as proven (Hegnauer, 1989, based on experiments with different species of *Swertia* and *Gentiana*). Within this biosynthetic route there are several side branches; from this point of view secologanin and sweroside are the most interesting nodes. One route from secologanin leads to the complex indole alkaloids (route I of Jensen, 1992). These are found in other families of Gentianales, but genera of Gentianaceae (including Potalieae) are not able to synthesize them; instead they produce pseudo-alkaloids from swertiamarin or gentiopicrine. Sweroside is interesting because this compound is also present in some Apocynaceae (Hegnauer, 1989) and in *Desfontainia* (Jensen, 1992); swertiamarin and gentiopicrine, however, occur only in Gentianaceae. So presence of only sweroside seems to be a plesiomorphic state within Gentianaceae; it is observed in the neotropical Gentianaceae *Irlbachia* and *Lisianthius* (Hamburger *et al.*, 1990; Shiolara *et al.*, 1994; Jensen & Schripsema, 2002 (Chapter 6, this volume)).

Sources and coding of xanthone compounds are described in Mészáros (1994) and Mészáros et al. (1996). These data were supplemented with new data for *Halenia* (Rodriguez et al., 1995), *Lomatogonium* (Khishgee & Pureb, 1993), *Schultesia* (Terreaux et al., 1995), and *Gentiana* sect. *Frigidae* (Butayarov et al., 1993). A new character was introduced to distinguish between taxa with no xanthones, taxa with only xanthone-C-glycosides (e.g., mangiferin), and taxa with also xanthone-O-glycosides. Data for flavonoids were also updated.

Massias et al. (1978) made a broad investigation of sugars. Simple sugars are widespread, but gentianose was found only in *Gentiana* and *Swertia*; it was not detected in nine other genera. Since that time another compound sugar, gentiobiose, has been documented for *Halenia* (Recio-Iglesias et al., 1992) and *Lomatogonium* (Schaufelberger & Hostettmann, 1984). Schilling (1976) detected L-(+)-bornesitol, a special sugar, in 20 of 24 investigated genera of Gentianaceae; its absence is documented for *Curtia*, *Exacum*, *Irlbachia*, and *Sebaea*. L-(+)-bornesitol also occurs in *Anthocleista* and in some Apocynaceae and Rubiaceae (Plouvier, 1990).

Methods

The data set was analyzed using parsimony analysis (Farris, 1970, 1983; Fitch, 1971) with equal a priori character weights and unordered characters. The analyses were performed with the computer program NONA (Goloboff, 1993). In all analyses we used subset coding for polymorphisms (see Mészáros et al., 1996, for some comments on polymorphisms and subset coding). Apart from the unordering of multistate characters and the total number of trees that can be stored in memory, all other default settings were retained in all analyses. By default, NONA collapses all branches that have no unambiguous synapomorphies (a character provides an unambiguous synapomorphy for a branch if a state transition occurs on that branch under every possible optimization of the character on the tree; Goloboff, 1993; see also Coddington & Scharff, 1994). The most-parsimonious cladograms and the cladograms that are one step longer were calculated using the instruction series "MULT*100; SUBOPTIMAL 1; MAX*". MULT*100 carries out 100 replications of randomizing the order of the taxa, creating a tree by means of stepwise addition, and submitting it to branch-swapping by means of tree bisection and reconnection. During each replicate a maximum of 20 trees was retained ("HOLD/20" setting, the default). The instruction "MAX*" is added to ensure full branch-swapping (i.e., unrestrained by the HOLD/20 setting) of the trees obtained with "MULT*100", also using tree bisection and reconnection. The "SUB-OPTIMAL 1" command, issued before "MAX*", instructs the program to keep all trees that are one step longer than the most-parsimonious trees. As descriptive measures of the fit between data and trees we calculated consistency and retention indices (C and R; Kluge & Farris, 1969; Farris, 1989). All consistency indices are calculated with autapomorphies included (see Yeates, 1992).

In order to evaluate the relative support of clades, we calculated branch support, i.e., the number of extra steps needed to lose a branch in the strict consensus of near-most-parsimonious trees (Bremer, 1994; also called Bremer support and, using an unfortunate terminology, decay index; see Källersjö et al., 1992). Because of the high number of near-most-parsimonious trees, we only calculated trees up to one step longer than the shortest. We also performed bootstrap (Felsenstein, 1985; but see Bremer, 1994) and jackknife analyses (Farris et al., 1996). These were performed with the aid of macros that are distributed together with NONA (Goloboff, 1993). For the bootstrap and jackknife analyses we ran 100 replicates each; in each replicate the best trees were obtained with a "MULT*10" command. In the jackknife analyses, we followed Farris et al.'s (1996) suggestion and randomly deleted 36% of the characters in each replicate. In Fig. 3.1, the reported value for a given clade is the percentage of replicates that support that clade (only values exceeding 50% are shown). In the bootstrap analysis we considered a replicate as supporting a clade when that clade is present in at least one tree for that replicate. In the jackknife analysis we used a stricter interpretation and considered a replicate as supporting a clade only when that clade is present in all trees for that replicate.

RESULTS

Standard parsimony analysis resulted in 100 most-parsimonious trees of 366 steps, with consistency index $C = 0.34$ and retention index $R = 0.56$. In all these trees, *Plumeria* and *Danais* grouped together, as did *Gelsemium*, *Exostema*, and *Rauwolfia*. The failure to group together the two representatives each of Rubiaceae and Apocynaceae indicates a problem in our taxon and/or character sampling at the level of the outgroups. To investigate if this had an influence on the ingroup relationships we performed a second analysis in which both Apocynaceae and Rubiaceae were constrained to be monophyletic. This analysis resulted in 100 trees of length 367 (one step longer) that, apart from the constrained families Apocynaceae and Rubiaceae, were identical to the trees of the unconstrained analysis. The strict consensus tree, arbitrarily rooted between Apocynaceae and the rest, is shown in Fig. 3.1. To check if the outgroups influenced branching within Gentianaceae, we also performed an analysis of ingroup taxa only. This yielded 50 most-parsimonious trees of 311 steps ($C=0.37$, $R=0.56$). The strict consensus of these trees (oriented as indicated by the previous analyses) is exactly the same as in Fig. 3.1.

The large polytomy near the base of the family arises partly because of the variable position of *Celiantha* and *Irlbachia*; in all most-parsimonious trees of all analyses this polytomy resolves as a clade comprising *Chorisepalum, Macrocarpaea, Symbolanthus,* and *Tachia,* with *Irlbachia* and *Celiantha* occupying various positions (see Fig. 3.2). A third genus with variable position is *Deianira*: it groups either with *Coutoubea* or as the sister to *Cotylanthera–Exacum*. By excluding these three genera with variable positions, more resolution is retained in the strict consensus tree (Fig. 3.2).

In all three cases (unconstrained outgroups, constrained outgroups, ingroup only) all trees up to one step longer than most parsimonious were calculated (yielding 4520, 3900, and 2885 trees, respectively). Within Gentianaceae, the groups with branch support >1 were identical but for one case: the *Blackstonia–Centaurium–Chironia–Eustoma–Orphium–Sabatia* clade has branch support $=1$ in the ingroup-only analysis, but branch support >1 in the two other cases (branches with branch support >1 are indicated by double bars in Fig. 3.1). Bootstrap and jackknife analyses were performed only for the ingroup-only analysis. In Fig. 3.1, bootstrap and jackknife values that exceed 50% are indicated above branches.

DISCUSSION

Considering that the strict consensus tree (Fig. 3.1) is not well resolved and that most branches that are present have low branch support, bootstrap, and jackknife values, the results of the cladistic analysis should not be overinterpreted. Therefore we will concentrate only on the most salient features.

Our analysis fails to corroborate tribe Exaceae (*Sebaea* is unresolved close to tribe Gentianeae to *Cotylanthera–Exacum*, while *Tachiadenus* appears as sister to *Voyria–Voyriella*; see Fig. 3.2). Nevertheless, the sister-group relationship between *Cotylanthera* and *Exacum*, hypothesized and discussed by Klackenberg in Struwe *et al.* (2002), is supported by both the jackknife and the bootstrap analyses.

Voyria and *Voyriella* appear as the sister group of *Tachiadenus* but only a single unambiguous synapomorphy supports this relationship: the very similar corolla fusion, resulting in a long corolla tube; within this tube the

Cladistics: a morphological approach 333

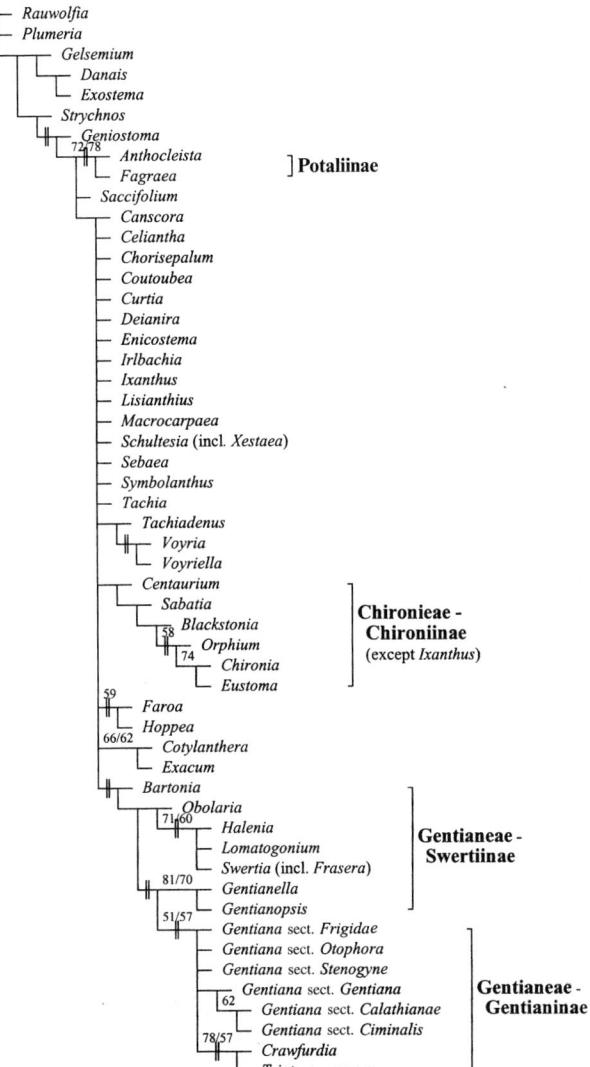

Figure 3.1. Summary of the parsimony analyses. Strict consensus tree of the 100 most-parsimonious trees that are obtained when Apocynaceae and Rubiaceae are both constrained to be monophyletic (367 steps, $C=0.34$, $R=0.56$), arbitrarily depicted with Apocynaceae basal. The strict consensus tree of the 100 trees of 366 steps that are obtained without constraints has identical relationships within Gentianaceae; the same result is also obtained with parsimony analysis of the ingroup only (50 trees of length 311, $C=0.37$, $R=0.56$). Double bars across branches indicate branches in Gentianaceae with branch support >1 (ingroup-only analysis); unmarked internal branches have branch support =1. Numbers above branches are bootstrap values (single numbers or numbers before slash) and jackknife values (numbers after slash) that exceed 50% (ingroup-only analysis).

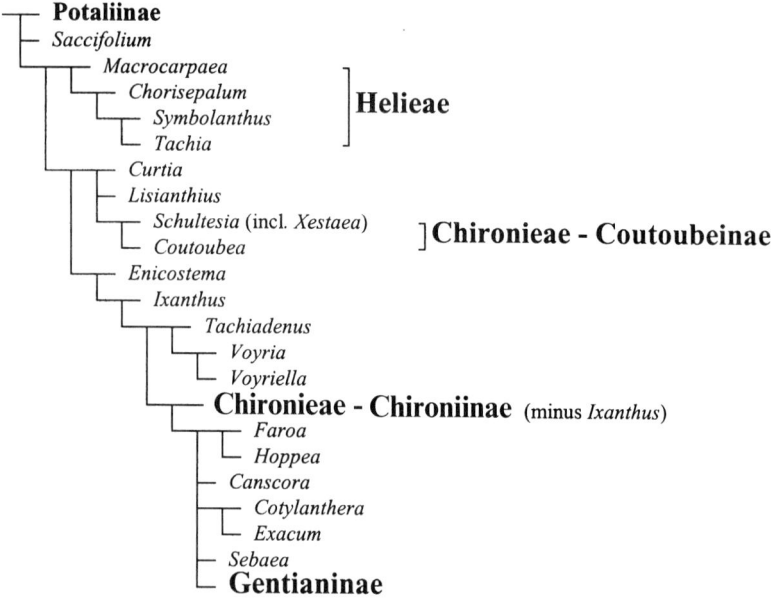

Figure 3.2. Strict consensus tree with exclusion of *Celiantha*, *Deianira*, and *Irlbachia*.

anthers, on very short filaments, are pressed against each other. Beside the characters that are related to the heterotrophic syndrome, *Voyria* and *Voyriella* share a very specific synapomorphy: pollen germination in thecae. However, this may also be the result of parallel evolution in the same tropical habitat. Molecular results, showing *Voyriella* to be closely related to *Curtia* and *Saccifolium* (Thiv et al., 1999a; Struwe et al., 2002), favor this latter interpretation.

Bartonia and *Obolaria* are nested within the monophyletic clade that represents tribe Gentianeae, which is in agreement with Grisebach's (1845) treatment of *Bartonia* and Holm's (1897) treatment of *Obolaria*; it is furthermore supported by the pollen morphological study of Nilsson and Skvarla (1969). However, contrary to the molecular analyses, subtribe Swertiinae, to which both *Bartonia* and *Obolaria* belong, is paraphyletic in our analysis. The monophyly of the Gentianeae clade, also obtained in Struwe et al.'s (2002) and Thiv et al.'s (1999a) analyses of more than 100 *mat*K sequences and over 150 *trn*L intron sequences of Gentianaceae and other Gentianales, confirms one of the results of our previous analysis (Mészáros et al., 1996).

Another result of this previous study that is confirmed in the current analysis is the close relationship between *Blackstonia*, *Eustoma*, *Orphium*, *Chironia*, and *Centaurium*. *Sabatia*, not included in the earlier analysis, is now added to this Chironiinae clade. These relationships are also obtained in Struwe *et al.*'s and Thiv *et al.*'s molecular analyses, in which *Bisgoeppertia*, *Cicendia*, *Exaculum*, *Geniostemon*, and possibly *Zygostigma* are also added to Chironiinae.

Basal in Gentianaceae according to our results is the trichotomy between the woody genera *Anthocleista* and *Fagraea*, the shrubby genus *Saccifolium*, and the rest of the family (Fig. 3.1). Within this last clade, Helieae (with the possible exclusion of *Irlbachia* and/or *Celiantha*) are sister to the rest of the family (Fig. 3.2). At the base of Helieae are the three typical woody genera *Chorisepalum*, *Macrocarpaea*, and *Symbolanthus*. In combination, this suggests a woody and (pan)tropical origin of Gentianaceae, supporting Carlquist's (1984) conclusion that *Symbolanthus* is primarily woody while *Ixanthus*, nested deeper in the family, is secondarily so. Woody ancestry has been argued for Apocynaceae (Sennblad, 1997: 11) and Rubiaceae (Carlquist, 1992: 319) as well, and all of this is in agreement with the general dominance of the trend from a woody to a herbaceous habit in other families (Anderberg & Ståhl, 1995: 1719) and in dicotyledons (Kremer & van Andel, 1995: 472). However, cladistic analyses of *mat*K and *trn*L intron sequences in Gentianaceae and other Gentianales (Thiv *et al.*, 1999a; Struwe *et al.*, 2002) contradict these hypotheses. On the basis of these data, the first two splits in Gentianaceae set apart tribes Saccifolieae and Exaceae, respectively, while both Potalieae and Helieae are well nested within the remainder of the family. While these results do not exclude a woody origin of the family – the herbaceous or suffrutescent state might be plesiomorphous – they contradict the primary woodiness in Potalieae and Helieae.

This discrepancy between the current analysis and the broader analysis based on *trn*L intron and *mat*K sequences leads to the obvious question; which of the results has more strength? While in general neither type of data is intrinsically superior for purposes of phylogenetic reconstruction, it seems that in this case results from the molecular data are more robust than those from our analysis. The question could be addressed formally by doing a combined analysis of molecular and morphological data. However, our data set has been conceived from the start as a genus-level data set (with the exception of the sections in *Gentiana*), thereby implicitly assuming monophyly of these supraspecific groups. This severely complicates combination with the molecular data, which are basically sequences of exemplar specimens of different species within genera. We tend to find weak support for several (sub)tribes that are also obtained with the molecular data (Potalieae, Helieae, and other examples below). However, the relationships among these groups are almost completely unresolved, and the little resolution we get at this level is very poorly supported. The molecular data, in contrast, yield better-supported relationships at this level. Given these results, and the technical problems of combining the data sets in this case, a formal combination of the two sets would seem to be of little use.

Several factors likely contribute to the poor results obtained with our morphological data set compared with those obtained with *trn*L and *mat*K sequences (Thiv *et al*., 1999a; Struwe *et al*., 2002). A first issue is the combined effect of limited taxon sampling and a limited number of informative characters relative to the molecular analyses. This may well explain our inability to retrieve both Rubiaceae (*Danais* + *Exostema*) and Apocynaceae (*Plumeria* + *Rauwolfia*) as monophyletic groups without imposing constraints upon the analysis. Next, there is a lack of good morphological studies for many tropical and subtropical representatives of Gentianaceae, which influences our data set in two ways. First, these taxa have a relatively high number of question marks, most simply representing missing information. Second, the lack of broader comparative studies often makes the primary homology statements that are expressed by the characters rather dubious. In contrast to this, the *mat*K and *trn*L sequences used by Struwe *et al.* (2002) and Thiv *et al.* (1999a) are both complete and easy to align, leading to higher information content and better hypotheses of primary homology. For these reasons, it could a priori be expected that the molecular data sets would give more and better-supported resolution, and this is precisely what is observed. An obvious way to proceed would be to increase research on poorly known characters (e.g., seed micromorphology and seed anatomy) and poorly known taxa, which often have ambiguous positions on different cladograms. At the same time, ontogenetic studies of flowers and inflorescences could help to detect pseudoconvergences (Kluge & Farris, 1969) in the floral region (see De Laet & Smets, 1996).

However, an additional problem for phylogenetic analysis of morphological data sets is posed by functional correlations among morphological traits. Given that only a limited number of morphological traits is available, this may well turn out to be a fundamental problem that is very difficult to overcome. In this particular case the heterotrophic genera provide a good example. As discussed above, they possess what we call the heterotrophic syndrome: co-occurrence of saprophytic or parasitic lifestyle, coralloid roots, reduced leaves, and loss of chlorophyll. It can be argued that within this syndrome the crucial characteristic is the capacity for a saprophytic or parasitic lifestyle; once this capacity has evolved, the coralloid roots, the reduced leaves, and the loss of chlorophyll may be simple adaptations to this new mode of living. Coding all characteristics of the syndrome as separate independent characters may then potentially lead to grouping according to correlated convergence rather than according to common descent. The same effect may explain the above-discussed discrepancy between morphological and molecular analyses when it comes to evolution of woodiness in Gentianaceae. All the outgroups in our analysis are trees or shrubs, which may then force the woody (sub)tribes Potalieae, *Saccifolium*, and Helieae to the base of the family.

ACKNOWLEDGMENTS

The authors thank Tim Motley for help in coding *Geniostoma*. This research was supported by the F.W.O., the Fund for Scientific Research – Flanders (Belgium) (project numbers 2.0038.91 and G.0143.95) and by the Research Council of the Katholieke Universiteit Leuven (grant OT/97/23). Jan De Laet was a postdoctoral fellow of the F.W.O., the Fund for Scientific Research – Flanders (Belgium).

Appendix 3.1
Data matrices

Numbers of characters and character states refer to Table 3.3. Polymorphisms are shown in square brackets; "?" indicates missing values and inapplicable characters.

Character numbers:

1	2	3	4	5	6	7	8	9	10	11
12	13	14	15	16	17	18	19	20	21	22
23	24	25	26	27	28	29	30	31	32	33
34	35	36	37	38	39	40	41	42	43	44
45	46	47	48	49	50	51	52	53	54	55
56	57	58	59	60	61	62	63	64	65	66
67	68	69	70	71	72	73	74	75	76	77
78	79	80	81	82	83	84				

Cladistics: a morphological approach

Outgroups:

Danais

0	0	?	?	0	?	0	1	0	1	?
0	0	1	?	0	1	?	0	[01]	1	?
0	?	0	?	0	?	?	0	1	2	?
?	2	0	?	0	0	0	0	0	0	?
?	0	2	1	0	?	?	2	?	1	0
1	0	?	?	?	?	?	1	1	[01]	0
?	0	0	1	?	[01]	?	?	?	?	?
0	?	?	?	?	?	?				

Exostema

0	0	0	?	1	?	0	0	0	1	?
0	0	1	0	0	[01]	1	[123]	[01]	[12]	0
?	?	0	?	0	?	?	1	[012]	[12]	1
?	2	0	?	0	0	0	0	0	0	1
?	0	0	1	0	?	?	0	?	[01]	0
?	?	?	?	?	?	?	[01]	[01]	?	?
1	0	0	1	1	1	0	?	?	?	?
0	?	?	?	?	?	?				

Gelsemium

0	0	0	0	0	0	?	[01]	0	1	?
0	0	0	0	0	1	?	[12]	0	1	0
0	0	0	1	0	?	?	1	1	[12]	2
?	2	1	0	0	0	0	0	0	0	0
?	0	0	0	0	1	0	4	?	0	0
?	?	?	?	?	?	?	2	[01]	?	?
3	0	0	1	1	0	0	?	0	3	?
0	?	?	?	?	?	?				

Geniostoma

0	0	0	0	0	0	0	0	1	[01]	1
0	0	0	0	0	0	?	0	0	1	0
1	0	0	0	0	?	?	[12]	1	[03]	1
0	2	2	0	0	0	0	0	0	0	0
?	0	1	0	0	[01]	0	1	?	0	0
0	0	?	?	?	?	?	0	0	?	?
2	0	0	2	2	0	1	?	?	?	?
0	?	?	?	?	?	?				

Plumeria

0	0	0	0	[01]	?	0	0	0	1	0
1	0	0	0	0	1	?	?	[01]	1	[01]
0	0	0	0	0	?	?	2	1	1	1
?	0	0	?	0	?	0	0	0	0	0
?	0	0	0	1	0	0	2	?	0	0
0	?	?	?	?	0	?	?	1	?	?
?	0	0	1	2	0	1	?	?	3	0
0	?	?	?	?	?	?				

Rauwolfia

0	0	[012]	?	0	?	0	0	0	[01]	?
1	0	0	0	0	1	?	0	0	1	0
[012]	0	0	1	0	0	?	2	1	[123]	[12]
?	2	[02]	0	0	?	0	0	0	0	0
0	0	2	0	1	[01]	0	1	0	?	2
0	0	0	1	0	0	0	?	0	?	?
?	0	0	1	2	0	[02]	?	?	3	0
0	?	?	?	?	?	?				

Strychnos

0	0	0	[01]	0	?	?	[01]	1	1	[01]
0	0	0	[01]	1	1	[01]	0	0	[12]	0
0	0	0	0	0	?	?	0	[12]	[012]	[012]
?	[02]	[02]	0	0	0	0	0	0	0	0
?	0	[12]	0	0	[01]	0	0	?	?	1
0	0	0	?	?	0	?	[12]	0	?	1
3	0	0	1	2	0	[012]	?	0	3	0
0	?	?	?	?	?	?				

Ingroups:
Anthocleista

0	0	0	0	1	1	0	0	1	1	1
0	1	0	0	0	0	?	0	[01]	2	0
0	0	0	0	0	?	?	2	0	1	2
0	?	1	0	0	0	0	0	0	0	1
0	0	2	0	0	1	0	1	?	2	1
?	?	?	?	?	?	?	[02]	[01]	0	?
?	0	0	2	2	0	2	1	?	1	?
2	0	0	0	[01]	[01]	1				

Cladistics: a morphological approach

Bartonia

[01]	2	[012]	?	2	?	0	[01]	?	?	?
0	0	0	?	1	?	?	[03]	[01]	2	0
0	[01]	0	?	0	0	0	1	2	3	1
0	0	[02]	0	1	0	[01]	0	0	0	0
?	0	2	0	0	1	[01]	3	0	0	0
?	?	?	?	?	?	?	2	0	[01]	0
?	?	0	1	[12]	0	1	?	?	?	?
?	?	?	?	?	?	?				

Blackstonia

0	2	0	1	2	?	0	0	?	1	?
0	0	0	1	1	?	1	1	0	0	0
0	0	0	?	0	?	0	2	0	0	0
0	0	0	0	0	0	[01]	0	0	0	0
0	0	2	0	0	1	0	2	?	0	0
?	0	0	1	?	0	?	2	0	[01]	0
[23]	0	0	1	1	0	[12]	1	0	2	0
2	0	0	0	0	1	1				

Canscora

0	2	[12]	?	?	?	0	0	?	?	?
0	0	0	1	1	0	1	0	[01]	2	0
2	[02]	0	0	0	0	1	2	2	[12]	1
0	0	0	0	0	0	[01]	0	1	0	0
0	0	2	0	0	1	0	2	0	0	0
0	0	0	0	0	0	0	[01]	0	[01]	0
?	0	0	1	1	0	[12]	?	0	?	[23]
[12]	0	0	1	1	1	1				

Celiantha

0	1	[01]	?	?	?	0	0	?	?	?
0	0	0	1	?	?	?	[02]	[01]	[12]	0
2	0	0	1	0	?	?	2	[12]	2	2
0	?	0	0	0	0	0	0	0	0	0
0	0	1	0	0	1	0	2	?	0	0
?	?	?	?	?	?	?	0	0	[01]	0
?	0	2	2	[13]	1	?	?	?	?	?
?	?	?	?	?	?	?				

Centaurium

0	[12]	[012]	1	2	0	0	0	[01]	1	?
0	0	0	1	1	[01]	1	0	[01]	[12]	0
[01]	1	0	0	0	0	[01]	2	1	[12]	1
0	0	0	1	0	0	1	0	0	0	0
0	0	2	0	0	2	0	2	0	0	0
0	0	0	1	0	0	0	[12]	0	[01]	1
[12]	0	0	1	1	0	[012]	1	0	2	0
[02]	[01]	0	[01]	[01]	[01]	[01]				

Chironia

0	[012]	[12]	1	1	?	0	0	?	[01]	?
0	0	0	?	0	0	?	1	1	1	0
[01]	[01]	0	1	0	0	0	2	1	[03]	0
0	0	0	1	0	0	1	0	0	0	0
0	0	2	0	0	0	0	[012]	0	0	[01]
?	?	?	?	?	?	?	[12]	0	0	0
[21]	0	0	1	1	0	2	1	0	2	?
2	0	0	1	1	1	1				

Chorisepalum

0	0	1	?	?	?	?	0	?	?	?
0	0	0	[01]	?	?	?	[03]	[01]	2	1
0	[12]	0	0	1	?	?	2	0	[01]	2
0	2	0	0	0	0	0	0	0	0	0
0	0	0	0	0	2	0	2	?	[01]	0
?	?	?	?	?	?	?	?	0	0	0
3	0	0	1	1	0	?	?	?	?	?
?	?	?	?	?	?	?				

Cotylanthera

1	2	1	?	?	?	0	0	?	?	?
0	0	0	?	?	?	?	3	0	2	0
1	1	0	?	0	0	0	2	2	0	0
0	0	[01]	1	0	1	0	0	0	0	0
1	0	0	0	0	1	0	1	1	0	0
[02]	1	0	0	0	0	0	3	0	1	0
?	0	0	1	[12]	0	1	?	?	?	?
?	?	?	?	?	?	?				

Cladistics: a morphological approach

Coutoubea

0	[12]	[01]	?	1	?	[01]	0	?	[01]	?
0	0	0	1	0	?	0	2	0	[01]	0
0	[01]	0	1	0	2	1	2	[01]	1	1
0	[02]	1	0	?	0	0	0	0	0	1
0	0	1	0	0	1	0	2	[12]	0	0
?	?	?	?	?	?	?	[01]	0	[01]	0
3	0	1	[12]	1	0	1	?	0	2	0
?	?	?	?	?	?	?				

Crawfurdia

0	1	0	1	?	?	0	1	?	?	?
0	0	0	1	?	?	1	[23]	[01]	1	0
2	0	1	?	0	0	0	3	1	2	2
?	2	?	0	?	0	0	0	0	0	0
0	0	1	0	0	2	1	2	0	0	0
?	?	?	?	?	?	?	[12]	1	?	?
?	0	0	1	1	0	?	?	?	?	?
?	?	?	?	?	?	?				

Curtia

?	2	1	?	?	?	0	0	?	?	?
0	0	0	1	0	?	1	0	[01]	[12]	0
0	1	0	1	0	?	0	2	[12]	[12]	2
0	?	[12]	0	0	0	0	0	0	0	0
0	1	1	0	0	1	0	1	?	0	0
?	?	?	?	?	?	?	[023]	0	[01]	[01]
0	?	0	1	[12]	[01]	?	0	?	2	?
?	?	?	?	?	?	?				

Deianira

0	1	0	?	1	?	0	0	?	?	?
0	0	0	[01]	0	[023]	1	2	1	2	0
[01]	[01]	0	?	0	2	0	2	2	1	1
0	2	0	1	0	1	0	0	0	0	0
?	0	2	0	0	1	0	2	[12]	0	0
?	?	?	?	?	?	?	[03]	0	0	0
?	0	1	[01]	1	0	0	?	?	?	?
?	?	?	?	?	?	?				

Enicostema

0	1	[012]	?	1	?	0	0	?	?	?
0	0	0	1	0	?	1	2	[01]	1	0
0	1	0	0	0	?	1	2	1	2	1
0	?	0	0	0	0	0	0	0	0	1
0	0	1	0	0	1	0	1	0	0	0
?	?	0	?	?	?	?	[12]	0	0	0
3	0	0	1	1	0	1	1	?	1	[12]
2	?	?	?	?	?	?				

Eustoma

0	[12]	?	?	?	?	0	0	?	?	?
0	0	0	1	0	?	1	1	[01]	1	0
0	1	0	?	0	0	0	2	1	3	0
0	0	0	1	1	0	1	0	0	0	0
0	0	2	0	0	1	0	1	0	0	0
0	0	0	?	?	0	?	1	0	0	0
[23]	0	0	1	1	0	2	?	0	2	0
2	0	0	1	1	1	1				

Exacum

0	[012]	[012]	1	2	?	0	0	?	1	?
0	0	0	1	1	[012]	1	0	[01]	[12]	0
[01]	2	0	0	0	2	0	2	[12]	0	0
0	0	[02]	1	0	1	0	0	0	0	[01]
1	0	2	0	0	0	0	[02]	1	0	0
0	0	0	[01]	0	0	0	0	0	0	[01]
?	0	0	1	[12]	0	[012]	0	?	2	1
0	?	?	?	?	?	?				

Fagraea

0	0	0	0	1	1	0	0	1	1	[01]
0	1	0	0	0	?	?	0	0	1	0
[12]	0	0	0	0	?	?	2	1	[12]	1
0	2	[01]	0	0	0	0	0	0	0	0
0	0	[01]	0	0	1	0	[12]	?	[12]	1
0	0	0	1	?	0	?	[023]	0	0	[01]
[23]	0	0	[12]	1	0	[02]	?	?	2	2
?	?	?	?	?	?	?				

Faroa

0	2	[012]	?	?	?	0	0	?	?	?
0	0	0	1	1	0	1	2	[01]	2	0
1	1	0	?	0	?	1	2	2	1	1
0	?	[12]	0	0	0	0	0	0	0	1
0	0	2	0	0	1	0	[012]	2	0	0
?	?	?	?	?	?	?	1	0	0	[01]
?	0	0	1	[12]	0	?	1	?	?	?
?	?	?	?	?	?	?				

Gentiana sect. *Calathianae*

0	[12]	1	?	2	0	1	0	?	?	?
0	0	0	1	[01]	1	0	3	[01]	1	0
2	[12]	1	?	0	0	0	3	1	1	1
0	2	0	0	0	0	0	0	0	0	0
0	0	0	0	0	2	1	2	0	0	0
?	?	?	0	?	?	?	3	0	1	0
3	0	0	1	1	0	[01]	1	1	2	2
[12]	0	0	0	0	1	1				

Gentiana sect. *Ciminalis*

0	1	1	?	2	0	1	0	?	?	?
0	0	0	1	?	1	1	3	1	1	0
1	0	1	?	0	0	0	3	1	2	2
1	2	0	0	0	0	0	1	0	0	0
0	0	0	0	0	2	1	2	0	0	0
?	?	?	?	?	?	?	3	0	1	0
?	0	0	1	1	0	1	1	1	2	2
2	0	0	0	0	1	1				

Gentiana sect. *Frigidae*

0	1	?	?	?	?	[01]	[01]	?	?	?
0	0	0	1	1	1	1	[23]	[01]	1	0
2	[02]	1	?	0	0	0	3	1	[23]	2
0	2	[02]	0	0	0	0	[01]	0	0	0
0	0	1	0	0	2	1	2	0	0	0
?	?	?	?	?	?	?	[23]	0	[01]	0
3	0	0	1	1	0	0	?	1	2	2
2	0	1	1	0	0	1				

Gentiana sect. *Gentiana*

0	1	2	1	2	1	0	0	?	1	0
0	0	0	1	1	0	1	2	[01]	1	[01]
2	0	1	?	0	0	0	3	1	[03]	2
1	2	0	0	0	0	0	[01]	0	0	0
0	0	0	0	0	1	1	2	0	0	0
0	0	0	0	1	0	1	2	1	[01]	0
[03]	0	0	1	1	0	1	1	1	2	[12]
[12]	[01]	0	0	0	1	0				

Gentiana sect. *Otophora*

0	1	1	?	?	?	0	0	?	?	?
0	0	0	1	0	?	1	[23]	[01]	1	0
1	0	1	?	0	0	0	3	1	0	[01]
0	2	0	0	1	0	0	0	0	0	0
0	0	[01]	0	0	2	1	2	0	0	0
?	?	?	?	?	?	?	2	1	?	?
?	0	0	1	1	0	?	?	?	?	?
?	?	?	?	?	?	?				

Gentiana sect. *Stenogyne*

0	1	1	?	?	?	0	0	?	?	?
0	0	0	1	?	?	1	3	[01]	1	0
[12]	[12]	1	?	0	0	0	3	1	2	2
?	2	0	0	?	0	0	0	0	0	0
0	0	[12]	0	0	2	1	2	0	0	0
?	?	?	?	?	?	?	[12]	[01]	0	0
?	0	0	1	1	0	1	?	?	2	?
?	?	?	?	?	?	?				

Gentianella

0	[12]	[012]	?	1	?	0	0	?	?	?
0	0	0	1	[01]	1	0	0	[01]	[12]	1
1	0	0	0	0	0	0	2	1	[02]	1
[01]	1	[12]	0	1	0	0	0	0	0	0
0	0	[012]	0	0	2	1	2	2	0	0
1	0	1	1	1	0	1	[12]	0	0	0
0	0	0	1	1	0	1	1	0	[02]	2
[12]	0	[01]	1	0	[01]	1				

Gentianopsis

0	[12]	?	0	?	?	0	0	?	?	?
0	0	0	1	1	1	0	1	1	2	1
1	[12]	1	0	0	2	0	2	2	2	1
0	1	1	0	1	0	0	0	0	0	0
0	0	1	0	0	2	1	2	1	0	0
0	0	0	1	1	0	1	[23]	0	0	0
?	0	0	[01]	1	0	[012]	?	0	2	[12]
[12]	0	0	0	0	1	1				

Halenia

0	[12]	1	?	?	?	0	0	?	1	?
0	0	0	1	0	?	[01]	1	[01]	2	0
0	0	0	0	0	0	0	2	2	3	0
0	1	[12]	0	1	0	0	0	0	0	0
0	0	[01]	0	0	[12]	0	2	0	0	0
2	0	0	1	1	0	1	[12]	0	0	0
0	0	0	1	1	0	0	1	[01]	?	1
2	1	1	1	0	1	0				

Hoppea

0	2	2	?	?	?	0	0	?	?	?
0	0	0	1	?	0	?	0	0	2	0
1	0	0	?	0	1	1	2	2	1	2
0	?	2	0	0	0	0	0	2	0	0
0	0	2	0	0	0	0	1	2	0	0
0	0	0	1	0	0	0	[01]	0	[01]	0
?	0	0	1	1	0	[01]	?	0	?	[23]
[12]	0	0	1	1	1	0				

Irlbachia

0	2	[012]	?	?	?	0	0	?	?	?
0	0	0	1	?	?	?	1	[01]	1	0
1	[01]	0	[10]	1	?	?	2	1	[123]	[12]
0	2	[01]	0	0	0	0	0	0	0	0
?	0	[01]	0	0	1	0	2	?	0	0
?	?	?	?	?	?	?	[01]	0	0	[01]
1	0	2	2	[12]	1	?	?	?	?	?
?	?	?	?	?	?	?				

Ixanthus

0	1	1	1	1	?	0	0	1	1	1
0	0	0	1	1	?	1	0	0	[12]	0
1	1	0	0	0	?	?	2	[12]	1	1
0	2	0	0	0	0	0	0	0	0	0
0	0	2	0	0	1	0	1	?	0	0
?	?	?	?	?	?	?	[12]	0	0	0
?	0	0	1	1	0	?	?	0	2	?
2	1	0	0	1	0	1				

Lisianthius

0	[012]	[01]	?	1	?	?	0	?	[01]	?
0	0	0	1	0	?	1	0	[01]	1	0
[01]	[012]	0	1	0	2	[01]	2	1	[12]	2
0	2	1	0	1	0	0	0	0	0	0
0	0	[012]	0	0	2	0	1	?	1	0
?	?	?	?	?	?	?	0	0	0	0
2	0	0	1	1	0	1	1	0	0	?
0	?	?	?	?	?	?				

Lomatogonium

0	[12]	[01]	?	?	?	0	0	?	?	?
0	0	0	1	0	?	0	[13]	[01]	[12]	0
0	0	0	?	0	?	0	2	[12]	0	0
0	1	2	0	?	0	0	0	0	0	0
0	0	0	0	0	1	0	3	?	0	0
?	0	?	?	?	?	?	[12]	0	0	0
?	0	0	1	1	[01]	0	[01]	[01]	1	[12]
2	0	0	1	0	1	1				

Macrocarpaea

0	[01]	[01]	?	1	?	0	0	?	?	?
0	0	0	[01]	0	?	1	0	[01]	1	0
[01]	0	0	[01]	1	0	0	2	1	[123]	2
0	2	0	0	0	0	0	0	0	0	0
0	0	1	0	0	1	0	2	?	0	0
?	?	?	?	?	?	?	[02]	0	[01]	0
3	0	0	1	1	[01]	1	1	?	?	?
2	1	0	0	1	1	1				

Obolaria

[01]	1	1	?	2	?	0	0	?	?	?
0	0	0	?	1	0	?	[023]	[01]	3	0
?	0	0	0	0	?	0	1	[12]	3	1
0	1	2	0	[01]	0	0	0	[01]	0	0
?	0	2	0	0	1	0	2	0	3	0
[02]	[01]	?	?	?	?	?	2	0	1	0
?	?	0	1	1	0	1	?	?	?	?
?	?	?	?	?	?	?				

Orphium

0	0	?	1	1	0	?	0	?	[01]	?
0	0	0	?	0	?	?	1	1	1	0
0	0	0	0	0	?	?	2	1	?	0
0	?	0	1	0	0	1	0	0	0	0
0	0	2	0	0	1	0	1	0	0	0
?	0	?	?	?	?	?	[12]	0	0	0
[21]	0	0	1	1	0	?	1	?	?	?
2	0	0	0	0	1	1				

Sabatia

0	[12]	[01]	?	?	?	[01]	0	?	?	?
0	0	0	1	1	?	1	0	[01]	[01]	0
0	0	0	?	0	?	0	2	[01]	0	0
0	0	0	1	0	0	1	0	0	0	0
0	0	2	0	0	1	0	2	?	0	0
?	?	?	?	?	?	?	1	0	0	0
[12]	0	0	1	[12]	0	[012]	1	?	2	?
?	?	?	?	?	?	?				

Saccifolium

0	0	0	0	2	?	0	0	1	0	?
0	0	0	1	0	2	0	3	0	1	0
0	0	0	1	0	0	?	1	1	?	2
0	0	1	0	0	0	0	0	0	0	0
?	?	1	0	0	0	0	1	?	?	0
0	0	?	?	?	?	?	0	0	0	0
?	?	0	1	2	0	?	?	?	?	?
?	?	?	?	?	?	?				

Schultesia–Xestaea

0	2	[01]	1	2	?	0	0	?	?	?
0	0	0	1	0	?	1	[03]	[01]	2	0
[012]	[12]	0	[10]	0	2	0	2	2	2	2
0	[02]	[01]	0	0	0	0	0	0	0	1
0	0	[01]	0	0	1	0	2	0	0	0
?	?	?	?	?	?	?	[01]	0	0	0
3	0	1	2	1	0	?	1	0	?	?
2	1	1	1	1	1	1				

Sebaea

0	[12]	[12]	?	?	?	0	0	?	?	?
0	0	0	?	1	0	1	[013]	[01]	[012]	0
0	[12]	0	0	0	0	0	2	[012]	[12]	[12]
0	?	[01]	0	0	0	0	0	0	1	0
0	0	[12]	0	0	[01]	0	[012]	[01]	0	0
?	0	?	?	?	?	?	2	0	[01]	0
4	0	0	1	1	0	0	0	?	?	?
?	?	?	?	?	?	?				

Swertia–Frasera

0	[12]	[012]	1	2	0	0	0	[01]	?	?
0	0	0	1	[01]	0	[01]	0	[01]	[012]	[01]
[01]	0	0	0	0	0	0	2	[012]	0	0
[01]	1	[012]	0	1	0	0	0	0	0	[01]
0	0	0	0	0	[01]	0	2	0	0	0
[03]	0	[01]	1	1	0	1	[12]	[01]	0	[01]
?	0	0	1	1	[01]	[012]	1	1	2	[012]
[12]	[01]	[01]	[01]	0	[01]	[01]				

Symbolanthus

0	0	[12]	0	0	?	?	0	1	1	1
0	0	0	[01]	?	?	?	[23]	[01]	1	0
0	[01]	0	0	1	?	?	2	1	2	2
0	2	[01]	0	0	0	0	0	0	0	1
0	0	0	0	0	1	0	2	?	0	[01]
?	?	?	?	?	?	?	0	0	?	?
?	0	1	1	1	0	2	1	?	?	?
?	?	?	?	?	?	?				

Tachia

0	1	[01]	?	?	?	0	0	?	?	?
0	0	0	[01]	?	?	?	3	0	1	0
[01]	[012]	0	0	1	?	?	2	1	2	2
0	2	[01]	0	0	0	0	0	0	0	0
0	0	0	0	0	1	0	2	?	0	0
?	?	?	?	?	?	?	[01]	0	0	0
1	0	0	1	[12]	1	?	1	?	[12]	?
?	?	?	?	?	?	?				

Tachiadenus

0	[012]	[012]	?	1	?	0	0	?	1	?
0	0	0	1	?	2	?	[01]	[01]	1	0
[01]	[02]	0	0	0	?	?	2	1	1	2
0	?	0	0	0	0	0	1	0	1	0
0	0	2	0	0	1	0	1	?	0	0
?	?	?	?	?	?	?	[01]	0	0	1
?	0	0	1	[12]	0	?	?	?	[12]	?
?	?	?	?	?	?	?				

Tripterospermum

0	1	0	1	?	?	0	1	?	?	?
0	0	0	1	0	?	1	[23]	[01]	1	0
1	[01]	?	?	0	0	?	3	1	[23]	2
?	2	?	0	?	0	0	0	0	0	0
0	0	[01]	0	0	2	1	2	0	0	[01]
?	?	?	?	?	?	?	[2]	1	[01]	0
?	0	0	1	[12]	0	1	?	0	?	?
[12]	[01]	[01]	[01]	1	1	[01]				

Voyria

1	1	0	?	?	?	0	0	?	?	?
0	0	0	?	?	0	?	[03]	[01]	1	0
[12]	0	0	[10]	0	?	?	2	1	[12]	2
0	?	[01]	0	0	0	0	[01]	0	0	0
0	0	[12]	0	0	[12]	[01]	1	?	[02]	0
[02]	[01]	0	0	0	[01]	1	[23]	0	[01]	[01]
2	1	0	2	0	0	1	?	?	?	?
?	?	?	?	?	?	?				

Voyriella

1	[12]	1	?	?	?	0	0	?	?	?
0	0	0	?	?	0	?	[02]	0	1	0
0	0	0	0	0	?	?	2	1	2	2
0	0	1	0	0	0	0	1	0	0	0
0	1	[02]	0	0	1	0	1	?	2	0
2	0	0	0	0	1	0	1	0	0	0
1	1	0	1	1	0	0	?	?	?	?
?	?	?	?	?	?	?				

LITERATURE CITED

Albert, V. A. & L. Struwe. 1997. Phylogeny and classification of *Voyria* (saprophytic Gentianaceae). *Brittonia* 49: 466–479.

Anderberg, A. & B. Ståhl. 1995. Phylogenetic relationships in the order Primulales, with special emphasis on the family circumscriptions. *Can. J. Bot.* 73: 1699–1730.

Arekal, D. 1961. Contribution to the embryology of *Hoppea dichotoma* Willd. (Gentianaceae). *Can. J. Bot.* 39: 1001–1006.

Barthlott, W. 1981. Epidermal and seed surface characters of plants: systematic applicability and some evolutionary aspects. *Nordic J. Bot.* 1: 345–355.

Baum, H. & W. Leinfellner. 1953. Die ontogenetischen Abänderungen des diplophyllen Grundbanes der Staubblätter. *Österr. Bot. Zeitschr.* 100: 91–135.

Beattie, A. J., D. E. Breedlove, & P. R. Ehrlich. 1973. The ecology of the pollinators and predators of *Frasera speciosa*. *Ecology* 54: 81–91.

Bentham, G. 1876. Gentianeae. Pages 799–820 in: G. Bentham & J. D. Hooker, eds. *Genera plantarum*, vol. 2, part 2. Reeve & Co., Williams & Norgate, London.

Bouman, F. & N. Devente. 1986. Seed micromorphology in *Voyria* and *Voyriella*. Pages 9–25 in: P. J. M. Maas & P. Ruyters, eds. *Voyria and* Voyriella *(saprophytic Gentianaceae)*. Flora Neotropica Monograph 41. The New York Botanical Garden, Bronx, NY.

Bouman, F. & S. Schrier. 1979. Ovule ontogeny and seed coat development in *Gentiana*, with a discussion on the evolutionary origin of the single integument. *Acta Bot. Neerl.* 28: 467–478.
Boutique, R. 1972. Gentianaceae. Pages 1–56 in: P. Bamps, ed. *Flore d'Afrique centrale (Zaire-Ruanda-Burundi)*. Jardin Botanique National de Belgique, Brussels.
Bremer, B. 1996. Phylogenetic studies within Rubiaceae and relationships to other families based on molecular data. *Opera Bot. Belg.* 7: 33–50.
Bremer, B., R. G. Olmstead, L. Struwe, & J. A. Sweere. 1994. *rbc*L sequences support exclusion of *Retzia*, *Desfontainia*, and *Nicodemia* from the Gentianales. *Pl. Syst. Evol.* 190: 213–230.
Bremer, K. 1994. Branch support and tree stability. *Cladistics* 10: 295–304.
Broome, C. R. 1978. Chromosome numbers and meiosis in North and Central American species of *Centaurium* (Gentianaceae). *Syst. Bot.* 3: 299–312.
Buchner, R. & C. Puff. 1993. The genus complex *Danais–Schizmatoclada–Payera* (Rubiaceae). Character states, generic delimitation and taxonomic position. *Adansonia* 15: 23–74.
Bureau, L.-E. 1856. *De la famille des Loganiacées, et des plantes qu'elle fournit à la médecine*. Thèse pour le doctorat en médecine. Faculté de Médecine de Paris, Paris.
Butayarov, A. V., E. K. H. Batirov, M. M. Tadzhibaev, E. E. Ibragimond, & V. M. Malikov. 1993. Xanthones from *Gentiana algida* and *G. karalinii*. *Khim. Prir. Soedin.* 901–902.
Carlquist, S. 1984. Wood anatomy of some Gentianaceae: systematic and ecological conclusions. *Aliso* 10: 573–582.
Carlquist, S. 1992. Wood anatomy of sympetalous dicotyledon families: a summary, with comments on systematic relationships and evolution of the woody habit. *Ann. Missouri Bot. Gard.* 79: 303–332.
Cobb, L. & P. J. M. Maas. 1983. Seed coat micromorphology in *Irlbachia* (Gentianaceae). *Proc. Kon. Ned. Akad. Wetensch.*, ser. C, 86: 127–136.
Coddington, J. & N. Scharff. 1994. Problems with zero-length branches. *Cladistics* 10: 415–423.
Conn, B. J. 1980. A taxonomic revision of *Geniostoma* subg. *Geniostoma* (Loganiaceae). *Blumea* 26: 245–364.
Corner, E. J. H. 1976. *The seeds of dicotyledons*, vols. 1 and 2. Cambridge University Press, Cambridge.
Cosner, M. E., R. K. Jansen, & T. G. Lammers. 1994. Phylogenetic relationships in the Campanulales based on *rbc*L sequences. *Pl. Syst. Evol.* 190: 79–95.
Cotthem, W. R., van. 1970. A classification of stomatal types. *Bot. J. Linn. Soc.* 63: 235–246.
De Laet, J. & E. Smets. 1996. A commentary on the circumscription and evolution of the order Gentianales, with special emphasis on the position of Rubiaceae. *Opera Bot. Belg.* 7: 11–18.
Demuth, K. 1993. *Morphologisch/anatomische sowie phytochemische Untersuchungen zur Symbiose von Gentianaceen mit vesikulär-arbusculären*

Mycorrhiza-Pilzen. Inaugural dissertation. Philipps-Universität Marburg, Fachbereich Biologie. Görich & Weiershäuser, Marburg.

Dickison, W. C. 1975. The bases of angiosperm phylogeny: vegetative anatomy. *Ann. Missouri Bot. Gard.* 62: 590–620.

Downie, S. R. & J. D. Palmer. 1992. Restriction site mapping of the chloroplast inverted repeat: a molecular phylogeny of the Asteridae. *Ann. Missouri Bot. Gard.* 79: 266–283.

Drexler, U. & M. J. Hakki. 1979. Embryologische und morphologische Untersuchungen an Pflanzen aus Westindies. 2. Zur Embryologie von *Eustoma exaltatum* (Gentianaceae) mit einer Bemerkung zum Phänomen der "instant pollen tube". *Willdenowia* 9: 131–147.

Dukas, R. & A. Dafni. 1990. Buzz-pollination in three nectariferous Boraginaceae and possible evolution of buzz-pollinated flowers. *Pl. Syst. Evol.* 169: 65–68.

Elias, T. S. & A. Robyns. 1975. Family 160. Gentianaceae. Pages 61–101 in: R. E. Woodson, Jr., R. W. Schery, and collaborators, eds. *Flora of Panama*, part 8. *Ann. Missouri Bot. Gard.* 62.

Endress, M. E., B. Sennblad, S. Nilsson, L. Civeyrel, M. W. Chase, S. Huysmans, E. Grafström *et al.* 1996. A phylogenetic analysis of Apocynaceae s. str. and some related taxa in Gentianales: a multidisciplinary approach. *Opera Bot. Belg.* 7: 59–102.

Erbar, C. 1997. Fieberklee und Seekanne – Enzian- oder Aster-verwandt? Zur Blütenentwicklung und systematischen Stellung der Menyanthaceae. *Bot. Jahrb. Syst.* 119: 115–135.

Farris, J. S. 1970. Methods for computing Wagner trees. *Syst. Zool.* 19: 83–92.

Farris, J. S. 1983. The logical basis of phylogenetic analysis. Pages 7–36 in: N. Platnick & V. A. Funk, eds. *Advances in cladistics*, vol. 2, *Proceedings of the second meeting of the Willi Hennig Society*. Columbia University Press, New York.

Farris, J. S. 1989. The retention index and the rescaled consistency index. *Cladistics* 5: 417–419.

Farris, J. S., V. A. Albert, M. Källersjö, D. Lipscomb, & A. G. Kluge. 1996 [1997]. Parsimony jackknifing outperforms neighbor-joining. *Cladistics* 12: 99–124.

Felsenstein, J. 1985. Confidence limits on phylogenies: an approach using the bootstrap. *Evolution* 39: 783–791.

Figdor, W. 1897. Ueber *Cotylanthera* Bl. *Ann. Jard. Bot. Buitenzorg* 14: 213–240.

Fitch, W. M. 1971. Towards defining the course of evolution: minimum change for a specified tree topology. *Syst. Zool.* 20: 406–416.

Fosberg, F. R. & M.-H. Sachet. 1980. Systematic studies of Micronesian plants. *Smithsonian Contr. Bot.* 45: 1–40.

Furman, T. E. & J. M. Trappe. 1971. Phylogeny and ecology of mycotrophic achlorophyllous angiosperms. *Quart. Rev. Biol.* 46: 219–225.

Garg, S. 1987. *Gentianaceae of the northwest Himalaya (a revision).* Today and Tomorrow's Printers and Publishers, New Delhi.

Gielly, L. & P. Taberlet. 1996. A phylogeny of the European gentians inferred from chloroplast *trn*L (UAA) intron sequences. *Bot. J. Linn. Soc.* 120: 57–75.

Gielly, L., Y.-M. Yuan, P. Küpfer, & P. Taberlet. 1996. Phylogenetic use of noncoding regions in the genus *Gentiana* L.: chloroplast *trn*L (UAA) intron versus nuclear ribosomal internal transcribed spacer sequences. *Mol. Phylogenet. Evol.* 5: 460–466.

Gilg, E. 1895. Gentianaceae. Pages 50–108 in: A. Engler & K. Prantl, eds. *Die natürlichen Pflanzenfamilien*, vol. 4(2). Verlag von Wilhelm Engelmann, Leipzig.

Gilg, E. 1897. Gentianaceae. Pages 282–283 in: A. Engler & K. Prantl, eds. *Die natürlichen Pflanzenfamilien*, Nachträge zum II–IV Teil. Engelmann, Leipzig.

Gilg, E. 1908. Gentianaceae. Pages 292–294 in: A. Engler & K. Prantl, eds. *Die natürlichen Pflanzenfamilien*, Nachträge II und III zum II–IV Teil. Engelmann, Leipzig.

Gill, L. S. & H. G. K. Nyawuame. 1990. Phylogenetic and systematic value of stomata in Bicarpellatae (Bentham et Hooker sensu stricto). *Feddes Repert.* 101: 453–498.

Gillett, J. M. 1959. A revision of *Bartonia* and *Obolaria* (Gentianaceae). *Rhodora* 61: 43–63.

Goethals, V. & E. Smets. 1995. Seed coat anatomy in the Gentianaceae: systematic significance. *Scripta Bot. Belg.* 11: 25.

Goloboff, P. A. 1993. *NONA version 1.8*. Program and documentation distributed by the author. Tucumán, Argentina.

Grisebach, A. H. R. 1839 [1838]. *Genera et species Gentianearum adjectis observationibus quibusdam phytogeographicis*. J. G. Cotta, Stuttgart.

Grisebach, A. H. R. 1845. Gentianaceae. Pages 39–141 in: A. de Candolle, ed. *Prodromus systematis naturalis regni vegetabilis*, vol. 9. Fortin, Masson, et Sociorum, Paris.

Grothe, E. H. M. & P. J. M. Maas. 1984. A scanning electron microscopic study of the seed coat structure of *Curtia* Chamisso & Schlechtendal and *Hockinia* Gardner (Gentianaceae). *Proc. Kon. Ned. Akad. Wetensch*, ser. C, 87: 33–42.

Guérin, P. 1904. Recherches sur le développement et la structure anatomique du tégument séminal des Gentianacées. *J. Bot. (Morot)* 18: 33–52, 83–88.

Guimarães, E. F. 1977. Revisão taxonômica do gênero *Deianira* Chamisso et Schlechtendal (Gentianaceae). *Arq. Jard. Bot. Rio de Janeiro* 21: 46–123.

Gustafsson, M. H. G., A. Backlund, & B. Bremer. 1996. Phylogeny of Asterales sensu lato based on *rbc*L sequences with particular reference to Goodeniaceae. *Pl. Syst. Evol.* 199: 217–242.

Gutsche, K., D. Glenny, & J. W. Kadereit. 1997. A contribution to the molecular phylogeny of the Gentianaceae-Gentianinae. *Scripta Bot. Belg.* 15: 73.

Hamburger, M., M. Hostettmann, H. Stoeckli-Evans, P. N. Solis, M. P. Gupta, & K. Hostettmann. 1990. A novel type of dimeric secoiridoid glycoside from *Lisianthius jefensis* Robyns & Elias. *Helv. Chim. Acta* 73: 1845–1852.

Hegnauer, R. 1989. *Chemotaxonomie der Pflanzen*, vol. 8. Birkhäuser Verlag, Basel.

Heusden, E. C. H., van. 1986. Floral anatomy of some neotropical Gentianaceae. *Proc. Kon. Ned. Akad. Wetensch.*, ser. C, 89: 45–59.

Hickey, L. J. 1979. A revised classification of the architecture of dicotyledonous leaves. Pages 26–39 in: C. R. Metcalfe & L. Chalk, eds. *Anatomy of the dicotyledons*, vol. 1, ed. 2. Clarendon Press, Oxford.

Hickey, L. J. & J. A. Wolfe. 1975. The bases of angiosperm phylogeny: vegetative morphology. *Ann. Missouri Bot. Gard.* 62: 538–589.
Ho, T.-N. & S.-W. Liu. 1990. The infrageneric classification of *Gentiana* (Gentianaceae). *Bull. Brit. Mus. (Nat. Hist.), Bot.* 20: 169–192.
Ho, T. N. & J. S. Pringle. 1995. Gentianaceae. Pages 1–139 in: Z.-Y. Wu & P. H. Raven, eds. *Flora of China*, vol. 16. Science Press, Beijing and Missouri Botanical Garden, St. Louis, MO.
Ho, T.-N., S.-W. Liu, & C. J. Wu. 1988. Gentianaceae. Pages 1–446 in: *Flora republicae popularis sinicae*, vol. 62. Science Press, Beijing.
Holm, T. 1897. *Obolaria virginica* L. A morphological and anatomical study. *Ann. Bot. (London)* 11: 369–383.
Holm, T. 1906. *Bartonia* Muehl. An anatomical study. *Ann. Bot. (London)* 20: 441–448.
Humbert, H. 1937. Un genre nouveau de Gentianacées-Chironiinées de Madagascar. *Bull. Soc. Bot. France* 84: 386–390.
Hungerer, B. K. & J. W. Kadereit. 1997. The biogeography of the "*Gentiana acaulis* group" examined with molecular markers. *Scripta Bot. Belg.* 15: 83.
Hutchinson, J. 1959. *The families of flowering plants*, vol. 1, ed. 2. Clarendon Press, Oxford.
Imhof, S., H. C. Weber, & L. D. Gomez. 1994. Ein Beitrag zur Biologie von *Voyria tenella* Hook. und *Voyria truncata* (Standley) Standley et Steyermark (Gentianaceae). *Beitr. Biol. Pflanzen.* 68: 113–123.
Jensen, S. R. 1992. Systematic implications of the distribution of iridoids and other chemical compounds in the Loganiaceae and other families of the Asteridae. *Ann. Missouri Bot. Gard.* 79: 284–302.
Jensen, S. R. & J. Schripsema. 2002. Chemotaxonomy and pharmacology of Gentianaceae. Pages 573–631 in: L. Struwe & V. A. Albert, eds. *Gentianaceae: systematics and natural history*. Cambridge University Press, Cambridge.
Källersjö, M., J. S. Farris, A. G. Kluge, & C. Bult. 1992. Skewness and permutation. *Cladistics* 8: 275–287.
Kaouadji, M. 1990. Flavonol diglycosides from *Blackstonia perfoliata*. *Phytochemistry* 29: 1345–1347.
Khishgee, D. & O. Pureb. 1993. Xanthones and flavonoids of *Lomatogonium rotatum*. *Khim. Prir. Soedin.* 761–762.
Khoshoo, T. N. & S. R. Tandon. 1963. Cytological, morphological and pollination studies on some Himalayan species of *Swertia*. *Caryologia (Pisa)* 16: 445–477.
Klackenberg, J. 1985. The genus *Exacum* (Gentianaceae). *Opera Bot.* 84: 1–144.
Klackenberg, J. 1987. Revision of the genus *Tachiadenus* (Gentianaceae). *Adansonia* 9: 43–80.
Klahr, A. 1993. *Vergleichende morphologisch-anatomische Untersuchungen an Apocynaceen-Wurzeln*. Diplomarbeit, Marburg. (Unpublished thesis)
Kluge, A. & J. S. Farris. 1969. Quantitative phyletics and the evolution of anurans. *Syst. Zool.* 18: 1–32.
Kremer, P. & J. van Andel. 1995. Evolutionary aspects of life forms in angiosperm families. *Acta Bot. Neerl.* 44: 469–479.

Krishna, G. G. & V. Puri. 1962. Morphology of the flower of some Gentianaceae with special reference to placentation. *Bot. Gaz. (Crawfordsville)* 124: 42–57.

Kusnezow, N. J. 1896–1904. Subgenus *Eugentiana* Kusnez. generis *Gentiana* Tournef. *Trudy Imp. S.-Peterburgsk. Bot. Sada* 15: 1–507.

Lammers, T. G. 1992. Circumscription and phylogeny of the Campanulales. *Ann. Missouri Bot. Gard.* 79: 388–413.

Leenhouts, P. W. 1962 [1963]. Loganiaceae. Pages 293–387 in: C. G. G. J. van Steenis, ed. *Flora Malesiana*, ser. 1, vol. 6(2). Wolters-Noordhoff, Groningen.

Leeuwenberg, A. J. M. & P. W. Leenhouts. 1980. Taxonomy. Pages 8–96 in: A. J. M. Leeuwenberg, ed. *Engler and Prantl's Die natürlichen Pflanzenfamilien, Angiospermae: Ordnung Gentianales, Fam. Loganiaceae*, vol. 28b (1). Duncker and Humblot, Berlin.

Leinfellner, W. 1950. Der Bauplan des synkarpen Gynözeums. *Österr. Bot. Zeitschr.* 103: 185–242.

Lindsey, A. A. 1940. Floral anatomy in the Gentianaceae. *Amer. J. Bot.* 27: 640–651.

Liu, S.-W. & T.-N. Ho. 1992. Systematic study on *Lomatogonium* A. Br. (Gentianaceae). *Acta Phytotax. Sin.* 30: 289–319.

Maas, P. J. M. & P. Ruyters, eds. 1986. Voyria *and* Voyriella *(saprophytic Gentianaceae)*. Flora Neotropica Monograph 41. The New York Botanical Garden, Bronx, NY.

Maas, P. J. M., S. Nilsson, A. M. C. Hollants, B. J. H. ter Welle, H. Persoon, & E. C. H. van Heusden. 1983. Systematic studies in neotropical Gentianaceae – the *Lisianthius* complex. *Acta Bot. Neerl.* 32: 371–374.

Maguire, B. 1981. Gentianaceae. The botany of the Guayana Highland – Part XI. *Mem. New York Bot. Gard.* 32: 330–388.

Maguire, B. & J. M. Pires. 1978. Saccifoliaceae – a new monotypic family of the Gentianales. Pages 230–245 in: B. Maguire and collaborators, eds. The botany of the Guayana Highland – Part X. *Mem. New York Bot. Gard.* 29.

Maheswari Devi, H. 1962. Embryological studies in Gentianaceae (Gentianoideae and Menyanthoideae). *Proc. Indian Acad. Sci.*, sect. B, 56: 195–216.

Marais, W. & I. C. Verdoorn. 1963. Gentianaceae. Pages 171–243 in: R. A. Dyer, L. E. Codd, & H. B. Rycroft, eds. *Flora of southern Africa*, vol. 26. Department of Agricultural Technical Services, Republic of South Africa, Pretoria.

Massias, M., J. Carbonnier, & D. Molho. 1978. Implications chimiotaxonomiques de la répartition de substances osidiques dans le genre *Gentiana*. *Bull. Mus. Nat. Hist. Nat.*, sér. 3, Écol., 504: 41–53.

McCoy, R. W. 1940. Floral organogenesis in *Frasera caroliniensis*. *Amer. J. Bot.* 27: 600–609.

Melderis, A. 1972. *Centaurium* Hill. Pages 56–59 in: T. G. Tutin, V. H. Heywood, N. A. Burges, D. M. Moore, D. H. Valentine, S. M. Walters, & D. A. Webb, eds. *Flora Europaea*, vol. 3. Cambridge University Press, Cambridge.

Mennega, A. M. W. 1980. Anatomy of the secondary xylem. Pages 112–161 in: A. J. M. Leeuwenberg, ed. *Engler and Prantl's Die natürlichen Pflanzenfamilien, Angiospermae: Ordnung Gentianales, Fam. Loganiaceae*, vol. 28b (1). Duncker and Humblot, Berlin.

Mészáros, S. 1994. Evolutionary significance of xanthones in Gentianaceae: a reappraisal. *Biochem. Syst. Ecol.* 22: 85–94.

Mészáros, S., J. De Laet, & E. Smets. 1996. Phylogeny of temperate Gentianaceae: a morphological approach. *Syst. Bot.* 21: 153–168.

Metcalfe, C. R. & L. Chalk. 1950. *Anatomy of the dicotyledons.* Clarendon Press, Oxford.

Miège, J. & J. Wüest. 1984. Les surfaces tégumentaires des graines de *Gentiana* et *Gentianella* vues au microscope électronique à balayage. *Ber. Schweiz. Bot. Ges.* 94: 41–59.

Mohan, J. S. S., M. Nataraj, & J. A. Inamdar. 1989. Foliar venation in some Gentianaceae and Menyanthaceae. *Ind. Bot. Contactor* 6: 77–81.

Müller, G. 1982. Contribution à la cytotaxonomie de la section *Cyclostigma* Griseb. du genre *Gentiana* L. *Feddes Repert.* 93: 625–722.

Nilsson, S. 1964. On the pollen morphology in *Lomatogonium* A. Br. *Grana Palynol.* 5: 298–329.

Nilsson, S. 1967a. Pollen morphological studies in the Gentianaceae-Gentianinae. *Grana Palynol.* 7: 46–145.

Nilsson, S. 1967b. Notes on pollen morphological variation in Gentianaceae-Gentianinae. *Pollen and Spores* 9: 49–58.

Nilsson, S. 1968. Pollen morphology in the genus *Macrocarpaea* (Gentianaceae) and its taxonomical significance. *Svensk Bot. Tidskr.* 62: 338–364.

Nilsson, S. 1970. Pollen morphological contributions to the taxonomy of *Lisianthus* L. s. lat. (Gentianaceae). *Svensk Bot. Tidskr.* 64: 1–43.

Nilsson, S. 2002. Gentianaceae: a review of palynology. Pages 377–497 in: L. Struwe & V. A. Albert, eds. *Gentianaceae: systematics and natural history.* Cambridge University Press, Cambridge.

Nilsson, S. & J. J. Skvarla. 1969. Pollen morphology of saprophytic taxa in the Gentianaceae. *Ann. Missouri Bot. Gard.* 56: 420–438.

Nixon, K. C. & J. I. Davis. 1991. Polymorphic taxa, missing values and cladistic analysis. *Cladistics* 7: 233–241.

Oehler, E. 1927. Entwicklungsgeschichtliche-zytologische Untersuchungen an einigen saprophytischen Gentianaceen. *Planta* 3: 641–733.

Olmstead, R. G., H. J. Michaels, K. M. Scott, & J. D. Palmer. 1992. Monophyly of the Asteridae and identification of their major lineages inferred from DNA sequences of *rbc*L. *Ann. Missouri Bot. Gard.* 79: 249–265.

Olmstead, R. G., B. Bremer, K. M. Scott, & J. D. Palmer. 1993. A parsimony analysis of the Asteridae sensu lato based on *rbc*L sequences. *Ann. Missouri Bot. Gard.* 80: 700–722.

Paiva, J. & I. Nogueira. 1990. Studies in African Gentianaceae. *Anales Jard. Bot. Madrid* 47: 87–103.

Pant, D. D. & P. F. Kidway. 1969. Ontogeny of stomata in some Gentianaceae. *Bot. J. Linn. Soc.* 62: 71–76.

Patel, R. C., J. A. Inamdar, & N. V. Rao. 1981. Structure and ontogeny of stomata in some Gentianaceae and Menyanthaceae complex. *Feddes Repert.* 92: 535–550.

Perrot, M. E. 1897 [1899]. Anatomie comparée des Gentianacées. *Ann. Sci. Nat. Bot.,* sér viii, VIII. 7: 105–292.

Perry, J. D. 1971. Biosystematic studies on the North American genus *Sabatia* (Gentianaceae). *Rhodora* 73: 309–369.
Philipson, W. R. 1972. The generic status of the southern hemisphere gentians. Pages 417–422 in: Y. S. Murty, B. M. Johri, H. Y. Mohan Ram, & T. M. Varghese, eds. *Advances in plant morphology*. Professor V. Puri commemorative volume. Sarita Prakashan, Meerut.
Pihlar, O., L. Struwe, & V. A. Albert. 1998. Neotropical Gentianaceae and whitesands: biogeography and character evolution. *Amer. J. Bot.* 85 (suppl.): 150–151.
Pilger, R. & K. Krause. 1915. Gentianaceae. Page 244 in: A. Engler & K. Prantl, eds. *Die natürlichen Pflanzenfamilien*, Ergänzungsheft III, Nachträge IV zum II–IV Teil. Engelmann, Leipzig.
Plouvier, V. 1990. Alditols et cyclitols: répartition et taxonomie chez les plantes supérieures. *Adansonia* 12: 209–223.
Pringle, J. S. 1990. Taxonomic notes on western American Gentianaceae. *Sida* 14: 179–187.
Punt, W. 1978. Evolutionary trends in the Potalieae (Loganiaceae). *Rev. Palaeobot. Palynol.* 26: 313–335.
Rao, K. S. & C. C. Chinnappa. 1983. Studies in Gentianaceae. Microsporangium and pollen. *Can. J. Bot.* 61: 324–336.
Rao, K. S. & M. Nagaraj. 1982. Studies in Gentianaceae. Embryology of *Swertia minor* (Gentianinae). *Can. J. Bot.* 60: 141–151.
Rath, C. 1993. Vergleichende Untersuchungen an den unterirdischen Organen von Rubiaceen. Diplomarbeit, Marburg. (Unpublished thesis)
Raynal, A. 1968. Les genres *Neurotheca* Benth. et Hook. et *Congolanthus* A. Raynal, gen. nov. (Gentianaceae). *Adansonia*, sér. 2, 8: 45–68.
Recio-Iglesias, M.-C., A. Marston, & K. Hostettmann. 1992. Xanthones and secoiridoid glycoside of *Halenia campanulata*. *Phytochemistry* 31: 1387–1389.
Rodriguez, S., J. L. Wolfender, G. Odontuya, O. Purev, & K. Hostettmann. 1995. Xanthones, secoiridoids and flavonoids from *Halenia corniculata*. *Phytochemistry* 40: 1265–1272.
Ronse Decraene, L.-P. & E. F. Smets. 1995. The distribution and systematic relevance of the androecial character oligomery. *Bot. J. Linn. Soc.* 118: 193–247.
Schaufelberger, D. & K. Hostettmann. 1984. Flavonoid glycosides and a bitter principle from *Lomatogonium carinthiacum*. *Phytochemistry* 23: 787–789.
Schilling, N. 1976. Distribution of L-(+)-bornesitol in the Gentianaceae and Menyanthaceae. *Phytochemistry* 15: 824–826.
Schinz, H. 1903. Versuch einer monographischen Übersicht der Gattung *Sebaea* R. Br. 1. Die Sektion *Eusebaea* Griseb. *Mitt. Geogr. Ges. Naturhist. Mus. Lübeck* 17: 3–55.
Schoch, E. 1903. Monographie der Gattung *Chironia* L. *Beih. Bot. Centralbl.* 14: 177–242; tabs. 15–16.
Sennblad, B. 1997. *Phylogeny of the Apocynaceae s. l.* Acta Universitatis Uppsaliensis, Comprehensive Summaries of Uppsala Dissertations from the Faculty of Science and Technology 295. HSC Uppsala University, Uppsala.
Shamrov, I. I. 1988. Ovule development and structural characteristics of the

embryo sac in some members of the Gentianaceae family. *Bot. Zhurn.* 73: 213–222. (In Russian)

Shamrov, I. I. 1991. The ovule of *Swertia iberica* (Gentianaceae): structural and functional aspects. *Phytomorphology* 41: 213–229.

Shamrov, I. I. 1996. Ovule development and significance of its features for Gentianaceae systematics. *Opera Bot. Belg.* 7: 113–118.

Shigenobu, Y. 1983. Karyomorphological studies in some genera of Gentianaceae. II. *Gentiana* and its allied four genera [Part 1]. *Bull. Coll. Child Devel., Kochi Women's University (Japan)* 7: 65–84.

Shigenobu, Y. 1984. Karyomorphological studies in some genera of Gentianaceae. II. *Gentiana* and its allied four genera [Part 2]. *Bull. Coll. Child Devel., Kochi Women's University (Japan)* 8: 55–104.

Shiolara, Y., K. Kato, Y. Ueda, K. Tamue, E. Syoha, N. Nishimoto, & F. de Oliveira. 1994. Secoiridoid glycosides from *Chelonanthus chelonoides*. *Phytochemistry* 37: 1649–1652.

Solereder, H. 1885. *Über den systematischen Wert der Holzstructur bei den Dicotyledonen*. Druck von R. Oldenbourg, München.

Solereder, H. 1899. *Systematische Anatomie der Dicotyledonen*. Verlag von Ferdinand Enke, Stuttgart.

Stebbins, G. L. 1974. *Flowering plants: evolution above the species level*. The Belknap Press of Harvard University Press, Cambridge, MA.

Stolt, K. A. H. 1921. Zur Embryologie der Gentianaceen und Menyanthaceen. *Kongl. Svenska Vetenskapsakad. Handl.* 61: 1–56.

Stolt, K. A. H. 1927. Über die Embryologie von *Gentiana prostrata* Hänk. und die Antipoden der Gentianaceen. *Bot. Not.* 80: 225–242.

Struwe, L. 1999. *Morphological and molecular phylogenetic studies in neotropical Gentianaceae*. Dissertation. Stockholm University, Stockholm.

Struwe, L. & V. A. Albert. 1997. Floristics, cladistics, and classification: three case studies in Gentianales. Pages 321–352 in: J. Dransfield, M. J. E. Coode, & D. A. Simpson, eds. *Plant diversity in Malesia*, III. Royal Botanic Gardens, Kew.

Struwe, L. & V. A. Albert. 1998. *Lisianthius* (Gentianaceae), its probable homonym *Lisyanthus*, and the priority of *Helia* over *Irlbachia* as its substitute. *Harvard Pap. Bot.* 3: 63–71.

Struwe, L., V. A. Albert, & B. Bremer. 1994 [1995]. Cladistics and family level classification of the Gentianales. *Cladistics* 10: 175–206.

Struwe, L., P. J. M. Maas, & V. A. Albert. 1997. *Aripuana cullmaniorum*, a new genus and species of Gentianaceae from white-sands of southeastern Amazonas, Brazil. *Harvard Pap Bot.* 2: 235–253.

Struwe, L., M. Thiv, J. W. Kadereit, A. S.-R. Pepper, T. J. Motley, P. J. White, J. H. E. Rova *et al.*. 1998. *Saccifolium* (Saccifoliaceae), an endemic of Sierra de la Neblina on the Brazilian–Venezuelan frontier, is related to a temperate-alpine lineage of Gentianaceae. *Harvard Pap. Bot.* 3: 199–214.

Struwe, L., J. W. Kadereit, J. Klackenberg, S. Nilsson, M. Thiv, K. B. von Hagen, & V. A. Albert. 2002. Systematics, character evolution, and biogeography of Gentianaceae, including a new tribal and subtribal classification. Pages 21–309 in: L. Struwe & V. A. Albert, eds. *Gentianaceae: systematics and natural history*. Cambridge University Press, Cambridge.

Sytsma, K. J. & B. A. Schaal. 1985. Phylogenetics of the *Lisianthius skinneri* (Gentianaceae) species complex in Panama utilizing DNA restriction fragment analysis. *Evolution* 39: 594–608.
Szujko-Lacza, J. & E. Gondar. 1983. Studies in the Gentianaceae II. Numerical evaluation of the two *Gentianella* species. *Feddes Repert.* 94: 473–491.
Szujko-Lacza, J. & S. Sen. 1979. Significance of anatomical features of the shoot in the systematics of Hungarian *Gentiana. Acta Bot. Acad. Sci. Hung.* 25: 365–403.
Takhtajan, A. 1991. *Evolutionary trends in flowering plants.* Columbia University Press, New York.
Taylor, P. 1973. A revision of the genus *Faroa* Welwitsch. *Garcia de Orta, Bot.* 1: 69–82.
ter Welle, B. J. H. 1986. Anatomy. Pages 25–29 in: P. J. M. Maas & P. Ruyters, eds. *Voyria and Voyriella (saprophytic Gentianaceae).* Flora Neotropica Monograph 41. The New York Botanical Garden, Bronx, NY.
Terreaux, C., M. Maillard, M. P. Gupta, & K. Hostettmann. 1995. Xanthones from *Schultesia lisianthoides. Phytochemistry* 40: 1791–1795.
Thiv, M. & J. W. Kadereit. 1997. The phylogeny of subtribe Erythraeinae of the Gentianaceae. *Scripta Bot. Belg.* 15: 148.
Thiv, M., L. Struwe, V. A. Albert, & J. W. Kadereit. 1999a. The phylogenetic relationships of *Saccifolium bandeirae* Maguire & Pires (Gentianaceae) reconsidered. *Harvard Pap. Bot.* 4: 519–526.
Thiv, M., L. Struwe, & J. W. Kadereit. 1999b [2000]. The phylogenetic relationships and evolution of the Canarian laurel forest endemic *Ixanthus viscosus* (Ait.) Griseb. (Gentianaceae): evidence from *mat*K and ITS sequence variation, and floral morphology and anatomy. *Pl. Syst. Evol.* 218: 299–317.
Tiemann, C., K. Demuth, & H. C. Weber. 1993. Zur VA-mycorrhiza von *Gelsemium rankinii* and *G. sempervirens* (Loganiaceae). *Beitr. Biol. Pflanzen* 68: 311–321.
Trivedi, B. S. & N. Upadhyay. 1983. Epidermal structure in *Gentiana* Linn. *J. Indian Bot. Soc.* 62: 124–132.
Tutin, T. G., 1972. Gentianaceae. Pages 56–67 in: T. G. Tutin, V. H. Heywood, N. A. Burges, D. M. Moore, D. H. Valentine, S. M. Walters, & D. A. Webb, eds. *Flora Europaea,* vol. 3. Cambridge University Press, Cambridge.
Ubsdell, R. A. E. 1979. Studies on variation and evolution in *Centaurium erythraea* Rafn. and *C. littorale* (D. Turner) Gilmour in the British Isles, 3. Breeding systems, floral biology and general discussion. *Watsonia* 12: 225–232.
Vasudevan, K. N. 1975. Contribution to the cytotaxonomy and cytogeography of the flora of the Western Himalayas (with an attempt to compare it with the flora of the Alps). Part I. *Ber. Schweiz. Bot. Ges.* 85: 57–84.
Vesque, M. J. 1875. Anatomie comparée de l'écorce. *Ann. Sci. Nat. Bot.,* sér. 6, 2: 82–198, pls. 9–11.
Vijayaraghavan, M. R. & U. Padmanaban. 1969. Morphology and embryology of *Centaurium ramosissimum* Druce and affinities of the family Gentianaceae. *Beitr. Biol. Pflanzen* 46: 15–37.
Vogel, S. 1958. Fledermausblumen in Südamerika. *Österr. Bot. Zeitschr.* 104: 491–530.

Vogel, S. 1969. Chiropterophilie in der neotropischen Flora. II. *Flora. Abt. B* 158: 185–222.
Vogel, S. 1978. Evolutionary shifts from reward to deception in pollen flowers. Pages 89–96 in: A. J. Richards, ed. *The pollination of flowers by insects.* Academic Press, London.
Vogel, S. 1993. Betrug bei Pflanzen: die Täuschblumen. *Abh. Math.-Naturw. Kl. Sächs. Akad. Wiss.* 1993(1): 1–48.
Wagenitz, G. 1964. Reihe Gentianales (Contortae, Loganiales, Apocynales). Pages 405–424 in: H. Melchior, ed. *Engler's Syllabus der Pflanzenfamilien*, vol. 2, ed. 12. Gebr. Börntraeger, Berlin.
Walker, J. W. & J. A. Doyle. 1975. The bases of angiosperm phylogeny: palynology. *Ann. Missouri Bot. Gard.* 62: 664–723.
Weaver, R. E., Jr. 1969. Cytotaxonomic notes on some neotropical Gentianaceae. *Ann. Missouri Bot. Gard.* 56: 439–443.
Weaver, R. E., Jr. 1972. A revision of the neotropical genus *Lisianthius* (Gentianaceae). *J. Arnold Arbor.* 53: 76–100, 234–272, 273–311.
Weaver, R. E., Jr. 1974. The reduction of *Rusbyanthus* and the tribe Rusbyantheae (Gentianaceae). *J. Arnold Arbor.* 55: 300–302.
Webb, C. J. & P. E. Pearson. 1993. The evolution of approach herkogamy from protandry in New Zealand *Gentiana* (Gentianaceae). *Pl. Syst. Evol.* 186: 187–191.
Weber, H. C. 1992. Abbreviationen von Wurzelsystemen als Erklärungsmöglichkeit für die phylogenetische Progression zum Endoparasitismus. Page 178 in: H.-P. Haschke & C. Schnarrenberger. *Botanikertagung 1992.* Akademie Verlag, Berlin.
Wilbur, R. L. 1955. A revision of the North American genus *Sabatia* (Gentianaceae). *Rhodora* 57: 1–33, 43–71, 78–104.
Wood, C. E., Jr. & R. E. Weaver, Jr. 1982. The genera of Gentianaceae in the southeastern United States. *J. Arnold Arbor.* 63: 441–487.
Yeates, D. 1992. Why remove autapomorphies? *Cladistics* 8: 387–389.
Yuan, Y.-M. 1993a. Karyological studies on *Gentiana* section *Cruciata* Gaudin (Gentianaceae) from China. *Caryologia* 46: 99–114.
Yuan, Y.-M. 1993b. Seed coat micromorphology and its systematic implications for Gentianaceae of western China. *Bot. Helv.* 103: 73–82.
Yuan, Y.-M. & P. Küpfer. 1993a. Karyological studies of *Gentianopsis* Ma and some related genera of Gentianaceae from China. *Cytologia* 58: 115–123.
Yuan, Y.-M. & P. Küpfer. 1993b. Karyological studies on *Gentiana* sect. *Frigida* s. l. and sect. *Stenogyne* (Gentianaceae) from China. *Bull. Soc. Neuchâteloise Sci. Nat.* 116: 65–78.
Yuan, Y.-M. & P. Küpfer. 1995. Molecular phylogenetics of the subtribe Gentianinae (Gentianaceae) inferred from the sequences of internal transcribed spacers (ITS) of nuclear ribosomal DNA. *Pl. Syst. Evol.* 196: 207–226.
Yuan, Y.-M. & P. Küpfer. 1997. The monophyly and rapid evolution of *Gentiana* sect. *Chondrophyllae* Bunge s. l. (Gentianaceae): evidence from the nucleotide sequences of the internal transcribed spacers (ITS) of nuclear ribosomal DNA. *Bot. J. Linn. Soc.* 123: 25–43.

Yuan, Y.-M., P. Küpfer, & J. J. Doyle. 1996. Infrageneric phylogeny of the genus *Gentiana* (Gentianaceae) inferred from nucleotide sequences of the internal transcribed spacers (ITS) of nuclear ribosomal DNA. *Amer. J. Bot.* 83: 641–652.

Zeltner, L. 1970. Recherches de biosystématique sur les genres *Blackstonia* Huds. et *Centaurium* Hill (Gentianacées). *Bull. Soc. Neuchâteloise Sci. Nat.* 93: 1–164.

4
Gentianaceae: a review of palynology

S. NILSSON

ABSTRACT

Palynology has traditionally been a major source of characters for taxonomic classification in the Gentianaceae. In this survey palynological data from the following genera are presented: *Adenolisianthus, Anthocleista, Aripuana, Calolisianthus, Celiantha, Chelonanthus, Chorisepalum, Congolanthus, Coutoubea, Curtia, Deianira, Fagraea, Gentianothamnus, Helia, Hockinia, Irlbachia, Lagenanthus, Lehmanniella, Lisianthius, Macrocarpaea, Neblinantha, Neurotheca, Potalia, Prepusa, Purdieanthus, Rogersonanthus, Saccifolium, Schultesia, Sebaea, Sipapoantha, Symbolanthus, Tachia, Tachiadenus, Tapeinostemon, Tetrapollinia, Wurdackanthus,* and *Zonanthus*. SEM micrographs of pollen from each genus are presented and palynological characters are evaluated from a phylogenetic standpoint. Gilg's (1895) tribal and subtribal division of the Gentianaceae, which was based mainly on pollen morphology, cannot be duly confirmed by the present study, and Struwe *et al.*'s (2002) classification is followed here.

The majority of taxa in Gilg's (1895) tribe Helieae have variously been referred to the genus "*Lisianthus*" as sections, to separate genera, or have been partly lumped into one genus. Pollen morphology suggests distinction of one West Indian–Central American group (*Lisianthius sensu stricto*) with monads and one neotropical group (mainly continental) with compound pollen grains (the *Irlbachia* complex). The generic name of *Lisianthius* should be restricted to West Indian and Central American taxa with single pollen grains. *Chorisepalum, Macrocarpaea,* and *Tachia* appear to belong to the same group. Gilg's (1895) generic concept is regarded as partly correct as is that of Maguire (1981) and Maguire and Boom (1989). Maas's (1985) broadly circumscribed *Irlbachia* is not in agreement with these palynological results. The genera *Celiantha, Neblinantha, Rogersonanthus, Sipapoantha,*

Symbolanthus, *Tetrapollinia*, and *Wurdackanthus* are associated within the *Irlbachia* complex. The genera *Lagenanthus*, *Lehmanniella*, and *Purdieanthus* are palynologically distinct, and pollen morphology does not support fusion into one genus, *Lehmanniella*. *Lehmanniella* and *Purdieanthus* are associated with the *Irlbachia* complex, while *Lagenanthus* appears palynologically dissimilar to the other two genera. The genera *Prepusa* and *Senaea* have tetrads similar to the *Irlbachia* complex and might be phylogenetically related. Pollen of *Sipapoantha* also shows similarity with *Prepusa* and *Senaea* pollen. *Coutoubea* and *Schultesia* have very similar types of tetrads and may be phylogenetically close.

Congolanthus and *Neurotheca* are supported as two separate genera with reference to pollen morphology. Some species of *Curtia* and *Hockinia montana* are heterostylous, with two similar types of pollen. The two genera are here regarded as closely allied and may be placed together in the same group.

Anthocleista and *Potalia* have pollen slightly resembling those of certain Apocynaceae or *Geniostoma* (Loganiaceae), not other Gentianaceae. The pollen grains of some species of *Fagraea* show similarity with those of some Gentianaceae taxa and *Gelsemium* (Gelsemiaceae). A transfer of *Anthocleista*, *Fagraea*, and *Potalia* to the Gentianaceae is not supported by palynological data, with the possible exception of *Fagraea* (Potalieae).

Keywords: Gentianaceae, palynology, taxonomy.

INTRODUCTION

In his classification of the Gentianaceae, Gilg (1895) primarily utilized pollen morphology to divide the subfamily Gentianoideae into five tribes: Gentianeae, Rusbyantheae, Helieae, Voyrieae, and Leiphaimeae (Table 4.1). The Gentianeae were further divided into five subtribes. Gilg mainly used size, shape, exine features, and single versus compound pollen grains to differentiate the various groups. Köhler (1905), following Gilg's classification, made a more detailed study of the pollen grains. His observations and illustrations largely confirmed Gilg's descriptions. In a series of publications Nilsson critically examined the pollen morphology of several taxa, i.e., the *Irlbachia* complex and *Lisianthius* (Nilsson, 1968, 1970; Robyns & Nilsson, 1970), Gentianeae-Gentianinae (Nilsson, 1967a,b), and Voyrieae (Nilsson & Skvarla, 1969). According to Gilg (1895) and later findings all genera of the Helieae *sensu* Gilg have pollen grains united in tetrads or polyads. The other tribes *sensu* Gilg have single, free pollen grains, the only exceptions being the genus *Macrocarpaea* as circumscribed by Ewan (1948a) which has

A review of palynology

Table 4.1. *Infrafamilial classification of Gentianaceae-Gentianoideae* sensu Gilg (1895)

I. Gentianeae
   a. Exacinae
   b. Erythraeinae
   c. Chironiinae
   d. Gentianinae
   e. Tachiinae
II. Rusbyantheae
III. Helieae
IV. Voyrieae
V. Leiphaimeae

both monads and tetrads (and *Potalia maguireorum* with tetrads, reported in this paper). The species of *Macrocarpaea* with tetrads have been segregated as the genus *Rogersonanthus* (Maguire & Boom, 1989). In the present work only major studies on Gentianaceae pollen are taken into account.

MATERIALS

In the present survey the following (primarily neotropical) genera are reviewed: *Curtia*, *Hockinia*, *Saccifolium*, and *Tapeinostemon* (Saccifolieae), *Gentianothamnus* and *Tachiadenus* (Exaceae), *Coutoubea*, *Deianira*, and *Schultesia* (Chironieae), *Adenolisianthus*, *Aripuana*, *Calolisianthus*, *Celiantha*, *Chelonanthus*, *Chorisepalum*, *Helia*, *Irlbachia* (including *Pagaea* and *Brachycodon*), *Lagenanthus*, *Lehmanniella*, *Macrocarpaea*, *Neblinantha*, *Prepusa*, *Purdieanthus*, *Rogersonanthus*, *Senaea*, *Sipapoantha*, *Symbolanthus*, *Tachia*, *Tetrapollinia*, *Wurdackanthus*, and *Zonanthus* (Helieae), and *Anthocleista*, *Congolanthus*, *Fagraea*, *Lisianthius*, *Neurotheca*, and *Potalia* (Potalieae). The generic concepts follow Gilg (1895), Maguire (1981), and Maguire and Boom (1989). Genera are arranged in tribes according to the new infrafamiliar classification of Struwe *et al.* (2002 (Chapter 2, this volume)).

The genus *Saccifolium* was earlier placed in the monotypic family Saccifoliaceae (Maguire & Pires, 1978). Pollen grains of *Anthocleista*, *Fagraea*, and *Potalia* in the Potalieae-Potaliinae (formerly in Loganiaceae; e.g., Leeuwenberg & Leenhouts, 1980; Punt, 1980) are also compared with those of Gentianaceae *sensu* Gilg. For a complete list of investigated species see Appendix 4.1.

METHODS

Pollen material collected from herbarium specimens was subjected to acetolysis (Erdtman, 1960). For preparation of slides for light microscopy (LM), the acetolyzed pollen grains were mounted in glycerine jelly and sealed with paraffin. The slides are deposited at the Palynological Laboratory, Swedish Museum of Natural History, Stockholm, Sweden.

Samples for scanning electron microscopy (SEM) were vacuum-coated with gold and examined with a Jeol JSM-6300. In some cases the pollen was freeze-sectioned (Ames Lab-Tek). The fractures were treated as above prior to examination under SEM.

For transmission electron microscopy (TEM), pollen material from herbarium specimens was dehydrated, treated with osmium tetroxide, and embedded in Spurr's (1969) resin. Ultrathin sections were cut with an LKB Ultrotome V using a diamond knife. The sections were post-stained in uranyl acetate and lead citrate prior to examination. The sections were examined in a Zeiss EM 10 electron microscope.

In general 10 pollen grains of each examined specimen were measured. Summaries of the results are presented in Tables 4.2–4.4. Terminology generally follows that of Punt *et al.* (1994; see also Struwe *et al.*, 2002: Appendix 2.2). Material was obtained from the following herbaria: BM, BR, C, CGE, COC, DUKE, E, F, G, GH, HAJB, HAL, K, L, LL, M, MO, NA, NY, P, PH, R, RB, S, U, UC, US, VEN, and W (abbreviations following Holmgren *et al.*, 1990).

POLLEN DESCRIPTIONS

The generic descriptions are arranged alphabetically below according to the classification of Struwe *et al.* (2002). For an overview of the pollen features of genera with pollen in monads or tetrads see Tables 4.3 and 4.4, respectively.

Most taxa in Gentianeae *sensu* Gilg with single grains have radially symmetrical, 3-colporate (rarely colpate), striate, striato-reticulate to reticulate pollen grains. *Rusbyanthus sensu* Gilg (1895; now *Macrocarpaea* spp.) has 3-colporate-porate, verrucose pollen grains and was classified in a tribe of its own, the Rusbyantheae. The genera with monads formerly placed in Gentianeae-Tachiinae *sensu* Gilg that are reviewed here generally have 3-colporate, coarsely to finely reticulate, striato-reticulate, or microreticulate to perforate pollen grains. Scabrate, microspinose, smooth, or verrucose-

Table 4.2. *Pollen types in the Gentianaceae (after Nilsson, 1970, and this paper)*

	Pollen types	Taxa
Monads	*Corymbosa*-type	*Macrocarpaea cinchonifolia*
	Glabra-type	*Macrocarpaea* spp.
	Fagraea berteroana-type	*Fagraea berteroana*
	Fagraea fragrans-type	*Fagraea fragrans, F. elliptica*
	Fagraea fragrans subtype *racemosa*	*Fagraea racemosa*
	Fagraea longiflora-type	*Fagraea longiflora*
	Fagraea ceilanica-type	*Fagraea ceilanica, F. involucrata*
	Longifolius-type	*Lisianthius, Macrocarpaea loranthoides*
	Potalia-type	*Anthocleista, Potalia*
Tetrads	*Aripuana*-type	*Aripuana*
	Caerulescens-type	*Tetrapollinia*
	Chelonoides-type	*Adenolisianthus, Chelonanthus* spp., *Rogersonanthus*
	Coutoubea-type	*Coutoubea, Schultesia, Xestaea*
	Deianira-type	*Deianira*
	Helia-type	*Helia*
	Lagenanthus-type	*Lagenanthus*
	Lehmanniella-type	*Lehmanniella*
	Neblinantha-type	*Neblinantha*
	Prepusa-type	*Prepusa, Senaea*
	Purdieanthus-type	*Purdieanthus*
	Sipapoantha-type	*Sipapoantha*
	Speciosus-type	*Calolisianthus* spp.
	Symbolanthus-type	*Symbolanthus, Wurdackanthus*
Polyads	*Amplissimus*-type	*Calolisianthus* spp.
	Imthurnianus-type	*Celiantha*
	Irlbachia-type	*Irlbachia*
	Uliginosus-type	*Chelonanthus* spp.

pilate pollen grains are less common. The size (polar axis) varies from less than 20 μm to up to 70 μm and the shape ranges from spheroidal (or even suboblate) to prolate.

Gilg (1895) distinguished 15 tetrad-bearing genera in his tribe Helieae. Maguire (1981, 1985) and Maguire and Boom (1989) created six new genera with pollen in tetrads or polyads, i.e., *Celiantha, Neblinantha, Rogersonanthus, Sipapoantha, Tetrapollinia,* and *Wurdackanthus*. To these may be added the recently described genus *Aripuana* (Struwe *et al.*, 1997).

Table 4.3. Pollen morphological data and vouchers for Gentianaceae taxa with monads (for footnotes see p. 393.)

Taxon	Country[a]	Vouchers[b]	P[c]	E[d]	Size range (μm)	P/E[e]	P/E range	Exine pattern[f]	Slide no.[g]	Pollen type
Anthocleista R. Br.										
A. amplexicaulis	Madagascar	*Pettersson 610* (UPS)	26	29	25–26×26–31	0.9	0.9–1.0	smooth, scabr. perf.	24160	*Potalia*
A. madagascariensis	Madagascar	*Schlieben 8083* (HBG)	28	29	25–30×24–33	1.0	0.9–1.2	smooth, scabr. perf.	24676	*Potalia*
A. microphylla	Equatorial Guinea	*Mildbraed 6434* (HBG)	37	42	36–37×40–44	0.8	0.8–0.9	smooth, scabr. perf.	24677	*Potalia*
Chorisepalum Gleason & Wodehouse										
C. carnosum	Venezuela	*Maguire et al. 53823* (NY)	33	25	28–41×20–33	1.32	1.2–1.7	ret.(–co. ret.)	24091	cf. *Glabra*
C. psychotrioides var. *psychotrioides*	Guyana	*Tillett et al. 43968* (NY)	39	35	34–42×32–41	1.11	1.0–1.2	co. ret.–ret.	12796	cf. *Glabra*
C. psychotrioides var. *psychotrioides*	Guyana	*Tillett et al. 44829* (NY)	47	42	45–50×40–46	1.12	1.0–1.2	co. ret.–ret.	12795	cf. *Glabra*
C. psychotrioides var. *psychotrioides*	Venezuela	*Steyermark 58668* (F)	29	26	28–32×24–28	1.12	1.0–1.2	co. ret.–ret.	7322	cf. *Glabra*
C. rotundifolium	Venezuela	*Steyermark 59730* (F)	35	33	33–36×31–35	1.06	1.0–1.1	ret.(–co. ret.)	7323	cf. *Glabra*
C. rotundifolium	Venezuela	*Steyermark 59730* (NY)	33	31	29–36×28–33	1.06	1.0–1.1	ret.(–co. ret.)	6164	cf. *Glabra*
C. sipapoanum	Venezuela	*Maguire & Politi 27921* (NY)	50	41	45–54×40–44	1.22	1.1–1.4	ret.–co. ret.	13803	cf. *Glabra*
Congolanthus A. Raynal										
C. longidens	Nigeria	*Daramola FHI 46923* (K)	38	40	36–40×39–42	0.95	0.9–1.0	fi. ret.	6575	
C. longidens	Congo	*Robyns 508* (NY)	37	40	34–41×33–44	0.93	0.8–1.1	fi. ret.	5871	
C. longidens	Congo	*Louis 1885* (S)	34	37	31–36×34–40	0.92	0.8–1.0	fi. ret.	24168	
Curtia Cham. & Schtldl.										
C. confusa	Brazil	*Smith & Klein 10617* (NY)	20	22	18–22×20–24	0.91	0.9–1.1	fi. ret., perf.	2256	

Species	Country	Collector						
C. confusa	Brazil	Dusén 17521 (MO)	20	20	19–23 × 17–22	1.00	0.9–1.4	fi. ret., perf.	3762
C. confusa	Venezuela	Huber et al. 5688 (U)	19	18	18–23 × 13–22	1.06	0.9–1.4	fi. ret., perf.	18340
C. diffusa	Brazil	Ynes Mexia 5750 (U)	21	22	21–22 × 18–22	0.95	1.0–1.3	fi. ret., perf.	18336
C. diffusa	Brazil	Duarte 2468 (U)	23	23	20–25 × 20–24	1.00	1.0–1.2	fi. ret., perf.	18337
C. diffusa	Brazil	Duarte 9030 (U)	26	29	24–30 × 26–30	0.90	0.8–1.1	fi. ret., perf.	18344
C. malmeana	Brazil	Dusén 8003 (S)	21	17	20–22 × 17–19	1.24	1.1–1.3	fi. ret., perf.	1330
C. obtusifolia	Brazil	Prance et al. 15500 (NY)	19	19	19–20 × 17–21	1.00	0.8–1.2	fi. ret., perf.	24095
C. obtusifolia	Venezuela	Huber 1578 (U)	21	18	20–23 × 14–22	1.17	1.0–1.5	fi. ret., perf.	18356
C. patula	Brazil	Glaziou 19656 (K)	20	15	20–21 × 13–15	1.33	1.3–1.6	fi. ret., perf.	6064
C. patula	Brazil	Irwin et al. 14124 (U)	23	20	22–24 × 17–21	1.15	1.0–1.4	fi. ret., perf.	4606
C. patula	Brazil	Barreto & Brade s.n. (R)	25	20	23–28 × 19–21	1.25	1.1–1.4	fi. ret., perf.	13919
C. pusilla	Brazil	Herb. R 15244 (R)	26	20	25–28 × 19–22	1.30	1.2–1.4	fi. ret., perf.	13920
C. quadrifolia	Venezuela	Steyermark & Bunting 102828 (NY)	22	22	21–23 × 19–24	1.00	1.0–1.2	fi. ret., perf.	17166
C. quadrifolia	Venezuela	Wurdack & Adderley 43626 (NY)	25	22	23–26 × 20–23	1.14	1.0–1.2	fi. ret., perf.	13795
C. stricta	Brazil	Glaziou 15245 (R 11665)	31	30	30–34 × 29–31	1.03	1.0–1.1	fi. ret., perf.	13921
C. tenella	Brazil	Lindeman & Haas 4995 (U)	18	18	17–22 × 17–20	1.00	0.8–1.2	fi. ret., perf.	6071
C. tenella	Brazil	Ratter & Fonseca 2803 (E)	19	19	18–20 × 17–20	1.00	0.9–1.1	fi. ret., perf.	17228
C. tenella	Mexico	Gr. Bitt. 5040 (US)	23	21	20–24 × 18–22	1.10	1.0–1.2	fi. ret., perf.	10436
C. tenella	Panama	Allen 994 (MO)	21	13	20–22 × 11–14	1.62	1.4–1.8	fi. ret., perf.	8765
C. tenuifolia	Guyana	Harrison 1245 (K)	18	18	15–20 × 17–20	1.00	0.9–1.3	fi. ret., perf.	6073
C. tenuifolia	Panama	Allen 994 (U)	20	17	18–22 × 15–19	1.18	1.1–1.4	fi. ret., perf.	6072
C. tenuifolia	Surinam	Lindeman 4377 (U)	16	15	14–19 × 13–15	1.07	0.9–1.3	fi. ret., perf.	3763
C. tenuifolia	Surinam	Forestry Bureau s.n. (U)	19	17	18–22 × 15–19	1.12	0.9–1.3	fi. ret., perf.	4413
C. tenuifolia	Venezuela	Maguire et al. 30910 (NY)	22	19	21–22 × 17–21	1.16	1.0–1.4	fi. ret., perf.	13797
C. tenuifolia ssp. tenella	Brazil	Malme s.n. (S)	21	19	19–22 × 19–21	1.11	0.9–1.1	fi. ret., perf.	24403

Table 4.3. (cont.)

Taxon	Country[a]	Vouchers[b]	P[c]	E[d]	Size range (µm)	P/E[e]	P/E range	Exine pattern[f]	Slide no.[g]	Pollen type
C. tenuifolia ssp. tenuifolia[h]	Brazil	Malme s.n. (S)	18	16	17–20×15–17	1.13	1.1–1.2	fi. ret., perf.	24402	
C. tenuifolia ssp. tenuifolia[i]	Brazil	Malme s.n. (S)	19	18	19–20×18–19	1.06	1.0–1.1	fi. ret. spi. fov.	24402	
C. verticillaris	Brazil	Plowman 9951 (U)	26	24	24–32×21–25	1.08	0.8–1.5	fi. ret., perf.	18335	
C. verticillaris	Brazil	Hatschbach 29978 (U)	28	25	22–31×21–26	1.12	0.8–1.3	fi. ret., perf.	18343	
C. verticillaris	Brazil	Hage et al. 296 (U)	30	28	25–32×25–30	1.07	0.9–1.1	fi. ret., perf.	18346	
C. verticillaris	Brazil	Harley 22534 (U)	28	23	24–31×22–24	1.22	1.1–1.4	fi. ret., perf.	18345	
Fagraea Thunb.										
F. berteroana	USA (Hawaii)	Motley s.n. (NY)	25	29	23–26×28–32	0.9	0.8–0.9	fi.ret.	24161	*Berteroana*
F. ceilanica	India	Saldanha 13586 (NY)	40	40	37–41×37–43	1.0	0.9–1.1	co. ret.	24180	*Ceilanica*
F. elliptica	Indonesia	Burley et al. 4335 (NY)	20	18	18–21×17–20	1.1	0.9–1.3	str. ret.	24178	*Fragrans*
F. fragrans	Philippines	Roque s.n. (NY)	20	18	19–22×17–20	1.1	1.0–1.3	str. ret.	24177	*Fragrans*
F. involucrata	Borneo	J. & Clemens 31661 (NY)	38	38	34–44×32–43	1.0	0.9–1.0	co. ret.	24181	*Ceilanica*
F. longiflora	Philippines	Edaño 708 (NY)	27	29	23–31×25–31	0.9	0.7–1.0	fi. ret.–fov.	24182	*Longiflora*
F. racemosa	Sumatra	Rahmat Si Toroes 3193 (NY)	29	27	28–31×23–29	1.1	1.0–1.2	str. ret.	24179	*Fragrans* subtype *Racemosa*
Gentianothamnus Humbert										
G. madagascariensis	Madagascar	Cours 3446 (P)	21	18	21–23×14–19	1.17	1.2–1.5	fi. ret.–perf. rug.	6931	
G. madagascariensis	Madagascar	Humbert & Cours 23715 (P)	22	18	20–24×14–21	1.22	0.9–1.6	fi. ret.–perf. rug.	6933	
G. madagascariensis	Madagascar	Humbert et al. 24817 (P)	21	17	19–24×13–20	1.24	0.9–1.6	fi. ret.–perf. rug.	6934	
G. madagascariensis	Madagascar	Humbert 24017 (MO)	17	17	14–20×17–20	1.00	0.9–1.1	fi. ret.–perf. rug.	24154	
G. madagascariensis	Madagascar	Humbert 23715 (MO)	21	17	18–24×14–20	1.24	1.0–1.7	fi. ret.–perf. rug.	24153	

Hockinia Gardner

Species	Country	Collection			Dimensions			Notes	Number	Section
H. montana[b]	Brazil	Glaziou 6899 (K)	28	24	25–30 × 22–26	1.17	1.1–1.3	fi. ret., perf.	4286	
H. montana[b]	Brazil	Herb. R 11688 (R)	34	30	33–35 × 29–33	1.13	1.0–1.2	fi. ret., perf.	13927	
H. montana[b]	Brazil	Brade 21208 (RB)	27	26	25–29 × 23–28	1.04	1.0–1.1	fi. ret., perf.	24654	
H. montana[i]	Brazil	Hatschbach 31428 (U)	39	38	37–41 × 37–40	1.03	0.9–1.1	fi. ret., perf.	17158	
H. montana[i]	Brazil	USNM 14718 (US)	31	27	30–33 × 24–29	1.15	1.0–1.3	fi. ret., perf. mi. spi.	5367	

Lisianthius P. Browne

Species	Country	Collection			Dimensions			Notes	Number	Section
L. acuminatus	Mexico	Breedlove 10062 (F)	41	28	40–44 × 25–31	1.46	1.4–1.7	ret.	6167	Longifolius
L. adamsii	Jamaica	Weaver 1293 (DUKE)	40	29	33–44 × 28–33	1.38	1.1–1.5	ret.	6067	Longifolius
L. auratus	Honduras	Williams & Williams 18382 (US)	42	30	40–44 × 28–33	1.40	1.2–1.5	ret.	4545	Longifolius
L. auratus	Nicaragua	Molina 15037 (F)	35	32	30–37 × 28–34	1.09	0.9–1.3	ret.–fi. ret.	6168	Longifolius
L. axillaris	Belize	Karling 54 (K)	36	28	35–39 × 24–35	1.29	1.0–1.5	ret.	4287	Longifolius
L. axillaris	Guatemala	Lundell 3159 (S)	32	23	31–34 × 22–24	1.39	1.3–1.6	ret.	633	Longifolius
L. brevidentatus	Guatemala	Türckheim 931 (L)	41	28	39–44 × 25–31	1.46	1.3–1.7	ret.	4260	Longifolius
L. capitatus	Jamaica	Harris 12030 (S)	45	30	42–51 × 28–32	1.50	1.4–1.7	ret.	3729	Longifolius
L. capitatus	Jamaica	Webster et al. 8362 (BM)	43	33	41–46 × 29–33	1.30	1.0–1.5	ret.	4303	Longifolius
L. collinus	Belize	Schipp 1205 (BM)	40	30	37–42 × 25–33	1.33	1.2–1.5	ret.	4400	Longifolius
L. congestus	Panama	Allen 783 (U)	35	26	33–37 × 24–29	1.35	1.2–1.5	fi. ret.–ret.	509	Longifolius
L. cordifolius	Jamaica	Harris 12315 (BM)	42	30	39–44 × 28–32	1.40	1.3–1.6	fi. ret., str. ret.	7356	Longifolius
L. cordifolius	Jamaica	Stearn 2 (BM)	35	33	33–40 × 31–35	1.06	0.9–1.3	ret., str.–ret.	4296	Longifolius
L. cordifolius	Jamaica	Webster & Wilson 4860 (BM)	36	30	34–39 × 28–33	1.20	1.1–1.3	ret., str.–ret.	24380	Longifolius
L. cordifolius	Jamaica	Yuncker 17374 (S)	38	32	35–41 × 29–35	1.19	1.1–1.4	ret.	24381	Longifolius
L. domingensis	Dominican Republic	Marcano 3658 (US)	47	30	44–50 × 29–34	1.57	1.3–1.7	ret.–co. ret.	6061	Longifolius
L. domingensis	Dominican Republic	Ekman 11516 (S)	42	32	39–46 × 29–36	1.31	1.2–1.5	ret.–co. ret.	3384	Longifolius
L. exsertus	Jamaica	Proctor 22114 (BM)	39	28	36–44 × 25–31	1.39	1.3–1.5	perf.(–fi. ret.)	7354	Longifolius
L. exsertus	Jamaica	Stearn 410 (K)	39	27	36–41 × 25–30	1.44	1.3–1.5	perf.(–fi. ret.)	4295	Longifolius

Table 4.3. (cont.)

Taxon	Country[a]	Vouchers[b]	P[c]	E[d]	Size range (μm)	P/E[e]	P/E range	Exine pattern[f]	Slide no.[g]	Pollen type
L. glandulosus	Cuba	Morton & Alain 9000 (US)	43	31	41–46×28–34	1.39	1.2–1.6	ret.	5982	Longifolius
L. glandulosus	Cuba	Ekman 1613 (K)	43	33	41–45×31–35	1.30	1.2–1.4	ret. (co. ret.)	4267	Longifolius
L. jefensis	Panama	Weaver & Foster 1481 (DUKE)	41	36	34–45×33–41	1.14	0.8–1.4	co. ret.	5978	Longifolius
L. latifolius	Jamaica	Hawkes et al. 2241 (K)	46	33	44–47×30–37	1.39	1.3–1.6	ret. (co. ret.)	4422	Longifolius
L. latifolius	Jamaica	Krug & Urban 662 (L)	48	32	46–50×30–34	1.50	1.4–1.6	ret.	4417	Longifolius
L. laxiflorus	Puerto Rico	Wagner 2 (U)	39	31	35–42×28–33	1.26	1.2–1.4	ret.	4435	Longifolius
L. laxiflorus	Puerto Rico	Sintensis 6077 (BM)	38	32	39–44×30–37	1.19	1.0–1.4	ret. (co. ret.)	4325	Longifolius
L. longifolius	Jamaica	Maxwell s.n. (BM)	44	29	42–46×28–31	1.52	1.4–1.6	ret.	7355	Longifolius
L. longifolius	Jamaica	Webster et al. 8432 (BM)	39	28	39–42×25–31	1.39	1.3–1.6	ret.	4399	Longifolius
L. nigrescens var. cuspidatus	Guatemala	Steyermark 50682 (F)	37	33	34–40×30–35	1.12	1.0–1.2	ret.	6068	Longifolius
L. nigrescens var. cuspidatus	Jamaica	Hort. Bot. Utrecht 0121118 (U)	44	29	40–54×25–33	1.52	1.2–1.9	ret.	4261	Longifolius
L. nigrescens var. cuspidatus	Mexico	Ghiesbreght 702 (K)	34	26	31–37×23–28	1.31	1.2–1.5	ret.	4326	Longifolius
L. peduncularis	Panama	Allen & Allen 4187 (MO)	41	32	41–43×30–34	1.28	1.2–1.4	ret.	4546	Longifolius
L. saponarioides	Guatemala	Heyde & Lux 2921 (K)	40	27	39–42×24–30	1.48	1.3–1.7	ret.	4300	Longifolius
L. scopulinus	Panama	Lewis et al. 2799 (MO)	34	34	31–39×29–39	1.00	0.8–1.3	ret.	1208	Longifolius
L. seemannii	Colombia	Haught 4652 (K)	38	26	36–41×24–30	1.46	1.3–1.6	fi. ret.	4420	Longifolius
L. seemannii	Costa Rica	Skutch 4109 (K)	40	28	34–44×25–31	1.43	1.3–1.6	ret.–fi. ret.	4301	Longifolius
L. silenifolius	Cuba	Alain & Acuna 1106 (US)	40	30	35–43×25–34	1.33	1.2–1.5	ret.	6065	Longifolius
L. silenifolius	Cuba	Ekman 12940 (K)	37	29	34–41×28–31	1.28	1.1–1.4	ret.	4408	Longifolius
L. silenifolius	Cuba	Bisse et al. 35146 (HAJB)	34	27	32–37×24–32	1.26	1.0–1.5	ret.	24141	Longifolius
L. skinneri	Costa Rica	Wilbur & Stone 10257 (DUKE)	41	34	39–47×31–39	1.21	1.0–1.4	ret.(–co. ret.)	6059	Longifolius

Species	Country	Collection								Type
L. skinnerii	Panama	*Weaver 1482* (DUKE)	42	32	41–46 × 31–34	1.31	1.2–1.5	ret.(–co. ret.)	6062	*Longifolius*
L. troyanus	Jamaica	*Weaver & Foster 1272* (DUKE)	39	29	35–46 × 26–32	1.34	1.2–1.6	fi. ret.–ret.	5983	*Longifolius*
L. umbellatus	Jamaica	*Perkins 1321* (GH)	40	32	39–43 × 29–35	1.25	1.2–1.4	ret.	6066	*Longifolius*
L. umbellatus	Jamaica	*Hort. Bot. Utrecht 012123* (U)	38	31	34–42 × 28–39	1.23	1.1–1.4	ret.	4328	*Longifolius*
L. viscidiflorus	Guatemala	*Contreras 4725* (LL)	38	27	35–40 × 25–30	1.41	1.3–1.5	ret.	5990	*Longifolius*
Macrocarpaea (Griseb.) Gilg										
M. affinis	Colombia	*Uribe 2432* (F)	29	33	26–32 × 31–35	0.88	0.8–0.9	ret.–co. ret.	700	*Glabra*
M. bangiana	Bolivia	*Bang 520* (US)	28	28	25–31 × 26–32	1.00	0.9–1.1	ret.–co. ret.	690	*Glabra*
M. bracteata	Venezuela	*Dorr et al. 4990* (NY)	38	37	33–44 × 29–42	1.03	0.8–1.2	ret.–co. ret.	24185	*Glabra*
M. browallioides	Panama	*Allen 4932* (MO)	43	49	41–45 × 46–53	0.88	0.8–0.9	ret.–co. ret.	7329	*Glabra*
M. cinchonifolia	Bolivia	*Rusby 1173* (NY)	32	33	29–34 × 31–35	0.97	0.9–1.0	verr., muri visible	4112	*Corymbosa*
M. cinchonifolia	Bolivia	*Rusby 1173* (BM)	32	33	30–35 × 29–39	0.97	0.9–1.2	verr., muri visible	4114	*Corymbosa*
M. cinchonifolia	Bolivia	*Rusby s.n.* (PH)	31	31	28–33 × 28–33	1.00	0.9–1.1	verr., muri visible	3347	*Corymbosa*
M. cinchonifolia	Bolivia	*Buchtien 5639* (NY)	31	31	28–33 × 29–34	1.00	0.9–1.1	verr., muri visible	2180	*Corymbosa*
M. cochabambensis	Bolivia	*Steinbach 8992* (F)	31	30	30–33 × 29–33	1.03	0.9–1.1	ret.–co. ret.	725	*Glabra*
M. densiflora	Colombia	*Lehmann 2087* (US)	33	31	30–36 × 28–33	1.06	1.0–1.2	ret.–co. ret.	1172	*Glabra*
M. domingensis	Dominican Republic	*Allard 18215* (S)	28	30	24–31 × 28–33	0.93	0.8–1.0	ret.–co. ret.	1164	*Glabra*
M. duquei	Colombia	*Cuatrecasas 21824* (F)	34	31	31–39 × 28–35	1.10	1.0–1.2	ret.–co. ret.	1752	*Glabra*
M. duquei	Colombia	*Cuatrecasas 21824* (US)	34	33	31–39 × 29–39	1.03	1.0–1.1	ret.–co. ret.	488	*Glabra*
M. glabra	Colombia	*Daniel 1450* (F)	31	32	26–35 × 29–34	0.97	0.8–1.2	ret.–co. ret.	1750	*Glabra*
M. glabra	Colombia	*Haught 5105* (US)	31	31	31–32 × 28–35	1.00	0.9–1.1	ret.–co. ret.	981	*Glabra*
M. glabra	Colombia	*Holton 470* (NY)	29	34	23–33 × 31–37	0.85	0.8–1.0	ret.–co. ret.	1912	*Glabra*
M. glabra	Colombia	*Killip 34074* (S)	33	34	29–35 × 31–40	0.97	0.8–1.1	ret.–co. ret.	1910	*Glabra*
M. glabra	Colombia	*Pennell 2195* (NY)	38	34	35–41 × 33–36	1.12	1.1–1.2	ret.–co. ret.	1911	*Glabra*
M. loranthoides	Peru	*Sandeman s.n.* (K)	41	31	36–44 × 30–34	1.32	1.1–1.4	ret.	4111	*Longifolius*
M. loranthoides	Peru	*Matthews 1315* (K)	41	32	36–43 × 24–34	1.28	0.9–1.7	ret.	7301	*Longifolius*
M. macrophylla	Colombia	*Killip & García 33650* (S)	28	33	26–31 × 29–36	0.85	0.8–0.9	ret.–co. ret.	1928	*Glabra*

Table 4.3. (cont.)

Taxon	Country[a]	Vouchers[b]	P[c]	E[d]	Size range (μm)	P/E[e]	P/E range	Exine pattern[f]	Slide no.[g]	Pollen type
M. macrophylla	Colombia	*Pennell 10670* (PH)	28	31	25–30 × 28–33	0.90	0.8–1.0	ret.–co. ret.	1853	*Glabra*
M. micrantha	Peru	*Klug 3138* (F)	29	33	26–32 × 31–34	0.88	0.8–0.9	ret.–co. ret.	652	*Glabra*
M. micrantha	Peru	*Klug 3138* (S)	28	31	24–30 × 28–33	0.90	0.8–1.0	ret.–co. ret.	730	*Glabra*
M. obtusifolia	Brazil	*Barreto 8845* (F)	34	38	33–35 × 33–41	0.89	0.8–1.0	ret.(–co. ret.)	729	*Glabra*
M. obtusifolia	Brazil	*Wilkes Exp. s.n.* (US)	29	32	25–33 × 28–35	0.91	0.8–1.0	ret.–co. ret.	653	*Glabra*
M. ovalis	Colombia	*Lehmann 5450* (F)	35	36	29–39 × 29–39	0.97	0.8–1.2	ret.–co. ret.	2718	*Glabra*
M. ovalis	Colombia	*Purdie s.n.* (K)	34	35	32–36 × 33–37	0.97	0.9–1.0	ret.–co. ret.	3346	*Glabra*
M. ovalis	Ecuador	*André 4549* (NY)	30	32	26–33 × 30–37	0.94	0.8–1.0	ret.–co. ret.	3183	*Glabra*
M. pachyphylla	Ecuador	*André 4513* (F)	38	39	35–43 × 32–42	0.97	0.9–1.1	ret.–co. ret.	2719	*Glabra*
M. pachyphylla	Colombia	*Cuatrecasas 11785* (US)	33	35	31–36 × 33–37	0.94	0.8–1.1	ret.–co. ret.	727	*Glabra*
M. revoluta	Peru	*Williams 7596* (F)	30	32	29–33 × 30–33	0.94	0.9–1.1	ret.–co. ret.	726	*Glabra*
M. rubra	Brazil	*Dusén 6965* (F)	28	28	25–32 × 23–32	1.00	0.8–1.4	ret.(–co. ret.)	1851	*Glabra*
M. rubra	Brazil	*Dusén 6965* (S)	28	32	26–32 × 31–33	0.88	0.8–1.0	ret.(–co. ret.)	1915	*Glabra*
M. rubra	Brazil	*Dusén 15790* (MO)	32	33	29–34 × 32–35	0.97	0.9–1.0	ret.(–co. ret.)	712	*Glabra*
M. rubra	Brazil	*Dusén 15790* (NY)	28	33	23–31 × 28–39	0.85	0.7–1.1	ret.	1913	*Glabra*
M. rubra	Brazil	*Dusén 15790* (U)	35	35	32–39 × 31–39	1.00	0.9–1.1	ret.(–co. ret.)	72	*Glabra*
M. stenophylla	Peru	*Williams 7582* (F)	30	31	28–35 × 28–35	0.97	0.9–1.2	ret.(–co. ret.)	724	*Glabra*
M. thamnoides	Jamaica	*Harris 5353* (US)	29	28	25–33 × 25–33	1.04	0.9–1.3	ret.–co. ret.	654	*Glabra*
M. thamnoides	Jamaica	*Harris 7718* (F)	32	33	30–33 × 32–36	0.97	0.8–1.0	ret.–co. ret.	2805	*Glabra*
M. valerii	Costa Rica	*Smith 1001* (F)	29	34	26–31 × 32–35	0.85	0.8–0.9	ret.–co. ret.	979	*Glabra*
M. sp. nov. 1[j]	Peru	*Weberbauer 7921* (F)	34	34	33–35 × 33–36	1.00	0.9–1.1	verr.	731	*Corymbosa*
M. sp. nov. 1[k]	Peru	*Vargas 4196* (US)	32	31	31–34 × 30–33	1.03	0.9–1.1	verr., muri visible	742	*Corymbosa*
M. sp. nov. 2[l]	Colombia	*Cuatrecasas 14986* (F)	29	31	26–33 × 29–33	0.94	0.8–1.2	ret.–co. ret.	1751	*Glabra*
M. sp. nov. 3[m]	Ecuador	*Steyermark 54805* (F)	32	33	30–34 × 31–36	0.97	0.9–1.1	ret.–co. ret.	952	*Glabra*
M. sp. nov. 4[n]	Peru	*Woytkowski 35417* (F)	34	35	33–36 × 33–36	0.97	0.9–1.0	verr.	2706	*Corymbosa*
M. sp. nov. 5[o]	Colombia	*Cuatrecasas et al. 12271* (F)	33	36	26–35 × 34–40	0.92	0.8–1.0	ret.–co. ret.	1161	*Glabra*

Species	Country	Collection								
M. sp. nov. 5º	Colombia	Cuatrecasas et al. 12271 (MO)	33	32	31–36×30–36	1.03	0.9–1.2	ret.–co. ret.	1930	Glabra
M. sp. nov. 6º	Peru	Weberbauer 6116 (F)	37	38	34–40×34–41	0.97	0.9–1.1	ret.–co. ret.	2720	Glabra
M. sp. nov. 7ᵠ	Ecuador	Steyermark 53548 (US)	30	35	26–33×29–35	0.86	0.8–1.0	ret.–co. ret.	699	Glabra
M. sp. nov. 7ᵠ	Ecuador	Steyermark 54812 (F)	30	34	28–33×33–36	0.88	0.8–1.0	ret.–co. ret.	951	Glabra
M. sp. nov. 8ᵠ	Peru	Stork & Horton 10118 (K)	34	36	31–36×33–39	0.94	0.9–1.1	ret.–co. ret.	3345	Glabra
M. sp. nov. 8ᵠ	Peru	Stork & Horton 10118 (NA)	35	31	33–39×30–33	1.13	1.1–1.2	ret.–co. ret.	1653	Glabra
M. sp. nov. 9ᵠ	Peru	Stork & Horton 9913 (K)	34	35	31–35×32–39	0.97	0.8–1.0	verr., muri visible	4110	Corymbosa
M. sp.ˢ	Peru	Ruiz & Pavón s.n. (F)	35	34	31–35×31–36	1.03	0.8–1.1	verr.	982	Corymbosa
Neurotheca Salisb. ex Benth. in Benth. & Hook.f.										
N. congolana	Angola	Baum 604 (K)	56	40	53–59×37–44	1.40	1.2–1.5	fi. ret.–ret.	6574	
N. congolana	Congo	Lebrun 6478 (BR)	55	39	50–61×35–43	1.41	1.3–1.5	fi. ret.–ret.	6745	
N. congolana	Congo	Bavicchi 412 (BR)	59	39	57–65×34–44	1.51	1.3–1.7	fi. ret.–ret.	6744	
N. congolana	Congo	Bequaert 7669 (BR)	58	38	56–62×33–40	1.53	1.4–1.7	fi. ret.–ret.	6743	
N. corymbosa	Gabon	Hallé & Villiers 5528 (P)	52	39	50–55×39	1.33	1.3–1.4	fi. ret.–ret.	6562	
N. corymbosa	Gabon	Thollon 43 (K)	59	43	55–64×39–48	1.37	1.2–1.5	fi. ret.–ret.	6567	
N. loeselioides	Brazil	Ule 8978 (L)	60	41	54–65×35–44	1.46	1.3–1.8	fi. ret.–ret.	1194	
N. loeselioides	Brazil	Nee 34765 (NY)	51	38	41–60×33–43	1.34	1.2–1.6	fi. ret.–ret.	24155	
N. loeselioides	Guyana	Tutin 618 (K)	55	39	52–62×39–42	1.41	1.3–1.5	fi. ret.–ret.	6568	
N. loeselioides	Guyana	Graham no. X.7 (K)	54	42	51–58×39–44	1.29	1.2–1.4	fi. ret.–ret.	6397	
N. loeselioides	Ghana	Harris s.n. (K)	55	36	51–59×33–39	1.53	1.3–1.7	fi. ret.–ret.	6399	
N. loeselioides	Ghana	Hall 865 (K)	62	47	58–66×40–54	1.32	1.2–1.6	fi. ret.–ret.	7037	
N. loeselioides	French Guinea	HLB 950.348–380 (L)	65	45	63–67×41–48	1.44	1.3–1.6	fi. ret.–ret.	1196	
N. loeselioides	Sierra Leone	Dawe WA 429 (K)	53	42	50–58×39–45	1.26	1.1–1.4	fi. ret.–ret.	6283	
N. loeselioides	Sierra Leone	W. H. & A. H. Brown s.n. (US)	67	49	64–69×44–55	1.37	1.2–1.6	fi. ret.–ret.	3205	
N. loeselioides	Surinam	J. & P. A. Florschütz 811 (U)	54	43	54–58×39–46	1.26	1.1–1.5	fi. ret.–ret.	3766	
N. loeselioides	Surinam	Lindeman 6520 (U)	58	41	52–64×35–50	1.41	1.3–1.6	fi. ret.–ret.	3767	

Table 4.3. (cont.)

Taxon	Country[a]	Vouchers[b]	P[c]	E[d]	Size range (μm)	P/E[e]	P/E range	Exine pattern[f]	Slide no.[g]	Pollen type
N. loeselioides	Surinam	*Lanjouw & Lindeman 157* (NY)	56	41	53–58×39–44	1.37	1.3–1.5	fi. ret.–ret.	5870	
N. loeselioides	Surinam	*van Donselaar 3700* (U)	50	40	48–55×33–50	1.25	1.0–1.6	fi. ret.–ret.	4812	
N. loeselioides ssp. *loeselioides*	Burundi	*Reekmans 9114* (S)	51	39	47–55×33–45	1.31	1.2–1.5	fi. ret.–ret.	24156	
N. loeselioides ssp. *loeselioides*	Senegal	*Berhaut 4411* (P)	52	39	47–57×35–45	1.33	1.3–1.6	fi. ret.–ret.	6557	
N. loeselioides ssp. *loeselioides*	Sierra Leone	*Melville & Hooker 281* (K)	65	45	63–69×40–46	1.44	1.4–1.6	fi. ret.–ret.	7211	
N. loeselioides ssp. *robusta*	French Guinea	*Schnell 2149* (P)	61	42	55–68×39–46	1.45	1.3–1.6	fi. ret.–ret.	6548	
Potalia Aubl.										
P. amara	Brazil	*Mori et al. 15785* (NY)	39	32	33–42×28–35	1.2	1.1–1.5	smooth, dist. protrud. marg.	24162	*Potalia*
P. "coronata", ined.	Peru	*van der Werff et al. 10254* (MO)	39	35	33–44×24–43	1.1	1.0–1.4	smooth, protrud. marg.	24163	*Potalia*
P. elegans	Venezuela	*Maguire et al. 29356* (NY)	35	32	33–41×28–39	1.1	1.0–1.3	smooth, perf. protrud. marg.	24164	*Potalia*
P. maguireorum[i]	Brazil	*Kubitzski et al. 79-216* (NY)	31	35	28–36×31–43	0.9	0.8–0.9	smooth, protrud. marg.	24165	*Potalia*
P. resinifera	Peru	*Vásquez & Jaramillo 7169* (MO)	36	33	33–39×29–37	1.1	1.0–1.3	smooth, protrud. marg.	24166	*Potalia*
P. "turbinata", ined.	Costa Rica	*Stevens et al. 25009* (MO)	38	34	34–43×30–43	1.1	1.0–1.3	smooth, protrud. marg.	24167	*Potalia*
Tachia Aubl.										
T. grandiflora	Brazil	*Prance et al. 5042* (S)	34	27	32–37×23–31	1.26	1.1–1.4	ret.–co. ret.	8050	cf. *Glabra*

T. grandifolia	Venezuela	*Maguire et al. 37496 (NY)*	45	40	43–47×37–44	1.13	1.0–1.2	co. ret.	13782	cf. *Glabra*
T. guianensis	Guyana	*Forest Dept. Guyana 3323* (K)	29	30	25–35×29–35	0.97	0.8–1.2	ret.	6209	cf. *Glabra*
T. guianensis	Guyana	*Martyn 207* (K)	34	30	32–35×28–32	1.13	1.0–1.2	ret.–co. ret.	3783	cf. *Glabra*
T. guianensis	Guyana	*Tutin 303* (BM)	32	26	31–34×23–28	1.23	1.1–1.4	ret.–co. ret.	442	cf. *Glabra*
T. guianensis	Guyana	*Sandwith 1111* (U)	32	26	30–33×24–29	1.23	1.0–1.4	ret., polar caps	6126	cf. *Glabra*
T. guianensis	Guyana	*Sandwith 1111* (K)	35	26	28–35×24–28	1.35	1.3–1.5	ret.–co. ret.	4819	cf. *Glabra*
T. guianensis	Surinam	*Wessels Boer 1540* (U)	35	30	31–37×26–32	1.17	1.0–1.3	ret.(–co. ret.)	4409	cf. *Glabra*
T. guianensis	Surinam	*Maguire 24611* (NY)	44	35	43–46×33–37	1.26	1.2–1.4	ret.–co. ret.	13781	cf. *Glabra*
T. guianensis	Surinam	*Maguire 24098* (U)	38	32	35–40×30–33	1.19	1.1–1.3	ret.–co. ret.	6274	cf. *Glabra*
T. guianensis	Surinam	*Maguire 24098* (NY)	43	39	42–45×30–42	1.10	1.0–1.3	ret.–co. ret.	13780	cf. *Glabra*
T. loretensis	Peru	*Klug 1277* (NY)	32	37	31–33×28–34	0.86	0.9–1.1	ret.–co. ret.	6400	cf. *Glabra*
T. occidentalis	Brazil	*Ducke 22389* (U)	29	32	24–33×29–35	0.91	0.9–1.0	smooth	6275	cf. *Glabra*
T. occidentalis	Brazil	*Maas et al. 6760* (NY)	26	29	24–29×26–31	0.90	0.9–1.0	smooth	24183	cf. *Glabra*
T. occidentalis	Colombia	*Schultes & Cabrera 14670* (U)	26	28	24–28×28–29	0.93	0.8–1.0	smooth	6127	cf. *Glabra*
T. occidentalis	Peru	*Killip & Smith 26163* (NY)	28	30	25–30×29–33	0.93	0.8–1.0	smooth	3275	cf. *Glabra*
T. schomburgkiana	Guyana	*Tillett & Tillett 45447* (NY)	39	35	37–41×33–36	1.11	1.0–1.2	ret.–co. ret.	13779	cf. *Glabra*
T. schomburgkiana	Guyana	*Steyermark & Pinkus 3* (NY)	30	26	28–33×23–28	1.15	1.1–1.4	ret.(–co. ret.)	3274	cf. *Glabra*
T. schomburgkiana	Guyana	*Pinkus 227* (US)	33	24	30–36×22–26	1.38	1.2–1.6	ret.(–co. ret.)	3273	cf. *Glabra*
T. schomburgkiana	Guyana	*Forest Dept. 2779* (K)	29	25	28–31×23–26	1.16	1.1–1.3	ret.(–co. ret.)	4397	cf. *Glabra*
T. schomburgkiana	Surinam	*Wessels Boer 1211* (U)	37	38	35–39×37–41	0.97	0.9–1.0	ret.	1351	cf. *Glabra*
Tachiadenus Griseb.										
T. antaisaka	Madagascar	*Humbert 13538* (S)	29	30	25–33×26–33	0.97	0.9–1.1	fi. ret.	18430	
T. carinatus	Madagascar	*Afzelius s.n.* (S)	28	29	23–26×24–29	0.97	0.8–1.1	fi. ret.	6063	
T. carinatus	Madagascar	*Humbert 20396* (S)	32	30	31–37×26–33	1.07	1.0–1.3	fi. ret.	18432	
T. carinatus	Madagascar	*Klackenberg 871124-5* (S)	30	27	26–34×19–31	1.11	0.9–1.6	fi. ret. rug.	24144	

Table 4.3. (cont.)

Taxon	Country[a]	Vouchers[b]	P[c]	E[d]	Size range (μm)	P/E[e]	P/E range	Exine pattern[f]	Slide no.[g]	Pollen type
T. gracilis	Madagascar	Catat 2528 (P)	28	26	24–31 × 22–28	1.08	1.0–1.3	fi. ret.	6533	
T. longiflorus	Madagascar	Humbert 11973 (S)	36	27	33–39 × 24–31	1.33	1.2–1.4	fi. ret.(–ret.)	18431	
T. longiflorus	Madagascar	Serv. Forest. Madag. 3240 (S)	33	31	30–37 × 28–34	1.06	1.0–1.4	fi. ret.(–ret.)	18433	
T. longiflorus	Madagascar	Elliot 1939 (K)	35	26	32–39 × 24–30	1.35	1.2–1.5	fi. ret.(–ret.)	6125	
T. longiflorus	Madagascar	Decary 6151 (P)	33	28	31–35 × 26–31	1.18	1.1–1.2	fi. ret.(–ret.)	6532	
T. platypterus ssp. platypterus	Madagascar	Cours 5008 (P)	31	27	29–33 × 25–29	1.15	1.0–1.2	fi. ret.	6531	
T. tubiflorus	Madagascar	Capuron s.n. (S)	35	25	30–37 × 21–30	1.40	1.2–1.7	perf. rug.	18434	
T. tubiflorus	Madagascar	Geay 8957 (P)	30	28	27–34 × 26–30	1.07	1.0–1.3	fi. ret. rug.	6935	
T. tubiflorus	Madagascar	Herb. J. Bot. TAN 3328 (P)	36	29	29–36 × 25–30	1.24	1.0–1.5	perf.–smooth	2761	
T. tubiflorus	Madagascar	Capuron s.n. (P)	29	23	25–33 × 20–26	1.26	1.1–1.5	perf. (fi. ret.)	6534	
Tapeinostemon Benth.										
T. longiflorum var. australe	Brazil–Venezuela	Maguire et al. 60507 (NY)	39	32	34–42 × 30–37	1.22	1.0–1.4	fi. ret.	13832	
T. longiflorum var. australe	Brazil–Venezuela	Steyermark 103937 (NY)	40	34	37–42 × 32–36	1.18	1.1–1.3	perf., fi. ret.	13777	
T. longiflorum var. longiflorum	Venezuela	Maguire et al. 42432 (NY)	40	33	40–44 × 30–34	1.21	1.0–1.3	fi. ret.	13778	
T. ptariense	Venezuela	Maguire et al. 42053 (NY)	28	26	25–33 × 23–29	1.08	1.0–1.3	perf., ret.	13833	
T. ptariense	Venezuela	Steyermark 59861 (F)	21	22	19–23 × 20–23	0.95	0.9–1.1	fi. ret.	37	
T. ptariense	Venezuela	Steyermark & Nilsson 171 (NY)	22	22	20–23 × 20–23	1.00	1.0–1.1	fi. ret.	1343	
T. sessiliflorum	Venezuela	Spruce 3293 (K)	21	22	20–22 × 20–24	0.95	1.0–1.0	fi. ret.	4414	

[392]

Taxon	Country	Voucher							Accession
T. spenneroides	Venezuela	*Steyermark 74801* (F)	20	23	19–22 × 22–24	0.87	0.8–1.0	fi. ret.	131
T. zamoranum	Ecuador	*André 4618* (F)	17	18	17–18 × 18–19	0.94	0.9–1.0	perf., fi. ret.	38
T. zamoranum	Ecuador	*Gaarol 74940* (QCA)	21	21	19–26 × 19–25	1.00	0.9–1.1	fi. ret., perf.	24042
Zonanthus Griseb.									
Z. cubensis	Cuba	*Wright 1346* (BM)	48	37	41–54 × 30–41	1.30	1.1–1.4	str. ret.(–fi. ret.)	4232
Z. cubensis	Cuba	*Wright 1346* (U)	48	36	43–54 × 31–41	1.33	1.3–1.5	str. ret.(–fi. ret.)	18672

Notes:

[a] Collection country for voucher.
[b] Herbarium abbreviations follow Holmgren *et al.* (1990).
[c] Polar axis.
[d] Equatorial diameter.
[e] Shape classes.
[f] Abbreviations for exine patterns are: co., coarsely; co. ret., coarsely reticulate (largest lumina exceeding 5 μm); dist., distinctly: fl., finely; fi. ret., finely reticulate (lumina, 2 μm); fov., foveolata; marg., pore margin; mi. spi., microspinulous; perf., perforate; protrud., protruding; ret., reticulate (lumina 2–5 μm); rug., rugulate; scabr., scabrate; spi., spinules; str. ret., striato-reticulate; verr., verrucose.
[g] Accession number in the pollen slide collection at (S).
[h] Longistylous.
[i] Brevistylous.
[j] Previously determined as *M. corymbosa*.
[k] Previously determined as *M. viscosa*.
[l] Previously determined as *M. macrophylla*.
[m] Previously determined as *M. ovalis*.
[n] Previously determined as *M. pachystyla*.
[o] Previously determined as *M. polyantha*.
[p] Previously determined as *M. revoluta*.
[q] Previously determined as *M. sodiroana*.
[r] Previously determined as *M. viscosa*.
[s] Previously determined as *M. pachystyla*.
[t] Pollen in tetrads 55 μm (50–65 μm).

Table 4.4. Pollen morphological data and vouchers for Gentianaceae taxa with pollen in tetrads or polyads (for footnotes see p. 407)

Taxon	Country[a]	Vouchers[b]	Diameter of tetrads / polyads (average)	Size range of tetrads (μm)	Size range of polyads (μm)	Single grains (μm) P[c]	Single grains (μm) E[d]	Exine pattern[e]	Slide no.[f]	Pollen type
Adenolisianthus (Progel) Gilg										
A. arboreus	Brazil	*Holt & Blake 582* (K)	62	55–72		33	43	diff. ret.	4430	*Chelonoides*
A. arboreus	Brazil	*Glaziou 9950* (P)	60	55–69		31	39	diff. ret.	7294	*Chelonoides*
A. arboreus	Venezuela	*Wurdack & Adderley 43722* (NY)	63	57–73		34	41	diff. ret.	24089	*Chelonoides*
Aripuana Struwe, Maas, & V. A. Albert										
A. cullmaniorum	Brazil	*Ferreira 5906* (NY)	83	81–91		42	41	ret. indist. diff.	24673	*Aripuana*
Calolisianthus Gilg										
C. acutangulus	Bolivia	*Cárdenas 4457* (US)	67/159	66–68	138–175	37	34	loop-like ret.	7314	*Amplissimus*
C. acutangulus	Brazil	*Malme 1378 B* (S)	70/131	70	108–158	40	43	loop-like ret.	2757	*Amplissimus*
C. acutangulus	Brazil	*Herb. Fl. Brazil 216* (BR)	68/146	64–72	133–163	44	55	loop-like ret.	7293	*Amplissimus*
C. amplissimus	Brazil	*Smith 6721* (S)	68/140	66–68	113–200	38	48	loop-like ret.	2756	*Amplissimus*
C. amplissimus	Brazil	*Jesus 56* (NY)	77/172	77	130–225	47	34	loop-like ret.	5363	*Amplissimus*
C. amplissimus	Brazil	*Dusén 16668* (NY)	66/143	66	125–188	35	44	loop-like ret.	7309	*Amplissimus*
C. macranthus	Brazil	*Vogel 640* (U)	68/123	52–74	103–148	38	48	loop-like ret.	3203	*Amplissimus*
C. macranthus	Brazil	*Irwin et al. 15272* (U)	70/147	66–74	123–158	39	47	loop-like ret.	4738	*Amplissimus*
C. pedunculatus	Brazil	*Dusén 9112* (S)	59	53–66		31	39	diff. ret. isl.	3730	*Speciosus*
C. pedunculatus	Brazil	*Irwin et al. 12829* (U)	57	52–61		30	39	diff. ret. isl.	4574	*Speciosus*
C. pedunculatus	Brazil	*Sellow s.n.* (HAL)	55	53–58		31	41	diff. ret. isl.	7345	*Speciosus*
C. pedunculatus var. *damazianus*	Brazil	*Barbosa s.n. (R 6719)*	68	57–75		34	46	diff. ret. isl.	13904	*Speciosus*
C. pendulus	Brazil	*Vogel 752* (U)	68	62–72		35	44	diff. ret. isl.	5465	*Speciosus*

Species	Country	Voucher								Subgenus
C. pendulus	Brazil	Brazil, São Paulo (M)	77	55–63		33	40	diff. ret. isl.	7346	Speciosus
C. pendulus	Brazil	Regnell I: 300 (S)	59	55–63		32	37	diff. ret. isl.	7349	Speciosus
C. pulcherrimus	Brazil	Barreto 2786 (F)	70	64–79		38	49	diff. ret. isl.	5788	Speciosus
C. pulcherrimus	Brazil	Barreto 2789 (F)	62	61–66		31	44	diff. ret. isl.	7312	Speciosus
C. speciosus	Brazil	Ynes Mexia 5842 (UC)	63	55–61		33	44	diff. ret. isl.	5781	Speciosus
C. speciosus	Brazil	Lindeman & Haas 6011 (U)	64	58–68		30	47	diff. ret. isl.	4548	Speciosus
C. speciosus	Brazil	Irwin et al. 14298 (U)	71	64–77		35	49	diff. ret. isl.	4604	Speciosus
C. speciosus	Brazil	Smith 6974 (US)	62	57–66		32	44	diff. ret. isl.	7315	Speciosus
C. speciosus	Brazil	Sellow s.n. (HAL)	58	55–61		28	42	diff. ret. isl.	7350	Speciosus
C. speciosus	Brazil	Macedo 3291 (S)	58	55–63		31	43	diff. ret. isl.	24157	Speciosus
C. speciosus	Brazil	Järneby s.n. (S)	62	61–66		32	45	diff. ret. isl.	24158	Speciosus
C. speciosus	Brazil	Joan de Alianca 5369 (S)	65	56–69		37	47	diff. ret. isl.	24159	Speciosus
C. speciosus	Brazil	Mart. Obs. 1153 (M)	50	52–66		29	41	diff. ret. isl.	7344	Speciosus
Celiantha Maguire										
C. bella	Venezuela	Silva & Brazão 60921 (NY)	134		120–148			glob. pila	24085	Imthurinanus
C. bella	Venezuela	Maguire et al. 42096 (NY)	189		180–213			glob. elong. proc.	17831	Imthurinanus
C. chimantensis	Venezuela	Steyermark 74983 (NY)	174		163–188			glob. pila	7307	Imthurinanus
C. imthurniana	Venezuela	Steyermark 58794 (F)	153		138–170			glob. pila	6585	Imthurinanus
C. imthurniana	Venezuela	Cardona 2664 (US)	124		110–140			glob. pila	5780	Imthurinanus
C. imthurniana	Venezuela	Quelch & McConnell 101 (K)	125		113–132			glob. pila	5791	Imthurinanus
Chelonanthus Gilg										
C. acutangulus	Bolivia	Herzog 1498 (L)	63	60–66		31	48	diff. ret.	4289	Chelonoides
C. acutangulus	Costa Rica	Skutch 2968 (NY)	73	68–78		37	53	diff. ret.	2200	Chelonoides
C. acutangulus	Peru	Melin 75 (S)	62	56–67		31	42	diff. ret.	1702	Chelonoides
C. acutangulus	Peru	Herb. Pavon s.n. (G)	51	48–55		26	36	diff. ret.	7310	Chelonoides
C. alatus	Brazil	Prance et al. 14376 (NY)	55	51–61		29	36	diff. ret.	24086	Chelonoides
C. alatus	Colombia	Killip 34482 (US)	67	66–68		36	48	diff. ret.	2409	Chelonoides
C. alatus	Grenada	Smith 10128 (NY)	76	62–84		40	59	diff. ret., elong. proc. pila	2404	Chelonoides

Table 4.4. (cont.)

Taxon	Country[a]	Vouchers[b]	Diameter of tetrads / polyads (average)	Size range of tetrads (μm)	Size range of polyads (μm)	Single grains (μm) P[c]	Single grains (μm) E[d]	Exine pattern[e]	Slide no.[f]	Pollen type
C. alatus	Mexico	Ynes Mexia 9144 (K)	71	66–77		37	50	diff. ret., pila	4551	Chelonoides
C. alatus	Surinam	L.P. 475 (U)	63	60–67		30	45	diff. ret., elong. proc. pila	556	Chelonoides
C. alatus	Venezuela	Steyermark 88754 (U)	72	66–78		31	50	diff. ret.	4571	Chelonoides
C. albus	Venezuela	Holt & Gehriger 327 (US)	52	46–54		26	35	diff. ret., elong. proc. pila	5776	Chelonoides
C. albus	Venezuela	Wurdack & Adderley 43593 (NY)	55	53–57		30	40	diff. ret., elong. proc. pila	24087	Chelonoides
C. angustifolius	Colombia	Schlim 72 (BR)	54	50–58		27	39	diff. ret., glob.	7287	Chelonoides
C. angustifolius	Colombia	Haught 2434 (US)	55	52–61		29	36	diff. ret., glob.	5777	Chelonoides
C. angustifolius	Colombia	Killip & Smith 14998 (BM)	54	50–61		28	36	diff. ret., glob.	4290	Chelonoides
C. angustifolius	Peru	Spruce s.n. (BR)	54	51–57		28	37	diff. ret., glob. pila	7299	Chelonoides
C. angustifolius	Venezuela	Steyermark 75290 (NY)	63	57–67		32	44	diff. ret., elong. proc. pila	6384	Chelonoides
C. angustifolius	Venezuela	Maguire 33712 (NY)	55	50–61		30	39	diff. ret., elong. proc. pila	24088	Chelonoides
C. angustifolius	Venezuela	Vareshi & Foldats 4681 (NY)	59	57–59		31	43	diff. ret., elong. proc. pila	6180	Chelonoides
C. angustifolius	Venezuela	Steyermark 58268 (US)	62	55–66		31	41	diff. ret., glob. pila	5778	Chelonoides
C. bifidus	Colombia	Pennell 1546 (NY)	55	50–58		28	36	diff. ret., elong. proc. pila	2219	Chelonoides

Species	Country	Collector								
C. bifidus	Guyana	Smith 3064 (MO)	56	51–62		28	38	diff. ret., glob. pila	632	Chelonoides
C. campanuloides	Brazil	Spruce 2005 (P)	138		138–163	42	47	loops, polar spi.	7296	Uligin.–Iribach.
C. campanuloides	Brazil	Spruce 2005 (K)	140		125–150	26	36	loops, polar spi.	7420	Uligin.–Iribach.
C. campanuloides	Colombia	Vogel 125 (U)	127		90–160	33	39	loops, polar spi.	5463	Uligin.–Iribach.
C. campanuloides	Venezuela	Holt & Blake 695 (US)	142		124–168	32	39	loops, polar spi.	5857	Uligin.–Iribach.
C. camporum	Peru	Melini 247 (S)	58	54–63				diff. ret., glob. pila	4988	Chelonoides
C. camporum	Peru	Melini 313 (S)	53	50–59		27	37	diff. ret., glob. pila	3415	Chelonoides
C. candidus	Brazil	Dusén 16538 (MO)	61	55–66		29	41	diff. ret., glob. pila	2201	Chelonoides
C. candidus	Brazil	Malme 3293 A (S)	58	53–62		30	41	diff. ret., glob. pila	3779	Chelonoides
C. chelonoides	Brazil	Irwin et al. 17270 (U)	70	55–63		32	44	diff. ret.	4603	Chelonoides
C. chelonoides	Grenada	Webster et al. 9551 (US)	65	61–72		32	42	diff. ret.	6583	Chelonoides
C. chelonoides	Guyana	Maguire & Fanshawe 22853 (U)	72	68–77		37	43	diff. ret.	4609	Chelonoides
C. chelonoides	Peru	Ynes Mexia 4153 (MO)	61	55–63		29	41	diff. ret.	2202	Chelonoides
C. chelonoides	Venezuela	Maguire et al. 36200 (NY)	126	113–138		70	96	diff. ret.	13755	Chelonoides
C. fistulosus	Guyana	Jenman 5435 (NY)	65	58–72		33	43	diff. ret., elong. pila	6383	Chelonoides
C. grandiflorus	Paraguay	Hassler 9986 (NY)	58	55–61		27	42	diff. ret., elong. pila	6166	Chelonoides
C. grandiflorus	Paraguay	Hassler 8824 (S)	63	57–72		33	43	diff. ret., elong. pila	5267	Chelonoides
C. grandiflorus	Surinam	Hulk 47a (U)	55	53–57		28	40	diff. ret., elong. pila	510	Chelonoides
C. purpurascens	French Guiana	Rova et al. 1902 (S)	70/115	61–77	88–125	44	42	diff. ret., polar loops	24650	Uliginosus

Table 4.4. (*cont.*)

Taxon	Country[a]	Vouchers[b]	Diameter of tetrads / polyads (average)	Size range of tetrads (μm)	Size range of polyads (μm)	Single grains (μm) P[c]	Single grains (μm) E[d]	Exine pattern[e]	Slide no.[f]	Pollen type
C. spruceanus	Brazil	*Melini 81* (S)	59/152	55–62	63–180	29	36	diff. ret., polar loops	5264	*Uliginosus*
C. spruceanus	Peru	*Klug 2345* (MO)	68/129	64–74	80–163	34	39	diff. ret., polar loops	2203	*Uliginosus*
C. uliginosus	Brazil	*Smith 1318* (US)	53/116	50–55	95–128	25	32	diff. ret., polar loops	570	*Uliginosus*
C. uliginosus	Colombia	*Vogel 266* (U)	62/125	58–68	95–148	34	36	diff. ret., polar loops	5468	*Uliginosus*
C. uliginosus	Colombia	*Herb. Lehmannianum 6046* (S)	58/116	59–70	108–168	35	38	diff. ret., polar loops	6253	*Uliginosus*
C. uliginosus	Colombia	*Schultes 3928* (K)	56/158	55–57	105–213	35	36	diff. ret., polar loops	4532	*Uliginosus*
C. uliginosus	Guyana	*Smith 3639* (K)	72/145	69–75	113–175	43	44	diff. ret., polar loops	4621	*Uliginosus*
C. uliginosus	Guyana	*Atkinson 14* (BM)	68/165	61–72	138–237	35	39	diff. ret., polar loops	4809	*Uliginosus*
C. uliginosus	Guyana	*Pinkus 287* (NY)	62/165	58–66	108–212	33	35	diff. ret., polar loops	6161	*Uliginosus*
C. uliginosus	Guyana	*Smith 3639* (NY)	64/119	61–66	108–133	33	40	diff. ret., polar loops	6178	*Uliginosus*
C. uliginosus	Guyana	*Cox & Hubbard 142* (NY)	57/155	52–75	120–198	32	34	diff. ret., polar loops	24090	*Uliginosus*
C. uliginosus	Surinam	*Kuyper 85* (U)	58/150	53–61	120–180	34	35	diff. ret., polar loops	555	*Uliginosus*

Species	Country	Collection									Genus
C. uliginosus	Venezuela	*Agostini 394* (NY)	58/159	53–61		100–225	28	35	diff. ret., polar loops	6162	*Uliginosus*
C. uliginosus var. *guianensis*	Colombia	*Vogel 0133* (U)	69/133	57–80		108–175	36	34	diff. ret., polar loops	4425	*Uliginosus*
C. uliginosus var. *guianensis*	Surinam	*Splitgerber 1090* (L)	80/188	77–83		125–255	39	42	diff. ret., polar loops	4610	*Uliginosus*
C. viridiflorus	Brazil	*Heringer 7883/77* (NY)	66	52–61			28	41	diff. ret.	5368	*Chelonoides*
C. viridiflorus	Colombia	*Vogel 20* (U)	65	62–67			31	46	diff. ret.	4521	*Chelonoides*
C. viridiflorus	Venezuela	*Breteler 4250* (U)	58	55–62			28	40	diff. ret.	4533	*Chelonoides*
Coutoubea Aubl.											
C. minor	Venezuela	*Kral et al. 71729* (NY)	54	51–56			28	38	homobro.	24092	*Coutoubea*
C. ramosa	Venezuela	*Tamayo 3460* (NY)	58	56–61			30	39	diff. ret.	24093	*Coutoubea*
C. reflexa	Venezuela	*Maguire et al. 36047* (NY)	63	61–65			34	41	diff. ret.	24094	*Coutoubea*
C. spicata	Panama	*Allen 1282* (U)	60	57–64			30	39	homobro.	2442	*Coutoubea*
Deianira Cham. & Schltdl.											
D. cordifolia	Brazil	*Malme 3323* (S)	39	36–42			21	29	fi. ret.	7455	*Deianira*
D. cyathifolia	Brazil	*Malme 3323 B* (S)	37	36–41			18	28	fi. ret.	1368	*Deianira*
D. cyathifolia	Brazil	*Malme 06746* (S)	39	36–41			21	29	fi. ret., segm.	24534	*Deianira*
D. erubescens	Brazil	*Smith 6962* (NY)	37	34–42			20	30	fi. ret.	2265	*Deianira*
D. erubescens	Brazil	*Ynes Mexia 5667* (MO)	38	33–43			19	28	fi. ret.–ret.	2266	*Deianira*
D. foliosa	Brazil	*Löfgren 315* (S)	33	30–34			17	23	fi. ret.	24535	*Deianira*
D. nervosa	Brazil	*Macedo 2404* (MO)	40	39–42			20	30	fi. ret.	2321	*Deianira*
D. nervosa	Brazil	*Smith 6964* (K)	52	50–55			26	38	ret.	4647	*Deianira*
D. nervosa	Brazil	*Smith 6964* (NY)	56	48–70			30	41	ret., segm.	2267	*Deianira*
D. pallescens	Brazil	*Macedo 2403* (MO)	37	35–39			18	37	fi. ret.	2322	*Deianira*
Helia Mart.											
H. brevifolia	Brazil	*Hatschbach 3617* (US)	61	53–58			28	39	diff. ret.	5850	*Helia*
H. brevifolia	Paraguay	*Jorgensen 4837* (MO)	60	55–66			29	39	diff. ret. perf.	2362	*Helia*
H. loesneriana	Venezuela	*Killip 37557* (US)	75	53–62			28	40	diff. ret.	5858	*Helia*

Table 4.4. (cont.)

Taxon	Country[a]	Vouchers[b]	Diameter of tetrads/polyads (average)	Size range of tetrads (μm)	Size range of polyads (μm)	Single grains (μm) P[c]	Single grains (μm) E[d]	Exine pattern[e]	Slide no.[f]	Pollen type
H. martii	Brazil	Glaziou 16264 (BR)	66	65–68		37	45	diff. ret.	18476	Helia
H. micrantha	Paraguay	Balansa 2141 (G)	64	58–70		33	42	diff. ret.	18479	Helia
H. oblongifolia	Brazil	Dusén 16537 (MO)	61	55–67		30	41	diff. ret. perf. fov.	2363	Helia
H. oblongifolia	Paraguay	Hassler 8875 (NY)	56	53–58		29	38	diff. ret.	5784	Helia
H. spathulata	Brazil	Emmerich 2147 & Andrade 1767 (R)	61	55–69		33	42	diff. ret. perf.	13926	Helia
Irlbachia Mart.										
I. cardonae	Guyana	Graham 129 (K)	168		125–200			polar spi.	4554	Irlbachia
I. cardonae	Venezuela	Steyermark 75730 (F)	159		133–193			polar spi.	6590	Irlbachia
I. cardonae	Venezuela	Tate 1360 (NY)	130		113–148			polar spi.	6217	Irlbachia
I. elegans	Brazil	Prance et al. 4866 (S)	90/158	83–99	140–175	53	51	polar spi.	8051	Irlbachia
I. elegans	Brazil	Prance et al. 3721 (S)	78/128	75–86	115–138	47	49	polar spi.	24584	Irlbachia
I. elegans	Venezuela	Maguire & Politi 27324 (US)	74/135	72–77	115–150	48	43	polar spi.	17209	Irlbachia
I. nemorosa	Venezuela	Morillo & Wood 4068 (VEN)	78/137	69–88	110–163	43	46	polar spi.	19288	Irlbachia
I. nemorosa	Venezuela	Steyermark & Dunsterville 113269 (VEN)	142		123–188	50	40	polar spi.	19290	Irlbachia
I. paruana	Venezuela	Cowan & Wurdack 31160 (NY)	185		165–203	60	55	polar spi.	13774	Irlbachia
I. phelpsiana	Venezuela	Cowan & Wurdack 31412 (NY)	154		138–180	50	53	polar spi.	13776	Irlbachia
I. plantaginifolia	Venezuela	Maguire et al. 42593 (NY)	78/127	73–86	113–138	39	46	polar spi.	13762	Irlbachia

Taxon	Country	Collector							Genus
I. poeppigii	Brazil	Poeppig 2854 (G)	119		95–140	42	44	polar spi.	7331	Irlbachia
I. pratensis	Venezuela	Maguire et al. 37563 (US)	81/117	73–94	90–125	45	45	polar blunt spi.	17204	Uligin.–Irlbach.
I. pratensis	Venezuela	Wurdack & Adderley 42949 (NY)	140		105–173			polar blunt spi.	13869	Uligin.–Irlbach.
I. pratensis	Venezuela	Maguire et al. 37563 (NY)	56/121	57–72	90–145	42	40	polar blunt spi.	24186	Uligin.–Irlbach.
I. pratensis	Venezuela	Wurdack & Monachino 39764 (NY)	119	108–128		66	83	polar blunt spi.	13754	Uligin.–Irlbach.
I. pumila	Venezuela	Williams 14362 (US)	123		113–130	37	33	polar spi.	2370	Irlbachia
I. pumila	Brazil	Spruce 2950 (NY)	112		100–130		32	polar spi.	7683	Irlbachia
I. ramosissima	Brazil	Schomburgk 989 (K)	83/126	82–84	100–148	40	34	polar spi.	5833	Irlbachia
I. recurva	Brazil	Schultes 8721 (US)	160		148–180	51	42	polar spi.	5860	Irlbachia
I. recurva	Colombia	Schultes 5863 (US)	147		123–175	61	58	polar spi.	2371	Irlbachia
I. recurva	Venezuela	Maguire & Politi 27324 (NY)	121		108–153	38	33	polar spi.	6216	Irlbachia
I. subcordata	Brazil	Vogel 311 (U)	158		138–193	42	33	polar spi.	5386	Irlbachia
I. tatei	Venezuela	Tate 743 (US)	83/157	75–88	138–213	48	43	polar spi.	17219	Irlbachia
I. tatei	Venezuela	Cardona 376 (US)	70/145	66–77	123–188	44	42	polar spi.	18387	Irlbachia
I. tatei	Venezuela	Steyermark 58167 (US)	79/142	76–83	130–173	48	42	polar spi.	18474	Irlbachia
Lagenanthus Gilg										
L. princeps	Colombia	Schlim 419 (BM)	73	70–77		36	60	diff. ret.	4828	Lagenanthus
L. princeps	Colombia	Murillo & Jaramillo s.n. (COL)	69	66–75		35	53	diff. ret.	24531	Lagenanthus
L. princeps	Venezuela	Cardona 119 (US)	57	53–58		27	43	diff. ret.	482	Lagenanthus
Lehmanniella Gilg										
L. splendens	Colombia	Sandeman 6046 (K)	65	61–72		36	50	co. ret.–exceed. co.	4810	Lehmanniella
L. splendens	Colombia	Uribe-Uribe 2605 (NY)	70	64–74		36	49	co. ret.–exceed. co.	6218	Lehmanniella
L. splendens	Colombia	John 20561 (NY)	77	72–81		38	57	co.ret.	6252	Lehmanniella
L. splendens	Colombia	Fonnegra & Roldan 2698 (NY)	70	66–75		38	50	muri distorted	24036	Lehmanniella

Table 4.4. (cont.)

Taxon	Country[a]	Vouchers[b]	Diameter of tetrads/polyads (average)	Size range of tetrads (μm)	Size range of polyads (μm)	Single grains (μm) P[c]	Single grains (μm) E[d]	Exine pattern[e]	Slide no.[f]	Pollen type
"*Lisianthus* L."[g]										
"*L.*" *acutilobus*	Venezuela	*Steyermark & Nilsson 732* (VEN)	83/147	73–94	120–163	47	46	polar spi.	17212	*Iribachia*
"*L.*" *acutilobus*	Venezuela	*Steyermark & Nilsson 732* (NY)	84/122	77–88	88–143	45	49	polar spi.	6382	*Iribachia*
"*L.*" *alpestris*	Brazil	*Martius 690* (M)	61	57–66		31	45	diff. ret. isl.	7254	*Speciosus*
"*L.*" *elegans*	Brazil	*Haas 15711* (U)	66	62–68		33	44	diff. ret. isl.	4436	*Speciosus*
"*L.*" *elegans*	Brazil	No collector (M)	60	55–62		31	39	diff. ret. isl.	7343	*Speciosus*
"*L.*" *scabridulus*	Venezuela	*Steyermark 57979* (F)	162		138–188			polar spi.	6584	*Iribachia*
"*L.*" *tenuifolius*	Brazil	*Mart. Obs. 1385* (M)	53	48–61		28	37	diff. ret. isl.	7327	*Speciosus*
Neblinantha Maguire										
N. neblinae	Brazil–Venezuela	*Steyermark 103862* (NY)	64	58–68		35	44	diff. ret. exceed co. ret.	24097	*Neblinantha*
N. parvifolia	Brazil	*Steyermark 103800* (NY)	89	83–96		47	59	diff. ret.–co. ret.	13817	*Neblinantha*
N. parvifolia	Brazil	*Maguire et al. 42419* (NY)	97	86–105		51	69	diff. ret.–co. ret.	13818	*Neblinantha*
Prepusa Mart.										
P. alata	Brazil	*Farney & Caruso 1195* (RB 276053)	86	78–90		29	37	pila, verr.	23895	*Prepusa*
P. connata	Brazil	*Glaziou 18972* (NY)	73	66–77		38	51	pila, verr.	5880	*Prepusa*
P. connata	Brazil	*Gardner 541* (CGE)	89	85–92		51	57	pila, verr.	17772	*Prepusa*
P. connata	Brazil	*Glaziou 18372* (K)	81	77–82		41	55	pila, verr.	5852	*Prepusa*
P. hookeriana	Brazil	*Vogel 1773* (U)	80	77–84		44	53	flat. proc., jigsaw	5881	*Prepusa*

Species	Country	Collection						Genus
P. hookeriana	Brazil	Lützenburg 1910 (NY)	71	68–75	38	45	flat. proc., jigsaw	Prepusa
P. hookeriana	Brazil	Vidal II-330 (R 107600)	71	65–78	40	50	flat. proc., jigsaw	Prepusa
P. montana	Brazil	Harley 22878 (U)	85	80–94	44	53	pila, verr.	Prepusa
P. viridiflora	Brazil	Duarte 10469 (RB)	75	69–79	42	47	pila, verr.	Prepusa
Purdieanthus Gilg								
P. pulcher	Colombia	Purdie s.n. (P)	60	56–66	29	49	diff. ret.	Purdieanthus
P. pulcher	Colombia	Purdie s.n. (K)	61	57–70	30	47	diff. ret.	Purdieanthus
P. pulcher	Colombia	Uribe 6473 (COL)	70	65–75	34	51	diff. ret.	Purdieanthus
Rogersonanthus Maguire & B. M. Boom								
R. arboreus	Trinidad	Broadway 6207 (S)	64	62–68	33	46	diff. ret.	Chelonoides
R. arboreus	Trinidad	Broadway 5648 (K)	55	52–61	28	36	diff. ret.	Chelonoides
R. quelchii	Venezuela	Steyermark & Wurdack 813 (F)	72	66–83	37	46	diff. ret.	Chelonoides
R. quelchii	Venezuela	Steyermark 93657 (VEN)	86	80–95	42	59	diff. ret.	Chelonoides
R. quelchii	Venezuela	Steyermark 58802 (F)	72	63–85	35	48	diff. ret.	Chelonoides
R. quelchii	Venezuela	Steyermark 58672 (F)	76	72–79	36	48	diff. ret.	Chelonoides
R. tepuiensis	Venezuela	Steyermark 59721 (F)	72	66–77	35	50	diff. ret.	Chelonoides
R. tepuiensis	Venezuela	Tate 1361 (F)	68	64–72	32	44	diff. ret.	Chelonoides
R. tepuiensis	Venezuela	Steyermark s.n. (VEN)	82	77–88	39	57	diff. ret.	Chelonoides
R. tepuiensis	Venezuela	Cardona 2249 (US)	77	73–80	37	49	diff. ret.	Chelonoides
Schultesia Mart.								
S. benthamiana	Guyana	Smith 2297 (U)	101	91–116	53	66	diff. ret.	Coutoubea
S. benthamiana	Guyana	Maas et al. 3661 (U)	78	74–83	40	52	diff. ret.	Coutoubea
S. benthamiana	Venezuela	Wurdack & Monachino 39942 (NY)	94	89–100	49	61	diff. ret.	Coutoubea
S. brachyptera	Guyana	Jansen-Jacobs et al. 569 (U)	73	70–77	37	49	diff. ret.	Coutoubea
S. brachyptera	Guyana	Maas et al. 7730 (U)	64	66–74	32	44	diff. ret.	Coutoubea

Table 4.4. (cont.)

Taxon	Country[a]	Vouchers[b]	Diameter of tetrads / polyads (average)	Size range of tetrads (μm)	Size range of polyads (μm)	Single grains (μm) P[c]	Single grains (μm) E[d]	Exine pattern[e]	Slide no.[f]	Pollen type
S. brachyptera	Panama	*Allen 1069* (U)	64	61–66		32	45	diff. ret.	2433	*Coutoubea*
S. brachyptera	Surinam	*Lindeman 4267* (U)	75	66–80		40	49	diff. ret.	2434	*Coutoubea*
S. brachyptera	Surinam	*Forest. Bureau s.n.* (U)	78	66–81		42	52	diff. ret.	4826	*Coutoubea*
S. guianensis	Brazil	*Brade 11372* (S)	95	92–100		48	61	diff. ret.	3618	*Coutoubea*
S. guianensis	Guinea Bissau	*Santo Expt. Bot. 1750* (K)	68	63–74		35	49	diff. ret.	24188	*Coutoubea*
S. guianensis	Guyana	*Jansen-Jacobs et al. 569* (U)	67	66–70		34	45	diff. ret.	22343	*Coutoubea*
S. guianensis	Guyana	*Poiteau s.n.* (K)	93	88–100		50	61	diff. ret.	5853	*Coutoubea*
S. guianensis var. *latifolia*	Sierra Leone	*Adams 157* (K)	69	65–74		37	49	diff. ret.	24187	*Coutoubea*
S. guianensis var. *latifolia*	Surinam	*Pons (L. B. B.) 12660* (U)	80	76–86		40	52	diff. ret.	22340	*Coutoubea*
S. guianensis var. *latifolia*	Venezuela	*Williams & Alston 373* (U)	92	88–99		46	65	diff. ret.	7737	*Coutoubea*
S. lisianthoides	Nicaragua	*Stevens & Krukoff 8224* (MO)	46	41–54		24	34	non-diff. ret., homobro.	24691	*Coutoubea*
S. lisianthoides	Nicaragua	*Pipoly 4526* (MO)	42	39–44		22	28	non-diff. ret., homobro.	24690	*Coutoubea*
S. lisianthoides	Panama	*Robyns 65-21* (MO)	49	44–46		22	33	non-diff. ret., homobro.	2380	*Coutoubea*
S. lisianthoides	Panama	*Greenm. & Greenm. 5203* (MO)	51	44–58		26	35	non-diff. ret., homobro.	2397	*Coutoubea*
S. pohliana	Brazil	*Hunt & Ramos 5828* (NY)	107	99–111		54	63	diff. ret., muri verm.	7316	*Coutoubea*

Species	Country	Collection						Genus
S. pohliana	Surinam	Irwin et al. 55237 (C)	75	73–80	37	50	diff. ret., muri verm.	22344	Coutoubea
S. subcrenata	Brazil	Malme 3179 a (S)	64	58–72	33	43	diff. ret.	7740	Coutoubea
S. subcrenata	Brazil	Weddell 2174 (P)	63	56–67	32	39	diff. ret.	22346	Coutoubea
Senaea Taub.									
S. caerulea	Brazil	Romariz 0127 (RB 59940)	74	70–79	38	50	pila, verr.	24051	Prepusa
S. janeirensis	Brazil	Martinelli et al. 12003 (RB 272215)	82	75–90	43	58	pila, verr.	21827	Prepusa
S. janeirensis	Brazil	Santos Lima 217 (RB)	91	87–98	53	61	pila, verr.	17774	Prepusa
Sipapoantha Maguire & B. M. Boom									
S. ostrina	Venezuela	Maguire & Politi 28092 (NY)	71	67–75	39	50	pila, verr.	24098	
Symbolanthus G. Don									
S. anomalus	Colombia	Uribe 2142 (US)	76	74–79	38	60	diff. ret.	2260	Symbolanthus
S. anomalus	Colombia	Lawrence 470 (NY)	75	70–79	36	58	diff. ret.	6392	Symbolanthus
S. anomalus	Colombia	H. L. B 912184-360 (L)	80	73–89	42	56	diff. ret.	7776	Symbolanthus
S. brittonianus	Bolivia	Bang 339 (K)	77	70–83	39	62	diff. ret.	4419	Symbolanthus
S. calygonus	Bolivia	Buchtien 2461 (L)	86	79–88	43	57	diff. ret.	4564	Symbolanthus
S. calygonus	Peru	Sandeman 187 (K)	86	81–88	42	63	diff. ret.	4484	Symbolanthus
S. calygonus	Peru	Allard 21095 (US)	93	88–95	48	70	diff. ret.	5848	Symbolanthus
S. daturoides	Peru	Mathews 1317 (K)	90	80–97	48	66	diff. ret.	7419	Symbolanthus
S. elisabethae	Brazil	Silva & Brazão 60653 (NY)	102	95–108	61	79	diff. ret.	13742	Symbolanthus
S. elisabethae	Guyana	Thurn 188 (K)	72	68–77	35	53	diff. ret.	5789	Symbolanthus
S. elisabethae	Venezuela	Pinkus 169 (NY)	77	73–80	37	57	diff. ret.	2262	Symbolanthus
S. gaultherioides	Colombia	Bro. Daniel 1693 (US)	70	66–77	35	52	diff. ret.	7306	Symbolanthus
S. latifolius	Bolivia	Steinback 9049 (U)	79	72–84	37	54	diff. ret.	4542	Symbolanthus
S. latifolius	Colombia	Vogel 116 (U)	96	94–101	48	67	diff. ret.	6057	Symbolanthus
S. macranthus	Ecuador	Espinosa 197 (NY)	87	80–88	39	60	diff. ret.	6391	Symbolanthus
S. magnificus	Colombia	Langenheim 3293 (US)	83	79–88	39	60	diff. ret.	5849	Symbolanthus
S. magnificus	Colombia	Uribe-Uribe 3660 (NY)	87	81–94	40	63	diff. ret.	6390	Symbolanthus
S. magnificus	Venezuela	Maguire 39441 (NY)	101	88–113	63	84	diff. ret.	13740	Symbolanthus

Table 4.4. (cont.)

Taxon	Country[a]	Vouchers[b]	Diameter of tetrads / polyads (average)	Size range of tetrads (µm)	Size range of polyads (µm)	Single grains (µm) P[c]	Single grains (µm) E[d]	Exine pattern[e]	Slide no.[f]	Pollen type
S. mathewsii	Bolivia	Pearce 640 (K)	70	62–76		37	51	diff. ret.	7777	Symbolanthus
S. mathewsii	Peru	Mathews 1836 (K)	85	79–90		43	59	diff. ret.	7303	Symbolanthus
S. nerioides	Peru	Davis s.n. (BM)	86	79–94		44	62	diff. ret.	16925	Symbolanthus
S. nerioides	Venezuela	Jahn 910 (US)	84	77–88		39	57	diff. ret.	5779	Symbolanthus
S. obscure-rosaceus	Peru	Woytkowski 8288 (US)	82	77–85		41	61	diff. ret.	5859	Symbolanthus
S. parvifolius	Venezuela	Maguire 35333 (NY)	108	93–115		56	77	diff. ret.	13738	Symbolanthus
S. pauciflorus	Peru	Spruce 4429 (BR)	72	68–74		35	52	diff. ret.	7324	Symbolanthus
S. pauciflorus	Peru	Spruce 4429 (NY)	75	69–80		37	52	diff. ret.	7325	Symbolanthus
S. pulcherrimus	Costa Rica	Weaver, Jr. 1406 (DUKE)	77	73–83		39	61	diff. ret.	6586	Symbolanthus
S. pulcherrimus	Panama	Pittier 3223 (US)	69	64–72		35	52	diff. ret.	5861	Symbolanthus
S. pulcherrimus	Panama	Allen 4946 (U)	86	84–96		44	63	diff. ret.	6058	Symbolanthus
S. rubro-violaceus	Costa Rica	Burger et al. 9355 (F)	88	84–90		43	65	diff. ret.	16920	Symbolanthus
S. rusbyanus	Peru	Wurdack 1068 (K)	80	77–86		40	60	diff. ret.	4563	Symbolanthus
S. stuebelii	Colombia	Scheider 1085 (S)	84	78–89		41	56	diff. ret.	16921	Symbolanthus
S. superbus	Colombia	Killip & Smith 15133 (US)	85	77–99		40	63	diff. ret.	2419	Symbolanthus
S. tricolor	Colombia	Haught 6148 (US)	78	72–86		38	55	diff. ret.	2264	Symbolanthus
S. tricolor	Colombia	Vogel 143 (U)	76	72–86		41	60	diff. ret.	4438	Symbolanthus
S. tricolor	Venezuela	Funck & Schlim 1412 (W)	88	84–91		44	65	diff. ret.	16923	Symbolanthus
S. vasculosus	Venezuela	Steyermark 89845 (NY)	81	77–94		39	59	diff. ret.	6165	Symbolanthus
Tetrapollinia Maguire & B. M. Boom										
T. caerulescens	Brazil	Dusén 2647 (MO)	67	62–72		35	42	non-polar blunt spi.	2372	Caerulescens

Taxon	Country	Voucher					Exine pattern	Accession	
T. caerulescens	Brazil	Irwin et al. 11661 (U)	64	56–66	33	41	non-polar blunt spi.	4633	Caerulescens
T. caerulescens	Guyana	Smith 2296 (NY)	66	63–70	36	42	non-polar blunt spi.	512	Caerulescens
T. caerulescens	French Guiana	Hallé 440 (U)	67	66–73	34	42	non-polar blunt spi.	4570	Caerulescens
T. caerulescens	Guyana	Jansen-Jacobs et al. 2764 (NY)	66	63–72	33	42	non-polar blunt spi.	24096	Caerulescens
T. caerulescens	Surinam	Maguire & Stahel 23682 (U)	68	61–77	34	42	non-polar blunt spi.	4424	Caerulescens
T. caerulescens	Venezuela	Steyermark 59353 (US)	69	62–76	37	45	non-polar blunt spi.	2192	Caerulescens
Wurdackanthus Maguire									
W. frigidus	Guadeloupe	Proctor 20339 (BM)	75	70–80	33	52	diff. ret.	4432	Symbolanthus
W. frigidus	Haiti	Torrey 248 (K)	69	55–83	38	50	diff. ret.	7553	Symbolanthus
W. frigidus	St. Vincent	Howard 11195 (BM)	73	65–78	36	50	diff. ret.	4271	Symbolanthus

Notes:
[a] Collection country for voucher.
[b] Herbarium abbreviations follow Holmgren et al. (1990).
[c] Polar axis.
[d] Equatorial diameter.
[e] Abbreviations for exine patterns are: co., coarsely; co. ret., coarsely reticulate (largest lumina exceeding 5 μm); diff., differentiated (reticulum); elong., elongate; exceed., exceedingly; fi., finely; fi. ret., finely reticulate (lumina <2 μm); flat., flattened; fov., foveolata; glob., globules; homobro., homobrochate; indist., indistinct; isl., islands; perf., perforate; proc., processes; ret., reticulate (lumina 2–5 μm); segm., segmented; spi., spinules; str. ret., striato-reticulate; verm., vermicular; verr., verrucose.
[f] Accession number in the pollen slide collection at (S).
[g] These taxa will eventually be assigned to valid genera.

Tribe Saccifolieae

Curtia CHAM. & SCHLTDL. (Fig. 4.1)

The pollen grains are released as monads, 3-colporate, 14–32×11–31 μm, subspheroidal to prolate, smooth, and perforate to foveolate, or microreticulate with minute spinules (microspines). In *Curtia tenuifolia* spp. *tenuifolia* two types of pollen are observed, smooth (Fig. 4.1A–B) and microspinose (Fig. 4.1C–D). The microspinose pollen grains appeared slightly larger than the smooth ones and could be correlated with brevistylous specimens (not always), while the smooth pollen grains could be associated with longi- or homostylous specimens. *Curtia obtusifolia* has smooth and perforate to foveolate (microreticulate) grains (Fig. 4.1E–F).

SPECIES EXAMINED: *Curtia conferta, C. diffusa, C. malmeana, C. obtusifolia, C. patula, C. pusilla, C. quadrifolia, C. stricta, C. tenella, C. tenuifolia,* and *C. verticillaris.*

PREVIOUS STUDIES: Gilg (1895) described the pollen grains of *Curtia* (and all Erythraeinae) as 3-colpate, smooth or finely "granular", while Köhler (1905) found them to be clearly 3-colporate and always reticulate. Maguire (1981) also reported microspinose grains in *Curtia tenuifolia*. In the same work LM micrographs of *Curtia obtusifolia* and *C. quadrifolia* were included and the latter was also illustrated by SEM micrographs.

Hockinia GARDNER (Fig. 4.2)

The pollen grains are released as monads, 3-colporate, 25–35× 22–33 μm, spheroidal to prolate, smooth, finely reticulate to perforate, foveolate, or microspinose. The colpus margin is distinct to indistinct. The smooth pollen grains observed are found in longistylous specimens (Fig. 4.2A–C), while the microspinose pollen occurred in a brevistylous specimen (Fig. 4.2D–F), a situation also found in *Curtia tenuifolia* ssp. *tenuifolia*.

SPECIES EXAMINED: *Hockinia montana.*

Saccifolium MAGUIRE & PIRES

PREVIOUS STUDIES: Maguire and Pires (1978) described the new family Saccifoliaceae, which included only the aberrant species *Saccifolium*

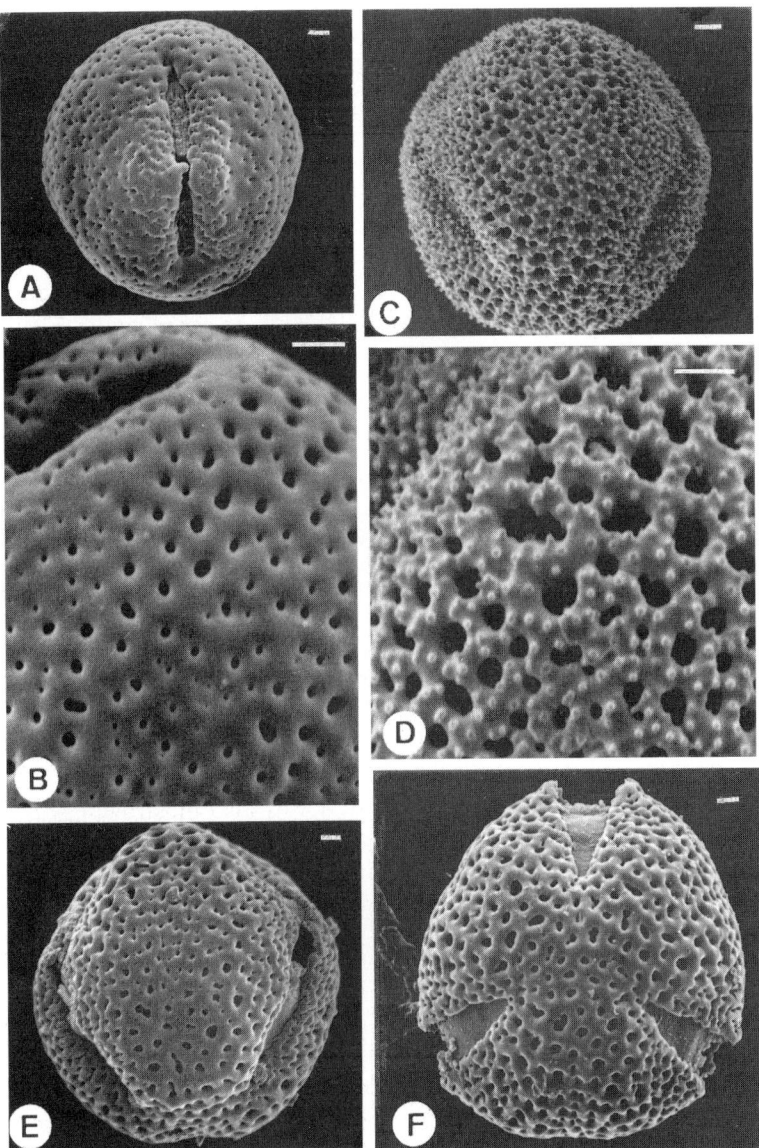

Figure 4.1. Pollen grains of *Curtia*. **A–B.** *C. tenuifolia* ssp. *tenuifolia*, longistylous (*Malme s.n.*). **A.** 3-colporate, perforate pollen in equatorial, apertural view, with equatorially constricted colpus. **B.** Detail of exine. **C–D.** *C. tenuifolia* ssp. *tenuifolia*, brevistylous (*Malme s.n.*). **C.** 3-colporate, perforate, and microspinose pollen in equatorial, mesocolpial view. **D.** Detail of exine. **E–F.** *C. obtusifolia* (*Prance et al. 15500*). **E.** 3-colporate, perforate to foveolate pollen in equatorial, mesocolpial view. **F.** The same; pollen in polar view. Scale bars = 1 μm.

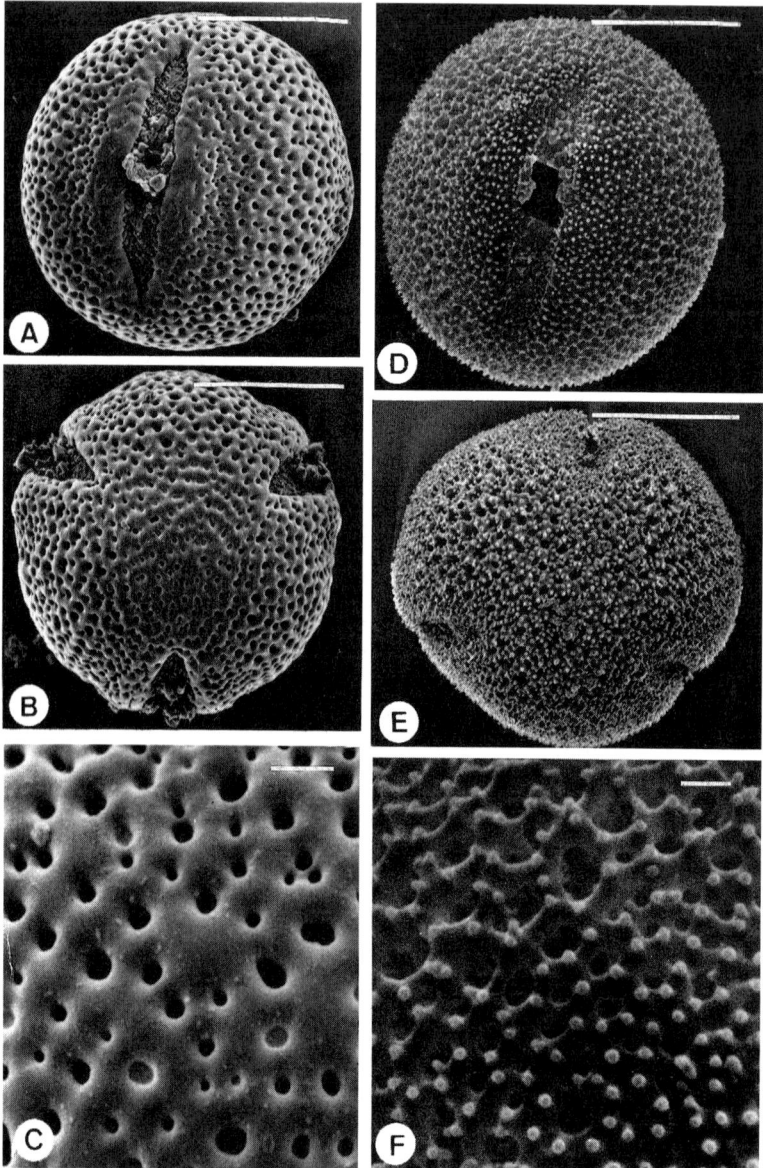

Figure 4.2. Pollen grains of *Hockinia montana*. **A–C.** Longistylous (*Brade 21208*). **A.** 3-colporate, perforate pollen in equatorial, apertural view; colpus margin distinct, smooth. **B.** The same; pollen in polar view. **C.** Detail of perforate-foveolate exine. **D–F.** Brevistylous (no collector, *2 Feb 1900*). **D.** 3-colporate, perforate to foveolate and microspinose pollen in apertural view; colpus margin indistinct. **E.** The same; pollen in polar view. **F.** Detail of exine. Scale bars: A, B, D, E = 10 μm; C, F = 1μm.

A review of palynology 411

bandeirae. The pollen grains were described as spheroidal, 3-colporate, and foveolate-perforate (Nilsson in Struwe *et al.*, 2002: Fig. 2.5). Judging from the LM and SEM micrographs published (Maguire & Pires, 1978: 241) this type of pollen appears similar to, for example, *Tapeinostemon* pollen (see Nilsson in Struwe *et al.*, 2002). The family Saccifoliaceae was regarded as associated with Gentianaceae, Menyanthaceae, and Loganiaceae by Maguire and Pires (1978), and has recently been shown to be a member of Gentianaceae (Thiv *et al.*, 1999).

Tapeinostemon BENTH. (Fig. 4.3)

The pollen grains are released as monads, 3-colporate, 17–40 × 18–37 μm, subspheroidal to prolate, reticulate to microreticulate, and the muri are smooth and relatively wide.

SPECIES EXAMINED: *Tapeinostemon capitatum*, *T. longiflorum*, *T. ptariense*, *T. spenneroides*, and *T. zamoranum*.

INTRASPECIFIC VARIATION: Considerable exine variation was found in two varieties of *Tapeinostemon longiflorum* with smooth or faintly rugulose polar areas respectively (Maguire, 1981). In two collections of *Tapeinostemon spenneroides* (Maguire, 1981) the polar areas were perforate or reticulate, respectively.

PREVIOUS STUDIES: Gilg (1895) reported the pollen grains of *Tapeinostemon* as 3-colpate and smooth, while Köhler (1905) observed 3-colporate grains with a finely reticulate exine pattern.

Tribe Exaceae

Gentianothamnus HUMBERT (Fig. 4.4)

The pollen grains are released as monads, 3-colporate, 14–24 × 13–21 μm, perforate and rugulate.

PREVIOUS STUDIES: The pollen morphology of *Gentianothamnus madagascariensis* was briefly described and illustrated in Lienau *et al.* (1986).

Tachiadenus GRISEB. (Fig. 4.5)

The pollen grains are released as monads, 3-colporate, 23–37 × 19–33 μm, finely reticulate to rugulate with unevenly thickened muri and rounded

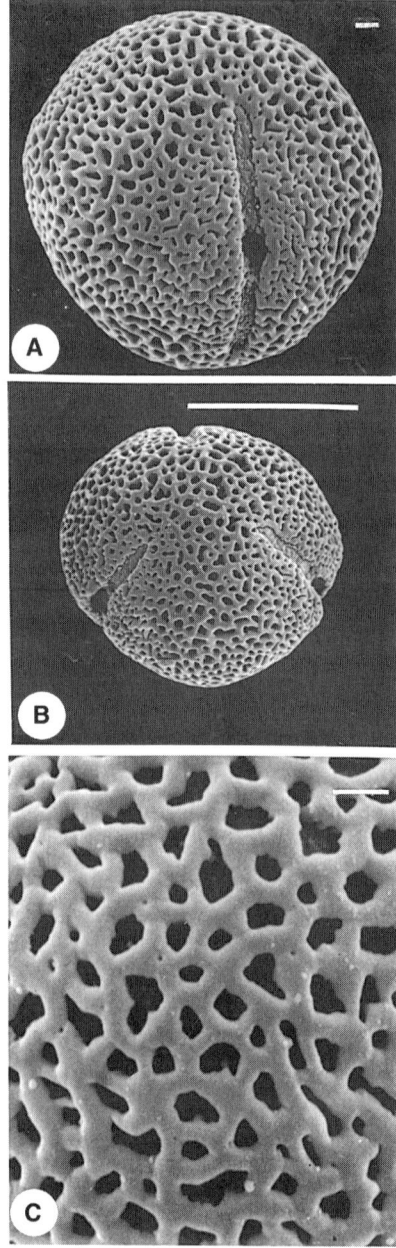

Figure 4.3. Pollen grains of *Tapeinostemon zamoranum* (*Gaarol 74940*). **A.** 3-colporate, finely reticulate pollen in equatorial, apertural view. **B.** The same; pollen in oblique polar view. **C.** Detail of finely reticulate exine. Scale bars: A, C = 1 μm; B = 10 μm.

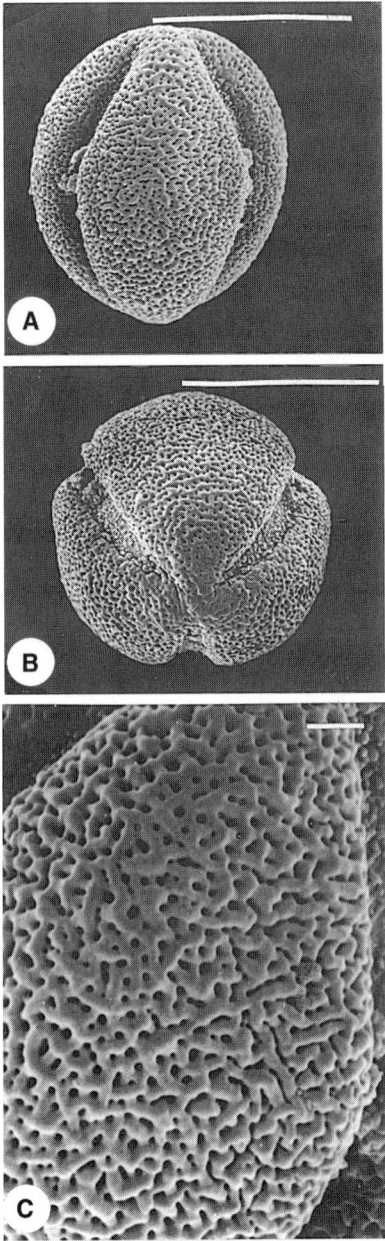

Figure 4.4. Pollen grains of *Gentianothamnus madagascariensis* (*Humbert & Cours 23715*). **A.** 3-colporate, perforate, microreticulate to rugulate pollen in equatorial, mesocolpial view. **B.** The same; pollen in slightly oblique polar view. **C.** Detail of exine. Scale bars: A, B = 10 μm; C = 1 μm.

lumina, or rugulate-perforate. The colpus margin is usually thickened and smooth.

SPECIES EXAMINED: *Tachiadenus carinatus* and *T. tubiflorus*.

INTERSPECIFIC VARIATION: The exine is reticulate in *Tachiadenus carinatus* (Fig. 4.5A–C), with a distinct and smooth colpus margin, while *T. tubiflorus* has an indistinct colpus margin and rugulate-perforate low exine relief (Fig. 4.5D–F).

PREVIOUS STUDIES: According to Gilg (1895) the pollen grains are clearly reticulate. Köhler (1905) described the pollen grains as 3-colporate and reticulate. Lienau et al. (1986) described and illustrated the reticulate pollen of *Tachiadenus longiflorus*. The size of *Tachiadenus tubiflorus* and *T.* sp. pollen was said to range between 32–37 μm (polar axis) and 27–30 μm (equatorial axis).

Tribe Chironieae

Coutoubea AUBL. (Fig. 4.6)

The pollen grains are united in tetrahedral tetrads that are 51–65 μm in diameter. The single grains are 3-colpate-colporate to porate, finely to coarsely reticulate, and in some species there are markedly larger meshes at the polar areas. The lumina are smooth with few to many granules or with elongated elements of different size and shape. The inside of the exine is densely granular. The aperture conditions vary from colpate (colpi of two contiguous grains concurrent) to porate (pores arranged pairwise) with the apertures situated in a distinct, finely reticulate to perforate or smooth area (*Coutoubea*-type).

SPECIES EXAMINED: *Coutoubea minor*, *C. ramosa*, *C. reflexa*, and *C. spicata*.

POLLEN TYPE: *Coutoubea* (defined here).

INTERSPECIFIC VARIATION: *Coutoubea minor* (Fig. 4.6A) and *C. spicata* have a relatively equal-sized reticulum while that of *C. ramosa* (Fig. 4.6B–C) and of *C. reflexa* (Fig. 4.6D–F) is more differentiated.

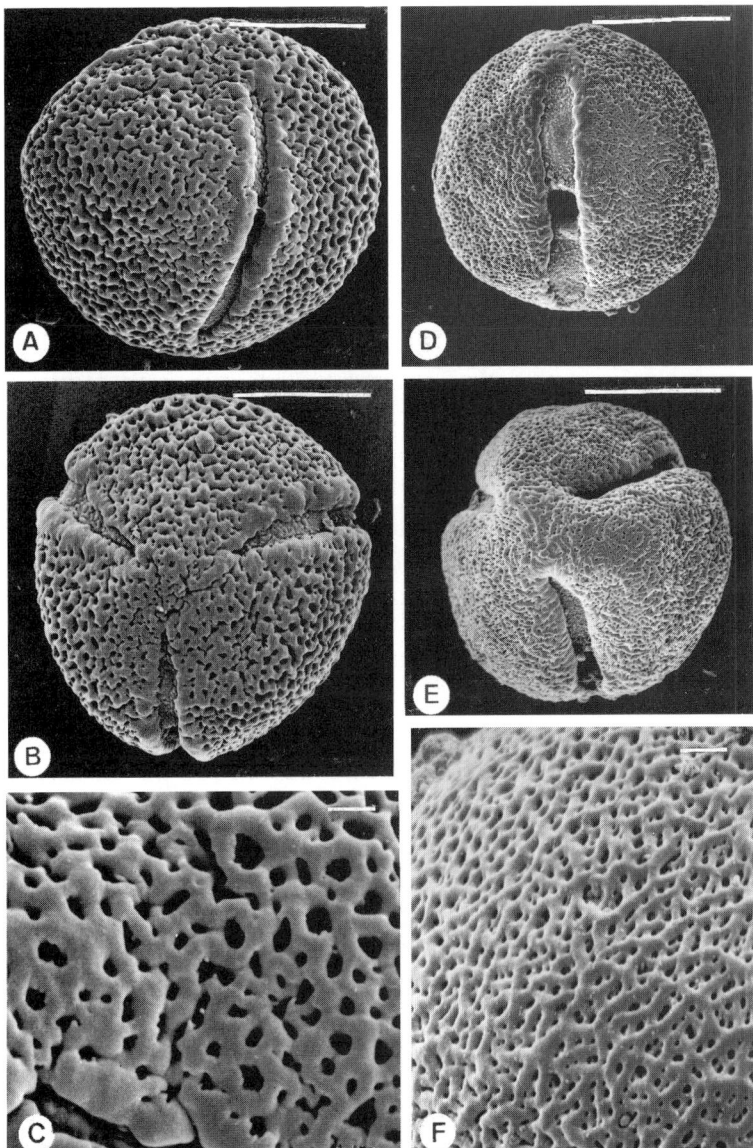

Figure 4.5. Pollen grains of *Tachiadenus*. **A–C.** *T. carinatus* (*Klackenberg 871124-5*). **A.** 3-colporate, perforate to foveolate (reticulate) pollen in oblique equatorial, apertural view; colpus bordered by a distinct, smooth colpus margin, os indistinct. **B.** The same; pollen in polar view. **C.** Detail of perforate to foveolate exine, near colpus. **D–F.** *T. tubiflorus* (*Capuron s.n.*). **D.** 3-colporate, perforate-rugulate pollen in equatorial, apertural view. **E.** The same; pollen in oblique polar view. **F.** Detail of perforate-rugulate exine. Scale bars: A, B, D, E = 10 μm; C, F = 1 μm.

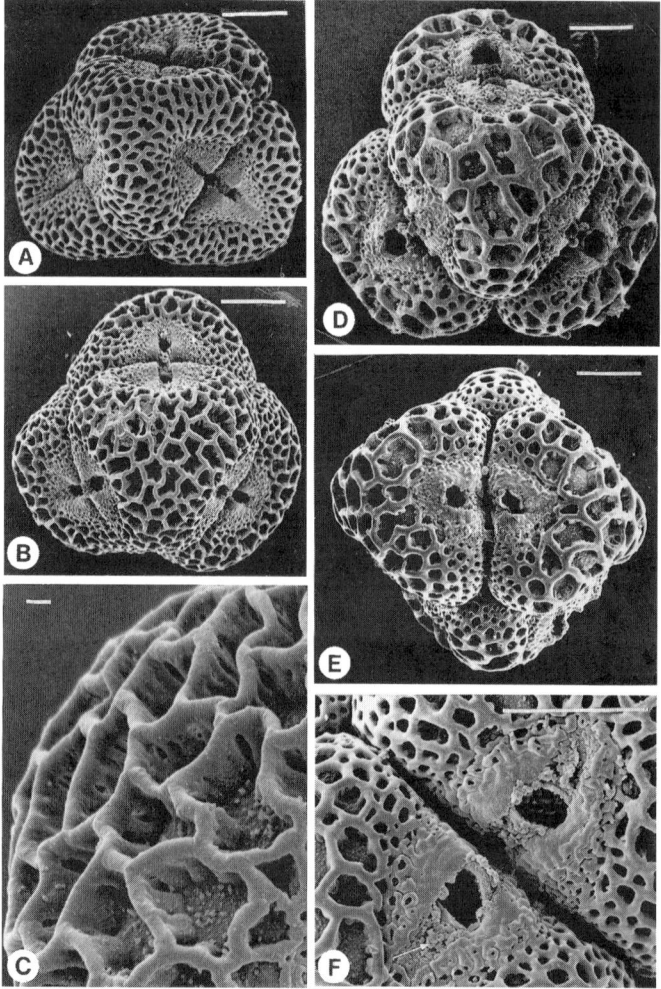

Figure 4.6. Pollen grains of *Coutoubea* (*Coutoubea*-type). **A.** *C. minor* (*Kral et al. 71729*). Tetrad in tetrahedral position; single grains 3-colporate, with equal-sized reticulum and smooth, perforated apertural areas. **B–C.** *C. ramosa* (*Tamayo 3460*). **B.** Tetrad in tetrahedral position; single grains 3-colporate, reticulate, with smooth apertural areas. The reticulum at the poles is relatively coarse and heterobrochate. **C.** Detail of exine showing sinuous muri supported by relatively tall columellae; the bottoms of lumina finely granular. **D–F.** *C. reflexa* (*Maguire et al. 36047*). **D.** Tetrad in tetrahedral position; single grains indistinctly 3-colporate to porate with smooth apertural areas. The reticulum is coarse at the polar areas, markedly diminishing in size toward the equator. **E.** The same; tetrad in decussate position with two distinct pores visible and smooth apertural areas. **F.** Detail of two smooth apertural areas beset with tiny, rounded structures at the edges (arrow), transgressing into the adjacent reticulate areas. The colpi are only faintly delimited. Scale bars: A, B, D–F = 10 μm; C = 1 μm.

PREVIOUS STUDIES: Gilg (1895) described the pollen of *Coutoubea* as having warts arranged in rows or indistinctly reticulate, while Köhler (1905) described and illustrated the pollen grains of *C. ramosa* as 3-porate, with relatively coarse reticulum, which is confirmed in the present study. Elias and Robyns (1975) described pollen of *Coutoubea spicata*, which was also illustrated by Agababyan and Tumanyan (1977). In addition, Maguire and Boom (1989) published several LM and SEM micrographs of *Coutoubea spicata* tetrads in different positions.

Deianira CHAM. & SCHLTDL. (Fig. 4.7)

The pollen grains are united in tetrahedral tetrads that are 33–55 μm in diameter. The single grains are 3-colpate and reticulate. Lumina are 1–2 μm in diameter and the muri appear suspended and slightly segmented due to thin, ring-like substructural features. The inside of the exine is smooth with the internal walls distinctly perforated (*Deianira*-type).

SPECIES EXAMINED: *Deianira cordifolia, D. cyathifolia, D. erubescens, D. foliosa, D. nervosa,* and *D. pallescens.*

POLLEN TYPE: *Deianira* (defined here).

PREVIOUS STUDIES: Gilg (1895) described the pollen of *Deianira* as finely reticulate to almost perforate, relating *Deianira* to *Helia* with reference to a similar exine pattern. Köhler (1905) did not observe any colpi (only pores) and the pollen grains were said to be reticulate to finely reticulate. Earlier observations agree with those of the present study.

Schultesia MART. (Figs. 4.8–4.9)

The pollen grains are united in tetrahedral tetrads that are 44–116 μm in diameter. The single grains are 3-porate and coarsely reticulate in the polar area with the reticulum diminishing in size toward the distinct, smooth to finely granular apertural area. The muri are relatively narrow and pointed and the lumina are filled with granules. The inside of the exine is finely granular.

SPECIES EXAMINED: *Schultesia benthamiana, S. brachyptera, S. guianensis, S. lisianthoides, S. pohliana, S. stenophylla,* and *S. subcrenata.*

POLLEN TYPE: *Coutoubea* (see *Coutoubea*).

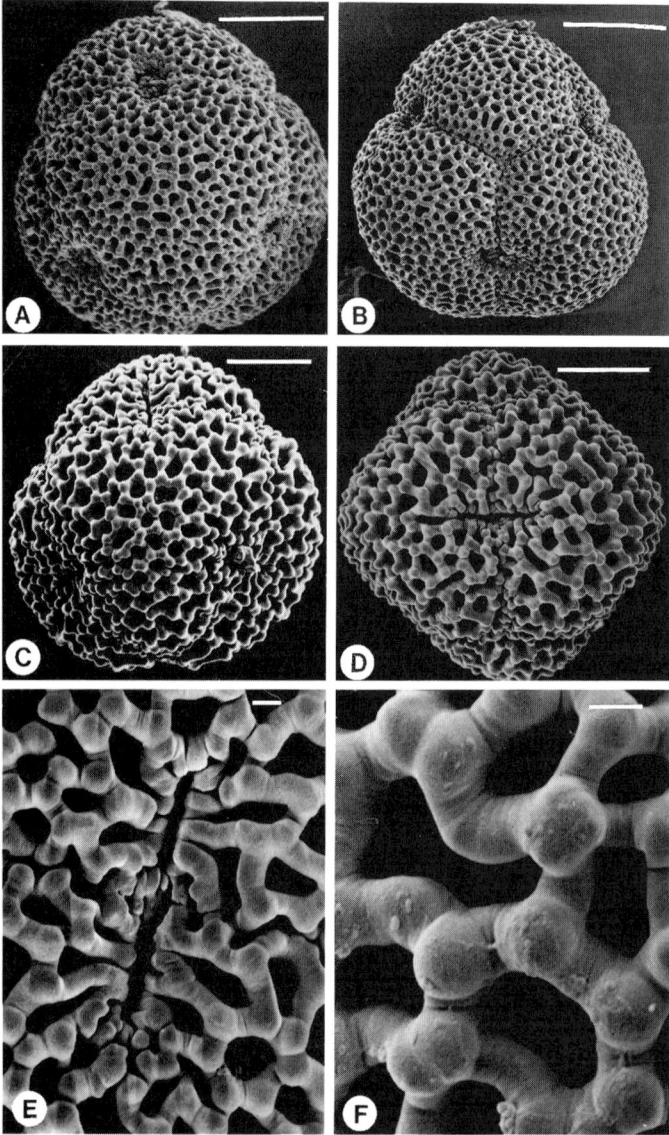

Figure 4.7. Pollen grains of *Deianira* (*Deianira*-type). **A–B.** *D. cyathifolia* (*Malme 3323 B*). **A.** Porate tetrad in tetrahedral position. **B.** The same; view of the back of tetrahedral tetrad. **C–F.** *D. nervosa* (*Smith 6964*). **C.** Colpate to colporate tetrad in tetrahedral position. **D.** The same, tetrad in decussate position showing a narrow colpus shared by two contiguous grains. **E.** Detail of exine at the colpus region. **F.** The same; detail of exine, muri with thin, ring-like substructures; the muri appear suspended between the columellae. Scale bars: A–D = 10 μm; E, F = 1 μm.

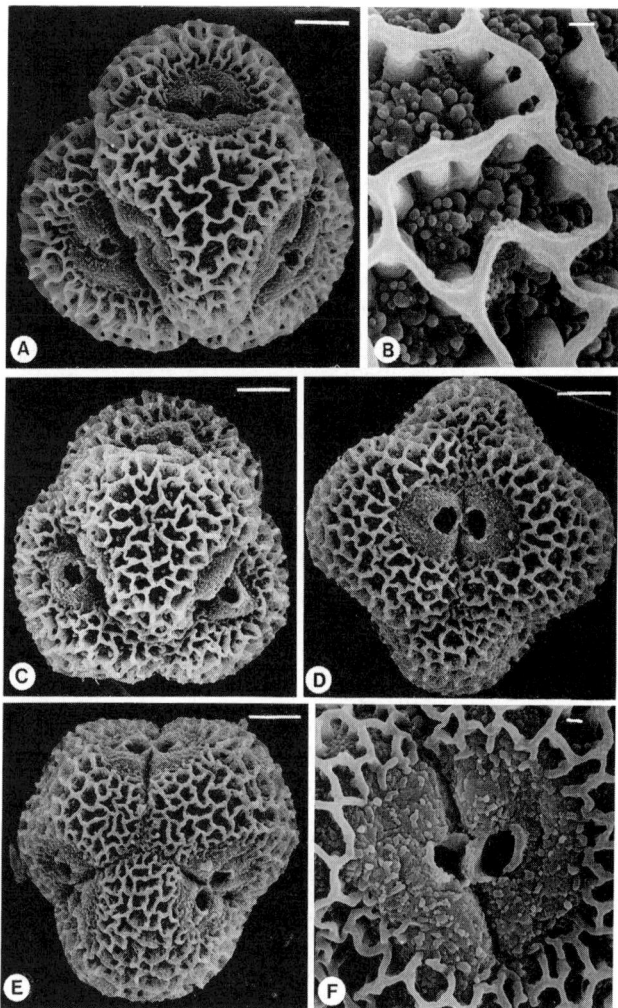

Figure 4.8. Pollen grains of *Schultesia* (*Coutoubea*-type). **A–B.** *S. benthamiana* (*Maas et al. 3661*). **A.** Tetrad in tetrahedral position; individual grains 3-porate, pores situated in smooth areas, exine reticulate with relatively coarse reticulum at the poles. **B.** Detail of reticulum with sinuous muri and lumina densely beset with granules or pila. **C–D.** *S. brachyptera* (*Maas et al. 7730*). **C.** Tetrad in tetrahedral position; individual grains 3-porate, pores located in smooth, apertural areas, reticulum of equal size all over. **D.** The same; tetrad in decussate position showing pores located in smooth, rounded areas, and an equal-sized reticulum. **E–F.** *S. guianensis* (*Jansen-Jacobs et al. s.n.*). **E.** Tetrahedral tetrads from below with porate, reticulate single grains. **F.** Detail of exine in the aperture region with pores from two adjacent grains located in finely granular areas surrounded by reticulum. Scale bars: A, C–E = 10 μm; B, F = 1 μm.

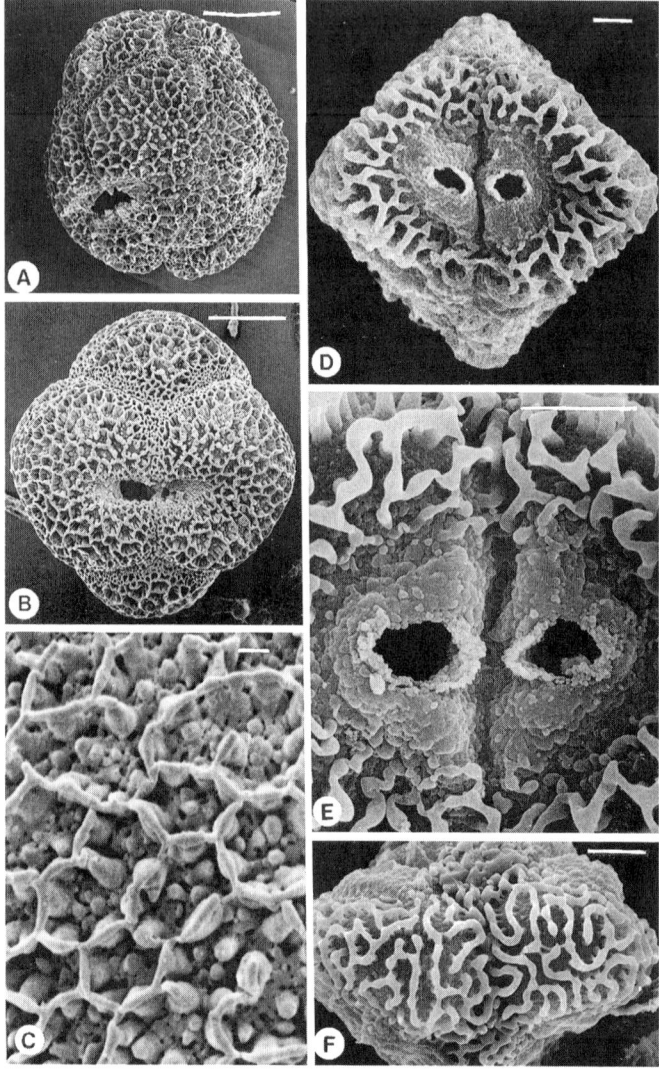

Figure 4.9. Pollen grains of *Schultesia* (*Coutoubea*-type). A–C. *S. lisianthoides* (= *Xestaea lisianthoides*; Stevens & Krukoff 8224). **A.** Tetrad in oblique tetrahedral view; individual grains 3-colporate, reticulate pattern relatively fine and equal-sized. **B.** The same; tetrad in decussate view with the pores/ora of two adjacent grains meeting. **C.** Detail of reticulate exine pattern; sinuous muri supported by relatively stout columellae, lumina with numerous granular processes. **D–F.** *S. pohliana* (*Irwin et al. 55237*). **D.** Tetrad in decussate position with square-shaped outline; the pores of adjacent grains located in smooth, apertural areas, are not fused. **E.** Detail of exine at the pores. **F.** The same; detail of reticulate pattern showing sinuous to vermicular muri. Scale bars: A, B, D–F = 10 μm; C = 1 μm.

A review of palynology 421

INTERSPECIFIC VARIATION: The shape of the tetrads in decussate position varies from relatively rounded to square-shaped due to the pointedness of the poles in side view. *Schultesia pohliana* (Fig. 4.9D–F) and *S. subcrenata* are markedly square-shaped in decussate view, i.e., with pointed poles. In *Schultesia pohliana* (Fig. 4.9F) the reticulum appears incomplete due to the fragmented muri.

PREVIOUS STUDIES: Gilg (1895) applied the same pollen description to *Schultesia* and *Coutoubea*, with both genera being described as having exine with aligned bumps and warts. Köhler (1905) also emphasized the similarity between the two genera, referring to the reticulate exine. In *Schultesia stenophylla* he observed curved muri with thickened nodes at their connection points, appearing as bumps, which explained Gilg's observations. *Schultesia benthamiana*, *S. brachyptera*, *S. pohliana*, and *S. subcrenata* were said to have straight rather than curved muri (cf. *S. stenophylla*) and *S. lisianthoides* was compared to *Coutoubea ramosa*. In the present pollen study the similarity between *Coutoubea* and *Schultesia* is confirmed. Possibly the muri are more keeled in *Schultesia* than in *Coutoubea* with reference to the studied taxa.

Schultesia lisianthoides (Fig. 4.9A–C), treated as the monotypic genus *Xestaea* in Struwe *et al.* (2002), deviates by the size of tetrads and their finer reticulum from all other *Schultesia* species (cf. Elias & Robyns, 1975). No differences as to muri (curved vs. straight) between any *Schultesia* taxa could be found. *Schultesia guianensis* was reported by Elias and Robyns (1975) as having markedly thick columellae, *S. heterophylla* as having (commonly) rhomboidal tetrads, and *S. lisianthoides* as having relatively small tetrads and fine reticulum. Elias and Robyns also pointed out the resemblance between *Coutoubea* and *Schultesia* pollen. *Schultesia guianensis* was illustrated by LM micrographs in Maguire and Boom (1989). Agababyan and Tumanyan (1977) described four taxa and presented illustrations of *Schultesia angustifolia*.

Tribe Helieae

Adenolisianthus (PROGEL) GILG (Fig. 4.10A)

The pollen grains are united in tetrahedral tetrads that are 55–73 μm in diameter. The single grains are indistinctly 3-colporate to 3-porate, finely to coarsely reticulate with enlarged meshes up to 10 μm in diameter around the equator. The muri are of markedly uneven width and thickness,

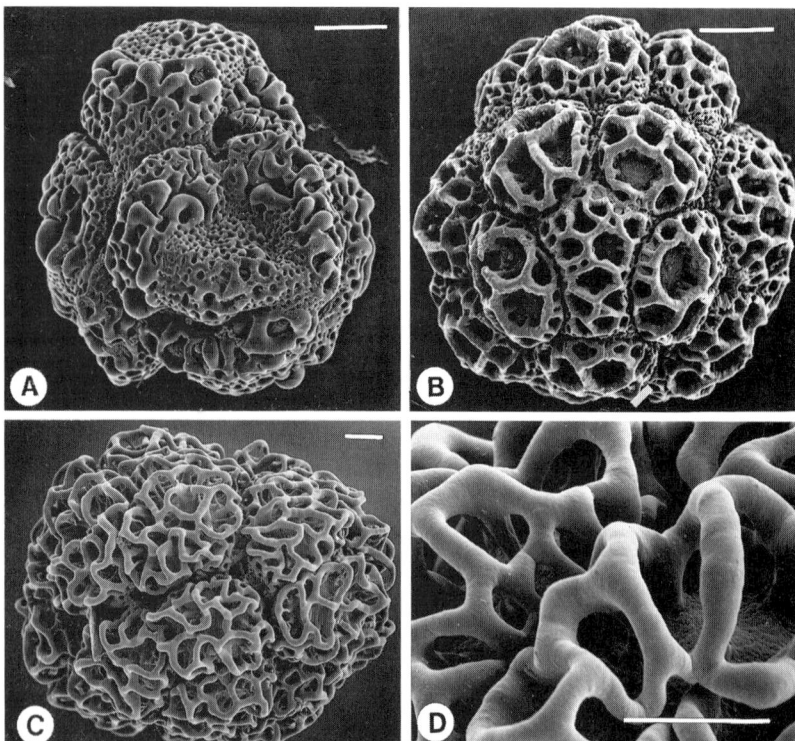

Figure 4.10. Pollen grains of *Adenolisianthus* (*Chelonoides*-type) and *Calolisianthus* (*Speciosus*-type and *Amplissimus*-type). **A.** *A. arboreus* (*Wurdack & Adderley 43722*). Tetrad in tetrahedral position, perforate-reticulate with coarse reticulum around the equator and at pole. **B.** *C. speciosus* (*Speciosus*-type; *Järneby s.n.*). Tetrad in tetrahedral position with islands of coarse, raised reticulum. **C.** *C. amplissimus* (*Amplissimus*-type; *Smith 6721*). Polyad, with coarse and raised, enveloping reticulum. **D.** *C. acutangulus* (*Amplissimus*-type; *Malme 1378 B*). Detail of raised, loop-like reticulum. Scale bars: A–C = 10 μm; D = 1 μm.

particularly at the equatorial or subequatorial zone but also at the poles, and sometimes are fragmented into rounded to elongated elements, 4–5 μm wide, or clavae-like processes, up to 4–5 μm in height. The inside of the exine is smooth.

SPECIES EXAMINED: *Adenolisianthus arboreus*.

POLLEN TYPE: *Chelonoides* (Nilsson, 1970).

A review of palynology 423

PREVIOUS STUDIES: Gilg (1895) noted that thickened muri were absent at the finely reticulate to almost perforate poles, while Köhler (1905) found that the 3-porate grains had widened meshes also at the poles, which is in full agreement with later published results (Nilsson, 1970) and the present study. Maguire and Boom (1989) provided a short description and illustrations (LM, SEM) of the tetrads (under the name *Chelonanthus fruticosus*). Note that the legend to their fig. 26 should refer to *Chelonanthus fruticosus* (Maguire & Boom, 1989: 40).

Aripuana STRUWE, MAAS, & V. A. ALBERT

The pollen grains are united in tetrahedral tetrads that are 81–91 μm in diameter. The single grains are 3-colp(or)ate and reticulate with only slightly coarser reticulum around the equator compared with the *Chelonoides*-type above (*Aripuana*-type).

SPECIES EXAMINED: *Aripuana cullmaniorum*.

POLLEN TYPE: *Aripuana* (defined here).

PREVIOUS STUDY: Struwe *et al.* (1997) provided SEM micrographs with the description of this genus.

Calolisianthus (GRISEB.) GILG (Fig. 4.10B–D)

The pollen grains are united either in tetrads or in polyads and represent two different pollen types. The tetrahedral, rarely rhomboidal, tetrads range from 52 to 77 μm in diameter. The single grains are generally 3-colporate to 3-porate, less commonly 3-colpate, finely to coarsely reticulate, with islands of coarse reticulum separated by grooves and spread over the distal face (Fig. 4.10B; *Speciosus*-type (Nilsson, 1970)). The polyads range in size from 108 to 225 μm and are coarsely reticulate with raised, loop-like muri (up to 17 μm in height) enveloping the firmly united tetrads (Fig. 4.10C–D; *Amplissimus*-type (Nilsson, 1970)). The apertures in the polyads are not always discernible. However, in cross-sections (TEM) the aperture margins are lamellate. In both tetrads and polyads the internal walls are perforated. *Calolisianthus tatei*, with spiny polyads (*Irlbachia*-type; Nilsson, 1970), has been transferred to the genus *Irlbachia* by Maguire (1981), and *C. imthurnianus*, with polar, isodiametric to elongated or clavae-like processes, was incorporated in the genus *Celiantha* (Maguire, 1981).

SPECIES EXAMINED: *Calolisianthus acutangulus, C. amplissimus, C. macranthus, C. pedunculatus, C. pendulus, C. pulcherrimus,* and *C. speciosus.* "*Lisianthus*" *alpestris*, "*L.*" *elegans*, and "*L.*" *tenuifolius* also have *Speciosus*-type of pollen.

POLLEN TYPES: *Amplissimus* and *Speciosus* (Nilsson, 1970).

PREVIOUS STUDIES: Gilg (1895) reported pollen tetrads with a regular, equal-sized reticulum (as in *Lagenanthus*). Neither polyads nor tetrads with differentiated reticulum were recognized by Gilg. Köhler (1905) also stressed the similarity between the 3-porate *Lagenanthus* pollen and the 3-colporate *Calolisianthus* pollen (*C. speciosus*). His observations and interpretation of displacement of the grooves in relation to the apertures in *Calolisianthus* are in full agreement with the present study, while similarity between *Lagenanthus* and *Calolisianthus* could not be confirmed. "*Lisianthus*" *alpestris*, "*L.*" *elegans*, and "*L.*" *tenuifolius* have pollen tetrads of *Speciosus*-type.

Celiantha MAGUIRE (Fig. 4.11)

The pollen grains are united in polyads that are 120–148 μm in diameter. Individual tetrads are not distinguishable. On top of each single grain there are some irregularly shaped, rounded to elongate pila-like processes surrounding a single globule, with or without supporting stilts rooted in the exine. The internal walls are perforated. The pore-like apertures are not always visible from the outside of the polyad, but are traceable in sections, and are clearly delimited by thickened, lamellar margins.

SPECIES EXAMINED: *Celiantha bella, C. chimantensis,* and *C. imthurnianus.*

POLLEN TYPE: *Imthurnianus* (Nilsson, 1970).

INTERSPECIFIC VARIATION: The two specimens examined of *Celiantha bella* are slightly different. In one collection (*Maguire et al. 42096*) there are rounded to elongated pieces of exine supported by columellae at the polar area (Fig. 4.11E), but in another collection (*Silva & Brazão 60921*) the single grains have globular processes only (Fig. 4.11A–B).

PREVIOUS STUDIES: The polyads of *Celiantha imthurniana* were described by Nilsson (1970). The polyads of *Celiantha bella* and *C. chimantensis* were illustrated by Maguire (1981).

A review of palynology 425

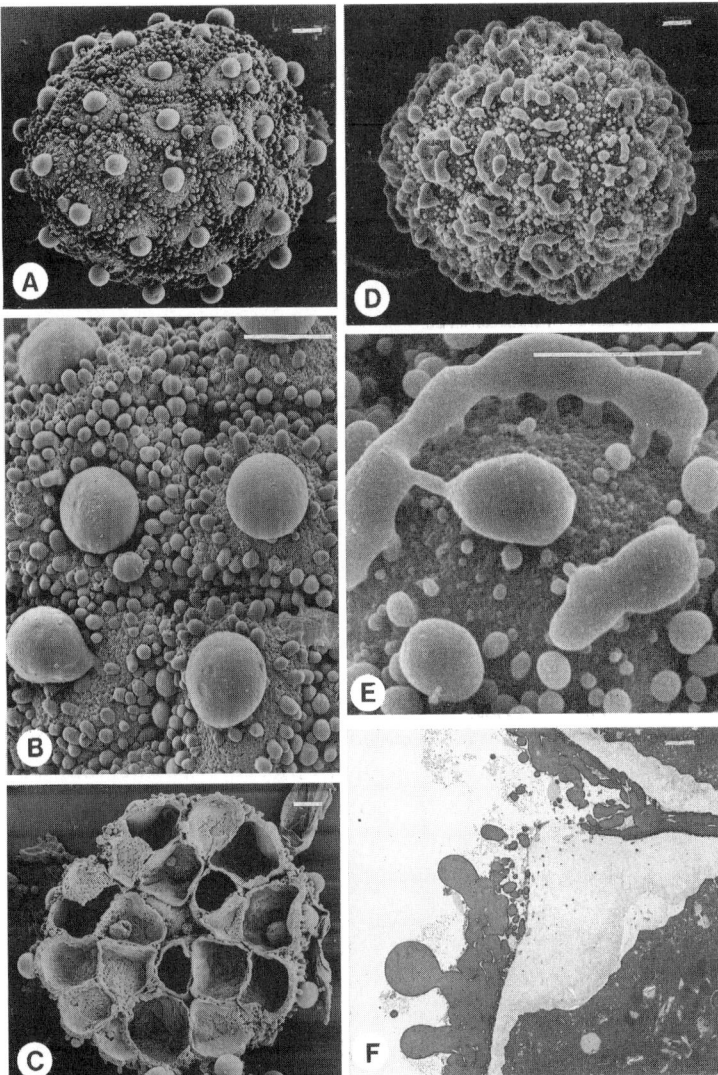

Figure 4.11. Pollen grains of *Celiantha bella* (*Imthurnianus*-type). **A–C.** (*Silva & Brazão 60921*). **A.** Polyad; single grains delimited from each other by numerous small globules; one large globule usually with columellar support on top of each grain (monad). **B.** Detail of the exine. **C.** Fractured polyad showing perforated inner walls between grains. **D–F.** (*Maguire et al. 42096*). **D.** Polyad; distal surface of the monads beset with elongated elements and pila. **E.** Detail of exine with elongated, mural fragments and globules (pila), supported by columellae. **F.** Section through an apertural area (pore) showing unstratified exine with pila and clava; the exine (foot layer) is lamellar at the pore margin. The intine is distinctly thickened (TEM). Scale bars: A–E = 10 μm; F = 1 μm.

Chelonanthus GILG (Figs. 4.12–4.13, 4.17A–C)

The pollen grains are united in tetrahedral tetrads that are 46–84 μm in diameter (*Chelonoides*-type), or are aggregated in polyads consisting of a complex of several loosely united, modified tetrads, and then 63–255 μm in diameter (*Uliginosus*-type). The single grains of the tetrads (*Chelonoides*-type; Fig. 4.12) are 3-colporate-porate. The equatorial zone and also polar areas consist of coarser reticulum with lumina 10 μm wide (or more) and irregularly thickened muri. The muri are often fragmented into elongate, irregular pieces or globules. The polar area is reticulate, rarely pilate. The inside of the exine is smooth to pitted and the internal walls are perforated.

The species included in the *Uliginosus*-type (Fig. 4.13) shed pollen in polyads, the tetrads of which have an equatorial-subequatorial zone and/or polar area with muri strongly projecting as loops and united apically in rings. *Chelonanthus campanuloides* (Fig. 4.17A–C) appears transitional to the *Irlbachia*-type, having a few polar spines that are free or united at their base. *Chelonanthus cardonae*, with spiny polyads, was transferred to *Irlbachia* by Maguire (1981).

SPECIES EXAMINED: *Chelonanthus acutangulus*, *C. alatus*, *C. albus*, *C. angustifolius*, *C. bifidus*, *C. campanuloides*, *C. camporum*, *C. candidus*, *C. chelonoides*, *C. grandiflorus*, *C. schomburgkii*, *C. spruceanus*, *C. uliginosus*, and *C. viridiflorus*. *Chelonanthus purpurascens*, with *Uliginosus*-type pollen, is synonymous with *C. uliginosus* (Struwe & Albert, 1998).

POLLEN TYPES: *Chelonoides* and *Uliginosus* (Nilsson, 1970).

INTER- AND INTERSPECIFIC VARIATION: *Chelonanthus alatus* (*Prance et al. 14376*; Fig. 4.12A–C) has a typical, differentiated reticulate pattern with unevenly wide, thickened muri; however, in other collections the muri are transformed to irregular, elongated pieces or rounded globules and pila. Muri fragmented into elongated pieces were also found in some collections of *Chelonanthus angustifolius* (Fig. 4.12F), *C. bifidus*, and *C. candidus*.

PREVIOUS STUDIES: Gilg (1895) reported the pollen grains of some *Chelonanthus* species to have thickened, mostly elongated exine processes. *Chelonanthus uliginosus* differed by having tetrads arranged in packages, and at the poles of the grains, the projecting exine was united into rings (Gilg, 1895). Gilg's description agrees largely with those of the *Chelonoides*- and *Uliginosus*-types by Nilsson (1970).

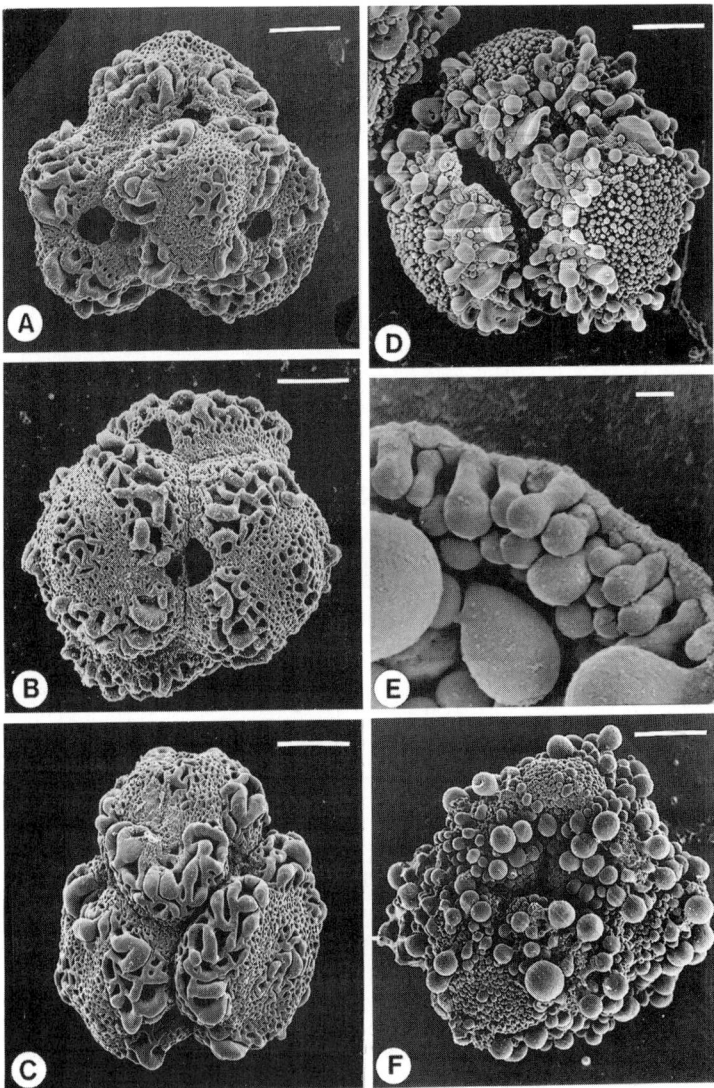

Figure 4.12. Pollen grains of *Chelonanthus* (*Chelonoides*-type). **A–C.** *C. alatus* (*Prance et al. 14376*). **A.** Porate tetrad in tetrahedral position with distinct reticulum of elongated, unevenly thickened muri around the equator and at the distal pole. **B.** The same; tetrad in decussate position; two grains sharing pore(s). **C.** Tetrad viewed from the back. **D–E.** *C. albus* (*Wurdack & Adderley 43593*). **D.** Back of tetrahedral tetrad showing an equatorial zone of elongated to rounded muri (processes). **E.** Detail of exine with isodiametric to slightly elongated macroprocesses, and numerous pila or skittle-like processes. **F.** *C. angustifolius* (*Maguire 33712*). Tetrad in decussate position with the equatorial reticulum fragmented into larger and smaller globules. Scale bars: A–D, F = 10 μm; E = 1 μm.

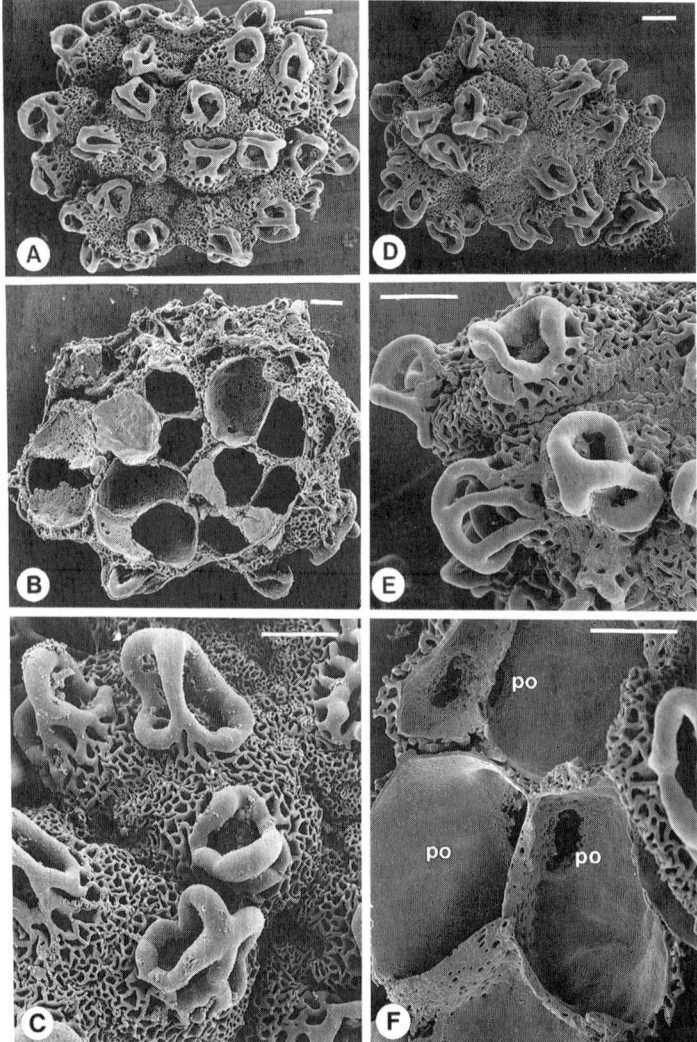

Figure 4.13. Pollen grains of *Chelonanthus* (*Uliginosus*-type). **A–C.** *C. uliginosus* (*Cox & Hubbard 142*). **A.** Polyad of loosely united tetrads with polar to subpolar loops (reticulum raised from the general surface of the grains). **B.** The same; fractured polyad with perforated inner walls between the monads. **C.** Detail of exine of some tetrads, view of back, which have prominently raised loops at the polar areas. **D–F.** *C. purpurascens* (*Rova et al. 1902*). **D.** Polyad finely reticulate with raised muri united at the apices to form loop-like structures at the poles of single grains (cf. *C. uliginosus*, Fig. 4.13A). **E.** Detail of finely reticulate exine and mural loop-like structures. **F.** Part of a fractured polyad showing some grains from the inside; the outer wall is smooth while the inner walls are perforate; three pores visible ("po"). Scale bars = 10 μm.

A review of palynology 429

Köhler (1905) pointed out the resemblance between pollen tetrads of *Adenolisianthus* and *Chelonanthus*. He noted that the exine was finely reticulate at the sides and that the muri at the polar area were thickened, raised, and apically fused to rings. In the same work he said that *Chelonanthus campanuloides* had pollen tetrads fused together in packages, as in *C. uliginosus* and *C. fistulosus*. Köhler's more detailed description agrees well with what was found in the present study. Maguire and Boom (1989) published LM and SEM micrographs of *Chelonanthus alatus*, *C. albus*, *C. angustifolius*, *C. chelonoides*, *C. schomburgkii*, and *C. uliginosus*.

Chorisepalum GLEASON & WODEHOUSE (Fig. 4.14)

The pollen grains are released as monads, 3-colporate, 28–54 × 20–46 μm, coarsely reticulate, and with densely granular lumina.

SPECIES EXAMINED: *Chorisepalum carnosum*, *C. ovatum*, *C. psychotrioides*, and *C. rotundifolium*.

PREVIOUS STUDIES: Maguire (1981) provided a pollen description of *Chorisepalum* and illustrated three species: *Chorisepalum carnosum*, *C. ovatum* var. *sipapoanum*, and *C. psychotrioides*. The pollen grains of the different species appear very similar, with coarsely reticulate exine and relatively wide muri.

Helia MART. (Fig. 4.15)

The pollen grains are united in tetrahedral tetrads that are 53–67 μm in diameter. The single grains are 3-colporate to porate, finely to coarsely reticulate, and have an equatorial zone of larger meshes. At the distal pole the exine is very finely reticulate to perforate.

SPECIES EXAMINED: *Helia brevifolia*, *H. loesneriana*, *H. martii*, *H. micrantha*, *H. oblongifolia*, and *H. spathulata*.

POLLEN TYPE: *Helia* (Nilsson, 1970).

INTERSPECIFIC VARIATION: In one collection of *Helia oblongifolia* the coarse reticulum extends toward the distal pole (cf. *Chelonoides*-type).

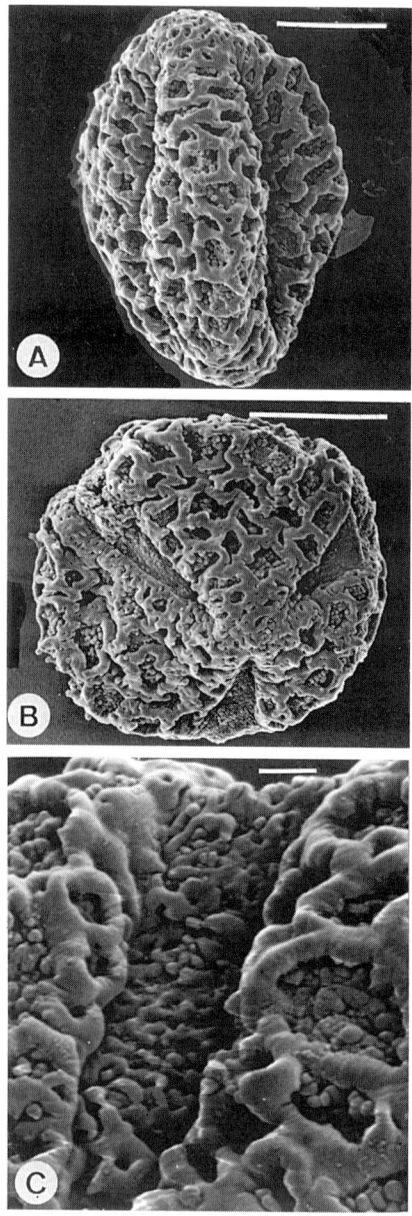

Figure 4.14. Pollen grains of *Chorisepalum carnosum* (aff. *Glabra*-type; *Maguire et al. 53823*). **A.** 3-colporate, reticulate pollen in equatorial, mesocolpial view. **B.** The same; pollen in oblique polar view. **C.** Detail of exine with relatively wide muri and densely granular lumina. Scale bars: A–B = 10 μm; C = 1 μm.

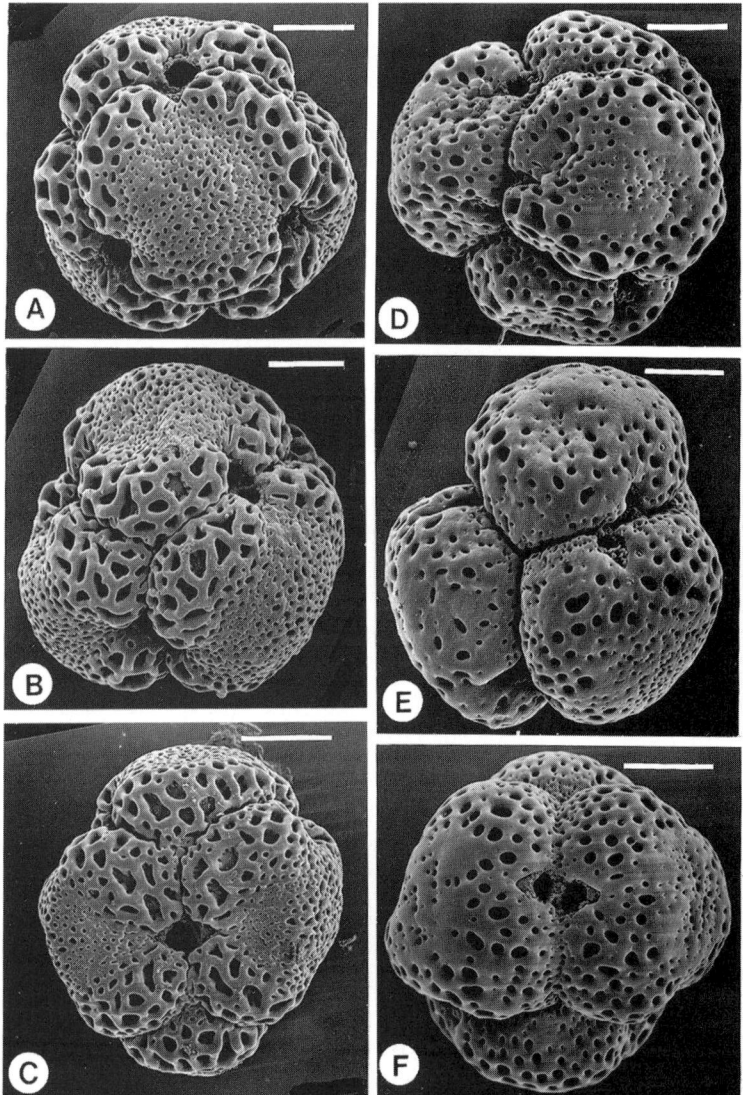

Figure 4.15. Pollen grains of *Helia* (*Helia*-type). **A–C.** *H. brevifolia* (*Jorgensen 4837*). **A.** Porate-colporate tetrad in tetrahedral position, with a prominently raised equatorial reticulum and perforate polar areas. **B.** The same; tetrad viewed from the back. **C.** Tetrad in decussate position showing two adjacent grains sharing a colpus with os. **D–F.** *H. oblongifolia* (*Dusén 16537*). **D.** Porate to colporate tetrad in tetrahedral position; the equatorial reticulum is only slightly raised, with rounded to oval-shaped lumina, the poles sparsely perforate. **E.** The same; tetrad viewed from the back. **F.** Tetrad in decussate position; adjacent grains have a common colpus. Scale bars = 10 μm.

PREVIOUS STUDIES: Gilg (1895) pointed out the microreticulate to perforate exine, which agrees with the present study. Erdtman (1952) published a palynogram of *Helia brevifolia*.

Irlbachia MART. (incl. *Pagaea* GRISEB. and *Brachycodon* BENTH.) (Figs. 4.16–4.17D–F)

The polyads consist of spiny, obscurely porate tetrads that are 110–203 μm in diameter. The single grains have small groups of spines at the distal polar area that are interconnected by a fine suspended network. The spines vary in shape and are either apically pointed, blunt, or obpyriform.

SPECIES EXAMINED: *Irlbachia cardonae, I. elegans, I. nemorosa, I. paruana, I. phelpsiana, I. plantaginifolia, I. poeppigii, I. pratensis, I. pumila, I. ramosissima, I. recurva,* and *I. subcordata*. "*Lisianthus*" *acutilobus* (Fig. 4.16D–E) and "*L.*" *scabridulus* also have *Irlbachia*-type of pollen. *Irlbachia pratensis* pollen seems transitional between the *Irlbachia*- and *Uliginosus*-types.

POLLEN TYPE: *Irlbachia* (Nilsson, 1970).

INTERSPECIFIC VARIATION: The shape and size of the spines varies between different collections of *Irlbachia cardonae*.

PREVIOUS STUDIES: Gilg (1895) investigated pollen of only one species of *Irlbachia, I. caerulescens* (now in *Tetrapollinia*). The tetrads of this species were said to have numerous spines, an observation in accordance with Köhler (1905) and the present study. *Pagaea sensu* Gilg (i.e., *Irlbachia*) was reported to have aggregated tetrads with 4–5 prominent spines at the poles. Köhler (1905) gave a more detailed description and illustration indicating that the spines are partly interconnected and surrounding a granular pole. These descriptions are confirmed by the present study.

Progel (1865) and Maguire (1981) included *Pagaea* in *Irlbachia* and illustrated the spiny polyads of *Irlbachia elegans, I. nemorosa, I. paruana, I. phelpsiana, I. plantaginifolia, I. pumila,* and *I. tatei* with LM and SEM micrographs (Maguire, 1981). The pollen grains of *Irlbachia plantaginifolia* do not have spines, but instead have a number of globules at the distal pole (cf. *Celiantha*). *Irlbachia caerulescens* was transferred to a new genus, *Tetrapollinia*, by Maguire (1981). *Irlbachia pratensis* has polyads intermediate between the *Uliginosus*- and *Irlbachia*-types (cf. *Chelonanthus campestris*) and was recently included in *Irlbachia* (Struwe et al., 1999).

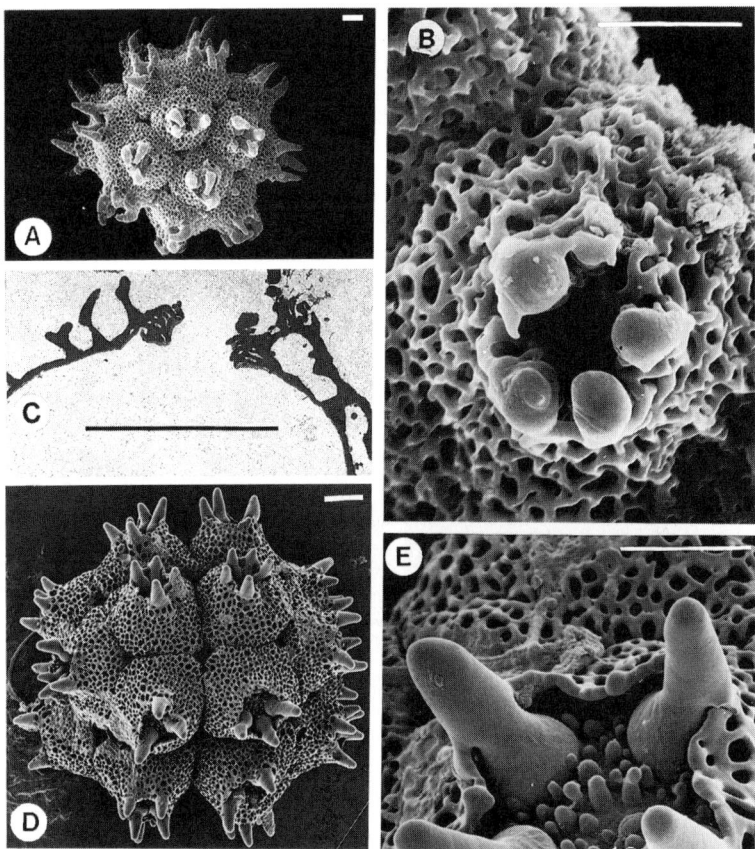

Figure 4.16. Pollen grains of *Irlbachia* (*Irlbachia*-type). **A–C.** *I. cardonae* (*Tate 1360*). **A.** Polyad with groups of pointed, polar spines; exine finely reticulate. **B.** Detail of exine at a pole with some spines interconnected by muri, forming a ring-like structure. **C.** Section through acetolyzed exine near an aperture. The ectexine (foot layer) is lamellated at the aperture margin and subtended by thin endexine which increases in thickness at the aperture (TEM). **D–E.** "*Lisianthus*" *acutilobus* (= *I. cardonae*; *Steyermark & Nilsson 732*). **D.** Polyad with groups of pointed, polar spines and delicate, reticulate exine pattern (cf. *I. cardonae*, Fig. 4.16A–C). **E.** Detail of finely reticulate exine attached to a spine; the smooth area inside the spines bears tiny clavae or pila. Scale bars = 10 μm.

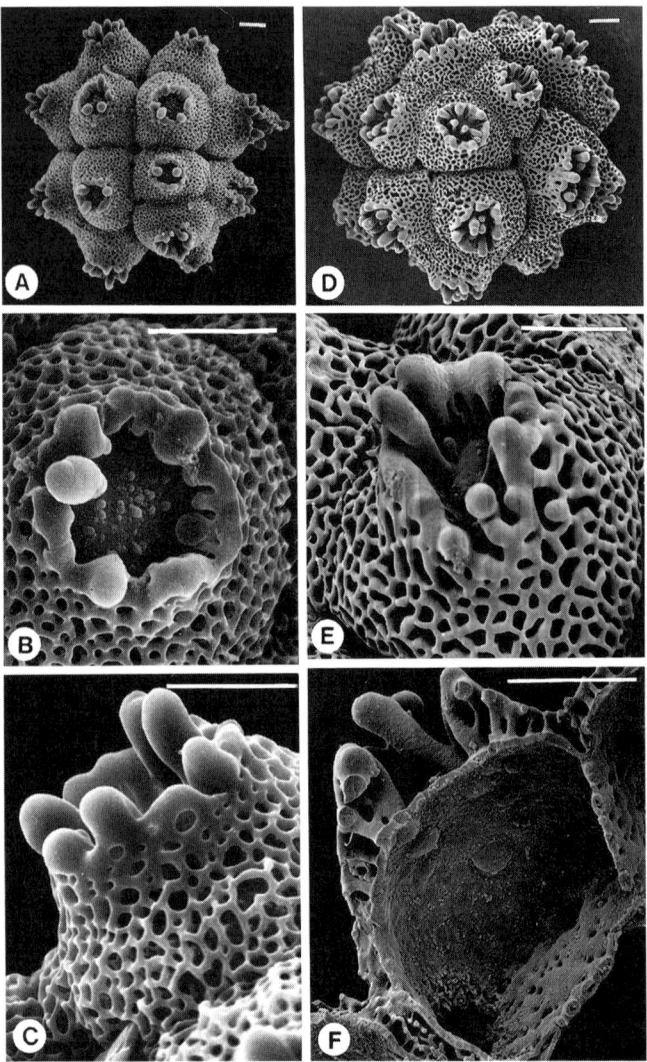

Figure 4.17. Pollen grains of *Chelonanthus* and *Irlbachia* (*Uliginosus–Irlbachia*-type). **A–C.** *Chelonanthus campanuloides* (= *I. pratensis*; Spruce 2005). **A.** Polyad of loosely united tetrads with groups of polar blunt spines. **B.** Detail of polar spines (reduced loops?) on top of a ring supported by columellae, enclosing a smooth area with few granules. **C.** The same; lateral view. **D–F.** *Irlbachia pratensis* (*Maguire et al. 37563*). **D.** Polyad with groups of relatively short and blunt spines at the poles of single grains. **E.** Detail near the polar area showing blunt processes on a mural ring or sometimes free-standing, clavae-like. **F.** The same; fractured polyad with part of a mural ring, a process and reticulum; the inside of the grain has a smooth to scabrate outer wall and perforate inner walls. Scale bars = 10 μm.

A review of palynology 435

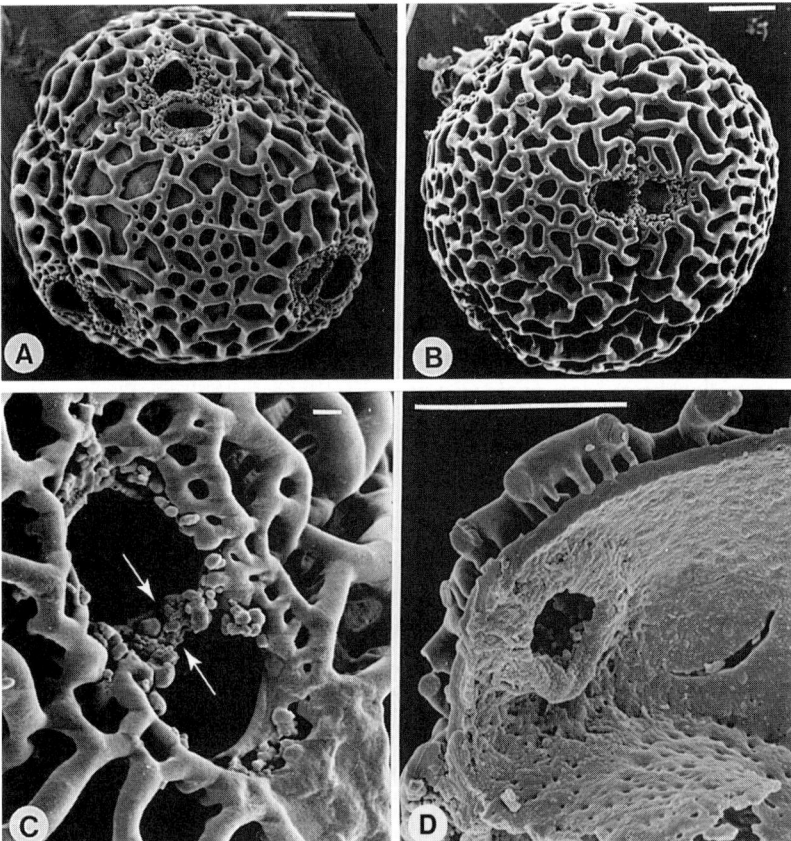

Figure 4.18. Pollen grains of *Lagenanthus princeps* (*Lagenanthus*-type). **A–C.** (*Murillo & Jaramillo s.n.*). **A.** Porate tetrad in tetrahedral position, with relatively equal-sized reticulum diminishing in size at the very distal pole. **B.** The same; tetrad in decussate position. **C.** Detail of exine at the pore area showing pores of two adjacent grains and their division wall (arrows). The pore margin is finely reticulate. **D.** (*Maas & Tillett 5242*). Fractured pollen grain showing exine with muri and columellae in cross-section and finely granular interior, a pore with thickened annular margin, and perforate inner wall. Scale bars: A, B, D = 10 μm; C = 1 μm.

Lagenanthus GILG (Fig. 4.18)

The pollen grains are united in tetrahedral tetrads that are 53–77 μm in diameter. The single grains are 3-porate with rounded to oval, well-defined pores provided with internal annuli. The exine is finely to coarsely reticulate and is only slightly differentiated. The internal walls are perforated.

SPECIES EXAMINED: *Lagenanthus princeps*.

POLLEN TYPE: *Lagenanthus* (Nilsson, 1970).

PREVIOUS STUDIES: Gilg (1895) described the pollen of *Lagenanthus* as having a coarse, equal-sized reticulum, which is not in accordance with the present study. However, Köhler (1905) described the pollen grains as clearly heterobrochate and with pores bordered by a finely reticulate pattern, which is in agreement with the present study.

Lehmanniella GILG (Fig. 4.19)

The pollen grains are united in tetrahedral tetrads that are 61–81 μm in diameter. The single grains are generally 3-porate, exceedingly coarsely reticulate, with lumina up to 40 μm in diameter. The reticulum is not always firmly attached to the nexine and is sometimes strongly deformed.

SPECIES EXAMINED: *Lehmanniella splendens*.

POLLEN TYPE: *Lehmanniella* (Nilsson, 1970).

INTERSPECIFIC VARIATION: In three collections (*Uribe-Uribe 2605*; *Fonnegra & Roldán 2698*; *Uribe 1135*) the exine appears folded and deformed (Fig. 4.19).

PREVIOUS STUDY: Gilg (1895) mentioned the broad, irregular, and strongly projecting exine bands, which agrees well with the present study.

Macrocarpaea (GRISEB.) GILG (Figs. 4.20–4.21)

Macrocarpaea bears single pollen grains. The monads of *Glabra*-type are 3-colporate, rarely porate, 23–44 × 26–42 μm, spheroidal to subspheroidal, with reticulate to verrucose-gemmate exine, and usually relatively wide muri. Pollen grains of *Corymbosa*-type (= *Rusbyanthus*-type; Nilsson, 1970) are released as monads, 3-porate to colporate, 28–35 × 28–39 μm, subspheroidal to spheroidal, and with warty exine (verrucae, gemmae, pila, and clavae are present); these are found only in *Macrocarpaea cinchonifolia*, *M.* sp. nov. 1, *M.* sp. nov. 4, and *M.* sp. nov. 9 (this material was previously determined as *M. corymbosa*, *M. pachystyla*, and *M. viscosa*; Table 4.3). *Macrocarpaea loranthoides* is exceptional in having pollen of *Longifolius*-type, otherwise characteristic of *Lisianthius*, and was previously included in that genus (Maas in Brako & Zarucchi, 1993: 1256). The

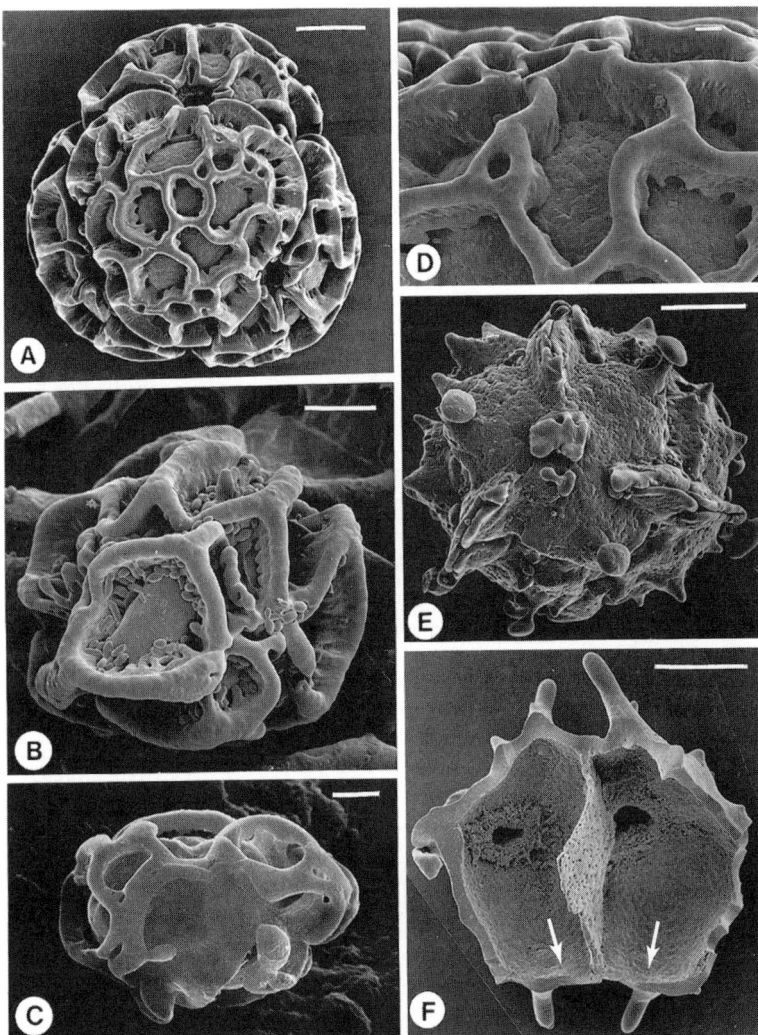

Figure 4.19. Pollen grains of *Lehmanniella splendens* (*Lehmanniella*-type). A–B, D. (*Sandeman 6046*). **A.** Tetrad in tetrahedral position with coarse reticulum, not different around the equator. **B.** A tetrad enveloped by markedly coarse reticulum. **C.** (*Uribe-Uribe 2605*). A smooth tetrad, the reticulum exceedingly coarse, and with strongly projecting muri. **D.** Detail of reticulum with relatively high, sinuous muri supported by short columellae, lumina smooth. **E.** (*Fonnegra & Roldan 2698*). Tetrad with distorted muri and uneven, sparsely perforate nexine. **F.** (*Uribe 1135*). The same; fragment of tetrad showing finely granular exine inside, two distinct pores with thickened margin, and two unopened pores (arrows). The inner wall is finely perforate. Scale bars: A–C, E, F = 10 μm; D = 1 μm.

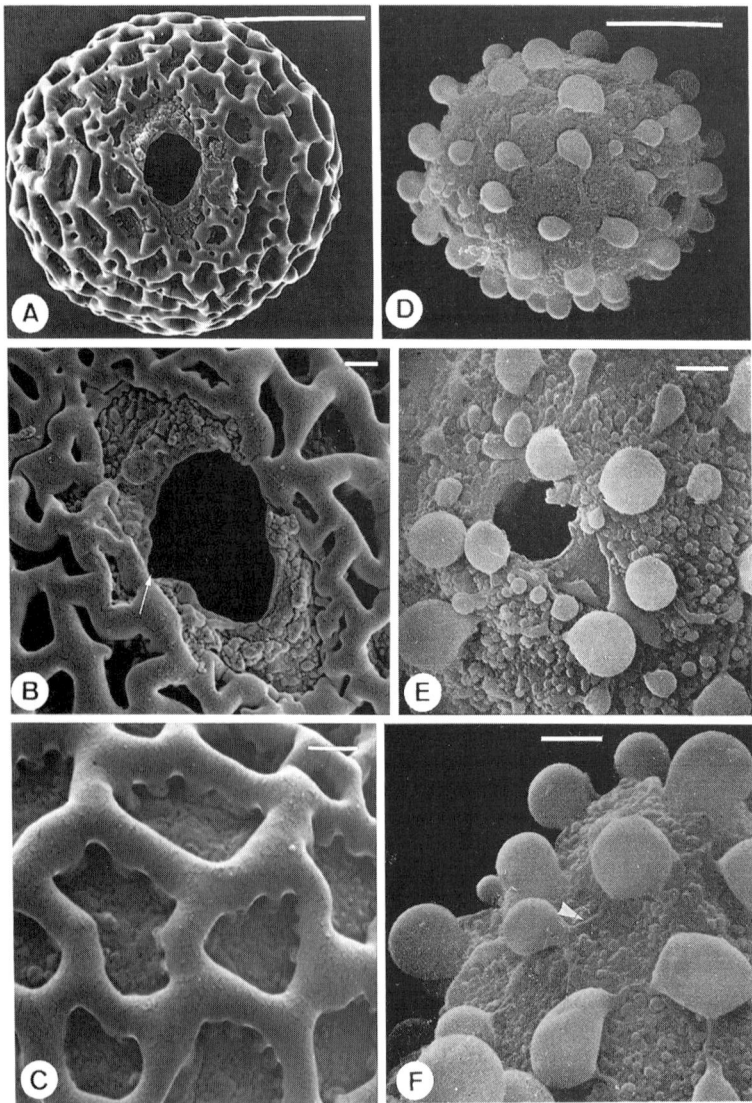

Figure 4.20. Pollen grains of *Macrocarpaea* (*Glabra*-type and *Corymbosa*-type). A–C. *M. bracteata* (*Glabra*-type; *Dorr et al. 4940*). **A.** 3-colporate, reticulate pollen in equatorial, apertural view. **B.** Detail of short colpus and os with lateral extensions. **C.** The same; detail of reticulate exine. D–F. *M. cinchonifolia* (*Corymbosa*-type; *Rusby 1173*). **D.** 3-porate-colporate, verrucose pollen in equatorial view. **E.** Detail of warty exine, and short, indistinct colpus with distinct pore. **F.** The same; detail of verrucose exine with remnants of tiny muri. Scale bars: A, D = 10 μm; B, C, E, F = 1 μm.

A review of palynology 439

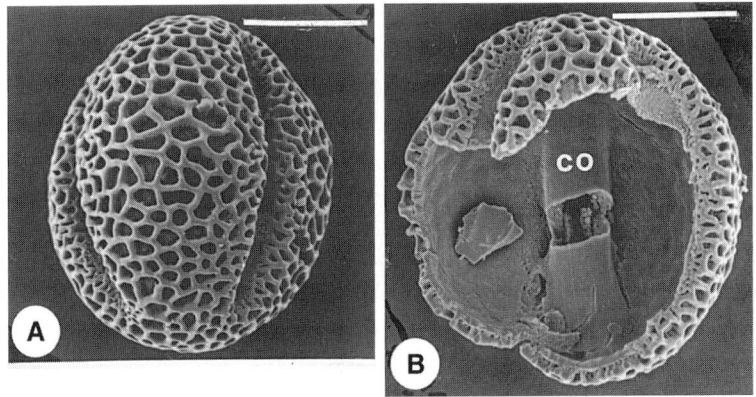

Figure 4.21. Pollen grains of *Macrocarpaea loranthoides* (*Longifolius*-type; Matthews 1315). **A.** 3-colporate, reticulate pollen in equatorial, mesocolpial view. **B.** The same; fractured pollen in equatorial view, showing smooth exine inside, and colpus floor ("co") with a central os. Scale bars = 10 μm.

genus name *Rogersonanthus* is now used (Maguire & Boom, 1989) for *Macrocarpaea* species with pollen in tetrads (= *Arborea*-type (Nilsson, 1968); *Chelonoides*-type (Nilsson, 1970)).

SPECIES EXAMINED: *Macrocarpaea affinis, M. bangiana, M. bracteata* (Fig. 4.20A–C), *M. browallioides, M. cinchonifolia* (Fig. 4.20D–F), *M. cochabambensis, M. densiflora, M. domingensis, M. duquei, M. glabra, M. loranthoides* (Fig. 4.21), *M. macrophylla, M. micrantha, M. obtusifolia, M. ovalis, M. pachyphylla, M. revoluta, M. rubra, M. stenophylla, M. thamnoides, M. valerii*, and the undescribed species nos. 1–10 (see Table 4.3).

POLLEN TYPES: *Glabra, Corymbosa*, and *Longifolius* (Nilsson, 1970).

INTERSPECIFIC VARIATION: Most species have 3-colporate, reticulate pollen grains, e.g., *Macrocarpaea bracteata* (Fig. 4.20A–C), or verrucose-gemmate pollen grains, e.g., *Macrocarpaea cinchonifolia* (Nilsson, 1968). Previous investigations of *Macrocarpaea viscosa*, which found both reticulate and verrucose pollen grains in the same species (Nilsson, 1968), are explained here by those collections representing three distinct, new species (Table 4.3; J. R. Grant, unpubl.).

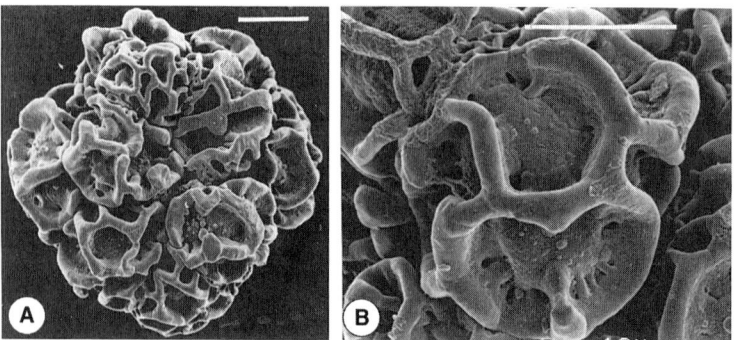

Figure 4.22. Pollen grains of *Neblinantha neblinae* (*Neblinantha*-type; Steyermark *103862*). **A.** View of the back of tetrad with coarsely reticulate pattern, with projecting muri in the equatorial region and smaller lumina toward the polar areas. **B.** Detail of exine showing coarse reticulum consisting of large lumina and projecting mural loop, supported by columellae on smooth exine floor. Scale bars = 10 μm.

PREVIOUS STUDIES: Gilg (1895) described the pollen of *Macrocarpaea* as distinctly reticulate, and Köhler (1905) described the pollen as 3-colporate and clearly reticulate with angular lumina. Maguire (1981) provided descriptions and LM and SEM micrographs of the new species *Macrocarpaea neblinae*, *M. piresii*, and *M. rugosa*, all confirming the present results (i.e., *Glabra*-type). Approximately 25 species were also studied by Nilsson (1968). Gilg (1895) found the *Rusbyanthus*-type (i.e., 3-porate) only in *Rusbyanthus cinchonifolius* (= *Macrocarpaea cinchonifolia*) and therefore classified the genus in a tribe of its own, Rusbyantheae. Köhler (1905) agreed with Gilg's observations. He found the exine to be basically smooth and compared the pollen of *Rusbyanthus* to that of *Prepusa connata*, with reference to the warty processes. However, the exine processes have slightly different shapes in *Prepusa*. There are vestiges of a reticulum in *Macrocarpaea cinchonifolia* and in some species of *Macrocarpaea*, and these are not found in *Prepusa*. Erdtman (1952) described and depicted *Rusbyanthus*-type pollen. However, the *Rusbyanthus*-type of pollen is identical to the *Corymbosa*-type of pollen (Nilsson, 1968, 1970).

Neblinantha MAGUIRE (Fig. 4.22)

The pollen grains are united in tetrahedral tetrads that are 58–105 μm in diameter. The single grains appear indistinctly porate. The exine is coarsely reticulate, consisting of relatively wide and strongly projecting, loop-like muri that are distally united to ring-like structures (*Neblinantha*-type).

SPECIES EXAMINED: *Neblinantha neblinae* and *N. parvifolia*.

A review of palynology 441

POLLEN TYPE: *Neblinantha* (defined here).

PREVIOUS STUDIES: In *Neblinantha parvifolia* the reticulum was reported to be of equal size (Maguire & Boom, 1989). In the generic description of *Neblinantha* the tetrads were said to be of *Helia*-type and with a coarse reticulum (Maguire & Boom, 1989). However, the pollen is more similar to the *Speciosus*-type (*Calolisianthus*) and even more to the *Lehmanniella*-type.

Prepusa MART. (Figs. 4.23, 4.24A–C)

The pollen grains are united in tetrahedral tetrads that are 66–94 μm in diameter. The single grains are 3-colporate to porate or colpate and pilate. The rounded to elongated pores (sometimes colpi) are arranged pairwise and are located in a narrow, smooth to granular zone. The processes are of different sizes, with small pila (1–2 μm high and wide) or verrucae (sometimes fused to irregular, flattened jigsaw-like pieces) with interspersed macropila, 5–6 μm high, 3–6 μm in diameter (*Prepusa*-type).

SPECIES EXAMINED: *Prepusa alata, P. connata, P. hookeriana,* and *P. montana.*

POLLEN TYPE: *Prepusa* (defined here).

INTERSPECIFIC VARIATION: *Prepusa hookeriana* (Fig. 4.24A–C) is lacking the pila- or verrucae-like processes and is covered only by a smooth, perforate surface layer (polar cap).

PREVIOUS STUDIES: Gilg (1895) reported that the larger processes were equally spaced, which is not always the case. Köhler (1905) observed that the single grains of *Prepusa connata* were 3-colporate, 50 μm in diameter, and with a basically smooth exine beset with "warts" ranging from 1 to 10 μm. *Prepusa hookeriana* was reported to differ from *P. connata* by having a finely reticulate exine (vs. smooth) with fewer and smaller processes.

Purdieanthus GILG (Fig. 4.24D–F)

The pollen grains are united in tetrahedral tetrads that are 56–70 μm in diameter. The single grains are 3-porate and reticulate. The reticulum is coarse and forms slightly projecting islands at the distal face. The curved to straight muri, provided with sparse perforations or holes, are not always closed around a lumen, but end abruptly. The limit between single grains is marked with aligned and rounded processes. The lumina have an uneven, perforate floor.

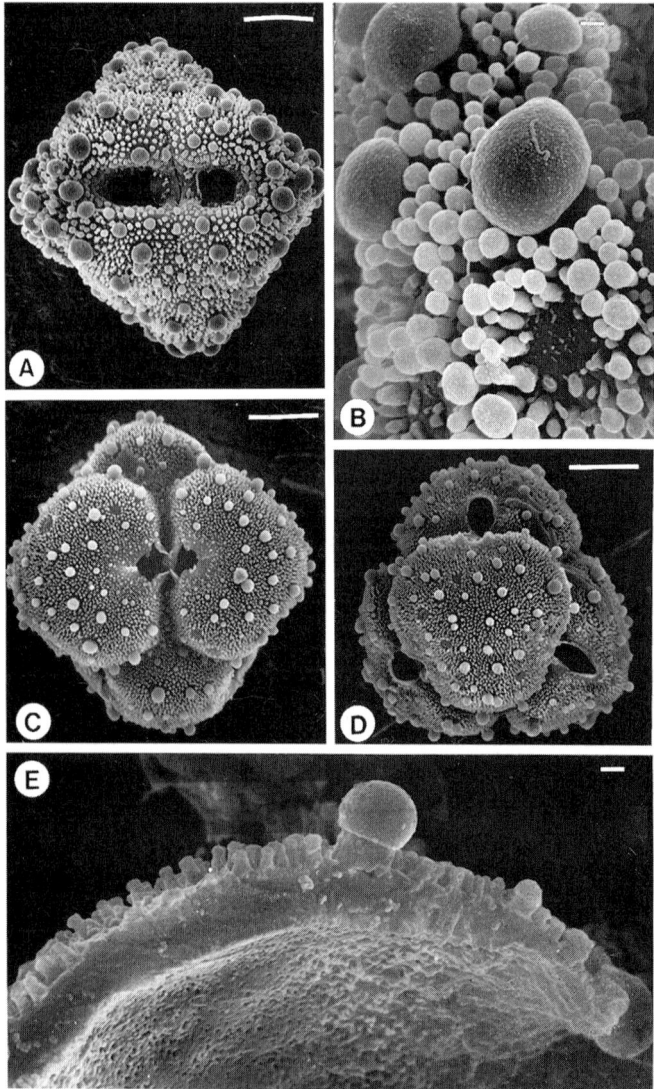

Figure 4.23. Pollen grains of *Prepusa* (*Prepusa*-type). **A–B.** *P. connata* (*Gardner 541*). **A.** Tetrad in decussate position; individual grains 3-colpate to porate, exine consisting of numerous pila (macroprocesses) and microprocesses. **B.** Detail of exine showing small pila and a few larger, rounded processes on top of a smooth nexine. **C–E.** *P. montana* (*Harley 22878*). **C.** Tetrad in decussate position, porate, exine with numerous micro- and fewer macroprocesses. **D.** Tetrad in tetrahedral position, colpate to porate with oval-shaped pores, exine beset with micro- and macroprocesses. **E.** The same; fragment of a grain with pila and one macroprocess (pila-like) with stout support. Scale bars: A, C, D, F = 10 μm; B, E = 1 μm.

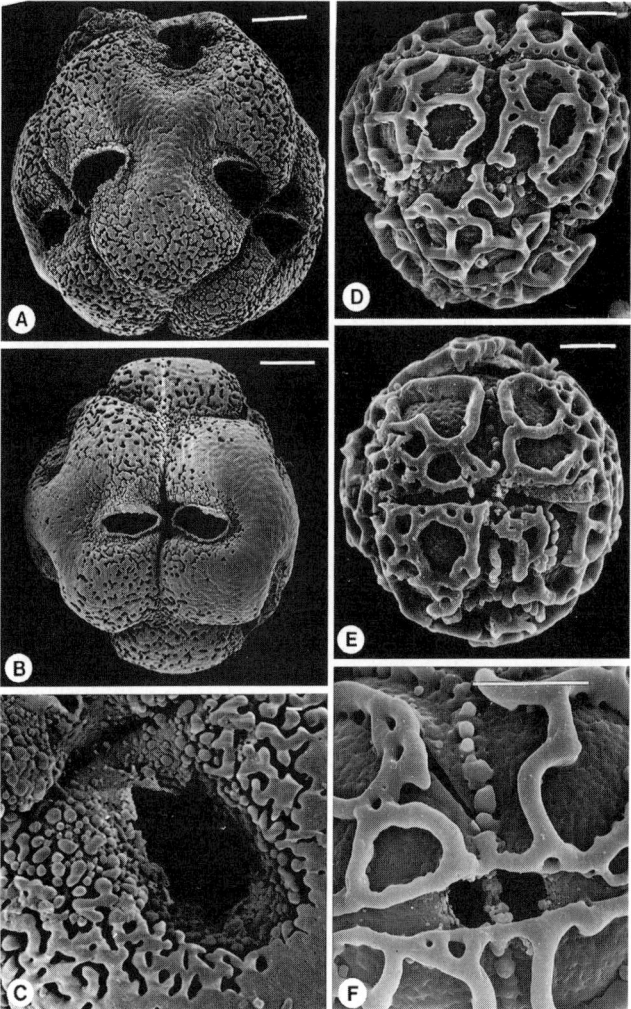

Figure 4.24. Pollen grains of *Prepusa* (*Prepusa*-type) and *Purdieanthus* (*Purdieanthus*-type). **A–C.** *Prepusa hookeriana* (*Vidal II-330*). **A.** Tetrad in tetrahedral position, individual grains 3-colpate, with a jigsaw-like, relatively smooth, exine pattern. **B.** The same; tetrad in decussate position with two relatively short and oval-shaped colpi; the exine has partly fused and flattened processes except at the markedly smooth polar areas. **C.** Detail of aperture, pore with lateral extensions. **D–F.** *Purdieanthus pulcher* (*Purdie s.n.*). **D.** Tetrad in tetrahedral position, individual grains 3-porate, reticulate with markedly large lumina in the equator areas; the border between grains is marked with smaller globules. **E.** The same; tetrad in decussate position. **F.** Detail of the coarsely reticulate exine at the aperture (poral) region; limit between grains marked by small, aligned globules. Scale bars: A, B, D–F = 10 μm; C = 1 μm.

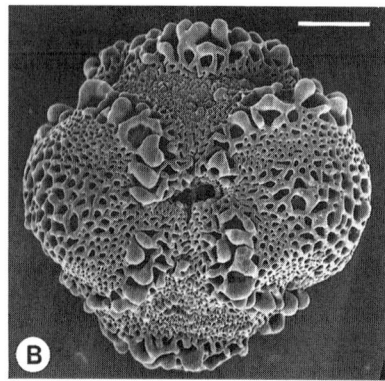
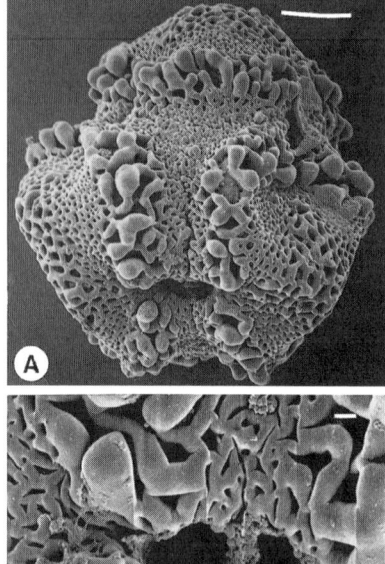

Figure 4.25. Pollen grains of *Rogersonanthus arboreus* (*Chelonoides*-type; Baksh & Nilsson 1984-3). **A.** Tetrad in oblique tetrahedral position, individual grains 3-porate, with a band around the equator consisting of muroid elements of different shape and size bordered by a fine reticulate pattern, which become coarser at the poles. **B.** The same; tetrad in decussate view with two pores from adjacent grains. **C.** Detail of exine near the two pores. Scale bars: A, B = 10 μm; C = 1 μm.

SPECIES EXAMINED: *Purdieanthus pulcher*.

POLLEN TYPE: *Purdieanthus* (Nilsson, 1970).

PREVIOUS STUDIES: According to Gilg (1895) the pollen grains are coarsely reticulate at "the sides", but perforate at the upper part of the projected areas, which largely agrees with the present study.

Rogersonanthus MAGUIRE (*Macrocarpaea pro parte sensu* Ewan)
(Fig. 4.25)

The pollen grains are united in tetrahedral tetrads that are 52–95 μm in diameter. The single grains are 3-colporate to porate. Equatorially or subequatorially there is a zone of coarser reticulum, which is occasionally also present at the very distal pole.

A review of palynology 445

SPECIES EXAMINED: *Rogersonanthus arboreus, R. quelchii*, and *R. tepuiensis* (*Macrocarpaea cerronis* and *M. salicifolia* were regarded as synonyms of *Rogersonanthus tepuiensis* by Maguire & Boom, 1989).

POLLEN TYPE: *Chelonoides* (Nilsson, 1970).

INTER- AND INTRASPECIFIC VARIATION: The polar and equatorial muri may be enlarged, fragmented, or transformed into globules.

PREVIOUS STUDIES: Neither Gilg (1895) nor Köhler (1905) mentioned any species of *Macrocarpaea* as having pollen grains in tetrads. Nilsson (1968, 1970) described both monads and tetrads in *Macrocarpaea*, naming the tetrads *Arborea*-type (later renamed *Chelonoides*-type). Maguire and Boom (1989) included LM micrographs of *Rogersonanthus quelchii* tetrads.

Senaea TAUB. (Fig. 4.26A–C)

The pollen grains are united in tetrahedral tetrads that are 70–90 μm in diameter. The single grains are 3-porate and pilate. The processes are of different sizes, and small, densely spaced pila (1–2 μm high, c. 1 μm in diameter) are intermixed with larger pila (3–6 μm high, 4–5 μm in diameter). The inside of the nexine is smooth. The apertures are rounded to oval-shaped and the pores are in pairs and are located in a smooth to granular zone.

SPECIES EXAMINED: *Senaea coerulea* and *S. janeirensis*.

POLLEN TYPE: *Prepusa* (see *Prepusa*).

PREVIOUS STUDIES: Gilg (1895) placed the two genera *Senaea* and *Prepusa* (the latter of which has exine with regularly spaced bumps) close to each other in his generic key to the Helieae. Köhler (1905) observed a finely reticulate exine pattern in *Prepusa hookeriana*, which could not be confirmed in the present study.

Sipapoantha MAGUIRE & BOOM (Fig. 4.26D–F)

The pollen grains are united in tetrahedral tetrads that are 67–75 μm in diameter. The single grains are 3-colporate to porate, pilate to verrucose, with smaller, densely spaced pila (1–2 μm) and larger, usually rounded, isodiametric pila, which are less often irregularly shaped in outline (4–5 μm) (*Sipapoantha*-type). They are spread out over the exine surface, and are

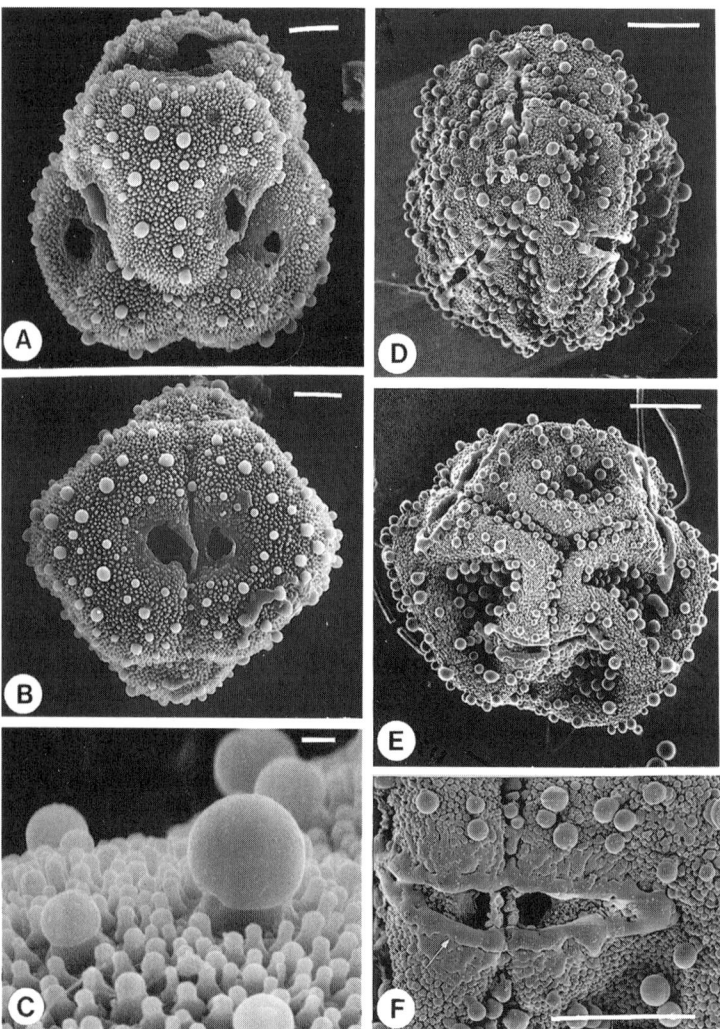

Figure 4.26. Pollen grains of *Senaea* (*Prepusa*-type) and *Sipapoantha* (*Sipapoantha*-type). **A–B.** *Senaea janeirensis* (*Martinelli et al. 12003*). **A.** Tetrad in tetrahedral position; individual grains are 3-porate, the exine with numerous microprocesses (pila) and fewer, unevenly large rounded macroprocesses. **B.** The same; tetrad in decussate position, pores surrounded by a smooth, narrow margin. **C.** *S. janeirensis* (*Santos Lima 217*). Detail of exine showing small pila and scattered macroprocesses (pila) with stout and solid support. **D–F.** *Sipapoantha ostrina* (*Maguire & Politi 28092*). **D.** Tetrad in tetrahedral position; individual grains are 3-colporate, with smaller and larger granular- to pila-like processes. **E.** The same; tetrad viewed from below with numerous scattered, partly aligned larger globules. **F.** Detail of exine at the aperture region with the colpi bordered by distinct, muroid elements. Scale bars: A, B, D–F = 10 μm; C = 1 μm.

A review of palynology

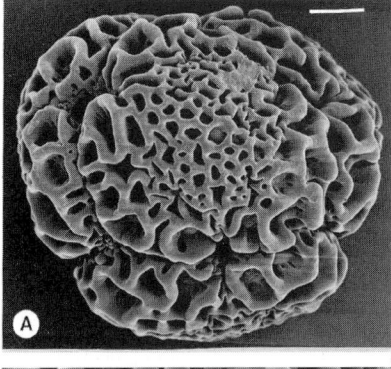
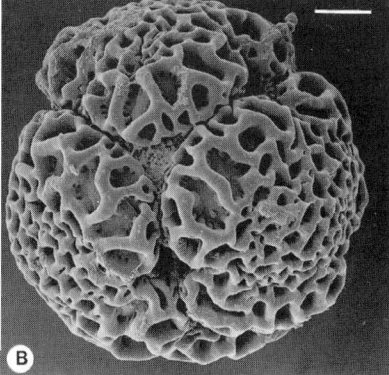
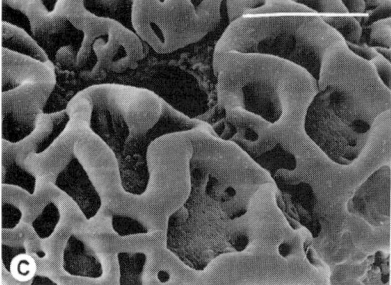

Figure 4.27. Pollen grains of *Symbolanthus pulcherrimus* (*Symbolanthus*-type; *Weaver 1406*). **A.** Tetrad in tetrahedral position; around the equator there is a band of coarse reticulum, which decreases in size at the pole. **B.** The same; tetrad, view of the back, with prominent, projecting reticulum. **C.** Detail of exine, muri unevenly thick and up-raised. Scale bars = 10 μm.

particularly dense on the demarcation line between the individual grains. The inside of the exine is smooth. The colpi are delimited by two elongated muri crossing the demarcation line.

SPECIES EXAMINED: *Sipapoantha ostrina*.

POLLEN TYPE: *Sipapoantha* (defined here).

PREVIOUS STUDY: The genus *Sipapoantha* was described by Maguire and Boom (1989) based on one species, *S. ostrina*. The tetrahedral tetrads were reported to be composed of 3-colporate grains with prominent globules of *Rusbyanthus*-type. However, the pollen tetrads of *Prepusa* and *Senaea* are more similar to *Sipapoantha* except for the distinct colpus margin in *Sipapoantha*.

Symbolanthus G. DON (Fig. 4.27)

The pollen grains are united in tetrahedral tetrads that are 62–115 μm in diameter. The single grains are 3-colporate (rarely 3-colpate, 3-porate) and reticulate, with a zone of larger meshes with distinctly projecting muri around the equator. At the polar area the muri are often thickened and slightly projecting.

SPECIES EXAMINED: *Symbolanthus anomalus*, *S. brittonianus*, *S. calygonus*, *S. daturoides*, *S. elisabethae*, *S. gaultherioides*, *S. latifolius*, *S. macranthus*, *S. magnificus*, *S. mathewsii*, *S. nerioides*, *S. obscure-rosaceus*, *S. pulcherrimus*, *S. rubro-violaceus*, *S. rusbyanus*, *S. stuebelii*, *S. superbus*, *S. tricolor*, *S. vasculosus*, and *S. yaviensis*.

POLLEN TYPE: *Symbolanthus* (Nilsson, 1970).

INTERSPECIFIC VARIATION: Only minor variations related to the width of muri and lumina shape are found.

PREVIOUS STUDIES: Gilg (1895) described the pollen tetrads of *Symbolanthus* as having large meshes at "the sides" and being almost perforate at the poles (cf. *Purdieanthus*). Köhler (1905) shared Gilg's views, which are also in agreement with the present study.

Tachia AUBL. (Fig. 4.28)

The pollen grains are released as monads, 3-colporate, 24–47 × 22–44 μm, spheroidal to ellipsoidal, coarsely reticulate or smooth, perforate and sparingly beset with globules, or reticulate-rugulate.

SPECIES EXAMINED: *Tachia grandifolia*, *T. guianensis*, *T. loretensis*, *T. occidentalis*, *T. parviflora*, and *T. schomburgkiana*.

INTERSPECIFIC VARIATION: The exine can be coarsely reticulate (*Tachia grandifolia* and *T. guianensis*; Fig. 4.28A–D), smooth (*T. schomburgkiana*), with globules (*T. occidentalis*; Fig. 4.28E–F), or reticulate-rugulate (*T. loretensis*). *Tachia guianensis* has variable pollen types (cf. Fig. 4.28A–B vs. C–D).

PREVIOUS STUDIES: Gilg (1895) reported the pollen grains of *Tachia* as clearly reticulate, and Köhler (1905) noted them to be 3-colporate and reticulate. Pollen of *Tachia* species was also described and illustrated by Maguire and Weaver (1975) and Agababyan and Tumanyan (1977).

A review of palynology

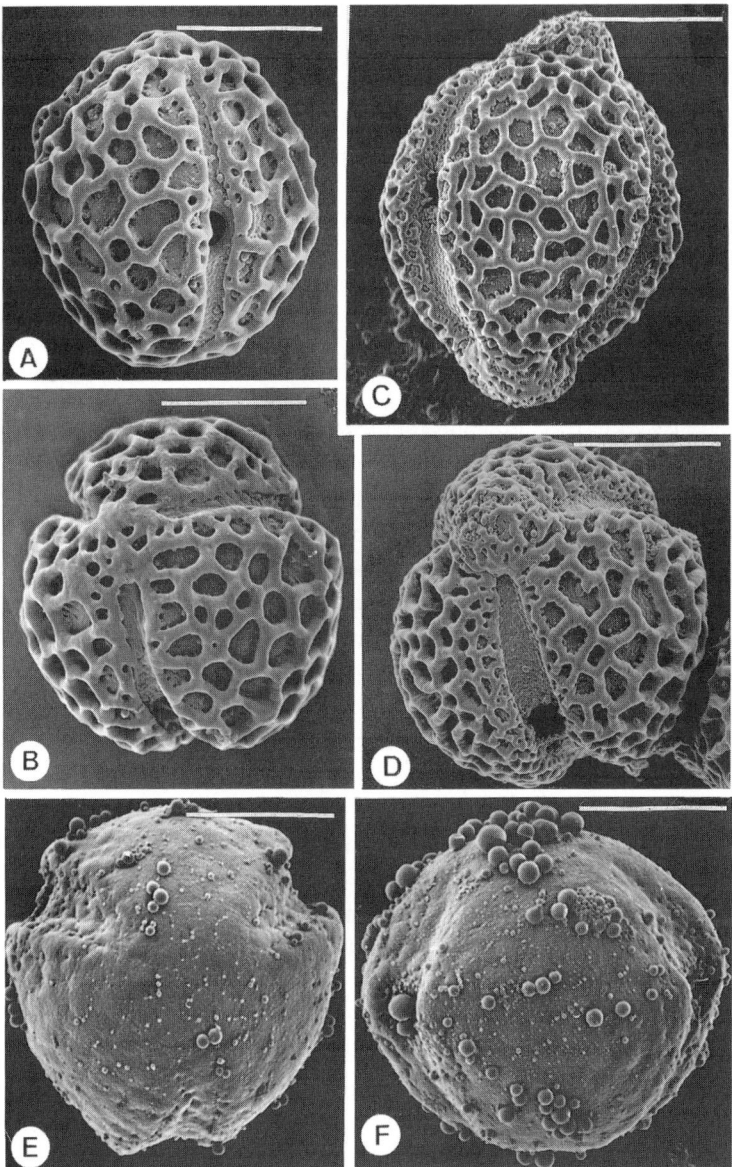

Figure 4.28. Pollen grains of *Tachia* (aff. *Glabra*-type). **A–B.** *T. guianensis* (*Rova et al. 1963*). **A.** 3-colporate, reticulate pollen in equatorial, apertural view. **B.** The same; pollen in oblique polar view. **C–D.** *T. guianensis* (*Sandwith 1111*). **C.** 3-colporate pollen in equatorial, mesocolpial view with polar caps. **D.** The same; pollen in oblique polar view. **E–F.** *T. occidentalis* (*Maas et al. 6760*). **E.** 3-colporate, smooth pollen in polar view. **F.** The same; pollen in equatorial, mesocolpial view; note granules of different sizes. Scale bars = 10 μm.

Tetrapollinia MAGUIRE & B. M. BOOM (Fig. 4.29A–C)

The pollen grains are united in tetrahedral tetrads that are 56–77 μm in diameter. The single grains of the spiny tetrads are 3-porate (less commonly 3-colporate), with a reticulate ground-pattern beset with spines that are up to 13 μm in height. The spines are usually concentrically arranged, pointed, varying in shape, and they are interconnected by a fine suspended network.

SPECIES EXAMINED: *Tetrapollinia caerulescens*.

POLLEN TYPE: *Caerulescens* (Nilsson, 1970).

PREVIOUS STUDIES: Gilg (1895) mentioned 3–4 species in *Irlbachia* but only *I. caerulescens* (= *Tetrapollinia caerulescens*) was described as having pollen grains united in tetrads with finely granular exine beset with pointed spines. Köhler (1905) described the tetrads as finely reticulate with long, pointed, and evenly distributed spines. Nilsson (1970) agreed with the above descriptions, although he regarded the variously shaped spines as concentrically arranged rather than evenly spaced. Thus, the pollen grains were different from those of typical *Irlbachia* species. Maguire and Boom (1989) created the monotypic genus *Tetrapollinia* for *Irlbachia caerulescens* and published both LM and SEM micrographs.

Wurdackanthus MAGUIRE (Fig. 4.29D–F)

The pollen grains are united in tetrads that range from 69 to 75 μm in diameter. The single grains are 3-colporate to 3-porate with a distinct zone of coarse reticulum around the equator and muri projecting like loops. At the pole the lumina are relatively large in diameter.

SPECIES EXAMINED: *Wurdackanthus frigidus*.

POLLEN TYPE: *Symbolanthus* (Nilsson, 1970).

PREVIOUS STUDIES: Gilg (1895) referred *Wurdackanthus frigidus* to the genus *Calolisianthus*, the pollen grains of which were described as tetrads with a regular, equal-sized reticulum, but this is in disagreement with later studies (Nilsson, 1970). Maguire (1985) described the new genus and species *Wurdackanthus argyreus*, which has coarsely reticulate pollen tetrads of *Symbolanthus*-type. Maguire and Boom (1989), who also published SEM micrographs of the two species, transferred *Wurdackanthus*

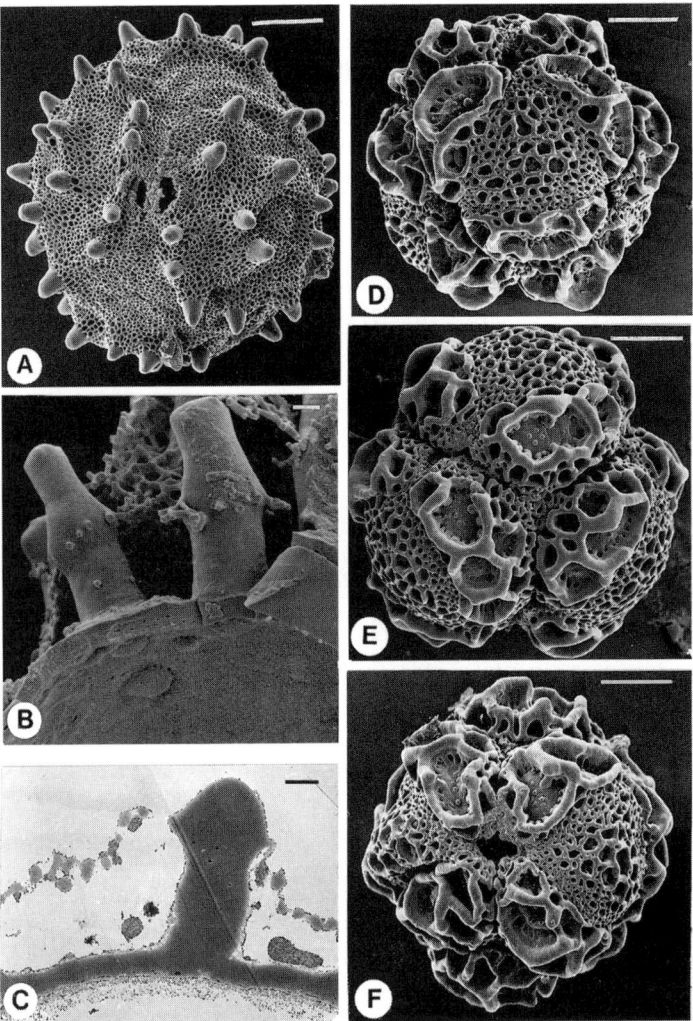

Figure 4.29. Pollen grains of *Tetrapollinia* (*Caerulescens*-type) and *Wurdackanthus* (*Symbolanthus*-type). **A–B.** *T. caerulescens* (*Jansen-Jacobs et al. 2764*). **A.** Tetrahedral tetrad in decussate position showing two pores. The delicately reticulate exine is beset with concentrically arranged, stout and blunt spines. **B.** The same; exine fracture showing two blunt spines with thickened girdle and delicate, attached exine suspended between the spines. The exine inside is smooth. **C.** *T. caerulescens* (*Irwin et al. 11661*). Section of exine with a cross-section of a spine and suspended reticulum; exine stratified into ectexine and thin endexine (TEM). **D–F.** *W. frigidus* (*Beard 17*). **D.** Tetrad in tetrahedral position, reticulate to coarsely reticulate with muri projecting around the equator. **E.** The same; tetrad from below with markedly raised, coarse reticulum at the equator. **F.** Tetrad in decussate position. Scale bars: A, D–F = 10 μm; B, C = 1 μm.

452   S. Nilsson

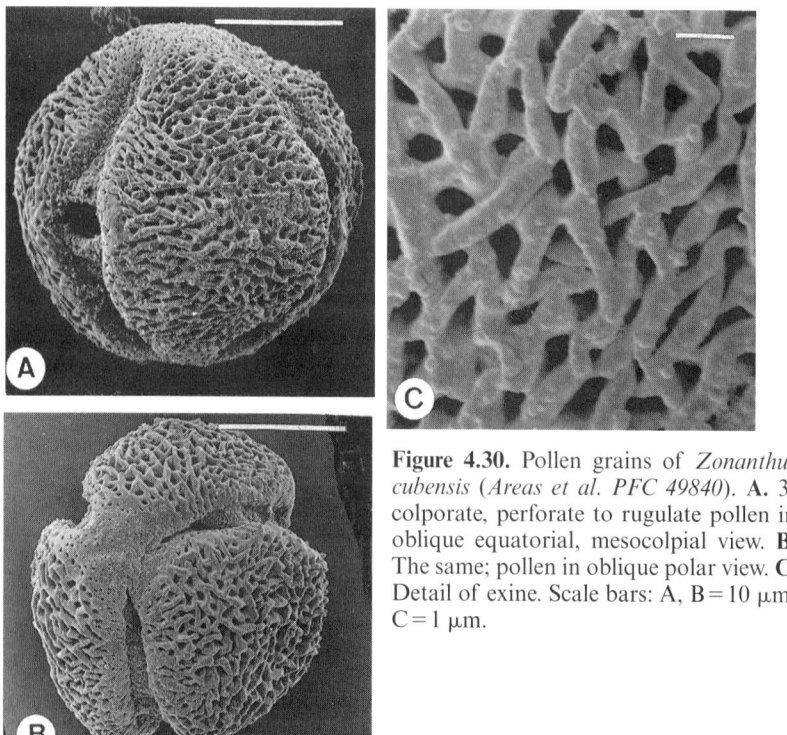

Figure 4.30. Pollen grains of *Zonanthus cubensis* (*Areas et al. PFC 49840*). **A.** 3-colporate, perforate to rugulate pollen in oblique equatorial, mesocolpial view. **B.** The same; pollen in oblique polar view. **C.** Detail of exine. Scale bars: A, B = 10 μm; C = 1 μm.

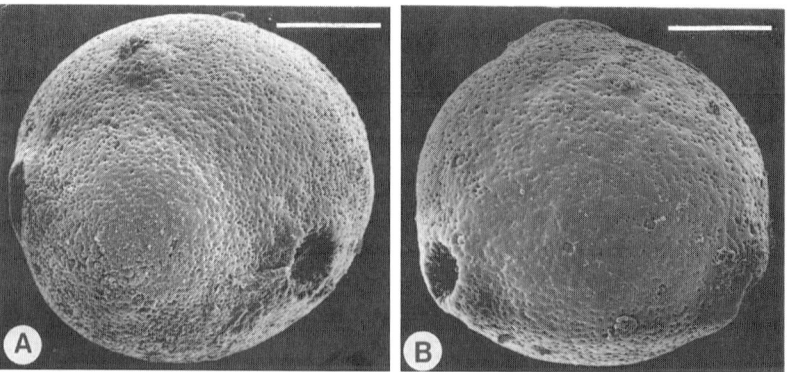

Figure 4.31. Pollen grains of *Anthocleista microphylla* (*Potalia*-type; *Mildbraed 6434*). **A.** Pollen in oblique, equatorial view, 3-porate, pore margins slightly elevated, exine smooth, densely perforate. **B.** The same; pollen in polar view. Scale bars: A, B = 10 μm.

frigidus from *Calolisianthus*. These previous results are confirmed in the present study.

Zonanthus GRISEB. (Fig. 4.30)

The pollen grains are released as monads, 3-colporate, 41–54×30–41 μm, subprolate, reticulate to striato-reticulate and usually with angular lumina.

SPECIES EXAMINED: *Zonanthus cubensis*.

PREVIOUS STUDIES: Gilg (1895) and Köhler (1905) described the pollen as reticulate.

Tribe Potalieae

Anthocleista R. BR. (Fig. 4.31)

The pollen grains are released as monads, 3(–4)-porate, 25–26×26–31 μm, oblate spheroidal to spheroidal, and smooth to scabrate.

SPECIES EXAMINED: *Anthocleista amplexicaulis*.

POLLEN TYPES: *Potalia* (Punt & Leenhouts, 1967); *Anthocleista nobilis* (Punt, 1978).

PREVIOUS STUDIES: Erdtman (1952) described the pollen of *Anthocleista parviflora* as 3-porate, delicately reticulate. Punt and Leenhouts (1967) investigated pollen of six *Anthocleista* species and reported pollen with an exine pattern varying from smooth, usually perforate, toward reticulate. Pores were said to be protruding with a distinct annulus (Punt, 1978) in the closely related taxon *Potalia amara* (Punt & Leenhouts, 1967).

Congolanthus A. RAYNAL (Fig. 4.32)

Neurotheca longidens was transferred to a new genus, *Congolanthus*, by Raynal (1968). *Congolanthus longidens* has relatively small grains (31–41 × 34–42 μm) compared with *Neurotheca*. The exine is thinner (4–5 μm) and more finely reticulate, with a stronger tendency toward striato-reticulate pattern compared with that of *Neurotheca*. The apertural features (lateral smooth os margins) are similar but the muri appear less keeled and the lumina more rounded in comparison with *Neurotheca*. The lateral extensions of the ora also appear more distinct than in *Neurotheca*.

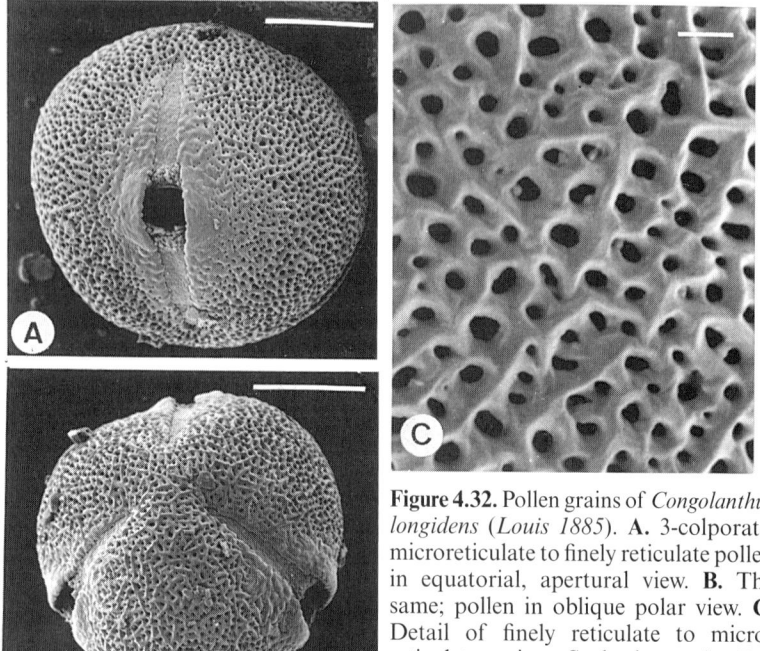

Figure 4.32. Pollen grains of *Congolanthus longidens* (*Louis 1885*). **A.** 3-colporate, microreticulate to finely reticulate pollen in equatorial, apertural view. **B.** The same; pollen in oblique polar view. **C.** Detail of finely reticulate to microreticulate exine. Scale bars: A, B = 10 μm; C = 1 μm.

SPECIES EXAMINED: *Congolanthus longidens*.

PREVIOUS STUDIES: Gilg (1895) described the pollen grains as perforate, but Köhler (1905) found them to be 3-colporate and clearly reticulate, which is in accordance with the present results. Raynal (1968) pointed out pollen morphological differences between *Congolanthus* and *Neurotheca*, i.e., subspheroidal and ovoidal pollen, respectively.

Fagraea THUNB. (Figs. 4.33–4.35)

The pollen grains are released as monads, either 3-colporate to porate and finely reticulate, 23–26×28–32 μm (*Fagraea berteroana*), 3-porate, reticulate to foveolate, 28–31×25–31 μm (*F. longiflora*), 3-colporate, striato-reticulate at mesocolpia and striate at poles, 18–22×17–20 μm (*F. elliptica*

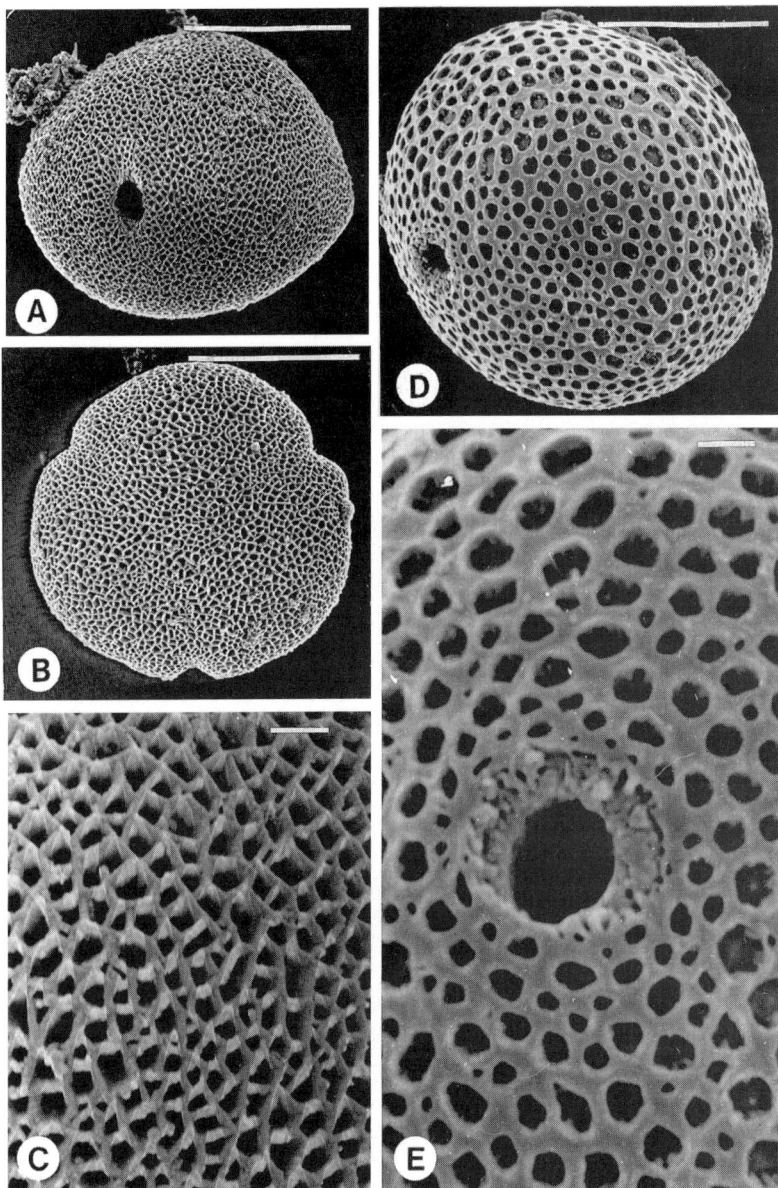

Figure 4.33. Pollen grains of *Fagraea*. **A–C.** *F. berteroana* (*F. berteroana*-type; *Motley s.n.*). **A.** Pollen in equatorial, apertural view, 3-colporate with short, vestigial colpi; exine microreticulate to finely reticulate. **B.** The same; pollen in polar view. **C.** Detail of reticulate exine. **D–E.** *F. longiflora* (*F. longiflora*-type; *Edano 708*). **D.** Pollen in equatorial view, 3-porate, exine foveolate-reticulate. **E.** Detail of the foveolate-reticulate exine. Scale bars: A, B, D = 10 μm; C, E = 1 μm.

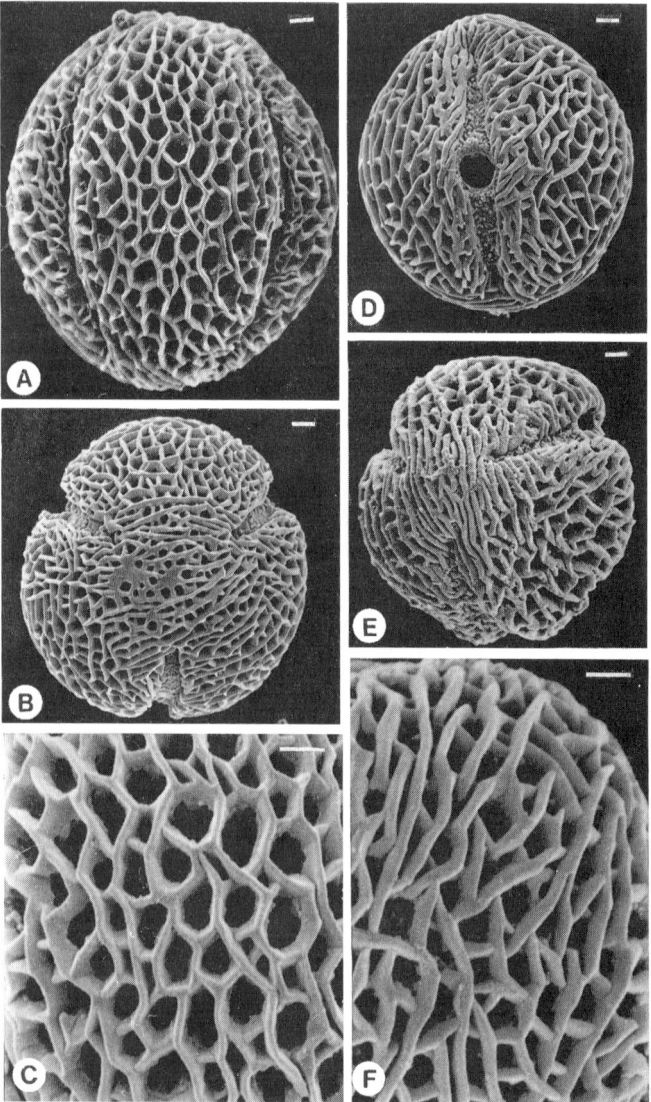

Figure 4.34. Pollen grains of *Fagraea* (*F. fragrans*-type). **A–C.** *F. fragrans* (*Roque s.n.*). **A.** Pollen in equatorial, mesocolpial view, 3-colporate, exine reticulate with lumina tending to be in rows. **B.** The same; pollen in polar view showing striate exine over the pole. **C.** Detail of reticulate exine with mostly straight muri and lumina in rows. **D–F.** *F. elliptica* (*Burley et al. 4335*). **D.** 3-colporate pollen in equatorial, apertural view; colpi long and narrow, exine striato-reticulate. **E.** The same; pollen in slightly oblique polar view, exine clearly striate at the pole. **F.** Detail of striato-reticulate exine, muri straight, branched, criss-crossing, lumina angular and of varying size. Scale bars: A, B, D, E = 10 μm; C, F = 1 μm.

A review of palynology 457

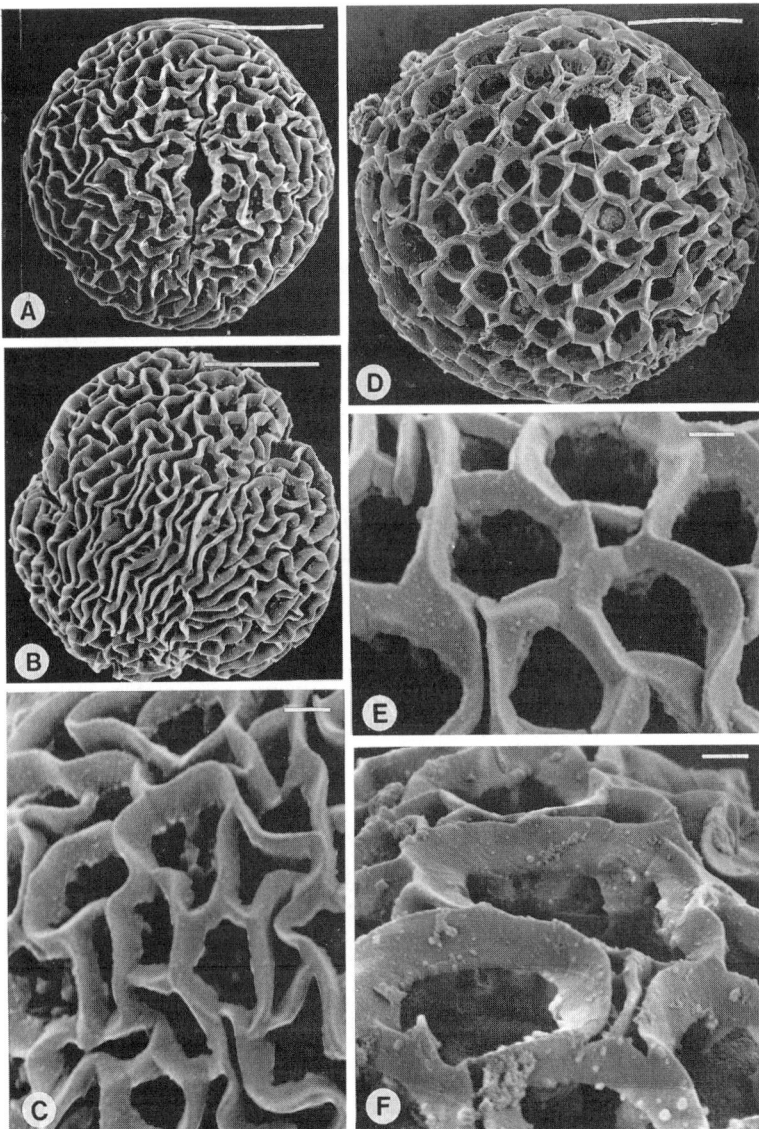

Figure 4.35. Pollen grains of *Fagraea*. **A–C.** *F. racemosa* (*F. fragrans*-type subtype *racemosa*; *Rahmat Si Toroes 3193*). **A.** Pollen in equatorial, apertural view, 3-colporate, with short indistinct colpi, exine striato-reticulate with relatively high, narrow, and sinuous muri. **B.** The same; pollen in polar view, striation distinct at the pole. **C.** Detail of reticulate-striate exine. **D–F.** *F. ceilanica* (*F. ceilanica*-type; *Saldanha 13586*). **D.** Pollen in oblique, equatorial view, 3-porate (one pore visible), distinctly reticulate. **E.** Detail of reticulate exine, muri high, narrow, keeled and inclined. **F.** The same; higher magnification. Scale bars: A, B, D = 10 μm; C, E, F = 1 μm.

and *F. fragrans*), 3-colporate to porate with markedly sinuous muri, 28–31 ×23–29 μm (*F. racemosa*), or 3-porate, coarsely reticulate with rounded to angular lumina enclosed by relatively high, narrow, and inclined muri, 37–44×32–43 μm (*F. ceilanica* and *F. involucrata*). For more detailed type descriptions of additional taxa see Punt and Leenhouts (1967) and Punt (1978).

SPECIES EXAMINED: *Fagraea berteroana, F. ceilanica, F. elliptica, F. fragrans*, and *F. racemosa*.

POLLEN TYPES: *Fagraea berteroana, F. ceilanica, F. fragrans*, and *F. longiflora* (defined here).

PREVIOUS STUDIES: Erdtman (1952) described two species of *Fagraea* as 3-colporate and striato-reticulate, and 3-porate and reticulate, respectively. Guinet (1962) described the pollen of *Fagraea obovata* (*F. ceilanica*) as 3-porate (with a hardly visible ectoaperture?) and reticulate, sometimes perforated. Rao and Lee (1970) described the pollen of *Fagraea fragrans* as colpate and coarsely reticulate, and that of *F. racemosa* (*F. morindaefolia*) as porate-colporate. Punt and Leenhouts (1967) and Punt (1978) published detailed accounts on the pollen grains of *Fagraea*. Pollen grains of *Fagraea ceilanica, F. elliptica*, and *F. racemosa* were illustrated with SEM micrographs (Punt, 1978, 1980).

It is confirmed in the present study that the *Fagraea berteroana*-type (*F. berteroana*) is well distinguished from all other pollen types by having finely reticulate pollen grains (Fig. 4.33A–C). In the *Fagraea ceilanica*-type (incl. *F. ceilanica, F. involucrata*, and *F. longiflora*) *sensu* Punt and Leenhouts (1962), *F. longiflora* (Fig. 4.33D–E) would be better omitted and placed in a separate type of its own. The *Fagraea ceilanica*-type would then include *F. ceilanica* (Fig. 4.35D–F) and *F. involucrata*. Punt (1978) put *Fagraea elliptica* and *F. fragrans* into different pollen types, the *F. elliptica*- and *F. fragrans*-types, respectively. The present results suggest that they more logically belong to the same pollen type. *Fagraea racemosa* is here regarded as a separate pollen subtype, although one closely related to the *F. fragrans*-type.

Lisianthius P. BROWNE (Fig. 4.36)

The pollen grains of *Lisianthius* are 3-colporate, 30–54×22–41 μm, subprolate to prolate, reticulate with a tendency toward a distinct, striato-reticulate exine pattern with markedly angular lumina, or striato-rugulate in low relief and perforate.

A review of palynology 459

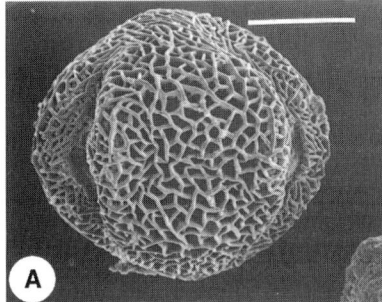
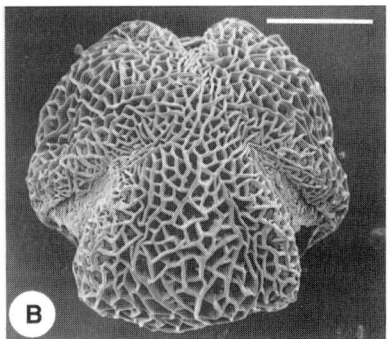
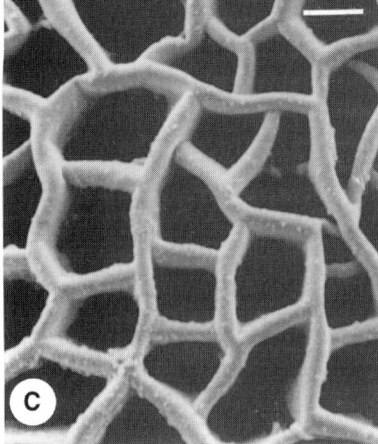

Figure 4.36. Pollen grains of *Lisianthius cordifolius* (*Longifolius*-type; Webster & Wilson 4860). **A.** 3-colporate, reticulate pollen in equatorial, mesocolpial view. **B.** The same; pollen in polar view. **C.** Detail of reticulate-striate exine. Scale bars: A, B = 10 μm; C = 1 μm.

SPECIES EXAMINED: *Lisianthius acuminatus, L. adamsii, L. auratus, L. axillaris, L. brevidentatus, L. capitatus, L. collinus, L. congestus, L. cordifolius* (Fig. 4.36), *L. domingensis, L. exsertus, L. glandulosus, L. jefensis, L. latifolius, L. laxiflorus, L. longifolius, L. nigrescens, L. peduncularis, L. saponarioides, L. scopulinus, L. seemannii, L. silenifolius, L. skinnerii, L. troyanus, L. umbellatus,* and *L. viscidiflorus*.

INTERSPECIFIC VARIATION: Exine varies from reticulate (e.g., *Lisianthius silenifolius*) to reticulate-striate (*L. cordifolius*), or striate with low relief (*L. exsertus*).

PREVIOUS STUDIES: Gilg (1895) described the pollen of *Lisianthius* as finely to coarsely reticulate (relating it to that of *Hockinia*). Köhler (1905) described the pollen grains as 3-colporate and reticulate. About 20 species were studied by Nilsson (1970). Agababyan and Tumanyan (1977) described and illustrated pollen of *Lisianthius cordifolius*.

Neurotheca SALISB. EX BENTH. IN BENTH & HOOK. F. (Fig. 4.37)

The pollen grains of *Neurotheca* are released as monads, 3-colporate, 41–69 × 33–55 μm, prolate to subprolate, with exine relatively thick (6–9 μm) and reticulate (lumina up to 3 μm in diameter, sinuous to rounded-angular). No morphological differences were found between pollen of *Neurotheca loeselioides* specimens from Africa (Fig. 4.37A–D) and from South America (Fig. 4.37E–F). A typical feature is the smooth, laterally thickened and raised, sparsely perforate exine at the os region. The lateral extensions of the ora are obscure.

SPECIES EXAMINED: *Neurotheca congolana*, *N. corymbosa*, and *N. loeselioides*.

PREVIOUS STUDIES: Gilg (1895) reported 3-colpate, smooth or finely "granular" pollen grains, while Köhler (1905) noticed 3-colporate, distinct reticulate grains, which is in accordance with the present study. Maguire (1981) published LM and SEM micrographs of *Neurotheca loeselioides*.

Potalia AUBL. (Figs. 4.38–4.39)

The pollen grains are single, 3-porate and 33–36 × 31–43 μm, or rarely loosely united in tetrahedral tetrads, and then *c.* 50 × 65 μm in diameter (*Potalia maguireorum*). The relatively large pores have a diameter of *c.* 5 μm with rims beset with numerous rounded to elongated protuberances. The pores have an inner fragmented or granular annular thickening. The exine is generally smooth and imperforate or foveolate-rugulate with perforations. The inside of the exine is smooth. In the tetrads (*Potalia maguireorum*; Fig. 4.39) the distal poles of the grains have a distinct, perforate to foveolate elevation supported by extended, elongated columellar elements.

SPECIES EXAMINED: *Potalia amara*, *P.* "coronata", ined., *P. elegans*, *P. maguireorum*, *P. resinifera*, and *P.* "turbinata", ined.

POLLEN TYPES: *Potalia* (Punt & Leenhouts, 1967); *Anthocleista nobilis* (Punt, 1978).

PREVIOUS STUDIES: Erdtman (1952) examined *Potalia amara* pollen, which he compared to pollen grains of *Anthocleista parviflora*. Punt and Leenhouts (1967) described pollen grains of *Potalia amara* as "3-porate and imperforate" and indicated that the pores were slightly more protruding than in *Anthocleista*. This is confirmed in the present study.

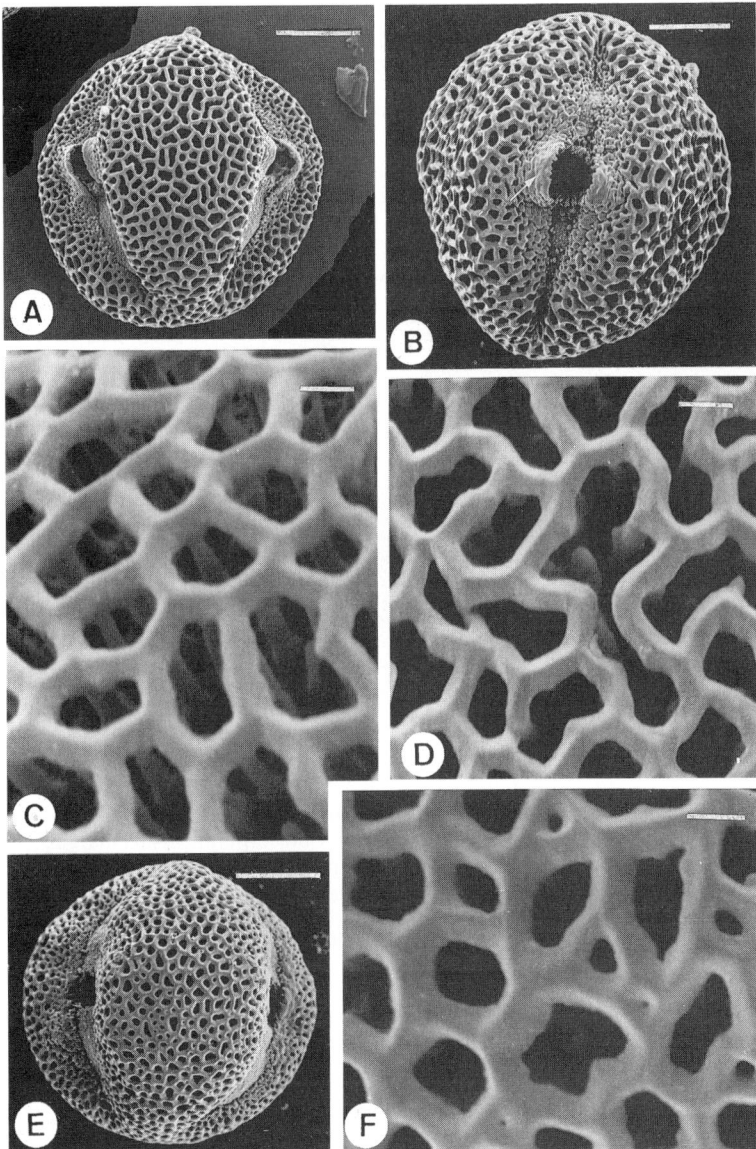

Figure 4.37. Pollen grains of *Neurotheca loeselioides*. **A–D.** (*Reekmans 9114*). **A.** 3-colporate, reticulate pollen in equatorial, mesocolpial view. **B.** Pollen in equatorial, apertural view, colpus and os with lateral elevations. **C.** Detail of exine, muri relatively straight. **D.** The same; detail of exine, muri sinuous. **E–F.** (*Nee 34765*). **E.** 3-colporate, reticulate pollen in equatorial, mesocolpial view. **F.** Detail of exine, muri relatively straight. Scale bars: A, B, E = 10 μm; C, D, F = 1 μm.

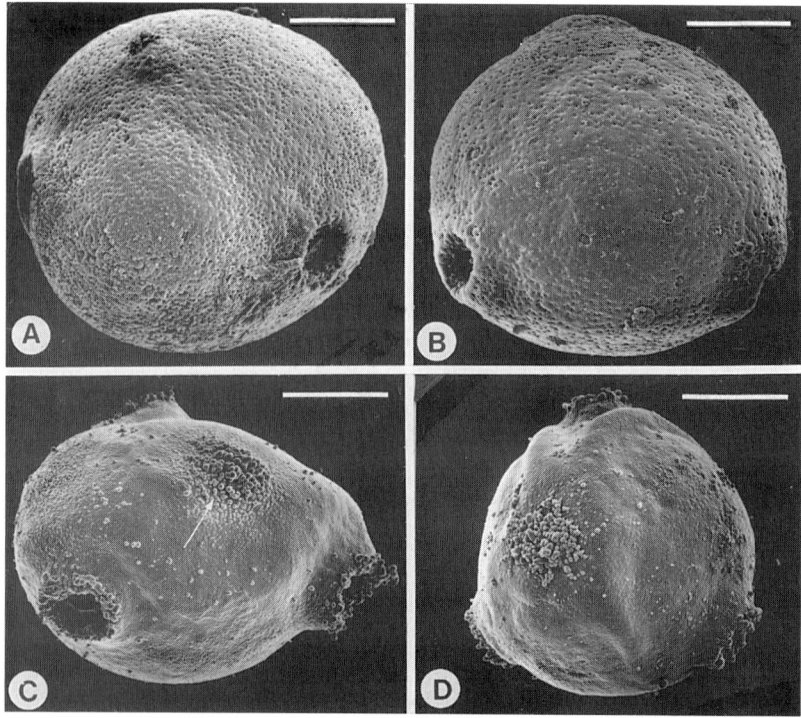

Figure 4.38. Pollen grains of *Potalia* (*Potalia*-type). **A–B.** *P. amara* (*Mori et al. 15785*). **A.** Pollen in slightly oblique equatorial view, 3-porate, with markedly protruding pore margins and granular rims, smooth, imperforate exine, with a rounded, distinctly patterned area at the pole. **B.** The same; polar view. **C.** *P. elegans* (*Maguire et al. 29356*). Pollen in equatorial view, 3-porate, pore margins only slightly protruding, exine smooth, distinctly and densely perforate. **D.** *P. resinifera* (*Vásquez & Jaramillo 7169*). Pollen in polar view, 3-porate, pore margins distinctly protruding with granular rims, exine smooth, imperforate; without marked polar area (cf. C and D). Scale bars = 10 μm.

DISCUSSION

The Saccifolieae

Curtia

The genus *Curtia* (Gentianeae-Erythraeinae) has different types of pollen varying from perforate-spinulose (*C. tenuifolia pro parte*) to finely reticulate or perforate, e.g., *C. obtusifolia*. Maguire (1981) pointed out the disparity in morphology and shape of *Curtia* pollen, suggesting a closer examination of the genus with reference to the differences observed. A correlation between

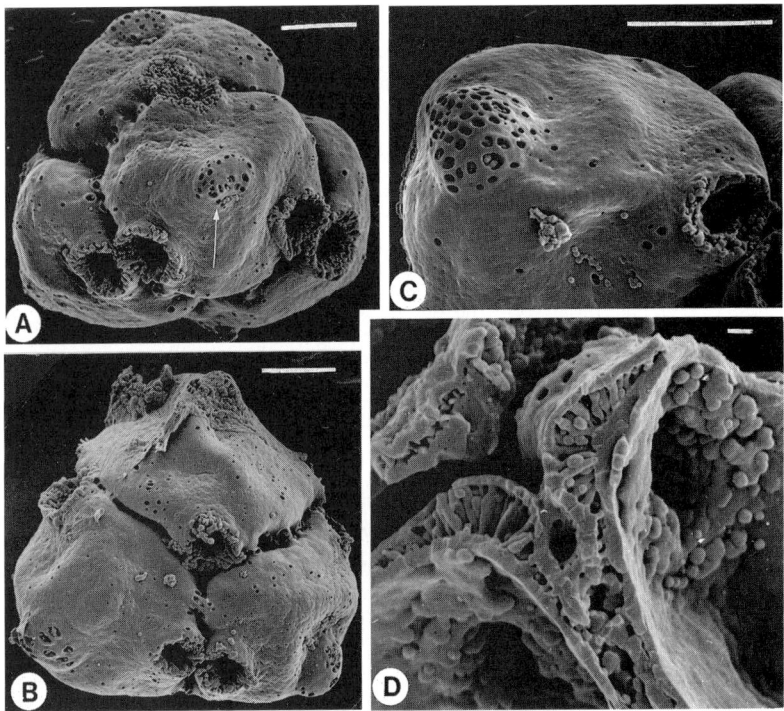

Figure 4.39. Pollen grains of *Potalia maguireorum* (*Potalia*-type; Kubitzski et al. 79-216). **A.** Pollen loosely united in tetrahedral tetrads, individual grains 3-porate, pores with markedly protruding margins and granular rims, exine smooth, sparsely perforate; at the poles there are dome-shaped structures with large perforations (cf. Fig. 4.38C and D). **B.** The same; view from the back. **C.** Close-up of a grain (in slightly oblique polar view) with a polar dome-shaped structure. **D.** Fracture showing part of the smooth exine inside two grains and thickened, granular pore margins; the dome-shaped structures are supported by columellae. Scale bars: A–C = 10 μm; D = 1 μm.

pollen morphology and heterostyly as observed here in *Curtia tenuifolia* appears plausible but is not always evident (cf. Fig. 4.1A–B vs. 4.1C–D). Grothe and Maas (1984) shared similar views based on seed micromorphology. A close relationship between *Curtia* and *Hockinia* is supported by pollen morphology.

Hockinia

The genus *Hockinia* was placed in Gentianeae-Tachiinae by Gilg (1895). Two types of pollen occur in the heterostylous *Hockinia montana*, one type with smooth or finely reticulate to perforate exine in longistylous specimens

(Fig. 4.2A–C), and another with microspinous pollen grains in brevistylous specimens (Fig. 4.2D–F). In terms of pollen morphology, *Hockinia* appears to be related to *Curtia*, and this has also been suggested by Grothe and Maas (1984) based on seed morphological similarities. The presence of two pollen types in *Hockinia* in relation to heterostyly is analogous to the case in *Curtia*. It is of interest to note that there is a pollen morphological similarity between *Hockinia montana*, *Curtia tenuifolia* ssp. *tenuifolia*, and the Cuban genus *Bisgoeppertia* (with microspinose grains). The last genus also belongs to the same subtribe in Gilg's classification, Gentianeae-Erythraeinae, but was moved to tribe Chironieae in the classification by Struwe *et al.* (2002).

Tapeinostemon

The genus *Tapeinostemon* (Gentianeae-Erythraeinae *sensu* Gilg) generally has small pollen grains (polar axis and equatorial diameter below 30 μm). Some species such as *Tapeinostemon longiflorum* have considerably larger grains with a polar axis reaching 40 μm or more. Maguire (1981) noticed clear differences in the exine structure (particularly at the polar areas) between two varieties of *Tapeinostemon longiflorum* and even more between two collections of *T. spenneroides*. One collection of the latter species (*Tillett et al. 45127*) has perforate pollen with an indistinct low-relief striation at the polar area, while another collection has pollen with distinctly reticulate to striato-reticulate polar areas (*Dunsterville & Dunsterville s.n.*). *Tapeinostemon zamoranum* (Fig. 4.3A–C) has reticulate grains that are rugulose-perforate at the colpus margin but are otherwise similar to those of *T. spenneroides*. Jonker (1948) placed *Tapeinostemon spenneroides* in his new genus *Stahelia*, which is not warranted palynologically. Steyermark (1951) did not agree that *Stahelia* should be regarded as a separate genus.

The Exaceae

Gentianothamnus

The monotypic Malagasy genus *Gentianothamnus* was described by Humbert (1937). The pollen grains were described as 3-colporate and scabrate with perforations (Lienau *et al.*, 1986), which is in accordance with the present findings. Humbert (1937) related *Gentianothamnus* with *Chironia* and *Orphium* (Gentianeae-Chironiinae *sensu* Gilg) or possibly to *Tachia* (Gentianeae-Tachiinae *sensu* Gilg), but the pollen grains of

Gentianothamnus show no similarity with any of these genera, but instead are slightly similar to those of *Tachiadenus*.

Tachiadenus

Tachiadenus (Gentianeae-Tachiinae *sensu* Gilg), a genus of 11 species endemic to Madagascar, was revised and palynologically examined by Klackenberg (1987, 1990). Two of the species were originally described as "*Lisianthus*", i.e., *Tachiadenus carinatus* and *T. tubiflorus*. According to Klackenberg (1990), *Tachiadenus tubiflorus* not only has a number of unique morphological features but the pollen grains also differ in being more finely reticulate to perforate with a vermicular (rugulate) surface pattern. Removal of *Tachiadenus* from the Gentianeae-Tachiinae *sensu* Gilg does not seem justified palynologically.

The Chironieae

Coutoubea and *Schultesia*

The two genera *Coutoubea* and *Schultesia* are very much alike in their pollen morphology. The outline of tetrads in decussate position varies in both genera due to the domed shape of the grains in side view. The reticulate pattern of the tetrads is differentiated into coarsely and distinctly heterobrochate polar areas bordered by a minutely reticulate pattern toward the smooth to granular apertural areas. *Coutoubea minor* has a less evident differentiation of the reticulum with the lumina relatively small and of equal size. The pollen grains are usually distinctly porate, although there are commonly transitions from colporate to porate grains. Gilg (1895) regarded the two genera to be closely related with reference to pollen although having different types of inflorescences. *Schultesia* has long, pedunculate inflorescences while *Coutoubea* usually has subsessile flowers in spikes or racemes.

Deianira

According to Gilg (1895) the genera *Deianira* and *Helia* are closely allied. However, *Deianira* differs from *Helia* in its exine reticulation. In *Helia* the exine is clearly differentiated into an equatorial-subequatorial zone with coarser reticulum and a perforate polar area. In *Deianira* the reticulum is undifferentiated with equally sized lumina and a special substructure of the muri.

The Helieae

The *Irlbachia* complex

The circumscription, taxonomic concepts, and treatments of the *Irlbachia* complex and "*Lisianthus*" *sensu lato* have varied considerably over the years (see Struwe & Albert, 1998). *Lisianthius sensu stricto* is now placed in Potalieae-Lisianthiinae, and is not closely related to the genera of the *Irlbachia* complex, which are placed in Helieae.

In Grisebach's (1839) neotropical tribe Lisyantheae the genus *Lisyanthus* was divided into five generic sections: *Macrocarpaea*, *Choriophyllum*, *Chelonanthus*, *Helia*, and *Calolisyanthus*, thereby including most species of the *Irlbachia* complex known at that time. Grisebach furthermore recognized the genera *Irlbachia* and *Leianthus* (the latter genus being synonymous with the true *Lisianthius*). *Leianthus* was divided into two sections: "*Lisianthus*" and *Omphalostigma*. It is noteworthy that Grisebach (1839) included the species from tropical South America in *Lisyanthus*, while the species of *Leianthus* (i.e., true *Lisianthius*) had a West Indian–Central American distribution. The genera *Irlbachia*, *Leiothamnus* (= *Symbolanthus*), and *Tachia* occur mainly in tropical South America. Bentham (1876) also kept *Leianthus* separate from *Lisyanthus*. The latter genus was subdivided into sections, i.e., *Adenolisianthus*, *Calolisyanthus*, *Chelonanthus*, *Choriophyllum*, *Helia*, *Irlbachia*, *Megacarpaea* (= *Macrocarpaea*), and *Symbolanthus*. *Pagaea* (= *Irlbachia*) was kept as a separate genus.

Gilg (1895) established several segregate genera from *Lisyanthus sensu* Grisebach. *Macrocarpaea* was placed in the same subtribe, together with some unrelated genera, e.g., *Hockinia*, *Tachia*, *Tachiadenus*, and *Zonanthus*. *Rusbyanthus* (*Macrocarpaea* spp.) was placed in a separate tribe, Rusbyantheae. The following "lisianthoid" genera segregated from *Lisyanthus* were placed in the tribe Helieae (pollen in tetrads): *Adenolisianthus*, *Calolisianthus*, *Chelonanthus*, *Helia*, *Irlbachia*, *Lagenanthus*, *Lehmanniella*, *Pagaea*, *Purdieanthus*, and *Symbolanthus*.

Steyermark *et al.* (1953), however, preferred to keep a sectional division of "*Lisianthus*" rather than following Gilg's (1895) classification and generic concept. He also disagreed with Perkins' (1902) idea to limit *Lisianthius* to taxa from Central America and the West Indies (and northern Colombia) only, a circumscription which, however, has strong palynological support. Rather, Steyermark shared the opinion of Bentham (1854) in recognizing one comprehensive genus, "*Lisianthus*", divided into sections with taxa interconnected by a series of intermediates.

A review of palynology 467

Weaver (1972) revised *Lisianthius* and defined it as a genus with West Indian–Central American distribution only, distinguishing the species from South American taxa by a number of morphological characters. *Lisianthius* (monads) was thereby separated from *Adenolisianthus*, *Calolisianthus*, *Chelonanthus*, *Helia*, *Irlbachia*, *Lagenanthus*, *Lehmanniella*, *Macrocarpaea* (monads), *Macrocarpaea* spp. (tetrads = *Rogersonanthus*), *Pagaea* (= *Irlbachia*), *Purdieanthus*, and *Symbolanthus*.

In the treatment of the tribe Lisyantheae by Maas *et al.* (1983) the genera *Adenolisianthus*, *Calolisianthus*, *Celiantha*, *Chelonanthus*, *Helia*, *Irlbachia*, *Lagenanthus*, *Lehmanniella*, *Leianthus*, *Leiothamnus*, *Lisianthius*, *Macrocarpaea*, *Pagaea*, *Purdieanthus*, and *Symbolanthus* were listed as members of their *Lisianthius* complex. In his survey of the Lisyantheae, Maas (1985) added *Prepusa* and *Senaea* and excluded *Leianthus* from the revised taxa. Furthermore the genera *Adenolisianthus*, *Brachycodon*, *Calolisianthus*, *Chelonanthus*, *Helia*, and *Pagaea* were reduced to *Irlbachia*, and *Lagenanthus* and *Purdieanthus* were merged into *Lehmanniella*. These taxonomic concepts remain unsupported from a palynological point of view. Maas (1985) proposed the generic name *Irlbachia* for *Adenolisianthus*, *Calolisianthus*, *Chelonanthus*, *Helia*, and *Pagaea*; however, the name *Helia* should have been used instead, following Kuntze's (1891) choice among names of equal priority (for details see Struwe & Albert, 1998).

It is evident that most of the above authors intended to split these species into two major groups, i.e., *Lisianthius* and another group with South American distribution, but the presence of three variants of genus names has caused considerable confusion. The latter group had been named the *Irlbachia* complex (e.g., Struwe *et al.*, 1997), which largely corresponds to tribe Helieae *sensu* Struwe *et al.* (2002). In the *Irlbachia* complex (*sensu* Struwe *et al.*, 1997) the following genera were included: *Adenolisianthus*, *Aripuana*, *Calolisianthus*, *Chelonanthus*, *Helia*, *Irlbachia*, *Macrocarpaea*, *Rogersonanthus*, *Symbolanthus*, *Tachia*, *Tetrapollinia*, and *Wurdackanthus*. Nilsson (1968, 1970) examined the pollen of most taxa of the *Irlbachia* complex. In the present study new data are presented from *Celiantha* and the presumably related genera *Aripuana*, *Neblinantha*, *Prepusa*, *Senaea*, and *Sipapoantha*. According to the present study the *Irlbachia* complex would contain the following genera (cf. Struwe *et al.*, 1997): *Adenolisianthus* (possibly congeneric with *Chelonanthus*), *Aripuana*, *Calolisianthus*, *Celiantha*, *Chelonanthus*, *Helia*, *Irlbachia*, *Rogersonanthus*, *Tetrapollinia*, *Symbolanthus*, and *Wurdackanthus*. *Neblinantha* and *Sipapoantha* should belong here, and probably *Prepusa* and *Senaea* (cf. Maas, 1985). All of these genera are also included in the tribe Helieae by Struwe *et al.* (2002).

The relationships among the genera included in the *Irlbachia* complex (tetrads, polyads) and the genera with pollen shed in monads remains obscure from a palynological standpoint. For example, palynologically, *Tachia* so far appears alien to the *Irlbachia* complex, but was included by Struwe *et al.* (1997). On the other hand, pollen grains of *Chorisepalum*, *Macrocarpaea*, and *Tachia* seem interrelated, and could, to a certain extent, be associated with *Lisianthius*. Further studies including, for example, pollen ontogeny, may eventually better elucidate relationships among taxa with single and compound pollen grains in the *Irlbachia* complex.

Adenolisianthus

Gilg's (1895) distinction between *Adenolisianthus* and *Chelonanthus* based on pollen is within the frame of variability of the *Chelonoides*-type. Bentham (1876) kept *Adenolisianthus* as a separate section of *"Lisianthus"*, but remarked on the similarity between the generic sections *Adenolisianthus*, *Chelonanthus*, and *Macrocarpaea* (only species with tetrads of *Arborea*-type, i.e., *Rogersonanthus*, should be considered in this context). Weaver (1972) placed *Adenolisianthus*, *Chelonanthus*, and *Rogersonanthus* close to each other on macromorphological grounds. Maas (1985) included *Adenolisianthus arboreus* in *Irlbachia* as *I. alata* ssp. *arborea* in his taxonomic survey. Maguire and Boom (1989) transferred this species to *Chelonanthus* under a new name, *C. fruticosus*. *Adenolisianthus* was again accepted by Struwe *et al.* (1999). Pollen morphological data suggest that *Adenolisianthus* could be reduced to a synonym of *Chelonanthus*.

Aripuana

Although the differentiation of reticulum (coarse at the equator and pole) is less pronounced, the *Aripuana*-type generally resembles the *Chelonoides*-type.

Calolisianthus and *Wurdackanthus*

Calolisianthus was regarded as a section of *Lisyanthus* by Grisebach (1839). Bentham (1876) also kept *Calolisianthus* as a separate section of his *"Lisianthus"*, while Gilg proposed its generic status and Maas (1985) included *Calolisianthus* in *Irlbachia*.

There are two, probably interrelated, pollen types in *Calolisianthus*, i.e., the *Speciosus*-type with tetrads, and the *Amplissimus*-type with polyads (Nilsson, 1970). *"Lisianthus" alpestris*, *"L." elegans*, and *"L." tenuifolius* have pollen of *Speciosus*-type and would better be included in the genus *Calolisianthus*. The transfer of *Calolisianthus tatei* to *Irlbachia* by Maguire

(1981) is fully supported. *Calolisianthus tatei* has no pollen morphological similarity to *C. amplissimus* (cf. Gleason & Wodehouse, 1931). In *Calolisianthus*, the *Speciosus*-type could be initially derived from the *Symbolanthus*-type and then further developed into the more aggregated *Amplissimus*-type. Maguire (1985) proposed a new genus, *Wurdackanthus*, and later added *Calolisianthus frigidus* (with *Symbolanthus*-type pollen) to this genus (Maguire & Boom, 1989). Maas (1985) included *Calolisianthus frigidus* in *Irlbachia* as *I. frigida*. *Wurdackanthus frigidus* could possibly be kept in the genus *Calolisianthus* based on pollen data, although it seems more closely related to the genus *Symbolanthus*. Thus, the creation of the new genus *Wurdackanthus* seems reasonable from a palynological standpoint.

Celiantha

The genus *Celiantha*, established by Maguire (1981), is distinctive through its polyads with polar globules. Transitions traceable from the *Uliginosus*- to the *Irlbachia*-types are observed (cf. *I. plantaginifolia*; Maguire, 1981). The transfer of *Calolisianthus imthurnianus* and "*Lisianthus*" *chimantensis* to *Celiantha* by Maguire (1981) is supported by palynological data. Steyermark *et al.*'s (1967) idea that "*Lisianthus*" *chimantensis* is allied to *Lisianthius collinus*, *L. corymbosus*, or *L. crenatus*, all taxa with single pollen grains, cannot be supported.

Chelonanthus

In terms of pollen morphology *Chelonanthus* contains two types, i.e., the variable *Chelonoides*-type (tetrads) and the *Uliginosus*-type (polyads) with polar loops united to rings. The relationships and possible developmental series among and between tetrads and polyads are of special interest. With the *Chelonoides*-type as a starting point, a trend from loosely united tetrads to polyads with reticulate to granular and finally spiny-globular exine ornamentation (as in *Irlbachia* and *Celiantha*) is proposed. Another tendency is fragmentation and transformation of muri into elongated pieces and eventually globular processes. Gilg (1895) considered the possibility of keeping the species with polyads (*Chelonanthus uliginosus* (= *C. purpurascens*) and *C. campanuloides* (= *Irlbachia pratensis*; Struwe *et al.*, 1999); see under *Irlbachia* below) apart from the rest of *Chelonanthus* with pollen in tetrads. However, with reference to the present studies it is suggested that the genus *Chelonanthus* not be split, but rather that the variation be interpreted as a morphological trend. The transfer of *Chelonanthus cardonae* to *Irlbachia* by Maguire (1981) is palynologically supported. Gilg's suggestion of a close affinity between *Chelonanthus camporum* and *C. angustifolius* is confirmed palynologically. *Chelonanthus arboreus* was transferred to *Macrocarpaea* by Ewan (1948a) and is now included in *Rogersonanthus*.

Great variability in the morphology of exine processes has been noted in *Chelonanthus*. Maguire and Boom (1989) also showed several examples of varying exine pattern in *Chelonanthus* with muri sometimes fragmented or hypertrophied into elongated to rounded processes (blunt spines, clavae, and granules). A superficial similarity to *Tetrapollinia caerulescens* may be noted.

Chorisepalum

The genus *Chorisepalum*, endemic to the Guayana Highlands, was erected by Gleason and Wodehouse (1931) based on *C. ovatum*. In his revision of *Chorisepalum*, Ewan (1947) added *Chorisepalum carnosum*, *C. psychotrioides*, and *C. rotundifolium*. Fifteen years later Steyermark (1962) described *Chorisepalum acuminatum*, which he regarded as closely allied to *C. rotundifolium*. One more species, *Chorisepalum breweri*, was later described by Steyermark and Maguire (1976).

The pollen of *Chorisepalum ovatum* was described and depicted by Gleason & Wodehouse (1931). It was said to be 30.8–34.2 μm in diameter, 3-colpate with a circular germinal aperture (i.e., 3-colporate), and with a prominent reticulum that is finer at the poles and at the colpus margin. This description is valid for all species examined. The lumina of the coarse reticulum may be exceedingly large.

The pollen of *Chorisepalum* resembles that of *Macrocarpaea* and *Tachia*. According to Gleason and Wodehouse (1931) the closest genera are *Zygostigma* or *Rusbyanthus* and possibly *Macrocarpaea*. *Zygostigma* is palynologically dissimilar, while the affinity to *Macrocarpaea* has palynological support. Ewan (1947) expressed a similar opinion and he even proposed that one species of *Macrocarpaea* (i.e., *M. calophylla*), not seen by either Ewan or the present author, may be a species of *Chorisepalum*. From the illustrations in Maguire (1981), and those included here, no important and consistent palynological differences could be seen among the species of *Chorisepalum* or between *Chorisepalum* and *Macrocarpaea*. With reference to the various opinions of relationships above and pollen morphological features, it is suggested that *Chorisepalum* be placed near *Macrocarpaea* and *Tachia* in Gentianeae-Tachiinae *sensu* Gilg or in Helieae *sensu* Struwe *et al.* (2002).

Helia

The pollen tetrads of *Helia* are similar to those of *Chelonanthus* (*Chelonoides*-type), although they are distinctive in having more finely reticulate and perforate polar areas. All tetrads studied were similar to each other. Grisebach (1839) classified some *Chelonanthus* species together with *Helia* species in his *Lisyanthus* sect. *Helia*, while Maas (1985) reduced *Helia* to *Irlbachia*, which was incorrect (cf. Struwe & Albert, 1998, regarding Kuntze, 1891). Kuntze's (1891) large generic concept of *Helia* is not supported palynologically (it is too heterogeneous).

Irlbachia (including *Brachycodon* and *Pagaea*)

Grisebach (1845) recognized both *Irlbachia* and *Pagaea* as separate genera. In his notes on north Brazilian Gentianaceae, Bentham (1854) put the species of *Pagaea* in *"Lisianthus"* sect. *Brachycodon*. *Irlbachia* was regarded as a section of *"Lisianthus"* close to sect. *Calolisyanthus*, while *Pagaea* was still kept as a separate genus by Bentham (1876). Gilg (1895) and Köhler (1905) also kept them as two separate genera. *Irlbachia caerulescens* (= *Tetrapollinia caerulescens*) was one of 3–4 species included in the genus *Irlbachia* by Gilg (1895).

It was Maguire (1981) who suggested *Irlbachia* and *Pagaea* to be congeneric under the name of *Irlbachia*. Maas (1985), in his taxonomic survey, included members of *Adenolisianthus, Calolisianthus, Chelonanthus, Helia,* and *Pagaea* in the broadly circumscribed genus *Irlbachia*. From a palynological point of view *Irlbachia* (*sensu* Maguire, 1981) is a homogeneous genus characterized by spiny polyads (*Irlbachia*-type; Nilsson, 1970). *Irlbachia sensu* Maas (1985) is more heterogeneous palynologically and therefore Maas's concept is not followed here. However, the reduction of *Pagaea* to *Irlbachia* has clear pollen morphological support.

The transfer of *Calolisianthus tatei* (as well as *Chelonanthus cardonae*) to *Irlbachia* has full palynological support (Maguire, 1981). Gleason's (1931) new species, *Calolisianthus tatei*, was regarded as close to *C. pulcherrimus*, but this has no pollen morphological support. Bentham (1876), who maintained *Pagaea* separate from *"Lisianthus"* sect. *Irlbachia*, thought that *"Lisianthus" breviflorus* should be referred to *Pagaea*, which agrees well with the spiny polyads. *"Lisianthus" acutilobus* from Venezuela was described by Steyermark (1966), including a drawing of a spiny polyad. It was regarded as closely allied to *"Lisianthus" breviflorus*, differing only in some minor flower morphological features from *Irlbachia cardonae* and *I. tatei*. Thus it seems best to include *"L." acutilobus* in the genus *Irlbachia* as well as *"Lisianthus" scabridulus*, which Steyermark *et al.* (1953) was inclined to refer to *Pagaea*. In fact, Struwe *et al.* (1999) recently treated *"Lisianthus" acutilobus* as synonymous with *Irlbachia cardonae* and *"L." scabridulus* as synonymous with *I. tatei*.

It is also interesting to note that Steyermark *et al.* (1953) held the opinion that *Irlbachia pratensis* and *"Lisianthus" campanuloides* were conspecific, which was also followed by Struwe *et al.* (1999), who treated them as *I. pratensis*. Both these species have polyads that are intermediate between the *Uliginosus*- and *Irlbachia*-types. *Irlbachia pratensis* has never been included in *Chelonanthus*. The pollen grains (polyads) are transitional between the *Uliginosus*- and *Irlbachia*-types of pollen, and therefore *Irlbachia pratensis* is best treated as a species of *Chelonanthus*.

Lagenanthus, *Lehmanniella*, and *Purdieanthus*

These three genera established by Gilg (1895) are easily separable with regard to pollen. In particular, *Purdieanthus* and *Lehmanniella* pollen seems more connected with the *Symbolanthus*- and *Speciosus*-types rather than with *Lagenanthus*, which Gilg associated with *Calolisianthus*. *Lehmanniella* pollen also has an extreme development of the reticulum that easily detaches from the smooth tetrad surface.

According to Ewan (1952) the genus *Lagenanthus* has two species, *L. princeps* and *L. parviflorus* (not seen). However, Maas (1981) concluded that *Lagenanthus parviflorus* did not belong to the Gentianaceae, but to the genus *Ravnia* or *Hillia* (Rubiaceae). In her treatment of *Hillia*, Taylor (1990) regarded *Lagenanthus parviflora* as a synonym of *Hillia triflora* var. *pittieri* (Rubiaceae), which also has palynological support. Ewan revised the monotypic genera *Lehmanniella* and *Purdieanthus* (Ewan, 1948b), separating the two genera based on leaf morphology. Maas (1985) lumped *Lagenanthus*, *Lehmanniella*, and *Purdieanthus* together under the generic name *Lehmanniella*. A third species, *Lehmanniella huanucensis*, was also added (Maas, 1985). From a pollen morphological point of view the genera are distinctive from each other and deserve to be maintained. *Purdieanthus* and *Lehmanniella* may be related to *Symbolanthus* and *Calolisianthus* but the relationships of *Lagenanthus* are unclear.

Macrocarpaea

Nilsson (1968) distinguished three types of pollen in the genus *Macrocarpaea*, i.e., reticulate monads (*Glabra*-type), warty monads (*Corymbosa*-type), and tetrads (*Arborea*-type). The majority of *Macrocarpaea* species have pollen of the *Glabra*-type. The pollen of two recently published species, *Macrocarpaea neblinae* and *M. piresii*, as well as *M. rugosa*, were described and illustrated by Maguire (1981). The *Corymbosa*-type (later called *Rusbyanthus*-type; Nilsson, 1970) contains several species: *Macrocarpaea cinchonifolia*, *M.* sp. nov. 1, *M.* sp. nov. 4, and *M.* sp. nov. 9 (Table 4.3). Gilg

(1897) suggested close affinity between *Macrocarpaea* species with verrucose pollen grains, and this also has pollen morphological support. Transitions between distinctly reticulate to verrucose pollen grains with remnants of muri between the "hypertrophied" meeting points of the muri were observed. In his review of *Macrocarpaea*, Ewan (1948a) suggested a close relationship between *M. pachystyla* (those collections here determined as *M.* sp. nov. 4) and the Bolivian genus *Rusbyanthus* based on corolla morphology, which has palynological support as well. Gilg's (1895) idea of creating a new genus, and even tribe (Rusbyantheae), for *Macrocarpaea cinchonifolia* was based primarily on the warty pollen grains. Ewan (1948a), however, believed that the genera *Rusbyanthus* and *Macrocarpaea* were closely allied. Similarly, Weaver (1972) placed *Rusbyanthus* and *Macrocarpaea* near to each other in his key. *Rusbyanthus cinchonifolius* was later combined to *Macrocarpaea cinchonifolia* by Weaver (1974). At the same time *Macrocarpaea pachystyla* was made a synonym of *M. cinchonifolia*. As a result not only the genus *Rusbyanthus* but also Gilg's entire tribe Rusbyantheae were reduced. This is completely supported by palynological data (Nilsson, 1968, 1970).

The *Macrocarpaea* species in the Guayana Highlands with tetrads (*Arborea*-type) were removed from *Macrocarpaea* to a new genus, *Rogersonanthus* (see below) by Maguire and Boom (1989). The tetrads cannot be differentiated from those of the *Chelonoides*-type (cf. *Chelonanthus* [Nilsson, 1970]; see *Rogersonanthus*). In conclusion, the genus *Macrocarpaea* would be best circumscribed by only those taxa having pollen of the *Glabra*- and *Corymbosa*-types (and with one species also having *Longifolius*-type pollen).

Neblinantha

The genus *Neblinantha* was erected by Maguire (1985). According to Maguire and Boom (1989) the two species included had pollen of *Helia*-type (cf. Nilsson, 1970). However, *Neblinantha* tetrads are more reminiscent of the tetrads of *Symbolanthus*- or *Speciosus*-type. Maguire and Boom (1989) described the pollen tetrads of the two *Neblinantha* species as slightly different. *Neblinantha neblinae* has a reticulum with thickened muri and *N. parvifolia* has a reticulum of uniform size. Although the pollen grains were said to be of *Helia*-type, no close relationship to any other Guayanan genera could be found (Maguire & Boom, 1989).

Prepusa and Senaea

Prepusa and *Senaea* are another genus pair (like *Coutoubea* and *Schultesia*) with tetrads of the same type. The exine bears different types of small processes (small pila, skittles, and globules) intermingled with larger rounded to pilate processes. Gilg (1895) pointed out the pollen morphological similarities between the two genera, which he separated on flower morphological differences. The tetrads show similarity to those of *Sipapoantha*. *Prepusa* and *Senaea* pollen may also be related to the *Chelonoides*-type of pollen (the *Irlbachia* complex) with reference to the globules and clavae of the exine.

Rogersonanthus

Rogersonanthus was described as a new genus by Maguire and Boom (1989) based on several *Macrocarpaea* species. Three species were included, namely *Rogersonanthus quelchii*, *R. tepuiensis* (including *Macrocarpaea cerronis* and *M. salicifolia*), and *R. arboreus*. *Macrocarpaea arborea* was originally described as *Chelonanthus arboreus* by Britton (1922) and *M. tepuiensis* as *Calolisianthus tepuiensis* (Gleason & Killip, 1939). In his revision of *Macrocarpaea*, Ewan (1948a) suggested a close relationship between *M. arborea*, *M. cerronis*, and *M. salicifolia*, an observation that has pollen morphological support despite minor inter- and intraspecific variation (Nilsson, 1968, 1970). *Macrocarpaea quelchii* was regarded as most closely allied to *M. ovalis* (*Glabra*-type) by Ewan, but this is contradicted by pollen morphology. *Macrocarpaea domingensis* was also postulated to be related to *M. quelchii* and associated species, but this is not palynologically supported. Maas (1985) lumped all species of the above taxa into *Irlbachia quelchii* (= *Rogersonanthus quelchii*), which is not contradicted by pollen morphology. Ewan's idea that *Macrocarpaea* and *Symbolanthus* are closely allied genera has no palynological support, while a relationship between *Rogersonanthus* and *Symbolanthus* could be possible.

The tetrads of *Rogersonanthus* are of typical *Chelonoides*-type. It is therefore in total agreement with palynology to remove these species from *Macrocarpaea*. However, the above five (or three) *Macrocarpaea* species may be better transferred to some other genus with the *Chelonoides*-type of pollen, e.g., *Chelonanthus*. Neither creation of a new genus, *Rogersonanthus* (Maguire & Boom, 1989), nor fusion into a heterogeneous genus, *Irlbachia* (Maas, 1985), seems relevant. It may be argued whether it was taxonomically sound or not to establish a new genus (cf. Maguire & Boom, 1989). If the *Chelonoides*-type of tetrads has arisen independently more than once and other characters support distinction from, for example, *Chelonanthus*, it seems justified to keep the above species in a separate genus (i.e., *Rogersonanthus*), which is provisionally followed in this study.

Sipapoantha

Sipapoantha tetrads (*Sipapoantha*-type) were compared to those of *Chelonanthus alatus* or *C. camporum* (Nilsson, 1970), which was interpreted to constitute a possible "remote ancestry" by Maguire & Boom (1989). *Rusbyanthus* was also mentioned with reference to the rounded exine processes. However, *Rusbyanthus* has single, usually porate, pollen grains, while *Sipapoantha* has tetrads. *Sipapoantha* pollen is more similar to *Prepusa* and *Senaea* tetrads although differs in having a prominent colpus margin.

Symbolanthus

The *Symbolanthus* pollen type can be derived from the basic *Chelonoides*-type whereby the equatorial-subequatorial reticulum tends to be more raised, forming coarsely reticulate islands that become gradually fused (i.e., *Speciosus*-type to *Amplissimus*-type). *Symbolanthus* (incl. *Leiothamnus*) was kept as a separate genus by Gilg (1895) and as a section of "*Lisianthus*" by Bentham (1876). Nilsson (1970) found the same type of tetrads in *Wurdackanthus frigidus*. Maas (1985) did not include *Symbolanthus* in his large circumscription of the genus *Irlbachia*.

Tachia

Most species of *Tachia* (Gentianeae-Tachiinae *sensu* Gilg) have reticulate to coarsely reticulate exine patterns. Exceptions are *Tachia occidentalis*, with a smooth, imperforate exine, *T. parviflora* with a smooth, perforate exine, and *T. loretensis* with a reticulate-corrugate exine (Maguire & Weaver, 1975). The division of *Tachia* into two sections is not supported by pollen morphology as smooth pollen grains occur in both sections. Maguire and Weaver (1975) also concluded that pollen morphology is of little phylogenetic significance in the genus *Tachia*. The reticulate pollen grains resemble those of *Macrocarpaea* and *Chorisepalum*.

Tetrapollinia

Grisebach (1839) recognized two species in the genus *Irlbachia*, i.e., *I. elegans* and *I. caerulescens*. The former species has typical *Irlbachia* polyads and is the type of *Irlbachia*, while *I. caerulescens* (= *Tetrapollinia caerulescens*) has spiny tetrads with an even distribution of the spines on the surface (*Caerulescens*-type). Despite this principal difference there is a certain resemblance between these taxa in exine details, e.g., a fine, delicate lace-like reticulum, attached to and suspended from the lower part of the spines. The tetrads also show superficial resemblance to certain

Chelonoides-tetrads with their acute to rounded processes. *Tetrapollinia caerulescens* was included in *Irlbachia* by Gilg (1895) and Köhler (1905). It was regarded as having a separate pollen type, the *Caerulescens*-type, by Nilsson (1970), and, as mentioned above, its relation to *Irlbachia* is still obscure from a palynological point of view. However, placing this taxon in a new genus, i.e., *Tetrapollinia*, following Maguire and Boom (1989), may be the best and ultimate solution.

Zonanthus

The genus *Zonanthus* (Gentianeae-Tachiinae) is monotypic and endemic to Cuba. The exine pattern is reticulate with striation at the apertural margin area. The taxonomic affinity to other genera is unknown. Gilg (1895) classified *Zonanthus* close to *Zygostigma* and *Macrocarpaea* in his key, but this does not fit palynologically. Pollen morphology does not, so far, provide suggestive evidence for the possible taxonomic relationships of *Zonanthus*.

The Potalieae

Anthocleista, *Fagraea*, and *Potalia*

The three genera *Anthocleista*, *Fagraea*, and *Potalia* have been classified as members of the Gentianaceae on several occasions, e.g., by Bureau (1856), Fosberg and Sachet (1980), Struwe *et al.* (1994, 2002), and Takhtajan (1997). This has primarily been due to increasing macromorphological and molecular evidence. Punt and Leenhouts (1967) concluded that *Anthocleista* and *Potalia* have pollen grains resembling *Geniostoma* (Loganiaceae), at least superficially. Affinity to Apocynaceae-Tabernaemontaneae (Leenhouts, 1962) remains palynologically unsupported, also in agreement with the present study (cf. also Endress *et al.*, 1996). Fosberg and Sachet (1980) suggested a transfer of the older tribe Potalieae (Loganiaceae) to the Gentianeae-Tachiinae (Gentianaceae). According to Fosberg and Sachet the Potalieae would then include *Anthocleista, Chorisepalum, Fagraea, Lisianthius, Macrocarpaea, Potalia, Rusbyanthus, Symbolanthus, Tachia,* and *Tachiadenus*. From a palynological point of view this is an unnatural grouping, as *Potalia* and *Anthocleista* have an overtly non-gentianaceous type of pollen with a possible similarity to certain Apocynaceae, but not to the Tabernaemontaneae (cf. Leenhouts, 1962), or to *Geniostoma* (Loganiaceae), which has similar apertures (pores) with distinctly protruding margins. On the other hand, the diverse genus *Fagraea*, e.g., *F. annulata* and *F. umbelliflora* (cf. Punt, 1978), has pollen similar to *Chorisepalum, Macro-*

A review of palynology 477

carpaea, and *Tachia*, whereas the pollen grains of *Gentianothamnus*, *Symbolanthus*, and *Tachiadenus* are totally different. Several of the 3-colporate *Fagraea* pollen types and subtypes distinguished by Punt and Leenhouts (1967) resemble gentianaceous pollen. The similarities and possible close relationships between *Fagraea elliptica*, *F. fragrans*, and *F. racemosa* (Punt, 1978) are also confirmed in the present study. Relationship between the above *Fagraea* species and Gentianaceae (and also *Gelsemium*, Gelsemiaceae) is supported by palynological data (cf. Punt, 1980; Endress *et al.*, 1996). Molecular data clearly place Potalieae in the Gentianaceae, despite the ambiguous pollen evidence.

Lisianthius

The generic name of *Lisianthius*/"*Lisianthus*"/*Lisyanthus* (Gentianaeae-Tachiinae *sensu* Gilg) has been used for taxa with pollen in monads as well as in tetrads or polyads (e.g., Bentham, 1876). In his monographic treatment of *Lisianthius*, Weaver (1972) separated *Lisianthius* from *Macrocarpaea*, and several other genera of the tribe Helieae *sensu* Struwe *et al.* (e.g., *Helia*, *Chelonanthus*, *Calolisianthus*, and *Symbolanthus*). Several macromorphological characters support this distinction, such as the presence of a capitate stigma and fibers in the corolla tube (vs. a bilamellate stigma and the absence of corolline fibers; Weaver, 1972). Steyermark *et al.* (1953) preferred to keep all the above genera together as sections under the generic name *Lisianthus*.

In accordance with Gilg (1895), Nilsson (1970), and Weaver (1972) the name *Lisianthius* should be used only for species distributed from Mexico to Central America down to Panama. *Lisyanthus* and "*Lisianthus*" are inappropriate generic names that should not be used (Struwe & Albert, 1998). Only one species of the true *Lisianthius*, *L. seemannii*, is found on the South American continent (Colombia). This is also one of the only two species in South America having monads of *Longifolius*-type (the other being *Macrocarpaea loranthoides*, previously placed in *Lisianthius*). Grisebach (1839) put this species into a section of its own, *Lisyanthus* sect. *Choriophyllum*, and mentioned its similarity to *Macrocarpaea*.

Neurotheca and *Congolanthus*

Neurotheca (Gentianeae-Erythraeinae *sensu* Gilg) includes one species (*N. loeselioides*) distributed in both northern South America and Africa as well as two other species confined to Africa. Raynal (1968) proposed a new generic name, *Congolanthus*, for the African species *Neurotheca longidens*. Both genera have reticulate pollen grains. *Congolanthus* has more spheroidal, more finely reticulate, and considerably smaller pollen grains than *Neurotheca* (polar axis c. 40 μm vs. 50–68 μm). No pollen morphological differences between the South American and African collections of *Neurotheca loeselioides* could be detected. The present study is in support of distinguishing two separate genera. Raynal's (1968) suggestion of a possible relationship between *Coutoubea* (Helieae *sensu* Gilg) and *Neurotheca* does not have any palynological support.

Some pollen morphological trends

Most pollen types appear non-variable and are confined to particular genera, while others show plasticity in morphological details, e.g., the *Chelonoides*-type, and the *Speciosus*-type to a minor degree. Some morphological trends among the distinguished pollen types seem evident (Nilsson, 1970: fig. 1), one trend being present in the *Chelonoides* → *Uliginosus* → *Irlbachia* → *Imthurnianus*-types, and another in the *Chelonoides* → *Symbolanthus* → *Speciosus* → *Amplissimus*-types. The morphological interrelationships among *Lagenanthus*, *Lehmanniella*, and *Purdieanthus* are less obvious, while the latter two resemble the *Symbolanthus*-type. The *Helia*-type much resembles the *Cheloniodes*-type, although it is still distinctive. It could be regarded as a variant within the *Chelonoides*-type (Nilsson, 1970). The *Glabra*- and *Corymbosa*-types (monads) appear related, while the *Longifolius*-type appears distinct although reminiscent of the *Glabra*-type to a minor degree.

Summary

Gilg's (1895) tribal and subtribal division of the Gentianaceae-Gentianoideae, which was based mainly on pollen morphology, cannot be duly confirmed by the present study.

Some species of *Curtia* and *Hockinia montana* (Saccifolieae) are heterostylous, with two similar types of pollen. The two genera are here regarded as closely allied and may be placed together in the same tribe (Saccifolieae), including *Saccifolium*, *Tapeinostemon*, and *Voyriella*. The positions of the genera *Gentianothamnus* and *Tachiadenus* (Exaceae) are not evident based on pollen morphology. *Coutoubea* and *Schultesia* (Chironieae-Coutoubeinae) differ in the morphology of the tetrads from the *Irlbachia* complex (Helieae), and these two genera are very similar palynologically and may be phylogenetically close.

Congolanthus and *Neurotheca* (Potalieae-Faroinae) are supported as two

A review of palynology 479

separate genera with reference to pollen morphology. The pollen grains of some species of *Fagraea* (Potalieae-Potaliinae) show similarity to those of some Gentianaceae taxa, e.g., *Lisianthius*, as well as to *Gelsemium* (Gelsemiaceae). A transfer of *Anthocleista, Fagraea*, and *Potalia* to the Gentianaceae from Loganiaceae, e.g., as subtribe Potalieae-Potaliinae, is not palynologically supported, *Fagraea* excepted. The genera *Anthocleista* and *Potalia* have very similar, 3–4-porate, usually smooth pollen grains with projecting pore margins, while those of *Fagraea* differ in aperture type and exine ornamentation. The pollen grains of *Fagraea* examined are 3-colporate-porate with reticulate, striato-reticulate, or foveolate exine pattern. *Anthocleista* and *Potalia* have pollen slightly resembling that of Loganiaceae, e.g., *Geniostoma*, but not other Gentianaceae.

The majority of taxa in Gilg's (1895) tribe Helieae have been variously referred to the genus "*Lisianthus*" as sections, to separate genera, or have been partly lumped into one genus. Pollen morphology suggests distinction of one West Indian–Central American group with monads (the genus *Lisianthius*, now in Potalieae) and one neotropical group (mainly continental) with compound pollen grains (the *Irlbachia* complex plus some additional genera, now in Helieae). The South American group is supported by pollen morphology, although still has distinguishable pollen types. Transitions are common. The taxonomic status of the *Irlbachia* complex cannot be decided based on pollen morphology alone. A transfer of "*Lisianthus*" *acutilobus* and "*L.*" *scabridulus* to *Irlbachia* is supported, as is a transfer of *I. pratensis*. Likewise, "*L.*" *alpestris*, "*L.*" *elegans*, and "*L.*" *tenuifolius* should be moved to *Calolisianthus*. In the present study Gilg's (1895) generic concept is regarded as partly correct, as is that of Maguire (1981) and Maguire and Boom (1989). Maas's (1985) classification and taxonomic concept (a broadly circumscribed *Irlbachia*) is not in agreement with the palynological results. The genera *Celiantha, Neblinantha, Rogersonanthus, Sipapoantha, Tetrapollinia*, and *Wurdackanthus* (Helieae) all seem associated within the *Irlbachia* complex.

The genus *Symbolanthus* (Helieae) has pollen tetrads reminiscent of those of the *Irlbachia* complex. The genera *Lagenanthus, Lehmanniella*, and *Purdieanthus* (Helieae) are palynologically distinct; thus, pollen morphology does not support their fusion into one genus, *Lehmanniella*, as proposed by Maas (1985). *Lehmanniella* and *Purdieanthus* are associable with the *Irlbachia* complex, but pollen of *Lagenanthus* appears dissimilar to the other two genera. The genera *Prepusa* and *Senaea* (Helieae) have similar tetrads to the genera of the *Irlbachia* complex and might be phylogenetically related. Pollen of *Sipapoantha* (Helieae) also shows similarity to

Prepusa and *Senaea* pollen. The new genus *Aripuana* (Helieae) has tetrads similar to the *Chelonoides*-type (*Chelonanthus* and *Rogersonanthus*).

Macrocarpaea should comprise only taxa with single, reticulate or verrucose pollen. The genera *Macrocarpaea* (Helieae) and *Lisianthius* (Potalieae-Lisianthiinae) are, in general, easily separable with reference to pollen. *Macrocarpaea*, *Tachia*, and *Chorisepalum* all appear to belong to the tribe Helieae. The phylogenetic position of *Zonanthus* (Helieae) is not evident based on pollen morphology.

ACKNOWLEDGMENTS

The author thanks Elisabeth Grafström, Magnus Hellbom, and Wieslaw Smolenski (Palynological Laboratory, Swedish Museum of Natural History) for skillful technical help. Anita Norrthon is acknowledged for help in preparing the manuscript. Jason R. Grant (University of Neuchâtel) is thanked for his help in determining material of *Macrocarpaea*.

Appendix 4.1
Specimens of Gentianaceae that were examined for this study and their voucher information

Herbarium abbreviations follow Holmgren et al. (1990). A superscript 'S' and a superscript 'T' indicate SEM and TEM, respectively. The slides are deposited at the Palynological Laboratory, Swedish Museum of Natural History, Stockholm. Synonyms (i.e., accepted species) follow Struwe and Albert (1998) and Struwe et al. (1999, 2002) and are indicated in parentheses after each taxon name, where applicable.

Adenolisianthus (Progel) Gilg
A. arboreus (Spruce ex Progel) Gilg. BRAZIL. Amazonas: Rio Negro, 1930, *Holt & Blake 582* (K); Rio de Janeiro, 1876, *Glaziou 9950* (P). VENEZUELA. Amazonas, 1959, *Wurdack & Adderley 43722* (NY)[S].

Anthocleista R. Br.
A. amplexicaulis Baker. MADAGASCAR. 1994, *Pettersson 610* (UPS).
A. madagascariensis Baker. MADAGASCAR. Moramanga, *Schlieben 8083* (HBG).
A. microphylla Gilg & Mildbr. EQUATORIAL GUINEA. Fernando Poo, 1911, *Mildbraed 6434* (HBG)[S].

Aripuana Struwe, Maas, & V. A. Albert
A. cullmaniorum Struwe, Maas, & V. A. Albert. BRAZIL. Amazonas, 1985, *Cid Ferreira 5906* (NY).

Calolisianthus Gilg
C. acutangulus (Mart.) Gilg. BOLIVIA. Santa Cruz, 1950, *Cárdenas 4457* (US). BRAZIL. Mato Grosso, 1894, *Malme 1378 B* (S)[S]; Cuiaba, *Herb. Fl. Brazil 216* (BR).
C. amplissimus (Mart.) Gilg. BRAZIL. D. F., Brasilia, 1964, *Jesus 56* (NY); Minas Gerais, Lagoa Santa, 1952, *Smith 6721* (S)[S]; Paraná, Morungava, 1915, *Dusén 16668* (NY).
C. macranthus Gilg. BRAZIL. Goiás, Serra do Morcégo, 1966, *Irwin et al. 15272* (U); São Paulo, Pirassununga, 1965, *Vogel 640* (U).
C. pedunculatus (Cham. & Schltdl.) Gilg. BRAZIL. Goiás, Chapada dos Veadeiros, 1966, *Irwin et al. 12829* (U); Paraná, Vila Velha, 1910, *Dusén 9112* (S); Regio inferior, *Sellow s.n.* (HAL).
C. pedunculatus var. *damazienus* Branverck. BRAZIL. Minas Gerais, Ouro Preto, *Barbosa s.n.* (R 6719).
C. pendulus (Mart.) Gilg. BRAZIL. Minas Gerais, Caldas, *Regnell I: 300* (S); Rio de Janeiro, Serra da Bocaina, 1965, *Vogel 752* (U); São Paulo (M).
C. pulcherrimus (Mart.) Gilg. BRAZIL. Minas Gerais, Serra da Piedade, Caeté, 1933, *Barreto 2786* (F); Belo Horizonte, 1933, *Barreto 2789* (F); Diamantina, 1931, *Mexia 5842* (UC).

C. speciosus (Cham. & Schltdl.) Gilg. BRAZIL. 1927, *Järneby s.n.* (S)[S]; 1974, *Joan de Aliança 5369* (S); Goiás Rio Paraná, 1966, *Irwin et al. 14298* (U); Mato Grosso, Corumbá, 1951, *Macedo 3291* (S); Minas Gerais, Morro, Vila Rica, *Mart. Obs. 1153* (M); Minas Gerais, 1958, *Smith 6974* (US); Paraná, 1967, *Lindeman & de Haas 6011* (U); Regio inferior, *Sellow s.n.* (HAL).

"*Lisianthus*" *alpestris* Mart. (species not yet combined into *Calolisianthus*). BRAZIL. Minas Gerais, Serra Morrado, *Martius 690* (M).

"*L.*" *elegans* Mart. (species not yet combined into *Calolisianthus*). BRAZIL. Serra dos Mulatos, 1967, *Haas 15711* (U); Minas Gerais, Mt. Itambé, *Anonymous s.n.* (M).

"*L.*" *tenuifolius* Spreng. (species not yet combined into *Calolisianthus*). BRAZIL. Minas Gerais, Serro Frio, *Mart. Obs. 1385* (M).

Celiantha Maguire

C. bella Maguire & Steyerm. VENEZUELA. Amazonas, Neblina, 1957, *Maguire et al. 42096* (NY)[S]; Amazonas, Neblina, 1966, *Silva & Brazão 60921* (NY)[S].

C. chimantensis (Steyerm. & Maguire) Maguire. VENEZUELA. Bolívar, Chimantá-tepui, 1953, *Steyermark 74983* (NY).

C. imthurniana (Oliver) Maguire. VENEZUELA. *Cardona 2664* (US); Bolívar, Mt. Roraima, 1944, *Steyermark 58794* (F); 1894, *Quelch & McConnell 101* (K).

Chelonanthus Gilg

C. acutangulus (Ruiz & Pav.) Gilg (= *C. alatus* (Pulle) Aubl.). BOLIVIA. Porongo, *Herzog 1498* (L). COSTA RICA. Prov. San José, 1936, *Skutch 2968* (NY). PERU. 1925, *Melin 75* (S); *Herb. Pavon s.n.* (G).

C. alatus (Aubl.) Pulle. BRAZIL. Amazonas, 1971, *Prance et al. 14376* (NY)[S]. COLOMBIA. El Meta, 1939, *Killip 34482* (US). GRENADA. 1956, *Smith 10128* (NY). MEXICO. Choapam., 1938, *Mexia 9144* (K). SURINAM. Lelydorplan, 1953, *L. P. 475* (U). VENEZUELA. Bolívar, 1961, *Steyermark 88754* (U).

C. albus (Spruce ex Progel) Badillo. VENEZUELA. *Holt & Gehriger 327* (US); Amazonas, 1959, *Wurdack & Adderley 43593* (NY)[S].

C. angustifolius (H. B. K.) Gilg. COLOMBIA. *Haught 2434* (US); Ocaña, 1846–1852, *Schlim 72* (BR); Santander, Mesa de los Santos, 1926, *Killip & Smith 14998* (BM). PERU. Tarapoto, 1855–56, *Spruce s.n.* (BR). VENEZUELA. *Steyermark 58268* (US); Bolívar, 1952, *Maguire 33712* (NY)[S]; 1953, *Steyermark 75290* (NY); Bolívar, 1956, *Vareshi & Foldats 4681* (NY).

C. bifidus (H. B. K.) Gilg (= *C. angustifolius* (H. B. K.) Gilg). COLOMBIA. Meta, Villavicencio, 1917, *Pennell 1546* (NY). GUYANA. Rupununi-Kuguwini Rivers, Feb 1938, *Smith 3064* (MO).

C. campanuloides (Spruce ex Progel) Gilg (= *Irlbachia pratensis* (H. B. K.) L. Cobb & Maas). BRAZIL. Amazonas, San Gabriel da Cachoeira, Rio Negro, 1851, *Spruce 2005* (K[S], P). COLOMBIA. Valle Anchicaya, 1956, *Vogel 125* (U). VENEZUELA. *Holt & Blake 695* (US).

A review of palynology 483

C. camporum Gilg. PERU. San Martin, Mt. Campana, 1925, *Melini 247* (S); 1925, *Melini 313* (S).
C. candidus Malme. BRAZIL. Mato Grosso, Serra da Chapada, 1903, *Malme 3293a* (S); Paraná, 1915, *Dusén 16538* (MO).
C. chelonoides (L.f.) Gilg (= *C. alatus* (Pulle) Aubl.). BRAZIL. Mato Grosso, Serra Azul, *Irwin et al. 17270* (U). GRENADA. Grand Etang Forest Preserve, *Webster et al. 9551* (US). GUYANA. Kamuni Creek, Essequibo River, 1944, *Maguire & Fanshawe 22853* (U). PERU. Huanuco, 1935, *Mexia 4153* (MO). VENEZUELA. Amazonas, Sanariapo, 1953, *Maguire et al. 36200* (NY).
C. fistulosus (Poir.) Gilg. GUYANA. 1889, *Jenman 5435* (NY).
C. grandiflorus (Aubl.) Hassler. PARAGUAY. *Hassler 8324* (S); Sierra de Amambay, 1907–08, *Hassler 9986* (NY). SURINAM. Upper Sipaliwini River, 1911, *Hulk 47a* (U).
C. purpurascens (Aubl.) Struwe, S. Nilsson, & V. A. Albert. FRENCH GUIANA. SW Rochambeae, 1994, *Rova et al. 1902* (S)[S].
C. schomburgkii (Griseb.) Gilg. BRAZIL. Acre, Rio Branco, 1909, *Ule 8448* (K). GUYANA. *Schomburgk 298* (K). VENEZUELA. Bolívar, Rio Pargueni, 1955, *Wurdack & Monachino 39764* (NY).
C. spruceanus (Benth.) Pilger. BRAZIL. Amazonas, Rio Uaupés, 1924, *Melini 81* (S). PERU. Loreto, 1931, *Klug 2345* (MO).
C. uliginosus (Griseb.) Gilg. (= *C. purpurascens* (Aubl.) Struwe, S. Nilsson, & V. Albert). BRAZIL. Rio de Janeiro, 1953, *Smith 1318* (US). COLOMBIA. Amazonas, Rio Igaraparoná, 1942, *Schultes 3928* (K); Dolores, Tolima, *Herb. Lehmannianum 6046* (S); Rio Zanza, Sierra Macarena Intendencia del Meta, 1956, *Vogel 266* (U). GUYANA. 1961, *Cox & Hubbard 142* (NY)[S]; Kanuku Mts., 1938, *Smith 3639* (K, NY); Manaka Railway, 1948, *Atkinson 14* (BM); Mazaruni, Kamarang River, 1938–Feb 1939, *Pinkus 287* (NY). SURINAM. Zanderij, 1912, *Kuyper 85* (U). VENEZUELA. Bolívar, S. E. Canaima, 1964, *Agostini 394* (NY).
C. uliginosus var. *guianensis* Griseb. (= *C. purpurascens* (Aubl.) Struwe, S. Nilsson, & V. Albert). COLOMBIA. Cundinamarca, 1955, *Vogel 0133* (U). SURINAM. Hanover, *Splitgerber 1090* (L).
C. viridiflorus (Mart.) Gilg. BRAZIL. D. F., Brasilia, 1961, *Heringer 7883/77* (NY). COLOMBIA. Huila, Altamira, 1956, *Vogel 20* (U). VENEZUELA. Merida, Barinas State, 1964, *Breteler 4250* (U).

Chorisepalum Gleason & Wodehouse
C. carnosum Ewan. VENEZUELA. Bolívar, Cerro Uroi, 1962, *Maguire et al. 53823* (NY)[S].
C. ovatum var. *sipapoanum* Maguire (= *C. sipapoanum* (Maguire) Struwe & V. A. Albert). VENEZUELA. Amazonas, Cerro Sipapo, Río Cuao, 1948, *Maguire & Politi 27921* (NY).
C. psychotrioides var. *psychotrioides* Ewan. GUYANA. Pakaraima Plateau, Merumé Mts., 1960, *Tillett et al. 43968* (NY); 1960, *Tillett et al. 44829* (NY). VENEZUELA. Bolívar, Mt. Roraima, 1944, *Steyermark 58668* (F).
C. rotundifolium Ewan. VENEZUELA. Bolívar, Mt. Ptari-tepuí, 1944, *Steyermark 59730* (F, NY).

Congolanthus A. Raynal
C. longidens (N. E. Br.) A. Raynal. NIGERIA. Bonny, 1962, *Daramola FHI 46923* (K). CONGO. Eala, 1925, *Robyns 508* (NY); 1936, *Louis 1885* (S)[S].

Coutoubea Aubl.
C. minor H. B. K. VENEZUELA. Amazonas, 1984, *Kral et al. 71729* (NY)[S].
C. ramosa Aubl. VENEZUELA. Bolívar, 1948, *Tamayo 3460* (NY)[S].
C. reflexa Benth. VENEZUELA. Amazonas, Puerto Ayacucho, 1953, *Maguire et al. 36047* (NY)[S].
C. spicata Aubl. PANAMA. Panama, Isla Taboga, 1938, *Allen 1282* (U).

Curtia Cham. & Schltdl.
C. confusa Grothe & Maas. BRAZIL. Paraná, 1916, *Dusén 17521* (MO); Santa Catarina, 1957, *Smith & Klein 10617* (NY). VENEZUELA. Amazonas, Río Negro, 1980, *Huber et al. 5688* (U).
C. diffusa (Mart.) Cham. BRAZIL. Minas Gerais, 1931, *Mexia 5750* (U); Minas Gerais, 1965, *Duarte 9030* (U); Minas Gerais, Serra do Cipo, 1950, *Duarte 2468* (U).
C. malmeana Gilg. BRAZIL. Paraná, 1909, *Dusén 8003a* (S).
C. obtusifolia (Benth.) Knobl. BRAZIL. Amazonas, 1971, *Prance et al. 15500* (NY)[S]. VENEZUELA. Amazonas, Atabapo, 1978, *Huber 1578* (U).
C. patula (Mart.) Knobl. BRAZIL. D. F., Brasília, 1966, *Irwin et al. 14124* (U); Minas Gerais, Serra do Cipó, *Barreto & Brade s.n.* (R); Minas Gerais, Serra do Cipó, near Congonhas da Serra, *Glaziou 19656* (K).
C. pusilla Griseb. BRAZIL. Minas Gerais, Ouro Preto (R 15244).
C. quadrifolia Maguire (= *C. conferta* (Mart.) Knobl.). VENEZUELA–BRAZIL. Amazonas, 1970, *Steyermark & Bunting 102828* (NY); Sabana Hechimoni, Río Siapa, 1959, *Wurdack & Adderley 43626* (NY).
C. stricta Mart. BRAZIL. Minas Gerais, Serra do Caraca, *Glaziou 15245* (R 11665).
C. tenella (Mart.) Knobl. (= *C. tenuifolia* (Aubl.) Knobl.). BRAZIL. D. F., 1976, *Ratter & Fonseca 2803* (E); *Lindeman & Haas 4995* (U). MEXICO. Gr. Bitt. 5040 (US). PANAMA. Panama, Pacora, 1938, *Allen 994* (MO).
C. tenuifolia (Aubl.) Knobl. GUYANA. Ebini Exper. Stat., *Harrison 1245* (K). PANAMA. Panama, 1938, *Allen 994* (U). SURINAM. E. of Kopi distr. Commewijne, 1953, *Lindeman 4377* (U); Zanderij, 1914, *Forestry Bureau s.n.* (U). VENEZUELA. Amazonas, Esmeralda, Cerro Moriche, 1951, *Maguire et al. 30910* (NY).
C. tenuifolia ssp. *tenuifolia* (Aubl.) Knobl. BRAZIL. Mato Grosso, Serra da Chapada, 1903, *Malme s.n.* (S)[S].
C. tenuifolia ssp. *tenella* (Mart.) Grothe & Maas. BRAZIL. Mato Grosso, Serra da Chapada, 1903, *Malme s.n.* (S).
C. verticillaris (Spreng.) Knobl. BRAZIL. Bahia, 1979, *Hage et al. 296* (U); 1980, *Harley 22534* (U); D. F., Chapada de Contagem, 1980, *Plowman 9951* (U); Minas Gerais, 1972, *Hatschbach 29978* (U).

Deianira Cham. & Schltdl.
D. cordifolia (Lhotzky ex Progel) Malme. BRAZIL. Mato Grosso, Santa Ana da Chapada, 1903, *Malme 3323* (S).

A review of palynology 485

D. cyathifolia B. Rodriguez. BRAZIL. Mato Grosso, Serra da Chapada, 1903, *Malme 3323B* (S)[S].
D. erubescens Cham. & Schltdl. BRAZIL. Minas Gerais, 1931, *Mexia 5667* (MO); 1952, *Smith 6962* (NY).
D. foliosa (Griseb.) Guimarães. BRAZIL. S. José dos Campos, 1909, *Löfgren 315* (S).
D. nervosa Cham. & Schltdl. BRAZIL. Minas Gerais, 1952, *Smith 6964* (K, NY[S], S); Mq S. Vincent, 1950, *Macedo 2404* (MO).
D. pallescens Cham. & Schltdl. BRAZIL. S. Vicente, *Macedo 2403* (MO).

Fagraea Thunb.
F. berteroana Gray. USA. Hawaii, Oahu, 1995, *Motley s.n.* (NY)[S].
F. ceilanica Thunb. INDIA. Hassan, 1969, *Saldanha 13586* (NY)[S].
F. elliptica Roxb. INDONESIA. 1990, *Burley et al. 4335* (NY)[S].
F. fragrans Roxb. PHILIPPINES. 1926, *Roque s.n.* (NY)[S].
F. involucrata Merr. BORNEO. 1933, *Clemens & Clemens 31661* (NY).
F. longiflora Merr. PHILIPPINES. Luzon, 1928, *Edano 708* (NY)[S].
F. racemosa Jack. SUMATRA. Sigamata, 1932, *Rahmat Si Toroes 3193* (NY)[S].

Gentianothamnus Humbert
G. madagascariensis Humbert. MADAGASCAR. Anjanaharibe, *Humbert 24017* (MO); 1951, *Humbert et al. 24817* (P); Marojejy, 1949, *Cours 3446* (P); Marojejy, 1949, *Humbert & Cours 23715* (MO, P[S], S).

Helia Mart.
H. brevifolia (Cham.) Gilg. BRAZIL. *Hatschbach 3617* (US). PARAGUAY. *Jorgensen 4837* (MO); *Jorgensen 4837* (S)[S].
H. loesneriana Gilg. VENEZUELA. *Killip 37557* (US).
H. martii (Griseb.) Mart. BRAZIL. *Glaziou 16264* (BR).
H. micrantha Gilg. PARAGUAY. Caaguazú, 1876, *Balansa 2141* (G).
H. oblongifolia Mart. BRAZIL. Paraná, Morungava, 1915, *Dusén 16537* (MO, S[S]). PARAGUAY. Near Caaguazú, 1905, *Hassler 8875* (NY).
H. spathulata (H. B. K.) Gilg. BRAZIL. Minas Gerais, Poços de Caldas, Morro do Ferro, 1964, *Emmerich 2147 & Andrade 1767* (R).

Hockinia Gardner
H. montana Gardner. BRAZIL. Minas Gerais, 1973, *Hatschbach 31428* (U); Rio de Janeiro, 1952, *Brade 21208* (RB)[S]; 1874, *Glaziou 6899* (K); (R 11688); 1900, *Anonymous s.n.* (US 14718)[S].

Irlbachia Mart.
I. cardonae (Gleason) Maguire. GUYANA. Imbaimadai, 1958, *Graham 129* (K). VENEZUELA. Bolívar, Auyan-tepui, 1937, *Tate 1360* (NY)[S,T]; Bolívar, Chimantá Massif, 1953, *Steyermark 75730* (F).
I. elegans Mart. BRAZIL. Amazonas, 1968, *Prance et al. 4866* (S); Manaus, 1966, *Prance et al. 3721* (S).
I. nemorosa (Willd. ex Roem. & Schult.) Merr. VENEZUELA. Amazonas, San Simon, *Morillo & Wood 4068* (VEN); Bolívar, 1977, *Steyermark & Dunsterville 113269* (VEN).

I. paruana Maguire (=*I. cardonae* (Gleason) Maguire). VENEZUELA. Amazonas, Cerro Parú, Río Ventuari, 1951, *Cowan & Wurdack 31160* (NY).
I. phelpsiana Maguire. VENEZUELA. Amazonas, Cerro Parú, 1951, *Cowan & Wurdack 31412* (NY).
I. plantaginifolia Maguire. VENEZUELA. Amazonas, Cerro Neblina, Río Yatua, 1958, *Maguire et al. 42593* (NY).
I. poeppigii (Griseb.) L. Cobb & Maas. SOUTH AMERICA. ("America merid."), 1895, *Poeppig 2854* (G).
I. pratensis (H. B. K.) L. Cobb & Maas. VENEZUELA. Amazonas, Casiquiare, Pacimoni, 1954, *Maguire et al. 37563* (NY[S], US); *Wurdack & Adderley 42949* (NY).
I. pumila (Benth.) Maguire. BRAZIL. Rio Negro, Casiquiari, Vasiva, Pacimoni, 1853–54, *Spruce 2950* (BM). VENEZUELA. 1942, *Williams 14362* (US).
I. ramosissima (Benth.) Maguire (=*I. poeppigii* (Griseb.) L. Cobb & Maas). BRAZIL. Amazonas, Rio Negro, *Schomburgk 989* (K).
I. recurva (Benth.) Progel (=*I. nemorosa* (Willd. ex Roem. & Schult.) Merr). BRAZIL. *Schultes 8721* (US). COLOMBIA. Caqueta, 1944, *Schultes 5863* (US). VENEZUELA. Amazonas, Danta Falls, 1948, *Maguire & Politi 27324* (NY).
I. subcordata (Benth.) Progel (=*I. nemorosa* (Willd. ex Roem. & Schult.) Merr). BRAZIL. Amazonas, Manaus, Rio Taramao, 1964, *Vogel 311* (U).
I. tatei (Gleason) Maguire. VENEZUELA. *Cardona 376* (US); Amazonas, Cerro Duida, 1929, *Tate 743* (US); 1944, *Steyermark 58167* (US).
"*Lisianthus*" *acutilobus* Steyerm. (=*Irlbachia cardonae* (Gleason) Maguire). VENEZUELA. Bolívar, Luepa, Cerro Venamo, 1960, *Steyermark & Nilsson 732* (NY[S], VEN).
"*L.*" *scabridulus* Steyerm. (=*Irlbachia tatei* (Gleason) Maguire). VENEZUELA. Amazonas, Cerro Duida, 1944, *Steyermark 57979* (F).

Lagenanthus Gilg

L. princeps (Lindl.) Gilg. COLOMBIA. Santander, 1969, *Murillo & Jaramillo s.n.* (COL)[S]; 1846–1852, *Schlim 419* (BM). VENEZUELA. Táchira, 1937, *Cardona 119* (US); 1980, *Maas & Tillett 5242* (NY)[S].

Lehmanniella Gilg

L. splendens (Hook.) Ewan. COLOMBIA. Antioquia, 1989, *Fonnegra & Roldan 2698* (NY[S], S); Antioquia, Narino, 1954, *Uribe-Uribe 2605* (NY)[S]; 1976, *Uribe 1135* (COL)[S]. Bucaramanga, 1948, *Sandeman 6046* (K)[S]; Santander, 1944, *John 20561* (NY).

Lisianthius P. Browne

L. acuminatus Perk. (=*Lisianthius* "perkinsiae", ined.). MEXICO. Chiapas, 1965, *Breedlove 10062* (F).
L. adamsii Weaver. JAMAICA. Parish of St. Anne, Douglas Castle, *Weaver 1293* (DUKE).
L. auratus Standl. HONDURAS. Comayaga, 1951, *Williams & Williams 18382* (US). NICARAGUA. Comarca de El Cabo, 1965, *Molina 15037* (F).
L. axillaris (Hemsl.) Kuntze. BELIZE. Tower Hill Estate, 1927, *Karling 54* (K). GUATEMALA. Petén, 4 May 1933, *Lundell 3159* (S).

A review of palynology 487

L. brevidentatus (Hemsl.) Kuntze. GUATEMALA. Alta Verapaz, *Türckheim 931* (L).

L. capitatus Urb. JAMAICA. St. Ann, 1915, *Harris 12030* (S); E. of Port Antonio, *Webster et al. 8362* (BM).

L. collinus Standl. BELIZE. Jacinto Hills, 1933, *Schipp 1205* (BM).

L. congestus Standl. PANAMA. Coclé, 1938, *Allen 783* (U).

L. cordifolius L. JAMAICA. St. Andrew, Cane River Valley, 1916, *Harris 12315* (BM); St. Andrew, 1955, *Stearn 2* (BM); St. Andrew, Long Mountain, 1954, *Webster & Wilson 4860* (BM)[S]; St. Andrew, Hope River, 1957, *Yuncker 17374* (S).

L. domingensis Urb. DOMINICAN REPUBLIC. Province of Los Guineos, 24 Mar 1958, *Marcano 3658* (US); Santo Domingo, Lomala Campona, 1929, *Ekman 11516* (S).

L. exsertus Sw. JAMAICA. Portland, 1961, *Proctor 22114* (BM); St. Andrews, Moresham River, 1956, *Stearn 410* (K).

L. glandulosus A. Rich. CUBA. Oriente, 1914, *Ekman 1613* (K); 1956, *Morton & Alain 9000* (US).

L. jefensis Robyns & Elias. PANAMA. Province of Panama, 1968, *Weaver 1481* (DUKE).

L. latifolius Sw. JAMAICA. St. Andrews Parish, 1958, *Hawkes et al. 2241* (K); *Krug & Urban 662* (L).

L. laxiflorus Urb. PUERTO RICO. Lares, 1887, *Sintensis 6077* (BM); 1963, *Wagner 2* (U).

L. longifolius L. JAMAICA. St. Andrew, 1927, *Maxwell s.n.* (BM); Trelawny, Cockpit Country, 1959, *Webster et al. 8432* (BM).

L. nigrescens Cham. & Schltdl. GUATEMALA. San Ildefanso Ixtahuacan – Cuilco, 1942, *Steyermark 50682* (F). JAMAICA. *Hort. Bot. Utrecht 0121118* (U). MEXICO. Chiapas, 1864–1870, *Ghiesbreght 702* (K).

L. peduncularis L. O. Williams. PANAMA. *Allen & Allen 4187* (MO).

L. saponarioides Cham. & Schltdl. GUATEMALA. Quiché, Rio Negro, 1892, *Heyde & Lux 2921* (K).

L. scopulinus A. Robyns & Elias. PANAMA. Veraguas, 1967, *Lewis et al. 2799* (MO).

L. seemannii (Griseb.) Kuntze. COLOMBIA. Antioquia, *Haught 4652* (K). COSTA RICA. San José, 1939, *Skutch 4109* (K).

L. silenifolius (Griseb.) Urb. CUBA. Pinar del Rio, 1949, *Alain & Acuna 1106* (US); 1921, *Ekman 12940* (K); Sierra del Rosario, 1977, *Bisse et al. 35146* (HAJB)[S].

L. skinnerii (Hemsl.) Kuntze. COSTA RICA. Alajuela, 1968, *Wilbur & Stone 10257* (DUKE). PANAMA. Province of Panama, 1968, *Weaver 1482* (DUKE).

L. troyanus Urb. JAMAICA. Parish of Westmoreland, 1967, *Weaver 1272* (DUKE).

L. umbellatus Sw. JAMAICA. Near Troy, 1917, *Perkins 1321* (GH); *Hort. Bot. Utrecht 012123* (U).

L. viscidiflorus Robins. GUATEMALA. Alta Verapaz, 1964, *Contreras 4725* (LL).

Macrocarpaea (Griseb.) Gilg
M. affinis Ewan. COLOMBIA. Santander, 1952, *Uribe 2432* (F).
M. bangiana Gilg. BOLIVIA. Yungas, *Bang 520* (US).
M. bracteata Ewan. VENEZUELA. Trujillo, *Dorr et al. 4990* (NY)[S].
M. browallioides (Ewan) Robyns & S. Nilsson. PANAMA. Boca del Toro, *Allen 4932* (MO).
M. cinchonifolia (Gilg) Weaver. BOLIVIA. Mapiri, 1886, *Rusby 1173* (BM, NY, K[S], S); Mapiri, 1886, *Rusby s.n.* (PH); Simaco, 1920, *Buchtien 5639* (NY).
M. cochabambensis Gilg. BOLIVIA. Chaparé, Cochabamba, 1929, *Steinbach 8992* (F).
M. densiflora (Benth.) Ewan. COLOMBIA. Cauca, Páramo de Mora, *Lehmann 2087* (US).
M. domingensis Urb. & Ekman. DOMINICAN REPUBLIC. La Vega, 1947, *Allard 18215* (S).
M. duquei Gilg. COLOMBIA. Valle del Cauca, 1946, *Cuatrecasas 21824* (F, US).
M. glabra (L.f.) Gilg. COLOMBIA. Cundinamarca, 1937, *Daniel 1450* (F); 1852, *Holton 470* (NY); 1939, *Killip 34074* (S); 1917, *Pennell 2195* (NY); *Haught 5105* (US).
M. loranthoides (Griseb.) Maas. PERU. Amazonas, Chachapoyas, 1835, *Matthews 1315* (K)[S]; Yurimaguas-Cajamarca, *Sandeman s.n.* (K).
M. macrophylla (H. B. K.) Gilg. COLOMBIA. Caldas, *Pennell 10670* (PH); 1939, *Killip & Garcia 33650* (S).
M. micrantha Gilg. PERU. Loreto, 1933, *Klug 3138* (F, S).
M. obtusifolia (Griseb.) Gilg. BRAZIL. Rio de Janeiro, 1838–1839, *Wilkes Exp. s.n.* (US); Santa Luzia, 1938, *Barreto 8845* (F).
M. ovalis (Ruiz & Pav.) Ewan. COLOMBIA. Cauca, *Lehmann 5450* (F); *Purdie s.n.* (K). ECUADOR. Loja-Zamora, 1876, *André 4549* (NY).
M. pachyphylla Gilg. COLOMBIA. Putumayo, *Cuatrecasas 11785* (US). ECUADOR. Loja, 1876, *André 4513* (F).
M. revoluta (Ruiz & Pav.) Gilg. PERU. Amazonas, *Williams 7596* (F).
M. rubra Malme. BRAZIL. Paraná, 1908, *Dusén 6965* (F, S); Paraná, Desiro, Ypiranga, 1914, *Dusén 15790* (MO, NY, U).
M. stenophylla Gilg. PERU. Amazonas, La Jalca, 1930, *Williams 7582* (F).
M. thamnoides (Griseb.) Gilg. JAMAICA. Near Vinegar Hill, *Harris 5353* (US); Road to St. Georges, 1899, *Harris 7718* (F).
M. valerii Standl. COSTA RICA. Alajuela, 1938, *Smith 1001* (F).
M. sp. nov. 1 (previously determined as *M. corymbosa* (Ruiz & Pav.) Ewan). PERU. Dept. Cusco, 1929, *Weberbauer 7921* (F).
M. sp. nov. 2 (previously determined as *M. macrophylla* (H. B. K.) Gilg). COLOMBIA. El Valle, 1943, *Cuatrecasas 14986* (F).
M. sp. nov. 3 (previously determined as *M. ovalis* (Ruiz & Pav.) Ewan). ECUADOR. Loja, 1943, *Steyermark 54805* (F).
M. sp. nov. 4 (previously determined as *M. pachystyla* Gilg). PERU. Dept. Junin, Tarma, 1948, *Woytkowski 35417* (F).
M. sp. nov. 5 (previously determined as *M. polyantha* Gilg). COLOMBIA. Cundinamarca, 1941, *Cuatrecasas et al. 12271* (F, MO).
M. sp. nov. 6 (previously determined as *M. revoluta* (Ruiz & Pav.) Gilg). PERU. 1909–1914, *Weberbauer 6116* (F).

M. sp. nov. 7 (previously determined as *M. sodiroana* Gilg). ECUADOR. Loja, Cachiyacu River, 1943, *Steyermark 54812* (F); Santiago-Zamora, Rio Tintas, *Steyermark 53548* (US).
M. sp. nov. 8 (previously determined as *M. viscosa* (Ruiz & Pav.) Gilg). PERU. Cajamarca, Prov. Cutervo, 1938, *Stork & Horton 10118* (K, NA).
M. sp. nov. 9 (previously determined as *M. viscosa* (Ruiz & Pav.) Gilg). PERU. Huanuco, 1938, *Stork & Horton 9913* (K).
M. sp. nov. 10 (previously determined as *M. viscosa* (Ruiz & Pav.) Gilg). PERU. Cusco, *Vargas 4196* (US).
M. sp. (previously determined as *M. pachystyla* Gilg). PERU. 1778–1788, *Ruiz & Pavón s.n.* (F).

Neblinantha Maguire
N. neblinae Maguire. BRAZIL–VENEZUELA. Serra de la Neblina, 1970, *Steyermark 103862* (NY)[S].
N. parvifolia Maguire. BRAZIL–VENEZUELA. Amazonas, Rio Titirica, 1970, *Steyermark 103800* (NY); Caño Grande, 1957, *Maguire et al. 42419* (NY).

Neurotheca Salisb. ex Benth. in Benth. & Hook.f.
N. congolana De Wild. & Th. Dur. ANGOLA. Longa, 1900, *Baum 604* (K). CONGO. Kinshasa, *Bavicchi 412* (BR); Kinshasa, *Bequaert 7669* (BR); Kinshasa, *Lebrun 6478* (BR).
N. corymbosa Hua. GABON. Cap Lopez, 1894–95, *Thollon 43* (K); Pointe Denis, 1968, *Hallé & Villiers 5528* (P).
N. loeselioides (Spruce ex Progel) Baill. BRAZIL. Amazonas, Rio Negro, 1910, *Ule 8978* (L); Amazonas, Rondonia, 1987, *Nee 34765* (NY)[S]. FRENCH GUINEA. Plateau du Benna, 1913, *HLB 950.348–380* (L). GHANA. Damongo Scarp., 1965, *Hall 865* (K); Kwalu Tafo, 1958, *Harris s.n.* (K). GUYANA. Kaieteur Plateau, 1958, *Graham no. X.7* (K); Kaietuk, 1933, *Tutin 618* (K). SENEGAL. Badi, 1954, *Berhaut 4411* (P). SIERRA LEONE. Free Town, *Brown & Brown s.n.* (US); 1922, *Dawe WA 429* (K). SURINAM. Sipaliwini Savanna, 1966, *van Donselaar 3700* (U); Zanderij, 22 Dec 1950, *Florschütz & Florschütz 811* (U); 1948, *Lanjouw & Lindeman 157* (NY); 1954, *Lindeman 6520* (U).
N. loeselioides ssp. *loeselioides* (Spruce ex Progel) Baill. BURUNDI. Bururi, 1980, *Reekmans 9114* (S)[S]. SENEGAL. Badi, 1954, *Berhaut 4411* (P).
N. loeselioides ssp. *robusta* (Hua) A. Raynal. FRENCH GUINEA. Plateau du Benna, 1944, *Schnell 2149* (P). SIERRA LEONE. Waterloo, 1958, *Melville & Hooker 281* (K).

Potalia Aubl.
P. amara Aubl. BRAZIL. Amapá, 1983, *Mori et al. 15785* (NY)[S].
P. "coronata", ined. (L. Struwe & V. A. Albert, unpubl.). PERU. Loreto, 1988, *van der Werff et al. 10254* (MO).
P. elegans Struwe & V. A. Albert. VENEZUELA. Amazonas, 1950, *Maguire et al. 29356* (NY)[S].
P. maguireorum Struwe & V. A. Albert. BRAZIL. Amazonas, 1979, *Kubitzki et al. 79-216* (NY)[S].

P. resinifera Mart. PERU. Loreto, 1986, *Vásquez & Jaramillo 7169* (MO)[S].
P. "turbinata", ined. (L. Struwe & V. A. Albert, unpubl.). COSTA RICA. Limon, 1987, *Stevens et al. 25009* (MO).

Prepusa Mart.
P. alata Porto & Brade. BRAZIL. Rio de Janeiro, Santa Maria Magdalena, 1986, *Farney & Caruso 1195* (RB 276053).
P. connata Gardner. BRAZIL. *Gardner 541* (CGE)[S]; Rio de Janeiro, 1891, *Glaziou 18972* (NY); Rio de Janeiro, Nova Friburgo, *Glaziou 18372* (K).
P. hookeriana Gardner. BRAZIL. 1965, *Vogel 773* (U); Rio de Janeiro, Serra dos Orgãos, 1910, *Lützenburg s.n.* (NY); 1952, *Vidal II-330* (R 107600)[S].
P. montana Mart. BRAZIL. *Harley 22878* (U)[S].
P. viridiflora Brade. BRAZIL. Espirito Santo, Serra Vitoria Minas, 1967, *Duarte 10469* (RB).

Purdieanthus Gilg
P. pulcher (Hook.) Gilg. COLOMBIA. Boyacà, Villa de Layva, 1970, *Uribe 6473* (COL); Mt. del Moro, *Purdie s.n.* (K[S], P).

Rogersonanthus Maguire & B. M. Boom
R. arboreus (Britton) Maguire & B. M. Boom. TRINIDAD. Morne Bleu, 1926, *Broadway 6207* (S); 1925, *Broadway 5648* (K); *Baksh & Nilsson 1984-3* (S)[S].
R. quelchii (N. E. Br.) Maguire & B. M. Boom. VENEZUELA. Bolívar, 1944, *Steyermark 58672* (F); Bolívar, Auyan-tepui, *Steyermark 93657* (VEN); Bolívar, Chimantá Massif, 1955, *Steyermark & Wurdack 813* (F); Bolívar, Mt. Roraima, 1944, *Steyermark 58802* (F).
R. tepuiensis (Gleason) Maguire & B. M. Boom (= *R. arboreus* (Britton) Maguire & B. M. Boom). VENEZUELA. Bolívar, Cave Camp, *Steyermark 59721* (F); 1937, Auyan-tepui, *Tate 1361* (F); *Steyermark s.n.* (VEN); Cardoni, *Cardona 2249* (US).

Schultesia Mart.
S. benthamiana Klotzsch ex Griseb. GUYANA. Rupununi, Manari, 1979, *Maas et al. 3661* (U)[S]; Rupununi River, 1937, *Smith 2297* (U). VENEZUELA. Amazonas, E. Río Orinoco, 1955, *Wurdack & Monachino 39942* (NY).
S. brachyptera Cham. GUYANA. Karanambo, 1988, *Maas et al. 7730* (U)[S]; Rupununi Savanna, Lethem, 1987, *Jansen-Jacobs et al. 569* (U). PANAMA. Veraguas, 24 Nov 1938, *Allen 1069* (U). SURINAM. Kopi, 1953, *Lindeman 4267* (U); Republick, Aug 1914, *Forest. Bureau s.n.* (U).
S. guianensis (Aubl.) Malme. BRAZIL. Rio de Janeiro, Jacarépaguá, 1932, *Brade 11372* (S). GUYANA. *Poiteau s.n.* (K); Rupununi Savanna, Nappi Village, 1987, *Jansen-Jacobs et al. s.n.* (U)[S]. SURINAM. Wia Wia reservation, 1970, *Pons (L. B. B.) 12660* (U). VENEZUELA. Carabobo, *Williams & Alston 373* (U). GUINEA BISSAU. *Santo Expt. Bot. 1750* (K). SIERRA LEONE. *Adames 157* (K).
S. lisianthoides (Griseb.) Benth. & Hook ex. Hemsl. (= *Xestaea lisianthoides* Griseb.). NICARAGUA. Zelaya, 1979, *Pipoly 4526* (MO 2912260); Zelaya, 1978, *Stevens & Krukoff 8224* (MO 3156547)[S]. PANAMA. 1922, *Greenm. & Greenm. 5203* (MO); Canal Zone, 1965, *Robyns 65-21* (MO).

S. pohliana Progel. BRAZIL. Mato Grosso, Xavantina, 1966, *Hunt & Ramos 5828* (NY). SURINAM. S. Kayser Airstrip, 1963, *Irwin et al. 55237* (C)[S].

S. subcrenata Klotzsch ex Griseb. BRAZIL. Mato Grosso, Cuiabá, 1903, *Malme 3179 a* (S); Salinas, 1844, *Weddell 2174* (P).

Senaea Taub.

S. caerulea Taub. BRAZIL. Minas Gerais–Diamantina, 1947, *Romariz 0127* (RB 59940).

S. janeirensis Brade. BRAZIL. *Santos Lima 217* (RB)[S]; Est. Rio de Janeiro, Mun. Santa Maria Magdalena, 1986, *Martinelli et al. 12003* (RB 272215)[S].

Sipapoantha Maguire & B. M. Boom

S. ostrina Maguire & B. M. Boom. VENEZUELA. Amazonas, Cerro Sipapo, 1949, *Maguire & Politi 28092* (NY)[S].

Symbolanthus G. Don

S. anomalus (H. B. K.) Gilg. COLOMBIA. Antioquia, 1951, *Uribe 2142* (US); Boyaca, N.W. of Bogotá, 1932, *Lawrence 470* (NY); Tequendama, near Bogotá, *H.L.B. 912184-360* (L).

S. brittonianus Gilg. BOLIVIA. Yungas, 1890, *Bang 339* (K).

S. calygonus (Ruiz & Pav.) Griseb. BOLIVIA. Beni, San Antonio, Mapira, 1907, *Buchtien 2461* (L). PERU. Above Canta, 1938, *Sandeman 187* (K); *Allard 21095* (US).

S. daturoides (Griseb.) Gilg. PERU. "Sambras-bamba", 1825, *Mathews 1317* (K).

S. elisabethae (Schomb.) Gilg. BRAZIL. Amazonas, Sierra de la Neblina, 1965, *Silva & Brazão 60653* (NY). GUYANA. Mt. Roraima, 1894, *Thurn 188* (K). VENEZUELA. Bolívar, 1939, *Pinkus 160* (NY).

S. gaultherioides Ewan. COLOMBIA. Antioquia, Cerro de la Vieja, 1938, *Bro. Daniel 1693* (US).

S. latifolius Gilg. BOLIVIA. Cochabamba, Chapare, 1929, *Steinback 9049* (U). COLOMBIA. Prov. Huila, Altamira, 1956, *Vogel 116* (U).

S. macranthus (Benth.) Ewan. ECUADOR. Loja, Namanda, 1946, *Espinosa 197* (NY).

S. magnificus Gilg. COLOMBIA. Santander, 1961, *Uribe-Uribe 3660* (NY); *Langenheim 3293* (US). VENEZUELA. *Maguire 39441* (NY).

S. mathewsii (Griseb.) Ewan. BOLIVIA. *Pearce 640* (K). PERU. Amazonas, Chachapoyas, *Mathews 1836* (K).

S. nerioides (Griseb.) Ewan. PERU. 1876, *Davis s.n.* (BM). VENEZUELA. *Jahn 910* (US).

S. obscure-rosaceus Gilg. PERU. *Woytkowski 8288* (US).

S. pauciflorus Gilg. PERU. Loreto, near Yurimaguas, 1855–56, *Spruce 4429* (BR, NY).

S. pulcherrimus Gilg. COSTA RICA. Cartago, SW El Empalme, 1968, *Weaver 1406* (DUKE). PANAMA. *Pittier 3223* (US); Bocas del Toro, Cerro Horqueta, 1947, *Allen 4946* (U).

S. rubro-violaceus Gilg. COSTA RICA. *Burger et al. 9355* (F).

S. rusbyanus Gilg. PERU. Amazonas, Bongará, 1962, *Wurdack 1068* (K).

S. stuebelii Gilg. COLOMBIA. *Schneider 1085* (S).

S. superbus Miers. COLOMBIA. *Killip & Smith 15133* (US).
S. tricolor Gilg. COLOMBIA. Cundinamarca, 1947, *Haught 6148* (US); 1956, *Vogel 143* (U). VENEZUELA. *Funck & Schlim 1412* (W).
S. vasculosus (Griseb.) Gilg. VENEZUELA. Aragua above El Limón, 1961, *Steyermark 89845* (NY).
S. yutajensis Maguire (= *S. yaviensis* Steyerm.). VENEZUELA. *Maguire 35333* (NY).

Tachia Aubl.

T. grandiflora Maguire & Weaver. BRAZIL. Amazonas, Rio Urubu, 1968, *Prance et al. 5042* (S). VENEZUELA. Amazonas, Piedra Tururumeri, *Maguire et al. 37496* (NY).
T. guianensis Aubl. FRENCH GUIANA. Mt. Mania, 1974, *Rova et al. 1963* (S)[S]; Bartica-Potaro Road, 1937, *Sandwith 1111* (K[S], U); Bartica-Potaro Road, 1933, *Tutin 303* (BM); Essequibo, *Martyn 207* (K); Makauria Creek, Essequibo River, 1940, *Forest Dept. Guyana 3323* (K). SURINAM. Arrowhead Basin, *Maguire 24611* (NY); Tafelberg-Saramacca River, 1963, *Wessels Boer 1540* (U); Tafelberg Creek, 1944, *Maguire 24098* (NY, U).
T. loretensis Maguire & Weaver. PERU. Loreto, Mishuyacu, 1930, *Klug 1277* (NY).
T. occidentalis Maguire & Weaver. BRAZIL. Amazonas, Tocantins, *Ducke 22389* (U); Amazonas, 1987, *Maas et al. 6760* (NY)[S]. COLOMBIA. Amazonas, Vaupés, Rio Apaporis, 1951, *Schultes & Cabrera 14670* (U). PERU. Huanuco, Junin, Santa Rosa, 1929, *Killip & Smith 26163* (NY).
T. schomburgkiana Benth. GUYANA. Karowtipu, Kako River, *Tillett & Tillett 45447* (NY); Kurupung Mts, Macreba Falls, 1938, *Forest Dept. 2779* (K); Kurupung River, Makreba Falls, 1939, *Steyermark & Pinkus 3* (NY); Membaru Creek, Mazaruni River, 1939, *Pinkus 227* (US). SURINAM. Paloemeu-Tapanahoni Rivers, 1963, *Wessels Boer 1211* (U).

Tachiadenus Griseb.

T. antaisaka Humbert. MADAGASCAR. Mt. Itrafanaomby, 1933, *Humbert 13538* (S).
T. carinatus (Desr.) Griseb. MADAGASCAR. Tamatave, 1912, *Afzelius s.n.* (S); Near Ambodimanga, 1987, *Klackenberg 871124-5* (S)[S]; Fort-Dauphin, 1947, *Humbert 20396* (S).
T. gracilis Griseb. MADAGASCAR. Tamatave, 1889, *Catat 2528* (P).
T. longiflorus Bojer ex Griseb. MADAGASCAR. Ambohimanga, 1928, *Decary 6151* (P); Andranokobaka, Moramanga, 1951, *Serv. Forest. Madag. 3240* (S); Kalambatitra, 1933, *Humbert 11973* (S); *Elliot 1939* (K).
T. platypterus ssp. *platypterus* Baker. MADAGASCAR. Distr. Ambalavao, 1955, *Cours 5008* (P).
T. tubiflorus (Roem. & Schult.) Griseb. MADAGASCAR. Tamatave, 1957, *Capuron s.n.* (S)[S]; Fènèrive, 1909, *Geay 8957* (P); Ivoloina, *Herb. J. Bot. TAN 3328* (P); Tampola, 1957, *Capuron s.n.* (P).

Tapeinostemon Benth.

T. longiflorum var. *longiflorum* Maguire & Steyerm. VENEZUELA. Amazonas, Cañon Grande, 1957, *Maguire et al. 42432* (NY).

T. longiflorum var. *australe* Maguire & Steyerm. BRAZIL–VENEZUELA. Sierra de la Neblina, 1965, *Maguire et al. 60507* (NY); 1970, *Steyermark 103937* (NY).
T. ptariense Steyerm. (= *T. spenneroides* Benth.). VENEZUELA. Amazonas, Sierra de la Neblina, 1957, *Maguire et al. 42053* (NY); Bolívar, Luepa-Cerro Venamo, 1960, *Steyermark & Nilsson 171* (NY); Bolívar, Ptari-tepui, 1944, *Steyermark 59861* (F).
T. sessiliflorum (Humb. & Bonpl. ex Schult.) Pruski & S. F. Sm. (formerly *T. capitatum* Benth.). VENEZUELA. Amazonas, Casiquiaré, 1853–54, *Spruce 3293* (K).
T. spenneroides Benth. VENEZUELA. Bolívar, Abácapa-tepui, 1953, *Steyermark 74801* (F).
T. zamoranum Steyerm. ECUADOR. Palanda, Santiago-Zamora, 1876, *André 4618* (F); Zamora, Chinchipe, 1988, *Gaarol 74940* (QCA)[S].

Tetrapollinia Maguire & B. M. Boom
T. caerulescens (Aubl.) Maguire & B. M. Boom. BRAZIL. D. F., Summit of Chapada da Contagem., 1966, *Irwin et al. 11661* (U)[T]; Paraná, 1903, *Dusén 2647* (MO). FRENCH GUIANA. Route Cayenne Rochambeau, 1962, *Hallé 440* (U). GUYANA. Rupununi River, 1937, *Smith 2296* (NY); Rupununi, 1992, *Jansen-Jacobs et al. 2764* (NY)[S]. SURINAM. Zanderij, 1944, *Maguire & Stahel 23682* (U). VENEZUELA. Bolívar, 1944, *Steyermark 59353* (US).

Wurdackanthus Maguire
W. frigidus (Sw.) Maguire & B. M. Boom. GUADELOUPE. Basse Terre, 1959, *Proctor 20339* (BM). HAITI. *Torrey 248* (K). ST. VINCENT. 1942, *Beard 17* (S)[S]; Soufrière Crater, 1950, *Howard 11195* (BM).

Zonanthus Griseb.
Z. cubensis Griseb. CUBA. Monte Cristo, *Areas et al. PFC 49840* (HAJB)[S]; 1860–1864, *Wright 1346* (BM, U).

LITERATURE CITED

Agababyan, V. S. & K. T. Tumanyan. 1977. Materialy k palinomorfologichkomu izocheniyu semeystva Gentianaceae 4. *Biol. Zh. Armenii* 30(8): 43–53. (In Russian)
Bentham, G. 1854. Notes on north Brazilian Gentianeae, from the collections of Mr. Spruce and Sir Robert Schomburgk. *J. Bot. Kew Gard. Misc.* 6: 193–204.
Bentham, G. 1876. Gentianeae. Pages 799–820 in: G. Bentham & J. Hooker, eds. *Genera plantarum*, vol. 2, part 2. L. Reeve & Co., Williams & Norgate, London.
Brako, L. & J. L. Zarucchi. 1993. Catalogue of the flowering plants and gymnosperms of Peru. *Monogr. Syst. Bot. Missouri Bot. Gard.* 45: 1–1286.
Britton, N. L. 1922. A Trinidad tree-gentian. *Bull. Dept. Agric. Trin. Tobago* 19: 230.
Bureau, L.-E. 1856. *De la famille des Loganiacées, et des plantes qu'elle fournit à la*

médicine. Thèse pour le doctorat en médecine. Faculté de Médicine de Paris, Paris.

Elias, T. S. & A. Robyns. 1975. Family 160. Gentianaceae. Pages 61–101 in: R. E. Woodson, Jr., R. W. Schery, and collaborators, eds. *Flora of Panama*, part 8. *Ann. Missouri Bot. Gard.* 62.

Endress, M. E., B. Sennblad, S. Nilsson, L. Civeyrel, M. W. Chase, S. Huysmans, E. Grafström et al. 1996. A phylogenetic analysis of Apocynaceae s. str. and some related taxa in Gentianales: a multidisciplinary approach. *Opera Bot. Belg.* 7: 59–102.

Erdtman, G. 1952. *Pollen morphology and plant taxonomy, Angiosperms.* Almqvist & Wiksell, Stockholm and Chronica Botanica Co., Waltham, MA.

Erdtman, G. 1960. The acetolysis method. A revised description. *Svensk Bot. Tidskr.* 54: 561–564.

Ewan, J. 1947. A revision of *Chorisepalum*, an endemic genus of Venezuelan Gentianaceae. *J. Wash. Acad. Sci.* 37: 392–396.

Ewan, J. 1948a. A revision of *Macrocarpaea*, a neotropical genus of shrubby gentians. *Contr. U. S. Natl. Herb.* 29: 209–251.

Ewan, J. 1948b. A review of *Purdieanthus* and *Lehmanniella*, two endemic Colombian genera of Gentianaceae, and biographical notes on Purdie and Lehmann. *Caldasia* 5: 85–98.

Ewan, J. 1952. A review of the neotropical lisianthoid genus *Lagenanthus* (Gentianaceae). *Mutisia* 4: 1–5.

Fosberg, F. R. & M.-H. Sachet. 1980. Gentianaceae, Loganiaceae. Pages 18–23 in: Systematic studies of Micronesian plants. *Smithsonian Contr. Bot.* 45: 1–40.

Gilg, E. 1895. Gentianaceae. Pages 50–108 in: A. Engler & K. Prantl, eds. *Die natürlichen Pflanzenfamilien*, vol. 4(2). Verlag von Wilhelm Engelmann, Leipzig.

Gilg, E. 1897. Beiträge zur Kenntnis der Gentianaceae, I. *Bot. Jahrb.* 22: 301–347.

Gleason, H. A. 1931. Botanical results of the Tyler–Duida Expedition. *Bull. Torrey Bot. Club* 58: 277–506.

Gleason, H. A. & E. P. Killip. 1939. The flora of Mount Auyan-tepui, Venezuela. *Brittonia* 3: 141–204.

Gleason, H. A. & R. P. Wodehouse. 1931. Gentianaceae. Pages 449–452 in: H. A. Gleason, ed. Botanical results of the Tyler–Duida Expedition. *Bull. Torrey Bot. Club* 58: 277–506.

Grisebach, A. H. R. 1839 [1838]. *Genera et species Gentianearum.* J. G. Cotta, Stuttgart and Tübingen.

Grisebach, A. H. R. 1845. Gentianaceae. Pages 39–141 in: A. de Candolle, ed. *Prodromus systematis naturalis regni vegetabilis*, vol. 9. Fortin, Masson, et Sociorum, Paris.

Grothe, E. H. M. & P. J. M. Maas. 1984. A scanning electron microscopic study of the seed coat structure of *Curtia* Chamisso & Schlechtendal and *Hockinia* Gardner (Gentianaceae). *Proc. Kon. Ned. Akad. Wetensch.*, Ser. C, 87: 33–42.

Guinet, P. 1962. Pollens d'Asie tropicale. *Inst. Fr. Pond. Trav. Sect. Scient. Techn.* V(1).

Holmgren, P. K., N. H. Holmgren, & L. C. Barnett. 1990. *Index Herbariorum*, part 1, *The herbaria of the world*, ed. 8. The New York Botanical Garden, Bronx, NY.

Humbert, H. 1937. *Gentianothamnus*, genre nouveau de Gentianacées de Madagascar. *Comptes Rend. Hebd. Séances Acad. Sci.* 204: 1747–1749.
Jonker, F. P. 1948. Remarks on the genera *Stahelia* and *Tapeinostemon* (Gentianaceae). *Rec. Trav. Néerl.* 41: 145–149.
Klackenberg, J. 1987. Revision of the genus *Tachiadenus* (Gentianaceae). *Bull. Mus. Nat. Hist. Nat., Paris*, sér. 4, 9, sect. B, *Adansonia* 1: 43–80.
Klackenberg, J. 1990. Famille 168. Gentianacées. Famille 168 bis. Menyanthacées. Pages 1–185 in: P. Morat, ed. *Flore de Madagascar et des Comores.* Association de botanique tropicale, Paris.
Köhler, A. 1905. Der Systematische Wert der Pollenbeschaffenheit bei den Gentianaceen. *Mitteil. Bot. Mus. Zürich* 25. (Dissertation)
Kuntze, C. E. O. 1891. *Revisio generum plantarum*, vol. 2. Arthur Felix, Leipzig.
Leenhouts, P. W. 1962 [1963]. Loganiaceae. Pages 293–387 in: C. G. G. J. van Steenis, ed. *Flora Malesiana*, ser. 1, vol. 6 (2). Wolters-Noordhoff, Groningen.
Leeuwenberg, A.J. M. & P. W. Leenhouts. 1980. Taxonomy. Pages 8–96 in: A. J. M. Leeuwenberg, ed. *Engler and Prantl's Die natürlichen Pflanzenfamilien, Angiospermae: Ordnung Gentianales, Fam. Loganiaceae*, vol. 28b (1). Duncker and Humblot, Berlin.
Lienau, K., H. Straka, & B. Friedrich. 1986. Fam. 168, Gentianaceae. Pages 9–15 in: H. Straka, ed. *Palynologia Madagassica et Mascarenica, Familien 167 bis 181.* Tropische und Subtropische Pflanzenwelt 55. Akademie der Wissenschaften und der Literatur, Franz Steiner Verlag, Wiesbaden, Mainz.
Maas, P. J. M. 1981. On the true identity of *Lagenanthus parviflorus* Ewan (Gentianaceae). *Ann. Missouri Bot. Gard.* 68: 685–688.
Maas, P. J. M. 1985. Nomenclatural notes on neotropical Lisyantheae (Gentianaceae). *Proc. Kon. Ned. Akad. Wetensch.*, ser. C, 88: 405–412.
Maas, P. J. M., S. Nilsson, A. M. C. Hollants, B. J. H. ter Welle, H. Persoon, & E. C. H. van Heusden. 1983. Systematic studies in neotropical Gentianaceae – the *Lisianthius* complex. *Acta Bot. Neerl.* 32: 371–374.
Maguire, B. 1981. Gentianaceae. Pages 330–388 in: B. Maguire and collaborators, eds. The botany of the Guayana Highland – Part XI. *Mem. New York Bot. Gard.* 32.
Maguire, B. 1985. Gentianaceae – part 2. *Phytologia* 57: 311–312.
Maguire, B. & B. M. Boom. 1989. Gentianaceae, part 3. Pages 2–56 in: B. Maguire and collaborators, eds. The botany of the Guayana Highland – Part XIII. *Mem. New York Bot. Gard.* 51.
Maguire, B. & J. M. Pires. 1978. Saccifoliaceae – a new monotypic family of the Gentianales. Pages 230–245 in: B. Maguire and collaborators, eds. The botany of the Guayana Highland – Part X. *Mem. New York Bot. Gard.* 29.
Maguire, B. & R. E. Weaver, Jr. 1975. The neotropical genus *Tachia* (Gentianaceae). *J. Arnold Arbor.* 56: 103–125.
Nilsson, S. 1967a. Pollen morphological studies in the Gentianaceae-Gentianinae. *Grana Palynol.* 7: 46–145.
Nilsson, S. 1967b. Notes on pollen morphological variation in Gentianaceae-Gentianinae. *Pollen Spores* 9: 49–58.
Nilsson, S. 1968. Pollen morphology in the genus *Macrocarpaea* (Gentianaceae) and its taxonomic significance. *Svensk Bot. Tidskr.* 62: 338–364.

Nilsson, S. 1970. Pollen morphological contributions to the taxonomy of *Lisianthus* L. s. lat. (Gentianaceae). *Svensk Bot. Tidskr.* 64: 1–43.
Nilsson, S. & J. J. Skvarla. 1969. Pollen morphology of saprophytic taxa in the Gentianaceae. *Ann. Missouri Bot. Gard.* 56: 420–438.
Perkins, J. 1902. Monographische Übersicht der Arten der Gattung *Lisianthus. Bot. Jahrb.* 31: 489–494.
Progel, A. 1865. Gentianaceae. Pages 197–248 in: C. F. P. von Martius, ed. *Flora Brasiliensis*, vol. 6(1). Frid. Fleischer, Munich.
Punt, W. 1978. Evolutionary trends in the Potalieae (Loganiaceae). *Rev. Palaeobot. Palynol.* 26: 313–335.
Punt, W. 1980. Pollen morphology. Pages 162–191 in: A. J. M. Leeuwenberg, ed. *Engler and Prantl's Die natürlichen Pflanzenfamilien, Angiospermae: Ordnung Gentianales, Fam. Loganiaceae*, vol. 28b (1). Duncker and Humblot, Berlin.
Punt, W. & P. W. Leenhouts. 1967. Pollen morphology and taxonomy in the Loganiaceae. *Grana Palynol.* 7: 469–516.
Punt, W., S. Blackmore, S. Nilsson, & A. Le Thomas. 1994. *Glossary of pollen and spore terminology*. LPP Contribution Series, No. 1. LPP Foundation, Utrecht.
Rao, A. N. & Y. K. Lee. 1970. Studies on Singapore pollen. *Pacific Sci.* 24: 255–268.
Raynal, A. 1968. Les genres *Neurotheca* Benth. et Hook. et *Congolanthus* A. Raynal, gen. nov. *Adansonia*, sér. 2, 8: 45–68.
Robyns, A. & S. Nilsson. 1970. *Macrocarpaea browallioides* (Ewan) A. Robyns & S. Nilsson, *comb. nov.* (Gentianaceae). *Bull. Jard. Bot. Natl. Belg.* 40: 13–15.
Spurr, A. R. 1969. A low-viscosity epoxy resin embedding medium for electron microscopy. *J. Ultrastruct. Res.* 26: 31–43.
Steyermark, J. A. 1951. The genus *Tapeinostemon* (Gentianaceae). *Lloydia* 14: 58–64.
Steyermark, J. A. 1962. Gentianaceae. *Bol. Soc. Venez. Cienc. Nat.* 23: 75–76.
Steyermark, J. A. 1966. Botanical novelties in the region of Sierra de Lema, Estado Bolivar, Venezuela – III. *Bol. Soc. Venez. Cienc. Nat.* 26: 411–452.
Steyermark, J. A. & B. Maguire. 1976. Gentianaceae, in: La vegetación en la cima del Macizo de Jaua. *Bol. Soc. Venez. Cienc. Nat.* 32: 392.
Steyermark, J. A. and collaborators. 1953. Contributions to the flora of Venezuela. Botanical exploration in Venezuela – III. *Fieldiana, Bot.* 28: 496–1071.
Steyermark, J. A., B. Maguire and collaborators. 1967. Botany of the Chimanta massif – Part II. *Mem. New York Bot. Gard.* 17: 440–464.
Struwe, L. & V. A. Albert. 1998. *Lisianthius* (Gentianaceae), its probable homonym *Lisyanthus*, and the priority of *Helia* over *Irlbachia* as its substitute. *Harvard Pap. Bot.* 3: 63–71.
Struwe, L., V. A. Albert, & B. Bremer. 1994 [1995]. Cladistics and family level classification of the Gentianales. *Cladistics* 10: 175–206.
Struwe, L., P. J. M. Maas, & V. A. Albert. 1997. *Aripuana cullmaniorum*, a new genus and species of Gentianaceae from white-sands of southeastern Amazonas, Brazil. *Harvard Pap. Bot.* 2: 235–253.
Struwe, L., P. J. M. Maas, O. Pihlar, & V. A. Albert. 1999. Gentianaceae. Pages

474–542 in: P. E. Berry, K. Yatskievych, & B. K. Holst, eds. *Flora of the Venezuelan Guayana*, vol. 5. Missouri Botanical Garden, St. Louis, MO.

Struwe, L., J. W. Kadereit, J. Klackenberg, S. Nilsson, M. Thiv, K. B. von Hagen, & V. A. Albert. 2002. Systematics, character evolution, and biogeography of Gentianaceae, including a new tribal and subtribal classification. Pages 21–309 in: L. Struwe & V. A. Albert, eds. *Gentianaceae: systematics and natural history*. Cambridge University Press, Cambridge.

Takhtajan, A. 1997. *Diversity and classification of flowering plants*. Columbia University Press, New York.

Taylor, C. M. 1990. Revision of *Hillia* subg. *Ravnia* (Rubiaceae: Cinchonioideae). *Selbyana* 11: 26–34.

Thiv, M., L. Struwe, V. A. Albert, & J. W. Kadereit. 1999. The phylogenetic relationships of *Saccifolium bandeirae* Maguire & Pires (Gentianaceae) reconsidered. *Harvard Pap. Bot.* 4: 519–526.

Weaver, R. E., Jr. 1972. A revision of the neotropical genus *Lisianthius* (Gentianaceae). *J. Arnold Arbor.* 53: 76–311.

Weaver, R. E., Jr. 1974. The reduction of *Rusbyanthus* and the tribe Rusbyantheae (Gentianaceae). *J. Arnold Arbor.* 55: 300–302.

5
The seeds of Gentianaceae

F. BOUMAN, L. COBB, N. DEVENTE, V. GOETHALS,
P. J. M. MAAS, AND E. SMETS

ABSTRACT

Seeds of representatives of 78 genera of the Gentianaceae were studied using scanning electron microscopy. Seeds of most taxa are relatively small, often between 0.3 mm and 0.7 mm in length. They develop from unitegmic, tenuinucellate ovules and are exotestal or sometimes have collapsed or reduced testas. In spite of their small size, the seeds show an extensive diversity in micromorphology.

Seed structure is often characteristic at the tribal, subtribal, or generic level. Testa cells vary in orientation, shape, undulation of the anticlinal walls and pitting of the inner walls. The outer wall is always thin and without a cuticular structure. Except for the inclusion of the Potaliinae, and the delimitation of the Saccifolieae, seed structure sustains the new classification of the Gentianaceae as proposed by Struwe et al. in Chapter 2 of this volume. Crystallized structures were found upon and under the exotestal cuticle of several genera of the Chironieae, Coutoubeinae, and Faroinae.

Seeds of the mycotrophic/saprophytic genera are all small and may be strongly reduced. Owing to adaptations to a saprophytic lifestyle, seed characters of saprophytes are of limited taxonomic use at the subgeneric level.

The seeds of the majority of species are dispersed by the wind. Most species are wind ballists. Endozoochorous dispersal is known in Potaliinae and in one species of *Chironia*. Seeds may belong to different seed bank types. Seed germination is epigeal. Seedlings are small and phanerocotylar.

Keywords: dispersal, embryology, Gentianaceae, germination, seed morphology, testa.

INTRODUCTION

Although the seeds of Gentianaceae are relatively small, mostly less than 0.7 mm, they reveal an impressive diversity in morphology when studied using scanning electron microscopy. Seed structure may supply an additional character set useful for taxonomic studies and is also relevant for functional aspects such as seed dispersal (Boesewinkel & Bouman, 1995). Until now, seeds of only a limited number of taxa of the Gentianaceae have been studied in detail.

As early as 1904, Guérin made a comparative light microscopical study of the seed anatomy of the Gentianaceae. He showed the diversity in exotestal structure of a number of representative taxa. In 1926, Netolitzky reviewed the literature on the seed anatomy of Gentianaceae. Since then, no serious review on this topic has been published. Seed micromorphology is increasingly being used in taxonomic studies at the generic level (Cobb & Maas, 1983; Klackenberg, 1983, 1985; Grothe & Maas, 1984; Miège & Wüest, 1984; Bouman & Devente, 1986; Yuan, 1993).

In this chapter, the results of a review of the existing literature and of a survey of the micromorphology of seeds of some 224 species and/or subspecies of 78 genera of Gentianaceae are presented.

DEVELOPMENT AND STRUCTURE OF OVULES AND SEEDS

Embryology and ovule development

The embryology of the Gentianaceae has been reviewed by Davis (1966) and more recently by Johri *et al.* (1992). Wall formation of the tetrasporangiate anthers conforms to the Dicotyledonous-type in *Swertia*, and to the Basic-type in *Eustoma* (Drexler & Hakki, 1979). The endothecium develops fibrous thickenings in most species. The tapetum is glandular, except in species of *Canscora* and *Gentiana*, where it is plasmodial. Tapetum cells remain uninucleate. Simultaneous cytokinesis of the microspore mother cells follows meiosis. The microspore tetrads are tetrahedral or isobilateral. Pollen grains are two- or three-celled when shed.

The ovules are tenuinucellate, usually unitegmic and anatropous. The archesporium is unicellular, sometimes multicellular. A single archesporial cell undergoes meiotic division and functions as the megaspore mother cell. The megaspore tetrad is linear and the chalazal megaspore develops into a *Polygonum*-type of embryo sac, with the exception of some saprophytic

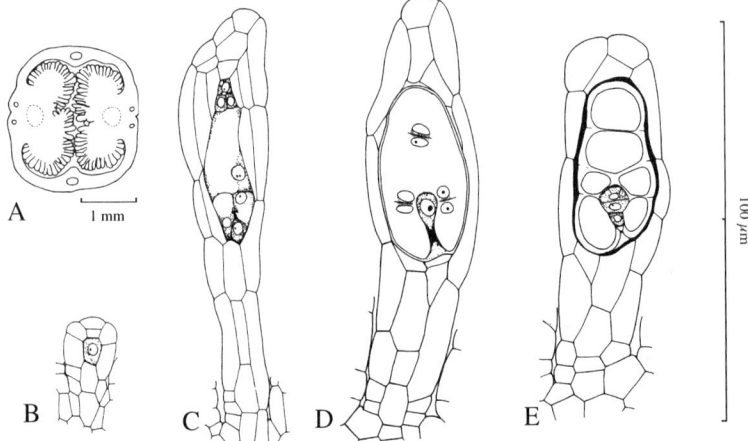

Figure 5.1. Seed development in *Voyria primuloides*, after Bouman and Louis (1990). **A.** Cross-section of ovary. **B.** Ovule primordium with archespore. **C.** Orthotropous, ategmic ovule with mature, inverted embryo sac. **D.** Developing seed with nuclear endosperm and zygote. **E.** Mature seed.

taxa. The synergids are beaked with tapering ends. The polar nuclei fuse before fertilization. The three antipodal cells remain mostly uninucleate and degenerate at embryo sac maturity. In some taxa the antipodals are persistent for some time and may become hypertrophied or bi- to multinucleate (*Halenia*). Secondary multiplication of antipodals, up to 16 cells, has been recorded in species of *Gentianella* and *Swertia* (Guérin, 1903; Stolt, 1921; McCoy, 1949). The polar nuclei fuse before fertilization. The nucellus is always resorbed at an early stage and does not contribute to the seed coat. The number of ovules per fruit varies from two to many. Few precise data are known for the mean number of ovules per fruit. Two ovules per fruit have been recorded for *Swertia piloglandulosa* ssp. *biovulata*. *Gentiana* and *Gentianella* species mostly have a mean number of 100 or fewer ovules per fruit (Luijten *et al.*, 1998), although up to 600 ovules per fruit have been counted in *Gentiana pneumonanthe* (Petanidou *et al.*, 1995).

The integument is usually thin and limited to a few layers, but may become up to 10 cells thick in the mature ovules of some *Gentiana* species (Bouman & Schrier, 1979). The inner layer of the integument does not develop into an endothelium (integumentary tapetum), except in *Exacum pumilum* (Johri *et al.*, 1992). The raphe is always small and the provascular bundle is always without differentiated xylem elements. In the smaller ovules, especially those of the saprophytic taxa, there is no sign of provascular tissue at all.

In most taxa, especially those with small to medium-sized seeds, the number of cells on the outer surface of the seed will not, or hardly not, exceed that of the ovule. During development, the integument may contain starch, which disappears during the cell wall thickening of the exotesta.

Hemitropous ovules have been observed in *Gentianella campestris* (V. Goethals, pers. obs.). Reduction of ovule structures is especially found in saprophytic taxa (Fig. 5.1). Orthotropous (atropous) ovules have been described in *Cotylanthera* (Figdor, 1897), and in *Voyriella* and a number of *Voyria* species (Oehler, 1927; Bouman & Louis, 1990). The report of orthotropy in *Halenia* (Stolt, 1921) needs confirmation. The integument is reduced in *Voyriella* and strongly reduced or absent and not forming a micropyle in *Cotylanthera* and a number of *Voyria* species (*V. aphylla*-group of Bouman & Devente, 1986). In orthotropous ovules, the embryo sac is inverted and directed toward the placenta, thus shortening the distance for the pollen tube.

Fertilization is porogamous and results in double fertilization. The division of the primary endosperm nucleus always precedes embryogenesis. Endosperm formation is usually of the nuclear type and later becomes cellular by centripetal wall formation. In *Voyriella* and some *Voyria* species the endosperm is *ab initio* of the cellular type, the first wall being transverse to the length of the embryo sac (Oehler, 1927). No cases of endosperm haustoria have been recorded in the family. In the mature seed the endosperm occupies the greater part of the seed and may be separated from the seed coat by a thick outer wall and a cuticle. The endosperm is rich in proteins and lipids and often has thick, pectinaceous walls.

In a number of taxa (*Lisianthius sensu stricto*) the endosperm has an uneven surface and may be called ruminate. According to the classification of Periasamy (1962), the rumination is of the *Verbascum*-type. In this type, the outer surface of the endosperm is impressed by the unequal radial elongation of the exotestal cells.

Embryo development conforms to the Solanad-type according to Davis (1966), and in some species to the Caryophyllad-type according to Johri *et al.* (1992). The differences in the embryology and seed structure between Gentianaceae and Menyanthaceae have been discussed by Maheswari Devi (1962).

Seed structure

The mature seeds of Gentianaceae are endospermous. The endosperm mostly contains oil and proteins and no starch. According to the classification of

Martin (1946), the seeds belong to the "Dwarf type" with embryos variable in relative size and with seed interiors from 0.3 to 2.0 mm long, or to the "Micro type" with seed interiors less than 0.2 mm long. The embryos are mostly small, rarely half or longer than half the seed length (*Frasera*), straight, and global, oval to elliptic, or oblong in shape. The cotyledons tend to be poorly developed. In the saprophytic genera *Voyria* and *Voyriella* the embryos are very small and without cotyledons. No mucilaginous seeds have been recorded in the family.

Seed shape

The seeds are diverse in shape and several distinct seed types may be discerned:

1 **Angular to polyhedral seeds.** Seeds of many Gentianaceae are basically (sub)conical to (sub)cubical with flattened, sometimes sunken sides and edges. The angular shape of the seeds is determined by the compression against neighboring seeds and against the fruit wall. Ovules and seeds are arranged in one layer. During development they may become flattened on their lateral sides by mutual compression, somewhat rounded at the chalazal end by contact with the fruit wall, and flat or concave at the hilar side by compression against the placenta. This type is found in the Exaceae, Helieae, and Potalieae-Faroinae.
2 **Globose to subglobose seeds** are found in the *Gentianella*-group and in *Coutoubea, Chironia, Lisianthius* spp., and *Sabatia*, while **subglobose to elliptic seeds** are present in many species of *Gentiana*.
3 **Winged or flattened seeds.** Seeds may become flattened and form wings. Such seeds are found in *Chorisepalum* and *Macrocarpaea rubra*, but occur especially in taxa of the Gentianeae, viz. in species of *Gentiana, Crawfurdia, Frasera, Swertia, Tripterospermum,* and *Veratrilla*. Wings are a generic character for *Chorisepalum, Crawfurdia,* and *Tripterospermum*.
4 **Fusiform seeds** have been recorded in some Asiatic species of *Gentiana* subg. Chondrophylleae (e.g., *G. thunbergii* and *G. nipponica*) and in *Gentianopsis ciliata*.
5 **Reduced seeds** are found in several saprophytic genera. **Scobiform seeds** are present in species of *Voyria* (*V. aphylla*-group).

In Gentianaceae, the raphe is never recognizable on the outside of the seed. Seeds of Gentianaceae are always exarillate: there are no records of arils, aril-like structures, or elaiosomes.

Seed size and weight

Seeds of most genera are usually relatively small (0.3–0.7 mm). Especially in the saprophytic taxa, seeds are reduced and may be very small. For example, seeds measuring less than 0.2 mm have been found in species of *Voyria* and *Bartonia*. Larger seeds are mostly found in the Gentianeae. Seeds of the *Gentianella*-group are about 1 mm. Seeds of *Gentiana* and *Tripterospermum* may measure up to 3.5 mm, and those of *Frasera* up to 7 mm.

Very few data are available on seed weight. The weight of seeds of Gentianaceae probably varies over a factor of almost 10^4. As far as is known, dry seed weight varies from 0.0002 mg in *Voyria aphylla* to c. 1.8 mg in *Tripterospermum sumatranum*. In the genus *Gentiana*, mean seed weight varies from c. 0.015 mg in *G. nivalis* to 0.15 mg in *G. purpurea* and c. 1 mg in *G. lutea* (Müller-Schneider, 1985).

Seed coat anatomy

As in the great majority of sympetalous taxa, the seed coat anatomy of the Gentianaceae is relatively simple (Boesewinkel & Bouman, 1984, 1995). Only the outer layer of the integument differentiates into an anatomically characteristic structure, the exotesta. The middle and inner layers of the integument become compressed during development and are later fully or partly resorbed. Knowledge on the seed coat anatomy of Gentianaceae was first reviewed and then extended by Guérin (1904). He showed the diversity in exotestal structure of a number of representative taxa (Fig. 5.2). Later, Netolitzky (1926) and Corner (1976) reviewed the literature on the seed structure of the entire family.

Bouman and Schrier (1979) described seed coat development in *Gentiana* in more detail. In *Gentianella* and related genera all integumentary layers remain thin-walled and the testa collapses at maturity. The mechanical functions of the seed coat are shifted to the outer layer of the endosperm, which is considerably thickened.

There are no differentiated vascular bundle(s) present in the raphe or in the seed coat. The seed anatomy of *Fagraea* (Potaliinae) has been described by Mohrbutter (1936) and Corner (1976).

Seed surface (exotesta)

For the description of exotestal characters we have used the terminology proposed by Barthlott and Ehler (1977), Barthlott (1981), and Stearn

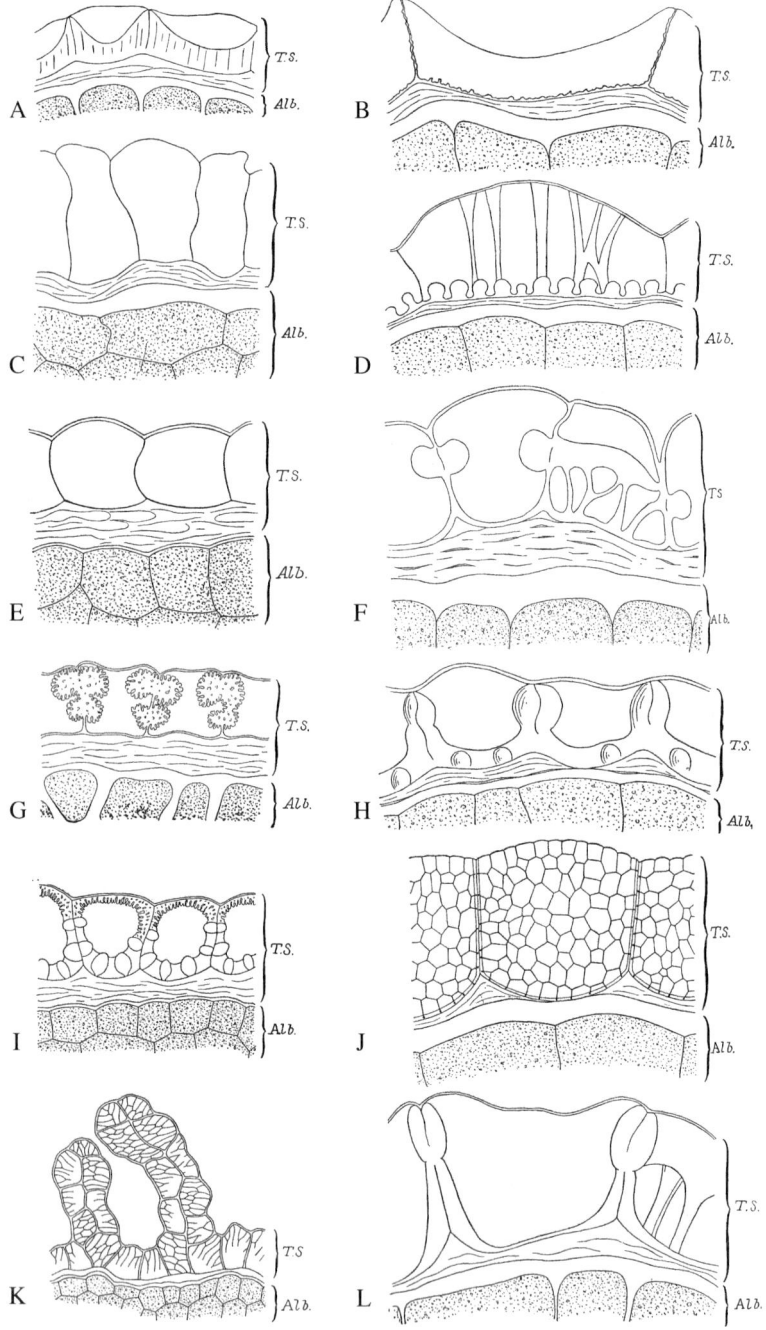

(1983). The testa is reticulate in most taxa. The structure of the seed surface is determined mainly by the pattern of the anticlinal walls and by the thickenings of the inner periclinal and anticlinal walls. The outer periclinal cell wall of the testa is usually mostly thin or very thin, and collapses during seed maturation. In the mature seed, the collapsed outer wall is pressed against the inner periclinal one and often reflects the structure of the inner wall. In some cases, especially when the testa cells are deep, the outer wall is tightened between the anticlinal walls and may tear, as in *Macrocarpaea rubra*. In taxa of the *Irlbachia* complex (Helieae), the outer wall may remain convex with band-like or reticulate cell wall thickenings. These thickenings are continuous with those of the anticlinal and inner walls. The testa of *Gentianopsis* species is completely covered with papillate to hair-like, blown-up cells. Testa cells are usually polygonal in surface view, sometimes more elongated as in *Gentiana*. In Exaceae, the testa cells may be star-shaped because of the V-undulating anticlinal walls. The anticlinal walls in exotestas of Gentianaceae may be straight, curved, waved, or undulating. The anticlinal boundaries are smooth, raised, or sunken. The cuticle is always smooth, sometimes with exudates (*Coutoubea* and *Orphium*). The thickenings of the exotestal cell walls may contain lignins. Non-reticulate seeds with undifferentiated, collapsed exotestas without distinct wall thickenings are known in *Gentianella*, *Halenia*, and related taxa.

SEED DISPERSAL

There have been relatively few studies on seed dispersal in Gentianaceae (Oostermeijer, 1996). The great majority of the fruits of the Gentianaceae are septicidal capsules that dehisce with two valves from the apex. The fruit wall is thin or hard. The fruit usually has numerous seeds, or rarely just a few. In *Chorisepalum* the pericarp splits from the base into four separate parts. In some *Bartonia* species the fruit is fenestrate, opening septicidally below the persistent style with the valves remaining connected at their tips. Fruits with lantern-like dehiscence or not opening and more or less rotting away are known in species of *Voyria*.

Figure 5.2. Seed coat anatomy in Gentianaceae, after Guérin (1904). The abbreviation "T.S." indicates seed coat (*tégument séminal*) and "Alb." indicates endosperm (*albumen*). All drawings are × 420. **A.** *Exacum metzianum*. **B.** *Sebaea ovata*. **C.** *Canscora decussata*. **D.** *Sabatia angularis*. **E.** *Gentiana serra*. **F.** *Gentiana lutea*. **G.** *Gentiana linoides*. **H.** *Gentiana septemfida*. **I.** *Gentiana bella*. **J.** *Gentiana yunnanensis*. **K.** *Gentiana stylophora*. **L.** *Gentiana nivalis*.

Wind is the prevalent dispersal vector of Gentianaceae seeds (Bouman & Devente, 1986). The majority of Gentianaceae grow in open or relatively open vegetation, in both temperate and tropical alpine ecosystems or tundras, where no trees restrict the wind velocity. Most species of Gentianaceae are wind ballists, also called rattleburs or semachores (Van der Pijl, 1982). Fruits stand on erect peduncles, often in stiff infructescences, although nodding peduncles are seen in many Helieae. Wings on the persistent calyx or fruit and/or the remains of a withered corolla may act as weather vanes and help in shaking the fruit and in determining the timing of seed release. In a number of temperate species the seeds are dispersed gradually and the period of seed dispersal may extend from late summer to winter, spring, and even the following summer (Threadgill et al., 1981).

Calyx wings are known in several tribes, especially in Exaceae and Gentianinae. The genus *Pterygocalyx*, for example, was named for this character. In species of *Exacum*, *Sebaea* (Klackenberg, 1985, 1987), and *Schultesia*, the persistent calyx lobes are usually furnished with a more or less prominent wing on the dorsal side of each lobe. The calyx wings of several other species enlarge in the fruit. In a number of Malagasy species the calyx is zygomorphic and furnished with two wings (*Exacum dipterum*) due to the reduction of three wings. *Tachiadenus* has a wing on the dorsal side of each sepal. The calyx of *Tachiadenus longiflorus* is unique because of its 10 wings. Wings on the fruit are also found in species of *Gentiana* subg. *Chondrophyllae*.

In fruit, the corolla either falls off soon after anthesis or is persistent. The marcescent corolla tube may narrowly enclose the fruit and is torn during fruit dehiscence. The persistent corolla may remain at the top of the fruit and prevent the full dehiscence of the apical parts of the two fruit valves. If this happens, the seeds are released gradually from the lateral slits of the fruit, as in species of *Centaurium* and in *Gentianopsis ciliata*.

Postfloral elongation of the peduncle, which raises the capsules above the leaves, is known for species of *Gentiana* (*G. clusii*, *G. kochiana*, *G. nivalis*, and *G. verna*). The formation of a carpo(gyno)phore to expose the fruit has also been described in species of *Tripterospermum* (Murata, 1989) and *Crawfurdia* (Smith, 1965).

In a number of taxa seeds are adapted by size and shape to further dispersal by wind. Seeds may be provided with ridges of elongated, air-filled cells (balloon-like cells) as in species of *Irlbachia*, be flattened, or have narrow to broad wings as in *Swertia*, *Frasera*, *Veratrilla*, *Macrocarpaea* and some sections of *Gentiana*. Seeds of *Crawfurdia* and *Tripterospermum* are three-winged, and those of *Urogentias* multiply winged. Seeds of *Gentianopsis ciliata* are fusiform with air-filled micropylar and chalazal ends. More extreme adaptations to wind dispersal are found in the *Voyria aphylla*-group, where fusi- to filiform seeds are released in groups, together with paraphyses (sterile, abortive seeds). Wind dispersal enables some *Voyria* species to reach tree trunks and branches up to 30 m high for growth as epiphytes (Groenendijk et al., 1997).

The number of reports on alternative types of dispersal is not great. In the genera *Gentiana* and *Gentianella* dispersal seems to be relatively more diverse. Müller (1955) and Müller-Schneider (1985) report dysochorous dispersal by snow finches in *Gentiana lutea* and endozoochorous dispersal by, among others, cattle, horse, deer, and chamois in *Gentianella amarella*, *G. campestris*, and *Gentiana cruciata*. Dispersal by running water has been described in *Gentiana brachyphylla* and *Lomatogonium carinthiacum*. Species of *Gentianopsis* often grow in wet or flooded habitats, and the exotestal papillae probably also promote flotation.

Endozoochorous dispersal, other than accidentally by herbivores, seems rare in Gentianaceae. The fruit is rarely a berry or a fleshy capsule. Fleshy, berry-like fruits have been reported, however, in species of *Chironia* and *Tripterospermum*. One species of *Chironia* (*C. baccifera*) has red, fleshy fruits when mature, drying black and splitting irregularly. In some species of *Tripterospermum* the fruits are red to dark purple berries with brown, dark purple, or black seeds. According to Pringle (1995), the fruit of *Symbolanthus* is also a nearly baccate capsule (a leathery, sometimes indehiscent capsule according to L. Struwe, pers. comm.). However, no observations on seed dispersal of these genera have yet been recorded.

In taxa of the Potalieae-Potaliinae, traditionally placed in the Loganiaceae, seeds are dispersed by animals. In the Potaliinae the fruit is a fleshy berry, and rarely dehiscent with four valves as in some large-fruited species of *Fagraea*. The fruit is green, white, yellow, or orange to bright red. *Fagraea littoralis* is ornithochorous (Docters van Leeuwen, 1934).

Dispersal is obscure in a number of *Voyria* species growing on the floor of humid tropical forest. These fruits are indehiscent and seeds are released after decay of the fruit wall. The small seeds are supposedly dispersed by rain wash or epizoochorously via mud on the fur or paws of animals (Bouman & Devente, 1986; Albert & Struwe, 1997). Imhof et al. (1994) suggest myrmecochorous dispersal; ants may bite off infructescences and transport fruits.

Sebaea oligantha, the only saprophyte of the genus *Sebaea*, represents a special case (Raynal, 1967). Next to heterostylous, chasmogamous flowers, which produce erect, dehiscent capsules, plants have short-styled, cleistogamous flowers, which are mostly hidden in the litter layer and produce geocarp capsules. Raynal (1967) suggests that soil insects may disperse the small, polyhedral seeds over short distances.

GERMINATION

With the exception of the mycotrophic/saprophytic taxa, germination is always epigeal and phanerocotylar. As far as described (Lubbock, 1892; Muller, 1978), the seedlings are small, like the seeds. The hypocotyl is short, mostly only a few millimeters in length, rarely about 1 cm as in *Anthocleista* (Miquel, 1985) and *Fagraea* (Burger, 1972). The cotyledons are always small, 1–5 mm long, various in shape but mostly oblong; those of *Erythraea* (= *Centaurium*) and *Orphium* are oblong-ovate and linear, respectively. They are sessile, nearly sessile, or with a petiole of only 2 mm. A distinct epicotyl is usually absent, especially in the rosette-forming species. The epicotyl is more distinct in *Fagraea* and *Orphium*.

Seed germination of gentian species is often very difficult under laboratory conditions. Favarger (1953) studied germination in 36 species of *Gentiana*, *Swertia*, *Halenia*, and *Lomatogonium*, while Threadgill *et al.* (1981) reviewed seed germination in Gentianaceae and studied dormancy in *Frasera caroliniensis*. Dormancy breaking and germination requirements proved quite diverse throughout the family. Seeds of a number of temperate gentians require a warm, moist period followed by a cold period before they are able to germinate. In temperate species, delayed dispersal coupled with the low temperature requirement for embryo growth means that the germination of seeds produced in a particular year will be potentially spread over two years (Threadgill *et al.*, 1981).

Thompson *et al.* (1997) presented data on the germination of 12 northwestern European species belonging to *Blackstonia*, *Centaurium*, *Gentiana*, and *Gentianella*. They may belong to one of all four seed bank types discerned: from transient ones with seeds present in the upper soil layer during part of the year only, to persistent ones with many seeds in lower soil layers and upper layers all year. An extreme of 50 000 seeds/m^2 was recorded in *Centaurium littorale*. Seeds of some temperate species may retain their viability for at least three years and germinate in spring or early summer. Germination may be promoted by gibberellic acid.

METHODS

The terminology of Barthlott and Ehler (1977), Barthlott (1981), and Stearn (1983) is used for the descriptions of seed structure. For reasons of uniformity, the nomenclature and classification of tribes and subtribes as proposed by Struwe *et al.* (2002 (Chapter 2, this volume)) is used in the present chapter. Almost all seed samples studied were collected from herbarium specimens (see Appendix 5.1 for material and voucher information).

Some seeds were pretreated with 10% ammonium solution and/or subsequent ultrasonic vibration to show the internal structure of the testa cells.

COMPARATIVE MORPHOLOGY OF THE SEEDS OF GENTIANACEAE

Tribe Exaceae

The seed structure of Exaceae has been studied by Klackenberg (1983, 1985, 1986, 1987). With the exception of *Sebaea sensu stricto* and the saprophytic genus *Cotylanthera*, the seeds of the Exaceae closely resemble one another and are discernible from those of the other tribes of Gentianaceae by the shape of the testa cells. The seeds are numerous, small and polyhedral, to angled by mutual compression. Inner and all lateral walls of the exotestal cells are strongly thickened and pitted, making the testa cells more or less horseshoe-shaped in cross-section. In most taxa the testa cells are star-shaped in surface view because of the undulating or V-undulating anticlinal walls, with mostly only one arm protruding in each neighboring cell. Outer cell walls are thin, collapsed, and lying on the inner one. Anticlinal walls are mostly undulating, sometimes curved or straight.

Exacum (Fig. 5.3A–B)

The seeds are numerous and minute to small, 140–700 μm, mostly about 250–500 μm in diameter, usually angular, polyhedral, subcubical to subconical in shape, rarely more rounded, or occasionally shallowly cup-shaped as in *Exacum radicans* and *E. subverticillatum* (Klackenberg, 1983, 1985). *Exacum sessile* has the smallest seeds, *c.* 140 μm in length, and only four testa cells thick from the top to the base of the seed. The testal structure varies between species, especially in the shape of the cells and the thickness of the anticlinal walls. Most species have star-shaped exotestal cells, sometimes with prominent arms. There are, however, some deviations. The

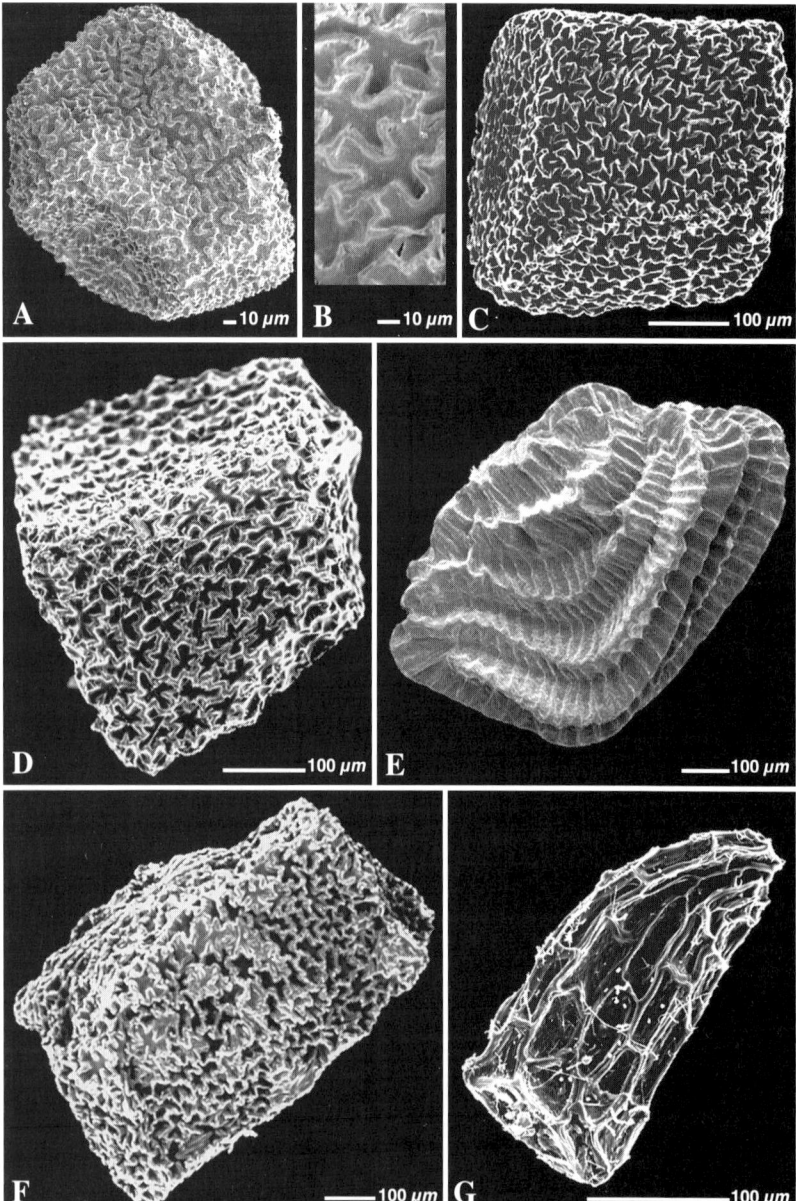

Figure 5.3. Scanning electron micrographs of seeds of Exaceae. **A–B.** *Exacum walkeri* (*Jayasuriya & Robyns 82*), seed and detail of the testa. **C.** *Ornichia madagascariensis* (*Phillipson et al. 3973*). **D.** *Tachiadenus carinatus* (*Lam & Meeuse 5498*). **E.** *Sebaea albidiflora* (*van Steenis 23468*). **F.** *Gentianothamnus madagascariensis* (*Malcomber et al. 869*). **G.** *Cotylanthera tenuis* (*Korthals 622*).

The seeds of Gentianaceae 511

star-shape is less pronounced in *Exacum lawii, E. pumilum, E. radicans, E. sessile, E. subverticillatum,* and *E. teres.* In many species of section *Africana* the anticlinal walls are straight or minutely wavy and the testa cells more or less polygonal as a consequence (Klackenberg, 1983, 1985). Anticlinal walls are more strongly thickened and more finely undulating in *Exacum oldenlandiodes.* In *Exacum teres* the anticlinal walls jut out at the corners to form a somewhat hook-like structure. The outer wall is thin and collapsed. The cuticle is smooth.

Ornichia (Fig. 5.3C)

The seeds are numerous, small, angular and polyhedral or subconical in shape. The seeds of *Ornichia madagascariensis* are about 10 cells high and measure about 450–580 μm in length. The exotestal cells are star-shaped, with prominent arms (Klackenberg, 1986). The outer wall is thin and collapsed in mature seeds. The cuticle is smooth.

Tachiadenus (Fig. 5.3D)

The seed structure of *Tachiadenus* has been described by Klackenberg (1987). The seeds are numerous, small, between 300 μm and 800 μm (those of *Tachiadenus carinatus* about 550 μm). The seeds are usually angular, polyhedral to subconical in shape. The seed coat is composed of many cells. Testa cells are anticlinally undulating in a star-shaped pattern.

In *Tachiadenus tubiflorus* the seeds are spherical, with the walls of the testa cells very protruding locally. This type of seed has not been found in any other species of Exaceae.

Sebaea (Fig. 5.3E)

The genus *Sebaea,* including *Belmontia, Lagenias,* and *Exochaenium,* is still poorly known and needs revision (J. Klackenberg, pers. comm.). Seed structure in *Sebaea sensu lato* is diverse. The seeds of species formerly attributed to *Belmontia* or *Exochaenium* resemble more those of the other genera of Exaceae (J. Klackenberg, pers. comm.; F. Bouman, pers. obs.). The seeds of species of *Sebaea sensu stricto,* including the Australian representative *S. albidiflora,* have their own type, resembling somewhat those of *Blackstonia.*

The seeds of *Sebaea sensu stricto* are often ridged or fringed and show a more bilateral symmetry, reflecting their anatropous nature. Seed size varies from 240 to 755 μm in length and from 175 to 480 μm in width. Testa cells are more or less rectangular, elongated perpendicular to the length of the seed, and arranged in regular longitudinal rows. Anticlinal walls are usually thin. The longitudinally-running ones are more pronounced, or strongly pronounced as in *Sebaea albidiflora* and *S. ambigua*, forming longitudinal ridges or fringes on the seeds. The outer periclinal walls are thin, collapsed at maturity, and may reflect the granular structure of the underlying wall.

The seeds of *Sebaea exacoides* are somewhat intermediate. The seed lacks ridges or fringes, the anticlinal walls are slightly undulated, but the testa cells are arranged in rows.

The seeds of *Sebaea oligantha* are small, angular to subconical, and $c.$ 250 μm in length. Testa cells are irregularly star-shaped with more raised, thicker anticlinal walls that are pronounced especially at the corners. The anticlinal boundaries are locally raised. Seeds of *Sebaea exigua* (*Exochaenium*) and *S. grandis* (*Belmontia*) also resemble the *Exacum*-type and measure about 270 and 160 μm, respectively.

Gentianothamnus (Fig. 5.3F)

Seeds of *Gentianothamnus madagascariensis* are angular in shape, mostly cubical to rectangular, and $c.$ 680×625 μm in size. Testa cells are puzzle piece-like (star-shaped) and oriented irregularly. The outer wall is collapsed, thin, and does not reflect the structure of the underlying inner wall. The anticlinal walls are V-undulating, prominent, and thick.

Cotylanthera (Fig. 5.3G)

The ovules and seeds of *Cotylanthera*, a small, Asiatic genus of saprophytes, are strongly reduced (Figdor, 1897; Oehler, 1927). The seeds develop from orthotropous, ategmic ovules and therefore do not have a testa in the proper sense, nor a micropyle. Partly depending on the length of the funicle, the seeds of *Cotylanthera tenuis* measure about 200–460× 90–160 μm. The seeds are saccate, acuminate toward the funicular end, and count 4–5 cells in length and 9–10 cells in circumference. The outer testal layer has a reticulate pattern and consists of elongated cells with thickened, lignified, straight to slightly curved anticlinal walls. The outer walls are thin and concave. The embryo sac is inverted, so the embryo lies near the funicular side. The embryo is small, of about 16 cells, and has a suspensor of three cells. Cotyledons are not differentiated in the mature seed. The embryo is surrounded by one layer of endosperm. Walls of the endosperm cells, especially the outer ones, are thickened and pectinaceous. The endosperm has crystalline protein bodies, but contains no starch (Figdor, 1897; Oehler, 1927). The seeds studied were all infected by fungi.

Tribe Chironieae

Seeds of taxa of the subtribes Canscorinae, Chironiinae, and Coutoubeinae resemble one another in several characters. The seeds are small and have a distinct reticulate pattern of polygonal cells with straight or undulated cell walls. The inner wall may have papillate thickenings of various length, intermingled with pits, or a more open reticulum with the papillae in the corners of the reticulum. The inner and anticlinal walls often differ in their wall thickenings. The cuticle is smooth or has flaky or granular exudates.

Crystallized structures are found upon and under the exotestal cuticle of several genera of the Chironieae, including *Chironia*, *Coutoubea*, and *Orphium*, and in the Faroinae, including *Neurotheca*. They resemble the exudates described on the leaves and stems of *Orphium* and are probably flavonoids. Seeds of *Cracosna* and *Exaculum* were not available for this study.

Subtribe Chironiinae

Chironia (Fig. 5.4A–D)

Seeds of *Chironia* are globose to elliptic. According to Marais and Verdoorn (1963), seeds of *Chironia serpyllifolia* are black to shiny brown, subglobose, reticulate with a pitted or deeply honeycombed testa. The mean seed size varies from 445 × 335 μm in *Chironia transvaalensis* to *c.* 620 × 480 μm in *C. densiflora*. The testa is reticulate, and about 5–8 cells high. The testa cells are polygonal in surface view. The anticlinal walls are straight or slightly bent. The outer walls are thin, collapsed, and reflect the papillate structure of the inner walls. The inner walls are pitted with papillae in the corners. The cuticle is smooth.

The seeds of fleshy-fruited *Chironia baccifera* are bigger, *c.* 1400 × 900 μm and about 10 cells high. Their seeds deviate from those of the non-baccate species in some structural details. The anticlinal walls are always straight, thicker, broader at the base, and locally hatched near the anticlinal boundaries. The inner wall is pitted with low papillae in the corners. The anticlinal walls deviate from the inner one by a coarser reticulate pitting. The anticlinal boundaries are locally raised. The outer walls are smooth and without any cuticular ornamentation.

Orphium (Fig. 5.4E–G)

Seeds of *Orphium fructescens* are globose to broadly elliptic and *c.* 870 × 780 μm in size. The testa is reticulate. The testa cells are polygonal in

Figure 5.4. Scanning electron micrographs of seeds of Chironieae-Chironiinae. A. *Chironia baccifera* (*Goldblatt 1397*). B. *Chironia transvaalensis* (*Wilms 974*), cross-section of seed coat. C–D. *Chironia baccifera* (*Coppejans 701*), seed coat after removal of the outer cell walls and detail of the anticlinal walls, respectively. E–G. *Orphium frutescens* (*Lotsy & Goddijn 1571*), seed, section of seed coat, and testa cell with exudates, respectively. H. *Eustoma grandiflorum* (*Thomas & Taylor 85509*). I. *Bisgoeppertia scandens* (*Zanoni et al. 40716*).

surface view, oriented irregularly, and are deeply concave. The anticlinal walls are distinct, straight to curved, and locally hatched near the anticlinal boundaries. The outer wall is thin, collapsed, and reflects the papillae of the inner wall. The inner wall is thickened and has numerous low papillae interspersed by pits. The anticlinal walls taper toward the periphery. The lower part has the same structure as the inner wall, while the upper part is thinner and more or less scalariform-pitted. The cuticle has irregular, flaky exudates, especially near the anticlinals. Seeds of *Bisgoeppertia*, *Eustoma*, and *Orphium* closely resemble each other.

Eustoma (Fig. 5.4H)

The seeds are widely elliptic to circular and $c.$ 330–450 × 270–370 μm in size. The hilum/micropyle is sunken. The testa is densely pitted by the concave, collapsed testa cells. Testa cells are polygonal. The anticlinal walls are straight to curved, and raised with a row of large pits. The outer wall is thin, collapsed, and reflects the papillae of the inner wall. The inner wall is reticulate with low papillae and small, rounded pits. The cuticle is smooth. The seeds of *Eustoma grandiflorum* resemble those of *Lisianthius* sect. *Lisianthius*.

Bisgoeppertia (Fig. 5.4I)

The seeds are elliptic to somewhat triangular in shape and $c.$ 500–680 × 450–550 μm in size. The testa is reticulate and the exotestal cells are polygonal and deeply concave. The anticlinal walls are straight, thick, and raised at the corners, making the seeds somewhat echinate. The inner wall is pusticulate. The cuticle is smooth.

Geniostemon (Fig. 5.5A)

The seeds are elliptic to ovate in shape. Seeds of *Geniostemon schaffneri* measure 310–360 × 230–305 μm. The testa cells are puzzle piece-like, with thick, sinusoidal anticlinal walls. The outer wall is collapsed and thin. The inner wall has not been observed. The cuticle is smooth. The seeds resemble those of *Hoppea dichotoma*.

Sabatia (Fig. 5.5B–C)

The seeds are globose to elliptic, or somewhat irregular in shape, and vary in size from about 340 × 220 μm in *Sabatia brevifolia*, 360 × 340 μm in *S. dodecandra*, to 480 × 355 μm in *S. gentianoides*. The testa is reticulate. Testa cells are irregular, more or less polygonal in shape, and taper toward the micropyle. The anticlinal walls are straight or curved. The outer walls are

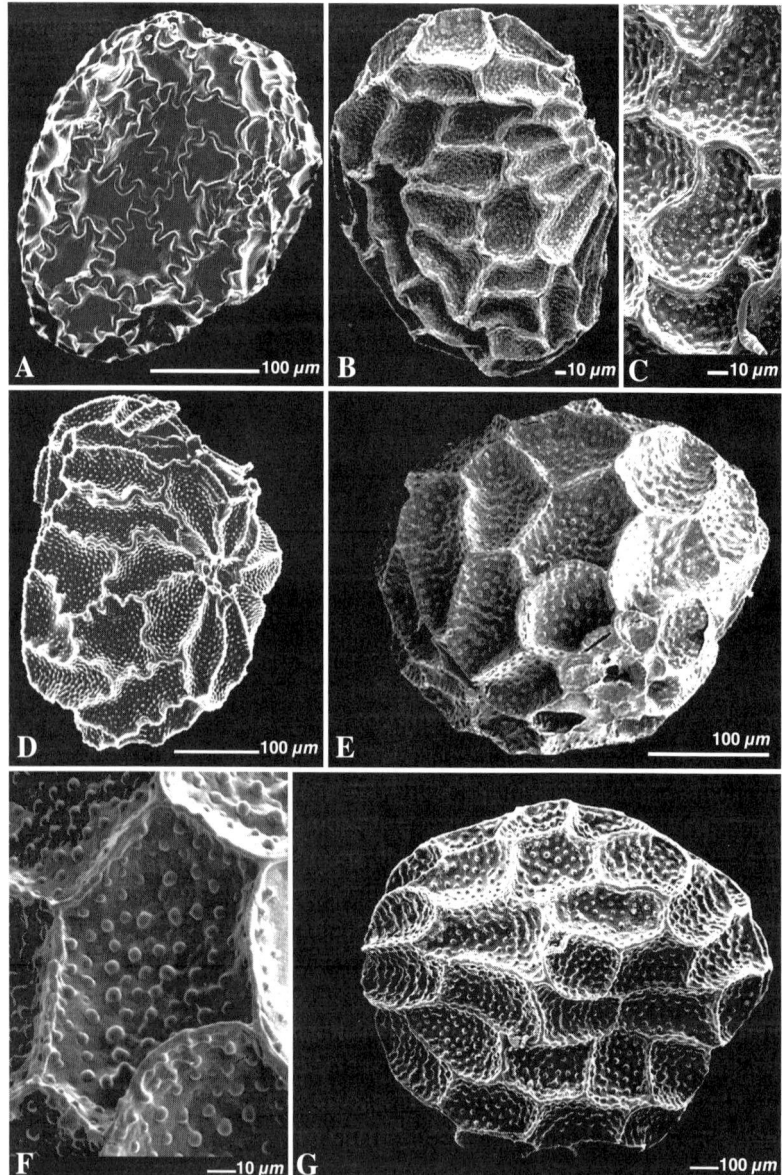

Figure 5.5. Scanning electron micrographs of seeds of Chironieae-Chironiinae. **A.** *Geniostemon schaffneri* (*Pringle 4938*). **B–C.** *Sabatia angularis* (*Frye s.n.*), seed and detail of testa. **D.** *Centaurium erythraea* (*Delvosalle s.n.*). **E–F.** *Centaurium arizonicum* (*Blumer 1792*), seed and detail of testa. **G.** *Zygostigma australe* (*Smith et al. 9375*).

thin, collapsed, and reflect the papillate thickenings of the underlying wall. The inner wall has small pits, with the anticlinal walls having a coarser reticulum. In *Sabatia brevifolia* the testa cells are shallower and the anticlinal walls are less distinct and show fewer, but larger, papillate thickenings per cell. The cuticle has sparse granular ornamentations, which are more profuse in *Sabatia dodecandra*.

Centaurium (Fig. 5.5D–F)

Seeds of *Centaurium* resemble those of *Sabatia* species, although the testa cells are mostly somewhat deeper and the reticulate pattern is more distinct. The seeds are globose to elliptic. Mean seed size varies from c. 335×265 μm in *Centaurium arizonicum*, 405×335 μm in *C. linariifolium*, to 420×365 μm in *C. littorale*. Testa cells are irregularly polygonal. The anticlinals are thin, straight, curved, or V-undulating. The anticlinal walls have larger meshes and the wall thickenings are continuous with the reticulum of the inner wall. The outer wall collapses at maturity and reflects papillae, and, when more closely appressed, the finely reticulately-thickened inner wall has papillae on the corners of the reticulum and pits between the reticulum. The cuticle is smooth.

Zygostigma (Fig. 5.5G)

The seeds of *Zygostigma australe* are globose to elliptic or somewhat irregular in shape, and are about 415×350 μm in size. Testa cells are polygonal, sometimes elongated and oriented irregularly. The outer wall is collapsed and shows the underlying papillae. The anticlinal walls are thin, straight, or slightly curved. The inner wall and anticlinal walls are pitted with papillae. The seeds closely resemble those of *Centaurium*.

Cicendia (Fig. 5.6A–B)

The seeds of *Cicendia filiformis* are broadly elliptic, c. 365×245 μm in size. The hilum is subapical. The testa is reticulate. The testa cells are irregular in shape and are not arranged in distinct rows. The anticlinal walls are raised, very thin, and are straight to curved. The inner wall has fine band-like thickenings running perpendicular to the length of the cell. The band-like thickenings are sometimes bifurcating toward the anticlinal walls or become faint with bead-like thickenings in the center of the wall. The cuticle is smooth.

Blackstonia (Fig. 5.6C–E)

The seeds are elliptic to broadly elliptic and vary in mean size from c. 350 ×260 μm in *Blackstonia intermedia* to 530×305 μm in *B. grandifolia*. The

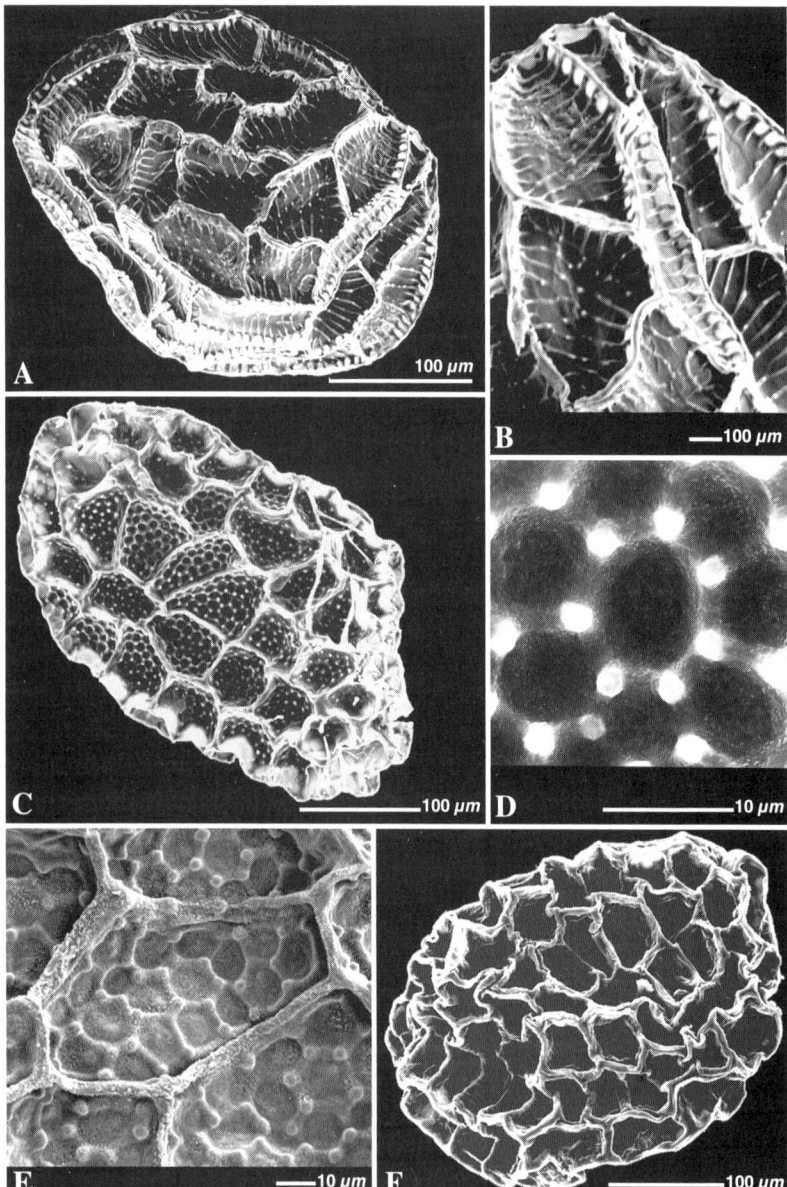

Figure 5.6. Scanning electron micrographs of seeds of Chironieae-Chironiinae. **A–B.** *Cicendia filiformis* (*Hubbeling s.n.*), seed and detail of testa. **C–D.** *Blackstonia perfoliata* (*Geerinck-Coutrez 7103*), seed and detail of periclinal wall. **E.** *Blackstonia intermedia* (*Guzzino 3547*), detail of testa. **F.** *Ixanthus viscosus* (*Bouharmont 25535*).

testa is reticulate. Testa cells are polygonal, isodiametric, or more rectangular, and are arranged in more or less longitudinal rows. The anticlinal walls are straight, membranous, and foraminately pitted. The inner wall has reticulate thickenings with polygonal pits and small papillae on the corners of the reticulum. The cuticle is smooth.

Ixanthus (Fig. 5.6F)

Seeds of *Ixanthus viscosus* are elliptic in shape and measure c. 335–430 × 200–295 μm in mean size. Testa cells are polygonal and irregularly oriented. The anticlinal walls are prominent and straight. The outer wall is thin and collapsed. The inner wall has reticulate thickenings. The cuticle is smooth. The seeds observed were not fully mature.

Subtribe Canscorinae

Canscora (Fig. 5.7A–B)

Seeds of *Canscora* are irregular in shape, angular, often cubical to rectangular and with shallowly sunken sides. Seeds of *Canscora diffusa* are about 300 × 230 μm and those of *C. heteroclita* 300 × 260 μm. The testal pattern is faintly reticulate with shallower cells. Testa cells are irregularly polygonal and oriented irregularly. In *Canscora heteroclita* the testa cells are elongated near the ridges to smaller and isodiametric at the sunken sides. The anticlinal walls are straight or curved. The outer wall is thin and somewhat tightened between the walls at the ridges, but mostly collapsed at the sides of the seed. The inner wall has a granular structure. The cuticle is smooth.

Schinziella (Fig. 5.7C–D)

Seeds of *Schinziella tetragona* are mostly cubical to rectangular, sometimes globose in shape. Seeds measure c. 370–465 × 260–420 μm. The testa is reticulate. Testa cells are polygonal, irregularly oriented, and locally differing in size and shape. The anticlinal walls are straight, prominent, and thick. The anticlinal boundary is sunken and may locally split. The outer walls are concave or convex, thin, and somewhat tightened between the walls. The cuticle is smooth. Seeds resemble those of *Canscora*.

Hoppea (Fig. 5.7E)

The seeds are elliptic, 230–290 × 140–210 μm in size in *Hoppea dichotoma*. The testa is reticulate, about six cells high. Testa cells are somewhat star-shaped because of the undulating anticlinal walls. The anticlinal walls are

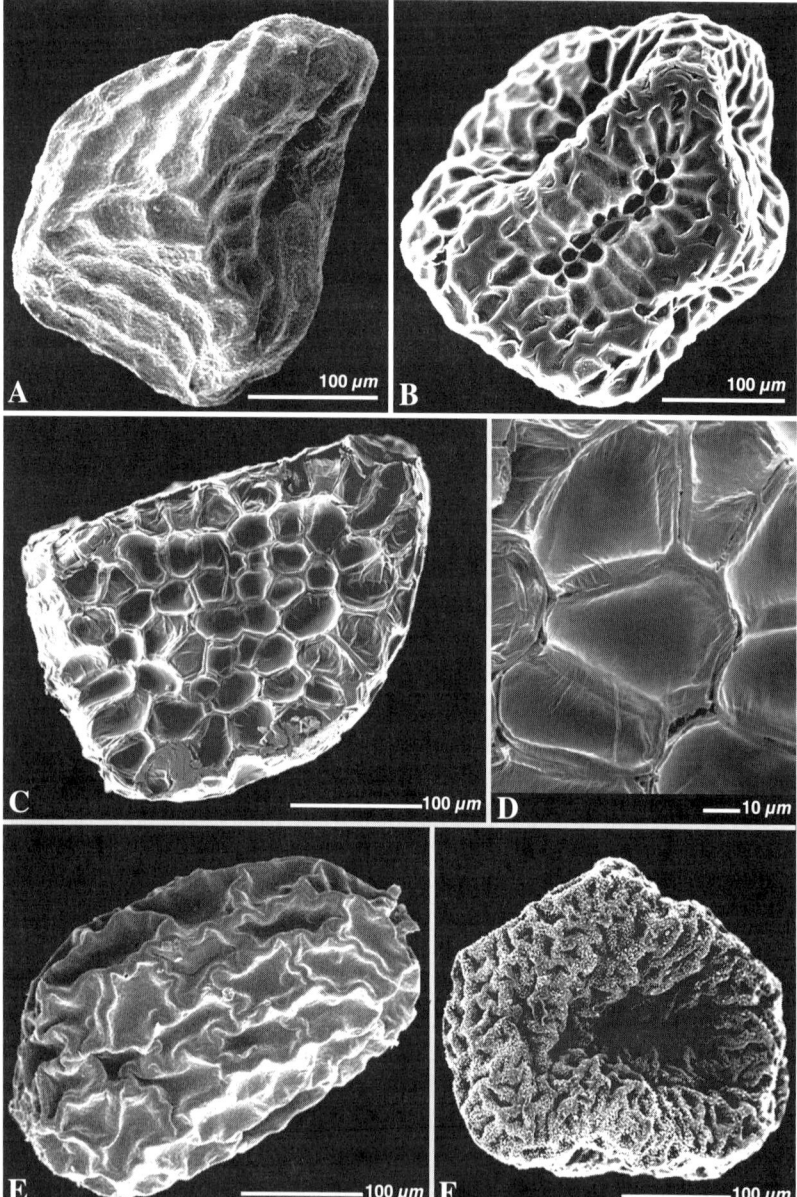

Figure 5.7. Scanning electron micrographs of seeds of Chironieae-Canscorinae. **A.** *Canscora diffusa* (*de Wilde et al. 4731*). **B.** *Canscora heteroclita* (*Robyns 7206*). **C–D.** *Schinziella tetragona* (*Carlier 349*), seed and detail of testa. **E.** *Hoppea dichotoma* (*Adam 26711*). **F.** *Microrphium pubescens* (*van Balgooy 2217*).

very thick. The outer wall is thin, collapsed, and does not reflect special structures. The anticlinal boundary and the cuticle are smooth.

Microrphium (Fig. 5.7F)

Seeds of *Microrphium pubescens* are irregular, globose to ovate in shape, and measure c. 350×270 μm. Testa cells are irregular in shape and orientation. The anticlinal walls are straight, prominent, and thick. The outer walls are concave and thin. The cuticular surface has scaly or scurfy exudates.

Subtribe Coutoubeinae

Coutoubea (Fig. 5.8A–D)

The seeds of *Coutoubea* are globose to almost globose and measure 450–580 × 350–500 μm in *Coutoubea humilis*, 435–630 × 425–535 μm in *C. reflexa*, 375–435 × 310–385 μm in *C. ramosa*, and 275–355 × 150–295 μm in *C. spicata*. The testa is reticulate, about 8–10 cells high. Testa cells are polygonal in surface view, not arranged in distinct rows, and become smaller toward the hilum/micropyle. The anticlinal walls are straight, distinct, and with bead-like thickenings as in *Coutoubea ramosa*, or undulating as in *C. spicata*. The anticlinal boundaries are flat, but may split in some seed samples. The outer wall is thin, but does not reflect the structure of the inner wall. The inner wall is reticulate with rounded meshes. The cuticle has distinct exudates of various structure, locally interspersed with smooth patches.

Deianira (Fig. 5.8E)

The seeds are globose to subglobose, about 430–480 μm in diameter in *Deianira pallescens*, elliptic, to sometimes angular as in *D. erubescens*. The testa is reticulate, the pattern and cell shape are somewhat less regular than in *Coutoubea*. The anticlinal walls are straight, or slightly curved with bead-like thickenings. The cuticle has distinct flaky exudates. Seeds of *Deianira* resemble those of *Coutoubea*.

Symphyllophyton

Seeds of *Symphyllophyton* cf. *caprifolioides* resemble those of *Coutoubea* and measure about 500–570 μm. The anticlinal walls are straight. The seeds observed were not fully mature.

Schultesia (Fig. 5.8F–G)

The seeds are variable in shape and in testal structure. They are elliptic, sometimes almost globose, often somewhat angular and/or pointed toward

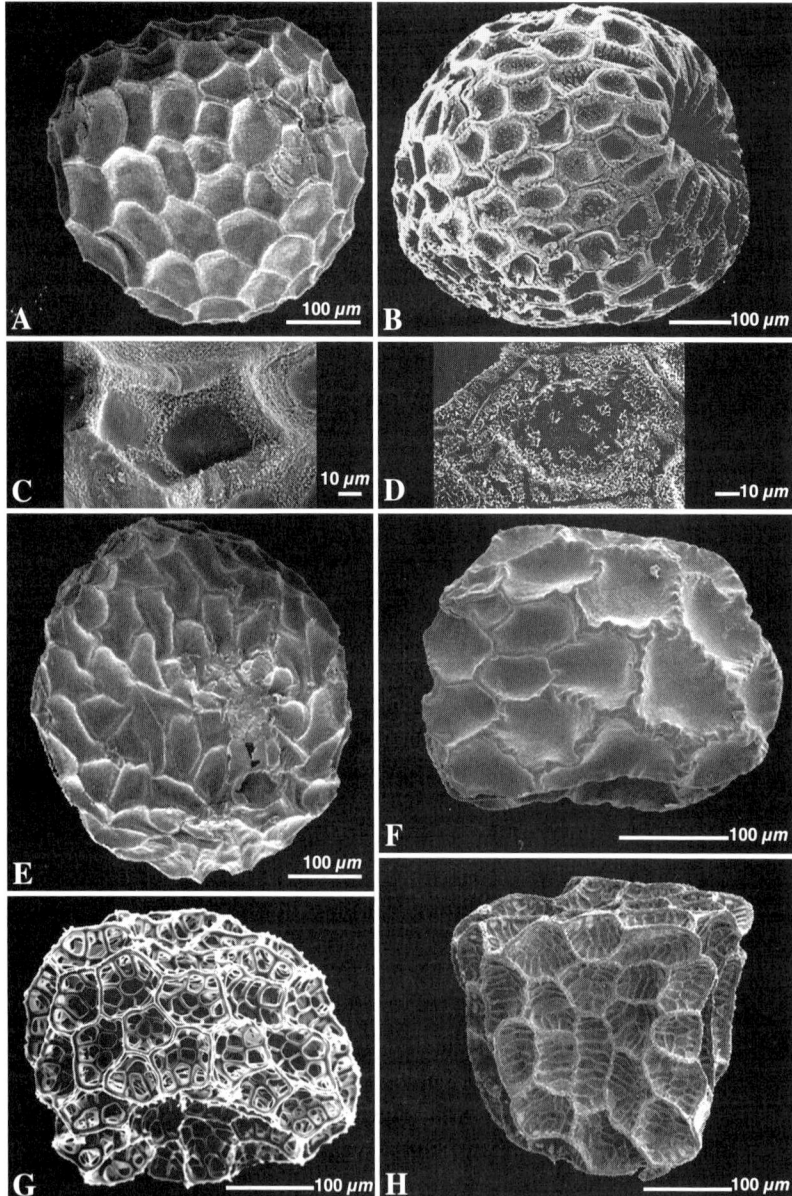

Figure 5.8. Scanning electron micrographs of seeds of Chironieae-Coutoubeinae. **A.** *Coutoubea ramosa* (*Croat 20178*). **B.** *Coutoubea humilis* (*Maguire & Fanshawe 32592*). **C.** *Coutoubea ramosa* (*Croat 20178*), detail of testa with exudates. **D.** *Coutoubea humilis* (*Maguire & Fanshawe 32592*), detail of testa with exudates. **E.** *Deianira pallescens* (*Irwin & Soderstrom 5920*). **F.** *Schultesia guianensis* (*Berg BG661*). **G.** *Schultesia crenuliflora* (*de Carvalho et al. 1068*), seed after removal of the outer testal walls. **H.** *Xestaea lisianthoides* (*Stevens 7103*).

The seeds of Gentianaceae 523

the micropyle. Seeds of *Schultesia stenophylla* var. *latifolia* measure c. 285 ×205 μm, those of *S. australis* 360–455×235–440 μm. The testa is distinctly reticulate. Testa cells are irregularly polygonal. The anticlinal walls are curved, straight as in *Schultesia crenuliflora*, or undulating as in *S. australis* and *S. guianensis*. The outer wall is thin, collapsed, and may reflect the structure of the inner wall. The inner wall is almost smooth in *Schultesia guianensis*, pitted in *S. stenophylla* var. *latifolia*, and reticulately thickened in *S. australis* and *S. crenuliflora*. Seeds of *Schultesia stenophylla* var. *latifolia* and *S. guianensis* resemble those of *Sabatia* species, but have thicker, undulating anticlinal walls and a pitted inner wall.

Xestaea (Fig. 5.8H)

Seeds of *Xestaea lisianthoides* are angular, elliptic to globose in shape, and measure 265–465×215–340 μm. The testa is reticulate. Testa cells are polygonal and distinctly sunken. The anticlinal walls are straight or slightly curved. The outer wall is thin, collapsed, and reflects the thickenings of the reticulate inner and anticlinal walls. The cuticle is smooth. Seeds of *Xestaea* resemble those of some *Schultesia* species.

Tribe Helieae

The *Irlbachia* complex (*Adenolisianthus, Calolisianthus, Celiantha, Chelonanthus, Helia, Irlbachia, Pagaea, Rogersonanthus, Tetrapollinia,* and *Wurdackanthus*)

Seed coat morphology of a number of taxa of Helieae has been studied by Cobb and Maas (1983). According to Maas *et al.* (1983) and contrary to Struwe *et al.* (1997), *Irlbachia sensu lato* should include the genera *Adenolisianthus, Calolisianthus, Chelonanthus, Helia, Pagaea, Rogersonanthus, Tetrapollinia,* and *Wurdackanthus*, as well as *Irlbachia sensu stricto*. This group of genera have also been known as the *Irlbachia* complex (including a few additional genera; Struwe *et al.*, 1997, 2002). The seeds of the above-mentioned genera are similar in general appearance and size, but very variable in shape and micromorphology, and show several overlapping characters between and sometimes within species. The taxa agree in medium-sized seeds, with mean seed sizes of 350–800×200–600 μm. The seeds vary in shape from almost globose to polyhedral or (sub)conical to (sub)cubical with rims, wing-like edges, or sharper ridges. The anticlinal walls of the testa cells are straight, curved, or sinusoidal to V-undulating, so that the testa cells vary accordingly from polygonal to star-shaped. The inner walls are pitted, smooth, or rough to warty with papillae either scattered over the entire surface or restricted to the lateral walls. The inner wall is rarely reticulately banded. The outer wall is always smooth without any cuticular ornamentation.

The testa of *Irlbachia* and its relatives may have different types of testa cells on the ridges and sides. Three types of testa cells may be discerned:

1 **Concave cells** with thin, collapsed outer walls, which are devoid of thickenings.
2 **Dome-like cells** with reticulate or band-like thickenings on the convex outer walls. The cells are filled with air.
3 **Collapsed domes**, cells with radiating band-like thickenings on their outer walls, collapsed in the center and often looking like collapsed domes (e.g., *Irlbachia plantaginifolia*, *Lehmanniella*, and *Neblinantha*). The cells are more or less intermediate between concave and dome-like cells.

The dome-like cells and collapsed domes seem to be unique within the Gentianaceae-Helieae.

The presence of concave or convex cells on the seed surface determines the shape of the seed. Roughly speaking, seeds of Helieae are polyhedral to angular and subconical to subcubical in shape when the testa is composed of concave cells only. Convex cells on the ridges form rims and round off the seeds, whereas seeds with convex cells on both the sides and the ridges tend to be more globose. Three seed types may be discerned in the *Irlbachia* complex (Figs. 5.9–5.10).

1 **Polyhedral or angular** seeds with only concave cells have been found in, for example, *Chelonanthus alatus*, *Irlbachia pratensis*, *I. tatei*, and *Tetrapollinia caerulescens*. The sides of the seeds may be sunken, making ridges more prominent and sometimes almost wing-like (as in specimens of *Irlbachia nemorosa*).
2 **Globose to almost globose seeds** with only dome-like cells are known in *Irlbachia poeppigii* and *I. pumila*. However, intermediate forms occur, viz. rimmed seeds with only dome-like cells and flat to shallow sunken sides in *Helia oblongifolia*, *Calolisianthus pedunculatus*, *C. pendulus*, and *C. speciosus*.
3 **Rimmed seeds** with concave cells on the sides and dome-like cells or collapsed domes on the rims are seen in *Celiantha imthurniana*, *Irlbachia plantaginifolia*, *Lehmanniella*, and *Rogersonanthus arboreus*.

The seeds of several species are heteromorphic in structure. Different specimens of one species may have seeds with dome-like, concave, or both dome-like and concave testa cells, as in *Chelonanthus purpurascens*, or a combination of two seed types as in *Calolisianthus amplissimus, C. angustifolia, C. pulcherrimus*, and *Irlbachia nemorosa*.

Seeds of *Adenolisianthus arboreus* are rimmed with shallow sunken sides. The testa has dome-like cells with band-like thickenings. The anticlinal walls are sinusoidal to V-undulating.

Seeds of *Calolisianthus* spp. (Fig. 5.9D) are angular, somewhat rectangular to cubic, and rimmed with shallow sunken sides. Seeds measure 330–500 ×215–400 μm. Seeds of *Calolisianthus pedunculatus, C. pendulus, C. pulcherrimus*, and *C. speciosus* have only dome-like cells. The inner walls are smooth, or rough to warty.

Seeds of *Celiantha* (Fig. 5.9F) are irregular in shape, globose to angular, subcubical to subconical and rimmed (those of *C. bella* are less rimmed). Seeds of *Celiantha imthurniana* are 290–400×360–450 μm, and those of *C. bella* 240–470×150–265 μm in size. Testa cells are concave, polygonal, and isodiametric in surface view. Dome-like cells with band-like thickenings are present only on the rims. The anticlinal walls are straight or curved. The inner wall is pitted and sometimes looks reticulate. The pits are 15–20 per cell and 2.0–3.5 μm wide. The seeds of *Celiantha* resemble those of *Irlbachia*.

Seeds of *Chelonanthus* spp. (Fig. 5.9E) are angular to rimmed and have sunken sides. Seeds are 395–550×370–490 μm in mean size. The testa is composed of dome-like cells and/or concave cells with band-like thickenings. The seeds of *Chelonanthus alatus, C. albus*, and *C. angustifolius* have both dome-like and concave cells with band-like thickenings. Seeds of *Chelonanthus longistylus* and *C. matogrossensis* have only concave testa cells with band-like thickenings. Seeds of *Chelonanthus purpurascens* and *C. viridiflorus* have dome-like cells and/or concave cells with band-like thickenings. The lateral walls are straight, sinusoidal, or V-undulating.

The seeds of *Helia oblongifolia* (Fig. 5.9A–B) are subglobose to slightly rimmed and measure 395–525×325–425 μm. Testa cells are polygonal and isodiametric in surface view. All testa cells are dome-like with convex outer walls. Those at the rims are somewhat more radially elongated. The outer wall of the domes has a coarse reticulum with polygonal to rounded meshes. The anticlinal walls are straight, also at their base. The inner walls are smooth.

The seeds of species of *Irlbachia* (Fig. 5.9C) may belong to one of the three seed types. Seeds of *Irlbachia poeppigii* and *I. pumila* have testas with only dome-like cells, those of *I. cardonae, I. pratensis*, and *I. tatei* have only concave cells, and the seeds of *I. plantaginifolia* have collapsed domes with

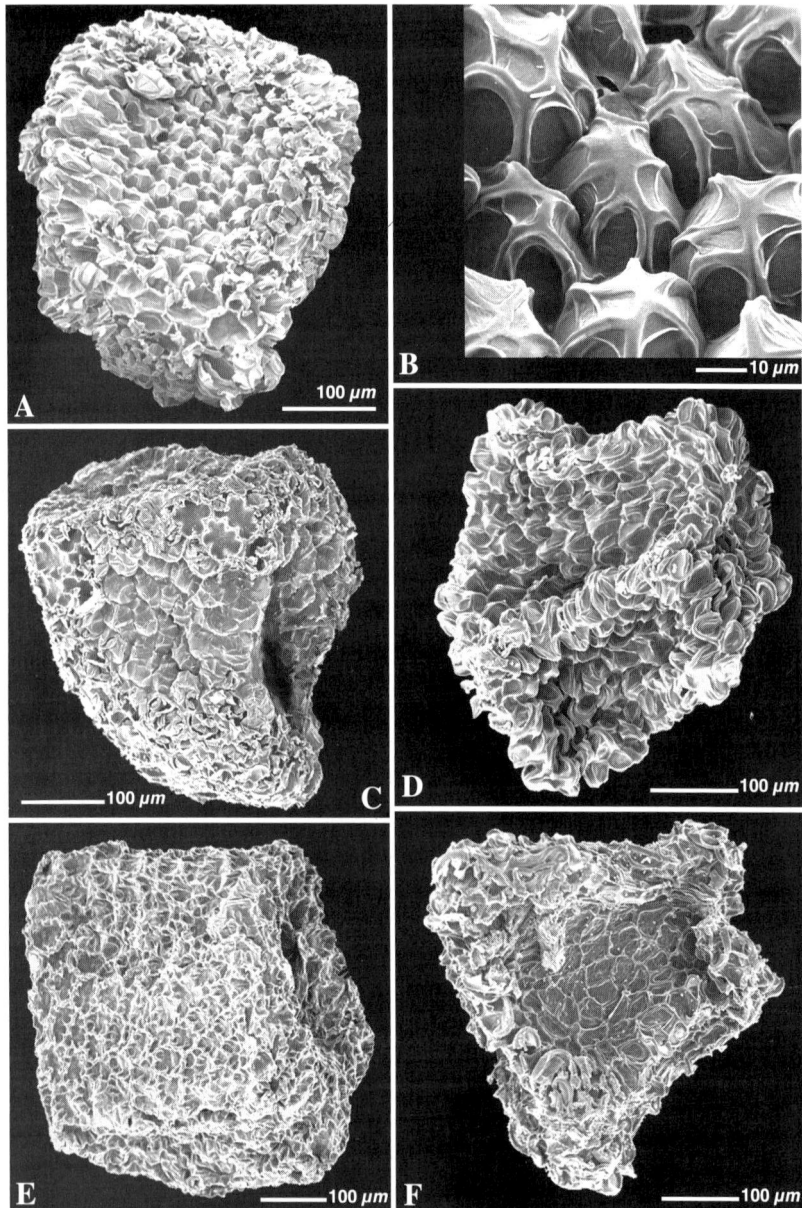

Figure 5.9. Scanning electron micrographs of seeds of Helieae. **A–B.** *Helia oblongifolia* (*Pedersen 12216*), seed and detail of testa. **C.** *Irlbachia poeppigii* (*Schomburgk 789*). **D.** *Calolisianthus speciosus* (*Mimura 395*). **E.** *Chelonanthus alatus* (*Martius s.n.*). **F.** *Celiantha imthurniana* (*im Thurn 306*).

The seeds of Gentianaceae 527

band-like thickenings tapering toward the center. The seeds of *Irlbachia nemorosa* have dome-like cells on the rims and concave cells on the sides. The testa cells are polygonal or puzzle piece-like because of sinusoidal anticlinal walls. *Irlbachia* seeds differ in the structure of the inner exotestal walls. *Irlbachia pratensis* has smooth walls, *I. poeppigii* and *I. pumila* have pitted inner walls, *I. nemorosa* has papillate walls, while in *I. plantaginifolia* the inner wall has both pits and papillae. *Irlbachia nemorosa* has two types of seeds. There is a tendency for seeds from high-altitude plants to have only concave testa cells and less pronounced papillae, while those from lowland plants often have both concave and dome-like cells with conical, V-shaped, or three-pronged papillae.

The seeds of *Rogersonanthus* (Fig. 5.10B–C) are angular, subcubical to subconical, and rimmed. They measure 410–545×325–545 μm. The testa of *Rogersonanthus arboreus* and *R. tepuiensis* has concave cells on the sides and domes, or collapsed dome-like cells at the ridges forming the rims. The rims are somewhat less pronounced in the specimens of *Rogersonanthus quelchii*. The concave testa cells are polygonal in shape, with inward-radiating wall thickenings, or devoid of wall thickenings. The general characters of the seeds agree with those of the Helieae.

Seeds of *Tetrapollinia caerulescens* (Fig. 5.10A) are polyhedral to angular. They are 300–410×250–350 μm in size. The testa consists only of concave cells. The outer wall is thin, collapsed, and often with "stretch marks" between the thickened angles of the walls. Testa cells are somewhat puzzle piece-like because of the sinusoidal to V-undulating anticlinal walls. The anticlinal walls have short papillae confined to the angles of the cell. The inner wall is warty and pitted.

The seeds of *Wurdackanthus frigidus* (Fig. 5.10D) are subglobose to subcubical in shape, rimmed, and sometimes with shallow sunken sides. The seeds measure 640–825×330–540 μm. The testa is reticulate. The rims have elongated, dome-like cells. The outer wall is reticulate with rounded meshes. Concave cells are small, polygonal, and have collapsed outer walls. The dome-like cells at the rims are more strongly radially elongated. The convex wall is reticulate with rounded meshes. The seeds resemble those of *Symbolanthus* in the elongated domes.

Neblinantha (Fig. 5.10E)

The seeds are irregular in shape, subcubical to subconical with distinct rims and sunken sides. Seeds of *Neblinantha neblinae* are relatively large, 655–725 ×400–590 μm in size. Testa cells are mostly dome-like with band-like thickenings on the rims, or collapsed domes with inward-radiating bands on the

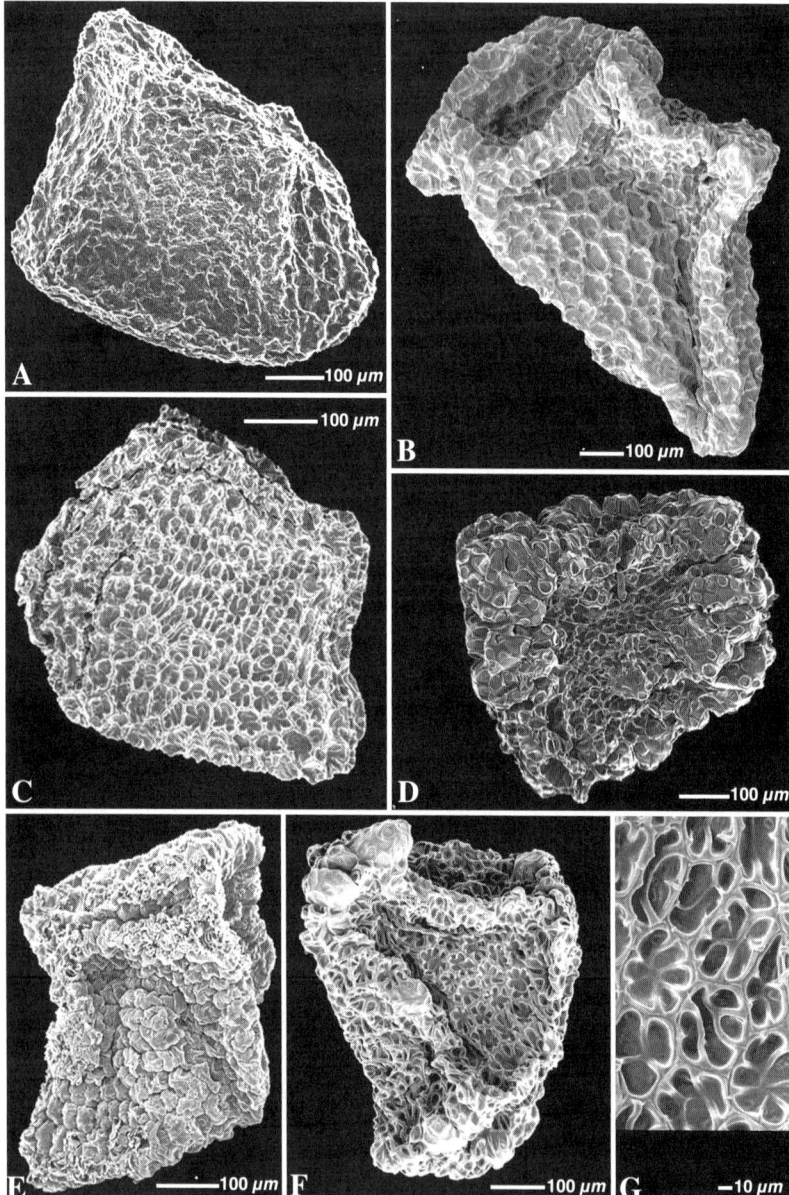

Figure 5.10. Scanning electron micrographs of seeds of Helieae. **A.** *Tetrapollinia caerulescens* (*Duarte 2558*). **B.** *Rogersonanthus tepuiensis* (*Steyermark & Wurdack 507*). **C.** *Rogersonanthus arboreus* (*Broadway s.n.*). **D.** *Wurdackanthus frigidus* (*Smith & Smith K18*). **E.** *Neblinantha neblinae* (*Plowman & Thomas 13645*). **F–G.** *Aripuana cullmaniorum*, seed (*Ramos et al. s.n.*) and detail of testa (*Cid Ferreira 5906*).

sides of the seed. Testa cells are polygonal with straight anticlinal walls. The cuticle is smooth. Seeds of *Neblinantha* resemble those of the *Irlbachia* complex. All seeds observed were dusted by fungi.

Aripuana (Fig. 5.10F–G)

Seeds of *Aripuana cullmaniorum* have been described by Struwe et al. (1997). The seeds are irregular in shape, often subconical or subcubical, angular, mostly rimmed, and 420–525×350–525 μm in size. The testa is reticulate. Testa cells are polygonal and of two types. Those on the rims are enlarged, dome-like, and have their outer walls thickened by a reticulum with rounded meshes or by band-like thickenings. Those on the lateral sides of the seed are concave and have band-like thickenings on the outer wall, looking like collapsed domes. The band-like thickenings often radiate toward the center. The anticlinal walls are straight, the anticlinal boundaries sunken. The inner wall has few pits, which are bordered by a rim. The cuticle is smooth. Seeds of *Aripuana* resemble those of *Rogersonanthus* because of the rims with dome-like cells with a reticulum of rounded meshes and concave cells with radiating band-like thickenings on the sides.

Lehmanniella (Fig. 5.11A–B)

The seeds of *Lehmanniella splendens* are irregular, subcubical to subconical in shape. Seeds have distinct rims and sunken sides, and measure 600–700×450–680 μm. The rims are formed by radially elongated, collapsed dome-like cells, which have about 10 bands radiating to the center with tiny perpendicular branches. The cells become shallower and increasingly collapsed toward the sides of the seed. Testa cells are polygonal to circular, collapsed domes. Their outer walls have band-like thickenings, radiating and tapering toward the center. Anticlinal walls are curved, and more undulating toward their base. The inner wall is rough to warty.

Seeds of the monotypic genera *Lagenanthus* and *Purdieanthus* resemble those of *Lehmanniella* in size, shape, and micromorphology. Seeds of *Lagenanthus princeps* are 620–710×580–730 μm in size. The outer walls of the collapsed domes have about 20, often branching, band-like thickenings. The seeds of *Purdieanthus pulcher* are 640–740×420–460 μm in size. The collapsed domes have about five, branched, band-like thickenings.

Sipapoantha (Fig. 5.11C)

The seeds are irregular in shape, subcubical to subconical, and with distinct rims and sunken sides. The sides are sometimes deeply sunken. Seeds of *Sipapoantha ostrina* are 515–825×555–775 μm in size. Testa cells are

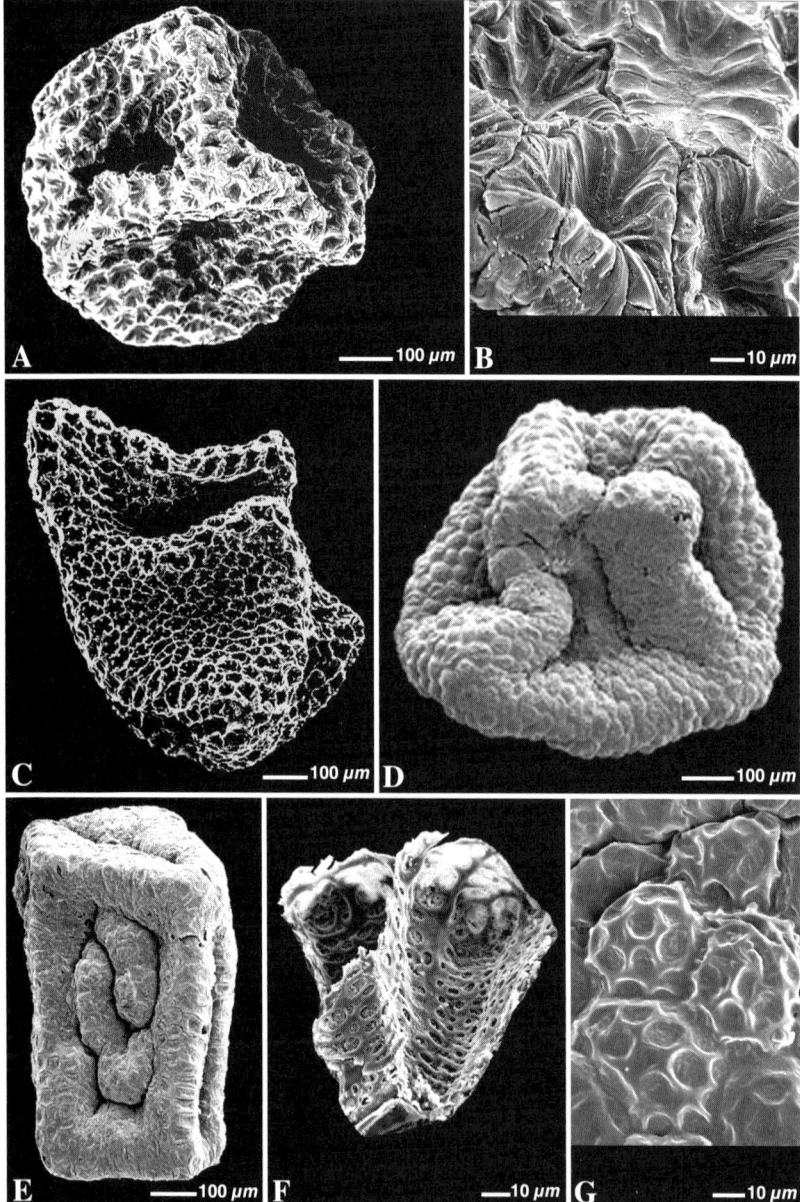

Figure 5.11. Scanning electron micrographs of seeds of Helieae. A–B. *Lehmanniella splendens* (*St. John 2056*), seed and detail of testa. C. *Sipapoantha ostrina* (*Steyermark et al. 124554*). D. "Roraimaea", ined. (*Cid Ferreira 9125*). E–F. *Symbolanthus calygonus*, seed (*Fosberg 19111*) and section of rim cells (*Maas et al. 5284*). G. "Roraimaea", ined. (*Cid Ferreira 9125*), detail of testa.

The seeds of Gentianaceae 531

polygonal and dome-shaped or concave, with thin, ring-like thickenings on the outer walls. The anticlinal walls have papillate thickenings. The inner wall is reticulate. The cuticle is smooth. Seeds of *Sipapoantha* resemble those of the *Irlbachia* complex, but differ in the ring-like thickenings of the testa cells.

"Roraimaea" (Fig. 5.11D,G)

Seeds of "Roraimaea", ined. (L. Struwe, S. Nilsson, & V. A. Albert, unpubl.) are somewhat irregular in shape with strongly swollen rims and sunken areas. Seeds measure 560–810 × 505–640 μm. Testa cells are polygonal, with those at the rims more radially elongated. All testa cells have convex outer walls. The outer wall is thickened by a foraminate reticulum with rounded meshes. The cuticle is smooth.

Symbolanthus (Fig. 5.11E–F)

Seeds of *Symbolanthus calygonus* are subcubical to subconical, and brain-like because of the broad, swollen, irregular rims and furrows. The seeds measure 700–1250 × 630–915 μm. Testa cells are polygonal, more rectangular at the rims. All testa cells are dome-like and radially stretched at the rims. The anticlinal walls have reticulate thickenings. The outer wall has band-like thickenings with rounded meshes or with bands radiating inwards. The anticlinal walls are straight to curved and undulating near their bases. The inner wall is pitted. The cuticle is smooth.

Macrocarpaea (Fig. 5.12A–C)

Seeds of *Macrocarpaea* show great variation in shape. Three main types may be discerned.

1 **Rimmed seeds** with concave cells at the sides and dome-like cells at the ridges forming rims are known in *Macrocarpaea ovalis* and *M. rugosa*. Seeds of *Macrocarpaea ovalis* are c. 560 × 360 μm. The concave testa cells are polygonal with inward-radiating wall thickenings. The seeds conform to the *Irlbachia* complex-type. Seeds of *Macrocarpaea glabra* and *M. luteynii* resemble the former ones, but deviate in having larger seeds with more pronounced rims. Their seeds are about 805 × 475 μm in size. Seeds of various collections of *Macrocarpaea glabra* differ in shape and in the elongation of the rim cells; those of *Maas & Plowman 1794* do not exceed 100 μm. Rim cells of *Macrocarpaea luteynii* (*Luteyn 7466*) are 80–250 μm long and form a kind of wing.

2 **Winged seeds**. These have radially elongated dome-like cells at the rims, forming the wing. The cells of the wing are up to 700 μm. The walls of

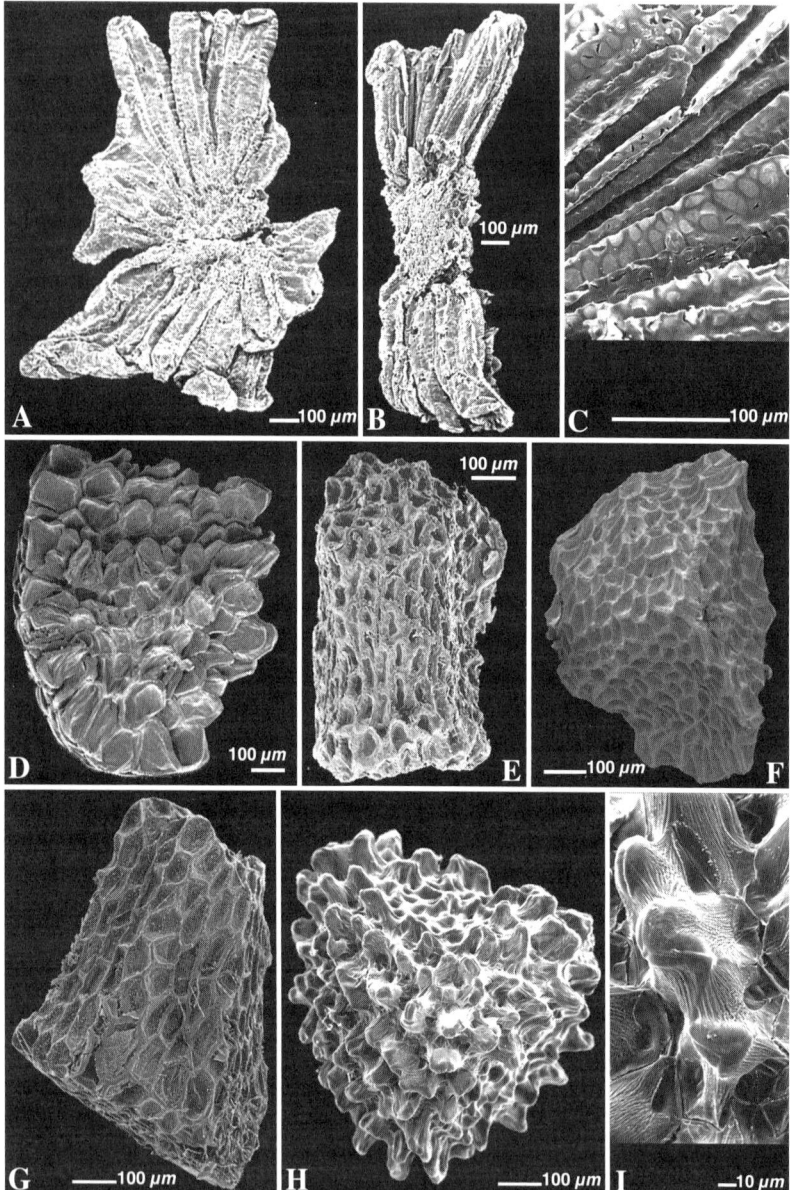

Figure 5.12. Scanning electron micrographs of seeds of Helieae. **A.** *Macrocarpaea corymbosa* (*Ruiz & Pavón s.n.*). **B–C.** *Macrocarpaea cinchonifolia* (*Bang s.n.*), seed and detail of wing. **D.** *Chorisepalum rotundifolium* (*MAC s.n.*). **E.** *Prepusa montana* (*Martius 2108*). **F.** *Senaea caerulea* (*Santos Lima & Brade 14215*). **G.** *Zonanthus cubensis* (*Wright 1346*). **H–I.** *Tachia gracilis* (*Tillett 752275*), seed and detail of testa.

the wing are reticulately to foraminately thickened. The wing is interrupted at the hilum/micropyle (*M. corymbosa*) and sometimes also interrupted at the chalaza, and then has two lateral wings (*M. cinchonifolia*). Seeds are about 1600×380 μm.

3 **Flattened seeds**. Seeds of *Macrocarpaea rubra* and *M. glaziovii* are irregular in shape, with extended chalazal and micropylar ends, and flattened, probably at the lateral sides. Seeds are $940–1290 \times 440–565$ μm and mostly have one more straight or concave side and one convex side. The testa is reticulate. Testa cells are polygonal, mostly elongated and more irregularly rectangular. Anticlinal walls and inner walls are finely reticulate to parallelly thickened. The outer wall is very thin, collapsed, and overlies the inner wall, or is tightened and then sometimes ripping. Anticlinal boundaries are raised. The seeds resemble those of *Senaea*.

Chorisepalum (Fig. 5.12D)

The seeds are laterally flattened, rimmed, and more or less triangular to elliptic in lateral view, mostly with the longer side convex and the other two sides straight or more irregular. Seeds of *Chorisepalum rotundifolium* are $1000–1540 \times 690–800$ μm, and those of *C. ovatum* $920–1055 \times 845–855$ μm in size. The rims have enlarged, papillate, air-filled, swollen cells. The testa is reticulate, with the cells polygonal to roundish and oriented irregularly. Testa cells at the lateral sides of the seed have thin outer walls and are collapsed; those at the rims are much larger, air-filled, and have non-collapsed outer walls with thin, branched, band-like thickenings. The inner wall is reticulate with few, large, rounded meshes. Seeds of *Chorisepalum* resemble those of *Macrocarpaea glabra*.

Prepusa (Fig. 5.12E)

The seeds vary in shape. Seeds are angular, triangular, rectangular to cubical, with ridges in *Prepusa alata*, *P. montana*, and *P. hookeriana*, and somewhat rimmed in *P. viridifolia*. The seeds are $500–790 \times 380–500$ μm in size. The testa is reticulate. Testa cells are polygonal to elliptic in surface view. The anticlinal walls are straight to curved. The anticlinal boundaries are raised. The outer wall is thin, collapsed, and does not reflect any underlying structure. The inner wall is reticulate with two to five large meshes of $10–20$ μm in diameter. In species with rims, testa cells are convex at the rims. The cuticle is smooth, granular in *Prepusa viridifolia*. Most seeds of *Prepusa* resemble *Senaea* seeds, but those of *P. viridiflora* resemble *Helia oblongifolia* and some *Irlbachia* species.

Senaea (Fig. 5.12F)

Seeds of *Senaea caerulea* are irregularly polygonal in shape, angular, slightly flattened and 570–715×385–520 μm in size. Seeds are often somewhat pointed at the chalaza. The testa is reticulate. Testa cells are polygonal or somewhat elongated, especially near the margins of the seed. All testa cells are concave with collapsed outer walls. The anticlinal walls are thickened, straight or curved. The anticlinal boundaries are raised. The inner wall is reticulate with two or three large meshes, or almost smooth. The cuticle is smooth. The testa of *Senaea* resembles that of *Prepusa hookeriana*.

Zonanthus (Fig. 5.12G)

Seeds of *Zonanthus cubensis* are irregularly polygonal, mostly rectangular to somewhat cubic, 670–700×390–670 μm in size. The testa is reticulate. Testa cells are polygonal and irregularly oriented. All testa cells are concave. The outer wall is thin, without thickenings, and collapsed in the mature seed. The anticlinal walls are thickened, straight or curved. The outer wall is thin, collapsed, and does not reflect the structure of the inner wall. The anticlinal boundaries are locally raised. The cuticle is smooth. The testa of *Zonanthus* resembles that of *Senaea* and *Prepusa*.

Tachia (Fig. 5.12H–I)

Seeds are echinate, more or less globose in outline. The anticlinal walls are locally strongly raised at some corners, giving the testa cells a papillate appearance. The seeds of *Tachia guianensis* are about 490–605 μm in diameter, those of *T. grandifolia* and *T. gracilis* about 750–805 μm. The central field of the testa cells is shifted to the side of the "papilla" and has a linear cuticular pattern, interspersed with granular structures. The anticlinal walls are thick and straight. The anticlinal boundaries are straight and raised. The inner wall is reticulate with rounded meshes to foraminate. The cuticle has a granular and linear ornamentation in *Tachia gracilis* and *T. grandifolia* and is almost smooth in *T. guianensis*.

Tribe Potalieae

Subtribe Faroinae

With the exception of those of *Urogentias*, the seeds of the Faroinae are relatively small; mean seed length does not exceed 500 μm. Many taxa have exudates on their outer surface. Seeds of *Urogentias* deviate in size, shape, and micromorphology.

Figure 5.13. Scanning electron micrographs of seeds of Potalieae-Faroinae. **A–B.** *Neurotheca loeselioides* (*vanden Berghen 1264*), seed and detail of testa with exudates. **C–D.** *Congolanthus longidens*, seed (*Coomans s.n.*) and detail of testa with exudates (*Couteaux 60*). **E.** *Pycnosphaera buchananii* (*Jacques-Felix 7416*) seed.

Neurotheca (Fig. 5.13A–B)

The seeds are irregular in shape, angular to triangular, mostly with flattened sides. Seed size is 325–480×230–395 μm in *Neurotheca loeselioides*. The testa is reticulate. Testa cells are polygonal, mostly isodiametric, tapering toward the micropyle. The anticlinal walls are thick and undulating. The anticlinal boundaries are smooth. The outer walls are collapsed. The cuticle has granular to verrucate exudates.

Congolanthus (Fig. 5.13C–D)

Seeds of *Congolanthus longidens* closely resemble those of *Neurotheca* in shape and microstructure. Seeds are triangular to almost globose. The seed size is 405–450×355–450 μm. Testa cells are polygonal. The anticlinal walls are undulating. The cuticle has granular to verrucate exudates, which become smooth near the corners of the cells.

Pycnosphaera (Fig. 5.13E)

Seeds of *Pycnosphaera buchananii* resemble those of *Neurotheca* in general morphology. Seeds are *c.* 330–460×300–380 μm in size. The testa is reticulate, about eight cells high. The anticlinal walls are slightly undulating to almost straight. The cuticle is tuberculate, and the granules fuse near the anticlinal boundaries and locally form smooth patches. The inner wall is pitted.

Djaloniella (Fig. 5.14A)

Seeds of *Djaloniella ypsilostyla* are broadly ellipsoid to elliptic. Seed size is *c.* 395–460×235–375 μm. The testa is reticulate, about eight cells high. Testa cells are polygonal and irregularly oriented. The anticlinal walls are thick and straight to undulating. The cuticle has granular to tuberculate exudates that fuse near the anticlinal boundaries.

Enicostema (Fig. 5.14B)

The seeds are globose to broadly elliptic, *c.* 405–550×435–515 μm in *Enicostema axillare*. The hilum is often indented. The testa is reticulate, about 16 cells high. Testa cells are polygonal and irregularly oriented. The anticlinal walls are straight, slightly undulating, or irregularly curved. The inner wall is reticulate. The cuticle has granular exudates that are fused near the anticlinal boundaries and locally form smooth patches.

Faroa (Fig. 5.14C–D)

Seeds of *Faroa* resemble those of the other taxa of the Faroinae in general characters. Seeds are globose in shape and measure *c.* 315×270 μm in *Faroa salutaris* and *c.* 380–485×350–445 μm in *F. acaulis*. The testa is reticulate. Testa cells are polygonal and irregularly oriented. The anticlinal walls are straight, or V- to U-undulating, prominent, and thickened. The outer wall is collapsed and thin. The cuticle has tuberculate to verrucate exudates in *Faroa acaulis* and a rosette of crystalloids in *F. salutaris*. The exudates fuse near the anticlinals.

The seeds of Gentianaceae 537

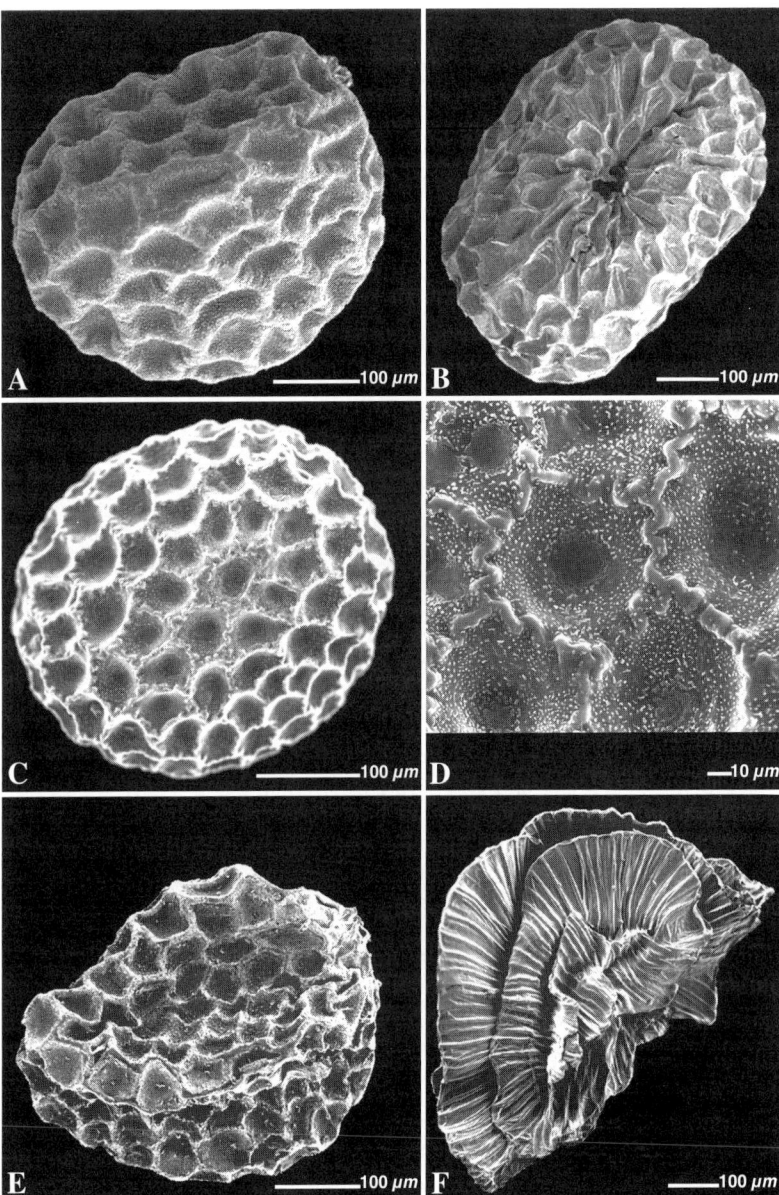

Figure 5.14. Scanning electron micrographs of seeds of Potalieae-Faroinae. A. *Djaloniella ypsilostyla* (*Morton 2442*). B. *Enicostema axillare* (*van Steenis 19537*). C–D. *Faroa acaulis* (*Goethals s.n.*), seed and detail of testa. E. *Karina tayloriana* (*Schmitz 4606*). F. *Urogentias ulugurensis* (*Schlieben 2786*).

Oreonesion

Seeds of *Oreonesion testui* resemble those of the other taxa of the *Faroinae* in general characters. Seeds are mostly triangular in shape and measure *c.* 400–555×350–470 μm. The testa is reticulate. Testa cells are isodiametric and irregularly oriented. The lateral walls are straight to undulating and thick. The outer wall is collapsed and thin. The cuticle is covered by granular to aculeate exudates. The seeds that were studied were not fully mature.

Karina (Fig. 5.14E)

Seeds of *Karina tayloriana* are angular, mostly triangular, or subglobose in shape. Seeds are *c.* 335–450×275–370 μm in size. The testa is reticulate. Testa cells are polygonal, isodiametric, and irregularly oriented. The anticlinal walls are straight, prominent, and thick. The outer wall is collapsed and thin. The cuticle is covered with verrucate exudates.

Urogentias (Fig. 5.14F)

The seeds of *Urogentias ulugurensis* are irregular, triangular to ovate in shape, and show a more bilateral symmetry, reflecting their anatropous nature. The seeds are flat, with several membranous, striate, longitudinal wings. The seeds measure 1085–1285×750–990 μm. Testa cells are polygonal, rectangular, and elongated perpendicular to the length of the seed, and are arranged in longitudinal rows. The anticlinal walls are straight, unequal, thin, and pitted. The longitudinally-running anticlinal walls are all strongly raised and form the wings. The outer wall is collapsed and thin. The inner wall is pitted. The cuticle is smooth.

Subtribe Lisianthiinae

Lisianthius (Fig. 5.15A–C)

Seeds of *Lisianthius* sect. *Lisianthius* are globose to ellipsoid, and sometimes pointed toward the micropyle. The seeds vary in size from *c.* 415× 335 μm in *Lisianthius skinneri*, 525×515 μm in *L. longifolius*, to 730× 545 μm in *L. glandulosus*. The testa is reticulate. The cells are polygonal in surface view. The testa surface is deeply concave and strongly affects the surface of the endosperm in, for example, *Lisianthius adamsii* and *L. glandulosus*, to more shallowly in *L. exsertus*. The anticlinal walls are sometimes raised at the corners, thereby making the seeds somewhat echinate. The outer walls are thin, collapsed, and reflect numerous granular or papillate cell wall thickenings of various length on the inner and anticlinal

The seeds of Gentianaceae 539

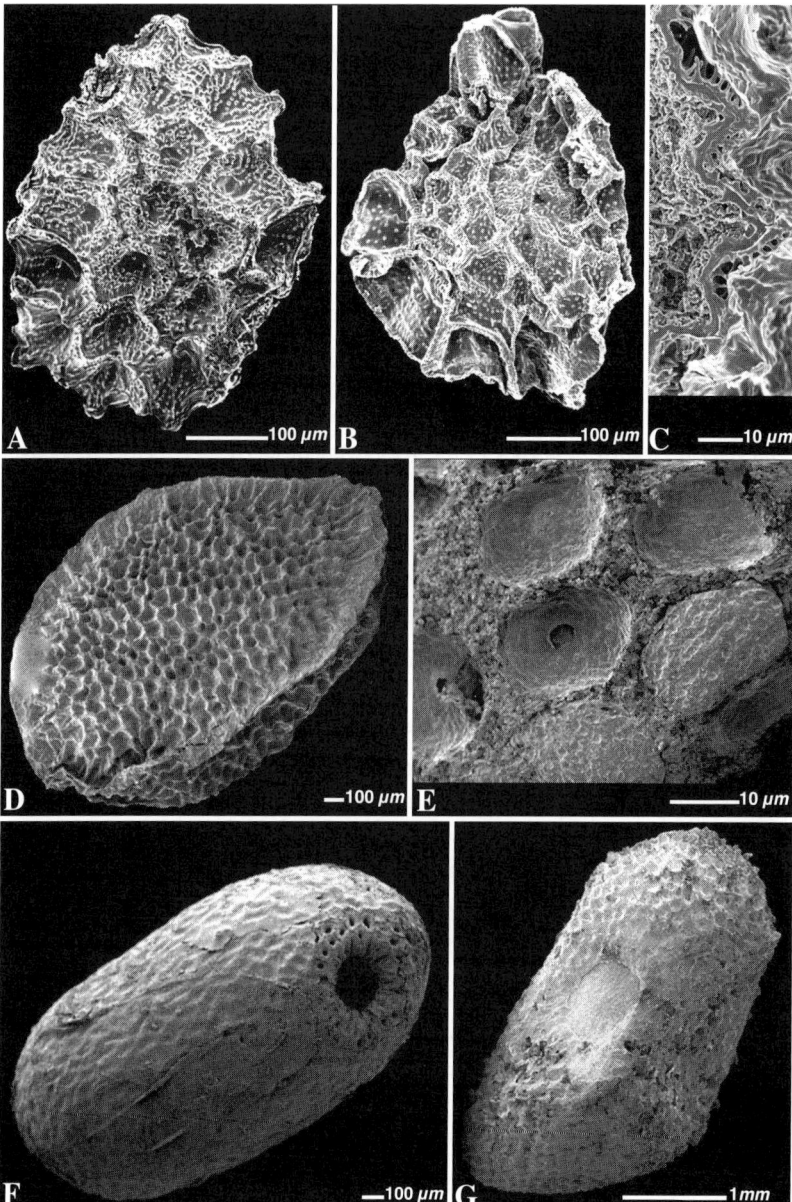

Figure 5.15. Scanning electron micrographs of seeds of Potalieae-Lisianthiinae and Potalieae-Potaliinae. **A.** *Lisianthius longifolius* (*Hart 1018*). **B.** *Lisianthius meianthus* (*Skutch 1568*). **C.** *Lisianthius adamsii* (*Webster & Wilson 5025*), section of testa. **D–E.** *Anthocleista amplexicaulis* (*Lam & Meeuse 5666*). **F.** *Fagraea obovata* (*Lörzing 14119*). **G.** *Potalia amara* (*Jansen-Jacobs et al. 2567*).

walls. The cuticle is smooth. Seed characters are useful at the genus level. Seeds of *Lisianthius* sect. *Omphalostigma* are somewhat more irregular in shape, and c. 570×350 μm in *L. meianthus*. Testa cells of *Lisianthius meianthus* are locally enlarged and form a kind of rim. The anticlinal walls are reticulately thickened near the anticlinals. The inner walls are papillate.

Subtribe Potaliinae

Anthocleista (Fig. 5.15D–E)

The seeds are obliquely ovoid to orbicular or irregularly polyhedral and laterally flattened. Seeds are c. 2200 μm in length. The hilum of *Anthocleista schweinfurthii* is slightly sunken and attached at the middle of the seed. The testa is reticulate with numerous testa cells. Testa cells are polygonal in *Anthocleista amplexicaulis*, but more circular in *A. schweinfurthii*. The anticlinal walls are thick, straight to curved. Testa cells have collapsed outer walls. The cuticle is smooth.

Fagraea (Fig. 5.15F)

Seeds of *Fagraea obovata* are irregularly elliptic to reniform in shape, c. 2270×1170 μm in size. The hilum is sunken and situated subapically at the concave side of the seed. The testa is faintly reticulate. Testa cells are polygonal in surface view, radially elongated, and more strongly elongated around the hilum. The anticlinal walls are straight and strongly thickened with fine, branching pits. The outer wall is slightly concave and detaches easily. The cuticle is smooth.

According to the recent revision by Wong and Sugau (1996), there are two seed forms in *Fagraea*. *Fagraea* sect. *Fagraea* consistently has ellipsoid-round seeds, whereas both *F.* sect. *Cyrtophyllum* and *F.* sect. *Racemosae* have only angular seeds.

Potalia (Fig. 5.15G)

Seeds of *Potalia amara* are irregular in shape, elliptic, and laterally flattened with the outer side more round. Seeds are red, brown, or black in color (L. Struwe & V. A. Albert, unpubl.). Seeds are c. 4150×2250 μm in size. The hilum is distinct, circular in shape, often lying in a shallow depression, at or near the middle of the seed. The testa is reticulate. Testa cells are polygonal in surface view. The outer wall is collapsed in the mature seed and detaches easily.

The seeds of Gentianaceae 541

Tribe Gentianeae

Seed structure in the Gentianeae has been studied by Ho and Liu (1990), Miège and Wüest (1984), Müller (1982), and Yuan (1993). Seed micromorphology provides a taxonomically significant character set useful in evaluating intrageneric relationships among the Gentianeae. Yuan (1993) presented data on the seed micromorphology of 46 species of seven genera of the Gentianeae from western China.

Subtribe Gentianinae

Gentiana (Fig. 5.16A–F)

Gentiana is the largest genus of the Gentianaceae and also shows the greatest diversity in seed morphology. Seeds of *Gentiana* are very diverse in size, shape, color, and micromorphology. Seed characters have been regularly used for the delimitation of subgenera and sections. Seeds of *Gentiana* are relatively large and may measure up to 4 mm. In *Gentiana* species the seed body is mostly elliptic to narrowly elliptic, and often flattened or winged. The seed color varies from yellowish to deep brown and black (Halda, 1996). The testa is reticulate or striato-reticulate. Testa cells are polygonal and mostly elongated. Anticlinal walls are thickened and straight, curved, or weakly undulating. The inner wall has granular or reticulate thickenings (*Gentiana pneumonanthe*) or is smooth (*G. cruciata*). The mean number of seeds per fruit varies from three to many (300). The seeds are wingless, variously winged, or multilamellate.

To date, the seeds of only a relatively small number of species have been studied in detail. Several seed types are found. According to Yuan (1993), the following types may be discerned.

1 **The reticulate-type.** The walls of the exotestal cells form a reticulate structure. Depending on the thickening of the anticlinal walls, the testa is finely or strongly reticulate. The meshes of the reticulum vary in the shape and size of the meshes. The reticulate-type is the most common in the *Gentiana* subgenera *Cruciata*, *Otophora*, *Microsperma*, and *Calathianae* and in species of subgenus *Chondrophyllae*.

The seeds of *Gentiana* subg. *Cruciata* (*Gentiana crassicaulis*, *G. cruciata*, *G. dahurica*, *G. fetissowii*, *G. macrophylla*, *G. officinalis*, and *G. straminea*) show a very uniform micromorphology with finely striate-reticulate sculpturing. The cell walls forming the reticulations are strongly thickened with annular bars in cross-section, leaving very narrow meshes and smooth inner walls (Bouman & Schrier, 1979). The

Figure 5.16. Scanning electron micrographs of seeds of Gentianeae-Gentianinae. **A.** *Gentiana cruciata* (*Hortus Botanicus Amsterdam*, no voucher). **B–C.** *Gentiana sedifolia* (*Hitcock 21075*), seed and detail of testa. **D–E.** *Gentiana pneumonanthe* (no voucher), seed and detail of testa. **F.** *Gentiana asclepiadea* (*Hortus Botanicus Amsterdam*, no voucher). **G.** *Tripterospermum luzonensis* (*Eijma 1326*). **H.** *Tripterospermum sumatranum* (*Rahmal Si Boeea 10126*). **I.** *Crawfurdia speciosa* (*Strachey & Winterbottom s.n.*).

The seeds of Gentianaceae   543

seeds are unwinged, mostly elliptic, and vary in mean length from 0.6 mm in *Gentiana olivieri* to 1.5 mm in *G. officinalis* (Halda, 1996). No distinct interspecific differences were observed by Yuan (1993).

The seeds of *Gentiana* subg. *Calathianae* (*Cyclostigma*) are more strongly reticulate, elliptic in shape, and vary in length from 0.5 mm in *Gentiana nivalis* to 1.1 mm in *G. schleicheri* (Miège and Wüest, 1984). Seeds are unwinged, or rarely with a wing at one end. The anticlinal walls are strongly thickened. The outer walls are collapsed and the inner walls have a coarse reticulum.

The seeds of *Gentiana* subg. *Chondrophyllae* are various in shape. Fusiform seeds have been recorded in the Asiatic species *Gentiana thunbergii* and *G. nipponica*. The seeds vary in mean length from 0.5 mm in *Gentiana piaszekii* to 2.0 mm in *G. flexicaulis* (Halda, 1996). Yuan (1993) observed 18 species in this subgenus and recognized three seed groups. Most species resemble the seeds of subgenus *Cruciata* (Yuan, 1993). In the first group, the testa is reticulate to striate-reticulate with rather narrow meshes and straight anticlinal walls. In the second group, the anticlinal walls have more or less pronounced undulations and the inner wall has a micropapillate secondary sculpture. The third group has rather rough striations with some fine, irregular, floccose structures.

2 **The undulate-type**. The testa shows a reticulate pattern, but the outer wall of the exotestal cells is convex and does not collapse. This type is found in species of *Gentiana* subg. *Stenogyne*.

3 **The honeycomb-type**. The testa is reticulate with hexagonal cells. The anticlinal walls are straight, raised, and form a honeycomb-like, reticulate (foveolate) structure. The outer wall is very thin, fully collapsed and lying against the reticulately thickened inner wall, and forms so-called hexagonal nets. This type is found in *Gentiana* subg. *Monopodiae* and in species of subgenus *Microsperma*. Seeds of subgenus *Monopodiae* are mostly ovate to elliptic and vary from 0.6 to 2.0 mm in length (Halda, 1996).

4 **The compound lamellar-type**. The testa resembles that of the honeycomb-type; however, the anticlinal walls are more irregular and locally more strongly raised, which may result in a spongy appearance. If mainly the longitudinal walls are raised, the seeds have membranous lamellae or irregular hyaline wings; if the anticlinal walls are raised more irregularly, the seeds are crinkle-winged or glistening white. This type is found in *Gentiana* subg. *Frigidae* (Yuan, 1993; Halda, 1996). The seeds are subglobose to elliptic, mostly between 1.0 mm and 1.5 mm long.

5 **The ribbed-type**. These seeds have longitudinal ribs formed by rows several testal cells wide. Ribs are sometimes bifurcating/fusing, or more

brain-like in *Gentiana clusii*. The ribs are often separated by deep furrows. The testa is reticulate. The testa cells are polygonal, and often somewhat elongated. The anticlinal walls are thickened, and the outer wall is thin and only slightly collapsing. This type is found in *Gentiana* subg. *Ciminalis* (sect. *Megalanthe/Thylacites*). The seeds are unwinged, mostly elliptic to narrowly elliptic in outline, often with a distinctly sunken micropyle surrounded by a collar, and vary in length from 0.7 to 1.8 mm (Miège & Wüest, 1984). *Gentiana alpina* deviates with its smaller, less ridged seeds, which resemble seeds of *Gentiana* subg. *Calathianae*.

6 **The winged-type.** The seeds are flattened and winged. The wing is mostly a single peripheral wing. Winged seeds are found in several *Gentiana* subgenera, viz. subg. *Gentiana*, *Stylophora*, *Newberryi*, and *Phyllocalyx*, and in species of subgenera *Pneumonanthe* and *Stenogyne*.

Species of *Gentiana* subg. *Gentiana* (subg. *Coelantha*) have relatively large, winged seeds. The seeds are mostly circular to elliptic in outline and laterally flattened. The wing is often tapering toward and absent at the hilum/micropyle. Seed length varies from 2.5–2.6 mm in *Gentiana purpurea* to 3.4–4.1 mm in *G. lutea* (Miège & Wüest, 1984). The testal pattern of the seed body is mostly linear. Testa cells are polygonal and mostly elongated. The exotestal cells of the seed body and wing are of the same type. The anticlinal walls are thickened and the outer wall is collapsed and reflects the reticulately thickened inner walls. The anticlinal boundary is raised. The cuticle is smooth.

Seeds of *Gentiana* sect. *Stylophora* (subg. *Megacodon*) are broadly winged and have a thickened, reticulate testa (Halda, 1996). Seeds of some species of subgenus *Pneumonanthe* are flattened and completely surrounded by a membranous wing. Seed length varies between 1.4 mm and 2.1 mm in *Gentiana asclepiadea*. The outer walls are thin and collapsed, the inner walls are reticulately thickened. Seeds of *Gentiana pneumonanthe* deviate in shape and are unwinged and narrowly elliptic, with extended chalazal and micropylar ends, and about 1.2–1.5 mm in length.

The seeds of *Gentiana* subg. *Stenogyne* are heterogeneous in shape. Most species have seeds that are triangular with narrow wings along the three edges (Yuan, 1993; Halda, 1996). The seeds are relatively small, up to 0.9 mm in *Gentiana striata*.

7 **The semi-winged-type.** The seeds resemble those of the winged-type; however, the narrow wing runs only along one side and/or at the chalazal side of the seed. This type is found in *Gentiana pudica* of subgenus

Dolichocarpa (Yuan, 1993) and in species of subgenus *Pneumonanthe* sect. *Septemfidae*.

8 **Gentiana subg. Isomeria**. No SEM studies on seeds of *Gentiana* subg. *Isomeria* are known to us. According to the descriptions and line drawings of Halda (1996), the seeds are globose to subglobose and about 1 mm long. The seeds are described as rugose, lamellate, and glistening white. The exotestal cells are elongated. Most probably the seeds belong to the honeycomb-type.

Tripterospermum (Fig. 5.16G–H)

The seeds of *Tripterospermum* have been described by Murata (1989) and Smith (1965). They resemble those of *Crawfurdia*. The seeds are basically triquetrous with three faces, and mostly laterally compressed. The species of section *Capsulifera* have capsular fruits and seeds that are compressed, discoid, and broadly three-winged. Species of section *Tripterospermum* have berries, and the seed shape varies. The seeds vary from compressed, broadly or narrowly winged to more triquetrous and ridged. Seeds of *Tripterospermum sumatranum* are c. 3670 × 1670 μm in size. Seeds of *Tripterospermum trinerve* measure c. 2400 × 1480 μm. Seeds of *Tripterospermum luzonense* measure c. 2170 × 1520 μm. The testa is finely reticulate. Testa cells are polygonal or irregular in shape, irregularly oriented, and similar on wing and seed body. The anticlinal walls are straight or curved and prominent. Outer walls are thin, collapsed, and do not show special structures. The cuticle is smooth.

Crawfurdia (Fig. 5.16I)

The seeds of *Crawfurdia* have been described by Smith (1965). The seeds of *Crawfurdia speciosa* are laterally compressed, discoid, three-winged, and measure c. 2400 × 1795 μm. The testa is finely reticulate with shallow cells. Testa cells are irregular in shape and orientation, and similar on wing and seed body. The anticlinal walls are mostly curved. Outer walls are thin, collapsed, and do not show special structures. The cuticle is smooth.

Subtribe Swertiinae

The unusual seed structure found in a number of taxa formerly belonging to *Gentiana* subg. *Gentianella* and some related taxa has been described by Netolitzky (1926). This type with smooth or almost smooth seed coats comprises the genera *Comastoma*, *Gentianella*, *Halenia*, *Lomatogonium*, and *Swertia* spp. The testa is very thin and fully collapsed. The exotesta has no distinct wall thickenings. The testal pattern may be visible as a thin reticulum. The surface structure is often determined by the underlying endosperm cells. The outer wall of the outermost endosperm layer is very thick and has taken over the protective function of the seed coat. Seeds of the genera *Jaeschkea, Latouchea, Megacodon,* and *Pterygocalyx* were not available for this study.

Gentianella (Fig. 5.17A)

Miège and Wüest (1984) studied the seeds of nine European species belonging to *Gentianella* sects. *Gentianella* and *Arctophila*. Yuan (1993) studied the Chinese *Gentianella azurea*. The seeds of all species closely resemble one another and show differences only in mean dimensions and tertiary structure. Seeds are brown in color, globose, elliptic, or ovate in shape, and sometimes have local impressions. Chalaza, hilum, and micropyle are mostly indistinct. Average seed size is between *c.* 600 μm and 900 μm. The seed surface is smooth, granular to irregularly ribbed, or wrinkled. The testa is undifferentiated, collapsed, fully crushed, and without a distinct testal pattern or faintly showing the collapsed layers. The walls of the exotestal cells are not thickened; the inner, outer, and anticlinal walls are very thin and may reflect the surface sculpture of the endosperm. The cuticle is without ornamentation.

Comastoma (Fig. 5.17B)

The seeds of three species of *Comastoma* have been described by Miège and Wüest (1984) and Yuan (1993). The seeds are elliptic, about 0.7–1.0 mm in length, and have a smooth or granular surface. The seeds are very similar to those of *Gentianella*.

Halenia (Fig. 5.17C–D)

Seeds of *Halenia* are of the same type as *Gentianella*. Seeds are globose to elliptic, sometimes slightly flattened, and brown, greenish-, or yellow-brown in color. Mean seed length varies from about 850 μm in *Halenia foliosa* to about 1250 μm in *H. dasyantha*. Seeds of *Halenia corniculata* are 1795 × 1360 μm. The testa has a granular to finely reticulate surface.

Lomatogonium (Fig. 5.17E)

The two Chinese species of *Lomatogonium* examined by Yuan (1993) have smooth seed coats and closely resemble those of *Gentianella*. Seeds of *Lomatogonium carinthiacum* are more irregular in shape, mostly elliptic with longitudinal ribs and sunken areas. The testa is smooth and very thin. The anticlinal walls are locally visible. The hilum is subapical. Seeds are 450–620 × 290–380 μm in size and have a length:width ratio of 1.7.

The seeds of Gentianaceae 547

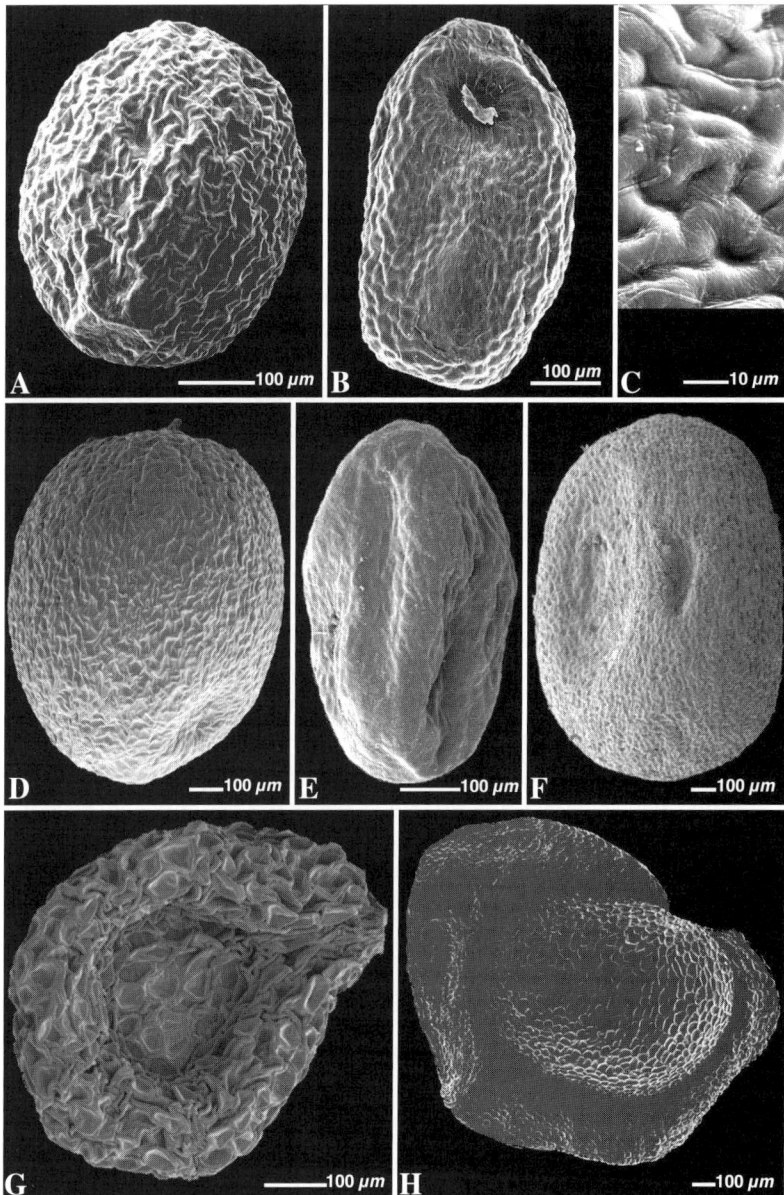

Figure 5.17. Scanning electron micrographs of seeds of Gentianeae-Swertiinae. **A.** *Gentianella germanica* (*Boom 36009*). **B.** *Comastoma tenellum* (*Kordeva & Sytina 1336*). **C.** *Halenia corniculata* (*Togasi 1227*), detail of testa. **D.** *Halenia dasyantha* (*Killip & Lehmann 38545*). **E.** *Lomatogonium rotatum* (*Braun-Blanquet 856*). **F.** *Swertia macrocarpa* (*Griffith 5829*). **G.** *Swertia bimaculata* (*Ohashi et al. 774316*). **H.** *Swertia perennis* (*Goethals s.n.*).

Swertia (Fig. 5.17F–H)

Seeds of *Swertia* have been described by Maiti and Banerji (1976) and Yuan (1993). Seeds are very variable in color, shape, size, and testal structure. Seeds are sometimes globose to elliptic-ovoid as in *Swertia dichotoma* and *S. tetraptera* and resemble those of *Comastoma* and *Halenia* (Yuan, 1993). However, seeds of many species are flattened, discoid, and provided with a peripheral margin or wing.

Seeds of *Swertia* subg. *Ophelia* are usually small and not winged, from about 420×370 μm in *Swertia gracilescens* to 2200×1700 μm in *S. macrosperma*. Seeds of the subgenera *Euswertia* and *Poephila* are often discoid, winged, and from 570×480 μm in *Swertia cuneata* to about 1900× 1700 μm in *S. hookeri* (Maiti & Banerji, 1976). The testa is variable, usually smooth or wrinkled, or rarely reticulate. The testa of *Swertia dilatata* is reticulate, with the cells arranged in rows. Testa cells are polygonal with straight anticlinal walls and a collapsed, thin outer wall. Seeds of *Swertia cordata* and *S. petiolata* have longitudinal cristae. The number of seeds in each capsule is as few as two in *Swertia piloglandulosa* ssp. *biovulata*, four in *S. macrosperma*, to as many as 300 in *S. alata*.

Veratrilla (Fig. 5.18A)

Seeds of *Veratrilla baillonii* are laterally compressed, elliptic in lateral view, discoid, and three-winged. Seeds measure c. 2000×1410 μm. The testa cells are isodiametric to elliptic and oriented irregularly. The anticlinal walls are straight. The outer wall is collapsed and thin, mostly with some striae. The cuticle is smooth.

Frasera (Fig. 5.18B)

Seeds of *Frasera caroliniensis* are elliptic in shape, flattened, and winged. The wings are regular and more or less equal all around the edge of the seed. The seeds measure 6525–7050×3600–3950 μm. Testa cells are isodiametric and oriented irregularly. The anticlinal walls are straight. The outer wall is collapsed and thin. The cuticle is smooth.

Gentianopsis (Fig. 5.18C–F)

Seeds of most species are elliptic to irregular in shape, and have a papillose testa, called "finger-like projections" by Yuan (1993). Seed size varies from 800–1080×480–525 μm in *Gentianopsis detonsa* to 1040–1285×715–860 μm in *G. crinita*. The exotestal cells in *Gentianopsis barbata*, *G. crinita*, *G. detonsa*, *G. paludosa*, and *G. procera* elongate radially and all differentiate into papillae

The seeds of Gentianaceae 549

Figure 5.18. Scanning electron micrographs of seeds of Gentianeae-Swertiinae.
A. *Veratrilla baillonii* (*Smith 4448*). **B.** *Frasera caroliniensis* (*Hill & Dorsey 24482*).
C. *Gentianopsis detonsa* (*Oosterveld 0246*). **D.** *Gentianopsis ciliata* (*Suringar 1865*).
E–F. *Gentianopsis crinita* (no voucher), section of testa and seed. **G.** *Obolaria virginica* (*Wilkens 2726*). **H.** *Bartonia tenella* (*Camden s.n.*).

or hair-like cells. The finger-like projections are usually longer along the periphery or at the chalaza and shorter on both the dorsal and ventral surfaces. Papillae and hairs may collapse in dry, mature seeds. The inner walls are warty or have short papillate thickenings. The cuticle is smooth.

The seeds of *Gentianopsis ciliata* deviate from the other species in shape and micromorphology. Seeds are narrowly elliptic or elliptic to oblanceolate to obovate because of the elongated micropylar and chalazal ends. Seeds are about 750×195 μm in size. The testa is reticulate. Testa cells are rectangular. The outer wall is thin, collapsed at the seed body. The cells at the micropyle and chalaza are more elongated, often air-filled, and more blown-up. The inner wall has a thin reticulum.

Obolaria (Fig. 5.18G)

The seeds of *Obolaria virginica* differ from those of all other Gentianaceae. Seeds are small, elliptic, c. 260×155 μm. The testa is minutely striate and consists of a few cells only, about three to four in the length direction of the seed. Testa cells are elongated and often irregular in shape. The anticlinal walls are straight or curved, strongly thickened; the anticlinal boundaries are locally raised and beaded. The outer wall is thin, collapsed, and does not reflect special structures. The cuticle is smooth.

Bartonia (Fig. 5.18H)

Seeds of *Bartonia* are very small, c. 135×85 μm in *B. tenella*, 160×90 μm in *B. virginica*, and 175×85 μm in *B. paniculata*. Seeds are irregular in shape, somewhat sac-like. The testal pattern is minutely reticulate, often indistinct because of the collapsed testa cells. The outer wall is thin, collapsed and wrinkled. The inner wall is without a distinct structure. The cuticle is smooth.

Tribe Saccifolieae

Saccifolium (Fig. 5.19A)

Seeds of *Saccifolium bandeirae* were described by Maguire and Pires (1978). The seeds are irregular in shape, angular with distinct rims and sunken sides, and about 450–700 μm in diameter. The testa is reticulate. Testa cells of the rims are enlarged, and the outer walls are partly collapsed. The anticlinal walls are straight to curved. The cuticle is smooth.

Curtia (Fig. 5.19B,E)

The seed structure of six species of *Curtia* was studied by Grothe and Maas (1984). The seeds are quite diverse in size and shape between and

The seeds of Gentianaceae 551

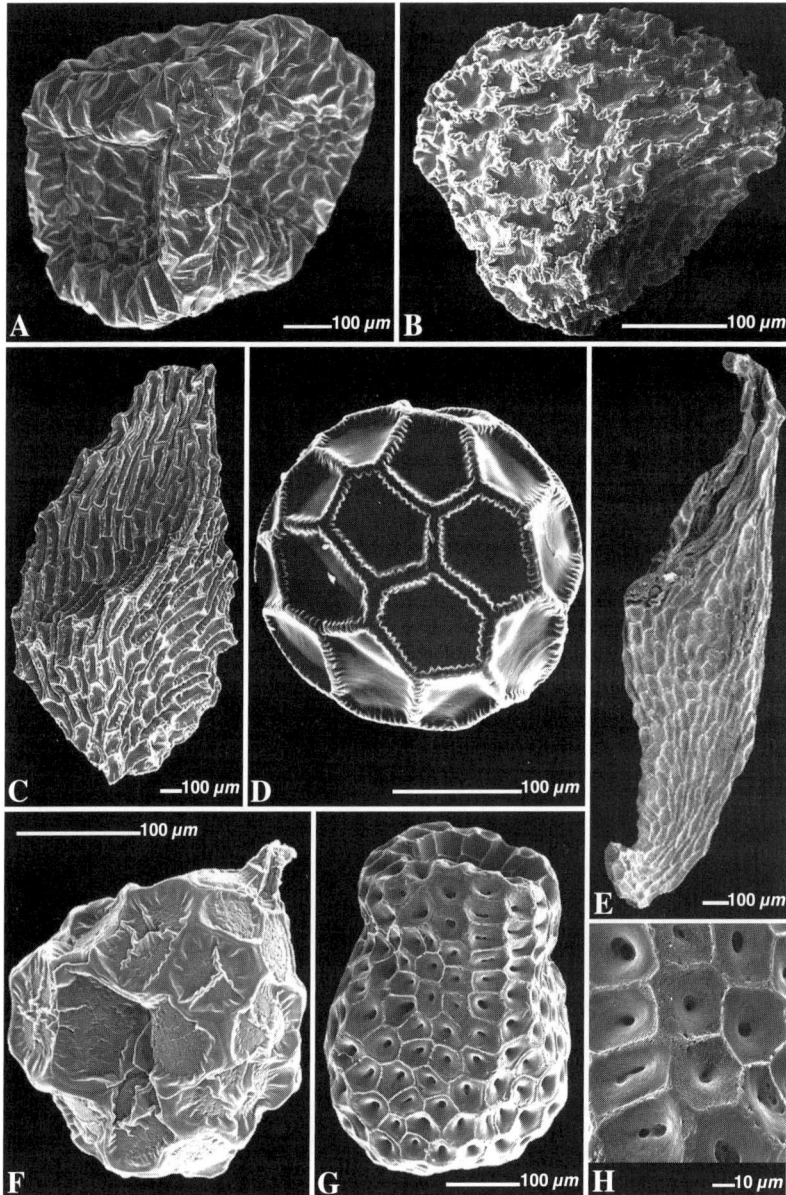

Figure 5.19. Scanning electron micrographs of seeds of Saccifolieae and *genus incertae sedis*. **A.** *Saccifolium bandeirae* (*Farney et al. 897*). **B.** *Curtia diffusa* (*Mexia 5750*). **C.** *Hockinia montana* (*Martinelli 227*). **D.** *Tapeinostemon spenneroides* (*Hoffman & Marco 2121*). **E.** *Curtia sp. nov.* (*Hoffman 3222*). **F.** *Voyriella parviflora* (*Maas et al. 2229*). **G–H.** *Voyria truncata* (*Steyermark 41610*), seed and detail of testa, outer cell walls removed.

within species. The shape varies from globose to elliptic, bowl-shaped, or more irregular with flattened sides. Seeds are mostly between 300 μm and 500 μm in mean length. The testa is reticulate or alveolate. Testa cells are polygonal or mostly with rectangular cells. The anticlinal walls are straight or undulating in *Curtia diffusa* and *C. tenuifolia*. The anticlinals have beaded thickenings. The anticlinal boundaries are smooth or locally raised or sunken. The outer wall is thin, collapsed in some species, and reflects pits of the inner wall. The cuticle is smooth or irregularly granular.

The seeds of *Curtia sp. nov.* (*Hoffman 3222*) deviate in shape and size from the above-mentioned species. The seeds are laterally flattened, have extended micropylar and chalazal ends, and measure 1990–2050 × 475–555 μm. The micropyle is slit-like and about one-third of the seed length. The testa is reticulate. Testa cells are polygonal or rectangular, those at the dorsal side of the micropyle more elongated. The anticlinal walls have beaded thickenings locally.

The *Curtia* species with smaller seeds and shallower testas resemble those of species of *Voyriella* and *Tapeinostemon*. The *Curtia* species with longer, flattened seeds resemble those of *Hockinia*.

Hockinia (Fig. 5.19C)

The seed structure of *Hockinia montana* was described in detail by Grothe and Maas (1984). The seeds are irregular in shape, often flattened and with distinct micropylar and chalazal ends. Seeds are 560–780 × 260–380 μm in size. The testa is reticulate, showing a regular, brick-wall-like arrangement, and is about 90 cells thick in surface view. Testa cells are elongated longitudinally. The anticlinal walls are thickened, straight near the anticlinal boundaries and undulating lower down, showing alternating lateral swellings with pronounced perforations between them. The outer periclinal wall is thin and collapsed. The inner wall has few large pits. The testal pattern resembles those of some *Curtia* species.

Tapeinostemon (Fig. 5.19D)

The seeds are minute, globose to elliptic in shape, and 180–300 × 200–290 μm in *Tapeinostemon spenneroides*. The testa is reticulate to alveolate, about five cells high. Testa cells are polygonal and irregularly oriented. The anticlinal walls are straight, and are striated or beaded. The outer wall is thin, concave, and at the periphery stretched and showing radial wrinkles. The cuticle is smooth.

Voyriella (Fig. 5.19F)

The seeds of *Voyriella parviflora* are almost globose, mostly acuminate toward the hilar end, and 180–245 × 165–215 μm in size. The testa is reticulate to alveolate, and up to five cells high. Testa cells are polygonal and isodiametric and have only slightly thickened walls. The anticlinal walls are straight and striated, or beaded. The outer periclinal walls are thin, collapsed, or collapsed only at the center, lying against the thickened inner wall, but tightened and showing radial wrinkles at the periphery. The anticlinal boundaries are raised. The inner wall has small pits. The cuticle is smooth. The embryo of *Voyriella* is better developed than in *Voyria* and is embedded in two layers of endosperm. The outermost cell walls of the endosperm are very thick and probably have a mechanical function.

Genus incertae sedis

Voyria (Figs. 5.19G–H, 5.20)

The seeds of *Voyria* were studied by Bouman and Devente (1986) and Bouman and Louis (1990). *Voyria* presents a striking example of extreme trends in reduction of ovule, embryo, and seed. The seeds show diversity in size, shape, and testal ornamentation. On the basis of seed characters, four different groups can be recognized.

1 **The *Voyria truncata*-group**. The seeds develop from unitegmic and anatropous ovules. The seeds are elliptic to urceolate, from 215–250 × 160–215 μm in *Voyria pittieri* to 430–495 × 350–380 μm in *V. clavata*. The mean length:width ratio is 1.3–1.4. The hilum is apical, distinct, and mostly sunken. The testa is reticulate. The cells are more or less arranged in a longitudinal pattern. Testa cells are strongly thickened, and U-shaped in cross-section. The anticlinal walls are straight or curved. The anticlinal boundaries are raised.
2 **The *Voyria tenuiflora*-group**. The ovules are ategmic and orthotropous with inverted embryo sacs. The seeds are saccate, acuminate toward the hilar end, and from 200–270 × 90–130 μm in *Voyria tenuiflora* to 390–600 × 145–200 μm in *V. spruceana*. The mean length:width ratio varies from 1.7 to 2.1. The anticlinal walls are not prominent and the cellular pattern of the testa is less distinct than in the other groups. The testa is only two to four cells in length and three to five cells in width. The inner walls of the testa have longitudinal rows of globose thickenings.
3 **The *Voyria rosea*-group**. The seeds develop from ategmic, orthotropous

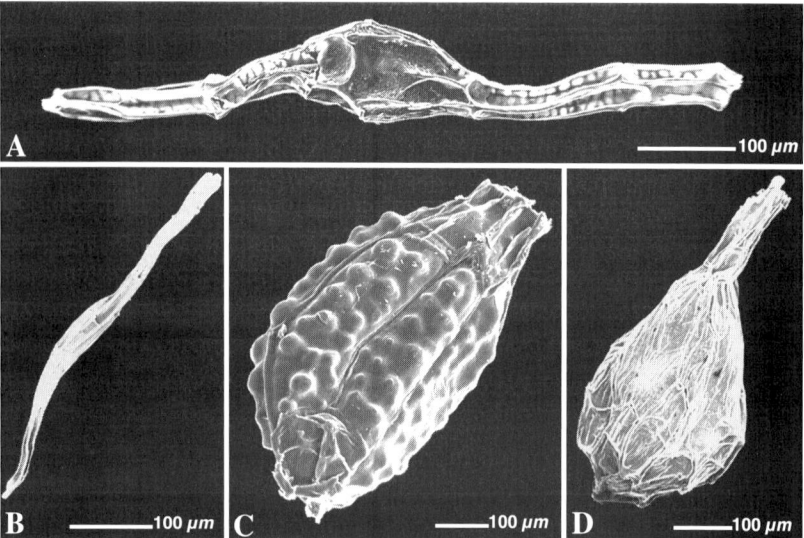

Figure 5.20. Scanning electron micrographs of seeds of *Voyria*, genus incertae sedis. **A.** *Voyria parasitica* (*Britton & Millspaugh 6109*). **B.** *Voyria aphylla* (*Teunissen LBB15200*), paraphyse. **C.** *Voyria tenuiflora* (*Lanjouw & Lindeman 2722*). **D.** *Voyria rosea* (*Maas LBB11045*).

ovules with inverted embryo sacs. The seeds are mostly saccate, and from 155–200×65–115 μm in *V. chionea* to 270–340×130–195 μm in *Voyria rosea*. The mean length:width ratio varies from 1.9 to 2.9. The group includes the single African representative *Voyria primuloides* (Bouman & Louis, 1990) and the Costa Rican endemic *V. kupperi*.

4 **The *Voyria aphylla*-group**. The ovules are ategmic and orthotropous with inverted embryo sacs. The mature seeds become fusi- to filiform by two projections, one formed by the outgrowth of the ovular apex, and the other representing the elongated funicle. The seeds measure from 550–780×35–60 μm in *Voyria tenella* to 1500–2000×75–105 μm in *V. obconica*, with a mean length of the seed body of 165 and 330 μm, respectively. The mean length:width ratio varies between species from 8 to 21. The testa cells show reticulate, foraminate, or linear thickenings on the anticlinal and inner walls. The outer wall is thin and collapsed. The seeds contain an undifferentiated embryo of three to four cells high and one layer or row of endosperm cells. In *Voyria parasitica* and close relatives, abortive seeds in the form of paraphyses are present in varying numbers. The paraphyses are the same length as, or somewhat shorter

The seeds of Gentianaceae 555

than, the fertile seeds. The seed body is less pronounced and with testal characters like those of the projections.

DISCUSSION

Like those of many other larger families, the seeds of Gentianaceae show an extensive diversity in size, shape, and micromorphology. Seeds of larger size are found mainly in the Gentianeae. Seed (micro)morphology provides an additional character set which may be of use at the tribal, subtribal, generic, and sometimes species level.

Delimitation of the family

Traditionally the Gentianaceae were divided into two subfamilies, the Gentianoideae and the Menyanthoideae. Current consensus is that the Menyanthaceae should be excluded from the Gentianaceae and even from the Gentianales (Nicholas & Baijnath, 1994). Menyanthaceae are now generally considered as a segregate family allied with the Asterales and not even closely related to the Gentianaceae. The families show considerable differences in embryology and seed structure (for a review on this topic see Johri *et al.*, 1992).

A close connection between the (sub)tribe Potaliinae, traditionally placed in the Loganiaceae as tribe Potalieae (Leeuwenberg, 1980), and the Gentianaceae was not much disputed until recently (Nicholas & Baijnath, 1994). In fact, increasing evidence, based on anatomical, chemical, and DNA studies (Struwe *et al.*, 2002), supports the transfer of Loganiaceae-Potalieae to the Gentianaceae as subtribe Potaliinae. However, seed morphology does not provide arguments for the inclusion of the Loganiaceae-Potalieae in the Gentianaceae. Seeds of Potaliinae differ markedly from the other Gentianaceae in their relatively large size (except Gentianeae), lateral funicular attachment, and strongly thickened exotestas.

With the description of the recently discovered, monotypic genus *Saccifolium* by Maguire and Pires (1978), a separate family Saccifoliaceae was created. The status of this family and its relationship with the Gentianaceae remain disputable (Nicholas & Baijnath, 1994). A close relationship between *Saccifolium* and *Gentiana* was initially found from DNA sequences of *trn*L intron based on herbarium material (Struwe *et al.*, 1998). However, new data from freshly collected material on *rbc*L, *trn*L intron, and *mat*K sequences now show that *Saccifolium* is in a sister clade to the rest of the Gentianaceae and should be classified in the tribe Saccifolieae

(Thiv et al., 1999; Struwe et al., 2002). The inclusion of *Curtia*, *Tapeinostemon*, and *Voyriella* in the Saccifolieae is not supported by seed micromorphology. The seeds of *Saccifolium* do not show a close resemblance to *Gentiana*, but fit well with those of the Gentianaceae in general.

Delimitation of tribes and subtribes

Seed structure provides additional arguments for the updated tribal and subtribal classification of the family proposed by Struwe et al. in Chapter 2 of this book. Seed morphological data confirm the current opinion that the conventional tribal and subtribal classification of the family (Gilg, 1895) must be updated. In particular, Gilg's subtribe Gentianineae appears artificial and there is no basis for the separate tribes Rusbyantheae, Voyrieae, and Leiphaimeae. Moreover, the composition of Gilg's subtribes Erythraeinae and Tachiinae and tribe Helieae needs rearrangement.

Exaceae

Seed morphology corroborates the Exaceae (Exacinae *sensu* Gilg) as a well-defined taxon. The seeds are relatively small, mostly less than 500 μm and not exceeding 800 μm in mean length. In most taxa the testa is characterized by star-shaped cells with smooth, thin, and collapsed outer walls, and smooth or pitted inner walls. The Malagasy genus *Tachiadenus* was originally placed in the tribe Exaceae, but was transferred to the Tachiinae by Gilg (1895). After a detailed study of the genus, Klackenberg (1985, 1987) concluded that *Tachiadenus* is a member of the Exaceae. This conclusion is sustained by seed morphological characters. The largest genus of the tribe, *Sebaea*, is heterogeneous in seed structure and shows a tendency toward the formation of longitudinal ridges or fringes. The seeds of the saprophytic genus *Cotylanthera* are relatively small and simple in testal structure.

Chironieae

The tribe Chironieae *sensu* Struwe et al. (2002) has three subtribes (Canscorinae, Chironiinae, and Coutoubeinae), all with relatively small seeds, in most species not exceeding 500 μm in mean length. The seeds in all three subtribes may have exudates on the cuticle.

The seeds of Canscorinae are always small and have a mean seed length of between 250 μm and 420 μm. Testa cells are mostly polygonal with straight or curved walls, but are sinusoidal in *Hoppea*. The inner wall is never papillate.

Chironiinae have small seeds that are mostly less than 500 μm and do not exceed 870 μm. The testa has a distinct reticulate pattern of polygonal, but never star-shaped, testa cells with often straight, sometimes curved or undulating anticlinal walls. The inner wall of the exotesta often has papillae intermingled with pits; sometimes the inner wall becomes reticulate. The seeds of *Bisgoeppertia*, *Eustoma*, and *Orphium* closely resemble one another in their deeply concave cells.

Mean seed length in Coutoubeinae does not exceed 550 μm. The testa cells are mostly polygonal with straight or curved walls, sinusoidal in some species of *Schultesia*. The inner wall is never papillate.

Helieae

Seed structure supports the current opinion that Helieae *sensu* Gilg needs revision. Seeds of *Coutoubea*, *Deianira*, and *Schultesia* do not correspond to those of the *Irlbachia* complex and relatives, resembling more those of Chironieae.

The differences of opinion of taxonomists over the last 150 years have muddled the nomenclature of the *Irlbachia* complex. Owing to the numerous synonyms it is very difficult for non-taxonomists to unravel all the existing names. The group needs a very careful revision based on interdisciplinary studies. Seed morphology does not support the very broad concept of "*Lisianthus*" widely adopted prior to 1970, but disputed on palynological evidence by Nilsson (1970; see Struwe *et al.*, 2002). Seeds of *Lisianthius sensu stricto* have their own seed type and differ from seeds of the *Irlbachia* complex and related genera.

The seeds of *Adenolisianthus*, *Calolisianthus*, *Celiantha*, *Chelonanthus*, *Helia*, *Irlbachia*, *Lagenanthus*, *Lehmanniella*, *Neblinantha*, *Pagaea*, *Purdieanthus*, *Sipapoantha*, *Symbolanthus*, *Tetrapollinia*, *Wurdackanthus*, and some species of *Macrocarpaea* all belong to the same basic type. Their seeds form a more or less continuous range from globose to angular seeds and from testas with only convex cells to testas with only concave cells. Some taxa show transitional stages or even di- to polymorphic seed types.

Seed characters do not support separate generic status for *Adenolisianthus*. The seeds resemble those of, for example, *Helia oblongifolia* and *Irlbachia pumila*, and have only dome-like testa cells. This is largely in accordance with pollen morphological data (Nilsson, 1970).

Seeds of *Tetrapollinia caerulescens* are angular and have a testa of only concave cells. These seeds resemble those of *Irlbachia pratensis* and *Chelonanthus alatus*. Seed characters support the close relationships between *Lagenanthus*, *Lehmanniella*, and *Purdieanthus*. Their seeds resemble one another and differ mainly by the number of band-like thickenings of the collapsed domes.

The brain-like appearance of the seeds of *Symbolanthus* is very characteristic. These seeds resemble somewhat those of *Wurdackanthus frigidus* in their more strongly elongated domes at the rims.

The genus *Macrocarpaea* is heterogeneous in seed structure. Based on pollen structure, Nilsson (1968) suggested the exclusion of the taxa with a *Chelonoides*-type of pollen. These species, included in the *Corymbosa* pollen type, now belong to the genus *Rogersonanthus* (Nilsson, 2002 (Chapter 4, this volume)). Seeds of *Aripuana* resemble those of *Rogersonanthus* in their rims with dome-like cells with a reticulum of rounded meshes and concave cells with radiating band-like thickenings on the sides.

The *Macrocarpaea glabra*-group has rimmed seeds resembling those of *Symbolanthus*, whereas the *M. corymbosa*-group has winged seeds of a type not found elsewhere in the Helieae. Meanwhile, the seeds of *Macrocarpaea rubra* resemble those of *Senaea*. The genus needs a critical revision. Seeds of *Chorisepalum* closely resemble those of certain *Macrocarpaea glabra* collections.

Seeds of *Lehmanniella*, "Roraimaea", *Sipapoantha*, and *Symbolanthus* resemble each other in the more strongly swollen rims and agree with the *Helia*-type in general structure.

The seeds of *Prepusa*, *Senaea*, and *Zonanthus* resemble each other in their polygonal testa cells with thick anticlinal walls and foraminate inner wall. *Tachia* has characteristic "echinate" seeds, but resembles the aforementioned genera in other testal characters.

Potalieae

The seeds of the three subtribes of Potalieae strongly differ from one another. Seeds of subtribe Potaliinae are clearly divergent from other Gentianaceae (see above).

With the exception of *Urogentias*, the seeds of subtribe Faroinae are characteristic at the subtribal level in shape, size, testal pattern, and presence of exudates. Seeds of *Congolanthus*, *Djaloniella*, *Enicostema*, *Faroa*, *Neurotheca*, *Oreonesion*, *Karina*, and *Pycnosphaera* closely resemble each other. Mean seed length varies from 310 µm in *Faroa* to 490 µm in *Enicostema*. *Urogentias* has multiply-winged seeds resembling those of certain *Sebaea* species.

Seeds of *Lisianthius* clearly differ from those of Helieae. The subdivision of the genus into sections *Lisianthius* and *Omphalostigma* is also supported by seed morphology. Seeds of *Lisianthius* resemble those of *Bisgoeppertia*,

Eustoma, and *Orphium* (Chironieae) in the deeply concave cells. However, cladistic analysis of morphological as well as molecular characters shows that *Lisianthius* is not related to these genera. In view of the increasing evidence, the resemblance in seed structure must be explained by convergent evolution.

Gentianeae

The seeds of the Gentianeae are distinguishable from those of the other tribes of the family. With the exception of the saprophytic genera *Bartonia* and *Obolaria*, the seeds are almost always larger than 0.5 mm. The seeds of the genus *Gentiana* are very diverse in size, shape, color, and micromorphology. The current review supports the theory that the sections may be characterized partly by their seed structure. Seed micromorphology may be of great help in the delimitation of the sections in the future.

The seeds of the genera of subtribe Swertiinae are of various types. Seeds of *Comastoma*, *Gentianella*, *Halenia*, *Lomatogonium*, and *Swertia* spp. have undifferentiated testas and smooth or almost smooth seed surfaces. The seed morphology of *Swertia* seems very diverse and only partly resembles that of *Gentianella*. Seed morphology largely agrees with the results from molecular studies.

Saprophytic taxa and *genus incertae sedis*

The saprophytic lifestyle has evolved several times independently within the Gentianaceae. The saprophytic taxa belong to three to four different (sub)tribes. The seeds of saprophytes may develop from ategmic ovules and are often reduced in structure, and are the smallest in size. Therefore, seed characters can generally be of limited taxonomic help, and only in elucidating evolutionary relationships at higher levels.

Bartonia and *Obolaria* were placed in the Erythraeinae by Gilg (1895). According to the recent phylogenetic studies by Struwe *et al*. (1998), these genera belong to the Swertiinae.

Ovules and seeds of *Voyria* and *Voyriella* differ in a number of characters, and the genera seem to represent different evolutionary lines. Seeds of *Voyriella* resemble those of some species of *Curtia* and *Tapeinostemon*. Based on pollen morphology, Nilsson and Skvarla (1969) also concluded that *Voyriella* would be better associated with *Curtia*. The seeds of *Curtia* species are diverse in shape and size. The species with the longer, flattened seeds resemble more those of *Hockinia*. According to the phylogenetic analysis of Struwe *et al*. (2002), *Voyriella* and *Curtia* are most closely related to *Saccifolium* and should be classified in the new tribe Saccifolieae.

The seeds of *Voyria* are quite diverse in structure, probably because of adaptation to different types of dispersal. *Voyria* subg. *Voyria* (Albert & Struwe, 1997) agrees with the *V. truncata*-group of Bouman and Devente (1986). This subgenus has many plesiomorphic character states, and its seeds develop from unitegmic, anatropous ovules. The seeds of *Voyria* subg. *Leiphaimos* develop from ategmic, atropous ovules but nevertheless achieve quite diverse morphologies.

Exudates

Crystallized structures are found on the surface and under the exotestal cuticle of several genera of the Chironieae, including *Chironia, Coutoubea, Microrphium,* and *Orphium,* and of the Potalieae-Faroinae. These exudates are of unknown chemical composition and ecological function. They resemble the exudate flavonoids found on leaves and stems of plants living in arid and semiarid areas. They have also been described for *Orphium frutescens* leaves (Wollenweber, 1989; Roitman et al., 1992) and are common on dried leaves and corollas of *Symbolanthus* (L. Struwe, pers. comm.).

ACKNOWLEDGMENTS

The authors thank Lena Struwe (Rutgers University), Jens Klackenberg (Swedish Museum of Natural History, Stockholm), Jan De Laet (American Museum of Natural History), Ton Leeuwenberg (Department of Plant Taxonomy, Wageningen), and Mike Thiv (Botanisches Museum, Berlin-Dahlem) for information and critical discussions on the taxonomy and phylogeny of the Gentianaceae. We acknowledge Eckhard Wollenweber (Institut für Botanik, der Technischen Hochschule, Darmstadt) for information on exudate flavonoids. Erik Grothe, Angelique Hollants, and Eric Simonis prepared some of the SEM photographs of the Helieae.

We also thank the curators of the following herbaria for allowing us to study their collections: A, AMD, BM, BOL, BR, C, CAY, F, G, GG, GH, GOET, IAN, INPA, K, L, LBB, LE, LV, M, MAC, MG, MO, MY, NY, P, PU, RB, S, SP, U, UC, UPS, US, VEN, W, WAG, and WIS.

Appendix 5.1
Material investigated for this study and its voucher information

Herbarium abbreviations follow Holmgren et al. (1990).

Adenolisianthus arboreus (Spruce ex Prog.) Gilg: *Garcìa-Barriga & Schultes 14147* (GH); *von Martius 893* (M); *Wurdack & Adderley 42901* (S, US).
Anthocleista amplexicaulis Baker: *Lam & Meeuse 5666* (WAG).
Anthocleista schweinfurthii Gilg: *Wieringa 1534* (WAG).
Aripuana cullmaniorum Struwe, Maas, & V. A. Albert: *Cid Ferreira 5906* (NY); *Ramos et al. s.n.* (INPA 62155).

Bartonia paniculata B. L. Rob.: *Boufford 15874* (U).
Bartonia tenella Willd.: *Camden s.n.* (L); *Wibbe s.n.* (BR).
Bartonia virginica (L.) Britton, Sterns, & Poggenb.: *Marie-Victorin & Rolland-Germain 49713* (WAG).
Bisgoeppertia scandens (Spreng.) Urb.: *Liogier 14835* (GH); *Zanoni et al. 40716* (U).
Blackstonia grandiflora Maire: *Pitard 720* (L).
Blackstonia intermedia (Ten.) Sennen ex Martinez: *Nicasio Guzzino 3547* (AMD).
Blackstonia perfoliata (L.) Huds.: *Excursion 1933* (AMD); *Geerinck-Coutrez 7103* (BR).

Calolisianthus amplissimus (Mart.) Gilg: *Anderson et al. 7462* (S); *Cardénas 4457* (US); *Dusén 16668* (F); *Irwin et al. 13871* (S), *24057* (RB); *Mosèn 4485* (S); *Plowman et al. 8962* (U); *Smith et al. 6721* (US).
Calolisianthus pedunculatus (Cham. & Schltdl.) Gilg: *Barbosa & da Silva 117* (RB); *Irwin et al. 29445* (SP); *Mello Barreto 2797* (F).
Calolisianthus pendulus (Mart.) Gilg: *Hatschbach 13519* (F), *15711* (MO); *Lindeman & de Haas 4224* (US); *Mosèn 1906* (S); *Vogel 611* (U).
Calolisianthus pulcherrimus (Mart.) Gilg: *Duarte 2435* (U); *Eiten 10989* (US); *Gehrt 150* (SP); *Irwin et al. 32414* (C); *Macedo 3056* (MO); *Mello Barreto 2802* (F); *Mexia 5842* (US); *Sazima & Semir 3885* (SP); *Williams & Assis 6347, 6552* (GH).
Calolisianthus speciosus (Cham. & Schltdl.) Gilg: *Irwin et al. 5271, 18135* (NY); *Mimura 395* (US); *Williams & Assis 7091* (GH).
Canscora diffusa (Vahl) R. Br. ex Roem. & Schult.: *Wilde et al. 4731* (WAG).
Canscora heteroclita (L.) Gilg: *Robyns 7206* (BR).
Celiantha bella Maguire & Steyerm.: *Liesner & Delascio 22126* (MO).
Celiantha imthurniana (Oliver) Maguire: *Bechine 114* (MY); *Steyermark 58794* (VEN); *im Thurn 306* (K).
Centaurium arizonicum (A. Gray) Heller: *Blumer 1792* (U).
Centaurium erythraea Raf.: *Delvosalle s.n.* (BR).

Centaurium linariifolium (Lam.) G. Beck: *Klinggräft s.n.* (AMD).
Centaurium littorale (Turner) Gilmour: *Hansen 6897* (BR).
Centaurium venustum Rob.: *Heller 6849* (L).
Chelonanthus alatus (Aubl.) Pulle: *Lanjouw & Lindeman 64* (U); *Leeuwenberg 11809* (WAG); *Maas et al. 3759* (U); *Maas & Steyermark 5349* (U); *Martius s.n.* (M); *Mexia 9144* (U); *Steyermark 88754* (U), *87994* (VEN); *Wilbur & Weaver 10869* (US).
Chelonanthus albus (Spruce ex Progel) Badillo: *Alencar 686* (U); *Pessoal do CPF 6188* (U); *Prance et al. 5002* (U); *Wurdack & Adderley 43593* (US).
Chelonanthus angustifolius H. B. K. (Gilg): *Maas et al. 5153, 5380, 5381* (U); *Soukup 5145* (US).
Chelonanthus longistylus (J. G. M. Pers. & Maas) Struwe & V. A. Albert: *Lanjouw & Lindeman H64* (U); *Leeuwenberg 11809* (WAG); *Maas & Steyermark 5349* (U); *Steyermark 88754* (U).
Chelonanthus matogrossensis (J. G. M. Pers. & Maas) Struwe & V. A. Albert: *Irwin et al. 17270* (U); *Kirkbride & Lleras 3016* (U); *Malme s.n.* (S).
Chelonanthus purpurascens (Aubl.) Struwe, S. Nilsson, & V. A. Albert: *Alencar 402* (U); *Coêlho 4* (INPA); *de la Cruz 1620* (NY); *Ducke RB25595* (RB); *Fosberg & Holdridge 19380* (US); *de Granville 1068* (CAY); *Harley 15551* (U): *Harrison 1345* (K); *Hitchcock 16955* (GH); *Klug 2345* (S).
Chelonanthus viridiflorus (Mart.) Gilg: *Irwin et al. 14609* (NY); *Mosèn 1479* (S); *Pedersen 12219* (C).
Chionogentias cf. *patula* (Kirk) L. Adams: *WRG 5-1947* (L).
Chironia baccifera L.: *Coppejans 701* (BR); *Goldblatt 1397* (WAG).
Chironia densiflora Scott-Elliot: *Wilms 975* (AMD).
Chironia gratissima S. Moore: *Dubois 1408* (WAG).
Chironia transvaalensis Gilg: *Wilms 974* (AMD).
Chorisepalum ovatum Gleason: *Liesner 24921* (NY).
Chorisepalum psychotrioides var. *acuminatum* (Steyerm.) Maguire: *Maguire et al. 46833* (MY).
Chorisepalum rotundifolium Ewan: *MAC s.n.* (U).
Cicendia filiformis (L.) Delarbre: *Hubbeling s.n.* (U 94076).
Cicendia quadrangularis (Lam.) Griseb.: *Buchtien s.n.* (L).
Comastoma tenellum (Rottb.) Toyok.: *Kordeva & Sytina 1336* (L).
Congolanthus longidens (N. E. Br.) A. Raynal: *Coomans s.n.* (AMD 022584); *Couteaux 60* (L); *Descoings 7810* (L).
Cotylanthera tenuis Blume: *Korthals 622* (L); *Carr 16396* (L); *Expedition Lundquist 549* (L).
Coutoubea humilis Sandwith: *Maguire & Fanshawe 32592* (U).
Coutoubea ramosa Aubl.: *Croat 20178* (L).
Coutoubea ramosa var. *ramosa* Aubl.: *Oldeman 2594* (U).
Coutoubea reflexa Benth.: *dos Santos & Coelho 708* (U).
Coutoubea spicata Aubl.: *Broadway 3846* (L); *Ginés 1236* (NY); *Cooper 333* (U).
Crawfurdia speciosa Wall.: *Strachey & Winterbottom s.n.* (BR).
Curtia conferta (Mart.) Knobl.: *Bunting et al. 4111* (MY); *Huber 5688* (U).
Curtia confusa Grothe & Maas: *Brade s.n.* (SP), *5891* (S, SP); *Dusén 4392* (S); *Sellow s.n.* (K).

Curtia diffusa (Mart.) Cham.: *Damazio 2048* (RB); *Mexia 5750* (NY).
Curtia sp. nov.: *Hoffman 3222* (NY, U, US).
Curtia tenuifolia ssp. *tenuifolia* (Aubl.) Knobl.: *Allen 994* (U); *Brade 13618, 17684* (RB); *Damazio 2057* (RB); *Dusén 8003a* (S); *Fiebrig 5250 = 4991* (L); *Heringer 8536/730* (NY); *Maas & Westra 4340* (U); *Koyama & Agostini 7334* (VEN); *Pedersen 8740* (C); *Pereira 1570* (RB); *Prance 4486* (U); *Progel 1618* (LE); *Schomburgk 167* (BM); *Silva 697* (MG).
Curtia tenuifolia ssp. *tenella* (Mart.) Grothe & Maas: *Brenes 220* (F); *Johnston 877* (GH); *Hinton 5040* (NY); *Koyama & Agostini 7390* (NY); *Pittier 12788* (VEN); *Pohl 5211* (W); *Standley 14548* (F); *Steyermark 13218* (F).
Curtia verticillaris (Spreng.) Knobl.: *Blanchet s.n.* (G); *Duarte 2148* (RB), *9210* (U); *Mexia 5817* (F); *Mori & Boom 14363* (NY).

Deianira erubescens Cham. & Schltdl.: *Riedel 1990* (BR).
Deianira pallescens Cham. & Schltdl.: *Irwin & Soderstrom 5920* (U).
Djaloniella ypsilostyla P. Taylor: *Morton 2442* (WAG); *Schnell 7375* (BR).

Enicostema axillare (Lam.) A. Raynal: *van Steenis 19537 (L)*.
Enicostema axillare (Lam.) A. Raynal ssp. *axillare*: *Sultan Abedin 3431* (NY); *Kotschy Iter Nubicum 224* (WAG).
Enicostema verticillatum (L.) Engl. ex Gilg: *Scheffler 219* (AMD).
Eustoma exaltatum (L.) Salisb. ex G. Don: *Ekman 11204* (K); *Lot et al. D 014130* (U).
Eustoma grandiflorum (Raf.) Shinners: *Thomas & Taylor 85509, 5010* (L).
Exacum hamiltonii G. Don: *Hooker s.n.* (U).
Exacum tetragonum Roxb.: *Clemens 10809* (BR).
Exacum walkeri Arn. ex Griseb.: *Jayasuriya & Robyns 82* (L).
Exochaenium exiguum A. W. Hill: *Wild 7960* (WAG).

Fagraea obovata Wall.: *Lörzing 14119 (L)*.
Faroa acaulis R. E. Fr.: *Goethals s.n.* (LV).
Faroa pusilla Baker: *Olaille 4926* (L).
Faroa salutaris Welw.: *B. Fritzche 291* (AMD); *Malaisse 12085* (BR).
Frasera caroliniensis Walter: *Hill & Dorsey 24482* (NY).
Frasera speciosa Dougl. ex Hook.: *Stanford et al. 2485* (U).

Geniostemon schaffneri Engelm. & Gray: *Pringle 4938 (BR)*.
Gentiana asclepiadea L.: *Hortus Botanicus Amsterdam*, no voucher.
Gentiana cruciata L.: *Hortus Botanicus Amsterdam*, no voucher.
Gentiana lutea L.: *Goethals s.n.* (LV).
Gentiana pneumonanthe L.: field collection in the Netherlands, no voucher.
Gentiana sedifolia H. B. K.: *Hitcock 21075* (US).
Gentianella campestris (L.) Harry Sm.: *Wittewaal s.n.* (U 61233).
Gentianella cunninghamii ssp. *cunninghamii*: *Adams 3323* (L).
Gentianella germanica ssp. *germanica* (Willd.) E. F. Warb.: *Boom 36009* (L).
Gentianopsis barbata (Froel.) Ma: *Martens s.n.* (BR).
Gentianopsis ciliata (L.) Ma: *Suringar 1865* (L).
Gentianopsis crinita (Froel.) Ma: coll. Marie Bouillé (Québec, Canada), no voucher.

Gentianopsis detonsa (Rottb.) Ma: *Oosterveld 0246* (U).
Gentianothamnus madagascariensis Humbert: *Malcomber et al. 869* (UPS).

Halenia corniculata (L.) Cornaz: *Togasi 1227* (BR).
Halenia dasyantha Gilg: *Killip & Lehmann 38545* (US).
Halenia foliosa Gilg: *Cuatrecasas 23108* (US).
Helia oblongifolia Mart.: *Brade 6132* (S); *Löfgren 1109* (C); *Pedersen 12216* (C); *Smith et al. 14566* (US).
Hockinia montana Gardner: *Martinelli 227* (U); *Martinelli & Simonis 8999* (U).
Hoppea dichotoma Willd.: *Adam 26711* (BR, WAG).
Hoppea fastigiata C. B. Clarke: *Hohenacker 581* (U).

Irlbachia nemorosa (Willd. ex Roem. & Schult.) Merr.: *Badillo 6144* (MY); *Bunting 4129* (MY); *Eavelo 15* (MY); *Cardona 2079, 2694, 2958, 2978* (US), *2986* (MO); *Maas et al. 4240, 5369, 5370* (U); *Maguire 27324* (S), *36396* (US); *von Martius s.n.* (M); *Moore 9663* (UC), *9755* (US); *Phelps 517* (US); *Prance 3721* (U, US), *4866, 18042* (U); *Schomburgk 147* (BM); *Schultes & Cabrera 162966* (GH), *16523* (US); *Silva & Santos 4704* (U); *Spruce 2524, 3055* (BR); *Steward Pr 20262* (U); *Steyermark 227, 345* (F), *732, 57929, 59087* (US), *97961* (G), *113399* (VEN); *Tate 1360* (US); *Trujillo 13567* (MY); *Ule 8731* (MG); *Williams 13937* (US); *Wurdack & Adderley 42926* (US).
Irlbachia plantaginifolia Maguire: *Bunting 3991* (U); *Maguire et al. 42593* (VEN); *Steyermark 102825* (US); *Williams 14202* (F).
Irlbachia poeppigii (Griseb.) L. Cobb & Maas: *Ducke RB 22386* (RB); *Holt & Blake 566* (US); *Lisboa 793* (INPA); *Nascimento 391* (INPA); *Pires et al. 13962* (U); *Poeppig 2854* (GOET); *Schomburgk 789* (K).
Irlbachia pratensis (H. B. K.) L. Cobb & Maas: *Huber 2665* (U); *Maguire 34694* (US), *37628* (S); *Pires & Leite 14803* (U); *Spruce 2005* (GOET); *Steyermark & Redmond 112805* (VEN); *Wurdack & Adderley 42949* (US).
Irlbachia pumila (Benth.) Maguire: *Level 53* (S); *Maguire 34502* (S): *Schultes & Lopez 9902* (US); *Spruce 2950* (BM).
Irlbachia tatei (Gleason) Maguire: *Cardona 376* (US); *Maguire 29556, 29625* (VEN); *Steyermark et al. 58167* (US), *58297, 108877* (F); *Tillett 751-106* (U).
Ixanthus viscosus (Aiton) Griseb.: *Bouharmont 25535* (BR); *D'Alleizette 4941* (L); *Bourgeau 1420* (WAG).

Karina tayloriana Boutique: *Schmitz 4606* (BR).

Lagenanthus princeps (Lindl.) Gilg: *Maas & Tillett 5242* (U); *Steyermark 57322* (F).
Lehmanniella splendens (Hook.) Ewan: *Cuatrecasas & Cowan 27218* (US); *Denslow 2358* (WIS); *St. John 20561* (US); *Zarucchi 3351* (NY).
Lisianthius adamsii Weaver: *Webster & Wilson 5025* (BM).
Lisianthius exsertus Sw.: *Webster & Wilson 5025* (BM); *Wiles 5188* (G).
Lisianthius glandulosus A. Rich.: *Linden 2014* (G).
Lisianthius longifolius L.: *Hart 1018* (G); *Hummel s.n.*, Apr 1958 (GB); *Linden 1672* (G).
Lisianthius meianthus Donn. Sm.: *Skutch 1568* (GH).
Lisianthius nigrescens var. *nigrescens* Cham. & Schltdl.: *Purpus s.n.* (U); *Bourgeau 2969* (U).

Lisianthius saponarioides Cham. & Schltdl.: *Contreras 6205* (G).
Lisianthius skinnerii (Hemsl.) Kuntze: *Dwyer 7771* (K); *Weaver & Wilbur 2249* (BM).
Lomatogonium carinthiacum (Wulf.) Rchb.: *Braun-Blanquet 856* (U); *van Lookeren 77* (L); *Martens s.n.* (BR).
Lomatogonium rotatum Fr. ex Nym.: *Buer s.n.* (L); *Herb. Hooker s.n.* (U).

Macrocarpaea cinchonifolia (Gilg) Weaver: *Bang s.n.* (NY).
Macrocarpaea corymbosa (Ruiz & Pav.) Ewan: *Ruiz & Pavón s.n.* (F 842844).
Macrocarpaea glabra (L.f.) Gilg: *Maas & Plowman 1794* (U).
Macrocarpaea glaziovii Gilg: *Gardner 539* (BM).
Macrocarpaea luteynii J. R. Grant & Struwe: *Luteyn 7466* (U).
Macrocarpaea macrophylla (H. B. K.) Gilg: *Uribe Uribe 1858* (U).
Macrocarpaea micrantha Gilg: *Wolfe 12269A* (F).
Macrocarpaea ovalis (Ruiz & Pav.) Ewan: *Camp 4795* (S).
Macrocarpaea rubra Malme: *Lindeman & de Haas 4071* (U).
Macrocarpaea rugosa Steyerm.: *Huber 8754* (K).
Macrocarpaea sp. nov., ined. (J. R. Grant, unpubl.): *Madison 3956* (NY).
Microrphium pubescens C. B. Clarke: *van Balgooy 2217* (L).

Neblinantha neblinae Maguire: *Plowman & Thomas 13645* (U).
Neurotheca corymbosa Hua: *Wieringa 1140* (WAG).
Neurotheca loeselioides (Spruce ex Progel) Baill.: *vanden Berghen 1264* (BR); *Chevalier 6128* (L).
Neurotheca loeselioides ssp. *loeselioides* (Spruce ex Progel) Baill.: *Maguire 24213* (U).

Obolaria virginica L.: *Curtiss s.n.* (NY); *Wilkens 2726* (L).
Oreonesion testui A. Raynal: *Jongkind 1811* (WAG); *Le Testu 8972* (BR).
Ornichia madagascariensis (Baker) Klack.: *Leeuwenberg 14194* (WAG).
Ornichia madagascariensis ssp. *pubescens* (Baker) Klack.: *Phillipson et al. 3973* (BR).
Orphium frutescens (L.) E. Mey.: *Lotsy & Goddijn 1571* (L); *Rehmann 1969* (BR); *Rodin 3295* (WAG).

Potalia amara Aubl.: *Jansen-Jacobs et al. 2567* (U).
Prepusa alata Porto & Brade: *Santos Lima 185* (RB).
Prepusa connata Gardner: *Glaziou 3813* (P).
Prepusa hookeriana Gardner: *von Lützelburg s.n.* (M).
Prepusa montana Mart.: *Furlan et al. CFCR1926* (U); *von Martius 2108* (M).
Prepusa viridiflora Brade: *Brade 19278* (RB).
Purdieanthus pulcher (Hook.) Gilg: *Linden 1217* (BM, K, W); *Purdie s.n.* (K); *Triana s.n.* (BM).
Pycnosphaera buchananii (Baker) N. E. Br.: *Jacques-Felix 7416* (L); *Richards 5696* (BR).

Rogersonanthus arboreus (Britton) Maguire & B. M. Boom: *Broadway s.n.* (US).
Rogersonanthus quelchii (N. E. Br.) Maguire & B. M. Boom: *Broadway 6207*

(BM), *s.n.* (US); *Maas & Steyermark 5363* (U); *Steyermark 93975, 117312B* (U); *Tate 1361* (US).
Rogersonanthus tepuiensis (Gleason) Maguire & B. M. Boom: *Steyermark & Wurdack 507* (NY).
"Roraimaea", ined. (L. Struwe, S. Nilsson, & V. A. Albert, unpubl.): *Cid Ferreira 9125* (U).

Sabatia angularis (L.) Pursh: *Holthuis 497* (L); *Frye s.n.* (U).
Sabatia brevifolia Raf.: *Genelle & Fleming 989* (U).
Sabatia dodecandra (L.) Britton, Sterns, & Poggenb.: *Bell 4896* (BR).
Sabatia gentianoides Ell.: *Bijhouwer 249* (WAG); *Webster & Wilbur 3212* (NY).
Sabatia stellaris Pursh: *Harriman 13482* (L).
Saccifolium bandeirae Maguire & Pires: *Farney et al. 897* (MO).
Schinziella tetragona (Schinz) Gilg: *Carlier 349* (BR).
Schultesia australis Griseb.: *Hatschbach 4518* (L); *Irwin & Soderstrom 6133* (U).
Schultesia brachyptera Cham.: *ex herb. P s.n.* (L 903251-217).
Schultesia crenuliflora Mart.: *de Carvalho et al. 1068* (U).
Schultesia guianensis (Aubl.) Malme: *Berg BG661* (NY, U).
Schultesia stenophylla var. *latifolia* Mart.: *Morton 1467* (WAG).
Sebaea albidiflora F. Muell.: *Meebold 7276* (BR); *Morrison s.n.* (AMD 087412); *van Steenis 23468* (L).
Sebaea brachyphylla Griseb.: *Bamps 2826* (WAG).
Sebaea cordata R. Br.: *Wright 1853–56* (L).
Sebaea grandis E. Mey.: *Dinter 5757* (AMD); *Wilms 969* (AMD).
Sebaea oreophila Gilg: *Stolz 728* (U).
Sebaea ovata (Labill.) R. Br.: unknown Australian collector, *s.n.* (U 12139).
Senaea caerulea Taub.: *Glaziou 19739* (K); *Santos Lima & Brade 14215* (U).
Sipapoantha ostrina Maguire & B. M. Boom: *Steyermark et al. 124554* (NY); *Steyermark et al. 105123* (U).
Swertia bimaculata Hook.f. & Thomson ex C. B. Clarke: *Ohashi et al. 774316* (L).
Swertia chirata Buch.-Ham. ex Wall: *Griffith 5831* (U).
Swertia macrocarpa (C. B. Clarke) C. B. Clarke: *Griffith 5829* (U).
Swertia perennis L.: *Goethals s.n.* (LV).
Swertia usambarensis Engl.: *Reekmans 10703* (WAG).
Symbolanthus calygonus s. lat. (Ruiz & Pav.) Griseb.: *Burger et al. 9355* (F); *Cuatrecasas 8024* (F); *Forero & Kirkbride 659* (U); *Fosberg 19111* (?); *Gillis & Plowman 10031* (A); *Killip & Smith 25433* (US); *Lawrance 140* (F); *Maas et al. 5009, 5284* (U); *Steyermark 55082* (F).
Symphyllophyton cf. *caprifolioides* Gilg: *Prance & Silva 58543* (U).

Tachia gracilis Benth.: *Tillett 752275* (U).
Tachia grandifolia Maguire & Weaver: *Prance et al. 21614* (U).
Tachia guianensis Aubl.: *de Granville 563* (P, U).
Tachiadenus carinatus (Desr.) Griseb.: *Gereau & Dumetz 3241* (WAG); *Lam & Meeuse 5498* (L).
Tachiadenus platypterus Baker: *d'Alleizette s.n.* (L 950341.515).
Tapeinostemon longiflorum Maguire & Steyerm.: *Funk 6723* (U).

Tapeinostemon spenneroides Benth.: *Hoffman & Marco 2121* (NY); *Steyermark et al. 130227* (U).
Tetrapollinia caerulescens (Aubl.) Maguire & B. M. Boom: *Duarte 2558* (RB); *Dusén 2647* (BM); *Harley et al. 15428* (U); *Heringer 11266* (S); *Hunt 6065* (SP); *Irwin et al. 15284* (RB), *17228* (F), *55242* (MO); *Löfgren 2055* (NY); *Maas et al. 3715, 5379* (U); *Maguire 30286, 33789* (VEN); *Malme 1964* (S); *Mori et al. 8027* (NY); *Pires & Cavalcante 52039* (S); *Rosa & Santos 1974* (NY); *Sastre 1333* (CAY); *Sidney et Onishi 1400* (S); *Silva 706* (MG); *Smith 2296* (F).
Tripterospermum fasciculatum (Wall.) Chater: *Rahmal Si Boeea 10126* (L).
Tripterospermum japonicum (Siebold & Zucc.) Maxim.: *Im & Kawahara 7813* (WAG); no collector, *s.n.* (AMD 020110).
Tripterospermum luzonensis (Vidal) Murata: *Eijma 1326* (L).
Tripterospermum sumatranum Murata: *Kessler 60* (L).
Tripterospermum trinerve Blume: *van Balgooy 2045* (L); *Du Mortier s.n.* (BR).

Urogentias ulugurensis Gilg & Gilg-Ben.: *Schlieben 2786* (BR).

Veratrilla baillonii Franch.: *Smith 4448* (UPS).
Voyria acuminata Benth.: *Holt & Blake 466* (US).
Voyria aphylla (Jacq.) Pers.: *Hammel 3439* (MO); *Nicolson 1929* (U); *Teunissen LBB15185, LBB15200* (U).
Voyria aurantiaca Splitg.: *Florschütz & Florschütz 1635* (U); *de Granville 933* (U); *Hijman & Weerdenburg 216* (U); *Maas et al. 2250* (U).
Voyria caerulea Aubl.: *Forest Dept. Brit. Guiana 2979* (K); *Kramer & Hekking 2812* (U); *Maas et al. 2307* (U); *Steyermark & Bunting 102554* (US).
Voyria chionea Benth.: *Poole 1863* (U).
Voyria clavata Splitg.: *Jenman s.n.* (K).
Voyria corymbosa ssp. *corymbosa* Splitg.: *Florschütz & Florschütz 1449* (U).
Voyria corymbosa ssp. *alba* (Standl.) Ruyters & Maas: *Maas et al. 2832* (U).
Voyria flavescens Griseb.: *Prance et al. 10370* (U); *Gehrt 32218* (SP).
Voyria kupperi (Suess.) Ruyters & Maas: *Liesner 1794* (U).
Voyria obconica Progel: *dos Santos & White 10384* (U).
Voyria parasitica (Schltdl. & Cham.) Ruyters & Maas: *Britton & Millspaugh 6109* (F); *Harris 9176* (NY); *Plowman 6339* (U).
Voyria pittieri (Standl.) L. O. Williams: *Duke 13624* (US); *Forest Dept. Brit. Guiana 4739* (K).
Voyria primuloides Baker: *de Wilde et al. s.n.* (WAG).
Voyria rosea Aubl.: *Maas LBB11045* (U); *Maas & Westra 4245* (U); *Lindeman 5769* (U); *Versteeg 9* (U); *de Goeie s.n.* (U).
Voyria spruceana Benth.: *van Donselaar 1777* (U); *Prance et al. 15754* (U).
Voyria tenella Hook.: *Lindeman 4431a* (U); *Maas LBB10793* (U); *Maas & Dressler 2699* (U).
Voyria tenuiflora Griseb.: *Florschütz & Maas 3156* (U); *Jonker & Jonker 618* (U); *Lanjouw & Lindeman 2722* (U).
Voyria truncata (Standl.) Standl. & Steyerm.: *Folsom et al. 6759* (MO); *Maas et al. 2834* (U); *Steyermark 41610* (F).

Voyriella parviflora (Miq.) Miq.: *Forest Dept. Brit. Guiana 3549* (K); *Maas et al. 2229, 2492, 2525, LBB 10767* (U); *Maas & Maas 2678* (U); *Oliveira 5436* (IAN).

Wurdackanthus frigidus (Sw.) Maguire & B. M. Boom: von Eggers 6904 (L, US); Smith & Smith K18 (K).

Xestaea lisianthoides Griseb.: *Stevens 7103, 8637* (U).

Zonanthus cubensis Griseb.: *Wright 1346* (GH); *Ekman 15030* (S).

Zygostigma australe (Cham. & Schltdl.) Griseb.: *Lindeman & Lindeman 9455* (U); *Smith et al. 9375* (US).

LITERATURE CITED

Albert, V. A. & L. Struwe. 1997. Phylogeny and classification of *Voyria* (saprophytic Gentianaceae). *Brittonia* 49: 466–479.

Barthlott, W. 1981. Epidermal and seed surface characters of plants: systematic applicability and some evolutionary aspects. *Nord. J. Bot.* 1: 345–355.

Barthlott, W. & N. Ehler. 1977. Raster-Elektronenmikroskopie der Epidermis-Oberflächen von Spermatophyten. *Trop. Subtrop. Pflanzenwelt.* 19: 1–105.

Boesewinkel, F. D. & F. Bouman. 1984. The seed: structure. Pages 567–610 in: B. M. Johri, ed. *Embryology of angiosperms.* Springer-Verlag, Berlin.

Boesewinkel, F. D. & F. Bouman. 1995. The seed: structure and function. Pages 1–24 in: J. Kigel and G. Galili, eds. *Seed development and germination.* Marcel Dekker, New York.

Bouman, F. & N. Devente. 1986. Seed micromorphology in *Voyria* and *Voyriella* (Gentianaceae). Pages 9–25 in: P. J. M. Maas & P. Ruyters, eds. Voyria *and* Voyriella *(saprophytic Gentianaceae).* Flora Neotropica Monograph 41. The New York Botanical Garden, Bronx, NY.

Bouman, F. & A. Louis. 1990. Seed structure in *Voyria primuloides* Baker (Gentianaceae): taxonomic and ecological implications. Pages 261–270 in: J. Paré, M. Bugnicourt, J. Mortier, M. Juguet, F. Vignon & J. Vignon, eds. *Some aspects and actual orientations in plant embryology.* Université de Picardie, Amiens.

Bouman, F. & S. Schrier. 1979. Ovule ontogeny and seed coat development in *Gentiana*, with a discussion on the evolutionary origin of the single integument. *Acta Bot. Neerl.* 28: 467–478.

Burger, H. D. 1972. *Seedlings of some tropical trees and shrubs, mainly southeast Asia.* Pudoc, Wageningen.

Cobb, L. & P. J. M. Maas. 1983. Seed coat micromorphology in *Irlbachia* (Gentianaceae). *Proc. Kon. Ned. Akad. Wetensch.*, ser. C, 86: 127–136.

Corner, E. J. H. 1976. *The seeds of dicotyledons.* Cambridge University Press, Cambridge.

Davis, G. L. 1966. *Systematic embryology of the angiosperms.* John Wiley & Sons, New York.

Docters van Leeuwen, W. M. 1934. Endozoische Samenverbreitung durch den Indischen Purpurstar: *Aplonis panayensis strigatus* Horsf. *Ber. Deutsch. Bot. Ges.* 52: 284–290.

Drexler, U. & M. I. Hakki. 1979. Embryologische und morphologische Untersuchungen an Pflanzen aus Westindien. 2. Zur Embryologie von *Eustoma exaltatum* (Gentianaceae), mit einer Bemerkung zum Phänomen der "instant pollen tubes". *Willdenowia* 9: 131–147.

Favarger, C. 1953. Sur la germination des gentianes. *Phyton* 4: 275–289.

Figdor, W. 1897. Ueber *Cotylanthera* Bl. Ein Beitrag zur Kenntnis tropischer Saprophyten. *Ann. Jard. Bot. Buitenzorg* 14: 213–247.

Gilg, E. 1895. Gentianaceae. Pages 50–108 in: A. Engler & K. Prantl, eds. *Die natürlichen Pflanzenfamilien*, vol. 4(2). Verlag von Wilhelm Engelmann, Leipzig.

Groenendijk, J. P., A. T. J. van Dulmen, & F. Bouman. 1997. The "forest floor" saprophytes *Voyria spruceana* and *V. aphylla* (Gentianaceae) growing as epiphytes in Colombian Amazonia. *Ecotropica* 3: 129–131.

Grothe, E. H. M. & P. J. M. Maas. 1984. A scanning electron microscopic study of the seed coat structure of *Curtia* Chamisso and Schlechtendal and *Hockinia* Gardner (Gentianaceae). *Proc. Kon. Ned. Akad. Wetensch.*, ser. C., 87: 33–42.

Guérin, P. 1903. Sur le sac embryonnaire et en particulier sur les antipodes des Gentianes. *J. Bot. (Morot)* 17: 101–108.

Guérin, P. 1904. Recherches sur le développement et la structure anatomique du tégument séminal des Gentianacées. *J. Bot. (Morot)* 18: 33–52, 83–88.

Halda, J. J. 1996. *The genus* Gentiana. SEN, Dobré.

Ho, T.-N. & S.-W. Liu. 1990. The infrageneric classification of *Gentiana* (Gentianaceae). *Bull. Brit. Mus. (Nat. Hist.), Bot.* 20: 169–192.

Holmgren, P. K., N. H. Holmgren, & L. C. Barnett. 1990. *Index Herbariorum*, part 1, *The herbaria of the world*, ed. 8. The New York Botanical Garden, Bronx, NY.

Imhof, S., H. C. Weber, & L. D. Gomez. 1994. Ein Beitrag zur Biologie von *Voyria tenella* Hook. und *Voyria truncata* (Standley) Standley et Steyermark (Gentianaceae). *Beitr. Biol. Pflanzen.* 68: 113–123.

Johri, B. M., K. B. Ambegaokar, & P. S. Srivastava. 1992. *Comparative embryology of angiosperms*, vol. 2. Springer-Verlag, Berlin.

Klackenberg, J. 1983. A reevaluation of the genus *Exacum* (Gentianaceae) in Ceylon. *Nord. J. Bot.* 3: 355–370.

Klackenberg, J. 1985. The genus *Exacum* (Gentianaceae). *Opera Bot.* 84: 1–144.

Klackenberg, J. 1986. The new genus *Ornichia* (Gentianaceae) from Madagascar. *Adansonia* 8: 195–206.

Klackenberg, J. 1987. Revision of the genus *Tachiadenus* (Gentianaceae). *Adansonia* 9: 43–80.

Leeuwenberg, A. J. M. 1980. Loganiaceae. Pages 1–225 in: *Engler and Prantl's Die natürlichen Pflanzenfamilien, Angiospermae: Ordnung Gentianales*, vol. 28b (1). Duncker and Humblot, Berlin.

Lubbock, J. 1892. *A contribution to our knowledge of seedlings*, vols. 1 and 2. Kegan Paul, Trench, Trübner and Co., London.

Luijten, S. H., J. G. B. Oostermeijer, A. C. Ellis-Adam, & H. C. M. den Nijs. 1998. Reproductive biology of the rare biennial *Gentianella germanica* compared with other gentians of different life history. *Acta Bot. Neerl.* 47: 325–336.

Maas, P. J. M., S. Nilsson, A. M. C. Hollants, B. J. H. ter Welle, H. Persoon, & E. C. H. van Heusden. 1983. Systematic studies in neotropical Gentianaceae – the *Lisianthius* complex. *Acta Bot. Neerl.* 32: 371–374.

Maguire, B. & J. M. Pires. 1978. Saccifolieaceae – a new monotypic family of the Gentianales. Pages 230–245 in: B. Maguire and collaborators, eds. The botany of the Guayana Highland – Part X. *Mem. New York Bot. Gard.* 29.

Maheswari Devi, H. 1962. Embryological studies in Gentianaceae (Gentianoideae and Menyanthoideae). *Proc. Indian Acad. Sci.*, sect. B, 56: 195–216.

Maiti, G. & M. L. Banerji. 1976. Exomorphic seed structure of the Himalayan species of *Swertia* Linn. (Gentianaceae). *Proc. Indian Acad. Sci.*, sect. B, 84: 231–237.

Marais, W. & I. C. Verdoorn. 1963. Gentianaceae. Pages 171–243 in: R. A. Dyer, L. E. Codd, & H. B. Rycroft, eds. *Flora of southern Africa*, vol. 26. Department of Agricultural Technical Services, Republic of South Africa, Pretoria.

Martin, A. C. 1946. The comparative internal morphology of seeds. *Amer. Midl. Naturalist* 36: 513–660.

McCoy, R. W. 1949. On the embryology of *Swertia caroliniensis*. *Bull. Torrey Bot. Club* 76: 430–439.

Miège, J. & J. Wüest. 1984. Les surfaces tégumentaires des graines de *Gentiana* et *Gentianella* vues au microscope électronique à balayage. *Bot. Helv.* 94: 41–59.

Miquel, S. 1985. *Plantules et premiers stades de croissance des espèces forestières du Gabon: potentialité d'utilisation en agroforesterie*. PhD thesis. Université Pierre et Marie Curie, Paris.

Mohrbutter, C. 1936. Embryologische Studien an Loganiaceen. *Planta* 26: 64–80.

Muller, F. M. 1978. *Seedlings of the north-western European lowland*. Junk, The Hague and Boston.

Müller, G. 1982. Contribution à la cytotaxonomie de la section *Cyclostigma* Griseb. de genre *Gentiana* L. *Feddes Repert.* 93: 625–722.

Müller, P. 1955. Verbreitungsbiologie der Blütenpflanzen. *Veröff. Geobot. Inst. ETH, Stiftung Rübel, Zürich* 30: 1–152.

Müller-Schneider, P. 1985. Verbreitungsbiologie der Blütenpflanzen Graubündens. *Veröff. Geobot. Inst. ETH, Stiftung Rübel, Zürich* 85: 1–263.

Murata, J. 1989. A synopsis of *Tripterospermum* (Gentianaceae). *J. Fac. Sci., Univ. Tokyo*, sect. 3, *Bot.* 14: 273–339.

Netolitzky, F. 1926. *Anatomie der Angiospermen-Samen. Handbuch der Pflanzenanatomie*, Band 10. Borntraeger, Berlin.

Nicholas, A. & H. Baijnath. 1994. A consensus classification for the order Gentianales with additional details on the suborder Apocynineae. *Bot. Rev.* 60: 440–482.

Nilsson, S. 1968. Pollen morphology in the genus *Macrocarpaea* (Gentianaceae) and its taxonomical significance. *Svensk Bot. Tidskr.* 62: 338–364.

Nilsson, S. 1970. Pollen morphological contributions to the taxonomy of *Lisianthus* L. s. lat. (Gentianaceae). *Svensk Bot. Tidskr.* 64: 1–43.

Nilsson, S. 2002. Gentianaceae: a review of palynology. Pages 377–497 in: L. Struwe & V. A. Albert, eds. *Gentianaceae: systematics and natural history*. Cambridge University Press, Cambridge.

Nilsson, S. & J. J. Skvarla. 1969. Pollen morphology of saprophytic taxa in the Gentianaceae. *Ann. Missouri Bot. Gard.* 56: 420–438.

Oehler, E. 1927. Entwicklungsgeschichtliche-zytologische Untersuchungen an einigen saprophytischen Gentianaceen. *Planta* 3: 641–733.

Oostermeijer, G. 1996. *Population viability of the rare* Gentiana pneumonanthe. PhD thesis. University of Amsterdam, Amsterdam.

Periasamy, K. 1962. The ruminate endosperm: development and types of rumination. Pages 62–74 in: *Plant embryology: a symposium*. CSRI, New Delhi.

Petanidou, T., H. C. M. den Nijs, J. G. B. Oostermeijer, & A. C. Ellis-Adam. 1995. Pollination ecology and patch-dependent reproductive success of the rare perennial *Gentiana pneumonanthe* in the Netherlands. *New Phytol.* 129: 155–163.

Pringle, J. S. 1995. Family 159A. Gentianaceae. Pages 1–131 in: G. Harling & L. Andersson, eds. *Flora of Ecuador*, vol. 53. Department of Systematic Botany, Gothenburg University, Göteborg.

Raynal, A. 1967. Sur un *Sebaea* Africain saprophyte (Gentianaceae). *Adansonia*, sér. 2, 7: 207–219.

Roitman, J. N., E. Wollenweber, & F. J. Arriaga-Giner. 1992. Xanthones and triterpene acids as leaf exudate constituents in *Orphium frutescens. J. Pl. Physiol.* 139: 632–634.

Smith, H. 1965. Notes on Gentianaceae. *Notes Roy. Bot. Gard. Edinburgh* 26: 237–258.

Stearn, W. T. 1983. *Botanical Latin*, rev. ed. 3. Nelson, London.

Stolt, K. A. H. 1921. Zur Embryologie der Gentianaceen und Menyanthaceen. *Kongl. Svenska Vetenskapsakad. Handl.* 61(14): 1–56.

Struwe, L., P. J. M. Maas, & V. A. Albert. 1997. *Aripuana cullmaniorum*, a new genus and species of Gentianaceae from white-sands of southeastern Amazonas, Brazil. *Harvard Pap. Bot.* 2: 235–253.

Struwe, L., M. Thiv, J. W. Kadereit, A. S.-R. Pepper, T. J. Motley, P. J. White *et al.* 1998. *Saccifolium* (Saccifoliaceae), an endemic of Sierra de la Neblina on the Brazilian–Venezuelan frontier, is related to a temperate-alpine lineage of Gentianaceae. *Harvard Pap. Bot.* 3: 199–214.

Struwe, L., J. W. Kadereit, J. Klackenberg, S. Nilsson, M. Thiv, K. B. von Hagen, & V. A. Albert. 2002. Systematics, character evolution, and biogeography of Gentianaceae, including a new tribal and subtribal classification. Pages 21–309 in: L. Struwe & V. A. Albert, eds. *Gentianaceae: systematics and natural history*. Cambridge University Press, Cambridge.

Thiv, M., L. Struwe, V. A. Albert, & J. W. Kadereit. 1999. The phylogenetic relationships of *Saccifolium bandeirae* Maguire & Pires (Gentianaceae) reconsidered. *Harvard Pap. Bot.* 4: 519–526.

Thompson, K., J. Bakker, & R. Bekker. 1997. *The soil seed banks of northwest Europe: methodology, density and longevity*. Cambridge University Press, Cambridge.

Threadgill, P. F., J. M. Baskin, & C. C. Baskin. 1981. Dormancy in seeds of *Frasera caroliniensis* (Gentianaceae). *Amer. J. Bot.* 68: 80–86.

Van der Pijl, L. 1982. *Principles of dispersal in higher plants*, rev. ed. 3. Springer-Verlag, Berlin, Heidelberg, and New York.
Wollenweber, E. 1989. Exkret-Flavonoide bei Blütenpflanzen und Farnen. *Naturwissenschaften* 76: 458–463.
Wong, K. M. & J. B. Sugau. 1996. A revision of *Fagraea* (Loganiaceae) in Borneo, with notes on related Malesian species and 21 new species. *Sandakania* 8: 1–93.
Yuan, Y. M. 1993. Seed coat micromorphology and its systematic implications for Gentianaceae of western China. *Bot. Helv.* 103: 73–82.

6
Chemotaxonomy and pharmacology of Gentianaceae

S. R. JENSEN AND J. SCHRIPSEMA

ABSTRACT

The occurrence of taxonomically informative types of compounds in the family Gentianaceae, namely iridoids, xanthones, mangiferin, and C-glucoflavones, has been recorded. The properties, biosynthesis, and distribution of each group of compounds are described. The iridoids (mainly secoiridoid glucosides) appear to be present in all species investigated, with a predominance of swertiamarin and/or gentiopicroside; c. 90 different compounds have been reported from 127 species in 24 genera. Xanthones are not universally present in Gentianaceae, but about 100 different compounds have been reported from 121 species in 21 genera. A coherent theory for the biosynthesis of xanthones, based partly on published biosynthetic results and partly on biosynthetic reasoning, is postulated and used to group the compounds into biosynthetic categories. Arranging the genera according to the xanthones present gives rise to four groups. Group 1 (*Anthocleista, Blackstonia, Gentianopsis, Macrocarpaea,* and *Orphium*) includes the taxa containing only few and biosynthetically primitive xanthones. Group 2 (*Comastoma, Gentiana, Gentianella, Lomatogonium, Swertia,* and, tentatively, *Tripterospermum*) contains xanthones with an intermediate degree of biosynthetic advancement. Group 3 (*Frasera, Halenia,* and *Veratrilla*) has the most advanced compounds, with the xanthones found in group 2 being the biosynthetic precursors. Group 4 (*Canscora, Centaurium, Chironia, Eustoma, Hoppea, Ixanthus,* and, with some reservation, *Schultesia*) contains another set of biosynthetically advanced compounds. A comparison of the above groups with strict consensus trees based on molecular data (*trn*L intron and *mat*K sequences) and the new classification proposed by Struwe *et al.* (2002)

shows very good correlation. On the evidence so far, members of the Exaceae do not contain xanthones. The taxa of group 1, with primitive xanthones, are found in several tribes (Chironieae, Gentianeae, Helieae, and Potalieae), while those of groups 2 and 3 comprise solely members of the Gentianeae. Finally, the taxa in group 4 are all members of the Chironieae. Mangiferin, a C-glucoxanthone with a biosynthesis different from the above xanthones, has been recorded from 42 species in seven genera. Of these, five belong to the Gentianeae and two to the Chironieae. The C-glucoflavones have been recorded from 78 species in nine genera. Three of these belong to the Potalieae, while the remaining six are members of the Gentianeae. Based on the above results, a tentative list of chemical characteristics for the tribes of the Gentianaceae is presented. Finally, some pharmacologically interesting properties of plant extracts or compounds from taxa within Gentianaceae are listed.

Keywords: biosynthesis, chemotaxonomy, Gentianaceae, pharmacology, secoiridoids, xanthones.

INTRODUCTION

The Gentianaceae contains many species with interesting phytochemical properties. They have been widely used in traditional medicine and also as constituents in bitters and similar concoctions. For this reason a considerable amount of chemical work has been performed on members of the family. In order to limit the information to the taxonomically most promising groups of compounds, we have concentrated our efforts on listing only the iridoids, the xanthones, mangiferin, and the C-glucoflavonoids, which are most characteristic for the family.

IRIDOIDS

The iridoids are a group of natural products belonging to the terpenoids, which are otherwise ubiquitous in plants. However, only a limited number of taxa possess the enzyme(s) that give rise to the cyclopentane ring that is characteristic for the carbocyclic iridoids. In plants, iridoids are usually found as glucosides and thus they are basically water soluble.

The distribution of iridoids has been shown to have considerable value as a systematic character (Jensen et al., 1975; Jensen, 1991, 1992) since they occur almost exclusively in the superorders Cornanae, Ericanae, Gentiananae, and Lamianae *sensu* Dahlgren (1989). Furthermore, two different principal

Chemotaxonomy and pharmacology 575

biosynthetic pathways exist (Jensen, 1991). One route leads from iridodial via deoxyloganic acid and loganin or loganic acid to secoiridoids and eventually to the complex indole alkaloids (Fig. 6.1). This route is found mainly in the Gentiananae and Cornanae, but never in the Lamianae. The other route, from the 8-epimeric precursor *epi*-iridodial via *epi*-deoxyloganic acid to aucubin and similar decarboxylated compounds, is found mainly in Lamianae, but never in Gentiananae.

Approximately 1200 different iridoids and secoiridoids (not counting the complex indole alkaloids) are known so far. At about the time we were finishing this chapter, a review on iridoids and secoiridoids in the Gentianaceae appeared (Rodriguez et al., 1998).

Biosynthesis

The iridoid glucosides of Gentianaceae are usually secoiridoids. The carbocyclic iridoids reported from the family are almost exclusively derivatives of loganin, the obligatory biosynthetic precursor of the secoiridoids. The few exceptions from this will be commented on later. Their biosynthesis (Fig. 6.1; bold numbers in the following text refer to the compounds drawn in the figures and are used synonymously with the compound names) was well investigated during the period 1967–1976 (see Takeda & Inouye, 1976), perhaps mainly because of the interest in the complex indole alkaloids found in related families within the order Gentianales. These compounds are derived from secologanin and are therefore, at least biosynthetically, also secoiridoids.

Early work established the terpenoid origin of the gentianaceous iridoid glucosides. Thus, feeding ^{14}C-labeled mevalonate (MVA; in Fig. 6.1 shown as mevalonolactone) to *Swertia caroliniensis* gave incorporation into gentiopicroside (**7**; Coscia & Guarnaccia, 1967). Inouye et al. (1967, 1970) simultaneously showed incorporation of mevalonolactone into **7** in *Gentiana triflora* and into sweroside (**5**) and swertiamarin (**6**) in *Swertia japonica*. Since the incorporation into **5** was ten times higher than that into **6**, the former was suggested to be a precursor of the latter. Further proof for the terpenoid pathway was found by incorporation of geranyl pyrophosphate into loganic acid (**2**), also a constituent of *Swertia caroliniensis* (Coscia & Guarnaccia, 1968), although it was not incorporated into gentiopicroside (**7**). This negative result was ascribed to dilution into the pool of **2** within the plant.

Next, it was established that **2** was an intermediate in the formation of **7** (Coscia et al., 1969, 1970; Guarnaccia et al., 1969). This was simultaneously

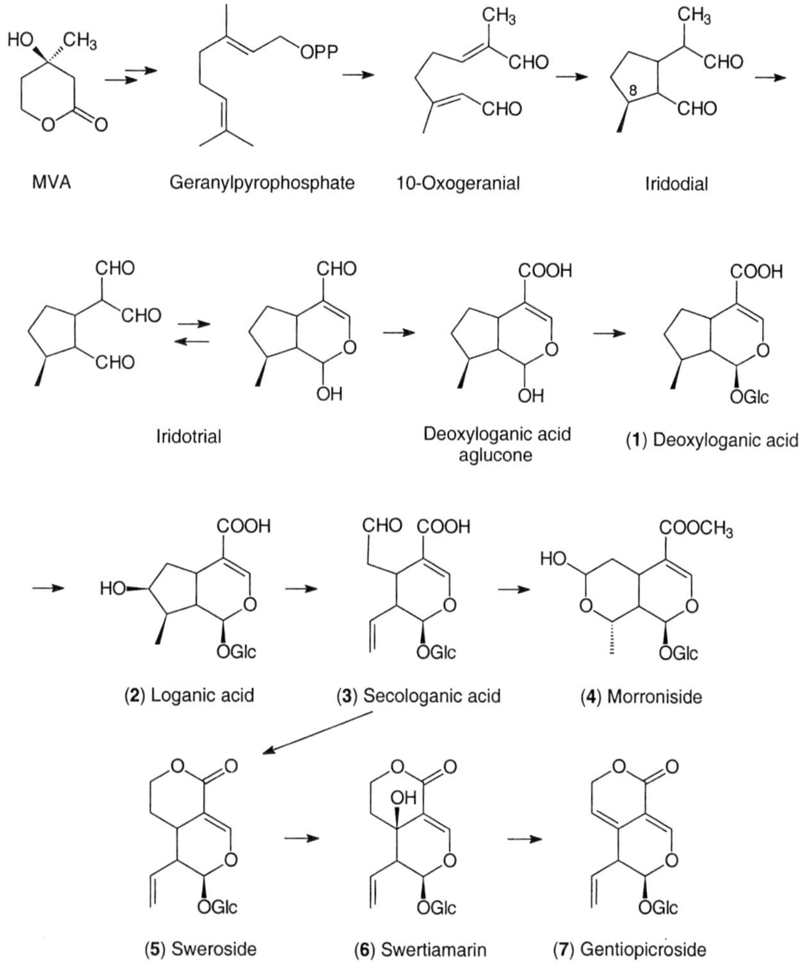

Figure 6.1. The biosynthetic pathway leading to the iridoid glucosides commonly found in Gentianaceae. MVA, mevalonolactone.

proved in *Gentiana triflora* by other groups (Inouye *et al.*, 1969, 1974a) and also in *Swertia petiolata* (Gröger & Simchen, 1969). Inouye's group later prepared labeled sweroside (**5**) and proved that this was also an intermediate for **7** in *Gentiana scabra* (Inouye *et al.*, 1971a).

The biosynthesis of morroniside (**4**), a constituent of *Gentiana thunbergii*, has also been investigated, and both loganic acid (**2**; Inouye *et al.*, 1974a) and deoxyloganic acid (**1**; Inouye *et al.*, 1976) have been shown to be precursors for **4** in this plant.

In several plant species from families other than the Gentianaceae, the pathway from MVA up to deoxyloganic acid has been established to be as shown in Fig. 6.1 (Inouye & Uesato, 1986; Jensen, 1991). Regarding the intermediacy of secologanic acid (**3**), this compound is a likely intermediate between **2** and **4**, but this has not yet been proved. Another uncertainty is the step between **2** and **3**, where the mechanism is still not understood.

The compound gentioside (**2u**; see Fig. 6.2), isolated from several species of *Gentiana* by Popov and Marekov (1971a), was also prepared in radioactive form by feeding ^{14}C-carbon dioxide to *G. asclepiadea* and isolating the compound. Using the labeled **2u** as a precursor, incorporation into gentiopicroside (**7**) was attained. However, the detailed mechanism for the conversion of deoxyloganic acid into the secoiridoids, with retention of both a C-7 and the C-8 proton, makes it highly unlikely that a compound such as **2u** without any proton at C-8 could be an intermediate in the biosynthesis (Takeda & Inouye, 1976). Furthermore, the structure given for gentioside is still somewhat doubtful because of the lack of complete NMR data for the compound.

Results

In the present work, we have classified the compounds according to biosynthetic complexity and displayed the derivatives of loganin (**2–2v**) in Fig. 6.2, those of secologanic acid and secologanol (**3–3r**) in Fig. 6.3, etc. The majority of the compounds are esters (at one or more hydroxyl groups) of the parent compound, signified by the presence of one or more (R^1–R^5)-groups in the derivatized oxygen atom(s). In each figure, the carboxylic acid moieties of the esters (R^n = A to S) are specified in a list next to the parent compound. Finally, the trivial name of each individual compound is listed in the figures.

The carbocyclic iridoids reported from the family are shown in Fig. 6.2. Some of the compounds are strictly derivatives not of loganin (**2**), but rather of its precursor deoxyloganin (**1**), so we have included three derivatives of 7-*epi*-loganin (**2p–2r**) together with gentioside (**2u**) and dihydrocornin (**2v**). Formally, the decarboxylated compounds reported from the roots of the Himalayan *Gentiana kurroo* (Sarg *et al.*, 1990, 1991) could also be included here. These compounds are aucubin and catalpol as well as a number of derivatives of the latter. However, it seems clear that this is a case of confusion with regard to the plant material used for the investigation, since other workers (Inouye & Nakamura, 1971b) isolated only the compounds morroniside (**4**) and gentiopicroside (**7**) from this species. Furthermore, *Picrorhiza*

578   S. R. Jensen & J. Schripsema

A = Acetyl
B = m-Hydroxybenzoyl
C = 2,3-Dihydroxybenzoyl
D = Sinapoyl
E = p-Coumaroyl
F = Caffeoyl
G = Feruloyl

(2) R1 = H; Loganic acid
(2a) R2 = C; Gentiournoside E
(2b) R4 = E; 4'-p-Coumaroylloganic acid
(2c) R1 = CH3; Loganin
(2d) R1 = CH3; R2 = E; 7-p-Coumaroylloganin
(2e) R1 = CH3; R3 = E; 2'-p-Coumaroylloganin
(2f) R1 = CH3; R3 = F; 2'-Caffeoylloganin
(2g) R1 = CH3; R3 = G; 2'-Feruloylloganin
(2h) R1 = CH3; R3 = E; 4'-p-Coumaroylloganin
(2i) R1 = CH3; R4 = B; 4'-m-Hydroxybenzoylloganin
(2j) R1 = CH3; R3 = B; R4 = A; 4'-Acetyl-2'-m-hydroxybenzoylloganin
(2k) R1 = CH3; R3 = B; R5 = A; 6'-Acetyl-2'-m-hydroxybenzoylloganin
(2l) R1 = CH3; R5 = B; 6'-m-Hydroxybenzoylloganin
(2m) R1 = CH3; R5 = H; 6'-[2(R)-Methyl-veratroyloxypropanoyl]-loganin
(2n) R1 = Glc; R2 = C; Gentiournoside D

(2p) R1 = B; Swertiaside
(2q) R1 = B; R2 = C; Senburiside I
(2r) R1 = J; Senburiside II

(2s) R = B; Depressoside

(2t) Loganetin   (2u) Gentioside   (2v) Dihydrocornin

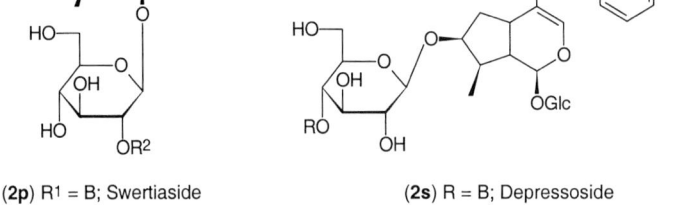

Figure 6.2. Carbocyclic iridoids of Gentianaceae, mainly loganic acid (**2**) and loganin (**2c**) and their ester derivatives.

kurrooa (Scrophulariaceae), also a Himalayan species the roots of which are much used in Chinese medicine, has been reported (Weinges *et al.*, 1972; Stuppner & Wagner, 1989) to contain the catalpol esters noted by Sarg *et al.* (1990). Finally, a report of harpagoside, an ester of the decarboxylated iridoid harpagide, from *Gentiana macrophylla* (Liu *et al.*, 1994a) seems rather unlikely, and other workers who investigated the plant (Wang & Lou, 1988; Kondo & Yoshida, 1993) did not report its presence.

The compounds derived from secologanic acid (**3**) and secologanol (**3g**) are listed in Fig. 6.3. Secologanic acid exists as an equilibrium mixture of the two forms shown, depending on the solvent. When a plant extract containing **3** is chromatographed on silica gel with solvent systems containing methanol, it is our experience that the two epimers of vogelioside (strictly C-7 epimers, but both depicted as **3a**) are readily formed. We therefore presume that vogelioside and *epi*-vogelioside are artifacts. The same is likely to be the case for the dioxolanylsecologanin (**3e**).

Iridoid glucosides derived from the hemiacetal morroniside (**4**) and the corresponding lactone, kingiside (**4c**), are only occasionally reported from Gentianaceae. Those known so far are listed in Fig. 6.4.

Compounds from the biosynthetic sequence sweroside (**5**), swertiamarin (**6**), and gentiopicroside (**7**), together with their derivatives, are the most characteristic iridoids from Gentianaceae. Also in this group is present a wealth of esters, in particular as derivatives of sweroside. The compounds derived from **5–7** are listed in Figs. 6.5–6.7.

Some monoterpene alkaloids have been reported from Gentianaceae and these deserve special mention. Alkaloids are relatively easy to isolate from plant material because of their alkaline properties. Thus, when partitioning the crude extract between dilute aqueous acid and an organic solvent, the alkaloids are found as salts only in the aqueous phase, together with other water-soluble constituents such as sugars, glycosides, etc., while most other organic compounds are removed in the organic solvent. The next preparative step is addition of a base to the aqueous phase to render the alkaloids insoluble in the alkaline water but now extractable by an organic solvent. After such a separation, a fraction consisting only of alkaloids is obtained – in principle, at least. However, by similar treatment of a plant extract containing iridoid glucosides, these compounds will to some degree be hydrolyzed in the acidic medium, particularly by standing, to yield dialdehydes. Until recently, ammonia was the preferred alkaline reagent since it is mild in action. But ammonia and dialdehydes react to form dihydropyridines, which, depending on the structure of the latter, may convert spontaneously to pyridines, which are more stable. These

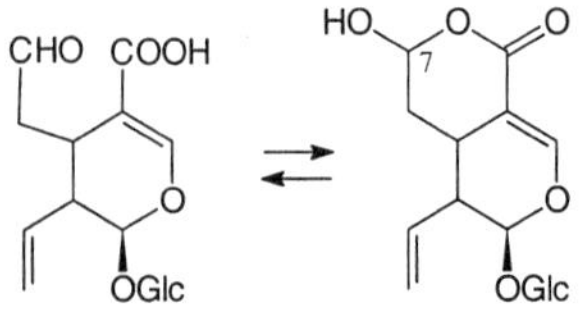

580   S. R. Jensen & J. Schripsema

(3) Secologanic acid 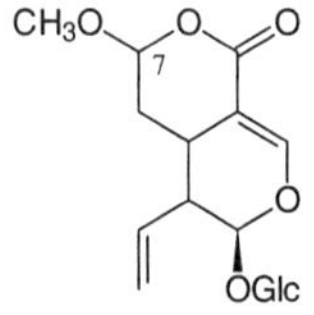 ⇌ (3a) Vogelioside 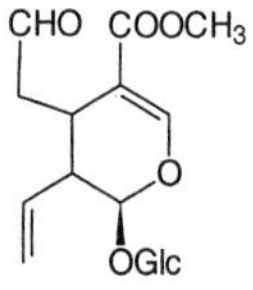   (3d) Secologanin

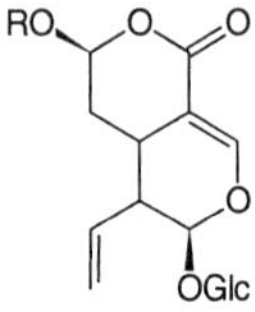

(3b) R = [tigloyl]   Chelonanthoside
(3c) R = [2-methylbutanoyl]   Dihydrochelonanthoside

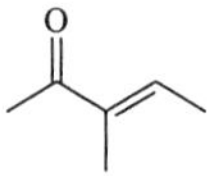

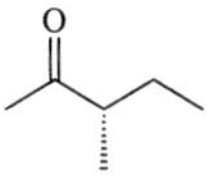

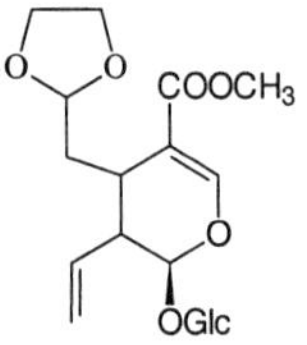

(3e) 7-Dioxolanylsecologanin

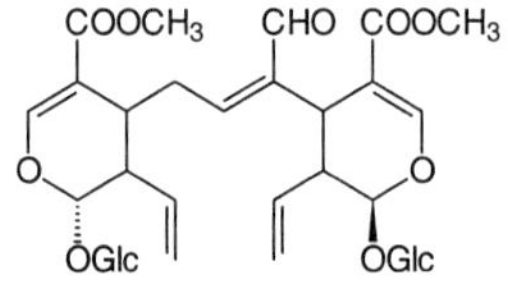

(3f) Centauroside

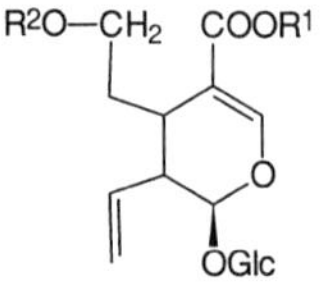

A = Acetyl
F = Caffeoyl
G = Feruloyl
K = 2,5-dihydroxybenzoyl
L = 2-hydroxy-3-glucosyloxybenzoyl

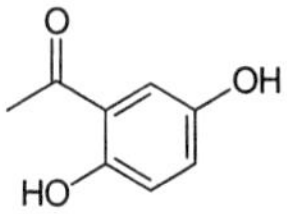

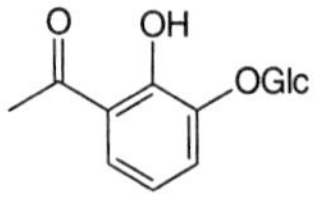

(3g) R¹ = CH₃; Secologanol
(3h) R¹ = CH₃; R² = A; 7-Acetylsecologanol
(3i) R¹ = CH₃; R² = K; 7-(2,5-Dihydroxybenzoyl)secologanol
(3j) R² = F; Grandifloroside
(3k) R² = G; Methylgrandifloroside
(3l) R¹ = CH₃; R² = G; Methyl methylgrandifloroside
(3m) R¹ = CH₃; R² = L; Depressin

Figure 6.3. Secoiridoids of Gentianaceae: glucosides derived from secologanic acid (**3**) and secologanol (**3g**).

Chemotaxonomy and pharmacology 581

(3n) Depresteroside (Gentiournoside A)
(3o) R1 = Glu; Gentiournoside B
(3p) R2 = Glu; Gentiournoside C

(3q) Rhodenthoside A

(3r) Lisianthoside

compounds are impossible to distinguish from genuine alkaloids of the plant (cf. Frederiksen & Stermitz, 1996). Swertiamarin (**6**) and gentiopicroside (**7**) are iridoids that are particularly susceptible to this treatment since they provide a good yield of the stable pyridine alkaloid gentianine (**7e**). Most other iridoids give less stable alkaloids and usually in poorer yield. The question of whether gentianine and the other *Gentiana* alkaloids are artifacts or inherent in the living plant has been discussed at some length by Hegnauer (1966), but we have preferred only to register the presence of gentianine in Table 6.1, where we consider that a report of this compound merely indicates that **6** and/or **7** are present in a particular plant species.

Finally, we have suggested an alternative structure for the compound gentioflavoside (**7c**) in Fig. 6.7, since the one originally proposed (Popov &

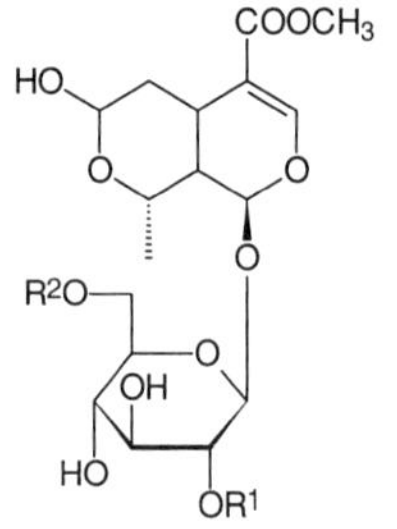

582   S. R. Jensen & J. Schripsema

(4) Morroniside
(4a) R1 = E; 4'-p-Coumaroylmorroniside
(4b) R2 = H; 6'-O-[2(R)-Methyl-3-veratroyl]-morroniside 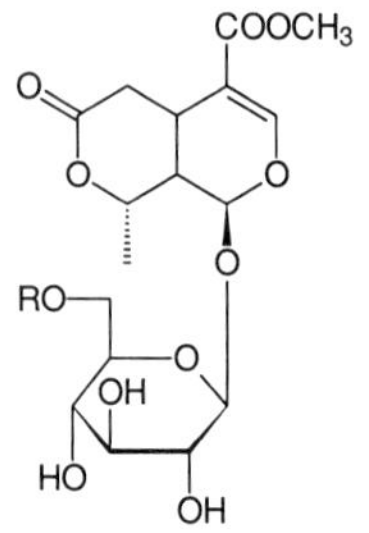
(4c) Kingiside
(4d) R = M; 6'-Vanilloylkingiside

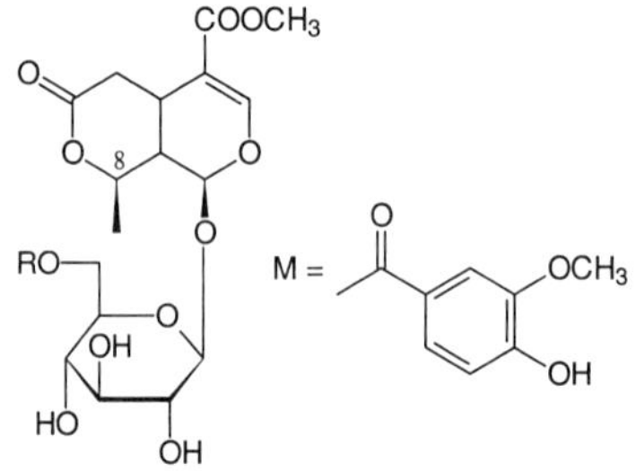

(4e) Epikingiside
(4f) R = M; 6'-Vanilloylepikingiside

Figure 6.4. Secoiridoids of Gentianaceae: morroniside (**4**) and kingiside (**4c**) and their derivatives (see the acyl moieties E and H in Fig. 6.2).

Marekov, 1971b) would be chemically unstable and would have lost the sugar moiety during chromatographic isolation.

The known occurrences of iridoids in the Gentianaceae are listed in Table 6.1. These compounds apparently have been found, or at least detected, in all cases in which they have been sought. An example of this is the large survey performed by van der Sluis (1985) on five populations of *Blackstonia perfoliata* and 99 populations of 11 species of *Centaurium*. Iridoids were found in all taxa investigated.

It appears that the biosynthetic pathway leading from sweroside (**5**) to swertiamarin (**6**) and gentiopicroside (**7**) is universally present in Gentianaceae. At least one of these compounds or their derivatives is found in practically all species investigated for iridoids. It should be noted that **5** is a fairly common compound in Gentianales and Cornales, but **6** and **7** have much more limited distributions, so far being recorded elsewhere only in a few species of Dipsacaceae (Jensen *et al.*, 1975; Jensen, 1992).

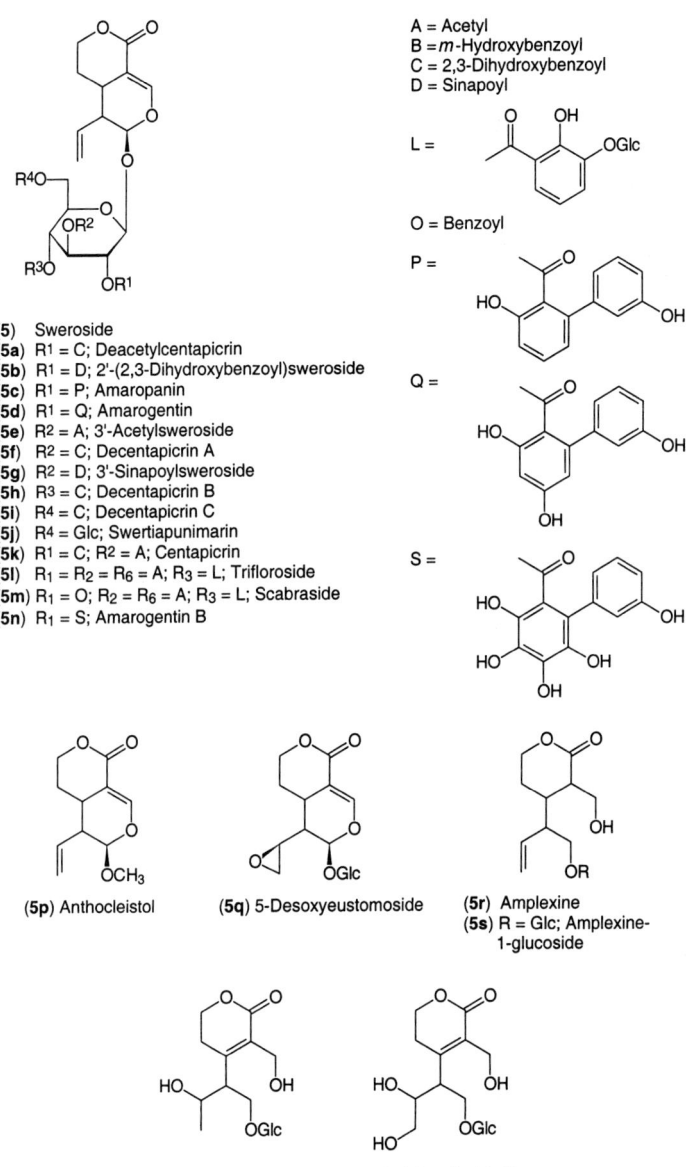

Figure 6.5. Secoiridoids of Gentianaceae: sweroside (5) and its derivatives.

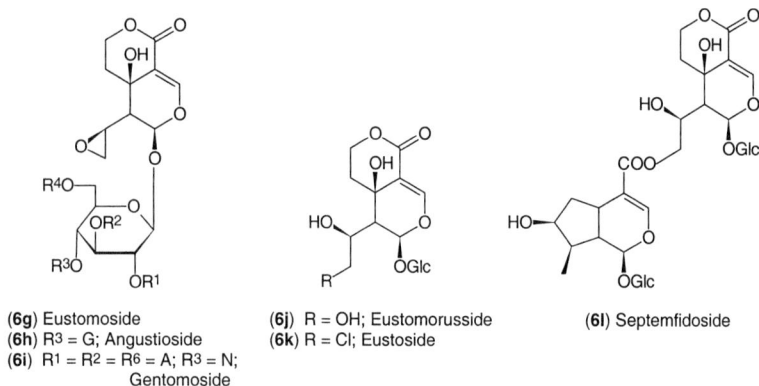

Figure 6.6. Secoiridoids of Gentianaceae: swertiamarin (6) and its derivatives.

XANTHONES

Xanthones (Figs. 6.8–6.9) have a rather restricted occurrence among higher plants, being found almost exclusively in Guttiferae and Gentianaceae (Carpenter et al., 1969; Sultanbava, 1980; Mandal et al., 1992a). Xanthones are sometimes found as the parent polyhydroxylated compounds but more often with a varying degree of substitution at the oxygen atoms. Most xanthones are mono- or poly-methyl ethers or are found as glycosides (Hostettmann & Wagner, 1977). Unlike iridoids, xanthones are apparently not present in all plant species investigated in the family Gentianaceae. This is documented by the systematic work of Hostettmann-Kaldas et al. (1981),

Chemotaxonomy and pharmacology 585

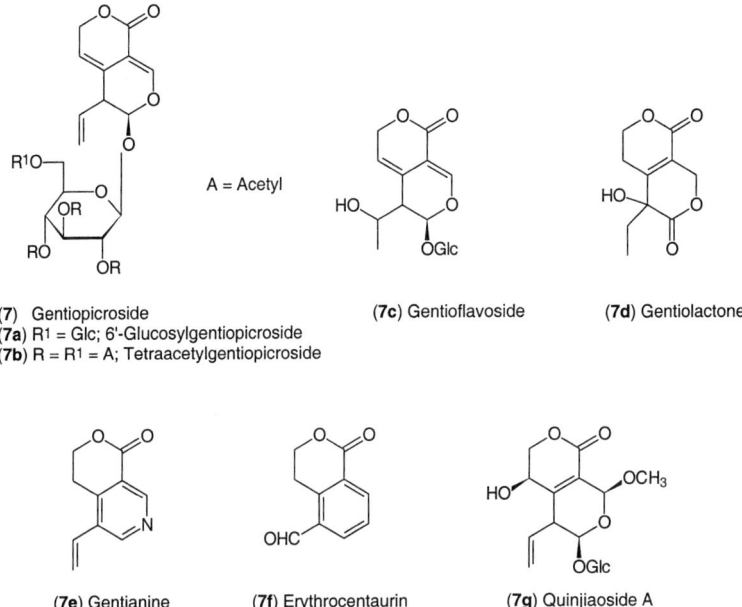

Figure 6.7. Secoiridoids of Gentianaceae; gentiopicroside (7) and its derivatives.

who found these compounds in only three out of six species of *Gentiana*, and van der Sluis (1985), who found that some species were apparently devoid of xanthones, while others consistently contained them and others again were variable in xanthone content.

The widespread occurrence and variation in oxidation pattern has been used by some workers to compute cladograms for taxonomic analysis. In their approach, Rezende and Gottlieb (1973) and Gottlieb (1982) consider the xanthones from Guttiferae (Clusiaceae) and Gentianaceae as a single entity with a common biosynthetic pathway to the compounds. From a frequency count, they come to the conclusion that the 1,3,5,6- and 1,3,6,7-tetraoxygenation patterns are the primitive state, and using these criteria they calculate scores for reduction and oxygenation in the A- and the B-ring. Thereby they assess the evolutionary advancement of six genera of Gentianaceae. This approach assumes (1) that the biosynthetic pathway is identical in Guttiferae and Gentianaceae, (2) that the "normal" xanthones and the C-glucoxanthone mangiferin (**10**) have the same biosynthesis, and, as a consequence of this, (3) that reduction is an important feature in the formation of xanthones in Gentianaceae. Below we will show that none of

Table 6.1. *The known occurrence of iridoids in Gentianaceae*

Species	Compounds reported[a]	References[b]
Anthocleista amplexicaulis	3j, 3k, 5r, 5s, 6	1, 2
Anthocleista djalonensis	5, 5r	3
Anthocleista grandiflora	3j, 3k	4
Anthocleista liebrechtsiana	6	5
Anthocleista nobilis	5p	6
Anthocleista procera	6, 7e	7
Anthocleista vogelii	3, 3a, 5	4, 8
Anthocleista zambesiaca	5, 7f	9
Blackstonia perfoliata	5, 6, 7	10
Centaurium chilense	5	11
Centaurium chloodes	5f, 6	11
Centaurium erythraea (= Erythraea centaurium)	2l, 2v, 3d, 3f, 5, 5a, 5k, 6, 7, 7c, 7e	12–17
Centaurium linarifolium	5f, 7f	18
Centaurium littorale	5a, 5f, 5h, 5i, 5k	13
Centaurium majus	5h, 6	11
Centaurium maritimum	5, 6, 7	11
Centaurium pulchellum	5, 6, 7	11
Centaurium quitense	5, 6, 7	11
Centaurium scilloides	5f, 6	11
Centaurium spicatum	5, 6, 7	11
Centaurium tenuiflorum	5, 6, 7	11
Chelonanthus chelonoides (= C. alatus)	3b, 3c, 5	19
Chironia baccifera	5, 6, 6e, 6g, 7	20
Chironia krebsii	5, 6, 6e, 6g, 7	20
Chironia palustris	5, 6, 6g, 6j, 7	20
Chironia purpurascens	5, 6, 6e, 6g, 6j, 7	20
Cicendia filiformis	7	21
Coutoubea spicata	6, 7	22
Curtia tenuifolia	2d, 3a, 3d, 3g, 5, 6, 7	23
Enicostema hyssopifolium	7f	24
Eustoma russellianum	5, 6, 6g, 6j, 6k, 7	25
Exacum affine	2e, 7	26
Exacum quinquenervium	7e	27
Exacum tetragonum	3k, 7	28
Fagraea fragrans	7e	29
Fagraea gracilipes	5, 5g	30
Fagraea obovata	5, 7	31
Faroa chalcophila	7	32
Faroa graveolens	7	33
Gentiana affinis	7	34
Gentiana algida	5, 5b, 5l, 5s, 6f, 7	35
Gentiana alpina	5, 6, 6d, 6g, 6j	36

Chemotaxonomy and pharmacology

Table 6.1. (cont.)

Species	Compounds reported[a]	References[b]
Gentiana asclepiadea	7c, 7e	37, 38
Gentiana atuntsiensis	7	39
Gentiana bulgarica	7	40
Gentiana burseri	5d, 7	41
Gentiana calycosa	7	34
Gentiana campestris	5q, 6, 6g, 6j, 6k, 7	42
Gentiana caucasica	7a	43
Gentiana cephalantha	7	39
Gentiana cephantha	7	44
Gentiana cerastioides	7	34
Gentiana depressa	2c, 2s, 3m, 3n, 3o	45–47
Gentiana detonsa	7	34
Gentiana farreri	5, 6, 7	48
Gentiana formosana	5, 5e	49
Gentiana gelida	5l, 6, 6f, 6g, 6i, 6j, 7	50
Gentiana kaufmanniana	7e	51
Gentiana kurroo	4, 7	52 (and see text)
Gentiana lactea	2, 5, 5a, 6, 7	55
Gentiana lutea	4, 5, 5d, 6, 7	52, 56
Gentiana macrophylla	5, 6, 7, 7g	57, 58
Gentiana manshurica	7	59
Gentiana olgae	7e	60
Gentiana olivieri	7e	61
Gentiana pannonica	5c	62
Gentiana pedicellata	2, 2b, 2c, 2e, 2f, 2g, 2h	63, 64
Gentiana punctata	2u, 5, 6, 7, 7c	21
Gentiana purpurea	5c, 5d, 6b, 7, 7d	65, 66
Gentiana pyrenaica	2, 2m, 4, 4a, 4b, 4c, 4d, 4e, 4f	67–70
Gentiana regescens	7e	71
Gentiana rhodantha	3q, 4c, 4e, 5, 6	72
Gentiana rigescens	7	39
Gentiana scabra	2, 5m, 6, 6f, 7, 7b	44, 73, 74
Gentiana septemfida	2, 5, 6, 6f, 6g, 6j, 6k, 6l, 7	75
Gentiana sino-ornata	5, 6, 7	48
Gentiana strictiflora	7	34
Gentiana suffrutescens	7	39
Gentiana thunbergii	4	76
Gentiana tianscanica	7e	60
Gentiana tibetica	7	77
Gentiana triflora	5l, 6, 7	78
Gentiana uchiyamana	6, 7	79
Gentiana urnula	2a, 2n, 3n, 3o, 3p	80
Gentiana verna	2, 2j, 2k, 3d, 3e, 3g, 3h, 3i, 5	81–83
Gentiana vvedenskyi	7e	60
Gentianella azurea	2, 6, 7	77
Gentianella bulgarica	2u, 5, 6, 6g, 7	21

Table 6.1. (cont.)

Species	Compounds reported[a]	References[b]
Halenia campanulata	3a	84
Halenia corniculata	3a, 5, 6	85
Halenia elliptica	5, 6	86
Ixanthus viscosus	2, 6, 7	87
Lisianthius jefensis	3r	88
Lisianthius seemanii	3r	16
Lomatogonium carinthiacum	6	89
Potalia amara	5t, 6	90
Potalia sp.	5u, 6	90
Sabatia angularis	7	91
Sabatia elliotii	5d, 7e	21, 91
Schultesia guianensis	6, 7e	92, 93
Swertia alata	6	94
Swertia angustifolia	5, 6, 6c, 6g, 6h	95
Swertia arisanensis	5	96
Swertia bimaculata	5, 6, 7	86
Swertia calycina	5, 7	97
Swertia caroliniensis	2c, 7	98
Swertia chirata	5, 6b	99
Swertia cincta	5, 6, 7	86
Swertia connata	7e	60
Swertia davidi	7e	100
Swertia elongata	5a, 6	101
Swertia erythrostricta	5, 6, 7	86
Swertia fasciculata	5, 6, 7	86
Swertia graciliflora	7e	102
Swertia japonica	2p, 2q, 2r, 5, 5d, 6, 6b, 7	103–106
Swertia lawii	7f	24
Swertia macrosperma	6, 7	86
Swertia marginata	7e	102
Swertia mileensis	5, 6, 6a	107, 108
Swertia mussotii	5d	109
Swertia nervosa	3a, 5, 6	110
Swertia patens	6	111
Swertia petiolata	7	112
Swertia pseudochinensis	6	113
Swertia pubescens	5, 6, 7	86
Swertia punicea	5, 5b, 6, 6d, 7, 7g	114, 115
Swertia randaiensis	7e	116
Swertia tetrapetala	6	117

Notes:
[a] Bold numbers refer to the compounds in Figs. 6.2–6.7.
[b] References: 1: Weber, 1974. 2: Rasoanaivo et al., 1994. 3: Onocha et al., 1995. 4: Chapelle, 1976a. 5: Cornelis & Chapelle, 1976. 6: Madubunyi et al., 1994. 7: Koch, 1965. 8: Chapelle,

Chemotaxonomy and pharmacology 589

1974. 9: Chapelle, 1976b. 10: Chapelle, 1973. 11: van der Sluis, 1985. 12: van der Sluis *et al.*, 1983. 13: van der Sluis & Labadie, 1981a. 14: Do *et al.*, 1987. 15: Rulko & Witkiewicz, 1972. 16: Sakina & Aota, 1976. 17: Tagaki *et al.*, 1982. 18: van der Sluis & Labadie, 1981b. 19: Shiobara *et al.*, 1994. 20: Wolfender *et al.*, 1993. 21: Hegnauer, 1966. 22: Schaufelberger *et al.*, 1987. 23: Kuwajima *et al.*, 1996a. 24: Ghosal *et al.*, 1974c. 25: Uesato *et al.*, 1979. 26: Kuwajima *et al.*, 1996b. 27: Delaude, 1984. 28: Das *et al.*, 1984. 29: Wan *et al.*, 1972. 30: Cambie *et al.*, 1990. 31: Jensen, 1992. 32: Delaude, 1985. 33: Delaude & Darimont, 1985. 34: Hostettmann-Kaldas *et al.*, 1981. 35: Tan *et al.*, 1996. 36: Mpondo *et al.*, 1990a. 37: Rulko & Nadler, 1970. 38: Mpondo & Chulia, 1988. 39: Luo & Lou, 1986. 40: Thanh *et al.*, 1987. 41: Stefanou *et al.*, 1976. 42: Mpondo *et al.*, 1990b. 43: Topuriya, 1978. 44: Yang *et al.*, 1985. 45: Chulia & Kaouadji, 1985. 46: Chulia *et al.*, 1994. 47: Chulia *et al.*, 1996. 48: Schauffelberger & Hostettmann, 1987. 49: Chung & Lin, 1993. 50: Calis *et al.*, 1990. 51: Sadykov, 1987. 52: Inouye & Nakamura, 1971b. 53: Sarg *et al.*, 1990. 54: Sarg *et al.*, 1991. 55: Schaufelberger & Hostettmann, 1988. 56: Bricout, 1974. 57: Wang & Lou, 1988. 58: Liu *et al.*, 1994a. 59: Zhang *et al.*, 1993. 60: Rakhmatullaev, 1971. 61: Rakhmatullaev *et al.*, 1969a. 62: Wagner & Vasirian, 1974. 63: Garcia & Chulia, 1987. 64: Chulia *et al.*, 1986. 65: Nyiredy *et al.*, 1986. 66: Suhr *et al.*, 1978. 67: Garcia *et al.*, 1989a. 68: Garcia *et al.*, 1989b. 69: Garcia *et al.*, 1989c. 70: Garcia *et al.*, 1990. 71: Sun & Xia, 1984. 72: Ma *et al.*, 1994. 73: Ikeshiro & Tomita, 1983. 74: Ikeshiro *et al.*, 1990. 75: Calis *et al.*, 1992. 76: Takeda & Inouye, 1976. 77: Zhang & Yang, 1994. 78: Inouye *et al.*, 1974b. 79: Chung & Lee, 1982. 80: Liu *et al.*, 1994b. 81: Mpondo *et al.*, 1989. 82: Mpondo & Garcia, 1990a. 83: Mpondo & Garcia, 1990b. 84: Recio-Iglecias *et al.*, 1992. 85: Rodriguez *et al.*, 1995b. 86: Gao *et al.*, 1994. 87: Ortega *et al.*, 1988. 88: Hamburger *et al.*, 1990. 89: Schaufelberger & Hostettmann, 1984. 90: S. R. Jensen & J. Schripsema, unpubl. 91: Korte, 1955. 92: Nobrega & Craveiro, 1988. 93: Terreaux *et al.*, 1995. 94: Khan *et al.*, 1979. 95: Luo & Nie, 1992. 96: Lin *et al.*, 1987. 97: Rodriguez *et al.*, 1995a. 98: Coscia *et al.*, 1969. 99: Chaudhuri & Daniewski, 1995. 100: Guo & Chen, 1980. 101: Kong *et al.*, 1995. 102: Rakhmatullaev *et al.*, 1969b. 103: Inouye *et al.*, 1971b. 104: Ikeshiro & Tomita, 1984. 105: Ikeshiro & Tomita, 1985. 106: Ikeshiro & Tomita, 1987. 107: Kikuzaki *et al.*, 1996. 108: He & Nie, 1980. 109: Sun *et al.*, 1991. 110: Luo & Nie, 1993a. 111: Liang *et al.*, 1984. 112: Gröger & Simchen, 1969. 113: Kitamura *et al.*, 1988. 114: Luo & Nie, 1993b. 115: Tan *et al.*, 1993. 116: Wu *et al.*, 1976. 117: Agata *et al.*, 1981.

these assumptions appears to be valid. Massias *et al.* (1982) carried out an analysis of four genera of Gentianaceae using the oxygenation patterns of the xanthones and the presence of some sugars and some C-glucoflavones to construct a phyletic scheme. In this scheme *Gentiana* comes out on one branch while *Comastoma*, *Gentianella*, and *Gentianopsis* appear on another. In that work it is assumed that the trisubstituted xanthones represent the primitive state, while the higher the oxidation state of the compounds, the more advanced the taxon. We agree with this assumption. Mészáros (1994) discusses the above two works and makes some unconventional evolutionary assumptions based on both the positions and the number of oxidations in the xanthone nuclei. He calculates advancement indices for 13 taxa, which are compared with those calculated on the basis of the other workers' hypotheses. One problem inherent in Mészáros's calculations is the use of an "outgroup", namely *Anthocleista*, which was considered a member of Loganiaceae (Struwe *et al.*, 2002 (Chapter 2, this volume)). In the present work, this taxon is a true member of Gentianaceae. Another problem is that there appears to be more than one line of xanthone advancement in the family (see below).

Biosynthesis

Biosynthetically, the xanthones are of mixed shikimate and acetate origin (Fig. 6.8). Thus phenylalanine, which is formed from shikimate, loses two carbon atoms from the side-chain and is oxidized to form *m*-hydroxybenzoic acid. This combines with three units of acetate (probably via malonate) to produce the intermediate shown. Proper folding and ring-closure gives a substituted benzophenone, which by an oxidative phenol coupling generates the central ring of the xanthone moiety.

Note that this oxidative coupling can in principle take place in two ways (see Fig. 6.8), depending on the folding of the benzophenone: either in *ortho*- or in *para*-position to the hydroxyl substituent in the potential B-ring to give 1,3,5-trihydroxyxanthone (**8**) or the 1,3,7-substituted analogue gentisein (**9**), respectively. Thus, depending on the orientation of the intermediate, two different hydroxylation patterns can be found. This will be discussed below.

Experimental proof for the overall pathway has been obtained from experiments performed on *Gentiana lutea* (Atkinson *et al.*, 1968; Gupta & Lewis, 1971): when plants were fed ^{14}C-labeled phenylalanine, the label was recovered solely in the B-ring (Fig. 6.8). Conversely, feeding of ^{14}C-labeled acetate gave incorporation of which the main part was found in the A-ring. Finally, administration of a ^{3}H-labeled analogue of the shown 2,3′,4,6-tetrahydroxybenzophenone proved that this was also a precursor and a likely intermediate. In the feeding experiments the compounds isolated from the plants were the monomethyl derivatives gentisin and isogentisin, obtained as a mixture and degraded to gentisein (**9**) in which the incorporation was measured. The alternative ring-closure to **8** has recently been shown to take place in cultured cells of *Centaurium erythraea*, where 2,3′,4,6-tetrahydroxybenzophenone is the precursor for 1,3,5-trihydroxyxanthone (**8**; Fig. 6.8) (Peters *et al.*, 1998). Furthermore, in these cell cultures compound **8** is selectively oxidized by a xanthone 6-hydroxylase to 1,3,5,6-tetrahydroxyxanthone (Schmidt *et al.*, 2000).

Interestingly, the iridoids amarogentin (**5d**) and amaroswerin (**6b**), first isolated from *Swertia japonica* (Inouye & Nakamura, 1971a), both contain

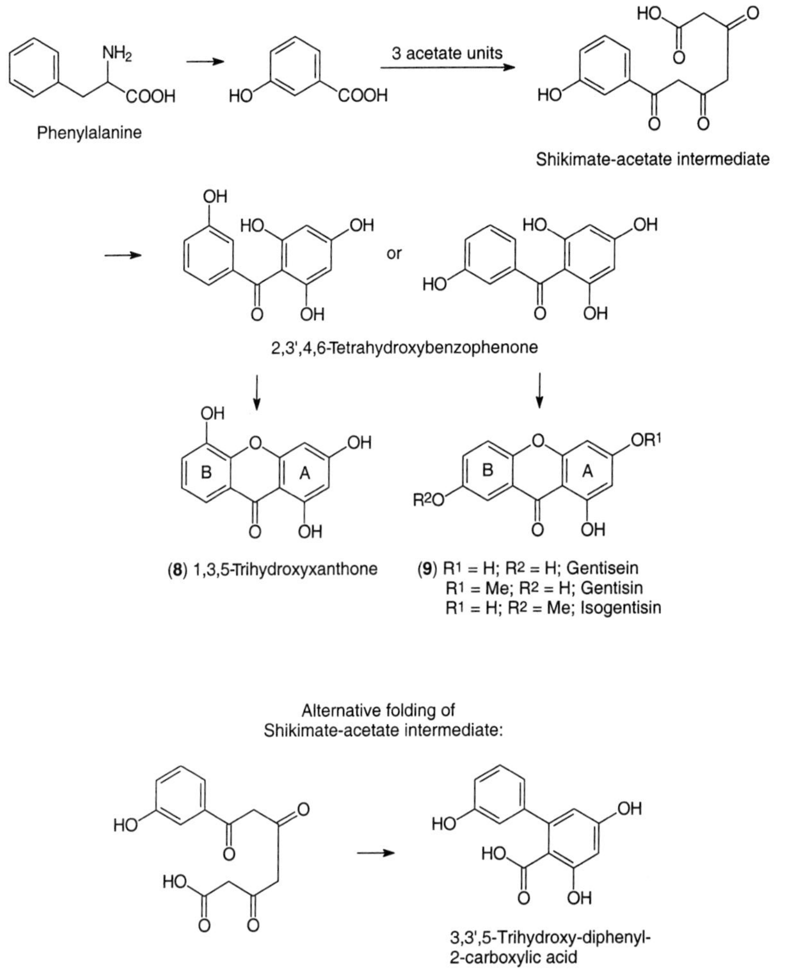

Chemotaxonomy and pharmacology 591

Figure 6.8. Biosynthetic pathways leading to the parent xanthones **8** and **9**, and to the *Swertia* acid.

the unique 3,3′,5-trihydroxydiphenyl-2-carboxylic acid moiety. The authors noted the similarity between the oxidation pattern of this acyl moiety and the above xanthones gentisin and isogentisin, which are common in Gentianaceae, and they suggested that they had a common biogenetic precursor, namely the shikimate–acetate intermediate discussed above (Fig. 6.8). The alternative folding pattern shown would, by ring-closure, produce the substituted diphenyl-2-carboxylic acid. Actual work on this idea was published only relatively recently by Inouye's group (Kuwajima *et al.*,

1990). Administration of radiolabeled acetate, phenylalanine, and benzoic acid to *Swertia japonica* showed that all were precursors and that the above hypothesis is very likely to be true.

Results

We have recorded approximately 600 occurrences of about 100 different xanthones in 110 species and 21 genera in the family (Table 6.2). Some plants are either very rich in these compounds or have been heavily worked on; an example is *Canscora decussata*, from which 31 different compounds have been isolated. In order to simplify the presentation of this large amount of data, we have preferred to demonstrate only the oxidation patterns of the parent xanthones, ignoring the variation in O-substitution with methyl, glycosyl, and other groups. This left 21 oxidation patterns (**8a–8n** and **9a–9n**) which could be presented in a presumed biosynthetic double tree- (or rather shrub-)like structure (Fig. 6.9). In the following discussion a number in bold will therefore represent not a single compound, but rather a group of compounds with the same oxidation pattern.

As discussed above, the biosynthetic pathway has been demonstrated in cultured cells of *Centaurium erythraea* for the formation of 1,3,5-trihydroxyxanthone (**8**), and alternatively, in *Gentiana lutea*, of 1,3,7-substituted gentisein (**9**). This should be sufficient evidence for the fact that **8** and **9** are actually the parent xanthones, and that these are not formed by reduction of tetrasubstituted precursors (**8j** and **9j**) in the plants, as presumed by Rezende and Gottlieb (1973). Also, no dioxygenated compounds have yet been discovered from the Gentianaceae. If reduction to less substituted compounds had been a valid mechanism in the biosynthetic pathway, we find it likely that at least a few representatives arising from this would have been found, particularly considering that such a large number of compounds are now known.

Almost all the compounds **9a–9n** may comfortably be derived from **9** by one or more (enzymatic) oxidations in positions *ortho* or *para* to an existing phenolic hydroxyl group (the oxidized positions in each case are indicated by the number above the arrow leading to the more oxidized compound in Fig. 6.9). The only position in **9** (or **9c**) which is not activated for enzymatic oxidation by one or more *ortho*- or *para*-hydroxyl groups is the C-5 carbon in the B-ring. This argument is supported by the fact that out of the 280 compounds in this part of the tree only eight (represented by **9e** and **9d**) have been formed by oxidation at C-5. Only after introduction of an oxygen at C-6 or C-8 is oxidation at C-5 likely. The remaining compounds may now easily be derived from a less oxidized compound

Chemotaxonomy and pharmacology 593

Table 6.2. *The known occurrence of xanthones in Gentianaceae*

Taxon	Compounds reported[a]	References[b]
Anthocleista vogelii	9j (3)	1, 2
Blackstonia perfoliata	9j (1)	3
Canscora decussata	8 (5), 8f (6), 8g (12), 8h (3), 9f (1), 9j (4), 9n (4)	4–12
Centaurium cachanlahuen	8h (1), 8j (1), 9j (2)	13
Centaurium erythraea	8f (1), 8g (1), 8h (1), 8j (1), 8n (1), 9j (1)	14
Centaurium linarifolium	8f (2), 8h (4), 8n (1), 9j (2)	15–17
Centaurium littorale	8h (2), 8j (2)	18
Centaurium pulchellum	8j (1), 9j (2)	19
Centaurium scilloides	8h (2), 8j (2)	19
Chironia krebsii	8 (6), 8f (1), 8h (5), 9 (1), 9j (4)	20
Comastoma pulmonarium	8j (2), 9j (4)	21
Eustoma grandiflorum	8 (1), 8g (1), 8h (1), 9 (1), 9j (1)	22
Frasera albicaulis	8 (1), 8a (1), 8b (1), 8c (1), 8j (1), 9 (2), 9a (2), 9b (2), 9c (2)	23
Frasera albomarginata	8 (1), 8a (1), 8c (1), 8j (1), 9 (1), 9b (1), 9c (1)	24
Frasera caroliniensis	8a (1), 8b (1), 8c (1), 8j (1), 9a (1), 9c (1)	25
Frasera speciosa	8a (1), 8b (1), 8c (1), 9a (1), 9c (2)	24
Frasera tetrapetala	8c (1), 9c (1)	26
Gentiana acaulis	9j (2)	27
Gentiana algida	8j (2), 8l (1)	28
Gentiana asclepiadea	9 (1)	29, 30
Gentiana barbata	9j (8)	31–34
Gentiana bavarica	9j (10)	35–37
Gentiana brachyphylla	9j (3)	38
Gentiana campestris	8j (5), 8l (2)	39–41
Gentiana cerastioides	8j (4)	42
Gentiana ciliata	9 (4), 9j (18)	43, 44
Gentiana corymbifera	8j (5), 8l (3)	45
Gentiana corymbosa	8j (4)	46
Gentiana dasyantha	8j (4)	46
Gentiana detonsa	9j (4)	42
Gentiana favrati	9j (3)	38
Gentiana karelinii	8j (1), 9j (4)	29
Gentiana kochiana	9j (6)	47, 48
Gentiana lactea	8j (5)	49
Gentiana lutea	9 (7), 9a (1), 9j (1)	50–55
Gentiana marcailhouana	8j (1), 9 (1)	56
Gentiana nivalis	9j (3)	38
Gentiana orbicularis	9j (3)	57
Gentiana pannonica	9 (1)	53
Gentiana punctata	9 (1)	53

Table 6.2. (cont.)

Taxon	Compounds reported[a]	References[b]
Gentiana purpurea	9 (1)	53
Gentiana rostani	9j (3)	38
Gentiana schleicheri	9j (2)	38
Gentiana strictiflora	8j (4)	42
Gentiana utriculosa	9j (2)	38
Gentiana verna	8j (1), 9j (8)	38, 58, 59
Gentianella azurea	8j (1)	60
Gentianella bellidifolia	8j (5), 8l (1), 9l (1)	61
Gentianella campestris	8j (5), 8l (2), 9j (1)	61
Gentianella germanica	8j (4), 8l (2)	61, 62
Gentianella ramosa	8j (5), 8l (2), 9j (1)	61, 62
Gentianella serotina	9l (3)	57
Gentianella stenocalyx	9 (1), 9j (5)	57
Gentianella tenella	8j (1), 9j (2)	61
Gentianopsis (12 species)	9 (2), 9j (5)	57
Halenia asclepiadea	8a (1), 8c (1), 9c (1)	46, 63
Halenia campanulata	8a (2), 8c (1)	64
Halenia corniculata	9a (2), 9c (2), 9d (5)	65
Halenia elliptica	8a (1), 8c (1), 8j (1), 9a (1), 9d (1)	66, 67
Hoppea dichotoma	8 (2), 8f (4), 8g (2)	68
Hoppea fastigiata	9e (2), 9j (1)	69–72
Ixanthus viscosus	9n (2)	73
Lomatogonium carinthiacum	8j (2), 9 (1), 9j (2)	74
Lomatogonium rotatum	8j (1), 9j (1)	75
Macrocarpaea glabra	9 (1), 9j (1)	76
Orphium frutescens	9j (3)	77
Schultesia lisianthioides (= Xestaea lisianthioides)	8 (2), 8a (2), 8f (1), 8j (1), 9j (3)	78
Swertia alata	8j (1)	79
Swertia angustifolia	8b (1), 8j (4), 8l (2), 9 (1), 9c (1), 9j (6), 9l (2)	80, 81
Swertia bimaculata	8b (3), 8c (1), 8d (1), 8j (3), 9c (3)	82, 83
Swertia calycina	8 (1), 9j (3)	65
Swertia chirata	8j (4), 9j (5)	84–86
Swertia chirayita	8j (2), 9j (1)	87, 88
Swertia connata	8j (1), 9j (3)	89, 90
Swertia cordata	9m (2)	91, 92
Swertia decussata	9j (1)	47
Swertia devidi	8j (1)	93
Swertia dilatata	9j (1)	94
Swertia elongata	8j (2)	95
Swertia erythrosticta	8j (6), 9j (3)	96
Swertia fasciculata	8j (1)	66

Chemotaxonomy and pharmacology 595

Table 6.2. (cont.)

Taxon	Compounds reported[a]	References[b]
Swertia franchetiana	**8** (1), **8j** (1)	97
Swertia gracilescens	**9j** (1)	94
Swertia hookeri	**8** (2), **8j** (7), **9j** (2)	98
Swertia iberica	**8j** (1), **9j** (6), **9k** (1)	35, 99, 100
Swertia japonica	**8j** (7), **9** (1), **9j** (3)	101–105
Swertia lawii	**8j** (2), **9j** (4), **9l** (1)	106
Swertia macrosperma	**8j** (3)	107
Swertia mileensis	**8a** (1), **8c** (1)	108, 109
Swertia mussotii	**8** (1), **8j** (5), **9j** (6), **9l** (1)	110–112
Swertia nervosa	**9j** (4)	94
Swertia paniculata	**8j** (5), **9j** (7), **9l** (1)	113, 114
Swertia patens	**8** (1), **8j** (1), **9j** (2)	115
Swertia perennis	**8j** (2), **9j** (7)	47, 59, 116, 117
Swertia perfoliata	**8j** (1), **9j** (1)	118
Swertia petiolata	**8j** (3), **9** (2), **9j** (2)	119–121
Swertia przewalskii	**9j** (7)	122
Swertia punicea	**8j** (1), **9j** (3), **9m** (1)	123, 124
Swertia purpurascens	**8j** (6), **9j** (2), **9l** (1), **9m** (1)	125–127
Swertia racemosa	**8j** (3), **9j** (3)	94
Swertia randaiensis	**8j** (2)	128, 129
Swertia speciosa	**8j** (4), **9** (3), **9j** (4)	130–132
Swertia swertopsis	**8j** (1), **9j** (3)	129
Swertia tetrapetala	**8a** (1), **8j** (1), **9c** (1)	133
Swertia tetraptera	**8a** (1), **9b** (1)	134
Swertia tosaensis	**8j** (1)	125
Swertia verticillifolia	**8j** (4), **9j** (2)	135
Tripterospermum lanceolatum	**8f** (1), **9b** (1), **9f** (1), **9l** (3)	136–138
Tripterospermum taiwanense	**9b** (1), **9f** (1)	139, 140
Veratrilla baillonii	**8a** (1), **8c** (1), **9a** (2), **9b** (2), **9c** (3)	141–143

Notes:
[a] Bold numbers refer to the compounds in Fig. 6.9 (number of different compounds found in parentheses)
[b] References: 1: Okorie, 1976. 2: Chapelle, 1974. 3: van der Sluis, 1985. 4: Ghosal et al., 1976. 5: Ghosal & Chaudhuri, 1975. 6: Ghosal & Chaudhuri, 1973. 7: Ghosal et al., 1973a. 8: Ghosal et al., 1974a. 9: Ghosal et al., 1977. 10: Ghosal & Biswas, 1979. 11: Ghosal et al., 1971. 12: Chaudhuri & Ghosal, 1971. 13: Versluys et al., 1982. 14: Kaouadji et al., 1986. 15: Parra et al., 1985. 16: Parra et al., 1984a. 17: Parra et al., 1984b. 18: van der Sluis & Labadie, 1985. 19: Miana & Al-Hazimi, 1984. 20: Wolfender et al., 1993. 21: Fan et al., 1988. 22: Sullivan et al., 1977. 23: Stout et al., 1969a. 24: Dreyer & Bourell, 1981. 25: Stout & Balkenhol, 1969. 26: Agata et al., 1984. 27: Plouvier et al., 1967. 28: Butayarov et al., 1993b. 29: Kitanov et al., 1991. 30: Goetz & Jacot-Guillarmod, 1977. 31: Pureb et al., 1991. 32: Nikolaeva et al., 1980a. 33: Nikolaeva et al., 1980b. 34: Nikolaeva et al., 1981. 35: Denisova et al., 1980b. 36: Hostettmann et al., 1974. 37: Hostettmann et al., 1976. 38: Hostettmann & Jacot-Guillarmod, 1977. 39: Kaldas et al., 1974. 40: Kaldas et al., 1975. (cont.)

Note b of Table 6.2 (cont.)
41: Kaldas et al., 1978. 42: Hostettmann-Kaldas et al., 1981. 43: Massias et al., 1976. 44: Goetz et al., 1978. 45: Massias et al., 1981. 46: Recio et al., 1990. 47: Rivaille et al., 1969. 48: Guyot et al., 1968. 49: Schaufelberger & Hostettmann, 1988. 50: Hayashi & Yamagishi, 1988. 51: Lubsandorzhieva et al., 1986. 52: Nikolaeva et al., 1983. 53: Verney & Debelmas, 1973. 54: Bellmann & Jacot-Guillarmod, 1973. 55: Atkinson et al., 1969. 56: Luong et al., 1980. 57: Massias et al., 1982. 58: Hostettmann & Jacot-Guillarmod, 1974. 59: Rivaille & Raulais, 1969. 60: Zhang & Yang, 1994. 61: Carbonnier et al., 1977. 62: Hostettmann-Kaldas & Jacot-Guillarmod, 1978. 63: Stout & Fries, 1970. 64: Recio-Iglesias et al., 1992. 65: Rodriguez et al., 1995b. 66: Gao et al., 1994. 67: Sun et al., 1983. 68: Ghosal et al., 1978a. 69: Mukherjee et al., 1995. 70: Mukherjee et al., 1990. 71: Mukherjee et al., 1991a. 72: Mukherjee et al., 1991b. 73: Ortega et al., 1988. 74: Sorig et al., 1977. 75: Khishgee & Pureb, 1993. 76: Stout et al., 1969b. 77: Roitman et al., 1992. 78: Terreaux et al., 1995. 79: Khan et al., 1979. 80: Dhoubhadel et al., 1980. 81: Ghosal et al., 1978b. 82: Inouye et al., 1971c. 83: Ghosal et al., 1975b. 84: Ghosal et al., 1973b. 85: Purushothaman et al., 1973. 86: Mandal & Chatterjee, 1987. 87: Saxena & Mukherjee, 1992. 88: Asthana et al., 1991. 89: Solov'eva et al., 1980a. 90: Solov'eva et al., 1980b. 91: Atta-ur-Rahman et al., 1994. 92: Khan & Haqqani, 1981. 93: Yu, 1984. 94: Tomimori et al., 1974. 95: Kong et al., 1995. 96: Hu et al., 1992. 97: Ding et al., 1988. 98: Ghosal et al., 1980. 99: Denisova et al., 1980a. 100: Denisova et al., 1980c. 101: Ashida et al., 1994. 102: Ishimaru et al., 1990. 103: Kanamori et al., 1984. 104: Sakamoto et al., 1982. 105: Basnet et al., 1994. 106: Ghosal et al., 1975a. 107: Zhou & Liu, 1990. 108: Liu & Huang, 1982. 109: He et al., 1982. 110: Sun & Ding, 1981. 111: Sun et al., 1991. 112: Ding & Sun, 1980. 113: Prakash et al., 1982. 114: Verma & Khetwal, 1985. 115: He et al., 1984. 116: Hostettmann & Miura, 1977. 117: Hostettmann & Jacot-Guillarmod, 1976. 118: Haqqani, 1981. 119: Khetwal et al., 1990. 120: Kulanthaivel et al., 1988. 121: Bhan & Kalla, 1982. 122: Hu et al., 1991. 123: Kanamori & Sakamoto, 1993. 124: Fukamiya et al., 1990. 125: Ghosal et al., 1974b. 126: Ghosal et al., 1975c. 127: Ahmad et al., 1973. 128: Chung et al., 1986. 129: Tomimori & Komatsu, 1969. 130: Massias et al., 1977. 131: Bisht et al., 1991. 132: Khetwal & Bisht, 1988. 133: Agata et al., 1981. 134: Neu et al., 1991. 135: Liao et al., 1991. 136: Lin et al., 1987. 137: Lin et al., 1982a. 138: Chen et al., 1992. 139: Lin et al., 1982b. 140: Lin et al., 1984. 141: Yang & Zhou, 1980a. 142: Yang & Zhou, 1980b. 143: Yang et al., 1995.

earlier in the tree, as shown by using a single oxidation at a time. Note that this postulated pathway does not require any reductions.

Similarly, the compounds **8a–8n** can be derived from **8**. In this case, however, it is position 7 in compound **8** which is inactive for oxidation. In analogy to the above, this position can be oxidized only after introduction of an oxygen at C-6 or C-8.

Using these principles, the double tree has been constructed. Only in a few cases can there be doubt about the placement of compounds with certain substitution patterns. Thus, **9d** could as well be derived from **8c** as from **9c** but it occurs in the same plants as the latter; similar arguments have been used for the other borderline cases.

Table. 6.3 lists the occurrences of different types of xanthones for the genera of Gentianaceae from which xanthones have been reported. Based on these data, four groups of genera can be distinguished, showing the following characteristics:

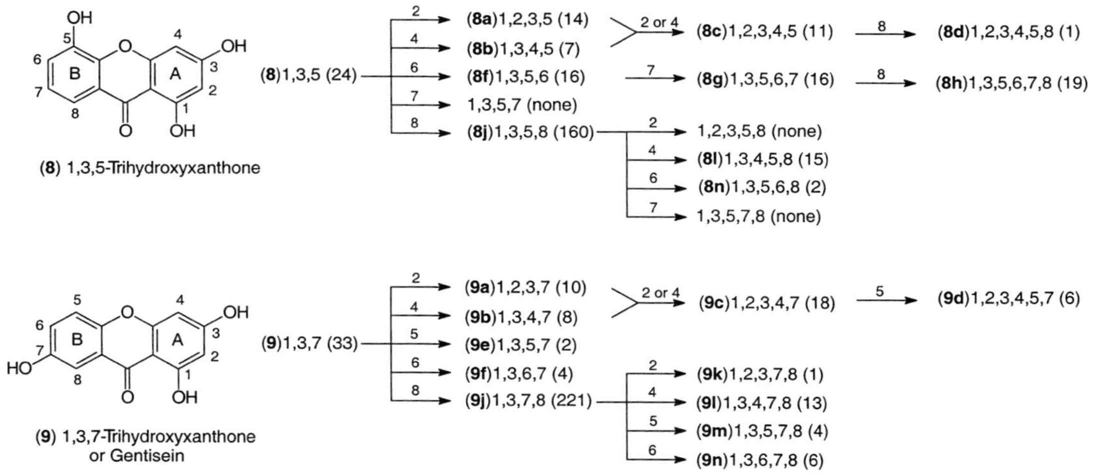

(8) 1,3,5-Trihydroxyxanthone

- (8a) 1,2,3,5 (14) → 2 or 4 → (8c) 1,2,3,4,5 (11) → 8 → (8d) 1,2,3,4,5,8 (1)
- (8b) 1,3,4,5 (7)
- (8f) 1,3,5,6 (16) → 7 → (8g) 1,3,5,6,7 (16) → 8 → (8h) 1,3,5,6,7,8 (19)
- 1,3,5,7 (none)
- (8j) 1,3,5,8 (160) → 2 → 1,2,3,5,8 (none)
  - → 4 → (8l) 1,3,4,5,8 (15)
  - → 6 → (8n) 1,3,5,6,8 (2)
  - → 7 → 1,3,5,7,8 (none)

(8) 1,3,5 (24)

(9) 1,3,7-Trihydroxyxanthone or Gentisein

- (9a) 1,2,3,7 (10) → 2 or 4 → (9c) 1,2,3,4,7 (18) → 5 → (9d) 1,2,3,4,5,7 (6)
- (9b) 1,3,4,7 (8)
- (9e) 1,3,5,7 (2)
- (9f) 1,3,6,7 (4) → 2 → (9k) 1,2,3,7,8 (1)
  - → 4 → (9l) 1,3,4,7,8 (13)
  - → 5 → (9m) 1,3,5,7,8 (4)
  - → 6 → (9n) 1,3,6,7,8 (6)
- (9j) 1,3,7,8 (221)

(9) 1,3,7 (33)

Figure 6.9. Presumed biosynthetic tree leading to substituted xanthones in Gentianaceae. Bold numbers refer to a parent compound with the oxidation pattern indicated. Each parent compound represents all derivatives (methyl ethers, glycosides, etc.) reported to have this oxidation pattern. The number given in parentheses shows how many times compounds with this oxidation pattern have been reported. Numbers above product arrows show the position in which oxidation takes place to give the next product.

Table 6.3. *The known occurrence of the different types of xanthones in genera of Gentianaceae*

Genus	8ᵃ	8a-8d	8f-8h	8j	8l	8n	9	9a-9d	9e	9f	9j	9k	9l	9m	9n
Anthocleista (1)ᵇ	—	—	—	—	—	—	—	—	—	—	3ᶜ	—	—	—	—
Blackstonia (1)	—	—	—	—	—	—	—	—	—	—	1	—	—	—	—
Canscora (1)	5	—	21	—	—	—	—	—	—	1	4	—	—	—	4
Centaurium (6)	—	—	14	7	—	2	—	—	—	—	7	—	—	—	—
Chironia (1)	6	—	6	—	—	—	1	—	—	—	4	—	—	—	—
Comastoma (1)	—	—	—	2	—	—	—	—	—	—	4	—	—	—	—
Eustoma (1)	1	—	2	—	—	—	1	—	—	—	1	—	—	—	—
Frasera (5)	2	12	—	3	—	—	3	14	—	—	—	—	—	—	—
Gentiana (29)	—	—	—	36	6	—	16	1	—	—	80	—	—	—	—
Gentianella (8)	—	—	—	21	7	—	1	—	—	—	9	—	4	—	—
Gentianopsis (12)	—	—	—	—	—	—	2	—	—	—	5	—	—	—	—
Halenia (4)	—	7	—	1	—	—	—	13	—	—	—	—	—	—	—
Hoppea (2)	2	—	6	—	—	—	—	—	2	—	1	—	—	—	—
Ixanthus (1)	—	—	—	—	—	—	—	—	—	—	—	—	—	—	2
Lomatogonium (2)	—	—	—	3	—	—	1	—	—	—	3	—	—	—	—
Macrocarpaea (1)	—	—	—	—	—	—	1	—	—	—	1	—	—	—	—
Orphium (1)	—	—	—	—	—	—	—	—	—	—	3	—	—	—	—
Schultesia (1)	2	2	1	1	—	—	—	—	—	—	3	—	—	—	—
Swertia (40)	6	10	—	86	2	—	7	6	—	—	92	1	6	4	—
Tripterospermum (2)	—	—	1	—	—	—	—	2	—	2	—	—	3	—	—
Veratrilla (1)	—	2	—	—	—	—	—	7	—	—	—	—	—	—	—

Notes:
ᵃ Bold numbers refer to the compounds in Fig. 6.9.
ᵇ Number of species investigated.
ᶜ The values correspond to the number of compounds reported from each genus with the actual substitution pattern. If the same compound occurs in more than one species of the same genus, each occurrence has been counted.

- **group 1**: only **9** and/or the derived **9j** are present;
- **group 2**: predominance of 8-substitution with compounds derived from both **8** and **9**;
- **group 3**: predominance of 2- and/or 4-substitution, but never 6-substitution;
- **group 4**: predominance of 6-substitution.

In **group 1**, with only 1,3,7- (**9**) and/or 1,3,7,8-tetrasubstituted xanthones (**9j**), we have five genera: *Anthocleista*, *Blackstonia*, *Gentianopsis*, *Macrocarpaea*, and *Orphium*. This could be considered the most primitive state. It should be noted that except for *Gentianopsis*, these genera have been poorly sampled, with only a single species of each. (Also note that many species of *Gentiana* and *Swertia* (Table 6.2) would belong here if they were the single example investigated from either genus!)

The genera in **group 2** could be seen as derived from those of **group 1** since they contain both the 1,3,5,8- and the 1,3,7,8-tetrasubstituted xanthones (**8j** and **9j**), these being formed by an oxidation at position 8 of either of the basic structures **8** or **9**. These types of xanthones occur sporadically in many genera, but they are not predominant in all. The genera in this group are *Comastoma*, *Gentiana*, *Gentianella*, *Lomatogonium*, and *Swertia*. This could be considered an intermediate degree of biosynthetic advancement. *Tripterospermum* may also belong in this group (see below).

In **group 3**, the 2- and/or 4-substituted xanthones (**8a–8d, 9a–d, 8l**, and **9l**) are the most common. The taxa of this group produce mainly the hexa-substituted compounds which demand the most oxidation steps, and they may be considered the most advanced. Group 3 includes the genera *Frasera*, *Halenia*, and *Veratrilla*. Nine of the 34 investigated *Swertia* species and five of the eight investigated *Gentianella* species also have some of these compounds present, but usually in admixture with the typical compounds from the group 2 taxa. Therefore, *Swertia* and *Gentianella* could be intermediates between these two groups. *Tripterospermum* is a special case. Two species have been investigated (Table 6.3), and in these were found the compounds **8f** and **9f** (6-substitution, three findings), which are otherwise characteristic for group 4. However, the remaining compounds in this genus, namely **9b** and **9l** (4-substitution, five findings), are elsewhere encountered mainly in *Swertia* and *Gentianella* on the one hand, and in group 3 on the other. Therefore, *Tripterospermum* is either an intermediate between groups 2 and 3, or could belong to group 4.

Compounds derived mainly by oxidation at the 6-position of **8** are the most common trait in the xanthones of **group 4** taxa. In these genera, the compounds **8f–8h, 8n, 9f**, and **9n** are present. This is another advanced group with many poly-substituted compounds. It includes *Canscora*, *Centaurium*, *Chironia*, *Eustoma*, *Hoppea*, and *Ixanthus*. With some reservations, we have also placed the genus *Schultesia* here. Although 6-substitution is not the predominant feature, it is still present, but the sampling is poor since only a single species has been investigated. As discussed above, *Tripterospermum* could also belong here.

It should be emphasized that the above division has been based on the reported isolations. For some genera only a single species has been investigated and only few xanthones have been isolated. For example, in *Blackstonia* only a single species was investigated and only a single compound was reported. A more thorough investigation of individual genera might lead to different conclusions. Despite this, a large number of compounds have now been isolated from many taxa, and we believe that the data can be given some credence. However, the inherent problem with chemical data is that lack of compounds is only rarely reported; usually this happens only in reports of systematic work.

Comparing the above data with the strict consensus tree presented by Struwe *et al.* (1998, 2002: Figs. 2.1–2.3) based on molecular data (*trn*L intron and *mat*K sequences), we see some interesting and strong correlations with the six major clades of Gentianaceae. To clarify the distribution patterns of the compounds, we have listed the sequenced genera within each tribe and shown the distribution of the compounds found in each genus (Table 6.4). For the xanthones, this is done indirectly because of the way they have been used to classify the genera above into four groups with different biosynthetic potential.

Notably, no xanthones have so far been reported from the Exaceae. However, it appears that only one genus of this clade, *Exacum*, has been investigated chemically (see Table 6.1), and further research may reveal the presence of xanthones also in this group.

The genera of **group 1**, with their poor sampling (usually a single species in each genus) and with only a few biosynthetically primitive compounds reported from each, are randomly found in the remaining clades (tribes). The genera of **group 2**, however, with xanthones that can be considered more advanced than those from group 1, all belong to the Gentianeae. *Tripterospermum*, which as discussed above has an ambiguous position with regard to chemistry, was not sampled in the work by Struwe *et al.* (1998). However, it is known from another nuclear DNA sequencing work (Yuan & Küpfer, 1995) that this genus belongs in tribe Gentianeae, and therefore it has been included in Table 6.4.

Within the Gentianeae further specialization in xanthone biosynthesis has taken place in the genera of **group 3**, which comprise *Frasera*, *Halenia*, and *Veratrilla*. The relationships of these genera with other genera of subtribe Swertiinae are unresolved in the strict consensus tree (Struwe *et al.*, 1998: fig. 1, 2002: Fig. 2.1), but the xanthones support a close relationship between the three genera *Frasera*, *Halenia*, and *Veratrilla* – with *Gentianella* and *Swertia* as probable intermediates to the remaining genera in tribe Gentianeae.

Finally, the taxa of **group 4** (including the chemotaxonomically more dubious *Schultesia*) have specialized in producing 6-substituted xanthones and are all members of the Chironieae.

In conclusion, the xanthone distribution data in Gentianaceae show a very

Chemotaxonomy and pharmacology

Table 6.4. *The distribution of chemical characters in the genera for which sequence data were presented by Struwe et al. (1998) and Thiv et al. (1999)*

Tribe[a]	Genus	Secoiridoids[b]	Xanthone data[c]	Mangiferin	C-Glucoflavones
Saccifolieae	*Saccifolium*				
Exaceae	*Exacum*	x			
Exaceae	*Ornichia*[d]				
Exaceae	*Sebaea*[d]				
Exaceae	*Tachiadenus*[d]				
Chironieae	*Blackstonia*	x	Group 1		
Chironieae	*Canscora*		Group 4	x	
Chironieae	*Centaurium*	x	Group 4		
Chironieae	*Chironia*	x	Group 4		
Chironieae	*Cicendia*	x			
Chironieae	*Coutoubea*	x			
Chironieae	*Deianira*[d]				
Chironieae	*Eustoma*	x	Group 4		
Chironieae	*Geniostemon*				
Chironieae	*Hoppea*		Group 4	x	
Chironieae	*Ixanthus*	x	Group 4		
Chironieae	*Orphium*		Group 1		
Chironieae	*Sabatia*	x			
Chironieae	*Schinziella*[d]				
Chironieae	*Schultesia*	x	Group 4		
Chironieae	*Symphyllophyton*[d]				
Chironieae	*Xestaea*[d]				
Helieae	*Calolisianthus*[d]				
Helieae	*Chelonanthus*	x			
Helieae	*Irlbachia*				
Helieae	*Macrocarpaea*		Group 1		
Helieae	*Neblinantha*[d]				
Helieae	*Symbolanthus*[d]				
Helieae	*Tachia*[d]				
Helieae	*Tetrapollinia*[d]				
Helieae	*Wurdackanthus*[d]				
Potalieae	*Anthocleista*	x	Group 1		
Potalieae	*Djaloniella*[d]				
Potalieae	*Enicostema*	x			x
Potalieae	*Fagraea*	x			x
Potalieae	*Faroa*	x			
Potalieae	*Lisianthius*	x			
Potalieae	*Neurotheca*[d]				
Potalieae	*Potalia*	x			x
Potalieae	*Pycnosphaera*[d]				
Potalieae	*Urogentias*[d]				
Gentianeae	*Bartonia*				
Gentianeae	*Comastoma*		Group 2		
Gentianeae	*Frasera*		Group 3		x
Gentianeae	*Gentiana*	x	Group 2	x	x
Gentianeae	*Gentianella*	x	Group 2	x	x
Gentianeae	*Gentianopsis*		Group 1	x	x
Gentianeae	*Halenia*	x	Group 3		

Table 6.4. (cont.)

Tribe[a]	Genus	Secoiridoids[b]	Xanthone data[c]	Mangiferin	C-Glucoflavones
Gentianeae	*Jaeschkea*[d]				
Gentianeae	*Lomatogonium*	x	Group 2		x
Gentianeae	*Megacodon*[d]				
Gentianeae	*Obolaria*[d]				
Gentianeae	*Swertia*	x	Group 2	x	x
Gentianeae	*Tripterospermum*[e]		Group 2(4)	x	
Gentianeae	*Veratrilla*		Group 3		

Notes:
[a] Classification according to Struwe *et al.* (2002).
[b] Mainly the secoiridoids swertiamarin (**6**) and gentiopicroside (**7**).
[c] The derived data; the groups are of genera with similar compounds present (see text).
[d] These genera have apparently not been chemically investigated.
[e] This genus was not sequenced by Struwe *et al.* (1998), but is known to belong to the same clade as *Gentiana* (Yuan & Küpfer, 1995).

good correlation with the phylogenetic results obtained from DNA sequence data. Further research on xanthones in genera uninvestigated thus far will probably prove to be very useful for chemosystematic work in the family.

MANGIFERIN

Mangiferin (**10**; Fig. 6.10) was first isolated from *Mangifera indica* (Anacardiaceae). In the figure, the structure is shown upside down compared with the xanthone structures in Figs. 6.8 and 6.9 in order to demonstrate the different biosynthetic relationship. Mangiferin is a so-called C-glucoxanthone since the sugar is attached to the xanthone nucleus by a carbon–carbon (C–C) bond. C-Glucosides are resistant to hydrolysis, in contrast to the more common O-glucosides, in which the sugar is attached by a carbon–oxygen (C–O) bond. The latter renders the compound hydrolyzable by enzymes or by acid catalysis to give the free sugar and the aglucone. Thus, C-glucosides are easily detected by chromatographic analysis of hydrolyzed plant extracts since they retain the sugar moiety after acid hydrolysis and consequently have special chromatographic properties compared with other phenolic plant constituents. Only a few compounds belong to this group, and mangiferin is by far the most widespread. In a review on the distribution of the C-glycoxanthones known at that time, Richardson (1983) found that mangiferin occurred more or less sporadically in 60 species, 18 genera, and 13 families of dicots (it is also present in monocots and in ferns). Of these,

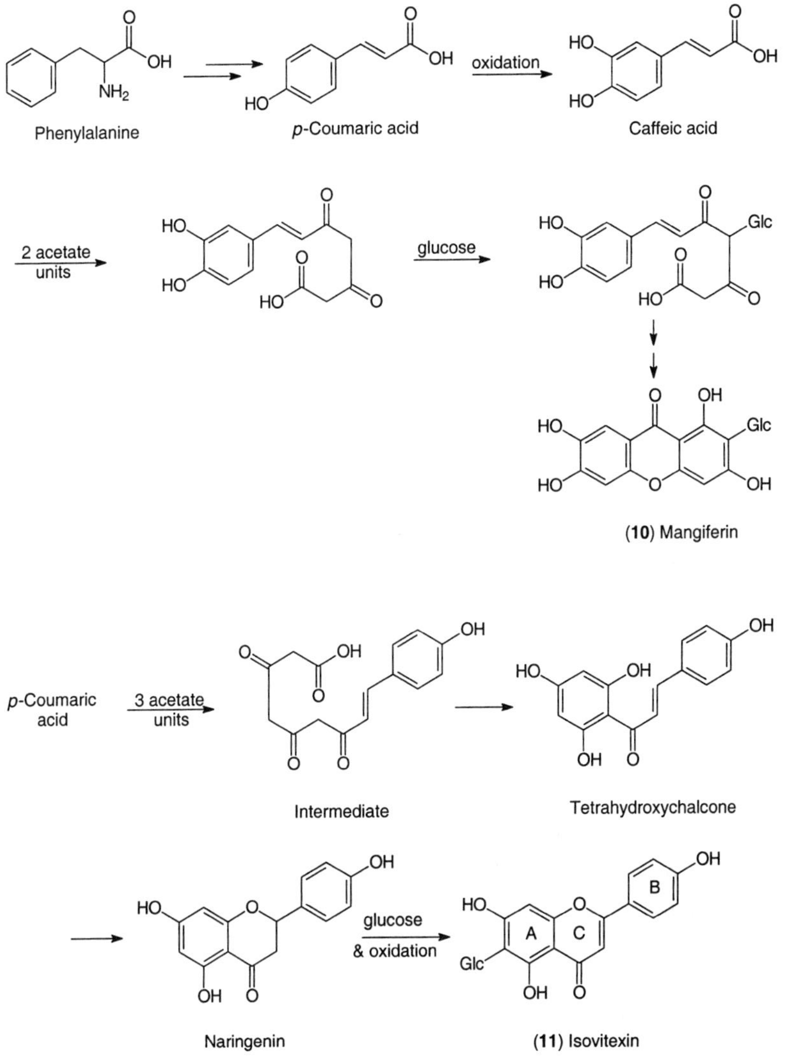

Chemotaxonomy and pharmacology 603

Figure 6.10. Biosynthetic pathways leading to mangiferin (**10**) and the C-glucoflavone isovitexin (**11**); note the common precursors.

Leguminosae (18 species and two genera) and Gentianaceae (23 species and three genera) appeared to be the main centers of distribution for this type of compound. Richardson noted that mangiferin was often found together with C-glucoflavones, but claimed that the compound was apparently of little chemotaxonomic significance.

Biosynthesis

The biosynthesis of the C-glucoside mangiferin (**10**) has been investigated in *Anemarrhena asphodeloides* (Liliaceae) with a result somewhat different from that found for the other xanthones (Fujita & Inoue, 1980). In this plant, all the carbon atoms of phenylalanine (Fig. 6.10) as well as of cinnamic and *p*-coumaric acid are incorporated into the xanthone nucleus, and benzoic acid is apparently not on the pathway. Thus we have a pathway for the formation of mangiferin which is distinct from that of the normal xanthones. This could of course be an example of a different pathway in a completely different taxon, but the oxidation pattern found in mangiferin is not the same as that seen in most of the other xanthones so far isolated from Gentianaceae, consistent with a different biosynthetic pathway as discussed below. As seen from Fig. 6.10, the pathway leading to **10** is more related to that found for the C-glucoflavones (e.g., isovitexin) than to that of the normal xanthones.

Results

Mangiferin (**10**) has so far been recorded in 42 species and seven genera within Gentianaceae (Table 6.5), almost a doubling since the review of Richardson (1983). The records in Table 6.5 include two O-glucosides of **10** in *Gentiana asclepiadea* (Goetz & Jacot-Guillarmod, 1977), but unlike the iridoids and the C-glucoflavones, **10** is not usually found in esterified form.

Also unlike the iridoids, mangiferin (**10**) is not consistently present in all the species of Gentianaceae investigated. Thus, in a chemotaxonomic investigation (Massias *et al.*, 1982), **10** was detected in 15 of the 18 species studied. In a similar study (Hostettmann-Kaldas *et al.*, 1981), **10** was found to occur in about half the 17 species studied.

The above data have also been listed in Table 6.4, and again we see some interesting correlations, despite the fact that only a limited number of genera have been reported to contain mangiferin. This compound has not so far been reported from the Exaceae. By far the largest number of reports are from *Gentiana*, *Gentianella*, *Gentianopsis*, *Swertia*, and *Tripterospermum*, all of which are well sampled and belong to the Gentianeae. The remaining two reports are from a single species each of *Canscora* and *Hoppea*, both of which belong to the Chironieae.

C-GLUCOFLAVONES

As in mangiferin (**10**), the sugar in C-glucoflavones is attached to the flavonoid nucleus by a C–C bond and these compounds are therefore much

Table 6.5. *The known occurrence of mangiferin in Gentianaceae*

Taxon	Reference
Canscora decussata	Ghosal & Chaudhuri, 1975
Fagraea blumei	Cuendet *et al.*, 1997
Gentiana asclepiadea	Goetz & Jacot-Guillarmod, 1977
Gentiana campestris	Kaldas *et al.*, 1975
Gentiana corymbifera	Massias *et al.*, 1981
Gentiana cruciata	Goetz *et al.*, 1976
Gentiana favrati	Hostettmann & Jacot-Guillarmod, 1977
Gentiana karelinii	Butayarov *et al.*, 1993a
Gentiana lactea	Schaufelberger & Hostettmann, 1988
Gentiana lutea	Bellmann & Jacot-Guillarmod, 1973
Gentiana marcailhouana	Luong *et al.*, 1980
Gentiana nivalis	Hostettmann & Jacot-Guillarmod, 1977
Gentiana orbicularis	Massias *et al.*, 1982
Gentiana pneumonanthe	Hostettmann-Kaldas *et al.*, 1981
Gentiana schistocalyx	Nikolaeva *et al.*, 1980c
Gentiana utriculosa	Hostettmann & Jacot-Guillarmod, 1977
Gentiana verna	Hostettmann & Jacot-Guillarmod, 1974
Gentianella campestris	Massias *et al.*, 1982
Gentianella caucasica	Lubsandorzhieva *et al.*, 1986
Gentianella germanica	Hostettmann-Kaldas & Jacot-Guillarmod, 1978
Gentianella ramosa	Hostettmann-Kaldas & Jacot-Guillarmod, 1978
Gentianella serotina	Massias *et al.*, 1982
Gentianella stenocalyx	Massias *et al.*, 1982
Gentianopsis (12 species)	Massias *et al.*, 1982
Hoppea dichotoma	Ghosal *et al.*, 1978a
Swertia calycina	Rodriguez *et al.*, 1995a
Swertia chirata	Ghosal *et al.*, 1973b
Swertia connata	Solov'eva *et al.*, 1980b
Swertia cordata	Atta-ur-Rahman *et al.*, 1994
Swertia dilatata	Tomimori *et al.*, 1974
Swertia elongata	Kong *et al.*, 1995
Swertia franchetiana	Ding *et al.*, 1988
Swertia gracilescens	Tomimori *et al.*, 1974
Swertia macrosperma	Zhou & Liu, 1990
Swertia mussotii	Ding & Sun, 1980
Swertia perennis	Hostettmann & Jacot-Guillarmod, 1976
Swertia perfoliata	Haqqani, 1981
Swertia randaiensis	Tomimori & Komatsu, 1969
Swertia speciosa	Massias *et al.*, 1977
Swertia swertopsis	Tomimori & Komatsu, 1969
Swertia tosaensis	Tomimori & Komatsu, 1969
Tripterospermum lanceolatum	Lin *et al.*, 1982a
Tripterospermum taiwanense	Lin *et al.*, 1984

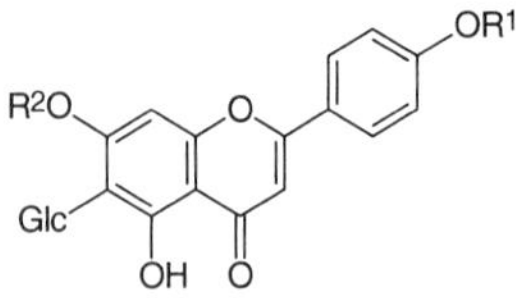

(11) Isovitexin
(11a) R1 = Me; Isocytisoside
(11b) R2 = Me; Swertisin

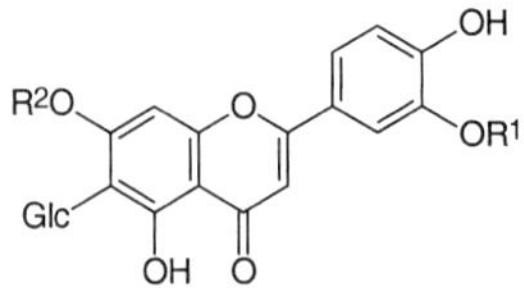

(11c) Isoorientin
(11d) R1 = Me; Isoscoparin
(11e) R2 = Me; Swertiajaponin

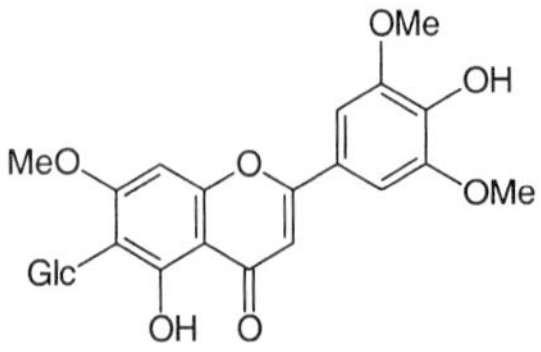

(11f) Isopyrenin

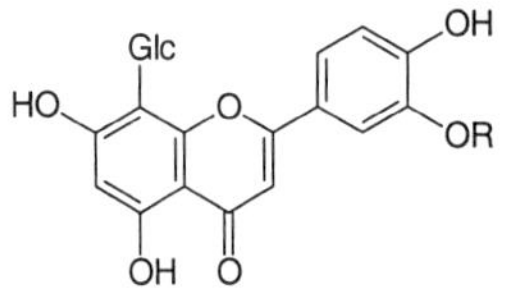

(11g) Orientin
(11h) R = Me; Scoparin

Figure 6.11. C-Glucoflavones reported from Gentianaceae.

more stable than the common O-glucosides. C-Glucoflavones are widely distributed in monocots (see Harborne et al., 1975), but are more scattered in dicots, being found mainly in Leguminosae, Gentianaceae, and Asteraceae. In Gentianaceae, the compounds were at that time (1975) known from six species of *Swertia* and eight species of *Gentiana*. C-Glucoflavones are usually found as such in the plants, but we have seen several reports of C-glucoflavones derivatized with additional O-glycosylation or as esters. However, these reports are very scattered, and we have merely recorded the presence of the parent compounds and their methyl ethers. The structures of the nine compounds so far isolated from the family are depicted in Fig. 6.11 (**11–11h**).

Biosynthesis

Flavonoids in plants are, like the normal xanthones, formed via a mixed shikimate–acetate pathway. However, the detailed pathway is entirely different since the complete carbon skeleton of phenylalanine is retained in the end-product. First, the phenylalanine is transformed to cinnamic acid and then further oxidized to *p*-coumaric acid as shown (Fig. 6.10). Condensation of *p*-coumaric acid with three acetate units gives first an intermediate which by ring-closure forms the A-ring to provide a tetrahydroxychalcone. Another ring-closure to form the central C-ring gives rise to the flavanone naringenin. This pathway is apparently also valid for the formation of C-glucoflavones in Gentianaceae. Thus, Fujita and Inoue (1979) fed radiolabeled precursors to *Swertia japonica* and got similar results in the biosynthesis of isovitexin (**11**), swertisin (**11b**), isoorientin (**11c**), and swertiajaponin (**11e**) – the route is given here only for the first compound. This shows that the sugar is probably attached at the flavanone stage, which is then transformed to the corresponding flavone found in the plant, namely **11**. The analogy to the biosynthesis of mangiferin is obvious, except that in the latter case only two acetate units are incorporated in the intermediate before cyclization. Note that in both cases all the hydroxy groups in the product stem from oxygen atoms originally present in the precursor.

Results

We have recorded C-glucoflavones in 78 species in nine genera (Table 6.6). Systematic surveys of *Gentiana*, *Gentianella*, *Gentianopsis*, and *Swertia*, which have been investigated most intensively, indicate that these compounds may be generally present, at least in parts of Gentianeae. Thus they were found in all of the 22 *Gentiana* species examined by Hostettmann *et al.* (1975) and Hostettmann-Kaldas *et al.* (1981), as well as in the 17 species from the first three of the above-mentioned genera examined by Massias *et al.* (1982). Similarly, they were found in all five species of *Swertia* investigated by Komatsu *et al.* (1968).

From the data in Table 6.6 it appears that the two biosynthetically primary compounds isovitexin (**11**) and isoorientin (**11c**) are by far the most common in the family. Increasing O-methylation has been linked to advancement of the plant taxa (Harborne, 1977), and presence of the methyl ethers **11a/11b** and **11d/11e** may thus be taken as an indication of this (Massias *et al.*, 1982).

When we combine the data from Table 6.6 into the table of sequenced taxa (Table 6.4) we again find an interesting correlation. C-Glucoflavones appear to be very common in the Gentianeae, where they have been reported from *Frasera*, *Gentiana*, *Gentianella*, *Gentianopsis*, *Lomatogonium*, and *Swertia*. The remaining reports are from a single species each of *Enicostema*, *Fagraea*, and *Potalia*, all of which belong to the Potalieae. Despite the poor sampling, it may be significant that no methyl ethers have been reported from this tribe.

Unfortunately, systematic investigations have been performed only in genera of the Gentianeae. It would be interesting to know if the compounds are consistently present in other genera.

608   S. R. Jensen & J. Schripsema

Table 6.6. *The known occurrence of C-glucoflavones in Gentianaceae*

Taxon	Compounds[a]					References[b]
Enicostema hyssopifolium	11	11b				1
Frasera tetrapetala		11b				2
Gentiana affinis			11c			3
Gentiana algida			11c		11g	3, 4
Gentiana argentea	11		11c			5
Gentiana arisanensis			11c			6
Gentiana asclepiadea	11		11c			7
Gentiana bavarica	11		11c			8
Gentiana brachyphylla	11		11c			8
Gentiana burseri	11		11c			9
Gentiana calycosa			11c			3
Gentiana campestris		11b	11c			10, 11
Gentiana cerastioides			11c			3
Gentiana ciliata	11		11c	11d		12
Gentiana corymbifera	11	11b	11c			13
Gentiana depressa	11		11c			14
Gentiana detonsa			11c			3
Gentiana elwesii	11		11c			14
Gentiana farreri			11c	11d		15
Gentiana favrati	11		11c			8
Gentiana karelinii			11c			16
Gentiana lactea	11		11c			17
Gentiana lutea	11		11c			18
Gentiana macrophylla	11		11c			19
Gentiana marcailhouana	11		11c			20, 21
Gentiana nivalis	11		11c			3
Gentiana olivieri			11c		11g	22
Gentiana orbicularis			11c			23
Gentiana pannonica	11		11c			9
Gentiana pedicellata	11		11c			14
Gentiana pneumonanthe	11			11d	11h	24
Gentiana prolata	11		11c			14
Gentiana punctata	11		11c	11d		25
Gentiana purpurea	11		11c			9
Gentiana pyrenaica	11	11a		11d	11f	26
Gentiana rostani	11		11c			8
Gentiana schistocalyx	11					27
Gentiana schleicheri	11		11c			8
Gentiana sikkimensis	11		11c			14
Gentiana sino-ornata			11c			15
Gentiana strictiflora			11c			3
Gentiana utriculosa	11		11c			8
Gentiana verna			11c			28
Gentiana villarsii	11		11c			9
Gentianella azurea			11c			29
Gentianella campestris		11b	11c			23

Chemotaxonomy and pharmacology 609

Table 6.6. (cont.)

Taxon	Compounds[a]				References[b]
Gentianella germanica		11b	11c		30
Gentianella ramosa		11b	11c		30
Gentianella serotina		11b	11c		23
Gentianella stenocalyx	11		11c		23
Gentianopsis (11 species)	11		11c	11d	23
Lomatogonium carinthiacum			11c		31
Lomatogonium rotatum			11c		32
Potalia amara		11b			33
Swertia alata		11b			34
Swertia bimaculata	11				35
Swertia diluta	11				35
Swertia franchetiana		11b			36
Swertia japonica	11	11b	11c	11e	35, 37
Swertia mussotii		11b			38
Swertia paniculata		11b	11c		39
Swertia perennis	11		11c		40
Swertia pseudochinensis	11				35
Swertia purpurascens		11b			41
Swertia randaiensis	11				35
Swertia swertopsis	11		11c		42

Notes:
[a] Bold numbers refer to the compounds in Fig. 6.11.
[b] References: 1: Ghosal & Jaiswal, 1980. 2: Agata et al., 1984. 3: Hostettmann-Kaldas et al., 1981. 4: Tan et al., 1996. 5: Chulia & Debelmas, 1977. 6: Kuo et al., 1996. 7: Goetz & Jacot-Guillarmod, 1977. 8: Hostettmann & Jacot-Guillarmod, 1977. 9: Hostettmann et al., 1975. 10: Kaldas et al., 1974. 11: Kaldas et al., 1975. 12: Goetz et al., 1978. 13: Massias et al., 1981. 14: Chulia & Debelmas, 1977. 15: Schaufelberger & Hostettmann, 1987. 16: Butayarov et al., 1993a. 17: Schaufelberger & Hostettmann, 1988. 18: Bellmann & Jacot-Guillarmod, 1973. 19: Tikhonova et al., 1989. 20: Luong et al., 1981. 21: Luong et al., 1980. 22: Ersoz & Calis, 1991. 23: Massias et al., 1982. 24: Burret et al., 1978. 25: Luong & Jacot-Guillarmod, 1977. 26: Marston et al., 1976. 27: Nikolaeva et al., 1980c. 28: Hostettmann & Jacot-Guillarmod, 1974. 29: Zhang & Yang, 1994. 30: Hostettmann-Kaldas & Jacot-Guillarmod, 1978. 31: Schaufelberger & Hostettmann, 1984. 32: Khishgee & Pureb, 1993. 33: S. R. Jensen & J. Schripsema, unpubl. 34: Khan et al., 1979. 35: Komatsu et al., 1968. 36: Ding et al., 1988. 37: Kubota et al., 1983. 38: Sun et al., 1991. 39: Verma & Khetwal, 1985. 40: Hostettmann & Jacot-Guillarmod, 1976. 41: Miana, 1973. 42: Tomimori & Komatsu, 1969.

PHYTOCHEMICAL CONCLUSIONS

The family Gentianaceae is apparently characterized by the universal occurrence of iridoid glucosides. The iridoids found are consistently secoiridoids; the occasional carbocyclic iridoids present are on the biosynthetic pathway leading to the secoiridoids (Fig. 6.1). Secoiridoids are characteristic for the whole order Gentianales and are hardly ever present in Scrophulariales/Lamiales, where iridoid glucosides of an alternative biosynthetic pathway are frequent. We have recorded c. 90 different iridoid compounds from the family (Figs. 6.2–6.7), from 127 species in 24 genera (Table 6.1). However, the most common iridoid glucosides present are those represented by the sequence sweroside (5), swertiamarin (6), and gentiopicroside (7). The last two are found almost exclusively in Gentianaceae, and are reported from 107 of the 127 species listed in Table 6.1. The iridoid glucosides are frequently found as the parent compounds above, but also often occur as derivatives, e.g., esterified with acid moieties, of which many are unique to the Gentianaceae.

Xanthones are also present in many species, although they are not universal in the family. About 100 different compounds are reported from 110 species in 21 genera. Using a biosynthetic scheme based partly on actual experiments (Fig. 6.8) and partly on deductions from structural features, the xanthones have been arranged according to biosynthetic complexity and to oxidation pattern (Fig. 6.9; Table 6.3). The genera have then been classified into groups according to the compounds present, and these groups provide a good fit with the clades indicated from phylogenetic results based on *trn*L intron and *mat*K sequences (Struwe *et al.*, 2002).

The C-glucoxanthone mangiferin (Table 6.5) has a more limited distribution than the iridoids and the normal xanthones. The compound has been reported from 42 species in seven genera, and it appears to be present only in two of the above groups.

The C-glucoflavones are much less variable than the iridoids and the xanthones; only nine different compounds have been reported so far from a total of 78 species in nine genera (Table 6.6). This group of compounds also has a limited distribution in the family, and is again found in only two of the above groups.

Combining the known chemical data and the sequence data (Table 6.4), we can now list the chemical characteristics of the tribes of Gentianaceae. A question mark indicates that data are available but scarce. The "absent" character is unfortunately very uncertain, since it may merely mean lack of data.

Saccifolieae:
Not investigated
Exaceae:
Secoiridoids (including **6** and **7**): present
Xanthones: absent
Mangiferin: absent
C-Glucoflavones: absent
Chironieae:
Secoiridoids (including **6** and **7**): present
Xanthones: 6-substituted but not 4-substituted are present
Mangiferin: present (?)
C-Glucoflavones: absent
Helieae:
Secoiridoids (including **6** and **7**): present
Xanthones: only primitive ones are present (?)
Mangiferin: absent
C-Glucoflavones: absent
Potalieae:
Secoiridoids (including **6** and **7**): present
Xanthones: only primitive ones are present (?)
Mangiferin: absent
C-Glucoflavones: present (?)
Gentianeae:
Secoiridoids (including **6** and **7**): present
Xanthones: 8-substituted or 2- and 4-substituted predominant
Mangiferin: present
C-Glucoflavones: present

SOME PHARMACOLOGICAL REMARKS ON THE GENTIANACEAE

Plants from the Gentianaceae are best known for their bitter taste, which can be related to their content of iridoids, such as amarogentin (**5d**), the most bitter compound known. Bitters have been traditional remedies for loss of appetite and fever, and are still included in many "tonic" formulations (Martindale, 1982). Various herbs from the Gentianaceae are common in pharmacopoeias: centaury (the dried flowering tops of the common centaury, *Centaurium erythraea*, and other species of *Centaurium*), chirata (the dried plant *Swertia chirata*, collected when in flower), and gentian (the dried, partially fermented rhizome and root of *Gentiana lutea*). Many other species from the family have been used for the same purpose in various regions of the world, e.g., *Chironia baccifera* in South Africa, *Erythraea chilensis* in Chile and Peru, *Coutoubea spicata* in Brazil, *Enicostema littorale* in India, *Frasera caroliniensis* in the USA, *Schultesia guianensis* in Central America, and *Centaurium australe* in Australia (Hegnauer, 1966).

Some specific activities have been reported for iridoids from Gentianaceae. Swertiamarin (**6**) should have anticholinergic properties (Bhattacharya *et al.*, 1976; Liang *et al.*, 1982; Yamahara *et al.*, 1991). For sweroside (**5**) and gentiopicroside (**7**) hepatoprotective activities have been reported (Zhou, 1991; Kondo *et al.*, 1994), and both compounds are being used as anti-hepatitis drugs.

Xanthones (especially mangiferin) are reported to give CNS stimulation (Bhattacharya *et al.*, 1972; Lin *et al.*, 1984), and this effect can be explained by their MAO inhibitory activity (Schaufelberger & Hostettmann, 1988). They should also have anti-inflammatory activity (Mandal *et al.*, 1992b). For bellidifolin and swerchirin a strong hypoglycemic activity has been reported (Saxena & Mukherjee, 1992; Saxena *et al.*, 1993; Basnet *et al.*, 1994, 1995). Norathyriol, a xanthone from *Tripterospermum*, was reported to have anti-inflammatory and analgesic activities. An investigation suggested that norathyriol might be a dual, but weak, cyclo-oxygenase and lipoxygenase pathway blocker (Wang *et al.*, 1994a). Norathyriol also suppressed inflammation-related edema, probably due partly to suppression of mast cell degranulation and hence reduction in the release of chemical mediators which increase vascular permeability, and partly, at least in higher doses, to protection of the vasculature from challenge by various mediators (Wang *et al.*, 1994b).

Flavan glucosides (from *Hoppea dichotoma*) are reported to have a sedative effect (Ghosal *et al.*, 1985). Antipsychotic activity has been reported for gentianine (Bhattacharya *et al.*, 1974), which antagonized amphetamine and decreased aggressiveness. Gentianadine decreased arterial pressure and had anti-inflammatory activity (Sadritdinov & Tulyaganov, 1972). An antifungal effect has been reported for anofinic acid and fomannoxin (Tan *et al.*, 1996) and for 2-methoxy-1,4-naphthoquinone (Rodriguez *et al.*, 1995a). Antifungal phosphocholine derivatives were extracted from *Irlbachia alata* (= *Chelonanthus alatus*; Tempesta *et al.*, 1994; Bierer *et al.*, 1995). In addition, a crude extract of *Swertia* with insect repellent activities has been reported (Okada, 1977).

Finally, swertifrancheside was found to be a potent inhibitor of the DNA polymerase activity of HIV-1 reverse transcriptase (Pengsuparp *et al.*, 1995), and five tetrahydroxyxanthones isolated from *Tripterospermum lanceolatum* were shown to have a strong inhibitory effect on the activity of

Moloney murine leukemia virus reverse transcriptase (Chang et al., 1992). The tetrahydroxyxanthones also inhibited angiotensin-1-converting-enzyme (ACE) activity in a dose-dependent manner (Chen et al., 1992).

ACKNOWLEDGMENTS

The authors thank the Commission of the European Union for a grant to J.S. (BIO2-CT94-8054).

LITERATURE CITED

Agata, I., H. Sekizaki, A. Sakushima, S. Nishibe, S. Hisada, & K. Kimura. 1981. Studies on the constituents of medicinal plants in Hokkaido. I. On the whole herb of *Swertia tetrapetala* Pall. 1. *Yakugaku Zasshi* 101: 1067–1071.
Agata, I., Y. Nakaya, S. Nishibe, S. Hisada, & K. Kimura. 1984. Studies on the constituents of medicinal plants in Hokkaido. II. On the whole herb of *Frasera tetrapetala* Pall. 2. *Yakugaku Zasshi* 104: 418–421.
Ahmad, S., M. Ikram, I. Khan, & N. M. Galbraith. 1973. Xanthones of *Swertia purpurescens*. *Phytochemistry* 12: 2542–2543.
Ashida, S., S. F. Noguchi, & T. Suzuki. 1994. Antioxidative components, xanthone derivatives, in *Swertia japonica* Makino. *J. Am. Oil Chem. Soc.* 71: 1095–1099.
Asthana, R. K., N. K. Sharma, D. K. Kulshreshtha, & S. K. Chatterjee. 1991. A xanthone from *Swertia chirayita*. *Phytochemistry* 30: 1037–1039.
Atkinson, J. E., P. Gupta, & J. R. Lewis. 1968. Benzophenone participation in xanthone biosynthesis (Gentianaceae). *J. Chem. Soc., Chem. Commun.*: 1386–1387.
Atkinson, J. E., P. Gupta, & J. R. Lewis. 1969. Phenolic constituents of *Gentiana lutea*. *Tetrahedron* 25: 1507–1511.
Atta-ur-Rahman, P. A., M. Feroz, M. I. Choudhary, M. M. Qureshi, S. Perveen, I. Mir, & M. I. Khan. 1994. Phytochemical studies on *Swertia cordata*. *J. Nat. Prod.* 57: 134–137.
Basnet, P., S. Kadota, M. Shimizu, & T. Namba. 1994. Bellidifolin: a potent hypoglycemic agent in streptozotocin (STZ)-induced diabetic rats from *Swertia japonica*. *Planta Med.* 60: 507–511.
Basnet, P., S. Kadota, M. Shimizu, Y. Takata, M. Kobayashi, & T. Namba. 1995. Bellidifolin stimulates glucose uptake in rat 1 fibroblasts and ameliorates hyperglycemia in streptozotocin (STZ)-induced diabetic rats. *Planta Med.* 61: 402–405.
Bellmann, G. & A. Jacot-Guillarmod. 1973. Phytochemistry of the genus *Gentiana*. I. Flavone and xanthone compounds in *Gentiana lutea* leaves. 1. *Helv. Chim. Acta* 56: 284–294.
Bhan, S. & A. K. Kalla. 1982. Chemical investigation of *Swertia petiolata* Royle. *Res. J. Fac. Sci., Kashmir Univ.* 1: 10–14.
Bhattacharya, S. K., P. K. S. P. Reddy, S. Ghosal, A. K. Singh, & P. Sharma.

1976. Chemical constituents of Gentianaceae. XIX. CNS-depressant effects of swertiamarin. *Indian J. Pharm. Sci.* 65: 1547–1549.
Bhattacharya, S. K., S. Ghosal, R. K. Chaudhuri, & A. K. Sanyal. 1972. *Canscora decussata* (Gentianaceae) xanthones. III. Pharmacological studies. *J. Pharm. Sci.* 61: 1838–1840.
Bhattacharya, S. K., S. Ghosal, R. K. Chaudhuri, A. K. Singh, & P. V. Sharma. 1974. Chemical constituents of Gentianaceae. XI. Antipsychotic activity of gentianine. *J. Pharm. Sci.* 63: 1341–1342.
Bierer, D. E., R. E. Gerber, S. D. Jolad, R. P. Ubillas, J. Randle, E. Nauka, J. Latour *et al.* 1995. Isolation, structure elucidation, and synthesis of Irlbacholine, 1,22-bis[[[2-(trimethylammonium)ethoxy]-phospinyl]oxy]docosane: a novel antifungal plant metabolite from *Irlbachia alata* and *Anthocleista djalonensis*. *J. Org. Chem.* 60: 7022–7026.
Bisht, R. S., G. P. Chamoli, K. N. Bijalwan, & G. C. Mishra. 1991. Xanthones and flavonoids from *Swertia speciosa*. *Himalayan Chem. Pharm. Bull.* 8: 31–34.
Bricout, J. 1974. Identification and determination of the bitter constituents of *Gentiana lutea* roots. *Phytochemistry* 13: 2819–2823.
Burret, F., A. J. Chulia, A. M. Debelmas, & K. Hostettmann. 1978. Presence of isoscoparin, isoscoparin-7–O-glucoside and saponarin in *Gentiana pneumonanthe* L. *Planta Med.* 34: 176–179.
Butayarov, A. V., E. K. Batirov, M. M. Tadzhibaev, & V. M. Malikov. 1993a. Xanthones and flavonoids of *Gentiana karelinii*. *Khim. Prir. Soedin*: 469–470.
Butayarov, A. V., E. K. Batirov, M. M. Tadzhibaev, E. E. Ibragimov, & V. M. Malikov. 1993b. Xanthones from *Gentiana algida* and *G. karelinii*. *Khim. Prir. Soedin*: 901–902.
Calis, I., H. Rueegger, Z. Chun, & O. Sticher. 1990. Secoiridoid glucosides isolated from *Gentiana gelida*. *Planta Med.* 56: 406–409.
Calis, I., T. Ersoz, A. J. Chulia, & P. Ruedi. 1992. Septemfidoside: a new bis-iridoid diglucoside from *Gentiana septemfida*. *J. Nat. Prod.* 55: 385–388.
Cambie, R. C., A. R. Lal, C. E. F. Rickard, & N. Tanaka. 1990. Chemistry of Fijian plants. Constituents of *Fagraea gracilipes* A. Gray. *Chem. Pharm. Bull.* 38: 1857–1861.
Carbonnier, J., M. Massias, & D. Molho. 1977. Taxonomic importance of the substitution scheme of xanthones in *Gentiana* L. *Bull. Mus. Natl. Hist. Nat., Sci. Phys.-Chim.* 13: 23–40.
Carpenter, I., H. D. Locksley, & F. Scheinmann. 1969. Xanthones in higher plants: biogenetic proposals and a chemotaxonomic survey. *Phytochemistry* 8: 2013–2026.
Chang, C. C., C. N. Lin, & J. Y. Lin. 1992. Inhibition of Moloney murine leukemia virus reverse transcriptase activity by tetrahydroxyxanthones isolated from the Chinese herb, *Tripterospermum lanceolatum* (Hayata). *Antiviral Res.* 19: 119–127.
Chapelle, J. P. 1973. Isolation of secoiridoid derivatives of *Anthocleista zambesiaca*. *Phytochemistry* 12: 1191–1192.
Chapelle, J. P. 1974. Chemical constituents of the leaves of *Anthocleista vogelii*. *Planta Med.* 26: 301–304.

Chapelle, J. P. 1976a. Grandifloroside and methylgrandifloroside, new iridoid glucosides from *Anthocleista grandiflora*. *Phytochemistry* 15: 1305–1307.
Chapelle, J. P. 1976b. Vogeloside and secologanic acid, secoiridoid glucosides from *Anthocleista vogelii*. *Planta Med.* 29: 268–274.
Chaudhuri, P. K. & W. M. Daniewski. 1995. Unambiguous assignments of the ^{1}H and ^{13}C chemical shifts of the major bitter principle of *Swertia chirata* by 2D NMR study and characterization of other constituents. *Pol. J. Chem.* 69: 1514–1519.
Chaudhuri, R. K. & S. Ghosal. 1971. Chemical constituents of the Gentianaceae. I. Xanthones of *Canscora decussata*. *Phytochemistry* 10: 2425–2432.
Chen, C. H., J. Y. Lin, C. N. Lin, & S. Y. Hsu. 1992. Inhibition of angiotensin-I-converting enzyme by tetrahydroxyxanthones isolated from *Tripterospermum lanceolatum*. *J. Nat. Prod.* 55: 691–695.
Chulia, A. J. & A. M. Debelmas. 1977. Contribution to the phytochemistry of the *Gentiana* genus. Study of flavonoids from Nepal *Gentiana*. *Plant. Med. Phytother.* 11: 112–118.
Chulia, A. J. & M. Kaouadji. 1985. Depressoside, a new iridoid isolated from *Gentiana depressa*. *J. Nat. Prod.* 48: 54–58.
Chulia, A. J., J. Garcia, E. M. Mpondo, & A. M. Mariotte. 1986. New glycosides in *Gentiana pedicellata* Wall. *Bull. Liaison – Groupe Polyphenols* 13: 50–55.
Chulia, A. J., J. Vercauteren, & M. Kaouadji. 1994. Depresteroside, a mixed iridoid-secoiridoid structure from *Gentiana depressa*. *Phytochemistry* 36: 377–382.
Chulia, A. J., J. Vercauteren, & A. M. Mariotte. 1996. Iridoids and flavones from *Gentiana depressa*. *Phytochemistry* 42: 139–143.
Chung, B. S. & Y. H. Lee. 1982. Secoiridoid glucosides from the root of *Gentiana uchiyamana* Nakai. *Saengyak Hakhoe Chi* (*Hanguk Saengyak Hakhoe*) 13: 1–6. (*Chem. Abstr.* 97: 141733)
Chung, M. I. & C. N. Lin. 1993. Studies on the constituents of Formosan gentianaceous plants. XV. A new acylated secoiridoid from *Gentiana formosana*. *J. Nat. Prod.* 56: 982–983.
Chung, M. I., K. H. Gan, C. N. Lin, & I. J. Chen. 1986. Studies on the constituents of Formosan gentianaceous plants. Part VII. Constituents of *Swertia randaiensis* Hayata and pharmacological activity of norswertianolin. *Kao-hsiung I Hsueh K'o Hsueh Tsa Chih* 2: 131–135.
Cornelis, A. & J. P. Chapelle. 1976. Carbon-13 nuclear magnetic resonance (^{13}C-NMR) in the systematic study of secoiridoid glucosides: structure confirmation of swertiamaroside. *Pharm. Acta Helv.* 51: 177–180.
Coscia, C. J. & R. Guarnaccia. 1967. Biosynthesis of gentiopicroside, a novel monoterpene. *J. Am. Chem. Soc.* 89: 1280–1281.
Coscia, C. J. & R. Guarnaccia. 1968. Natural occurrence and biosynthesis of a cyclopentanoid monoterpene carboxylic acid. *Chem. Commun*: 138–140.
Coscia, C. J., L. Botta, & R. Guarnaccia. 1969. Monoterpene biosynthesis. I. Occurrence and mevalonoid origin of gentiopicroside and loganic acid in *Swertia caroliniensis*. *Biochemistry* 8: 5036–5043.
Coscia, C. J., L. Botta, & R. Guarnaccia. 1970. On the mechanism of iridoid and secoiridoid monoterpene biosynthesis. *Arch. Biochem. Biophys.* 136: 498–506.
Cuendet, M., K. Hostettmann, O. Potterat, & W. Dyatmico. 1997. Iridoid

glucosides with free radical scavenging properties from *Fagraea blumei*.
Helv. Chim. Acta 80: 1144–1152.
Dahlgren, G. 1989. The last Dahlgrenogram, a system of classification of the
dicotyledons. Pages 237–260 in: K. Tan, ed. *Plant taxonomy, phytogeography
and related subjects: The Davis and Hedge Festschrift*. Edinburgh University
Press, Edinburgh.
Das, S., R. P. Sharma, J. N. Baruah, P. Kulanthaivel, & W. Herz. 1984.
Secoiridoids from *Exacum tetragonum*. *Phytochemistry* 23: 908–909.
Delaude, C. 1984. Gentianine, alkaloid from *Exacum quinquenervium* Griseb.
(Gentianaceae). *Bull. Soc. R. Sci. Liege* 53: 54–56.
Delaude, C. 1985. Gentianin, an alkaloid from *Faroa chalcophila* P. Taylor
(Gentianaceae). *Bull. Soc. R. Sci. Liege* 54: 18.
Delaude, C. & E. Darimont. 1985. Studies on the alkaloidal material from *Faroa
graveolens* Baker (Gentianaceae). *Bull. Soc. R. Sci. Liege* 54: 127–128.
Denisova, O. A., V. I. Glyzin, A. V. Patudin, & D. A. Fesenko. 1980a. Xanthones
from *Swertia iberica* roots. *Khim. Prir. Soedin.*: 190–195.
Denisova, O. A., V. I. Glyzin, & A. V. Patudin. 1980b. Xanthone glycosides of
Swertia iberica. *Khim. Prir. Soedin.*: 569–570.
Denisova, O. A., E. V. Solov'eva, V. I. Glyzin, & A. V. Patudin. 1980c. Xanthone
glycosides from the aboveground part of *Swertia iberica*. *Khim. Prir. Soedin.*:
724–725.
Dhoubhadel, S. P., P. P. Wagley, & S. D. Pradhan. 1980. A new xanthone
glycoside from *Swertia angustifolia* Buch.-Ham. ex D. Don. *Indian J. Chem.*,
sect. B, 19B: 929–930.
Ding, J., S. Fan, B. Hu, & H. Sun. 1988. On the xanthone glycosides and flavonoid
glucoside from *Swertia franchetiana* H. Smith. *Zhiwu Xuebao* 30: 414–419.
Ding, J.-Y. & H.-F. Sun. 1980. Studies on the antihepatitis constituents of
zangyinchen (*Swertia mussotii*). I. Isolation and identification of mangiferin
and oleanolic acid. *Chung Ts'ao Yao* 11: 391–392.
Do, T., S. Popov, N. Marekov, & A. Trifonov. 1987. Iridoids from Gentianaceae
plants growing in Bulgaria. *Planta Med.* 53: 580.
Dreyer, D. L. & J. H. Bourell. 1981. Xanthones of the Gentianaceae. Part VI.
Xanthones from *Frasera albomarginata* and *F. speciosa*. *Phytochemistry* 20:
493–495.
Ersoz, T. & I. Calis. 1991. C-glucosylflavones from *Gentiana olivieri*. *Hacettepe
Univ. Eczacilik Fak. Derg.* 11: 29–36.
Fan, S., B. Hu, J. Ding, & H. Sun. 1988. Chemical constituents of *Comastoma
pulmonarium* (Toucz.) Toyohuni. *Zhiwu Xuebao* 30: 303–307. (*Chem. Abstr.*
110: 21067)
Frederiksen, S. & F. R. Stermitz. 1996. Pyridine monoterpenes alkaloid formation
from iridoid glycosides. A novel PMTA dimer from geniposide. *J. Nat. Prod.*
59: 41–56.
Fujita, M. & T. Inoue. 1979. Biosynthesis of C-glucosylflavones in *Swertia
japonica*. *Yakugaku Zasshi* 99: 165–171.
Fujita, M. & T. Inoue. 1980. Biosynthesis of mangiferin in *Anemarrhena
asphodeloides* Bunge. I. The origin of the xanthone nucleus. *Chem. Pharm.
Bull.* 28: 2476–2481.

Fukamiya, N., M. Okano, K. Kondo, & K. Tagahara. 1990. Xanthones from *Swertia punicea. J. Nat. Prod.* 53: 1543–1547.
Gao, G. Y., M. Li, Y. X. Feng, & P. Tan. 1994. Determination of effective constituents in 11 *Swertia* and related plants by HPLC. *Acta Pharm. Sin.* 29: 910–914.
Garcia, J. & A. J. Chulia. 1987. 4'-*p*-Coumaroyl iridoid glucosides from *Gentiana pedicellata. Planta Med.* 53: 101–103.
Garcia, J., S. Lavaitte, & C. Gey. 1989a. 8–Epikingiside and its vanillate ester, isolated from *Gentiana pyrenaica. Phytochemistry* 28: 2199–2201.
Garcia, J., E. M. Mpondo, G. Cartier, & C. Gey. 1989b. Secoiridoids and a phenolic glucoside from *Gentiana pyrenaica. J. Nat. Prod.* 52: 996–1002.
Garcia, J., E. M. Mpondo, & R. Nardin. 1989c. Loganin and a new iridoid glucoside from *Gentiana pyrenaica. J. Nat. Prod.* 52: 423–425.
Garcia, J., E. M. Mpondo, & M. Kaouadji. 1990. Kingiside and derivative from *Gentiana pyrenaica. Phytochemistry* 29: 3353–3355.
Ghosal, S. & K. Biswas. 1979. Chemical constituents of Gentianaceae. Part XXVI. Two new 1,3,5,6,7-pentaoxygenated xanthones from *Canscora decussata. Phytochemistry* 18: 1029–1031.
Ghosal, S. & R. K. Chaudhuri. 1973. Chemical constituents of the Gentianaceae. VI. New tetraoxygenated xanthones of *Canscora decussata. Phytochemistry* 12: 2035–2038.
Ghosal, S. & R. K. Chaudhuri. 1975. Chemical constituents of Gentianaceae. XVI. Antitubercular activity of xanthones of *Canscora decussata. J. Pharm. Sci.* 64: 888–889.
Ghosal, S. & D. K. Jaiswal. 1980. Chemical constituents of Gentianaceae. XXVIII: Flavonoids of *Enicostemma hyssopifolium* (Willd.) Verd. *J. Pharm. Sci.* 69: 53–56.
Ghosal, S., R. K. Chaudhuri, & A. Nath. 1971. Chemical constituents of the roots of *Canscora decussata.* II. *J. Indian Chem. Soc.* 48: 589–590.
Ghosal, S., R. K. Chaudhuri, & A. Nath. 1973a. Chemical constituents of Gentianaceae. IV. New xanthones of *Canscora decussata. J. Pharm. Sci.* 62: 137–139.
Ghosal, S., P. V. Sharma, R. K. Chaudhuri, & S. K. Bhattacharya. 1973b. Chemical constituents of the Gentianaceae. V. Tetraoxygenated xanthones of *Swertia chirata. J. Pharm. Sci.* 62: 926–930.
Ghosal, S., R. K. Chaudhuri, & K. R. Markham. 1974a. Chemical constituents of the Gentianaceae. XII. Structure of the pentaoxygenated xanthones of *Canscora decussata. J. Chem. Soc., Perkin* I: 2538–2541.
Ghosal, S., P. V. Sharma, & R. K. Chaudhuri. 1974b. Chemical constituents of Gentianaceae. X. Xanthone-O-glucosides of *Swertia purpurascens. J. Pharm. Sci.* 63: 1286–1289.
Ghosal, S., A. K. Singh, P. V. Sharma, & R. H. Chaudhuri. 1974c. Chemical constituents of Gentianaceae. IX. Natural occurrence of erythrocentaurin in *Enicostemma hyssopifolium* and *Swertia lawii. J. Pharm. Sci.* 63: 944–945.
Ghosal, S., P. V. Sharma, & R. K. Chaudhuri. 1975a. Tetra- and pentaoxygenated xanthones of *Swertia lawii. Phytochemistry* 14: 1393–1396.
Ghosal, S., P. V. Sharma, & R. K. Chaudhuri. 1975b. Chemical constituents of

Gentianaceae. 18. Xanthones of *Swertia bimaculata*. *Phytochemistry* 14: 2671–2675.
Ghosal, S., P. V. Sharma, R. K. Chaudhuri, & S. K. Bhattacharya. 1975c. Chemical constituents of Gentianaceae. XIV. Tetraoxygenated and pentaoxygenated xanthones of *Swertia purpurascens*. *J. Pharm. Sci.* 64: 80–83.
Ghosal, S., R. Ballava, P. S. Chauhan, K. Biswas, & R. K. Chaudhuri. 1976. Chemical constituents of Gentianaceae. Part 21. New 1,3,5-trioxygenated xanthones in *Canscora decussata*. *Phytochemistry* 15: 1041–1043.
Ghosal, S., K. Biswas, & R. K. Chaudhuri. 1977. Chemical constituents of Gentianaceae. Part 22. Structures of new 1,3,5-tri- and 1,3,5,6,7-pentaoxygenated xanthones of *Canscora decussata* Schult. *J. Chem. Soc., Perkin* I: 1597–1601.
Ghosal, S., D. K. Jaiswal, & K. Biswas. 1978a. Chemical constituents of Gentianaceae. Part XXV. New glycoxanthones and flavanone glycosides of *Hoppea dichotoma*. *Phytochemistry* 17: 2119–2123.
Ghosal, S., P. V. Sharma, & D. K. Jaiswal. 1978b. Chemical constituents of Gentianaceae. XXIII: Tetraoxygenated and pentaoxygenated xanthones and xanthone O-glucosides of *Swertia angustifolia* Buch.-Ham. *J. Pharm. Sci.* 67: 55–60.
Ghosal, S., K. Biswas, & D. K. Jaiswal. 1980. Chemical constituents of Gentianaceae. Part 27. Xanthone and flavonol constituents of *Swertia hookeri*. *Phytochemistry* 19: 123–126.
Ghosal, S., D. K. Jaiswal, S. K. Singh, & R. S. Srivastava. 1985. Chemical constituents of Gentianaceae. Part 32. Dichotosin and dichotosinin, two adaptogenic glucosyloxy flavans from *Hoppea dichotoma*. *Phytochemistry* 24: 831–833.
Goetz, M. & A. Jacot-Guillarmod. 1977. Contribution to the phytochemistry of the genus *Gentiana*. XXII. Identification of new O-glucosides of mangiferin in *Gentiana asclepiadea* L. *Helv. Chim. Acta* 60: 2104–2106.
Goetz, M., K. Hostettmann, & A. Jacot-Guillarmod. 1976. Contribution to the phytochemistry of the genus *Gentiana*. Part 15. Flavonic and xanthonic C-glucosides from *Gentiana cruciata*. *Phytochemistry* 15: 2015.
Goetz, M., F. Maniliho, & A. Jacot-Guillarmod. 1978. Study of the flavonic and xanthonic compounds in the leaves of *Gentiana ciliata* L. *Helv. Chim. Acta* 61: 1549–1554.
Gottlieb, O. R. 1982. Pages 89–95 in: *Micromolecular evolution, systematics and ecology*. Springer-Verlag, Berlin.
Gröger, D. & P. Simchen. 1969. Über den Einbau von Loganin in Gentiopicrosid. *Z. Naturforsch.* 24b: 356–357.
Guarnaccia, R., L. Botta, & C. J. Coscia. 1969. Mechanism of secoiridoid monoterpene biosynthesis. *J. Am. Chem. Soc.* 91: 204–206.
Guo, X.-F. & C.-P. Chen. 1980. Isolation and identification of gentianine from *Swertia davidi* Franch. *Chung Ts'ao Yao* 11: 200. (*Chem. Abstr.* 94: 71293)
Gupta, P. & J. R. Lewis. 1971. Biogenesis of xanthones in *Gentiana lutea*. *J. Chem. Soc* (C): 629–631.
Guyot, M., J. Massicot, & P. Rivaille. 1968. New xanthones extracted from *Gentiana kochiana*. *Comptes Rend. Acad. Sci., Paris*, sér. C, 267: 423–425.

Hamburger, M., M. Hostettmann, H. Stoeckli-Evans, P. N. Solis, M. P. Gupta, & K. Hostettmann. 1990. A novel type of dimeric secoiridoid glycoside from *Lisianthius jefensis* Robyns et Elias. *Chim. Acta* 73: 1845–1852.

Haqqani, M. H. 1981. Chemical investigation of *Swertia perfoliata*. *Fitoterapia* 52: 5–6.

Harborne, J. B. 1977. Flavonoids and the evolution of angiosperms. *Biochem. Syst. and Ecol.* 5: 7–22.

Harborne, J. B., T. J. Mabry, & H. Mabry. 1975. *The flavonoids*. Chapman & Hall, London.

Hayashi, T. & T. Yamagishi. 1988. Two xanthone glycosides from *Gentiana lutea*. *Phytochemistry* 27: 3696–3699.

He, R.-Y. & R.-L. Nie. 1980. Studies on bitter principles from *Swertia mileensis* T. N. He et W. L. Shi. *Yunnan Chih Wu Yen Chiu* 2: 480–482.

He, R., S. Feng, & R. Nie. 1982. Isolation and identification of xanthones from *Swertia mileensis*. *Yunnan Zhiwu Yanjiu* 4: 68–76.

He, R., S. Feng, & R. Nie. 1984. Isolation and identification of xanthones from *Swertia patens*. *Yunnan Zhiwu Yanjiu* 6: 341–343.

Hegnauer, R. 1966. Pages 176–192 in: *Chemotaxonomie der Pflanzen. Dicotyledonae: Daphniphyllaceae–Lythraceae* (vol. 4). Birkhäuser Verlag, Basel.

Hostettmann, K. & A. Jacot-Guillarmod. 1974. Phytochemistry of *Gentiana*. VII. Flavonic and xanthonic compounds in *Gentiana verna* leaves. I. *Helv. Chim. Acta* 57: 1155–1158.

Hostettmann, K. & A. Jacot-Guillarmod. 1976. Identification of xanthones and new arabinosides of flavone C-glucosides from *Swertia perennis* L. *Helv. Chim. Acta* 59: 1584–1591.

Hostettmann, K. & A. Jacot-Guillarmod. 1977. Xanthones et C-glucosides flavoniques du genre *Gentiana* section *Cyclostigma*. *Phytochemistry* 16: 481–482.

Hostettmann, K. & I. Miura. 1977. A new xanthone diglucoside from *Swertia perennis* L. *Helv. Chim. Acta* 60: 262–264.

Hostettmann, K. & H. Wagner. 1977. Xanthone glycosides. *Phytochemistry* 16: 821–829.

Hostettmann, K., R. Tabacchi, & A. Jacot-Guillarmod. 1974. Phytochemistry of the *Gentiana* genus. VI. Xanthones in the leaves of *Gentiana bavarica*. *Helv. Chim. Acta* 57: 294–301.

Hostettmann, K., M. D. Luong, M. Goetz, & A. Jacot-Guillarmod. 1975. Identification of flavone C-glucoside in the *Gentiana* species (section *Coelanthe*). *Phytochemistry* 14: 499–500.

Hostettmann, K., A. Jacot-Guillarmod, & V. M. Chari. 1976. Contribution to the phytochemistry of genus *Gentiana*, XVIII: Structure of gentiabavarutinoside, a new acylated xanthone glycoside from *Gentiana bavarica* L. *Helv. Chim. Acta* 59: 2592–2595.

Hostettmann-Kaldas, M. & A. Jacot-Guillarmod. 1978. Contribution to the phytochemistry of the genus *Gentiana*. Part XXIII. Xanthones and flavone C-glucosides of the genus *Gentiana* (subgenus *Gentianella*). *Phytochemistry* 17: 2083–2086.

Hostettmann-Kaldas, M., K. Hostettmann, & O. Sticher. 1981. Xanthones, flavones, and secoiridoids of American *Gentiana* species. *Phytochemistry* 20: 443–446.
Hu, B., J. Ding, H. Sun, & S. Fan. 1991. The chemical constituents of *Swertia przewalskii* Pissjauk. *Zhiwu Xuebao* 33: 507–510.
Hu, B., H. Sun, S. Fan, & J. Ding. 1992. The xanthones of *Swertia erythrosticta* Maxim. *Zhiwu Xuebao* 34: 886–888.
Ikeshiro, Y. & Y. Tomita. 1983. A new bitter secoiridoid glucoside from *Gentiana scabra* var. *buergeri*. *Planta Med.* 48: 169–173.
Ikeshiro, Y. & Y. Tomita. 1984. A new iridoid glucoside of *Swertia japonica*. *Planta Med.* 50: 485–488.
Ikeshiro, Y. & Y. Tomita. 1985. Iridoid glucoside of *Swertia japonica*. *Planta Med.* 51: 390–393.
Ikeshiro, Y. & Y. Tomita. 1987. Senburiside II, a new iridoid glucoside from *Swertia japonica*. *Planta Med.* 53: 158–161.
Ikeshiro, Y., I. Mase, & Y. Tomita. 1990. A secoiridoid glucoside from *Gentiana scabra* var. *buergeri*. *Planta Med.* 56: 101–103.
Inouye, H. & Y. Nakamura. 1971a. Über die Monoterpenglucoside und Verwandte Naturstoffe – XIV. Die Struktur der beiden stark bitter schmeckenden Glucoside Amarogentin und Amaroswerin aus *Swertia japonica*. *Tetrahedron* 27: 1951–1966.
Inouye, H. & Y. Nakamura. 1971b. Monoterpene glucosides and related natural products. XVI. Occurrence of secoiridoid glucosides in gentianaceous plants especially in the genera *Gentiana* and *Swertia*. *Yakugaku Zasshi* 91: 755–759.
Inouye, H. & S. Uesato. 1986. Biosynthesis of iridoids and secoiridoids. *Prog. Chem. Org. Nat. Prod.* 50: 169–236.
Inouye, H., S. Ueda, & Y. Nakamura. 1967. Zur Biosynthese der bitteren Glucoside der Gentianazeen, des Gentiopicrosids, des Swertiamarins und des swerosids. *Tetrahedron Lett.* 3221–3226.
Inouye, H., S. Ueda, Y. Aoki, & Y. Takeda. 1969. Zur Biosynthese der Iridoidglucoside. *Tetrahedron Lett.*: 2351–2354.
Inouye, H., S. Ueda, & Y. Nakamura. 1970. Studies on monoterpene glucosides. XII. Biosynthesis of gentianaceous secoiridoid glucosides. *Chem. Pharm. Bull.* 18: 2043–2049.
Inouye, H., S. Ueda, & Y. Takeda. 1971a. Studies on monoterpene glucosides and related natural products. XIII. Incorporation of [10–^{14}C]-sweroside into gentiopicroside and the alkaloids in *Vinca* and *Cinchona* plants. *Chem. Pharm. Bull.* 19: 587–594.
Inouye, H., S. Ueda, & Y. Nakamura. 1971b. Monoterpene glucosides. X. Secoiridoide-glucosides from *Swertia japonica*. Isolation of five secoiridoid-glucosides as well as the structural clarification of sweroside, swertiamarin, and gentiopicroside. *Chem. Pharm. Bull.* 18: 1856–1865.
Inouye, H., S. Ueda, M. Inada, & M. Tsujii. 1971c. Xanthones of *Swertia bimaculata*. *Yakugaku Zasshi* 91: 1022–1026.
Inouye, H., S. Ueda, K. Inoue, & Y. Takeda. 1974a. Studies on monoterpene glucosides and related natural products. XXIII. Biosynthesis of the

secoiridoid glucosides, gentiopicroside, morroniside, oleuropein, and jasminin. *Chem. Pharm. Bull.* 22: 676–686.
Inouye, H., S. Ueda, Y. Nakamura, K. Inoue, T. Hayano, & H. Matsumura. 1974b. Monoterpene glucosides and related natural products. XXIV. Trifloroside, a new secoiridoid glucoside from *Gentiana triflora*. *Tetrahedron* 30: 571–577.
Inouye, H., S. Ueda, & Y. Takeda. 1976. Biosynthesis of secoiridoid compounds. *Heterocycles* 4: 527–565.
Ishimaru, K., H. Sudo, M. Satake, Y. Matsunaga, Y. Hasegawa, S. Takemoto, & K. Shimomura. 1990. Amarogentin, amaroswerin and four xanthones from hairy root cultures of *Swertia japonica*. *Phytochemistry* 29: 1563–1565.
Jensen, S. R. 1991. Plant iridoids, their biosynthesis and distribution in angiosperms. Pages 133–158 in: J. B. Harborne & F. A. Tomas-Barbaran, eds. *Ecological chemistry and biochemistry of plant terpenoids*. Clarendon Press, Oxford.
Jensen, S. R. 1992. Systematic implications of the distribution of iridoids and other chemical compounds in the Loganiaceae and other families of the Asteridae. *Ann. Missouri Bot. Gard.* 79: 284–302.
Jensen, S. R., B. J. Nielsen, & R. Dahlgren. 1975. Iridoid compounds, their occurrence and systematic importance. *Bot. Notis. (Lund)* 128: 148–180.
Kaldas, M., K. Hostettmann, & A. Jacot-Guillarmod. 1974. Contribution à la phytochimie de genre *Gentiana* IX. Etude de composes flavoniques et xanthoniques dans les feuilles de *Gentiana campestris* L. 1. *Helv. Chim. Acta* 57: 2557–2561.
Kaldas, M., K. Hostettmann, & A. Jacot-Guillarmod. 1975. Phytochemistry of the genus *Gentiana*. XIII. Flavone and xanthone compounds in the leaves of *Gentiana campestris*. 2. *Helv. Chim. Acta* 58: 2188–2192.
Kaldas, M., I. Miura, & K. Hostettmann. 1978. Campestroside, a new tetrahydroxyxanthone glucoside from *Gentiana campestris*. *Phytochemistry* 17: 295–297.
Kanamori, H. & I. Sakamoto. 1993. The components of *Swertia punicea*. *Hiroshima-ken Hoken Kankyo Senta Kenkyu Hokoku* 1: 13–16.
Kanamori, H., I. Sakamoto, M. Mizuta, K. Hashimoto, & O. Tanaka. 1984. Studies on the mutagenicity of Swertiae Herba. I. Identification of the mutagenic components. *Chem. Pharm. Bull.* 32: 2290–2295.
Kaouadji, M., I. Vaillant, & A.-M. Mariotte. 1986. Polyoxygenated xanthones from *Centaurium erythrea* roots. *J. Nat. Prod.* 49: 359.
Khan, M. I. & M. H. Haqqani. 1981. Chemical investigation of *Swertia cordata*. *Fitoterapia* 52: 165–166.
Khan, T. A., M. H. Haqqani, & N. M. Nisar. 1979. Chemical investigation of *Swertia alata*. *Planta Med.* 37: 180–181.
Khetwal, K. S. & R. S. Bisht. 1988. A xanthone glycoside from *Swertia speciosa*. *Phytochemistry* 27: 1910–1911.
Khetwal, K. S., B. Joshi, & R. S. Bisht. 1990. Constituents of high-altitude Himalayan herbs. Part 4. Tri- and tetraoxygenated xanthones from *Swertia petiolata*. *Phytochemistry* 29: 1265–1267.
Khishgee, D. & O. Pureb. 1993. Xanthones and flavonoids of *Lomatogonium rotatum*. *Khim. Pir. Soedin*: 761–762.

Kikuzaki, H., Y. Kawasaki, S. Kitamura, & N. Nakatani. 1996. Secoiridoid glucosides from *Swertia mileensis. Planta Med.* 62: 35–38.
Kitamura, Y., M. Dono, H. Miura, & M. Sugii. 1988. Production of swertiamarin in cultured tissues of *Swertia pseudochinensis. Chem. Pharm. Bull.* 36: 1575–1576.
Kitanov, G. M., T. V. Dam, & I. Asenov. 1991. Chemical composition of *Gentiana asclepiadea* root. *Khim. Prir. Soedin.*: 425–427.
Koch, M. 1965. Gentianine and swertiamarin from *Anthocleista procera. Trav. Lab. Matière Med. Pharm. Galenique Fac. Pharm.* Paris: 50–94.
Komatsu, M., T. Tomimori, Y. Makiguchi, & K. Asano. 1968. Constituents of *Swertia japonica.* III. Flavonoid constituents of the plants of *Swertia* species. *Yakugaku Zasshi* 88: 832–837.
Kondo, Y. & K. Yoshida. 1993. Constituents of roots of *Gentiana macrophylla. Shoyakugaku Zasshi* 47: 942–943. (*Chem. Abstr.* 120: 294090)
Kondo, Y., F. Takano, & H. Hojo. 1994. Suppression of chemically and immunologically induced hepatic injuries by gentiopicroside in mice. *Planta Med.* 60: 414–416.
Kong, D., Y. Jiang, Y. Yao, S. Luo, & H. Li. 1995. Glucoside constituents of hengengzhangyacai (*Swertia elongata*). *Zhongcaoyao* 26: 7–10.
Korte, F. 1955. Amarogentin, ein neuer Bitterstoff aus Gentianaceen. Charakteristische Pflanzeninhaltstoffe, IX. *Mitteil. Chem. Ber.* 88: 704–707.
Kubota, M., M. Hattori, & T. Namba. 1983. Studies on the evaluation of cultivated *Swertia japonica* Makino. (I) Variation in the contents of various components during the efflorescence. *Shoyakugaku Zasshi* 37: 229–236.
Kulanthaivel, P., S. W. Pelletier, K. S. Khetwal, & D. L. Verma. 1988. Isolation of a new xanthone and 2-hydroxydimethylterephthalate from *Swertia petiolata. J. Nat. Prod.* 51: 379–381.
Kuo, S.-H., M.-H. Yen, M.-I. Chung, & C.-N. Lin. 1996. A flavone C-glycoside and an aromatic glucoside from *Gentiana* species. *Phytochemistry* 41: 309–312.
Kuwajima, H., N. Hayashi, K. Takaishi, K. Inoue, Y. Takeda, & H. Inouye. 1990. Studies on monoterpene glucosides and related compounds. LXV. Biosynthesis of the biphenylcarboxylic acid moiety of amarogentin and amaroswerin. *Yakugaku Zasshi* 110: 484–489.
Kuwajima, H., S. Hagiwara, E. Fujino, K. Takaishi, Y. Tachibana, & K. Inoue. 1996a. Iridoid glucosides from *Curtia tenuifolia. Planta Med.* 62: 91–92.
Kuwajima, H., N. Shibano, K. Takaishi, K. Inoue, & T. Shingu. 1996b. An acetophenone glycoside from *Exacum affine. Phytochemistry* 41: 289–292.
Liang, J., D. Han, H. Li, & X. Yuan. 1982. Isolation and identification of swertiamarin, the active principle in *Swertia patens* Burk. *Yaoxue Tongbao* 17: 242–243.
Liang, J., D. Han, & H. Li. 1984. Studies on the active constituents of *Swertia patens. Zhongyao Tongbao* 9: 226–228. (*Chem. Abstr.* 102: 12243)
Liao, Z., B. Hu, L. Ji, S. Fan, J. Ding, & H. Sun. 1991. Chemical constituents of *Swertia verticillifolia. Zhiwu Xuebao* 33: 968–970.
Lin, C.-N., C.-H. Chang, M. Arisawa, M. Shimizu, & N. Morita. 1982a. Studies on the constituents of Formosan Gentianaceae. Part IV. Two new xanthone glycosides from *Tripterospermum lanceolatum. Phytochemistry* 21: 205–208.

Lin, C.-N., C.-H. Chang, M. Arisawa, M. Shimizu, & N. Morita. 1982b. Studies on the constituents of Formosan gentianaceous plants. Part III. A xanthone glycoside from *Tripterospermum taiwanense* and rutin from *Gentiana flavomaculata*. Phytochemistry 21: 948–949.

Lin, C.-N., M. I. Chung, M. Arisawa, M. Shimizu, & N. Morita. 1984. The constituents of *Tripterospermum taiwanense* Satake var. *alpinum* Satake and pharmacological activity of some xanthone derivatives. Shoyakugaku Zasshi 38: 80–82.

Lin, C. N., M. I. Chung, K. H. Gan, & J. R. Chiang. 1987. Studies on the constituents of Formosan gentianaceous plants. Part IX. Xanthones from Formosan gentianaceous plants. Phytochemistry 26: 2381–2384.

Liu, J. & M. Huang. 1982. Isolation and identification of xanthones from Qing Ye Dan (*Swertia mileensis*). Zhongcaoyao 13: 433–434.

Liu, Y., X. Li, Y. Liu, & C. Yang. 1994a. Iridoid glycosides from *Gentiana macrophylla*. Yunnan Zhiwu Yanjiu 16: 85–89. (*Chem. Abstr.* 121: 175127)

Liu, Y., X. Li, Y. Liu, & C. Yang. 1994b. Five iridoidal glycosides from *Gentiana urnula*. Yunnan Zhiwu Yanjiu 16: 417–423. (*Chem. Abstr.* 123: 79604)

Lubsandorzhieva, P. B., G. G. Nikolaeva, V. I. Glyzin, A. V. Patudin, T. D. Dargaeva, & A. D. Bakuridze. 1986. Content of mangiferin in species of the family Gentianaceae. Rastit. Resur. 22: 233–236.

Luo, J. & Z. Lou. 1986. TLC-densitometry determination of bitter glycosides in the Chinese drug Longdan, radix Gentianae, and its quality evaluation. Yaoxue Xuebao 21: 40–46. (*Chem. Abstr.* 104: 213344)

Luo, L. H. & R. L. Nie. 1992. Iridoid glycosides from *Swertia angustifolia*. Yaoxue Xuebao 27: 125–129.

Luo, Y. & R. Nie. 1993a. The secoiridoids isolated from *Swertia nervosa* (G. Don) Wall. ex C. B. Clarke. Zhiwu Xuebao 35: 307–310. (*Chem. Abstr.* 120: 101911)

Luo, Y. & R. Nie. 1993b. The secoiridoids isolated from *Swertia punicea*. Yunnan Zhiwu Yanjiu 15: 97–100. (*Chem. Abstr.* 119: 113374)

Luong, M. D. & A. Jacot-Guillarmod. 1977. Contribution to the phytochemistry of the genus *Gentiana*. XXI: The cinnamoyl-C-glucosylflavones and their O-glucosides in *Gentiana punctata* L. Helv. Chim. Acta 60: 2099–2103.

Luong, M. D., P. Fombasso, & A. Jacot-Guillarmod. 1980. Contribution to the phytochemistry of genus *Gentiana*. XXV. Study on the flavonic and xanthonic compounds in leaves of *Gentiana* × *marcailhouana* Ry. New flavonic cinnamoyl-C-glucosides. Helv. Chim. Acta 63: 244–249.

Luong, M. D., J. Saeby, P. Fombasso, & A. Jacot-Guillarmod. 1981. Phytochemistry of genus *Gentiana*. XXVI. Identification of a new di-O-glucosyl cinnamoyl-C-glucosylflavone in the leaves of *Gentiana* × *marcailhouana* Ry. Helv. Chim. Acta 64: 2741–2745.

Ma, W.-G., N. Fuzzati, J.-L. Wolfender, K. Hostettmann, & C.-R. Yang. 1994. Rhodenthoside A, a new type of acylated secoiridoid glycoside from *Gentiana rhodentha*. Helv. Chim. Acta 77: 1660–1671.

Madubunyi, I. I., K. P. Adam, & H. Becker. 1994. Anthocleistol, a new secoiridoid from *Anthocleista nobilis*. Z. Naturforsch. (Biosci.) 49C: 271–272.

Mandal, S. & A. Chatterjee. 1987. Structure of chiratanin, a novel dimeric xanthone. Tetrahedron Lett. 28: 1309–1310.

Mandal, S., P. C. Das, & P. C. Joshi. 1992a. Naturally occurring xanthones from terrestrial flora. *J. Indian Chem. Soc.* 69: 611–636.

Mandal, S., P. C. Das, P. C. Joshi, A. Chatterjee, C. N. Islam, M. K. Dutta, B. B. Patra *et al.* 1992b. Anti-inflammatory action of *Swertia chirata*. *Fitoterapia* 63: 122–128.

Marston, A., K. Hostettmann, & A. Jacot-Guillarmod. 1976. Contribution to the phytochemistry of genus *Gentiana*, XIX: Identification of new C-glycosylflavones in *Gentiana pyrenaica* L. *Helv. Chim. Acta* 59: 2596–2600.

Martindale. 1982. Bitters, in: J. E. F. Reynolds, ed. *The Extra Pharmacopoeia*, ed. 28. The Pharmaceutical Press, London.

Massias, M., J. Carbonnier, & D. Molho. 1976. 1,3,7-Trisubstituted xanthones of *Gentiana ciliata* L. Chromatographic separation of the two isomers: gentisin and isogentisin. *Bull. Mus. Natl. Hist. Nat., Sci. Phys.-Chim.* 10: 45–52.

Massias, M., J. Carbonnier, & D. Molho. 1977. Xanthones of *Swertia speciosa* Wall. Contribution to the chemotaxonomy of the genus. *Bull. Mus. Natl. Hist. Nat., Sci. Phys.-Chim.* 13: 55–61.

Massias, M., J. Carbonnier, & D. Molho. 1981. Xanthones and C-glucosyl-flavones from *Gentiana corymbifera*. *Phytochemistry* 20: 1577–1578.

Massias, M., J. Carbonnier, & D. Molho. 1982. Chemotaxonomy of *Gentianopsis*: xanthones, C-glycosylflavonoids and carbohydrates. *Biochem. Syst. Ecol.* 10: 319–327.

Mészáros, S. 1994. Evolutionary significance of xanthones in Gentianaceae: a reappraisal. *Biochem. Syst. Ecol.* 22: 85–94.

Miana, G. A. 1973. Flavonoids of *Swertia purpurascens*. *Phytochemistry* 12: 728–729.

Miana, G. A. & M. G. Al-Hazimi. 1984. Xanthones of *Centaurium pulchellum*. *Phytochemistry* 23: 1637–1638.

Mpondo, E. M. & A. J. Chulia. 1988. 6′-O-β-D-glucosyl gentiopicroside: a new secoiridoid from *Gentiana asclepiadea*. *Planta Med.* 54: 185–186.

Mpondo, E. M. & J. Garcia. 1990a. Secologanin and derivatives from *Gentiana verna*. *Planta Med.* 56: 125.

Mpondo, E. M. & J. Garcia. 1990b. Two iridoid glucosides from *Gentiana verna*. *Phytochemistry* 29: 643–644.

Mpondo, E. M., J. Garcia, & J. Lestani. 1989. New secoiridoid glucosides from *Gentiana verna*. *J. Nat. Prod.* 52: 1146–1149.

Mpondo, E. M., J. Garcia, G. Cartier, & G. Pellet. 1990a. 6′-O-β-D-glucosyl-swertiamarine: a new secoiridoid from *Gentiana alpina*. *Planta Med.* 56: 334.

Mpondo, E. M., J. Garcia, & A. J. Chulia. 1990b. Secoiridoid glucosides from *Gentiana campestris*. *Phytochemistry* 29: 1687–1688.

Mukherjee, K. S., C. K. Chakraborty, S. Laha, D. Bhattacharya, & T. P. Chatterjee. 1990. A new xanthone of *Hoppea fastigiata*. *J. Indian Chem. Soc.* 67: 1003–1004.

Mukherjee, K. S., C. K. Chakraborty, T. P. Chatterjee, D. Bhattacharya, & S. Laha. 1991a. 1,5,7-Trihydroxy-3-methoxyxanthone from *Hoppea fastigiata*. *Phytochemistry* 30: 1036–1037.

Mukherjee, K. S., C. K. Chakraborty, S. Laha, D. Bhattacharya, & T. P. Chatterjee. 1991b. 1,7-Dihydroxy-3,5-dimethoxyxanthone from *Hoppea fastigiata*. *Int. J. Pharmacogn.* 29: 225–227.

Mukherjee, K. S., S. Laha, T. K. Manna, & S. C. Roy. 1995. Further work on *Limnophila rugosa* (Roth.) Merrill and *Hoppea fastigiata* Clarke (Gentianaceae). *J. Indian Chem. Soc.* 72: 63–64.
Neu, B., J. Guo, J. Chen, & J. Ma. 1991. Chemical constituents of *Swertia tetraptera* Maxim. var. *xinglongensis* Ji Ma et R. N. Zhao. *Zhongguo Zhongyao Zazhi* 16: 549–550.
Nikolaeva, G. G., V. I. Glyzin, D. A. Fesenko, & A. V. Patudin. 1980a. Xanthone glycosides of *Gentiana barbata*. *Khim. Prir. Soedin.*: 841–842.
Nikolaeva, G. G., V. I. Glyzin, D. A. Fesenko, & A. V. Patudin. 1980b. Xanthone compounds of *Gentiana barbata*. *Khim. Prir. Soedin.*: 255.
Nikolaeva, G. G., V. I. Glyzin, B. A. Krivut, A. Silla, & A. V. Patudin. 1980c. Glycosides – γ-pyrone derivatives from *Gentiana schistocalyx*. *Khim. Prir. Soedin.*: 833–834.
Nikolaeva, G. G., V. I. Glyzin, A. V. Patudin, & D. A. Fesenko. 1981. *Gentiana barbata* xanthoglycosides. II. *Khim. Prir. Soedin.*: 392–393.
Nikolaeva, G. G., V. I. Glyzin, M. S. Mladentseva, V. I. Sheichenko, & A. V. Patudin. 1983. Xanthones from *Gentiana lutea*. *Khim. Prir. Soedin.*: 107–108.
Nobrega, E. M. & A. A. Craveiro. 1988. New alkaloid from *Schultesia guianensis*. *J. Nat. Prod.* 51: 962–965.
Nyiredy, S., C. A. J. Erdelmeier, K. Dallenbach-Toelke, K. Nyiredy-Mikita, & O. Sticher. 1986. Preparative on-line overpressure layer chromatography (OPLC): a new separation technique for natural products. *J. Nat. Prod.* 49: 885–891.
Okada, T. 1977. Insect repellant. Japan. Patent 78 75,327. (*Chem. Abstr.* (1978) 89: 192516)
Okorie, D. A. 1976. A new phthalide and xanthones from *Anthocleista djalonensis* and *Anthocleista vogelii*. *Phytochemistry* 15: 1799–1800.
Onocha, P. A., D. A. Okorie, J. D. Connolly, & D. S. Roycroft. 1995. Monoterpene diol, iridoid glucoside and dibenzo-α-pyrone from *Anthocleista djalonensis*. *Phytochemistry* 40: 1183–1189.
Ortega, E. P., R. E. Lopez-Garcia, R. M. Rabanal, V. Darias, & S. Valverde. 1988. Two xanthones from *Ixanthus viscosus*. *Phytochemistry* 27: 1912–1913.
Parra, M., M. T. Picher, E. Seoane, & A. Tortajada. 1984a. New xanthones isolated from *Centaurium linarifolium*. *J. Nat. Prod.* 47: 123–126.
Parra, M., E. Seoane, & A. Tortajada. 1984b. Additional new xanthones isolated from *Centaurium linarifolium*. *J. Nat. Prod.* 47: 868–871.
Parra, M., M. T. Picher, E. Seoane, & A. Tortjada. 1985. Xanthones and secoiridoids isolated from methanolic extract of *Centaurium linarifolium*. *J. Nat. Prod.* 48: 998–999.
Pengsuparp, T., L. Cai, H. Constant, H. S. Fong, L.-Z. Lin, A. D. Kinghorn, J. M. Pezzuto *et al.* 1995. Mechanistic evaluation of new plant-derived compounds that inhibit HIV-1 reverse transcriptase. *J. Nat. Prod.* 58: 1024–1031.
Peters, S., W. Schmidt, & L. Beerhues. 1998. Regioselective oxidative phenol couplings of 2,3′,4,6-tetrahydroxybenzophenone in cell cultures of *Centaurium erythraea* Rafn. and *Hypericum androsaemum* L. *Planta* 204: 64–69.

Plouvier, V., J. Massicot, & P. Rivaille. 1967. Gentiacaulein, a new tetrasubstituted xanthone, the aglycon of gentiacauloside from *Gentiana acaulis*. *Comptes Rend. Hebd. Seances Acad. Sci.*, Sér. D, 264: 1219–1222.
Popov, S. & N. Marekov. 1971a. A new iridoid precursor of gentiopicroside. *Phytochemistry* 10: 3077–3079.
Popov, S. S. & N. L. Marekov. 1971b. Gentioflavoside, a new secoiridoid found in some *Gentiana* species. *Chem. Ind. (London)*: 655.
Prakash, A., P. C. Basumatary, S. Ghosal, & S. S. Handa. 1982. Chemical constituents of *Swertia paniculata*. *Planta Med.* 45: 61–62.
Pureb, O., Y. Zham'yansan, & K. Oyuun. 1991. Xanthones and flavonoids of *Gentiana barbata*. *Khim. Prir. Soedin.*: 284–285.
Purushothaman, K. K., A. Sarada, & V. Narayanaswami. 1973. Chemical examination of *Swertia chirata*. *Leather Sci. (Madras)* 20: 132–134.
Rakhmatullaev, T. U. 1971. Alkaloids from *Gentiana olgae*, *Gentiana vvedenskyi*, *Gentiana tianscanica*, and *Swertia connata*. *Khim. Prir. Soedin.* 7: 128.
Rakhmatullaev, T. U., S. T. Akramov, & S. Y. Yunusov. 1969a. Alkaloids of *Gentiana olivieri*. *Khim. Prir. Soedin.* 5: 608.
Rakhmatullaev, T. U., S. T. Akramov, & S. Y. Yunusov. 1969b. Alkaloids from *Swertia marginata*, *S. graciliflora*, and *Dipsacus azureus*. *Khim. Prir. Soedin.* 5: 64–65.
Rasoanaivo, P., M. Nicoletti, G. Multari, G. Palazzino, & C. Galeffi. 1994. Research on African medicinal plants. Part XXXIII. Secoiridoids and related monoterpenes of *Anthocleista amplexicaulis*. *Fitoterapia* 65: 38–43.
Recio, M. C., I. Slacanin, M. Hostettmann, A. Marston, & K. Hostettmann. 1990. Phytochemical investigation of species of the genera *Gentiana* and *Halenia* from South America. *Bull. Liaison – Groupe Polyphenols* 15: 215–218.
Recio-Iglecias, M.-C., A. Marston, & K. Hostettmann. 1992. Xanthones and secoiridoid glucosides of *Halenia campanulata*. *Phytochemistry* 31: 1387–1389.
Rezende, C. M. A. & O. R. Gottlieb. 1973. Xanthones as systematic markers. *Biochem. Syst. Ecol.* 3: 63–70.
Richardson, P. M. 1983. The taxonomic significance of C-glycosylxanthones in flowering plants. *Biochem. Syst. Ecol.* 11: 371–375.
Rivaille, P. & D. Raulais. 1969. Xanthones and other constituents of *Gentiana* and *Swertia*. Presence of a new triterpene in *Gentiana verna*. *Comptes Rend. Acad. Sci.*, sér. D, 269: 1121–1124.
Rivaille, P., J. Massicot, M. Guyot, V. Plouvier, & M. Massias. 1969. Xanthones from *Gentiana kochiana*, *Swertia decussata* and *S. perennis*. *Phytochemistry* 8: 1533–1541.
Rodriguez, S., J.-L. Wolfender, E. Hakizamungu, & K. Hostettmann. 1995a. An antifungal naphthoquinone, xanthones and secoiridoids from *Swertia calycina*. *Planta Med.* 61: 362–364.
Rodriguez, S., J.-L. Wolfender, G. Odontuya, O. Purev, & K. Hostettmann. 1995b. Xanthones, secoiridoids and flavonoids from *Halenia corniculata*. *Phytochemistry* 40: 1265–1272.
Rodriguez, S., A. Marston, J.-L. Wolfender, & K. Hostettmann. 1998. Iridoids and secoiridoids in the Gentianaceae. *Curr. Org. Chem.* 2: 627–648.

Roitman, J. N., E. Wollenweber, & F. J. Arriaga-Giner. 1992. Xanthones and triterpene acids as leaf exudate constituents in *Orphium frutescens*. *J. Plant Physiol.* 139: 632–634.
Rulko, F. & K. Nadler. 1970. Alkaloids of Gentianaceae. VI. Alkaloids of *Gentiana asclepiadea*. *Pharm. Pharmacol.* 22: 329–332.
Rulko, F. & K. Witkiewicz. 1972. Gentiana alkaloids. VII. Alkaloids of centaury (*Erythraea centaurium*) Diss. *Pharm. Pharmacol.* 24: 73–77.
Sadritdinov, F. & N. Tulyaganov. 1972. Pharmacology of the new alkaloid gentianadine. Pages 152–154 in: M. B. Sultanov, ed. *Farmakol. Alkaloidov Ikh Proizvod.* "Fan", Tashkent.
Sadykov, Y. D. 1987. Alkaloids of some plants present in the flora of Tadzhikistan. *Izv. Akad. Nauk Tadzh. SSR, Otd. Biol. Nauk*: 48–52.
Sakamoto, I., T. Tanaka, O. Tanaka, & T. Tomimori. 1982. Xanthone glucosides of *Swertia japonica* Makino and a related plant: structure of a new glucoside, isoswertianolin, and structure revision of swertianolin and norswertianolin. *Chem. Pharm. Bull.* 30: 4088–4091.
Sakina, K. & K. Aota. 1976. Studies on the constituents of *Erythraea centaurium* (Linne) Persoon. I. The structure of centapicrin, a new bitter secoiridoid glucoside. *Yakugaku Zasshi* 96: 683–688.
Sarg, M., M. El-Domiaty, O. M. Salama, M. M. Bishr, & A. R. El-Gindy. 1990. Pharmacognostical study of the rhizomes and roots of *Gentiana kurroo* Royle. *Mansoura J. Pharm. Sci.* 6: 49–72.
Sarg, T., O. Salama, M. El-Domiaty, M. Bishr, S. El S. S. Mansour, & E. Weight. 1991. Iridoid glucosides from *Gentiana kurroo* Royle. *Alexandria J. Pharm. Sci.* 5: 82–86.
Saxena, A. M. & S. K. Mukherjee. 1992. Mechanism of blood sugar lowering action of *Swertia chirayita*: effect of impure swerchirin (SWI) on insulin release from isolated beta cells of the pancreas. *J. Microb. Biotechnol.* 7: 27–29.
Saxena, A. M., M. B. Bajpai, P. S. R. Murthy, & S. K. Mukherjee. 1993. Mechanism of blood sugar lowering by a swerchirin-containing hexane fraction (SWI) of *Swertia chirayita*. *Indian J. Exp. Biol.* 31: 178–181.
Schaufelberger, D. & K. Hostettmann. 1984. Flavonoid glycosides and a bitter principle from *Lomatogonium carinthiacum*. *Phytochemistry* 23: 787–789.
Schaufelberger, D. & K. Hostettmann. 1987. High-performance liquid chromatographic analysis of secoiridoid and flavone glycosides in closely related *Gentiana* species. *J. Chromatogr.* 389: 450–455.
Schaufelberger, D. & K. Hostettmann. 1988. Chemistry and pharmacology of *Gentiana lactea*. *Planta Med.* 54: 219–221.
Schaufelberger, D., M. P. Gupta, & K. Hostettmann. 1987. Flavonol and secoiridoid glycosides from *Coutoubea spicata*. *Phytochemistry* 26: 2377–2379.
Schmidt, W., S. Peters, & L. Beerhues. 2000. Xanthone 6-hydroxylase from cell cultures of *Centaurium erythraea* Rafn. and *Hypericum androsaemum* L. *Phytochemistry* 53: 427–431.
Shiobara, Y., K. Kato, Y. Ueda, K. Taniue, Y. Syoha, N. Nishimoto, F. de Oliveira *et al.* 1994. Secoiridoid glucosides from *Chelonanthus chelonoides*. *Phytochemistry* 37: 1649–1652.

Sluis, W. G., van der. 1985. Chemotaxonomic investigations of the genera *Blackstonia* and *Centaurium* (Gentianaceae). *Plant Syst. Evol.* 149: 253–286.
Sluis, W. G., van der & R. P. Labadie. 1981a. Secoiridoids and xanthones in the genus *Centaurium*. Part II: Secoiridoid glucosides in *Centaurium spicatum*. *Planta Med.* 41: 221–231.
Sluis, W. G., van der & R. P. Labadie. 1981b. Secoiridoids and xanthones in the genus *Centaurium*. Part III. Decentapicrins A, B and C. New *m*-hydroxybenzoyl esters of sweroside from *Centaurium littorale*. *Planta Med.* 41: 150–160.
Sluis, W. G., van der & R. P. Labadie. 1985. Polyoxygenated xanthones of *Centaurium littorale*. *Phytochemistry* 24: 2601–2605.
Sluis, W. G., van der, J. M. Van der Nat, Á. L. Spek, Y. Ikeshiro, & R. P. Labadie. 1983. Secoiridoids and xanthones in the genus *Centaurium*. VI. Gentiogenal, a conversion product of gentiopicrin (gentiopicroside). *Planta Med.* 49: 211–215.
Solov'eva, E. V., V. I. Glyzin, & A. V. Patudin. 1980a. Gentiabavaroside from *Swertia connata*. *Khim. Prir. Soedin.*: 570–571.
Solov'eva, E. V., O. A. Denisova, V. I. Glyzin, & A. V. Patudin. 1980b. Xanthone compounds of *Swertia connata*. *Khim. Prir. Soedin.*: 840–841.
Sorig, T., L. Toth, & G. Bujtas. 1977. Isolation of xanthones from *Lomatogonium carinthiacum* (Wulfen), Rchb. *Pharmazie* 32: 803.
Stefanou, E., K. Hostettmann, & A. Jacot-Guillarmod. 1976. Contribution to the phytochemistry of the genus *Gentiana*. Part 12. Secoiridoids and xanthones of *Gentiana burseri*. *Phytochemistry* 15: 330–331.
Stout, G. H. & W. J. Balkenhol. 1969. Xanthones of the Gentianaceae. I. *Frasera caroliniensis*. *Tetrahedron* 25: 1947–1960.
Stout, G. H. & J. L. Fries. 1970. Xanthones of the Gentianaceae. V. The xanthones of a *Halenia* species. *Phytochemistry* 9: 235–236.
Stout, G. H., E. N. Christensen, W. J. Balkenhol, & K. L. Stevens. 1969a. Xanthones of the Gentianaceae. II. *Frasera albicaulis*. *Tetrahedron* 25: 1961–1973.
Stout, G. H., B. J. Reid, & G. D. Breck. 1969b. Xanthones of the Gentianaceae. IV. Xanthones of *Macrocarpaea glabra*. *Phytochemistry* 8: 2417–2419.
Struwe, L., M. Thiv, J. W. Kadereit, A. S.-R. Pepper, T. J. Motley, P. J. White, J. H. E. Rova *et al.* 1998. *Saccifolium* (Saccifoliaceae), an endemic of Sierra de la Neblina on the Brazilian–Venezuelan frontier, is related to a temperate-alpine lineage of Gentianaceae. *Harvard Pap. Bot.* 3: 199–214.
Struwe, L., J. W. Kadereit, J. Klackenberg, S. Nilsson, M. Thiv, K. B. von Hagen, & V. A. Albert. 2002. Systematics, character evolution, and biogeography of Gentianaceae, including a new tribal and subtribal classification. Pages 21–309 in: L. Struwe & V. A. Albert, eds. *Gentianaceae: systematics and natural history*. Cambridge University Press, Cambridge.
Stuppner, H. & H. Wagner. 1989. Minor iridoids and phenol glucosides of *Picrorhiza kurrooa*. *Planta Med.* 55: 467–469.
Suhr, I. H., P. Arends, & B. Jensen. 1978. Gentiolactone, a secoiridoid dilactone from *Gentiana purpurea*. *Phytochemistry* 17: 135–138.

Sullivan, G., F. D. Stiles, & K.-H. A. Rosler. 1977. Phytochemical investigation of *Eustoma grandiflorum* (Raf.) Shinners. *J. Pharm. Sci.* 66: 828–831.
Sultanbava, M. U. S. 1980. Xanthonoids of tropical plants. *Tetrahedron* 36: 1465–1506.
Sun, H. & J. Ding. 1981. Isolation and identification of the xanthone constituents from *Swertia mussotii* Franch. *Zhiwu Xuebao* 23: 464–469.
Sun, H. & C. Xia. 1984. Chemical constituents in *Gentiana regescens*. *Zhongyao Tongbao* 9: 33–34.
Sun, H., B. Hu, S. Fen, & J. Ding. 1983. Three new xanthones from *Halenia elliptica* D. Don. *Zhiwu Xuebao* 25: 460–467. (*Chem. Abstr.* 100: 117831)
Sun, H., B. Hu, J. Ding, & S. Fan. 1991. The glucosides from *Swertia mussotii* Franch. *Zhiwu Xuebao* 33: 31–37.
Tagaki, S., M. Yamaki, E. Yumioka, T. Nishimura, & K. Sakina. 1982. Studies on the constituents of *Erythraea centaurium* (Linne) Persoon. II. The structure of centauroside, a new bis-secoiridoid glucoside. *Yakugaku Zasshi* 102: 313–317.
Takeda, Y. & H. Inouye. 1976. Studies on monoterpene glucosides and related natural products. XXX. The fate of the C-8 proton of 7-deoxyloganic acid in the biosynthesis of secoiridoid glucosides. *Chem. Pharm. Bull.* 24: 79–84.
Tan, P., Y. L. Liu, & C. Y. Hou. 1993. Swertiapunimarin from *Swertia punicea* Hemsl. *Yaoxue Xuebao* 28: 522–525. (*Chem. Abstr.* 119: 245535)
Tan, R. X., J.-L. Wolfender, W. G. Ma, L. X. Zhang, & K. Hostettmann. 1996. Secoiridoids and antifungal aromatic acids from *Gentiana algida*. *Phytochemistry* 41: 111–116.
Tempesta, M., S. D. Jolad, S. King, G. Mao, R. C. Bruening, J. E. Kuo, T. V. Truong et al. 1994. Phosphocholine derivatives having antifungal activity. Int. Patent Appl. WO 94 08,563. (*Chem. Abstr.* 121: 103986)
Terreaux, C., M. Maillard, M. P. Gupta, & K. Hostettmann. 1995. Xanthones from *Schultesia lisianthoides*. *Phytochemistry* 40: 1791–1795.
Thanh, D., S. Popov, N. Handjieva, A. Dyulogerov, & A. Trifonov. 1987. Chemical ionization mass spectrometry of silylated secoiridoids. *F. E. C. S. Int. Conf. Chem. Biotechnol. Biol. Act. Nat. Prod., [Proc.]* 5: 110–114. VCH, Weinheim, Germany.
Thiv, M., L. Struwe, V. A. Albert, & J. W. Kadereit. 1999. The phylogenetic relationships of *Saccifolium bandeirae* Maguire & Pires (Gentianaceae) reconsidered. *Harvard Pap. Bot.* 4: 519–526.
Tikhonova, L. A., N. F. Komissarenko, & T. P. Berezovskaya. 1989. Flavone C-glycosides from *Gentiana macrophylla*. *Khim. Prir. Soedin.*: 287–288.
Tomimori, T. & M. Komatsu. 1969. Constituents of *Swertia japonica*. VI. Flavonoid and xanthone constituents of *S. randaiensis* and *S. swertopsis*. *Yakugaku Zasshi* 89: 1276–1282.
Tomimori, T., M. Yoshizaki, & T. Namba. 1974. Nepalese crude drugs. II. Xanthone constituents of the plants of *Swertia* species. *Yakugaku Zasshi* 94: 647–651.
Topuriya, L. I. 1978. Components of *Gentiana caucasica* and *Gentiana schistocalyx*. *Khim. Prir. Soedin.*: 413.
Uesato, S., T. Hashimoto, & H. Inouye. 1979. Three new secoiridoid glucosides from *Eustoma russellianum*. *Phytochemistry* 18: 1981–1986.

Verma, D. L. & K. S. Khetwal. 1985. Phenolics in the roots of *Swertia paniculata* Wall. *Sci. Cult.* 51: 305–306.
Verney, A. M. & A. M. Debelmas. 1973. Xanthones of *Gentiana lutea, G. purpurea, G. punctata, G. pannonica. Ann. Pharm. Fr.* 31: 415–420.
Versluys, C., M. Cortés, J. T. López, J. R. Sierra, & I. Razmilic. 1982. A novel xanthone as secondary metabolite from *Centaurium cachanlahuen. Experientia* 38: 771–772.
Wagner, H. & K. Vasirian. 1974. Deoxyamarogentin. New bitter principle from *Gentiana pannonica. Phytochemistry* 13: 615–617.
Wan, A. S. C., E. Macko, & B. Douglas. 1972. Pharmacological investigations of gentianine from *Fagraea fragrans. Asian J. Med.* 8: 334–335.
Wang, J.-P., T.-F. Ho, C.-N. Lin, & C.-M. Teng. 1994a. Effect of norathyriol, isolated from *Tripterospermum lanceolatum*, on A23187-induced pleurisy and analgesia in mice. *Naunyn-Schmiedeberg's Arch. Pharmacol.* 350: 90–95.
Wang, J. P., S. L. Raung, C. N. Lin, & C. M. Teng. 1994b. Inhibitory effect of norathyriol, a xanthone from *Tripterospermum lanceolatum*, on cutaneous plasma extravasation. *Eur. J. Pharmacol.* 251: 35–42.
Wang, Y. & Z. Lou. 1988. TLC identification of Chinese traditional drug "Qinjiu", roots of *Gentiana macrophylla* and allied species. *Yaowu Fenxi Zazhi* 8: 348–349. (*Chem. Abstr.* 110: 160443)
Weber, N. 1974. Plants from Madagascar. 4. Terpenoid and other constituents of *Hernandia voyroni* and *Anthocleista amplexicaulis. Phytochemistry* 13: 2006–2007.
Weinges, K., P. Kloss, & H.-D. Henkels. 1972. Naturstoffe aus Arzneipflanzen, XVII. Picrosid-II, ein neues 6-Vanilloyl-catalpol aus *Picrorhiza kurrooa* Royle und Benth. *Liebigs Annalen* 759: 173–182.
Wolfender, J.-L., M. Hamburger, K. Hostettmann, J. D. Msonthi, & S. Mavi. 1993. Search for bitter principles in *Chironia* species by LC-MS and isolation of a new secoiridoid diglycoside from *Chironia krebsii. J. Nat. Prod.* 56: 682–689.
Wu, T.-S., H.-J. Tien, & C.-N. Lin. 1976. Studies on the constituents of Formosan gentianaceous plants. Part I. Constituents of *Swertia randaiensis* Hayata. *J. Chin. Chem. Soc. (Taipei)* 23: 53–55.
Yamahara, J., M. Kobayashi, H. Matsuda, & S. Aoki. 1991. Anticholinergic action of *Swertia japonica* and an active constituent. *J. Ethnopharmacol.* 33: 31–35.
Yang, Y. B. & J. Zhou. 1980a. Studies on the xanthones of *Veratrilla baillonii* Franch. II. A new xanthone. *Yunnan Chih Wu Yen Chiu* 2: 468–472.
Yang, Y. B. & J. Zhou. 1980b. Studies on the xanthones of *Veratrilla baillonii* Franch. I. Structures of veratriloside and veratrilogenin. *Yao Hsueh Hsueh Pao* 15: 625–629.
Yang, Y. B., X. Y. Pu, X. Peng, & H. Yin. 1995. Xanthones from *Veratrilla baillonii* Franch. III. Structure elucidation of a new xanthone glycoside. *Yaoxue Xuebao* 30: 440–442.
Yang, Z., Y. Zhang, Z. Yang, & C. Yang. 1985. Determination of contents of gentiopicroside in various parts of *Gentiana cephantha* Fr. *Baiqiuen Yike Daxue Xuebao* 11: 488–491.

Yu, R. 1984. Studies on the constituents of *Swertia devidi* Franch. *Zhiwu Xuebao* 26: 675–676.

Yuan, Y.-M. & P. Küpfer. 1995. Molecular phylogenetics of the subtribe Gentianinae (Gentianaceae) inferred from the sequences of internal transcribed spacers (ITS) of nuclear ribosomal DNA. *Pl. Syst. Evol.* 196: 207–226.

Zhang, Y.-J. & C.-R. Yang. 1994. Two triterpenoids from *Gentiana tibetica*. *Phytochemistry* 36: 997–999.

Zhang, Z., X. Han, Z. Cai, H. Liu, L. Wang, & X. Hu. 1993. Gentiopicroside production from root cultures of *Gentiana manshurica*. *Zhiwu Shengli Xuebao* 19: 66–70.

Zhou, H. M. & Y. L. Liu. 1990. Structure of swertiamacroside from *Swertia macrosperma* C. B. Clarke. *Yaoxue Xuebao* 25: 123–126.

Zhou, J. 1991. Bioactive glycosides from Chinese medicines. *Mem. Inst. Oswaldo Cruz, Rio de Janeiro* 86 (suppl. 2): 231–234.

Index

Page numbers in **bold** indicate the main description of the tribe, subtribe, or genus in Chapter 2. Taxonomic rank for all groups except families, genera, and species, as well as some generic synonyms, are indicated in parentheses.

Adenolisianthus **141**, 525
  distribution 137–8
  number of species 24, 138
  palynology 170–1, 377, 379, 381, 394, 421–3, 468
  phylogeny 468, 557
  seed anatomy 523, 557
Africa
  floras 270–1
  generic diversity 14–15
Agathotes 235 (= *Swertia*)
Aliopsis 229, 246 (= *Gentianella*)
Aloitis 229, 246 (= *Gentianella*)
Anagallidiinae (subtribe) 54
Anagallidium 54, 235 (= *Swertia*)
Analgesics 612
Anatomy
  flower *see* Floral anatomy *and* Floral anatomy under specific groups
  pollen *see* Palynology
  seed *see* Seed anatomy under specific groups *and* Seeds
Androecium *see* Stamens *and* Flower morphology under specific groups
Anemarrhena 604
Anthers *see* Stamens
Anthocleista 4, 6, **195–6**, 539–40
  biogeography 196–7
  chromosome number 329
  dispersal 9
  distribution 190, 193, 195
  floral anatomy 323
  flower morphology 192, 195
  fruit morphology 8, 195–6, 325
  germination 508
  habit 10, 195
  number of species 25, 190, 195
  palynology 213–14, 216, 324, 377, 379, 381–2, 452–3, 476–7, 479
  phylogeny
    molecular data 29, 197

  morphological data 196, 315, 333, 335, 353
  palynological data 378, 379, 476–7, 479
  phytochemical data 598
  phytochemistry 330, 586, 593, 598, 601
  seed anatomy 539–40
  vegetative morphology 195, 320, 321
Anti-inflammatory activity 612
Anti-psychotic activity 612
Ants 161, 210
APG (Angiosperm Phylogeny Group) 194
Apiales (order) 3
Apocynaceae 2, 3, 4, 6, 311
  chromosome number 329
  evolution of habit 335
  flower morphology 7
  fruit morphology 9
  habit 10
  mycorrhiza type 319
  number of species 11
  palynology 378, 379
  phylogeny 40, 42, 44, 331, 332
    molecular data 9, 29–37, 30, 31, 35–7, 40, 336
    morphological data 316, 331, 333, 335–6
  phytochemistry 329, 330
  pollination 7
Aripuana 137–8, **141–4**, 164, 165, 528–9
  distribution 138
  fruits 8
  number of species 24, 138
  palynology 171, 377, 379, 381, 394, 408, 423, 468, 480
  phylogeny
    molecular data 149
  pollination 7
  seed anatomy 529, 558
Asclepiadaceae 3

633

Asia
  floras 271–2
  generic diversity 14–15
Asterales (order) 2, 3, 555
Asteranthe 224
Asteridae (subclass) 3–4, 7
Auricles *see* Vegetative morphology
Australia
  floras 270–1
  generic diversity 14–15

Bark 59–60
Bartonia 5, 222–3, **224–5**
  biogeography 263–5
  dispersal 505
  distribution 224
  floral anatomy 322–3
  number of species 25, 224
  palynology 65, 249
  phylogeny 559
    molecular data 29, 245
    morphological data 314, 333–4, 354
  phytochemistry 601
  seed anatomy 503, 549–50, 559
  trophy 320
  vegetative morphology 321
Belmontia 85, 511; *see Sebaea*
Bentham, G. 1, 56–7, 76, 110, 112–3, 194, 224
Biogeography
  Chironieae 135–7
  Exaceae 103–8
  Faroinae 13, 190, 193, 200, 219
  Gentianaceae 14, 15, 22, 23, 28
  Gentianeae 262–6
  Helieae 13, 138–9, 186–9, 191
  Lisianthiinae 219
  Potalieae 219–21
  Saccifolieae 66
  Swertiinae 263–4
  Africa 103–6, 137, 196–7, 265
  Andes 154, 182, 184, 189, 264–5
  Asia, temperate 262–6
  Asia, tropical 103–7, 136, 197
  Australia 103, 106, 265
  Brazilian Shield 66, 136, 154, 182–3, 186–90
  Canary Islands 137
  Caribbean 154, 182–3
  Central America 197, 203, 220–1
  Europe 262–6
  Guayana Shield 66, 136, 182–90, 221
  Indian Ocean basin 103–7, 136, 196–7
  Madagascar 103–6, 196–7
  New Zealand 265
  North America 137, 262–6
  Panama isthmus 136, 220–1, 262–3
  South America 66, 136, 182–90, 191, 196–7, 220–1, 264–5
    highland/lowland taxa 151, 187–90
    refugia hypothesis 186–7
    white-sand species 151, 157, 187, 221
  austral origin 16, 103–7, 182
  Boreotropics hypothesis 220–1, 265, 268
  climatic changes 186–7, 191
  Cretaceous 104, 135
  endemicity 188–9
  Gondwanic patterns 16, 103–7, 135, 182–3, 206, 220–1
  long-distance dispersal 105, 136, 220
  neotropical patterns 181–2
  Neotropics 136, 182–90, 219, 220–1
  paleotropics 103, 136, 219
  pantropical patterns 103–8, 193, 196–8, 203, 206, 219–21
  Quaternary 186–90, 262–6
  temperate origin 262–6
  temperate patterns 262–6
  tepui 66, 182, 184, 187–90
  Tertiary 104, 262–6
  transatlantic disjuncts 136–7, 219–21, 266
  white-sand species 157, 182, 187–90, 212
Bisgoeppertia 108, **110**, 113
  classification 112
  distribution 110
  number of species 24
  palynology 65
  phylogeny 335
  seed anatomy 514–15, 557, 559
Bitters 574, 611; *see also* Pharmacology
Blackstonia 108, **111**, 113, 126, 136, 137, 224, 517–18
  chromosome number 329
  classification 112
  distribution 111
  embryology 326
  flower morphology 322–3
  germination 508
  number of species 24
  phylogeny 119, 206, 314, 332–3, 335, 354
  phytochemistry 586, 593, 598, 599, 601
  seed anatomy 511, 517–19
  vegetative morphology 320
Boraginaceae 31
Boreotropics hypothesis 220, 265
Brachycodon 379, 472–3 (= *Irlbachia*)
Bracts *see* Inflorescence
Buddlejaceae 4

C-Glucoflavones 574, 604–9, 610
Calathiana 54 (= *Gentiana*)
Calathiinae (subtribe) 54
Calolisianthus 137–8, **144–5**, 164–5, 524–6
  biogeography 182

distribution 138
number of species 24, 138
palynology 171, 377, 379, 381, 394–5, 423–4, 468–9, 479
phylogeny
  molecular data 29, 149
  morphological data 468–9, 479
phytochemistry 601
pollination 8
seed anatomy 523–5, 557
Calyx
  Potalieae 211
  aestivation 39
  fusion 39, 148, 168, 192, 211, 223
  glands 5, 39
  homeotic mutation 169
  intracalycine membrane 223, 238–9
  keels 39, 68–70, 168, 211
  persistence 140, 142, 169
  raised veins 192, 198
  reduced sepal number 234
  wings 39, 68–70, 73, 168
  zygomorphy 68, 223
Campanulaceae 10
Canscora 108, **111**, **114**, 117, 125–6, 224, 519–20, 592–3
  classification 112
  distribution 111
  embryology 326, 499
  flower morphology 111, 114, 321–3
  number of species 24
  phylogeny 114, 314, 333–5
  phytochemistry 592–3, 598–9, 601, 604–5
  seed anatomy 504, 519–20
  subgenus
    Canscora 111
    Heterocanscora 111
    Pentanthera 114, 117
    Phyllocyclus 114, 120
  vegetative morphology 111, 114, 321
Canscorinae (subtribe) 108, **125–6**
  classification 50
  distribution 125
  key 47
  number of genera 24, 50
  number of species 50
  phylogeny 135–6, 556
    molecular data 125–6
  seed anatomy 513, 519–21, 556
Caryophyllaceae 7
Celiantha 138–9, **145**, 524–6
  biogeography 182
  distribution 138
  flower morphology 321–2
  number of species 24, 138, 145
  palynology 171–2, 324, 377, 379, 381, 395–9, 424–6, 469, 479

phylogeny
  morphological data 315, 332–3, 335, 355
  palynological data 377, 469, 479
  pollination 8
  seed anatomy 524–5, 557
Centaurium 108, **114–15**, 123, 126, 508, 516–17, 524–7
  biogeography 137
  chromosome number 328, 329
  classification 112
  dispersal 5–6
  distribution 114
  embryology 326
  floral anatomy 321–3
  germination 508
  number of species 13, 15, 24
  paraphyly 13, 127
  pharmacology 611
  phylogeny 314, 332–3, 335–6
  phytochemistry 582, 586, 590, 592–3, 598, 601
  pollination 7, 324
  section
    Gyrandra 127
  seed anatomy 328, 516–17
  vegetative morphology 320–1
Central America
  floras 275
  generic diversity 14–15
Chelonanthus 137–8, **145–7**, 164–5, 524–6, 557
  distribution 138, 146, 182, 184
  nomenclature 146
  number of species 24, 138
  palynology 172, 377, 379, 381, 426–9, 434, 469–70, 480
  phylogeny 468–9
    molecular data 149, 197
  phytochemistry 586, 601
  pioneer species 187
  pollination 8
  seed anatomy 523–6, 557
  wood anatomy 320
Chemistry *see* Phytochemistry
Chinese medicine 579; *see also* Pharmacology *and* Ethnobotany
Chionogentias 229, 246 (= *Gentianella*)
Chironia 108, **115**, 123, 126–7, 513–14, 612
  biogeography 137
  classification 112
  dispersal 507
  distribution 115
  exudates on seeds 560
  floral anatomy 323
  fruit morphology 325
  number of species 24

Chironia (cont.)
palynology 129
phylogeny 314, 332–3, 335, 356
phytochemistry 586, 593, 598–9, 601
pollination 324
seed anatomy 502, 513
vegetative morphology 320–1
Chironiaceae 48
Chironieae (tribe) 23, 67, **108–10**
biogeography 135–7
chromosome number 127
classification 49–51, 112–13
exudates on seeds 560
flower morphology 7
key 47
number of genera 12, 24, 49, 108
number of species 11, 12, 49, 108
number of subtribes 24
palynology 128–35, 379, 414–21, 465
phylogeny 108, 119, 123–8, 333, 334, 465, 556–7
  molecular data 40, 42, 44, 49, 124
phytochemistry 601, 611
seed anatomy 513–23, 556–7
taxonomic history 110
Chironiinae (subtribe) 108
classification 49
key 47
number of genera 12, 24, 50
number of species 12, 50
phylogeny 126–8, 556
  molecular data 126–8
  morphological data 333–5
seed anatomy 327, 513–19, 556, 559
Chlora 49 (= *Blackstonia*)
Chloreae (tribe) 49
Chlorinae (subtribe) 49
Chorisepalum 138–9, **147–8**, 164, 532–3
biogeography 182–4
dispersal 505
distribution 138, 147
floral anatomy 322
number of species 24, 138
palynology 172, 377, 379, 382, 429–30, 470
phylogeny 148, 480
  molecular data 148–9
  morphological data 314, 332–5, 357
  palynological data 148, 377, 470
seed anatomy 502, 532–3, 558
vegetative morphology 321
Chromosome number 328–9; *see also*
  Chromosome number under specific groups
Cicendia 108, **115–16**, 118, 126, 517–18, 586
classification 112
distribution 115

number of species 24
phylogeny 335
phytochemistry 601
seed anatomy 517
Cladistics *see* Phylogeny under specific
  groups *and* Molecular systematics
Classification 24–7, 40–4, 48–55
as diversity assessment 10
historical 56
Chironieae 110–13
Exaceae 66–7, 71–7
Gentianaceae 1–2, 5, 311–16, 377–9, 478–80, 555
Gentianeae 224
Helieae 140–1, 377, 466–76, 479
Lisianthius vs. *Lisyanthus* 146–7, 377
Potalieae 194–5
Saccifolieae 56–7, 379
molecular data 29–37, 335–6; *see also*
  Molecular systematics
morphological data 310–75
overview of new classification 24–7, 48–54
palynological data 377–97
phytochemical data 574, 589–90, 610
seed data 498–571
Cleistogamy 39, 68, 86, 90
Clusiaceae 584, 585
Colla, L. A. 71
Colleters 4, 39, 59–60, 128, 140, 161, 166, 168
Comastoma 222–3, **225**, 246–7, 547
biogeography 263
distribution 225
flower morphology 230
nomenclature 225
number of species 25, 225
palynology 249–51
phylogeny 245–7, 589, 599
phytochemistry 589, 593, 598–9, 601
seed anatomy 545–8, 559
Congolanthus **196**, **198**, 535–6
distribution 190, 196
number of species 25, 190, 196
palynology 214, 216, 377, 379, 382, 453–4, 477–8
phylogeny 378, 477–8
seed anatomy 535–6, 558
Cornanae (superorder) 574–5
Corolla
Gentianeae 230, 239, 245
Potalieae 212–13
aestivation 5–6, 39, 144, 169, 192, 223, 239
color 212–13
fimbriae (hairs) 223, 229–30, 235, 243–7
fusion 4, 39, 169, 192

plicae 223, 238–9
spurs 223, 231, 249
zygomorphy 39, 142
Corona 142; *see also* Floral anatomy
Cotylanthera 5, 66, **77–9**, 102–4
  distribution 77
  embryology 326, 501
  number of species 24
  phylogeny 107, 556
    morphological data 314, 332–4, 357
  pollination 325
  seed anatomy 509–10, 512, 556
  staminal morphology 324
  trophy 320
Coutoubea 108, **116**, 125, 521–2, 612
  chromosome number 116
  classification 112
  distribution 116
  exudates on seeds 560
  floral anatomy 321–2
  number of species 24
  palynology 129, 132, 324, 377, 379, 381, 399, 414, 416–17, 465
  phylogeny 378, 465, 557
    molecular data 31
    morphological data 315, 332–4, 358
  phytochemistry 586, 601
  seed anatomy 502, 505, 513, 521–2, 557
  vegetative morphology 320
Coutoubeaceae 48
Coutoubeinae (subtribe) 108
  classification 51
  key 47
  number of genera 24, 51
  number of species 51
  phylogeny 123, 135–6, 334, 556
    molecular data 125
  seed anatomy 513, 521–3, 556
Cracosna 108, **116**, 123, 125
  classification 112
  distribution 116
  number of species 24
  seed anatomy 513
Crawfurdia 222, **225–6**, 237–8, 542, 545
  biogeography 262
  dispersal 506
  distribution 226
  floral anatomy 321–2
  fruit morphology 8, 325
  nomenclature 226
  number of species 25, 226
  palynology 251
  phylogeny 314, 333, 358
    molecular data 197
  section
    Crawfurdia 226 (= *Crawfurdia*)
    Dipterospermum 225 (= *Crawfurdia*)
  Protocrawfurdia 226, 238
    (= *Crawfurdia*)
  seed anatomy 327, 502, 542, 545
  vegetative morphology 321
Curtia **58**, 62–3, 551–2
  biogeography 66
  distribution 58
  floral anatomy 321–2
  number of species 24
  palynology 63, 65, 377, 379, 382–4, 408–9, 462–3, 478
  phylogeny 62–3, 378, 556
    molecular data 29
    morphological data 314, 333–4, 359
  palynological data 462–3, 478
  phytochemistry 330, 586
  seed anatomy 328, 550–2, 556, 559
  taxonomic history 55–6
  vegetative morphology 321
Cyrtophyllum 200 (= *Fagraea*)
Cytology *see* Chromosome number under specific groups

Danais 316, 331, 333, 336
Deianira 108, **116–17**, 125, 224, 521–2
  classification 112
  distribution 116
  number of species 24
  palynology 132, 324, 377, 379, 381, 399, 417–18, 465
  phylogeny 465, 557
    molecular data 29
    morphological data 315, 332–3, 359
  phytochemistry 601
  pollination 324–5
  seed anatomy 521–2, 557
Desfontainiaceae 4, 329
Development
  flowers 169–70
  homeotic mutation 169
  stamens 93
Dipsacales (order) 3, 4
Disk *see* Glands
Dispersal 5–6, 498, 560
  ants 217
  bat 9, 217
  bird 9, 101, 107, 217, 507
  earthworms 101, 107
  long-distance *see* Biogeography
  mammals 217
  on animals 101
  rain 101, 266
  water 507
  wind 101, 266, 506–7
Diversity
  Andes 139, 153, 159, 182, 229
  endemic genera 15

Diversity (*cont.*)
  generic 11–12
  generic, by continent 14–15
  species 11–12
Djaloniella **198**
  distribution 190, 198
  number of species 25, 190, 198
  phylogeny 201
  phytochemistry 601
  seed anatomy 536–7, 558
DNA sequencing *see* Molecular systematics
Don, G. 56, 74, 110, 112–13, 224
"Duplipetala" 51, 108, 114, **117**, 123, 125, 126
  classification 112
  distribution 117
  number of species 24

Ellipandrae (series) 50–1
Emblingia 55
Endlicher, S. L. 56, 74–5, 110, 112–13
Engler, A. 76
Enicostema **198–9**, 224, 333, 536–7
  distribution 190, 193, 198, 220
  flower morphology 192
  number of species 25, 190, 198
  phylogeny 314, 334, 360
    molecular data 197
  phytochemistry 586, 601, 607–8, 612
  seed anatomy 536–7, 558
  vegetative morphology 321
Epigyny 3–4
Ericanae (superorder) 574
Erythraea 49–50, 612 (= *Centaurium*)
Erythraeeae (series) 49–50
Erythraeinae (subtribe) 49–50, 313, 322–3, 327, 508, 556, 559
Erythreae (tribe) 49–50
Ethnobotany; *see also* Pharmacology
  Exaceae 101
  Helieae 181
  Potalieae 218
  Africa 218
  Asia 101, 218
  Latin America 181, 218
  Madagascar 101
  traditional uses 101
Europe
  floras 272–3
  generic diversity 14–15
Eustoma 108, **117–18**, 123, 126, 514–15
  biogeography 137
  classification 112
  distribution 117
  embryology 499
  floral anatomy 323
  horticulture 117

number of species 24
palynology 132–3
phylogeny 314, 332–3, 335, 360
phytochemistry 586, 593, 598–9, 601
pollination 324
seed anatomy 514–15, 557, 559
Evolution
  bird dispersal 9
  calyx 90, 124, 197
  calyx wings 91, 168
  calyx zygomorphy 90–1
  chromosome numbers 127, 208
  colleters 128
  convergent 2, 559
  corolla 91–2, 169, 197
    color 91–2, 124, 160, 169, 196
    fimbriae (hairs) 229–30, 235, 245
    glands 229–30, 245
    spurs 249
  dispersal 5–6, 9, 101, 217, 498, 505–8, 560
    bird 507
    water 507
    wind 506–7
  epigyny 3–4
  floral anatomy 167–8
  fruit types 6, 8–10, 197
  fusion of carpels 6, 8–9, 196, 216
  habit 10–11, 128, 139, 165, 205, 208–10, 218, 320, 335, 337
  hairs 128, 222
  inflorescence 89, 167, 210
  leaf morphology 124, 128, 166
  morphology, Chironieae 123–8
  palynology; *see also* Palynology
    Chironieae 135
    exine 62, 124–5, 179
    monads 149, 178
    polyads 149, 178–9
    tetrads 124–5, 135, 149, 178–9, 208
  phytochemistry, xanthones 237
  placentation 98, 143, 180, 196–7, 216, 237
  pollen types 380–1, 408
  pollination
    bat 9, 139, 180
    syndromes 7–9, 139, 180
  saprophytes 63, 89
  seed *see* Seeds
  stamens 92–4, 124–7, 170, 196–7, 201, 213, 230
  stem morphology 124
  stigma 42, 125, 140, 142, 179–80
  supermerous flowers 124, 128, 197, 206, 211
Exacaceae 49
Exaceae (tribe) 23, **67–71**
  biogeography 103–8
  character synopsis 101–3

Index 639

chromosome number 100
classification 49
dispersal 101, 506
distribution 103–4
flower morphology 7, 67, 70, 72, 89–94, 98–9
fruit morphology 71, 98–100
key 46
number of genera 12, 24, 49, 66, 88
number of species 12, 13, 49
palynology 94–8, 379, 411–14, 464–5
pharmacology 101
phylogeny 40, 42, 44, 67, 88–94, 101–2, 556
  molecular data 40, 102
  morphological data 102, 332, 335, 464–5
  palynological data 464–5
  phytochemical data 101, 600
phytochemistry 101, 601, 611
pollination 7, 99
seed anatomy 67, 71, 74, 99–100, 502, 509–12, 556
taxonomic history 66–7, 71–7
vegetative structures 67–8, 88–9
Exacinae (subtribe) 49, 322, 323
Exaculum 108, **118**, 123, 126, 137
  classification 112
  distribution 118
  number of species 24
  phylogeny 335
  seed anatomy 513
Exacum 66–7, **79–81**, 102, 103
  biogeography 105–7
  character evolution 319
  dispersal 506
  distribution 79
  embryology 326
  floral anatomy 324
  number of species 24
  palynology 94–6
  phylogeny 313, 314, 332–4, 361
  phytochemistry 330, 586, 600, 601
  pollination 324–5
  section
    Africana 81
    Exacum 81
  seed anatomy 327, 328, 504, 509–11
  vegetative morphology 320–1
Exadenus 231 (= *Halenia*)
Exochaenium 85; see *Sebaea*
Exostema 316, 322, 331
  phylogeny 333, 336, 350
Exudates see Exudates on seeds under specific groups

Fagraea 2–3, 6, **199–200**, 539–40
  chromosome number 329
  dispersal 9, 507–8
  distribution 190–1, 193, 199
  floral anatomy 322–3
  flower morphology 192, 197, 199
  fruit morphology 8–9, 197, 200, 325
  germination 508
  habit 10, 199
  nomenclature 200
  number of species 13, 15, 25, 190, 199
  palynology 2, 197, 214–15, 216, 324, 377–9, 381, 384, 454–8, 476–7, 479
  phylogeny 4, 200, 378
    molecular data 32, 197
    morphological data 315, 333, 335, 361
    palynological data 476–7, 479
  phytochemistry 586, 601, 605, 607
  pollination 9
  section
    Cyrtophyllum 540
    Fagraea 540
    Racemosae 540
  seed anatomy 503, 507, 539–40
  vegetative morphology 320–1
Fagraeeae (tribe) 53
Fagraeinae (subtribe) 53
Faroa **200–1**, 536–7
  distribution 190, 199
  floral anatomy 321–2
  number of species 25, 190, 199
  phylogeny 201, 314, 333–4, 362
  phytochemistry 586, 601
  seed anatomy 328, 536, 558
Faroinae (subtribe) **193**
  classification 52, 194–5
  distribution 13, 193, 219
  exudates on seeds 560
  key 47
  number of genera 25, 53
  number of species 53
  phylogeny 195, 201, 207–8
    molecular data 197
  seed anatomy 502, 534–8, 558
Filaments see Stamens
Floral anatomy 73
  Helieae 167–8
  Potalieae 211–12
  anthers 78, 93
  corona 142
  fibers in flowers 211
  gynoecium 73, 99, 168
  petal epidermis 67
  postgenital fusion of carpels 1–2, 6, 216
  sclereids 168
  staminal appendages 192–3, 201, 213
  supermerous flowers 211; see also Flower morphology
  vascularization 99, 167–8, 208, 211–12, 248, 240

640 Index

Flower development *see* Development
Flower morphology
  apocarpy 1
  bauplan 1, 3, 6–7
  calyx *see* Calyx
  cleistogamy 39, 68, 86, 90
  colleters *see* Colleters
  corolla *see* Corolla
  disk *see* Glands
  double stigma 68, 99, 101–2
  glands *see* Glands
  number of floral parts 39, 90, 108, 124, 148, 169, 192–3, 211, 239
  spurs on petals 231
  stamens *see* Stamens
  supermerous flowers 39, 108, 120, 128, 192–3, 195–7, 206, 211
Fossils 104, 221
Frasera 222–3, **226–7**, 235, 248, 548–9
  biogeography 263–5
  classification 226, 247
  distribution 226
  number of species 25, 226
  phylogeny 245, 247, 264, 373
  phytochemistry 593, 598, 599–601, 607–8
  seed anatomy 327, 502–3, 548
  seed dormancy 508
Fruit dispersal 505–8; *see also* Dispersal
Fruit morphology 6, 8–10, 43, 115, 143, 180, 193, 216–17, 223
  dehiscence 43, 143, 180, 223
  fleshy fruits 6, 8–10, 43, 115, 143, 193, 195–7, 216, 223, 236, 240

Gardneria 8
Gelsemiaceae 3, 4, 6
  number of species 11
  palynology 378, 479
  phylogeny 40, 479
    molecular data 32, 40
Gelsemium 4, 316
  floral anatomy 321–2
  mycorrhiza type 319
  phylogeny 331, 333, 351
  palynology 479
Generic skewness 11–12
Geniostemon 108, **118**, 126, 601
  classification 113
  distribution 118
  number of species 24
  phylogeny 335
  seed anatomy 515, 516
Geniostoma 10, 316, 320, 322, 324
  palynology 216, 378
  phylogeny 333, 335, 351
Gentiana 62, 222–3, **227–8**, 237–8, 503, 542–5
  biogeography 262, 265

  chromosome number 228, 242, 329
  classification 227–8, 242
  dispersal 506–7
  distribution 227
  embryology 499–500
  floral anatomy 321–3
  fruit morphology 325
  germination 508
  nomenclature 227–8
  number of species 13, 25, 227–8
  palynology 251–4
  pharmacology 611
  phylogeny 222, 241–2, 313–14
    molecular data 197, 241–2, 555
    morphological data 240
    phytochemical data 589, 599
  phytochemistry 329, 330, 575–9, 581, 585–7, 592–4, 598–9, 601, 604–8
  pollination 7–8
  section
    Amarella 225 (= *Comastoma*)
    Calathianae 228, 241–2 (= *Gentiana*)
    Chondrophyllae 228, 241–2 (= *Gentiana*)
    Ciminalis 228, 241–2
    Comastoma 225 (= *Comastoma*)
    Crossopetalum 229, 256 (= *Gentianopsis*)
    Cruciata 228, 241 (= *Gentiana*)
    Dipterospermum 225 (= *Crawfurdia*)
    Dolichocarpa 228 (= *Gentiana*)
    Fimbricorona 228 (= *Gentiana*)
    Frigidae 228, 241–2 (= *Gentiana*)
    Gentiana 228, 241–2 (= *Gentiana*)
    Imaicola 229 (= *Gentianopsis*)
    Isomeria 228 (= *Gentiana*)
    Kudoa 228 (= *Gentiana*)
    Megacodon 233 (= *Megacodon*)
    Microsperma 228 (= *Gentiana*)
    Monopodiae 228 (= *Gentiana*)
    Otophora 228 (= *Gentiana*)
    Phyllocalyx 228 (= *Gentiana*)
    Pneumonanthe 228, 241–2 (= *Gentiana*)
    Stenogyne 225, 228, 241–2 (= *Gentiana*)
    Stylophora 233 (= *Megacodon*)
  seed anatomy 502–5, 541–5, 559
  subgenus
    Calathianae 323, 328–9, 333, 362, 541, 543–4
    Chondrophyllae 502, 506, 541, 543
    Ciminalis 263, 322–3, 363, 544
    Cruciata 329, 507, 541, 543
    Dolichocarpa 545
    Frigidae 329, 330, 333, 363, 543
    Gentiana 322–3, 327, 333, 364, 544
Gentianella 225, 227, 244 (= *Gentianella*)

Isomeria 545
Microsperma 541, 543
Monopodiae 543
Newberryi 544
Otophora 323, 327, 333, 364, 541
Phyllocalyx 544
Pneumonanthe 262, 323, 327, 544, 545
Stenogyne 238, 262, 327, 333, 365, 543, 544
Stylophora 544 (= *Megacodon*)
Gentianaceae
 age 104
 bauplan 3, 4, 6, 10
 biogeography 14, 15, 22, 23, 28; see also Biogeography *and* Biogeography under specific groups
 chromosome number 328–9; see also Chromosome number under specific groups
 classification 23, 24–8, 40–3, 48–54
  alphabetical list 24–7
  circumscription 555–60
  Gilg's system 28, 313–16
  monophyletic 3, 5, 6, 10, 28, 48
  taxonomic history 1–2, 5, 311–16, 377–9, 478–80, 555
 conservation 28
 description **39**, **42–3**, **45**
 dispersal 9, 498, 505–8
 distribution 15, 23
 diversity 11–16, 28
 embryology 326–7, 499–501
 endemicity 14–15, 188
 floral anatomy *see* Floral anatomy under specific groups
 floristics 23, 270–6
 flower morphology 7, 8, 39, 42, 230, 321–4
 fruit morphology 8–10, 39, 43, 325–6
 geographic origin 103–7
 germination 508
 habit 10–11, 22, 319–20, 337
 key 45–7
 life cycle 319–20
 mycorrhiza type 319
 number of genera 11–12, 23, 48, 311
 number of species 11–14, 23, 28, 48, 311
 palynology 45, 324, 377–497
 phylogeny
  molecular data 2, 4, 6, 7, 23, 29–37, 40–4, 197, 313, 334–6
  morphological data 310–76, 378–9
  palynological data 382–407, 462–78
  phytochemical data 600
  seed data 498, 555–60
 phytochemistry *see* Phytochemistry
 pollen types 381
 pollination 7–8, 324–5
 saprophytes 5, 39, 559–60
 seed anatomy 327–8, 498, 501–5, 509–55
 species radiation 15
 stomata 321
 trophy 39, 319–20, 559–60
 vegetative morphology 39, 320–1
Gentianales (order) 3–5, 311
 flower morphology 7
 fruit morphology 9
 minimum age 136
 number of species 11
 phylogeny 4, 40–3, 555
 phytochemistry 329, 610
 pollination 7–8
 seed anatomy 7
 vegetative morphology 4
Gentiananae (superorder) 574–5
Gentianeae (tribe) 23, **222–3**, 224–66
 biogeography 262–6
 classification 53–5
 distribution 14, 222
 floral anatomy 321–3
 flower morphology 222–3, 239–40, 245
 fruit morphology 223, 237, 240
 generic circumscriptions 222
 key 46
 number of genera 12, 53
 number of species 12, 53
 number of subtribes 25, 237
 palynology 65, 237, 249–62, 378, 380–1
 phylogeny 41, 43–4, 53, 237–49, 266, 559
  molecular data 41, 43–4, 237, 241
  morphological data 239, 332–4
  phytochemical data 600
 phytochemistry 237, 601–2, 611
 seed anatomy 327, 502–3, 541–50, 559
 species diversity 14
 taxonomic history 224
 vegetative morphology 222, 242
Gentianella 222–3, **228–9**, 230, 501, 506–7, 546–7
 biogeography 263–5
 chromosome number 229, 329
 circumscription 244, 246
 classification 229
 dispersal 507
 distribution 228
 embryology 326, 500
 floral anatomy 323
 flower morphology 229–30
 fruit morphology 229, 325
 germination 508
 nomenclature 228
 number of species 13, 15, 25, 229
 palynology 254–6
 paraphyly 13, 15

Gentianella (cont.)
  phylogeny 222, 242, 245–7, 264–5
    molecular data 33, 245
    morphological data 314, 333, 365
    phytochemical data 589, 599–600
  phytochemistry 589, 594, 598, 601, 604–5, 607–9
  pollination 324–5
  section
    Comastoma 250–1 (= *Comastoma*)
    Crossopetalum 256 (= *Gentianopsis*)
  seed anatomy 502–3, 505, 545–6, 559
  subgenus
    Eublephis 229 (= *Gentianopsis*)
Gentianinae (subtribe) 222
  biogeography 262–3
  chromosome number 242
  classification 54
  dispersal 506
  embryology 326
  flower morphology 239
  fruit morphology 240
  habit 240
  key 46
  number of genera 25, 54
  number of species 54
  palynology 250, 252, 378
  phylogeny 54, 237–42, 265, 333–4, 556
    molecular data 238, 241
    morphological data 238, 313
  seed anatomy 541–5, 556
  vegetative morphology 238, 240
Gentianoideae (subfamily) 48
Gentianopsis 222, 223, **229–31**, 238, 249, 502, 506–7, 548–50
  biogeography 263, 265
  chromosome number 231, 329
  dispersal 507
  distribution 231
  floral anatomy 321–3
  flower morphology 230, 244
  fruit morphology 244, 325
  nomenclature 229, 231
  number of species 25, 231
  palynology 256–7
  phylogeny 230, 244–5, 249, 264
    molecular data 33, 245
    morphological data 230, 244, 314, 333, 366
    phytochemical data 589, 598
  phytochemistry 589, 598, 601, 604, 607, 609
  pollination 324
  seed anatomy 244, 505, 548–50
Gentianothamnus 66–7, **81–2**, 102–3, 108
  distribution 81
  number of species 24

  palynology 96, 377, 379, 384, 411, 413, 464–5
  phylogeny 82, 464–5
  seed anatomy 510, 512
Geology *see* Biogeography
Gesneriaceae 4
Gilg, C. 56, 112–13
Gilg, E. 28, 48, 56–7, 76, 110, 112–13, 140, 194, 224, 312, 314–15
  palynology 134–5, 177–8
Glands
  disk at base of ovary 42, 68, 99, 142, 180, 192–3, 208, 223, 239
  disk at inner base of calyx 5, 59–60
  on leaves 59
  on outside of calyx 5, 39, 168
  on petals (corolline) 39, 223, 229–30, 233, 235, 239
  on stamens (staminal) 68, 72, 92
Gondwana *see* Biogeography
Gray, A. 194
Grisebach, A. H. R. 56, 75–6, 110, 112–13, 140, 224, 312
Guttiferae 584–5
Gynoecium
  number of carpels 42
  placentation 42, 67, 71, 75, 98, 143, 180, 195–6, 212, 216, 223
  position 39, 192
  semi-inferior ovary 78
  stigma decurrent 223, 225
  stigma shapes 42, 140, 142, 180, 192, 217
  style 42, 142, 179–80, 192

Habit 10, 39, 139, 165, 208–9
Hairs *see* Vegetative morphology
Halenia 222–3, **231–2**, 248–9, 546–7
  biogeography 262–5
  chromosome number 232, 328
  distribution 231
  embryology 326, 500
  flower morphology 231, 321–3
  germination 508
  number of species 13, 15, 25, 231
  palynology 256–8
  phylogeny 248, 264
    molecular data 33, 245
    morphological data 314, 333, 366
    phytochemical data 599
  phytochemistry 330, 588, 594, 598, 600–1
  pollination 324
  seed anatomy 505, 545–8, 559
  trophy 501
Helia 137–8, **148**, 150, 524–6, 557
  biogeography 182
  distribution 138, 148
  number of species 24, 138, 148

palynology 172, 377, 379, 381, 399–400, 429, 431–2, 470–1
phylogeny 148, 470–1
seed anatomy 523–6, 557
Helieae (tribe) 23, **137–40**
biogeography 181–90
character synopsis 181
circumscription 137–9
classification 51–2
dispersal 506
distribution 13, 138–9, 181–2, 184–9, 191
evolution of woody habit 337
floral anatomy 167–8
flower morphology 142, 168–70, 179–80
fruit morphology 143, 179–80
habit 10, 139, 337
highland/lowland genera 138–9
inflorescence 167
key 46
nomenclature 150
number of genera 12, 24–5, 52, 138–9
number of species 11–12, 52, 138–9, 189
palynology 170–9, 378, 380–1, 421–53, 466–76, 479–80
pharmacology 181
phylogeny 41, 43–4, 125, 149, 163–5, 336, 556–7
 molecular data 41, 149, 163–5
 morphological data 334–6
 palynological data 170–9, 466–76, 479–80
phytochemistry 601, 611
pollination 7, 8, 139, 180–1
seed anatomy 502, 523–34, 556, 558
taxonomic history 140–1, 377, 466–76, 479
vegetative morphology 165–7, 181
Herbaceous habit *see* Habit *and* Evolution, habit
Herbal medicines *see* Pharmacology *and* Ethnobotany
Heterostyly 5, 57–9, 61–3
palynology 65
Hippieae (tribe) 52
Hippion 52 (= *Enicostema*)
Hockinia **58–9**, 63, 478, 551–2
biogeography 66
distribution 58
number of species 24
palynology 63, 65, 377, 379, 385, 408, 410, 463–4, 478
phylogeny 62–3, 378, 463–4, 478
seed anatomy 551–2, 559
taxonomic history 55–6
Homeotic mutation 169
Hoppea 108, **118–19**, 125–6, 515, 519–520, 612

classification 113
distribution 118
embryology 326
floral anatomy 321–3
number of species 24
pharmacology 612
phylogeny 314, 333–4, 367
phytochemistry 594, 598–9, 601, 604–5
seed anatomy 519–21, 556
vegetative morphology 321
Horticulture
Eustoma 117
Lagenanthus 151–2
ornamentals 117
Hutchinson, J. 194
Hydrangeaceae 33

Incertae sedis 25, 45, 266–8, 559–60; *see also Voyria*
classification 54
Indumentum *see* Vegetative morphology, hairs
Inflorescence 39, 89, 167, 210
bracts 167, 234
Interpetiolar stipules and sheaths 4, 6, 39, 166, 210; *see also* Vegetative morphology
Intraxylary phloem 4
Introduced species [*see Centaurium*]
Iridoids 329, 574–84, 586–8, 590, 611–12
Irlbachia 137–9, **150–1**, 379, 471–2, 524–7, 557
biogeography 151, 182, 184–5
dispersal 506
distribution 138, 150
generic complex 138–9, 150
nomenclature 150
number of species 24, 150
palynology 173, 324, 377–9, 381, 400–1, 432–4, 466–8, 471–2, 479
phylogeny 164–5, 377–8, 466–8, 557
 molecular data 34, 149, 151
 morphological data 315, 332–3, 335, 367
 palynological data 151, 471–2, 479
phytochemistry 329–30, 601
pollination 324
seed anatomy 328, 505, 523–7, 529, 531, 557
taxonomic history 150
vegetative morphology 321
Irlbachia clade (subgroup in Helieae)
biogeography 184
phylogeny 149, 164
Ixanthus 108, **119**, 126, 136–7, 224, 518–19
biogeography 119

Ixanthus (cont.)
  classification 113
  number of species 24
  palynology 133
  phylogeny 119, 314, 333–5, 368
  phytochemistry 588, 594, 598–9, 601
  seed anatomy 518–19
  vegetative morphology 320–1
  wood anatomy 320, 335

Jaeschkea 222, **232**
  biogeography 263
  chromosome number 232
  distribution 232
  number of species 25, 232
  palynology 258–9
  phylogeny 245
  phytochemistry 602
Jussieu, A. L. de 1–3, 6, 22, 194

Karina 195, **201–2**, 537–8
  distribution 191, 201
  number of species 25, 191
  phylogeny 201
  seed anatomy 201, 537–8, 558
Key to tribes and subtribes 45–7
Kingdon-Wardia 235 (= *Swertia*)
Knoblauch, E. 56–7, 110, 112–13
Kuhlia 200 (= *Fagraea*)
Kuntze, O. 146
Kurramiana 232 (= cf. *Jaeschkea*)

Labordia 8
Lagenanthus 138–9, **151–2**
  biogeography 182
  distribution 138, 152
  number of species 24, 138, 152
  palynology 173, 377, 379, 381, 401, 435–6, 472, 479
  phylogeny 378, 472, 479, 557
  pollination 8
  seed anatomy 529, 557
Lagenias 85, 511; see *Sebaea*
Lamiales (order) 3–4, 610
Lamianae (superorder) 574–5
Latex 39
Latouchea 222, **232**, 243
  distribution 232, 263–4
  number of species 25, 232
Leaves *see* Vegetative morphology
Leeuwenberg, A. J. M. 194
Lehmanniella 138–9, **152**, 529–30
  distribution 138, 152
  number of species 25, 138, 152
  palynology 173, 377, 379, 381, 436–7, 472, 479
  phylogeny 378, 472, 479, 557
  seed anatomy 524, 529–30, 557–8
  Leiphaimeae (tribe) 54, 556
  Leiphaimos 266 (= *Voyria*)
  Lisianthiinae (subtribe) **193**; see also *Lisianthius*
  biogeography 202, 219
  classification 53, 194, 202–3
  distribution 193
  key 47
  number of genera 25, 53
  number of species 53
  palynology 480
  phylogeny 203, 208
    molecular data 197, 203
    morphological data 219
  pollination 203
  seed anatomy 538–40
Lisianthius 145–6, 152, 159, 194, **202–3**, 322, 502, 515, 538–40 ("Lisianthus", see also *Lisyanthus*)
  biogeography 219
  chromosome number 202, 328
  classification 139, 202
  distribution 191, 193, 202–3
  embryology 501
  floral anatomy 321–2
  flower morphology 192, 202, 210
  fruit morphology 202, 325
  number of species 25, 191, 202
  palynology 215–16, 377–9, 381, 385–7, 402, 458–9, 477, 479–80
  phylogeny 203, 377, 477, 479, 557, 559
  molecular data 34, 197
  morphological data 219, 313–14, 333–4, 368
  phytochemistry 329, 588, 601
  pollination 203, 324
  section
    Lisianthius 202, 515, 538
    Omphalostigma 202, 210, 540
  seed anatomy 502, 538–40, 557–9
  taxonomic history 202, 377
  vegetative morphology 202, 320
Lisianthus see *Eustoma*
Lisyantheae (tribe) 51
Lisyanthinae (subtribe) 51
Lisyanthus 51, 145–6 ("Lisianthus", see also *Lisianthius*)
  section
    Adenolisianthus 141 (= *Adenolisianthus*)
    Calolisyanthus 144 (= *Calolisianthus*)
    Chelonanthus 145 (= *Chelonanthus*)
    Macrocarpaea 152 (= *Macrocarpaea*)
    Symbolanthus 159 (= *Symbolanthus*)
Loganiaceae 1, 3–6, 23, 193, 311
  dispersal 507
  fruit morphology 8

habit 10
number of species 11
palynology 378–9, 479
phylogeny 40, 42, 44, 479, 555
  molecular data 29–37, 32, 34–7, 40, 42, 44
  morphological data 316
pollination 8
Lomatogonium 222–3, **232–3**, 246–7, 546–7
  biogeography 263, 265
  classification 233
  dispersal 507
  distribution 233
  floral anatomy 321–2
  flower morphology 233
  fruit morphology 233, 325
  germination 508
  nomenclature 233
  number of species 25, 233
  palynology 258, 260
  phylogeny 230, 245–7, 313–14, 333, 369
  morphological data 230
  phytochemistry 330, 588, 594, 598–9, 602, 607, 609
  pollination 324
  section
    Pleurogynella 233 (= *Lomatogonium*)
  seed anatomy 545–7, 559

Maas, P. J. M. 146
Macrocarpaea 138–9, **152–4**, 502, 505, 508, 531–3, 558
  biogeography 154, 182–4
  dispersal 506
  distribution 138, 152–3
  habit 10
  number of species 13, 15, 25, 138–9, 153
  palynology 173–4, 377, 379, 381, 387–9, 436, 438–40, 472–3, 480
  phylogeny 153, 164, 377, 472–3, 480, 558
  molecular data 35, 149, 197
  morphological data 314, 332–5, 369
  phytochemical data 598
  phytochemistry 594, 598, 601
  pollination 324
  seed anatomy 531–3, 557–8
  species diversity 153
  taxonomic history 152–3
  vegetative morphology 321
  widespread species 186–7
Macrocarpaea clade (subgroup in Helieae)
  biogeography 183–4
  phylogeny 149, 167
Madagascar
  generic diversity 14–15
Mangifera 602
Mangiferin 574, 585, 602–4, 605, 610, 612

MAO inhibitors 612
*mat*K data 28–37; *see also* Molecular systematics
Medicines *see* Pharmacology
Megacodon 222, **233–4**, 243
  biogeography 263–4
  distribution 233
  number of species 25, 233
  palynology 260
  phylogeny 244–5
  phytochemistry 602
Menyanthaceae 2, 3, 555
Mesomelitae (series) 224
Microcala 74, 115; *see Cicendia*
Microrphium 108, **119–20**, 123, 125–6
  classification 113
  distribution 119
  exudates on seeds 560
  number of species 24
  seed anatomy 520–1
Mitrasacme 1
Mitreola 1–3
Molecular systematics 2, 21, 23, 28–37, 334–5
  Boraginaceae 31
  Canscorinae 125–6
  Chironieae 123–8
  Chironiinae 126–8
  Coutoubeinae 123–5
  Gelsemiaceae 32
  Gentianeae 555
  Gentianinae 238, 555
  Helieae 149, 163–4
  Hydrangeaceae 33
  Loganiaceae 29, 32, 34–7
  Oleaceae 32, 34
  Potalieae 196–7
  Rubiaceae 29–36, 336
  Saccifolieae 62, 555
  Solanaceae 35
  Strychnaceae 32
  Valerianaceae 37
  5S-NTS data 160, 184
  18S data 267
  classification 48–55
  Genbank numbers 29–37
  ITS (internal transcribed spacer) data
    Chironieae 123, 126–7
    Gentianeae 237–8, 241–5
    Helieae 149, 163–5
    Potalieae 196–7
  jackknife analysis 38
  *mat*K data
    alignment 38
    result 38, 42–4
    methodology 28, 38
    parsimony analysis 38

Molecular systematics (cont.)
  rbcL data 62, 221
  restriction enzyme data 203
  trnL intron data
    alignment 38
    result 38, 40–1, 44
    vouchers 29–37
Monophyly of genera 15, 28
Monotypic genera 14
Morphology, flower see Flower morphology
  and Flower morphology under
  specific groups
Morphology, pollen see Palynology
Morphology, vegetative see Vegetative
  morphology and Vegetative
  morphology under specific groups
Mostuea 4
Mycorrhiza 222, 266, 319

Neblinantha 138–9, **154**, 527–8
  biogeography 182
  distribution 138, 154
  number of species 25, 138, 154
  palynology 174–5, 377, 379, 381, 402,
    440–1, 473, 479
  phylogeny 377, 479
  phytochemistry 601
  pollination 8
  seed anatomy 524, 527–9, 557
Nectaries see Glands
Neurotheca **203–4**, 535
  distribution 191, 193, 203, 220
  flower morphology 192
  number of species 25, 191, 203
  palynology 215–16, 377, 379, 389–90,
    460, 461, 477–8
  phylogeny 201, 378, 477–8
  molecular data 197
  phytochemistry 601
  seed anatomy 201, 513, 535–6, 558
New classification see Classification
New Zealand see Australia
North America
  floras 273–5
  generic diversity 14–15

Obolaria 5, 222–4, **234**, 243, 549–50
  biogeography 263–5
  chromosome number 234
  distribution 234
  floral anatomy 322–3
  number of species 25, 234
  palynology 260
  phylogeny 559
    molecular data 35, 245
    morphological data 314, 333–4, 370
  phytochemistry 602

seed anatomy 549–50, 559
  trophy 320
  vegetative morphology 321
Obolariaceae 48
Octopleura 203 (= Neurotheca)
Oleaceae 7, 32, 34
Oleales (order) 3
Ontogeny see Development
Ophelia 235 (= Swertia)
Oreonesion **204**
  distribution 191, 204
  number of species 25, 191, 204
  phylogeny 201, 204
  seed anatomy 538, 558
Ornamental plants 117
Ornichia 66, **83–4**, 102–3
  distribution 83
  number of species 24
  palynology 96
  phytochemistry 601
  seed anatomy 510–11
Orphium 108, **120**, 126, 513–15, 560, 594
  biogeography 137
  classification 113
  distribution 120
  exudates on seeds 560
  floral anatomy 323
  germination 508
  number of species 24, 120
  palynology 133
  phylogeny 314, 332–3, 335, 370
  phytochemistry 594, 598, 601
  pollination 324
  seed anatomy 505, 513–15, 557, 559
  vegetative morphology 320–1
Ovary see Flower morphology under
  specific groups

Pacific region see Australia
Pagaea (= Irlbachia)
Palynology 45
  Chironieae 128–35, 379, 414–21, 465
  Exaceae 94–8, 379, 411–14, 464–5
  Gelsemiaceae 479
  Gentianaceae 377–497
  Gentianeae 65, 249–62, 378, 380, 381
  Helieae 170–9, 421–53, 466–76, 479–80
  Lisianthiinae 480
  Loganiaceae 378–9, 479
  Potalieae 213–16, 379, 453–63, 476–9
  Saccifolieae 63–5, 379, 408–11, 462–4,
    478
  evolutionary trends 478; see also
    Evolution
  exine 45
  glossary of terms 277–8
  methodology 38, 380

monads 45, 381
pollen types 380–1, 408
polyads 45, 381
tetrads 45, 381
vouchers 38; see also Palynology under specific groups
Paraphyletic genera 13, 127
Perimelitae (series) 224
Petal see Flower morphology under specific groups
Pharmacology 574, 579, 611–13; see also Phytochemistry and Ethnobotany
Exaceae 101
Helieae 181
Potalieae 218
Phyllocyclus 108, 114, **120**, 123, 125
classification 113
distribution 120
number of species 24
Phylogeny; see also Phylogeny under specific groups and Molecular systematics
Chironieae 123–6
Saccifolieae 62–3
Phytochemistry 2, 4, 6, 311, 329–30, 574–613
Apocynaceae 329, 330
Chironieae 611
Cornales 582
Dipsacaceae 582
Exaceae 101, 586, 601, 611
Gentianales 582
Gentianeae 611
Helieae 611
Potalieae 329, 588, 601, 611
Rubiaceae 330
Saccifolieae 601
alkaloids 579, 581
amarogentin 611
aucubin 577
C-glucoflavones 574, 604–9, 610
carbohydrates 329
catalpol 577
convergent evolution 2
decussatin 194
flavone-O-glycosides 101
flavonoids 329, 560
gentianacaulin 194
gentianine 194
gentianose 330
gentioflavoside 581
gentiopicrine 329
gentiopicroside 194, 577, 579, 581, 582, 610, 612
glycoflavones 101
indole alkaloids 4, 311
iridoids 329, 574–84, 586–8, 590, 611, 612
L-(+)-bornesitol 101

loganin 577, 578
mangiferin 574, 585, 602–4, 605, 610, 612
methyl-swertianin 194
morroniside 577
pseudo-alkaloids 329
secoiridoids 2, 6, 194, 218, 329, 575, 580, 584; see also Iridoids
secologanic acid 579, 580
secologanin 329
sugars 330
swerosides 329, 579, 582–3, 610, 612
swertiamarin 194, 329, 579, 581–2, 610
triterpenes 329
xanthones 2, 6, 101, 194, 329–30, 574, 584–5, 589–602, 610, 612
Picrophloeus 200 (= *Fagraea*)
Picrorhiza 577, 579
Pioneer species 187
Pistil see Flower morphology and Gynoecium
Pistillipollenites 104
Pitygentias 229, 246 (= *Gentianella*)
Plant–animal interactions; see also Dispersal and Pollination
ants 160, 210
Pleurogyne 233 (= *Lomatogonium*)
Pleurogynella 233 (= *Lomatogonium*)
Plicae 223, 238–9
Plumeria 316, 321–2, 331, 333, 336, 352
Pollen see Palynology
Pollination
Apocynaceae 7
Exaceae 7, 99
Gentianaceae 7–8, 324–5
Helieae 7–8, 139, 180–1
Loganiaceae 8
Potalieae 217
Rubiaceae 7, 8
bat 8–9, 139, 180–1, 217
bee 8, 217
bird 8–9, 139, 181, 217
butterfly 7, 217
cross-pollination 99
evolution of 7–8
hawkmoth 7, 9, 139, 181
insect 139, 217
self-pollination 99
thrips 92
Polymerous flowers 211; see also Flower morphology
Population studies 203
Potalia 1–3, 6, **205–6**, 379, 539–40
dispersal 9
distribution 191, 193, 205
floral anatomy 323
fruit morphology 8–9, 205
habit 10, 205

Potalia (cont.)
number of species 15, 25, 191, 205
palynology 2, 215–16, 377, 379, 381, 390, 460, 462, 476–7, 479
phylogeny 4, 205–6, 378–9, 476–7, 479
  molecular data 35
  morphological data 219, 313
phytochemistry 588, 601, 607, 609
pollination 9
seed anatomy 539–40
Potaliaceae 48, 53
Potalieae (tribe) 23, **191–3**, 194–207
biogeography 219–21
character diagnosis 218
classification 52–3
dispersal 9, 217, 507
distribution 13, 190–1
exudates on seeds 560
floral anatomy 192, 210–11
flower morphology 211–13, 216–17
fruit morphology 8–9, 194, 216–17
habit 10, 193, 208–10, 218–19, 337
key 45, 47
nomenclature 53
number of genera 12, 25, 53, 191
number of species 12, 53, 190–1
number of subtribes 25
palynology 213–16, 379, 453–63, 476–80
pharmacology 218
phylogeny 194, 207–8, 336, 476–9, 555
  molecular data 41, 43–44, 197
  morphological data 194, 218, 313, 335–6
  phytochemical data 194
phytochemistry 329, 601, 611
pollination 9, 217
seed anatomy 327, 502, 533–40, 558–9
taxonomic history 194–5
vegetative morphology 210
Potaliinae (subtribe) 6, **193**
classification 53, 194
dispersal 9, 507
distribution 193, 219
flower morphology 192, 208
fruit morphology 8–9
habit 10
key 45
number of genera 25, 53
number of species 53
palynology 379, 479
phylogeny 208, 218, 333–4, 479, 555
  molecular data 197
pollination 9
seed anatomy 539–40, 555, 558
vegetative morphology 165–6
Potalioideae (subfamily) 53
Prepusa 138–9, **154–5**, 157, 224, 533–4

biogeography 182
distribution 138, 154
number of species 25, 138, 154
palynology 175, 377, 379, 381, 402–3, 441–3, 473–4, 479, 480
phylogeny 378, 479–80
seed anatomy 533–4, 558
Primulaceae 7
Pterygocalyx 222–3, **234–5**, 238, 249
biogeography 263–5
dispersal 506
distribution 234
number of species 25, 234
phylogeny 245, 249, 264
Ptychanthe 224
Purdieanthus 138–9, **155**, 164–5
biogeography 182
distribution 138
number of species 25, 138
palynology 175, 377, 379, 381, 403, 441, 444, 472, 479
phylogeny 378, 472, 479, 557
  molecular data 149
seed anatomy 529, 557
Pycnosphaera **206**, 535–6
distribution 191, 206
number of species 25, 191, 206
phylogeny 206
phytochemistry 601
seed anatomy 535–6, 558

Rauwolfia 316, 320–2, 331, 333, 336, 352
Refugia hypothesis (South America) 186–7
Reichenbach, H. G. L. 52, 71, 74–5
Restriction enzyme data 203
Retzia 4
Rogersonanthus 137–8, 140, **155–6**, 524, 527–8
biogeography 182
distribution 138
number of species 25, 138
palynology 175, 377, 379, 381, 403, 408, 444–5, 474, 479–80
phylogeny 156, 377, 474, 479, 558
pollination 8
seed anatomy 523–4, 527–8
taxonomic history 155
"Roraimaea" 138–9, **156–7**
classification 52
distribution 138, 156
number of species 25, 138
phylogeny 157
pollination 8
seed anatomy 531, 558
Rubiaceae 3–5, 311
chromosome number 329
evolution, woody habit 335

flower morphology 7
fruit morphology 9
mycorrhiza type 319
number of species 11
phylogeny 40, 331, 333, 336
  molecular data 29–36, 40, 336
phytochemistry 330
pollination 7, 8
Rusbyantheae (tribe) 51, 556
Rusbyanthus 51, 152 (= *Macrocarpaea*)
  palynology 175

Sabatia 108, **120–1**, 126, 137, 325, 515–17
  chromosome number 328–9
  classification 113
  distribution 120–1
  floral anatomy 323
  number of species 24
  phylogeny 314, 332–3, 335, 371
  phytochemistry 588, 601
  pollination 324
  section
    Campestria 120
    Dodecandrae 120
    Sabatia 120
  seed anatomy 502, 504, 515–17
Sabatieae (tribe) 49
Saccifoliaceae 23, 48, 311, 555
Saccifolieae (tribe) 11, 23, 55, **57**
  biogeography 66
  classification 48, 56
  distribution 13, 66
  generic synopses 58–62
  key 46
  morphology 62
  number of genera 12, 24, 49
  number of species 12, 49, 55
  palynology 62, 63–5, 379, 408–11, 462–4, 478
  phylogeny 40, 42, 44, 55, 62–3, 379, 462–4, 478, 555, 560
    molecular data 40, 555
    morphological data 35
    palynological data 62
  phytochemistry 601, 611
  seed anatomy 550–3
  taxonomic history 56–7, 379
Saccifolioideae (subfamily) 57
Saccifolium 5, **59**, 62–3, 311, 550–1
  biogeography 66
  distribution 59
  flower morphology 5
  habit 337
  number of species 24
  palynology 64–5, 377, 379, 411, 478
  phylogeny 62–3, 379, 478, 555, 560
    molecular data 27, 36, 40, 555

Index 649

  morphological data 315, 334–5, 371
  phytochemistry 601
  seed anatomy 550–1
  taxonomic history 56
  vegetative morphology 5, 320
Sanango 4
Saprophytes 5, 39, 63, 67, 70, 78, 225, 559–60
  distribution 107
Scanning electron microscopy (SEM), methodology 38, 380
Schinziella 108, **121**, 125–6, 519–20
  classification 113
  distribution 121
  floral anatomy 323
  number of species 24
  phytochemistry 601
  seed anatomy 519–20
Schultesia 108, **121–2**, 123, 125, 224, 522–3, 612
  classification 113
  dispersal 506
  distribution 121
  floral anatomy 321–2
  number of species 24
  palynology 133–4, 324, 377, 379, 381, 403–6, 417, 419–21, 465
  phylogeny 378, 465, 557
    molecular data 36
    morphological data 315, 333–4, 372
    phytochemical data 600
  phytochemistry 330, 588, 594, 598–9, 601
  seed anatomy 521–3, 557
  vegetative morphology 321
Sczukinia 235 (= *Swertia*)
Sebaea 5, 66, **84–7**, 101, 103, 511–12
  dispersal 506–7
  distribution 84
  floral anatomy 323
  number of species 13, 15, 24, 84
  palynology 96–7, 377
  phylogeny 86, 101, 556
    molecular data 36
    morphological data 314, 332–4, 372
  phytochemistry 330, 601
  seed morphology 504, 509, 510, 511–12, 556
  trophy 39, 67, 512
Sebaeeae (tribe) 49
Sebaeinae (subtribe) 49
Secondary compounds *see* Phytochemistry
Sedatives 612
Seed dormancy 508
Seed testa *see* Seeds
Seeds
  Chironieae 513–23
  Exaceae 99–100, 509–12

Seeds (*cont.*)
  Gentianeae 541–50
  Helieae 523–34
  Potalieae 534–40
  Saccifolieae 550–5
  Voyria 43, 553–5
  anatomy 501–5
  development 499–500
  dispersal 505–8
  exudates *see* Exudates on seeds under specific groups
  germination 508–9
  phylogenetic information 555–60
  shape 43, 502
  size 503
  star-shaped testa cells 67, 71, 74, 100
  types 541–5
Senaea 138–9, **157**, 532
  biogeography 182
  distribution 138, 157
  number of species 25, 138, 157
  palynology 175–6, 379, 381, 445–6, 473–4, 479–80
  phylogeny 157, 378, 479–80, 558
  seed anatomy 532–4, 558
Sepal *see* Calyx
Sipapoantha 138–9, **157–8**, 529–30
  biogeography 182
  distribution 138, 157
  number of species 25, 138
  palynology 176, 377, 379, 381, 405, 408, 445, 447, 475, 479
  phylogeny 157–8, 377–8, 475, 479
  seed anatomy 529–31, 557–8
Solanaceae 4, 35
Solanales (order) 3
Solanum 325
South America 186–9, 191, 221
  climatic changes 186–7, 191
  endemicity 188–9
  floras 275–6
  generic diversity 14–15, 183
  glaciation 186–7
  highland/lowland taxa 187–9
  refugia hypothesis 186–7
  species diversity 229
  white-sand species 151, 157, 187, 221
Species number *see* Number of species under specific groups
Species skewness 11–12
Spigelia 1–3, 8
Spur on petal 231, 249; *see also* Corolla
Stamens
  Gentianeae 230
  Potalieae 213
  anther dehiscence 42, 192
  apical glandular appendage 68, 72, 92

apical sterile appendage 140, 142, 170, 192
asymmetric 238, 240
basal appendage/gland on thecae 68, 72, 92
curved filaments 140, 142, 240
elongate connectives 62, 92
fused anthers 42, 62, 68, 70, 72, 94, 223
fused filaments 39, 192, 195, 213
glands *see* Glands
helically twisted anthers 109–10, 115, 120, 121, 124, 126–128
insertion 39, 72, 140, 192, 213, 230, 239
porate anthers 68, 72, 78, 92
recurved anthers 68, 72, 109, 140, 142, 170, 228
reduced number 39, 108–9
staminal appendages 192–3, 201, 213
winged filaments 240
zygomorphy 39, 68, 70, 99, 170, 228
Stem anatomy 266
Stigma *see* Flower morphology *and* Flower morphology under specific groups
Stilbaceae 4
Stipules *see* Vegetative morphology
Strychnaceae 32
Strychnos 8, 316, 320–1, 333, 353
Style *see* Flower morphology under specific groups
Supermerous flowers *see* Flower morphology
Swertia 222, 226, 233, **235**, 247, 547–8, 576
  biogeography 245, 262–3, 265
  chromosome number 235, 328–9
  classification 198, 247–8
  dispersal 506
  distribution 235
  embryology 326, 499–500
  floral anatomy 321–3
  flower morphology 235, 247
  germination 508
  number of species 13, 15, 25, 235
  palynology 260–1
  paraphyly 13, 245, 248
  pharmacology 611
  phylogeny 222, 235, 245, 247–8, 265
    molecular data 36, 245
    morphological data 314, 333, 373
    phytochemical data 590–2, 594–5, 598–9, 602, 604–7, 609
  pollination 324–5
  section
    Ophelia 248 (= *Swertia*)
  seed anatomy 327, 502, 545, 547–8, 559
  vegetative morphology 321
Swertia subgenus *Lomatogonium* 233 (= *Lomatogonium*)
Swertieae (tribe) 54
Swertiinae (subtribe) 222, 235, 238

Index 651

biogeography 245, 263–4
classification 54
flower morphology 239, 243
number of genera 25, 54, 246
number of species 25, 54
palynology 255, 257, 259
phylogeny 243–9, 263, 265, 333, 559
molecular data 243–5
morphological data 242–3, 246
seed anatomy 545–50, 559
Symbolanthus 125, 137–9, **158–60**, 530–1
biogeography 182, 185
chromosome number 329
dispersal 507
distribution 138, 159
exudates on leaves 560
fruit morphology 9, 326
number of species 13, 15, 25, 138, 159–60
palynology 176, 324, 377, 379, 381,
   405–6, 447–8, 475, 479
phylogeny 160, 164–5, 378, 475, 479
molecular data 36–7, 149
morphological data 315, 332–5, 373
phytochemistry 601
pollination 8, 324
seed anatomy 527, 531, 557–8
vegetative morphology 320–1
wood anatomy 320
Symbolanthus clade (subgroup in Helieae)
phylogeny 149, 164–5
Sympetaly *see* Corolla
Symphyllophyton 108, **122**, 125
classification 113
distribution 122
number of species 24
palynology 134
phytochemistry 601
seed anatomy 521

Tachia 137–8, **160–1**, 164, 224, 532, 534
biogeography 183–4
distribution 138, 160
number of species 25, 138
palynology 176–7, 377, 379, 390–3,
   448–9, 475
phylogeny 377, 480
molecular data 37, 149, 197
morphological data 314, 332–4, 374
phytochemistry 601
pollination 8
section
   Schomburgkiana 160
   Tachia 160, 166
seed anatomy 532, 534, 558
Tachiadenus 66–8, **87–8**, 102–3, 506, 511
dispersal 506
distribution 87

floral anatomy 323
habit 319
number of species 24
palynology 97, 377, 379, 414–15, 465
phylogeny 465, 556
molecular data 37
morphological data 313–14, 332–4, 374
phytochemistry 601
pollination 7
seed anatomy 328, 510–11, 556
trophy 319
Tachiinae (subtribe) 51, 556
Takhtajan, A. 57, 77, 194
Tapeinostemon 57, **59–61**, 62–4, 551–2
biogeography 66
distribution 59
number of species 24, 59
palynology 64–5, 377, 379, 411–12, 464,
   478
phylogeny 62–3, 464, 478, 556
seed anatomy 551–2, 556, 559
taxonomic history 55–6
Taxonomy *see* Classification
Teleiandrae (series) 49
Testa *see* Seeds
Tetragonanthus 231 (= *Halenia*)
Tetrapollinia 137, 139, **161–2**, 164–5, 524,
   527–8, 557
distribution 139, 161
number of species 25, 139, 161
palynology 177, 377, 379, 381, 406–8,
   450–1, 475–6, 479
phylogeny 378, 475–6, 479
molecular data 149
phytochemistry 601
seed anatomy 523, 527, 557
Thorne, R. 57
Transmission electron microscopy (TEM)
   38, 380
Trees *see* Habit
Tripterospermum 222–3, **236**, 542, 545
biogeography 262
chromosome number 236
dispersal 506–7
distribution 236
floral anatomy 322
fruit morphology 8, 325
nomenclature 236
number of species 25, 236
palynology 261
pharmacology 612
phylogeny 314, 333, 375, 600
phytochemistry 595, 598–600, 602, 604–5
seed anatomy 502–3, 542, 545
vegetative morphology 321
*trn*L intron *see* Molecular systematics
Twining habit 238

Unisexual flowers 39, 223
Urogentias 195, **207**, 537–8
  dispersal 507
  distribution 191, 207
  number of species 25, 191
  phylogeny 207
  phytochemistry 601
  seed anatomy 327–8, 533, 538, 558
Utania 200 (= *Fagraea*)

Valerianaceae 37
Vegetative morphology
  auricles 210
  bark 59–60
  colleters *see* Colleters
  hairs 84, 127, 153, 238, 240
  interpetiolar structures 4, 6, 39, 166, 210
  latex 39
  leaf venation 39, 89, 160, 166
  rugose leaves 148, 166
  saccate leaves 5, 59–60
  twining habit 238
  whorled leaves 89, 235
Veratrilla 222–3, 235, **236**, 248, 548–9, 595
  biogeography 263
  classification 236
  dispersal 506
  distribution 236
  number of species 25, 236
  palynology 261–2
  phylogeny 245, 248
  phytochemistry 595, 598–600, 602
  pollination 325
  seed anatomy 502, 548
Verbenaceae 4
Vicariance *see* Biogeography
Voyria 5, 224, **266–8**, 328, 503, 507, 553–4, 559–60
  biogeography 220
  dispersal 107, 266, 505, 507, 560
  distribution 266
  embryology 326, 501
  floral anatomy 7, 321–2
  flower morphology 266–8
  fruit morphology 326
  number of species 25, 55, 266
  palynology 65, 268, 324, 378
  phylogeny 54, 267, 556
    molecular data 42, 267
    morphological data 313, 315, 332–4, 375
  pollination 7
  seed anatomy 328, 500, 502–4, 507, 550, 553–5, 559–60
  subgenus
    Leiphaimos 267, 560
    Voyria 267

  trophy 320
  vegetative anatomy 266
Voyrieae (tribe) 54–5, 378, 556
Voyriella 5, **61–2**, 63, 266, 551, 553
  distribution 61
  embryology 326, 501
  fruit morphology 326
  number of species 24, 61
  palynology 64–5, 478
  phylogeny 62–5, 478, 556
    molecular data 37
    morphological data 315, 332–4, 375
  seed anatomy 502, 551–3, 556, 559
  taxonomic history 55–6
  trophy 320

Wood anatomy 209, 266
  intraxylary phloem 4
Woodiness *see* Evolution, habit, Habit *and* Habit under specific groups
Woody habit *see* Habit
Wurdackanthus 137, 139, **162**, 164–5, 527–8
  biogeography 182
  distribution 139, 162
  number of species 25, 139, 162
  palynology 177, 377, 379, 381, 407–8, 450–2, 468–9, 479
  phylogeny 160, 162, 378, 468–9, 479
    molecular data 149
  phytochemistry 601
  seed anatomy 523, 557, 558

Xanthones 2, 6, 101, 194, 329–30, 574, 584–5, 589–602, 610, 612; *see also* Phytochemistry
Xestaea 108, **122**, 125, 522–3
  classification 113
  distribution 122
  number of species 24
  palynology 134
  phylogeny 334, 372
  phytochemistry 601
  seed anatomy 523

Zonanthus 139, **163**, 177
  biogeography 182–4
  distribution 139, 163
  number of species 25, 139
  palynology 377, 379, 393, 452–3, 476, 480
  phylogeny 163
  seed anatomy 534, 558
Zygostigma 108, **122–3**, 516–17
  classification 113
  distribution 122
  number of species 24
  phylogeny 335
  seed anatomy 517